Separation and Purification Technologies in Biorefineries

Separation and Purification Technologies in Biorefineries

SHRI RAMASWAMY

Department of Bioproducts and Biosystems Engineering, University of Minnesota, USA

HUA-JIANG HUANG

Department of Bioproducts and Biosystems Engineering, University of Minnesota, USA

BANDARU V. RAMARAO

Department of Paper and Bioprocess Engineering,
State University of New York College of Environmental Science and Forestry, New York, USA

WILEY

A John Wiley & Sons, Ltd., Publication

This edition first published 2013

Registered office
John Wiley & Sons Ltd, The Atrium, Southern Gate, Chichester, West Sussex, PO19 8SQ, United Kingdom

For details of our global editorial offices, for customer services and for information about how to apply for permission to reuse the copyright material in this book please see our website at www.wiley.com.

Cover Acknowledgement

Cover photos and design advice courtesy of David Hansen, Minnesota Agricultural Experiment Station; Sara Specht, Graphic Designer, College of Food, Agricultural, and Natural Resource Sciences, University of Minnesota, USA

Library of Congress Cataloguing-in-Publication Data

Ramaswamy, Shri, 1957-
Separation and purification technologies in biorefineries / Shri Ramaswamy, Hua-Jiang Huang, Bandaru V. Ramarao.
pages cm
Includes index.
ISBN 978-0-470-97796-5 (cloth)
1. Biomass conversion. 2. Biomass energy. I. Title.
TP248.B55R36 2013
333.95'39–dc23

2012035282

A catalogue record for this book is available from the British Library.

ISBN: 9780470977965

Set in 10pt/12pt Times by Laserwords Private Limited, Chennai, India

MIX
Paper from responsible sources
FSC® C013604

Contents

List of Contributors

Zurina Zainal Abidin, Department of Chemical and Environmental Engineering, Faculty of Engineering, Universiti Putra Malaysia, Serdang, Selangor, Malaysia

Sara M. Badenes, Department of Bioengineering and Institute for Biotechnology and Bioengineering, Centre for Biological and Chemical Engineering, Instituto Superior Técnico, Technical University of Lisbon, Lisbon, Portugal

Prashan P. Barve, Chemical Engineering and Process Development Division, CSIR-National Chemical Laboratory, Pune, India

M. Berrios, Department of Inorganic Chemistry and Chemical Engineering, University of Cordoba, Cordoba, Spain

Bhavin V. Bhayani, Department of Paper and Bioprocess Engineering, Empire State Paper Research Institute, State University of New York College of Environmental Science and Forestry, Syracuse, New York, USA

Dayang Radiah Awang Biak, Department of Chemical and Environmental Engineering, Faculty of Engineering, Universiti Putra Malaysia, Serdang, Selangor, Malaysia

Costin Sorin Bildea, University "Politehnica" of Bucharest, Department of Chemical Engineering, Bucharest, Romania

Joaquim M. S. Cabral, Department of Bioengineering and Institute for Biotechnology and Bioengineering, Centre for Biological and Chemical Engineering, Instituto Superior Técnico, Technical University of Lisbon, Lisbon, Portugal

Lourdes Casas, Chemical Engineering and Food Technology Department, University of Cadiz, Cádiz, Spain

Biaohua Chen, State Key Laboratory of Chemical Resource Engineering, Beijing University of Chemical Technology, Beijing, China

Chim Yong Chin, PureVision Technology, Inc., Ft. Lupton, Colorado, USA

Tai-Shung Chung, Department of Chemical and Biomolecular Engineering, National University of Singapore, Singapore

Michael John Cooney, University of Hawaii at Manoa, Hawaii Natural Energy Institute, Honolulu, Hawaii, USA

Frederico Castelo Ferreira, Department of Bioengineering and Institute for Biotechnology and Bioengineering, Centre for Biological and Chemical Engineering, Instituto Superior Técnico, Technical University of Lisbon, Lisbon, Portugal

Mats Galbe, Department of Chemical Engineering, Lund University, Lund, Sweden

Mohd Yusof Harun, Department of Chemical and Environmental Engineering, Faculty of Engineering, Universiti Putra Malaysia, Serdang, Selangor, Malaysia

Bo Hu, Department of Bioproducts and Biosystems Engineering, University of Minnesota, Saint Paul, Minnesota, USA

Hua-Jiang Huang, Department of Bioproducts and Biosystems Engineering, University of Minnesota, Saint Paul, Minnesota, USA

M. A. Izquierdo-Gil, Department of Applied Physics I, Faculty of Physics, University Complutense of Madrid, Madrid, Spain

Ann-Sofi Jönsson, Department of Chemical Engineering, Lund University, Lund, Sweden

Sanjay P. Kamble, Chemical Engineering and Process Development Division, CSIR-National Chemical Laboratory, Pune, India

Anton A. Kiss, Arnhem, The Netherlands

Aspi K. Kolah, Department of Chemical Engineering and Materials Science, Michigan State University, East Lansing, Michigan, USA

Bhaskar D. Kulkarni, Chemical Engineering and Process Development Division, CSIR-National Chemical Laboratory, Pune, India

Zhigang Lei, State Key Laboratory of Chemical Resource Engineering, Beijing University of Chemical Technology, Beijing, China

Carl T. Lira, Department of Chemical Engineering and Materials Science, Michigan State University, East Lansing, Michigan, USA

Congcong Lu, Coatings Technology Center, DCM, The Down Chemical Company, Midland, Michigan, USA

Casimiro Mantell, Chemical Engineering and Food Technology Department, University of Cadiz, Cádiz, Spain

Mika Mänttäri, Lappeenranta University of Technology, Department of Chemical Technology, Laboratory of Membrane Technology and Technical Polymer Chemistry, Lappeenranta, Finland

A. Martín, Department of Inorganic Chemistry and Chemical Engineering, University of Cordoba, Cordoba, Spain

M. A. Martín, Department of Inorganic Chemistry and Chemical Engineering, University of Cordoba, Cordoba, Spain

Enrique Martínez de la Ossa, Chemical Engineering and Food Technology Department, University of Cadiz, Cádiz, Spain

Dennis J. Miller, Department of Chemical Engineering and Materials Science, Michigan State University, East Lansing, Michigan, USA

Marianne Nyström, Lappeenranta University of Technology, Department of Chemical Technology, Laboratory of Membrane Technology and Technical Polymer Chemistry, Lappeenranta, Finland

Imran Rahman, Chemical Engineering and Process Development Division, CSIR-National Chemical Laboratory, Pune, India

Bandaru V. Ramarao, Department of Paper and Bioprocess Engineering, Empire State Paper Research Institute, State University of New York College of Environmental Science and Forestry, Syracuse, New York, USA

Shri Ramaswamy, Department of Bioproducts and Biosystems Engineering, University of Minnesota, Saint Paul, Minnesota, USA

Miguel Rodríguez, Chemical Engineering and Food Technology Department, University of Cadiz, Cádiz, Spain

J. A. Siles, Department of Inorganic Chemistry and Chemical Engineering, University of Cordoba, Cordoba, Spain

Marian Simo, Praxair Technology Center, Tonawanda, New York, USA

Panu Sukitpaneenit, Department of Chemical and Biomolecular Engineering, National University of Singapore, Singapore

Bart Van der Bruggen, K.U.Leuven Department of Chemical Engineering, Laboratory for Applied Physical Chemistry and Environmental Technology, Leuven, Belgium

G. Peter van Walsum, Forest Bioproducts Research Institute, Department of Chemical and Biological Engineering, University of Maine, Orono, Maine, USA

Saravanan Venkatesan, Shell Global Solutions International B.V., Department of Innovation Biodomain, Amsterdam, The Netherlands. Present Address: Shell Technology Centre Bangalore, India

Ola Wallberg, Department of Chemical Engineering, Lund University, Lund, Sweden

Nien-Hwa Linda Wang, School of Chemical Engineering, Purdue University, West Lafayette, Indiana, USA

Yan Wang, Department of Chemical and Biomolecular Engineering, National University of Singapore, Singapore

Natalia Widjojo, Department of Chemical and Biomolecular Engineering, National University of Singapore, Singapore

Shang-Tian Yang, William G. Lowrie Department of Chemical and Biomolecular Engineering, The Ohio State University, Columbus, Ohio, USA

Hamdan Mohamed Yusoff, Department of Chemical and Environmental Engineering, Faculty of Engineering, Universiti Putra Malaysia, Serdang, Selangor, Malaysia

Guido Zacchi, Department of Chemical Engineering, Lund University, Lund, Sweden

Jianguo Zhang, Department of Bioproducts and Biosystems Engineering, University of Minnesota, Saint Paul, Minnesota, USA

Preface

The depletion of fossil resources, global climate change, and a growing world population all make it imperative that we find alternative, renewable sources of materials, chemicals, transportation fuels, and energy to address increasing global demand. Biorefineries will be an integral part of the future sustainable bioeconomy. In addition to sustainable biomass resources and effective biomass conversion technologies, separation and purification technologies will play a very important role in the successful development and commercial implementation of biorefineries. Due to the widely varying characteristics and composition of biomass, and the varying associated potential conversion technologies, biorefineries offer very interesting challenges and opportunities associated with the separation and purification of complex biomass components and the manufacture of valuable products and co-products. Generally, separation and purification processes can account for a large fraction (about 20–50%) of the total capital and operating costs of biorefineries. Significant improvement in separation and purification technologies can greatly reduce overall production costs and improve economic viability and environmental sustainability.

Examples of separation and purification needs in biorefineries include pre-extraction of value-added phytochemicals from lignocellulosic biomass, separation of biomass components (including cellulose, hemicellulose, lignin and extractives), extraction and purification of hemicellulose prior to pulping, separation of valuable chemicals from biomass hydrolyzate, removal of fermentation inhibitors enabling improved conversion efficiency and yield, concentrating process streams for varying end products and applications, integration of separation and purification technologies with bioprocessing, as well as downstream product separation and purification, syngas clean-up, purification of reactants, purification of glycerol from biodiesel production for production of intermediates such as succinic acid, and separation and purification of products such as ethanol, butanol, and lactic acid (there are many more examples).

In this book, technical experts from around the world offer their perspectives on the different separation and purification technologies that pertain to biorefineries. They provide basic principles, engineering design and specific applications in biorefineries, and also highlight the immense challenges and opportunities. There are significant opportunities for developing totally new approaches to separation and purification especially suitable for biorefineries and their full integration in the overall biorefineries. For example, adsorption with a molecular sieve is efficient in breaking the ethanol–water or butanol-water azeotrope for biofuel dehydration. Membrane separation, especially ultrafiltration and nanofiltration, represents a promising procedure for recovery of hemicelluloses from hydrolyzates and lignin from spent liquor. Hybrid separation systems such as extractive-fermentation and fermentation-membrane pervaporation are promising approaches to the removal of product inhibition, and hence to the improvement of process performance. Fermentation, bipolar membrane electrodialysis, reactive distillation, and reactive absorption are suitable for separation of products obtained by esterification, as in biodiesel production. Integrated bioprocessing—consolidated bioprocessing integrating pre-treatment, bioprocessing, separation, and purification—offers tremendously exciting new opportunities in future biorefineries.

The editors are grateful to all the contributors for making this very timely book possible. We hope that it will serve as a good resource for industrial and academic researchers, scientists, and engineers as we all work together to address the challenges, develop innovative solutions, and contribute to the development of sustainable biorefineries.

Shri Ramaswamy
Hua-Jiang Huang
Bandaru V. Ramarao

Part I
Introduction

1

Overview of Biomass Conversion Processes and Separation and Purification Technologies in Biorefineries

Hua-Jiang Huang and Shri Ramaswamy

Department of Bioproducts and Biosystems Engineering, University of Minnesota, USA

1.1 Introduction

There has been an increasing interest in conversion of biomass to biofuels, energy and chemicals due to increase in global demand, price and decrease in potential availability of crude oil, the need for energy independence and energy security, and the need for reduction in greenhouse gases emission from fossil fuel contributing to global climate change, and so forth.

Biomass feedstock suitable for producing biofuels, energy and co-products can be starchy biomass (e.g., corn/wheat kernel, cassava), sugarcane and sugar beet, ligocellulosic biomass including agricultural residues (e.g., corn stover, crop residues such as wheat straw and barley straw, and sugar cane bagasse), forest wastes, fast-growing trees such as hybrid poplar and willow, fast-growing herbaceous crops such as switchgrass and alfalfa, oily plants such as soybean and rapeseed, microalgae, waste cooking oil, animal manure, as well as municipal solid waste. The total amount of biomass feedstock available is huge. In the United States, based on the estimation by U.S. Department of Energy (U.S. Department of Energy 2011), total potential biomass resource is about 258 (baseline)–340 (high-yield scenario) million dry tons in 2012. Potential supplies at a forest roadside or farmgate price of \$60 per dry ton range from 602 to 1009 million dry tons by 2022 and from about 767 to 1305 million dry tons by 2030, depending on the assumptions for energy crop productivity (1% to 4% annual increase over current yields). This estimate excludes resources that are currently being used, such as corn grain and woody biomass used in the forest products industry. Worldwide, the biomass availability is also significantly high of the order of 5.0 billion tons per year (Bauen *et al*. 2009; U.S. Department of Energy 2011).

Separation and Purification Technologies in Biorefineries, First Edition.
Edited by Shri Ramaswamy, Hua-Jiang Huang, and Bandaru V. Ramarao.

Biofuels made from starchy crops, sugar plants as well as vegetable oils are usually called first-generation biofuels; for example, bioethanol produced from maize, starch, or sugar via fermentation, biodiesel from soybean oil, rapeseed oil, palm oil, or other plant oil by transesterification. Biogas from anaerobic digestion of waste streams also belongs to the first-generation biofuels. As the first-generation biofuels produced from food crops competes with food production and supply, and biogas can only be produced in small quantities, the first-generation biofuels alone generally cannot meet our energy requirements. Biofuels such as cellulosic ethanol made from lignocellulosic biomass such as woody crops, fast-growing trees and herbaceous crops, agricultural residues and forestry waste are referred to as the second-generation biofuels. The focus for second-generation biofuels was primarily ethanol. Unlike the first-generation biofuels, the second-generation biofuels are based on non-food crops and other lignocellulosic biomass; it can also bring about significant reduction in greenhouse gas emissions as well as reduction in fossil fuel use. The third-generation biofuels are made from genetically modified energy crops that may be carbon-neutral, biofuels from algae, or biofuels directly produced from microorganisms or using advances in biochemistry. Fourth-generation biofuels have also been suggested, which are carbon negative—they consume more carbon than they generate during their entire life cycle. Examples of this could be carbon-fixing plants such as low input high-diversity perennial grasses (Tilman, Hill, and Lehman 2006).

A biorefinery is a facility to convert biomass to bioproducts including bioenergy (fuels, heat and power) and diverse array of co-products (including materials and chemicals) (Huang *et al*. 2008; Huang and Ramaswamy 2012). The biorefinery concept is similar to today's petroleum refinery, which produces multiple fuels and products from petroleum (http://www.nrel.gov/biomass/biorefinery.html). Biorefinery can be divided into two basic conversion platforms: biochemical conversions, and thermo-chemical conversions. A biorefinery can also be a combination of both biochemical and thermo-chemical conversion approaches. Biochemical conversions of biomass using enzymes and microorganisms (yeast and bacteria) are often referred to as "sugar-platform" conversions, where biomass is firstly pretreated and hydrolyzed to mono-sugars: glucose, xylose, arabinose, galactose, and mannose, and so forth. The mono-sugars are then fermented or digested to biofuels such as bioethanol and biobutanol, or chemicals such as lactic acid and succinic acid, depending on the biocatalysts used. Thermo-chemical conversion of biomass includes biomass combustion for heat and power, pyrolysis for bio-oil and biochar, hydrothermal liquefaction to bio-oils as major product, and biomass gasification to syngas. Syngas (mainly CO and H_2) from biomass gasification can be further synthesized into a wide range of different fuels and chemicals under different catalysts and operating conditions; biomass gasification or "syngas platform" represents the major thermo-chemical platform. In addition to these basic thermo-chemical conversions, there are a variety of other chemical conversion processes such as conversion of oil-containing biomass such as soybean and microalgae for biodiesel, and the conversion of building block chemicals such as lactic acid to its corresponding commodities, chemicals, polymers and materials.

This chapter provides an overview of the separation and purification technologies in biorefineries for producing bioproducts including biofuels, bioenergy, biochemicals and materials, with more emphasis on lignocelluose biorefineries.

1.2 Biochemical conversion biorefineries

In the biochemical conversion biorefineries or "sugar platforms," biomass is subjected to hydrolysis and saccharification and then the resulting sugars, including hexoses (glucose, mannose, and galactose) and pentoses (xylose, arabinose) are converted to biofuels such as ethanol and butanol, chemicals, and materials.

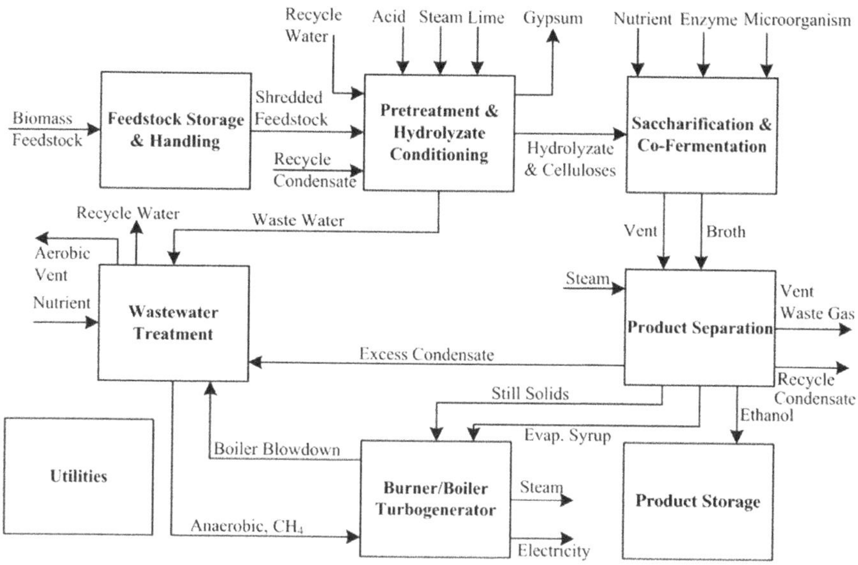

Figure 1.1 *Simplified process block diagram of basic lignocellulose to ethanol biorefinery (Aden et al. 2002; Huang et al. 2008)*

As an example, the basic process for conversion of cellulosic biomass to fuel ethanol is shown in Figure 1.1, which mainly consists of the following eight major process areas (Aden *et al*. 2002):

1. Feedstock handling including biomass storage and size reduction (shredding).
2. Pretreatment and hydrolyzate conditioning or detoxification. Here, the shredded biomass is pretreated with dilute sulfuric acid at a high temperature (using steam), and thus most of the hemicellulose is hydrolyzed to fermentable monosugars (mainly xylose, mannose, arabinose, and galactose) while glucan in the hemicellulose and a small fraction of the cellulose are converted to glucose. In addition, the hydrolysis reaction produces acetic acid liberated from acetate in biomass, furfural and hydroxymethyl furfural (HMF) from degradation of pentose and hexose sugars respectively. These compounds are inhibitory to the subsequent fermentation so, following the pretreatment, the prehydrolysys slurry is flashed to remove a portion of the acetic acid, and most of the furfural and HMF. The hydrolyzate, after being separated from the solids, is then overlimed to pH 10 by adding lime to remove the remaining inhibitors, followed by neutralization and precipitation of gypsum. After filtering out the gypsum, the detoxified hydrolyzate and the solids (cellulose) are sent to the saccharification and co-fermentation area. This step also solubilizes some of the lignin in the feedstock and make the cellulose accessible to subsequent enzymatic hydrolysis.
3. Saccharification and co-fermentation. The cellulose is biochemically hydrolyzed or saccharified to glucose by cellulase enzyme in the continuous hydrolysis tanks. The co-fermentation of the detoxified hydrolyzate slurry is carried out in anaerobic fermentation tanks in series using the microorganism *Zymomonas mobilis*. With several days of separate and combined saccharification and cofermentation, most of the cellulose and xylose are converted to ethanol.
4. Product separation and purification. Beer is firstly preconcentrated by distillation, followed by vapor-phase molecular sieve separation for ethanol dehydration. The postdistillation slurry from the

distillation bottom is separated into the solids and liquid. The liquid is then evaporated and separated into the concentrated syrup, and the condensed water is recycled in the process. The solids and the syrup obtained are sent to the combustor.

5. Wastewater treatment. Part of the evaporator condensate, together with the wastewater from pretreatment area, is treated by anaerobic digestion. The biogas (rich in methane) from anaerobic digestion is sent to the combustor for energy recovery. The treated water is recycled for use in the process.
6. Product storage.
7. Combustion of solids (lignin) for heat (steam) and power. The solids from distillation, the concentrated syrup from the evaporator, and biogas from anaerobic and aerobic digestion are combusted in a fluidized bed combustor to produce high-pressure steam for electricity production and process heat. Generally, the process produces excess steam that is converted to electricity by steam turbines for use in the plant and for sale to the grid.
8. Utilities.

This process involves a number of separation tasks as follows:

- removal of inhibitors from hydrolyzate prior to fermentation;
- liquid–solid separation such as separation of prehydrolyzate slurry and postdistillation slurry;
- ethanol recovery from beer by distillation and its dehydration using molecular sieve adsorption;
- water scrubbing of fermentation vents for recovering of the ethanol;
- water recovery by multiple effect evaporation;
- gas-solid (particles) separation from combustion flue gas.

The capital and operating costs of all the above separation processes account for a large fraction of the total capital and operation costs of the whole process.

The lignocellulose bioethanol process described above is only one case of "sugar-platform" biorefineries. Other bioconversion processes have similar steps in preparation of fermentable mono-sugars from biomass feedstock. In other words, in addition to bioethanol the biomass-derived mono-sugars including pentose and hexose can be fermented to other biofuels such as butanol, and biochemicals such as carboxylic acids (including succinic, fumaric, malic, itaconic, glutamic, lactic, 3-hydroxypropionic, citric, and butyric acids) (Yang *et al*. 2006), other chemicals (e.g., 1,3-propanediol), and materials, depending on the microorganism used. Among the carboxylic acids, succinic, fumaric, malic, itaconic, glutamic acids, and 3-hydroxypropionic acids are the major building block chemicals that can subsequently be converted to a number of high-value bio-based chemicals and materials. Building-block chemicals are molecules with multiple functional groups that have the potential to be transformed into new families of useful molecules. Biological transformations account for the majority of routes from plant feedstocks to building blocks, but chemical transformations predominate in the conversion of building blocks to molecular derivatives and intermediates (U.S. Department of Energy 2004). In addition, xylitol, and arabinitol are also important building-block chemicals. They can be employed to produce commodity and specialty chemicals such as xylaric acid, glycerol, propylene glycol, ethylene glycol, and lactic acid. Xylitol and arabinitol can be produced by hydrogenation of sugars or extraction from biomass pretreatment (U.S. Department of Energy 2004). In the following section, some important biofuel and building block chemicals including biobutanol, succinic acid, itaconic acid, 3-Hydroxypropionic acid, 1,3-propanediol, and lactic acid will be briefly introduced.

Biobutanol (C_4H_9OH) can be used as a chemical solvent in the food and pharmaceutical industries, and as a fuel. Biobutanol as a fuel is superior to ethanol in that it has higher energy content, lower vapor pressure, lower hygroscopy and hence causes less corrosion to pipelines and equipment. It has a higher octane rating, and is more safe. Butanol can be produced by ABE (Acetic acid, Butanol and Ethanol) fermentation of biomass carbohydrates using *C. acetobutylicum*, *C. beijerinckii*, or *C. saccharobutylicum*. The ABE fermentation broth is very dilute, with total ABE concentration of less than 20 g/L (A:B:E = 3:6:1 (molar)), and the butanol yield is low. This makes product separation a big challenge (Green 2011).

Succinic acid ($HOOCCH_2CH_2COOH$), also called amber acid or butanedioic acid, is primarily used as a sweetener in the food industry. In addition, it is a key building block for deriving both commodity and specialty chemicals such as 1,4-butanediol (BDO), tetrahydrofuran (THF), γ-butyrolactone (GBL), pyrrolidinones, and N-Methylpyrrolidone (NMP) (U.S. Department of Energy 2004; Cukalovic and Stevens 2008). Succinic acid is produced by fermentation of glucose using an engineered form of the organism *A. succiniciproducens* and, most recently, via an engineered *Eschericia coli* strain. Currently, highly efficient microorganism for production of succinic acid are *A. succinogenes*, *A. succiniciproducens*, and *M. succiniciproducens* (Cheng *et al*. 2012). The process also has the benefit of carbon dioxide fixation, as seen in its reaction formula (Zeikus, Jain and Elankovan 1999):

$$C_6H_{12}O_6 + CO_2 = HOOCCH_2CH_2COOH + CH_3COOH + HCOOH$$

In addition to glucose, glycerol can also be the carbon source for succinic acid fermentation. This provides a good opportunity to produce a value-added chemical from glycerol, the relatively cheap co-product of biodiesel production.

Itaconic acid, or methylsuccinic acid ($HO_2CCH_2CH(CH_3)CO_2H$), is used in polymers, paints, coatings, medicines, and cosmetics (Bressler and Braun 1999). As a value-added building block chemical, itaconic acid has the potential to be used for deriving both commodity and specialty chemicals such as 2-methyl-1,4-BDO, 3-methyl THF, 3-&4-methyl-GBL, 2-methyl-1,4-butanediamine, and other value-added chemicals (U.S. Department of Energy 2004). It is produced commercially by the fungal fermentation of carbohydrates. The most commonly used organism for itaconic acid production is *Aspergillus terreus*, grown under phosphate-limited conditions (Willke and Vorlop 2001).

3-Hydroxypropionic acid (3-HPA), as an important C3 building block, has the potential to derive several commodity and specialty chemicals such as 1,3-propanediol (1,3-PDO), acrylic acid, methyl acrylate, acrylamide, and other valuable chemicals (U.S. Department of Energy 2004). 3-HPA can be produced from glycerol using a recombinant strain *E. coli* (Raj *et al*. 2008), *Klebsiella pneumoniae* (Luo *et al*. 2010a; Huang *et al*. 2012), or from glucose using a recombinant strain *E. coli* (Rathnasingh *et al*. 2010). When cultivated aerobically on a glycerol medium containing yeast extract, the recombinant *E. coli* SH254 produced 3-HPA at a maximum of 6.5 mmol l^{-1} (0.58 g l^{-1}). The highest specific rate and yield of 3-HPA production were estimated as 6.6 mmol g^{-1} cdw h^{-1} and 0.48 mol mol^{-1} glycerol, respectively (Raj *et al*. 2008). The engineered *K. pneumoniae* can effectively produce 3-HPA and 1,3-PDO from glycerol under anaerobic conditions (Huang *et al*. 2012).

1,3-propanediol (1,3-PDO) is used in manufacturing polymers, medicines, cosmetics, food, and lubricants (Drożdżyńska, Leja and Czaczyk 2011). It can be produced from glycerol using pathogenic microorganisms such as *Klebsiella pneumoniae* and non-pathogenic microorganisms such as *Clostridium butyricum*, *Clostridium acetobutylicum*, and *Lactobacillus diolivorans*. *C. butyricum* has been reported to produce 1,3-PDO with a titer of 94 g/l when using glycerol as the carbon source (Wilkens *et al*. 2012). A recombinant strain of *C. acetobutylicum* produces up to 84 g/l in fed-batch cultivation (González-Pajuelo *et al*. 2005). The 1,3-PDO concentration obtained was 73.7 g/l in a fed-batch co-feeding glucose and glycerol with a molar ratio of 0.1. *L. diolivorans* proves to be a top candidate microorganism for industrial production of 1,3-PDO from glycerol. The wild-type strain produces up to 0.85 g 1,3-PDO/l h and product concentrations up to 85.4 g/l (Pflügl *et al*. 2012). 1,3-PDO can also be produced from glucose and molasses in a two-step process using two recombinant microorganisms. The first step is the conversion of glucose or other sugar into glycerol by the metabolic engineered *S. cerevisiae* strain HC42 adapted to high (>200 g l^{-1})

glucose concentrations. The second step is to convert glycerol to 1,3-PDO in the same bioreactor using the engineered strain *C. acetobutylicum* DG1 (pSPD5). The best results were obtained with an initial glucose concentration of 103 $g\,l^{-1}$, leading to a final 1,3-PDO concentration of 25.5 $g\,l^{-1}$, a productivity of 0.16 $g\,l^{-1}\,h^{-1}$ and 1,3-PDO yields of 0.56 $g\,g^{-1}$ glycerol and 0.24 $g\,g^{-1}$ sugar (Mendes *et al*. 2011). Recently, 1,3-PDO production by microorganisms were reviewed (Saxena *et al*. 2009; Drożdżyńska, Leja, and Czaczyk 2011).

Lactic acid is widely used in the food industry (Zhang, Jin, and Kelly 2007), and as a building-block chemical (Lee *et al*. 2011). It can be used for the production of biodegradable and biocompatible polymers such as polylactic acid (PLA), lactate esters, propylene glycol, acrylic acid and esters (Adsul *et al*. 2011). The current status of the production of potentially valuable chemicals from lactic acid via biotechnological routes has been reviewed recently (Gao, Ma and Xu 2011). Lactic acid can be produced from lignocellulose-derived sugars using microorganisms such as recombinant *Escherichia coli* (Dien, Nichols and Bothast 2001), *Bacillus coagulans* (Maas *et al*. 2008), *Lactobacillus sp*. (Wee and Ryu 2009), and *Lactococcus lactis* (Laopaiboon *et al*. 2010). There has been a recent overview of the lactic acid production (Vijayakumar, Aravindan, and Viruthagiri 2008; Abdel-Rahman, Tashiro, and Sonomoto 2011).

Biofuels (ethanol and butanol) and valued-added building-block chemicals (e.g., succinic acid, 3-HPA, and 1,3-PDO) derived from lignocellulosic carbohydates by biochemical conversion as described earlier, are often very dilute in their fermentation broths. This usually causes high production costs. In addition to improving microbial biocatalysts to increase substrate and hence product concentrations, yields, and productivities, development of efficient separation and purification processes with low costs are much needed.

1.3 Thermo-chemical and other chemical conversion biorefineries

1.3.1 Thermo-chemical conversion biorefineries

The major thermo-chemical conversion biorefineries involve combustion, hydrothermal liquefaction, pyrolysis, and gasification of biomass into heat (steam) and power, biofuels and chemicals.

Biomass combustion, the complete oxidation process, is a simple way to recover energy from biomass. As the steam turbine used in the process for generating power is not efficient, combustion of biomass, especially the whole biomass, is not the best option. Owing to the simplicity and the maturity of the combustion technology, combustion of the whole biomass, including non-fermentable residues, is commercially common. Combustion of biomass solid residues from distillation for steam and power for process use, as part of Figure 1.1, is a typical example. The carbon dioxide produced from biomass combustion was originally absorbed by the biomass plant during growth from environment via photosynthesis; so it is assumed to be carbon-neutral. In terms of separation, postcombustion capturing and sequestration of CO_2 from flue gases produced by the biomass combustion is very important and interesting.

Biomass pyrolysis is a thermal conversion process converting biomass to liquid (bio-oil), solid (char) and gas in the absence of oxygen. Based on different reaction rates and product distributions, pyrolysis can be classified as four categories: torrefaction, carbonization, intermediate pyrolysis, and fast pyrolysis. Table 1.1 shows the typical product yields for pyrolysis of wood using different modes and conditions.

The pyrolysis bio-oil can be used as feedstock of gasification for producing syngas, which can then be synthesized into fuels and chemicals. In addition, bio-oil can be used to produce transportation fuels. Fast pyrolysis liquid has a higher heating value of about 17 MJ/kg as produced with about 25 wt.% water that cannot easily be separated. Besides, pyrolysis bio-oil has a high oxygen content of around 35–40 wt% (Bridgwater 2012), leading to instability and relatively low heating value. Thus, pyrolysis bio-oil needs to be catalytically upgraded to transportation fuels and fuel additives by hydrotreating, cracking and decarboxylation, or esterification of bio-oil with alcohols followed by water separation to reduce their oxygen content and improve their thermal stability (Bulushev and Ross 2011). Bio-oil upgrading technologies have

Table 1.1 Typical product weight yields (dry wood basis) for different pyrolysis of wood. Adapted from Bridgwater, A. V., © 2012 with permission from Elsevier

Pyrolysis mode	Temperature (°C)	Residence time	Yields (%)		
			Liquid	*Solid*	*Gas*
Torrefaction (slow)	~290	~10–60 min	0	80	20
Carbonization (slow)	~400	hours to days	30	35 (char)	35
Intermediate	~500	~10–30 s	50	25 (char)	25
Fast	~500	~1 s	75	12 (char)	13

been recently reviewed (Huber and Corma 2007; Bulushev and Ross 2011; Bridgwater 2012). Furthermore, the separation of some chemicals such as acids and phenolics from bio-oil is another alternative option. Bio-oil is a complex mixture of several hundreds of organic compounds including hydroxyaldehydes, hydroxyketones, sugars, carboxylic acids, phenolics (phenols, guaiacols, catechols, syringols, isoeugenol) and other oligomeric lignin derivatives, along with around 25% water. About 35–50% of the bio-oil constituents are non-volatile (Czernik and Bridgewater 2004). Separation of value-added compounds from bio-oil becomes significantly important.

Hydrothermal liquefaction (HTL) is the process where the reaction of biomass is carried out in water media at high temperature and pressure with or without added catalyst. Its products include a bio-oil fraction, a water fraction containing some polar organic compounds, a gaseous fraction and a solid residue fraction (Biller and Ross 2011). Generally, HTL operates at 280–370 °C and 10–25 MPa (Behrendt *et al.* 2008). As HTL operates in water media, it can process directly the wet biomass feedstock such as wet microalgae (Wu, DeLuca and Payne 2010; Zou *et al.* 2010; Anastasakis and Ross 2011; Vardon *et al.* 2011; Vardon *et al.* 2012), animal manure (Yin *et al.* 2010; Vardon *et al.* 2011; Theegala and Midgett 2012), and digested anaerobic sludge (Vardon *et al.* 2011) without the need for predrying the biomass. Thus, the HTL process has energy-saving potential and it is a promising conversion process. There has been a recent overview of HTL of biomass for bio-oil (Akhtar and Amin 2011; Toor, Rosendahl and Rudolf 2011). The Hydro Thermal Upgrading (HTU®) process is one example of HTL. The HTU process, carried out at 300–350 °C, 100–180 bar and a residence time of 5–20 min, produces bio-oil (or biocrude) having a heating value of 30–35 MJ/kg (Goudriaan and Naber 2008; Toor *et al.* 2011). Due to the low oxygen content (10–18%wt), this bio-oil can be upgraded by hydrodeoxygenation (HDO) to premium quality diesel fuel. The thermal efficiency of the HTU process is 70–90% (Goudriaan and Naber 2008).

Biomass gasification is a partial oxidation process operating at a temperature in the range of 700–850 °C and a pressure of 0.1–3 MPa using steam, air or oxygen as oxidant. For gasification of black liquor from pulp mills can be conducted at conditions of 900–1200 °C and 2–3 MPa. It is one of the prominent thermochemical conversion methods to produce renewable fuels, energy, chemicals and materials. In addition to producing heat and power, synthesis gas from biomass gasification can be subsequently converted into liquid transportation fuels such as diesel and gasoline, alternative fuels such as methanol, dimethyl ether (DME) and ethanol, and other chemicals under different catalysts and operating conditions (Huang and Ramaswamy 2009). Synthetic diesel can be produced by the Fischer–Tropsch (FT) synthesis of syngas over iron or cobalt-based or hybrid (composite) catalysts (Khodakov, Chu, and Fongarland 2007). Methanol, which is also a material for fuel cell in addition to being an alternative fuel, can be synthesized from syngas over the Cu/ZnO catalyst (Zhang *et al.* 2009). Dimethyl ether can be produced by dehydration of methanol. It can also be manufactured directly from syngas by a single-step process using the hybrid catalyst composed of CuO, ZnO, Al_2O_3, and/or Cr_2O_3) for methanol synthesis and an acid function catalyst (such as γ-Al_2O_3, H-ZSM-5 or HY zeolites) for conversion of methanol into DME (Bae *et al.* 2008). In addition, mixed alcohols can be synthesized from syngas. Mixed alcohols synthesis from syngas is an important process for the production of oxygenated fuels, fuel additives, and other intermediates for value-added chemical feedstock for applications in medicine, cosmetics, as lubricants, as detergents, and for polyester (Fang *et al*. 2009). The potential catalysts for mixed alcohols synthesis from syngas include Cu-based catalysts and Mo-based catalysts. The synthesis of mixed alcohols from syngas over Cu-Fe based catalyst consists of alcohol formation (major reaction), hydrocarbon formation, and water–gas shift reaction are the side reactions (Fang *et al*. 2009). Methanol can also be synthesized to gasoline over zeolites. Hydrogen can be produced from syngas for fuel cell or power generation, or synthesis of ammonia for fertilizer. Table 1.2 shows the reactions of these important biofuels.

1.3.1.1 Example: Biomass to gasoline process

Biomass can be converted to gasoline via methanol synthesis and methanol-to-gasoline (MTG) technologies, as illustrated in Figure 1.2. In this process, biomass feedstock, after shredding and drying, is sent

Table 1.2 *Reactions of common syngas-based fuel synthesis*

Product	Reactants	Main reactions	Catalyst	Ref.
Methanol	$CO + H_2$	$CO + 2H_2 \rightarrow CH_3OH$ $CO_2 + 3H_2 \rightarrow CH_3OH + H_2O$ $CO + H_2O \rightarrow CO_2 + H_2$	Cu-ZnO-Al2O3 hybrid	(Zhang, Xiao and Shen 2009)
Diesel and waxes	$CO + H_2$	$nCO + 2nH_2 \rightarrow C_nH_{2n} + nH_2O$ $nCO + (2n + 1) H_2 \rightarrow C_nH_{2n+2} + nH_2O$	iron or cobalt	(Khodakov, Chu and Fongarland 2007)
Mixed alcohols	$CO + H_2$	alcohol formation: $nCO + 2nH_2 = C_nH_{2n+1}OH + (n-1)H_2O$; hydrocarbon formation: $nCO + 2nH_2 = C_nH_{2n} + nH_2O$; $nCO + (2n + 1)H_2 = C_nH_{2n+2} + nH_2O$ water–gas-shift reaction equilibrium: $CO + H_2O = CO_2 + H_2$	Cu-Fe	(Fang *et al.* 2009)
DME	Methanol	$2CH_3OH = CH_3OCH_3 + H_2O$	γ-Al_2O_3 or modified ZSM-5 zeolite	(Fu *et al.* 2005; Kim *et al.* 2006)
	$CO + H_2$	Two overall reaction routes: (1) $3CO + 3H_2 = CH_3OCH_3 + CO_2$ (2) $2CO + 4H_2 = CH_3OCH_3 + H_2O$		(Ogawa *et al.* 2003)
Gasoline	Methanol	$nCH_3OH \Leftrightarrow \frac{n}{2}CH_3OCH_3 + \frac{n}{2}H_2O \xrightarrow{-nH_2O} C_nH_{2n} \rightarrow n[CH_2]$	zeolite ZSM-5	(Chang 1992)

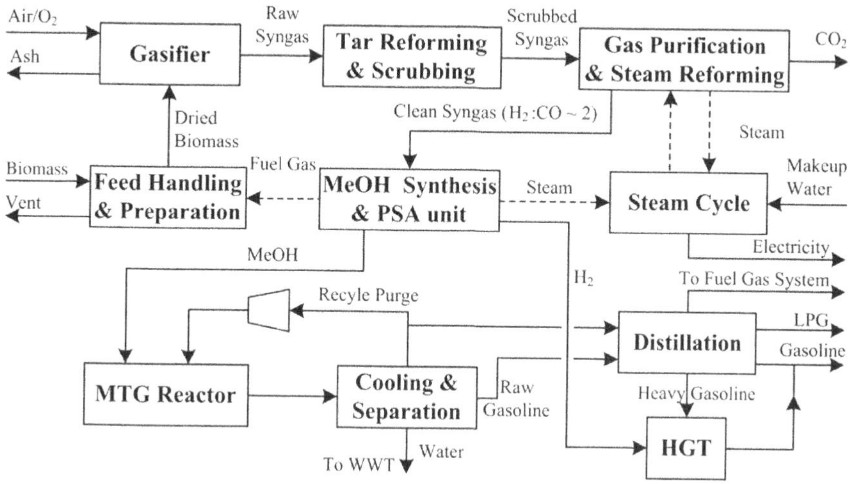

Figure 1.2 *Block diagram of biomass-to-gasoline process (Jones and Zhu 2009)*

to the gasifier for producing syngas. The raw syngas is sent to a tar reformer, a particulate scrubber, and finally a sulfur removal unit. Then the syngas enters a steam reformer where CH_4 is converted to H_2 and CO and the H_2/CO ratio is adjusted to that required by methanol synthesis. Excess CO_2 is removed by amine absorption. The clean syngas is then compressed and sent to the methanol synthesis. Part of the purge gas from methanol synthesis is used to produce hydrogen by a pressure swing adsorption (PSA) unit; the remaining purge gas is used as fuel for drying the feedstock. Raw methanol is converted to hydrocarbons and water in the MTG reactors. The raw gasoline isolated from water by phase separation, is distilled to produce fuel gas, liquefied petroleum gas (LPG), light gasoline, and heavy gasoline. The heavy gasoline is hydrotreated with hydrogen from the PSA to meet the final gasoline specifications. Steam generated in the process is collected and sent to the steam cycle for power generation. Some steam is used in steam reforming and other processes (Jones and Zhu 2009).

1.3.2 Other chemical conversion biorefineries

In addition to the major thermo-chemical conversion approaches mentioned above, biorefineries may also involve various other chemical conversion processes. For instance, production of value-added building block chemicals such as levulinic acid and sorbitol, the conversion of oil-containing biomass for biodiesel, and conversion of those building block chemicals described above to commodity, chemicals and materials. Next, some important value-added building block chemicals including levulinic acid, glycerol, sorbitol, and xylitol/arabinitol are briefly introduced, followed by an example of chemical conversion process.

1.3.2.1 Levulinic acid

Levulinic acid is an important platform molecule that can be used to produce a wide range of compounds such as γ-valerolactone (GVL), 2-methyltetrahydrofuran, δ-aminolevulinic acid, β-acetylacrylic acid, diphenolic acid, and 1,4-pentanediol (U.S. Department of Energy 2004). Levulinic acid can be catalytically converted to fuel additives through intermediates such as γ-valerolactone and valeric acid, and this has been recently highlighted (Lange *et al*. 2010; Bond *et al*. 2010; Bozell 2010). Also, 2-methyltetrahydrofuran and various levulinate esters derived from levulinic acid can be used as gasoline and biodiesel additives, respectively (U.S. Department of Energy 2004). Different from biofuels production via fermentation of biomass-derived sugars, levulinic acid is produced by acid catalyzed hydrolysis of biomass-derived sugars, a conventional chemical processing approach. This presents another promising route for biofuels.

1.3.2.2 Glycerol

Glycerol can be used as raw material for the cosmetics, pharmaceutical, and food industries (Leoneti, Aragão-Leoneti, and de Oliveira 2012). It is the major co-product of biodiesel production by transesterification of oils, with a weight ratio of 1/10 (glycerol/biodiesel). Glycerol can be considered a renewable building block for producing value-added products obtained by chemical (syn-gas, acrolein, and 1,2-propanediol) or bio-chemical (ethanol, 1,3-propanediol, D-lactic acid, succinic acid, propionic acid, and poly-3-hydroxybutyrate) routes (Posada *et al*. 2012). The wide use of glycerol in producing so many chemical building blocks plus its low price due to the fast growth of biodiesel industry and the surplus of glycerol makes it an excellent renewable feedstock and important building block for producing multiple products in biorefineries. Moreover, glycerol can be utilized to produce triacetin (or 1,2,3-triacetoxypropane), a biofuel additive, by esterification of glycerol with acetic acid. However, the glycerol from biodiesel production as a by-product must be purified before it is used in these industries (Leoneti, Aragão-Leoneti and de Oliveira 2012). Distillation, solvent extraction, ionic exchange, electrodialysis, and simulated moving bed (SMB) can be used for separation and purification of glycerol.

1.3.2.3 Sorbitol

Sorbitol is a potential key chemical intermediate from biomass resources for deriving a number of intermediates and chemicals such as propylene glycol, ethylene glycol, glycerol, lactic acid, and isosorbide (U.S. Department of Energy 2004). Sorbitol is commercially produced by the hydrogenation of glucose.

1.3.2.4 Xylitol/Arabinitol

Xylitol and arabinitol, the sugar alcohols, can be produced by hydrogenation of 5-carbon sugars xylose and arabinose from biomass. There is no major technical barrier associated with the production of xylitol and arabinitol (U.S. Department of Energy 2004). Separation and purification of the pentoses, xylose and arabinose, is important for production of xylitol and arabinitol. In addition, xylitol, and arabinitol can be produced by direct extraction from biomass pretreatment processes. Efficient separation and purification approaches such as ion exchange and nanofiltration are also necessary for this route.

1.3.2.5 Example: Conversion of oil-containing biomass for biodiesel

As an example, the conventional process of the plant oil to biodiesel conversion is shown in Figure 1.3. In this process, fatty acid methyl ester (FAME, biodiesel) is synthesized by esterification of oil with methanol over an alkali catalyst (NaOH). The resultant liquid mixture enters the methanol distillation column where methanol is removed and recycled for use as the reactant. The bottom liquid out of the distillation column is then washed and separated into the oil phase (raw FAME) and the aqueous phase (mainly glycerol). The raw FAME is purified by distillation, while the aqueous solution is neutralized with H_3PO_4, followed by filtering out the solid Na_3PO_4, and the distillation for glycerol concentration.

This homogeneous process using liquid catalyst (NaOH) has many disadvantages: requirement of alkali and acid chemicals and their handling, large separation burden and hence high separation capital and

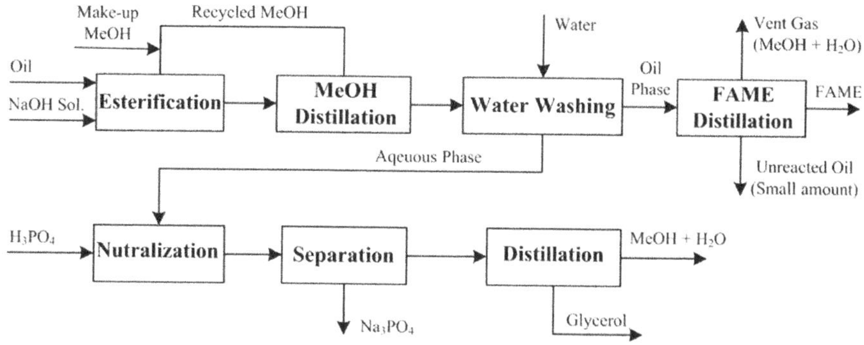

Figure 1.3 *Simplified block diagram of conventional biodiesel production process (Zhang, Dube and McLean 2003)*

operation costs. In addition, dehydrated vegetable oil with less than 0.5 wt.% free fatty acids, an anhydrous alkali catalyst and anhydrous alcohol are necessary for commercially viable alkali-catalyzed systems, and thus the low-cost waste cooking oil is not suitable as feedstock for this process; otherwise, soap occurs during the biodiesel production and this requires additional soap related separation, making the system more costly (Zhang, Dube, and McLean 2003). To overcome these disadvantages of the conventional biodiesel process, heterogeneous biodiesel process using solid catalyst can be applied. Figure 1.4 shows the simplified block diagram of the Esterfip–H biodiesel process (Axens-IFP Group Technologies).

In this continuous system, oil reacts with methanol in two fixed-bed reactors packed with a non-noble metal solid catalyst supplied by Axens. Excess methanol is removed after each of the two reactors by a partial flash vaporization. Esters and glycerol are then separated in a settler. Glycerol phases from each reactor, after being separated from settlers, are combined and the last traces of methanol are removed by vaporization. Biodiesel is produced after final recovery of methanol by full vaporization under vacuum (Bacovsky *et al*. 2007). This process has many advantages: high biodiesel yield (close to theoretical); high purity glyderol without the need for further purification; no soap formation and no low-value fatty acids; no handling of hazardous acid and base chemicals; much lower catalytic cost as compared to other processes (Bacovsky *et al*. 2007).

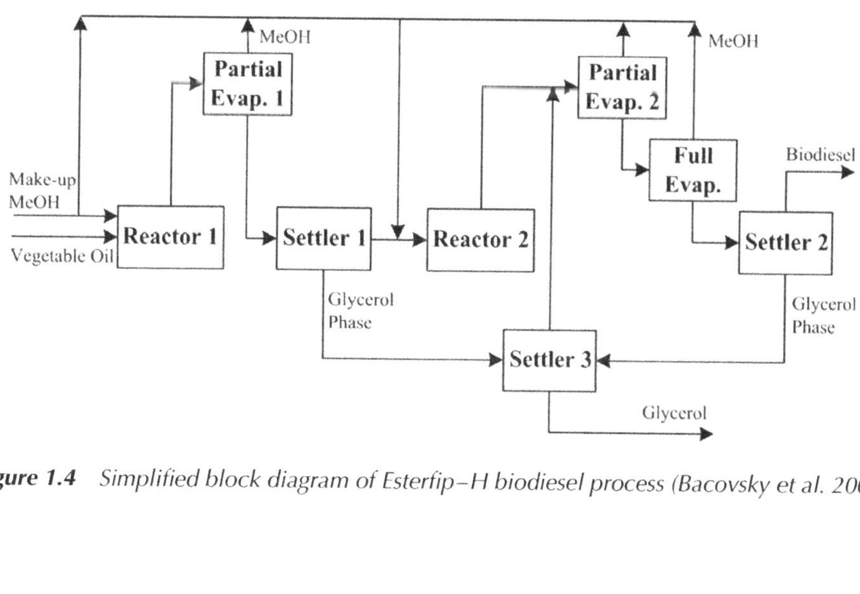

Figure 1.4 *Simplified block diagram of Esterfip–H biodiesel process (Bacovsky et al. 2007)*

1.4 Integrated lignocellulose biorefineries

Integrated lignocellulose biorefineries (ILCB) or integrated forest biorefineries (IFBR) are comprehensive approaches that make full use of all the components of biomass feedstock to produce heat (steam) and power, biofuels, cellulose fibers for pulp and paper, and multiple products (chemicals, polymers or materials). Figure 1.5 below is the general ILCB, modified from the diagram of the advanced pulp mill-based integrated forest biorefinery (IFBR) (Huang *et al.* 2010). The ILCB include not only the pulping process for pulp and paper, but also the following processes that could make value-added coproducts:

- separation of phytochemicals from woody biomass at mild conditions (optional);
- extraction of hemicellulose prior to pulping for biofuels and chemicals;
- extraction of lignin and chemicals (e.g., acetic acid) from spent pulping liquors;
- gasification of biomass including spent pulping liquor and forest residues and agricultural residues, for heat and power, syngas production, and syngas synthesis into fuels an chemicals such as methanol, DME, diesel, gasoline, and mixed alcohols;
- the extracted hemicellulose, combined with isolated short fiber, is hydrolyzed to monosugars, which are then fermented to sugar-based biofuels (e.g., ethanol, butanol), building blocks (e.g., lactic acid, succinic), and chemicals, depending on the microorganism used.

For changing a current pulp mill to an ILCB, the additional incremental costs for realizing a commercial biorefinery can be minimized by fully utilizing the existing infrastructure. Modification of the modern day pulp mills into ILCB presents an excellent opportunity to produce, in addition to valuable cellulose fiber, co-products include fuel grade ethanol/butanol and additional energy, thus resulting in increased revenue streams and profitability and potentially lower the greenhouse gas emissions (Huang *et al.* 2010).

Separation and purification technologies also play a significant role in the ILCB. Pre-extraction of value-added chemicals such as phytochemicals and extraction of hemicellulose prior to pulping,

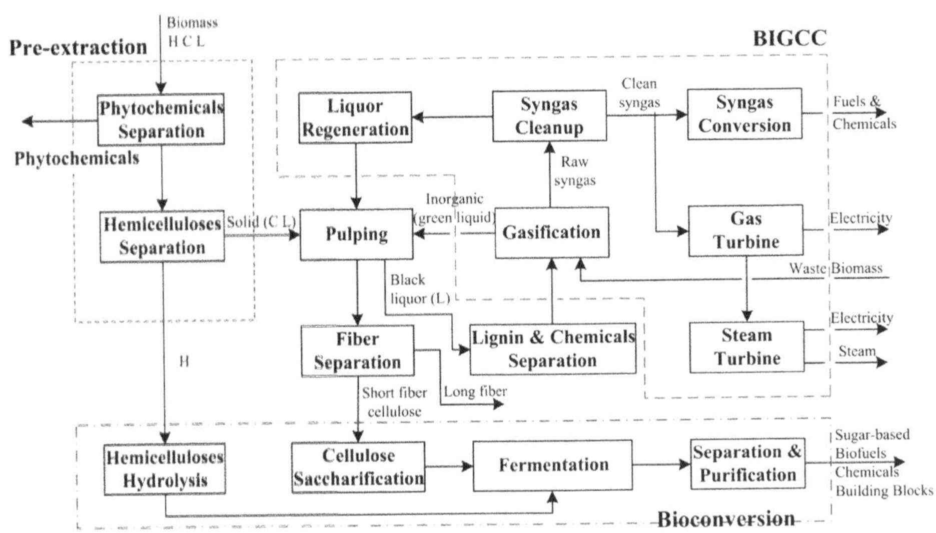

Figure 1.5 *Block diagram of the general ILCB*

separation of valuable chemicals from biomass prehydrolysis liquor, syngas cleanup, purification of reactants, for example purification of glycerol from biodiesel production for production of intermediates such as succinic acid, and separation and purification of products (ethanol, butanol, lactic acids etc.) are only some of the examples. Generally, the capital and operating costs of separation and purification processes usually account for a large fraction (about 20–50%) of the total capital and operating costs of biorefineries. Significant improvement in of separation and purification technologies can significantly reduce the overall production costs.

1.5 Separation and purification processes

As discussed earlier, in each of the multitude of lignocellulose based biorefinery applications, in addition to the biomass conversion processes, separation and purification of the biomass components and the products streams and their full integration with the overall process is of utmost importance. In many instances this can be the single biggest factor influencing the overall success and commercialization of biorefineries. Given the significance and importance of this area, separation and purifications technologies and their applications in biorefineries is the focus of this book.

The following section presents a brief introduction and outlines the challenges and opportunities in many of the plausible separation and purification technologies in biorefineries. Each of the separation and purification technologies is then the focus of the remainder of the book and they are dealt in greater detail in each of the following chapters.

1.5.1 Equilibrium-based separation processes

1.5.1.1 Absorption

Absorption is often used for separation of particles or desired gas components from a gas mixture into a liquid solvent phase. In biorefineries, absorption is commonly used for removal of acid gases such as H_2S and CO_2 from syngas prior to synthesis of syngas into methanol and diesel, and so forth. There are two major type of absorption: physical and chemical absorption. Physical absorption is commercially used to remove acid gas such as CO_2 and H_2S from syngas in the production of hydrogen, ammonia and methanol. The most well-known physical absorption processes are the Selexol process using the dimethyl ethers of polyethylene glycol at relatively high pressure (2.07–13.8 MPa) and the Rectisol process using cold methanol at −40 °C and 2.76–6.89 MPa for separating H_2S and CO_2 (Kohl and Nielsen 1997). Other major absorption processes include the Purisol process using N-methyl-2-pyrollidone, and the FLUOR process using propylene carbonate (Olajire 2010).

Currently, both the chemical absorption based on aqueous methyldiethanolamine (MDEA) and the Selexol process are selected in commercial IGCC (Integrated Gasification Combined Cycle) facilities for removal of acid gases. While physical absorption processes can meet the stringent sulfur cleanup required by catalytic synthesis of syngas, they are more expensive than the MDEA-based chemical absorption. On the other hand, although the Selexol process by itself is more expensive than an MDEA process, the total acid gas removal (AGR), sulfur recovery process, and tailgas treating process system, based on Selexol, could be more cost effective than the system based on MDEA, especially if the syngas pressure is high and deep sulfur removal (e.g., to 10–20 ppmv) is required. The Rectisol process is capable of deep sulfur removal, but it is the most expensive AGR process. Hence, Rectisol is generally used for chemical synthesis of syngas where very pure syngas is required (Korens, Simbeck and Wilhelm 2002). An overview of CO_2 separation has recently been presented elsewhere (Olajire 2010).

1.5.1.2 Distillation

Distillation is a commonly used separation method in chemical and biochemical industries. There are different distillation processes for liquid mixture separation: ordinary distillation, azeotropic distillation, extractive distillation. For separation and dehydration of ethanol from fermentation broth, it is impossible to separate ethanol–water in a single distillation column because ethanol forms an azeotropic mixture or azeotrope, at 95.6% by weight with water at a temperature of 78.15 °C. The separation and dehydration of ethanol usually consists of two steps: the ordinary distillation is firstly used to obtain approximately 92.4 wt% ethanol from the dilute broth, azeotropic distillation, extractive distillation, liquid–liquid extraction, and adsorption and so forth are then applied for further dehydration. The major distillation processes including ordinary distillation, azeotropic distillation, and extractive distillation potentially used in biorefineries has been reviewed taking ethanol separation and dehydration as example (Huang *et al*. 2008).

Molecular distillation (MD) is a special distillation process that is carried out under high-vacuum conditions and is suitable for the fractionation and separation of chemicals from pyrolysis bio-oils (Wang *et al*. 2009; Guo *et al*. 2009, 2010). Under these conditions the mean free path length of the molecules to be separated is generally longer than the distance between the evaporation surface and the condenser surface. It can also be used for purification of biodiesel obtained by esterification of cooking oil with methanol (Wang *et al*. 2010), and isolating heat sensitive phytochemicals from biomass or biomass extract (Huang and Ramaswamy 2012). As described before, the properties of pyrolysis liquid can be improved by hydrogenation and/or HDO. On the other hand, pyrolysis bio-oil is a valuable source for the production of chemicals, such as alcohols, aldehydes, ketones, acids, phenolics and sugars. Separation of these chemicals, for example the acid compounds for refining pyrolysis oil (Guo *et al*. 2009) and phenolic fraction for production of pharmaceuticals, adhesives, and specialty polymers (Žilnik and Jazbinšek 2011) from bio-oil, is an alternative option. Wang *et al*. (2010) explored the purification of crude biodiesel with molecular distillation and showed that it resulted in the high yield of FAME (up to 98.32%). In order to enhance the condensation efficiency of molecular distillation, traditional vacuum distillation was firstly used to remove most of the water in the crude bio-oil. The resulting bio-oil was then fractionated by molecular distillation. Results indicated that the distilled fractions were rich in low molecular weight carboxylic acids and ketones; the residual fraction hardly contains water and it has improved heating values of 21.29 MJ/kg and 22.34 MJ/kg for two operating conditions (80 °C, 1600 Pa and 80 °C, 340 Pa), respectively.

Steam distillation is a conventional commercially utilized process for isolating volatile organic compounds such as essential oils that are sensitive to high heat from plant material. Different from the earlier separation methods, steam distillation is used for direct separation of the desirable components from solid biomass feedstock, not liquid mixture. In this method, steam is introduced by heating water, and passed through the oil-containing plant material. With the addition of steam, the oil–water mixture boils at a lower temperature ($<$100 °C at 1 atm) allowing heat-sensitive compounds to be separated with less decomposition. Steam distillation is suitable for extracting light components whose vapour pressures are relatively high ($\geq$1.33 kPa at 100 °C). For components whose vapour pressures at 100 °C are between 0.67 kPa and 1.33 kPa, superheated steam is used for the distillation. Steam distillation can be used to separate light components of essential oils and bioactive compounds from biomass (Huang and Ramaswamy 2012), and this could bring value-added co-products for biorefineries.

Chapter 2 by Lei *et al*. provides additional details on distillation and its applications in biorefineries.

1.5.1.3 Liquid-liquid extraction

Liquid-liquid extraction (LLE), or solvent extraction, is a conventional separation process where one or more mixed solvents are used to extract desirable component from the feed liquid phase to the solvent phase. Liquid-liquid extraction can be used for separating biofuels and chemicals from dilute liquid mixtures—for example, extracting bioalcohols (Simoni *et al*. 2010) and carboxylic acids (Bressler and Braun 1999; Aşçi and İnci 2012; Oliveira *et al*. 2012) from their fermentation broths, extracting inhibitors (compounds toxic to microorganisms used for fermentation) from biomass hydrolyzates (Grzenia, Schell, and Wickramasinghe 2011), and removing impurities (soap, methanol, and glycerol) in biodiesel from used cooking oils (Berrios *et al*. 2011). For example, Chapeaux *et al*. (2008) and Simoni *et al*. (2010) studied the LLE of 1-butanol from water using ionic liquids (ILs) as solvents. Experimental results show that some ILs have high distribution coefficients and selectivities of 30 to 300. 1-hexyl-3-methylimidazolium tris(pentafluoroethyl) trifluorophosphate shows especially good extraction capability with the distribution coefficient of 5 and the selectivity of 300 for 5 wt% 1-butanol aqueous mixture.

Organic acids such as succinic, maleic, lactic, and itaconic acids can be extracted from their fermentation broths by amine extractants, which is based on reactive extraction. For instance, extraction of itaconic acid from aqueous solutions has been studied by six different solutions of trioctylamine (TOA)–tridodecylamine (TDA) mixtures and one of the following diluents: dimethyl phthalate (DMP), methyl isobutyl ketone (MIBK), 2-octanone, 1-octanol, cyclohexyleacetate (CHA), and 1-decanol. The maximum itaconic acid recovery was 98.39% with DMP and 3.14 mol L^{-1} initial concentration of the TOA–TDA mixture (Aşçi and İnci 2012). In addition, organic acids, particularly acetic acid, are reported from the aqueous fraction of the pyrolysis liquid using a long chain aliphatic tertiary amine. The best results were obtained with TOA in 2-ethyl-hexanol (40 wt%, as diluent) with 84% acetic acid recovery at equilibrium conditions (room temperature). Formic acid and glycolic acid present in the feed were also co-extracted with 92% and 69% extraction efficiencies respectively, as well as relatively non-polar compounds such as substituted phenolics and ketones (Rasrendra *et al*. 2011). Furthermore, the extraction of succinic acids, l-lactic, and l-malic from fermentation broths and dilute waste water using ionic liquid as extractant was investigated, and the results show that phosphonium-based ILs can be better extractants than the organic solvents traditionally used (Oliveira *et al*. 2012).

Extraction of acetic acid from biomass hydrolysates using mixed solvent consisting of 85% octanol and 15% Alamine 336 (w/w) for the purpose of inhibitor removal or detoxification, extraction of 5-hydroxymethylfurfural (HMF) from an aqueous reaction solution obtained by acid dehydration of six carbon sugars for production of HMF, using MIBK as extractant, and the extraction of glycerol from 2-butanol into an aqueous phase during the manufacture of biodiesel have also been studied (Grzenia *et al*. 2011).

Liquid-liquid extraction of the key chemicals from bio-oils have been investigated (Vitasari, Meindersma, and de Haan 2011; Žilnik and Jazbinšek 2011). For instance, different aqueous extractions and extraction with combined use of a hydrophobic-polar solvent and antisolvent for extraction of fast pyrolysis bio-oils were studied. Results show that alkali solution was more efficient than water or aqueous NaHSO3 solution; MIBK was shown to be the most efficient solvent for extraction of phenolics from bio-oil in combination with 0.1 M or 0.5 M aqueous NaOH solution, followed by butyl acetate (Žilnik and Jazbinšek 2011).

Chapter 3 by Hu *et al*. provides additional details on liquid-liquid extraction and its applications in biorefineries.

1.5.1.4 Supercritical fluid extraction

In the supercritical fluid extraction (SFE) process, a supercritical fluid is used to extract the valuable solutes from a solid matrix or a liquid mixture at its supercritical condition. $ScCO_2$ is the most commonly used supercritical fluid in the food, pharmaceutical, and chemical industries. Being non-polar, or hydrophobic, $ScCO_2$ is very suitable for extracting hydrophobic constituents from biomass (Huang and Ramaswamy 2012). For example, some value-added phytochemicals such as pigments, phenolics, and carotenoids can be recovered from microalgae with $ScCO_2$ extraction. Phytochemicals from plants including other plants such as switchgrass and alfalfa have the potential to be used in pharmaceuticals, cosmetics, nutritional, and consumer products. Extraction of phytochemicals at mild conditions prior to biomass pretreatment could bring value-added co-products in addition to using biomass for producing biofuels, chemicals, and materials. This could help lower the overall production cost of the major products of biorefineries. In addition, lipid in microalgae can be extracted via ScCO2 extraction for biodiesel production (Halim *et al.* 2011; Soh and Zimmerman 2011). The extracted lipid in this case had a suitable fatty acid composition for biodiesel (Halim *et al.* 2011). Besides, the $ScCO_2$ extraction has a comparable efficiency in extracting lipids compared to the conventional solvent extraction such as hexane extraction, indicating potential energy benefits by avoiding conventional algal mass dehydration prior to extraction. In other words, $ScCO_2$ extraction is a promising procedure for extracting algae oil for biodiesel production (Soh and Zimmerman 2011). A brief review on $ScCO_2$ of phytochemials from biomass has been recently published (Huang and Ramaswamy 2012).

Chapter 4 by Mantell *et al.* provides additional details on super critical fluid extraction and its applications in biorefineries.

1.5.2 Affinity-based separation

Adsorption, ion exchange, and chromatography are the three conventional sorption processes where certain adsorbates are selectively transferred from the fluid phase to the surface of insoluble, rigid particles suspended in liquid in a vessel or packed in a column.

Both adsorption and ion exchange can be used for efficient removal of inhibitors from biomass hydrolysate. For illustration, the detoxification of sugarcane bagasse hydrolyzate to improve ethanol production by *Candida shehatae* NCIM 3501 was studied and comparisons were made between five detoxification methods: neutralization, overliming, activated charcoal, ion-exchange resins (IER), and enzymatic detoxification using laccase. Results show that ion exchange was most efficient in removing furans (63.4%), total phenolics (75.8%), and acetic acid (85.2%); activated carbon is the second best with 38.7, 57 and 46.8% removal of furans, phenolics and acetic acid, respectively (Chandel *et al.* 2007). In addition, adsorption and ion exchange can be used for product separation and purification. The adsorption for ethanol-water separation was previously reviewed (Huang *et al.* 2008). Obviously adsorption with different adsorbents can also be applied for separation and purification of other biofuels and chemicals, for example the dehydration of biobutanol with molecular sieve, which is similar to ethanol dehydration. Here are some examples of the application of adsorption and ion exchange in separation and purification of biofuels and chemicals. The raw biodiesel from esterification of used cooking oils contains several impurities: free glycerol, methanol, free fatty acids (FFA), soap, catalyst, metals, water and glycerides (Berrios and Skelton 2008). These impurities should be removed to improve the biodiesel quality to its standard specification. Biodiesel is traditionally purified by water washing, which introduces additional water leading to increased cost and production time. One alternative commercial process uses adsorption with magnesium silicate as adsorbent (Magnesol®). Using the Magnesol process, methanol can be efficiently removed (Berrios and Skelton 2008). Other research showed that adsorption (magnesium silicate and bentonite) can remove soap, methanol, and glycerol effectively (Berrios *et al.* 2011). Glycerol and free fatty acids (FFA) can also be removed efficiently from biodiesel with the adsorption process using silica gel as adsorbent (Yori *et al.* 2007; Manuale *et al.* 2011). Like adsorption, ion-exchange resin is a commercial process that can be used for purification of biodiesel (Berrios and Skelton 2008; Berrios *et al.* 2011), separation of carboxylic acids such as succinic acid (Zeikus *et al.* 1999), as well as purification of xylose from biomass prehydrolyzates (Vegas *et al.* 2005). Using the ion exchange process, glycerol and free fatty acids (FFA) can be efficiently removed from biodiesel (Berrios and Skelton 2008). Ion-exchange resin (Lewatit® GF202) was also applied for purification of used cooking oil biodiesel. Soap, methanol and glycerol removal were 52.2%, 98.8% and 20.2%, respectively. This resin has the advantage in that

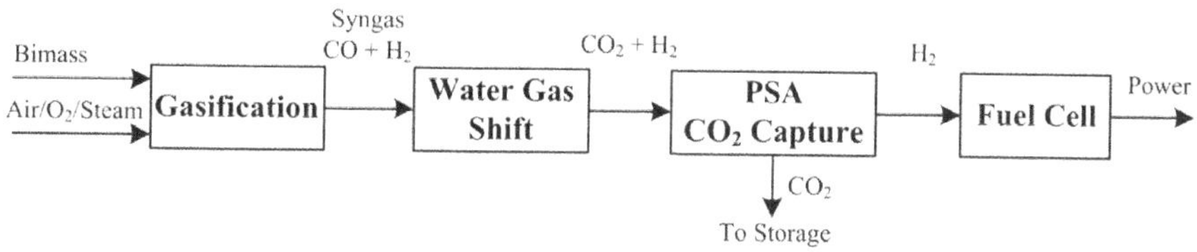

Figure 1.6 *Hydrogen production for fuel cell*

it can be regenerated for reuse, while other resins can be used only once (Berrios *et al*. 2011). Another application of ion exchange in lignocellulosic biorefineries is the purification of succinic acid where the ion exchangers are used for simultaneous acidification and crystallization (Zeikus *et al*. 1999).

Chapter 5 by Venkatesan provides additional details on adsorption and its applications in biorefineries. Chapter 6 by Berrios *et al*. provides additional details on ion exchange and its applications in biorefineries.

Pressure swing adsorption (PSA) can be used for hydrogen purification (Majlan *et al*. 2009; Lopes, Grande and Rodrigues 2011) and for capturing CO_2 (Ribeiro, Santos and Rodrigues 2010). One of the primary applications of PSA is for removal of CO_2 as the final step in the production and purification of hydrogen for use in biorefineries and in the production of ammonia, or the separation of CO_2 from biogas to increase the methane content. Figure 1.6 shows the block diagram of hydrogen production for fuel cell where PSA is used to capture CO_2 and purify H_2 for fuel cell.

Chapter 19 by Simo provides additional details on pressure swing adsorption and its applications in biorefineries especially using dehydration of ethanol as a case study.

1.5.2.1 Simulated moving-bed chromatography

Simulated moving-bed (SMB) chromatography is a continuous separation and purification technique that has better performance (higher throughput and less solvent requirement) than traditional batch chromatography. In the SMB process, a flow of liquid (mobile phase) moves countercurrent to a constant flow of solid (stationary phase), resulting in more efficient separation. SMB technology has been commercially used in pharmaceutical and specialty chemical manufacturing. It can also be applied for biofuels and chemicals separation in biorefineries. For instance, SMB can be used for purification of glycerol from biodiesel production. The sequential SMB chromatography, using the Ambersep BD50 resin, can extract glycerol with 99.5% purity in the extract stream. The raffinate stream contains the salts and other organic impurities including free fatty acids (Lancrenon and Fedders 2008). Similarly, a commercial SMB process using gel-type acidic ion-exchange resin beads was introduced to separate fatty acid salts and inorganic salts from the crude glycerol byproduct of the biodiesel production (Rezkallah 2010). Besides, the SMB chromatography has been proposed for purification of oligosaccharides made up of xylose and arabinose units (Ohsaki, Tamura and Yamaura 2003). In addition, a four-zone SMB chromatography was studied for isolating lactic acid from acetic acid, a major impurity in the fermentation broth of *L. rhamnosus* resulting in 99.9% purity and over 93% yield of lactic acid (Lee *et al*. 2004). More recently, the four-zone SMB system was investigated to separate sugars (glucose and xylose) and 1-ethyl-3-methylimidazolium acetate (EmimAc) from the biomass hydrolyzate where EmimAc, an ionic liquid, was used as the biomass pretreating agent for biomass hydrolysis. Glucose, xylose, and EmimAc were recovered at the yields of 71.38, 99.37 and 98.92% respectively (Mai *et al*. 2012). In summary, SMB chromatography could be efficiently applied for separation and purification of chemicals in biorefineries.

Chapter 7 by Wang *et al*. provides additional details on simulated moving bed and its applications in biorefineries.

1.5.3 Membrane separation

Membrane separation technologies have been widely researched for biofuel separation in biorefineries (Huang *et al*. 2008; He *et al*. 2012).

Electrodialysis (ED) is a process used to extract ions selectively from one solution through ion-exchange membranes to another solution based on electric potential difference. It can remove low molecular weight ionic components efficiently from a liquid mixture. Its applications include seawater desalination and salt production, drinking water production, desalting of glycol, glycerol purification, and organic acid production, and so forth.

Electrodialysis is commonly used for the separation of organic acids or carboxylic acids such as acetic acid and oxalic acid (Wang *et al*. 2011), citric acid (Wang, Wen and Zhou 2000; Wang *et al*. 2011), gluconic acid (Wang, Huang and Xu 2011), and succinic acid (Groot 2011) from their fermentation broths. An overview on the application of electrodialysis for production of organic acids has been presented (Huang *et al*. 2007). As an example, lactic acid can be produced by continuous fermentation with an integrated product recovery process based on bipolar membrane electrodialysis, as illustrated in Figure 1.7. In this process, conventional electrodialysis is used to concentrate the lactate salt, and then bipolar membrane electrodialysis is applied for the conversion of the lactate into lactic acid and base. The resulting lactic acid is purified by ion exchange, while the resulting base is recycled to the fermenter to control the pH-value (Strathmann 2010). This system requires a much smaller amount of ion-exchange resin in a final purification step compared to the conventional lactic acid production in a batch process where the lactic acid is isolated and purified mainly by ion-exchange resulting in a large volume of waste water with regeneration salts (Strathmann 2010).

Other potential similar applications of bipolar membrane electrodialysis include the recovery of gluconic acid from sodium gluconate, ascorbic acid from sodium ascorbate, and succinic acid from sodium succinate.

Recently, membrane technologies have been widely studied for biorefineries. Microfiltration (0.050–10 μm), ultrafiltration (1–100 nm), or nanofiltration (<2 nm) can be selected for separation of biofuels and chemicals, depending on the molecules to be separated.

Membrane can be used for removal of inhibitors such as acetic acid. The bioconversion of lignocellulosic biomass usually involves conversion (hydrolysis) of cellulose and hemicellulose to monosugars, followed by fermentation of the monosugars into the desired products. Acetic acid is liberated from acetate in biomass during biomass pretreatment or hemicellulose hydrolysis. As acetic acid is an inhibitor to the subsequent fermentation, it must be removed from the hydrolyzate prior to fermentation. Wickramasinghe and Grzenia (2008) showed that anion exchange membrane can efficiently remove acetic acid from biomass

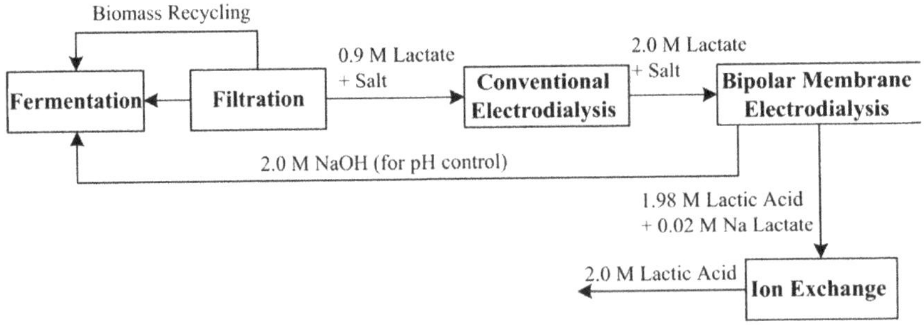

Figure 1.7 Block flow diagram of the lactic acid production process with integrated bipolar membrane electrodialysis (Strathmann 2010)

hydrolysates, and it exhibited better separation performance in terms of throughput and product loss compared to anion-exchange resin.

Membrane technologies can be applied for algal biomass harvesting. Algal biomass harvesting is a key step and a big challenge for microalgae biodiesel production because the cells are small (3–30 μm) and fragile, their density is close to water leading to difficulty in separation by gravity, and it is a highly diluted aqueous slurry (Ríos *et al.* 2012). Microfiltration and ultrafiltration can be applied for harvesting algal biomass, offering several advantages such as mild operating conditions without using additional chemicals (Rossignol *et al.* 1999; Rossi *et al.* 2004; Rossi *et al.* 2005). Ríos *et al.* (2012) used a pH-induced flocculation-sedimentation as preconcentration for antifouling, followed by dynamic microfiltration. The preconcentration step concentrated about ten times at a relatively low cost and enlarged the particle size for dynamic microfiltration. The pilot experiments at optimized conditions resulted in concentration factor up to 200 and permeability up to 600 $L/h/m^2/bar$ (Ríos *et al.* 2012).

Chapter 21 by Cooney provides additional details on oil extraction from algae as a case study in biorefinery applications.

Membrane processes can be used for separating hemicelluloses from biomass hydrolyzates or process water of pulp mills. For example, nanofiltration (NF) is suitable for separating hemicelluloses of small molecular weights from hydrolyzates. Biomass pretreatments such as alkaline process usually produce hemicelluloses with smaller molecular weights, compared to other pretreatments such as hot water pretreatment. In this case, nonfiltration, is much better than ultrafiltration for separating hemicelluloses from hydrolyzates (Schlesinger *et al.* 2006). For isolating hemicelluloses from alkaline process liquors containing 200 g/l NaOH, for instance, the hemicelluloses of molar mass over 1000 g/mol are almost retained. In addition, two of the membranes with the nominal molecular weight cutoff (MWCO) of 200–300 and 200–250 g/mol, respectively can retain up to 90% of hemicelluloses, while the tight ultrafiltration membrane with MWCO of 2000 g/mol retain less than 70% hemicelluloses (Schlesinger *et al.* 2006). Ali *et al.* patented an alkaline treatment system for recovering hemicelluloses where prefiltration units with a screen size of 400–650 mesh, followed by one NF membrane was able to retain compounds with a molecular weight of about 200 and higher (Ali *et al.* 2005). Besides, ultrafiltration (UF) can be used for isolating the hemicelluloses or the hemicellulose galactoglucomannan from process water from a thermomechanical pulp mill (Persson, Jönsson, and Zacchi 2005; Persson and Jönsson 2010). Different hydrophobic and hydrophilic UF membranes with 1–5 kDa cutoff were studied and compared for separating hemicelluloses from the process water of the thermo-mechanical pulping of spruce. Results show that the hydrophilic membrane C005F, from Microdyn Nadir GmbH with cut-off 5 kDa, had the highest flux and the most efficient separation of product and contaminants (salts and monosugars). The flux was 140 $L\,m^{-2}{\cdot}h^{-1}$ at 0.8 MPa and 40 °C. The retention of hemicelluloses and monosugars were 90% and 3% respectively (Persson, Jönsson and Zacchi 2005). In addition, hydrophobic membranes were fouled by hydrophobic molecules such as lignin and resins, while hydrophilic membranes had no fouling (Persson, Jönsson, and Zacchi 2005).

Membrane can be applied for lignin recovery from pulp mill waste liquors (Jönsson, Nordin and Wallberg 2008; Jönsson and Wallberg 2009) *and biomass prehydrolysis liquor* (Alriols *et al.* 2010). Lignin constitutes up to 30% of biomass. Effective use of lignin is critically important for biorefineries. There are three categories of opportunities for lignin use. First, power—fuel—syngas, i.e., for power by combustion, and for fuel and syngas via gasification (near term); Second, macromolecules such as carbon fiber, polymer modifiers, adhesives and resins (medium-term opportunities), and, third, aromatic chemicals such as BTX chemicals (benzene, toluene, and xylene), phenol, lignin monomer molecules, and oxidized lignin monomers including vanillin and vaillic acid (long term) (Holladay *et al.* 2007). Lignin recovery is necessary for the second and the third categories of lignin use. Lignosulphonates have long been separated by UF from spent liquor in sulfite pulp mills. The isolation of lignin from kraft black liquor has often been extracted by precipitation. This requires changing the pH or the liquor temperature, which could be less cost effective. For this reason, the membrane method has been studied for lignin recovery (Jönsson, Nordin, and Wallberg 2008; Jönsson and Wallberg 2009). For instance, a hybrid UF/NF process was used for separating lignin from the black liquor withdrawn before the evaporation unit. UF was firstly used to retain most hemicelluloses and large molecules. The resulting permeate having 100 g/l lignin with lean or poor hemicelluloses was then concentrated by NF, leading to the product stream (retentate) of 165 g/l lignin (Jönsson, Nordin and Wallberg 2008). In addition, the ethanol organosolv pre-treatment coupled with membrane UF was utilized for fractionation and separation of lignin and other fractions from non-woody biomass, *Miscanthus sinensis*. The organosolv process allowed fractionation of the biomass feedstock into different fractions of products: cellulose hemicellulose-derived sugars and lignin. Ultrafiltration using tubular ceramic membranes with different cutoffs (5, 10 and 15 kDa) was used to obtain specific molecular weight lignin fractions (Alriols *et al*. 2010). Ultrafiltration with similar membranes was applied for recovering lignin from black liquor from the alkaline pulping of the *Miscanthus sinensis* (7.5% NaOH, 90 min and 90 °C) (Toledano *et al*. 2010a). In comparison with selective precipitation, UF has the advantages in that its lignin has higher purity (contains less contaminants such as hemicelluloses), and the UF process allowed controlling the molecular weight of the obtained fractions by selecting the right cutoff of the membrane (Toledano *et al*. 2010b).

Chapter 18 by Zacchi *et al*. provides additional details on cellulosic bioethanol production as a case study in biorefineries.

Chapter 20 by van Walsum provides additional details on separation and purification processes pertaining to lignocellulose hydrolyzates and their applications in biorefineries.

Membrane techniques can be utilized for biodiesel separation and purification. Conventional technologies used for biodiesel separation, such as gravitational settling, decantation, filtration, and biodiesel purification such as water washing, acid washing, and washing with ether and absorbents, have proven to be inefficient and less cost effective. The membrane technology shows great promise for the separation and purification of biodiesel (Atadashi, Aroua, and Aziz 2011).

Membrane techniques can be used for separation of liquid mixtures, for example, carboxylic acids from dilute solutions. Lactic acid is widely used in food and chemical industries. It can be manufactured by either chemical synthesis or carbohydrate fermentation. The high cost of the traditional lactic acid production by lactose fermentation is associated with the separation steps required for food-grade lactic acid. In order to reduce costs, different separation techniques such as reactive extraction, membrane technology, ion exchange, electrodialysis and distillation have been studied for lactic acid separation (Gonzalez *et al*. 2008; Pal *et al*. 2009). Some researches have shown that NF can be used to remove lactic acid from the fermentation broths for improving the fermentation yield (Gonzalez *et al*. 2008; Umpuch *et al*. 2010). Nanofiltration and reverse osmosis membranes can also be applied for separation of carboxylic acids from aqueous fraction of fast pyrolysis bio-oils (Teella 2011). Another example is the application of membrane in separation and purification of ionic liquid solvents by NF (Abels *et al*. 2012).

Chapter 22 by Kamble *et al*. provides additional details on separation and purification technologies in biopolymer production processes.

Membrane technologies can be used for gas separation and purification. Separation of hydrogen, a clean energy carrier, is a good example. Hydrogen can be combusted in fuel cells and gas turbines with zero or near-zero emissions at a high efficiency (Berchtold *et al*. 2012). H_2 is also widely used in chemical industry, for example, for upgrading bio-oil via hydrotreating, and for ammonia synthesis for fertilizer. Hydrogen can be separated from syngas produced by biomass gasification (National Academy of Science 2004; U.S. Department of Energy 2007; Huang and Ramaswamy 2011) or biogas produced by dark fermentation of biomass carbohydrate using anaerobic bacteria in the dark (National Academy of Science 2004; Kovacs *et al*. 2006). Membrane gas separation technology are widely used to separate hydrogen from syngas or the biogas produced, to provide a high purity H_2 product (Ji, Feng and Chen 2009; Sánchez, Barreiro, and

Maroño 2011). For instance, a robust industrially viable polybenzimidazole (PBI)/stainless steel composite membrane was developed and evaluated for H_2 separation at elevated temperatures. The PBI composite membrane demonstrated exceptional long-term thermo-chemical stability and excellent separation performance for H_2 over the other syngas components. The H_2 permeance and H_2/CO_2 selectivity of the composite membrane for simulated dry syngas were 7 GPU (~88 barrer) and 47, respectively (Berchtold *et al.* 2012). Among the microporous membranes, the X-ray amorphous metal oxide membranes, mainly silica, and zeolite membranes, especially the MFI-type membranes (silicalite-1 and ZSM-5), are the most common ones (Caro and Noack 2010). In addition, membrane technology can also be utilized for CO_2 separation from synthesis gas, natural gas or biogas (Zhao *et al.* 2008; Park *et al.* 2010; Sandström, Sjöberg, and Hedlund 2011).

Chapter 8 by Jonsson *et al.* provides additional details on membrane separation processes of microfiltration, ultrafiltration, and diafiltration and their applications in biorefineries.

Chapter 9 by Nisstrom *et al.* provides additional details on membrane separation processes of nanofiltration and its applications in biorefineries.

Membrane pervaporation is one of the most promising technologies for molecular-scale liquid/liquid separations in biorefinery, petrochemical, pharmaceutical industries, and so forth. It is highly selective, economical, safe and ecofriendly (Jiang *et al.* 2009). It has been widely studied for removal of inhibitory products from fermentation broth (Huang *et al.* 2008). For instance, a continuous cultivation of *Clostridium acetobutylicum* ATCC 824 is described using a two-stage design to mimic the two phases of batch culture growth of the organism. A hydrophobic pervaporation unit was coupled to the second fermentor containing the highest solvent titers. This *in situ* product recovery technology efficiently decreased butanol toxicity in the fermentor while the permeate was enriched to 57–195 $g\,L^{-1}$ total solvents depending on the solvent concentrations in the fermentor. By the alleviation of product inhibition, the glucose concentration could be increased from 60 to 126 $g\,L^{-1}$ while the productivity increased concomitantly from 0.13 to 0.30 $g\,L^{-1}\,h^{-1}$. The continuous fermentation was conducted for 1172 h during which the pervaporation was coupled to the second fermentor for 475 h with an average flux of 367 $g\,m^{-2}\,h^{-1}$. The energy consumption was calculated for a 2 wt.% *n*-butanol fermentation broth and compared with the conventional process (Hecke *et al.* 2012).

Chapter 10 by Chung *et al.* provides additional details on membrane pervaporation and its applications in biorefineries.

1.5.4 Solid–liquid separation

1.5.4.1 Conventional filtration

Conventional filtration is a mature, commercially available solid–liquid separation technology. It has been widely used in chemical and biochemical industries. With the solid biomass as starting feedstock for producing biofuels, chemicals and materials, the biorefineries involves a numberof solid–liquid separation tasks, such as separation of prehydrolyzate slurry and post-distillation slurry. Therefore, selection and/or design of efficient, cost-efffective filtration processes are equally important for improving the overall process performance.

Chapter 12 by Ramarao *et al.* provides additional details on conventional filtration and its applications in biorefineries.

1.5.4.2 Solid–liquid extraction

Solid–liquid extraction (SLE) is the process where a solvent or solvent mixture is used to extract valuable compounds from the solid matrix of feedstock. The SLE technologies mainly include conventional solid–liquid extraction, ultrasound-assisted extraction, microwave-assisted extraction, and pressurized subcritical liquid extraction. Biomass feedstock, such as woody and perennial plant materials, and microalgae, usually contains significant phytochemicals such as phenolics, terpenes, sterols, enzymes, polysaccharides, alkaloids, toxins, and pigments, depending on the biomass species. These phytochemicals are value-added co-products, which can be used in neutraceutical and pharmaceutical industries. In order to reduce the overall production cost of biofuels and chemicals from biomass, it is necessary to extract and separate bioactive compounds or phytochemicals as value-added co-products prior to or during biomass conversion (Huang and Ramaswamy 2012). Phytochemicals from plants are usually present in very dilute concentrations. The heat-sensitive properties of phytochemicals and the increased difficulty of SLE over LLE bring a great challenge for efficient separation of phytochemicals from such a dilute biomass matrix. Based on the recent review on the phytochemicals separation (Huang and Ramaswamy 2012), the SLE technologies can be effectively applied for selective isolation of phytochemicals from biomass feedstock.

Chapter 13 by Abidin *et al*. provides additional details on solid–liquid extraction and its applications in biorefineries.

1.5.4.3 Precipitation and crystallization

Pre-extraction of hemicelluloses from wood chips prior to pulping for production of value-added products has gained interests in development of integrated forest biorefinery (Huang *et al*. 2008; Al-Dajani and Tschirner 2008; Mao, Genco and Yoon 2008). Ethanol precipitation can be used for recovery of hemicelluloses (polysaccharides) from the pretreated hydrolyzates (pre-hydrolysis liquor) (Liu *et al*. 2011b) and the spent liquors from pulp mills (Liu *et al*. 2011a). Precipitation by acidification using carbon dioxide, or sulfuric acid can be applied for extracting lignin from kraft black liquor (Öhman *et al*. 2007a; Öhman *et al*. 2007b; Öhman *et al*. 2007c; Wallmo *et al*. 2009; Minu *et al*. 2012). For the separation of hemicelluloses and lignin from pre-hydrolysis liquor (PHL), lignin was firstly removed by acidification of PHL to a pH of 2, resulting in 47% lignin precipitation. The lignin precipitation could be further improved by adding polyethylene oxide and poly aluminum chloride, or ethyl acetate. The hemicelluloses was then precipitated and isolated by adding ethanol to the acidified PHL, at a volumetric ratio (ethanol/PHL) of 4 to 1 (Liu *et al*. 2011a).

Precipitation and crystallization can also be used for separation and purification of succinic acid from its fermentation broth. As an example, the broth liquor in a fed batch bioreactor using *Actinobacillus succinogenes* BE-1 as biocatalyst has the concentrations of succinic acid, formic acid, lactic acid and acetic acid of 97.8 g/L, 23.5 g/L, 5.1 g/L and 17.4 g/L respectively. By controlling the fermentation broth at 4 °C and pH <2, succinic acid was easily and selectively crystallized and isolated with 70% yield and 90% purity, while the by-products lactic acid, acetic acid and formic acids were miscible in the solution. In comparison, succinic acid isolated by the traditional calcium precipitation coupled with ion exchange adsorption, had 52% yield and 92% purity (Li et al. 2010). Huh *et al*. studied the production of the highly purified succinic acid from the fermentation broth by recombinant microorganism, *Mannheimia succiniciproducens*. In their proposed method, the preseparation process such as reactive extraction and vacuum distillation was firstly used to effectively remove acid byproducts. The crystallization was then applied without adding any salts to produce highly purified succinic acid, with 99.8% purity and 73.1% yield (Huh *et al*. 2006).

1.5.5 Reaction-separation systems for process intensification

For reversible reactions such as esterification of plant oil with methanol for biodiesel production, the conversion rate and the product yield are limited by the reaction equilibrium. Integration of a reversible reaction with separation into a reaction-separation unit allows removing the byproduct (water) simultaneously with reaction, shifting the reaction equilibrium toward the key product (ester). Thus, the reaction performance increases. For many fermentation processes the products themselves also inhibit the fermentation processes. Thus, the product concentration and yield, as well as the substrate concentration, are limited to the low levels, leading to low fermentation performance. To get higher product concentration and yield and allow using higher concentration of the substrate, the fermentation can be integrated with separation into a system so that the product can be removed simultaneously with fermentation. This can eliminate or reduce significantly the product inhibition and hence increase the product yield and concentration.

The common reaction-separation systems include hybrid reaction-membrane separation (membrane filtration, membrane electrodialysis, membrane pervaporation, or membrane distillation), fermentation-extraction (extractive fermentation), reactive distillation, and absorptive distillation systems.

1.5.5.1 Reaction–membrane separation systems

Reaction–membrane separation systems include membrane bioreactors, (chemical) membrane reactors, bioreactor-membrane pervaporation, bioreactor-membrane distillation, and so forth. Membrane bioreactors can be used for biodiesel production. For example, Dube *et al*. (2007) developed a membrane reactor that removed unreacted vegetable oil from the FAMEs product after transesterification, yielding high-purity biodiesel. A related review has been presented by Atadashi, Aroua and Aziz (2011).

In addition, fermentation-membrane pervaporation system has been widely investigated. Taking a hybrid reaction–membrane pervaporation system for butanol production such as two stage continuous cultivation of *C. acetobutylicum* ATCC 824 coupled with a hydrophobic pervaporation unit using PDMS composite membranes was employed for ABE production. With this *in situ* product-recovery technology, the product (butanol) inhibition decreased significantly in the fermentor. Correspondingly, the glucose concentration increased from 60 to 126 $g\,L^{-1}$, the productivity increased from 0.13 to 0.30 $g\,L^{-1}\,h^{-1}$, and the permeate was enriched to 57–195 $g\,L^{-1}$ total solvents depending on the solvent concentrations in the fermentor (Hecke *et al*. 2012).

The fermentation–bipolar membrane electrodialysis system for succinic acid production is another case of a reaction–membrane separation system. In this process, the fermentation is neutralized with sodium hydroxide, forming soluble sodium succinate during the fermentation. The whole broth is filtered with a microfiltration unit to separate the cells and large insoluble particles from the succinate broth. The filtered sodium succinate is fed to a batch desalting electrodialysis unit, where the ionic species are separated from the non-ionic ones (sugars) and molecules with large molecular masses. The sodium succinate solution is then fed to a batch bipolar membrane electrodialysis unit where the ionic species are converted to their equivalent acid and base forms and separated. Sodium ions are transported across the cation membrane and associate with the hydroxyl ions to form sodium hydroxide, which is reused for fermenter neutralization. After succinic acid is purified (99.5%), a further distillation is required to purify acetic acid to 99.9% (Luo *et al*. 2010b).

Chapter 11 by Izquierdo-Gil provides additional details on membrane distillation and their applications in biorefineries.

Chapter 14 by Cabral *et al*. provides additional details on membrane bioreactors and their applications in biorefineries.

1.5.5.2 Extractive fermentation (Reaction–LLE systems)

Extractive fermentation, a combination of fermentation and liquid-liquid extraction, has been widely investigated. As an example, five non-ionic surfactants (Triton X 114, L64, L62, L62LF, and L61) were examined as extractants for extracting butanol from the fermentation broth in extractive acetone-butanol (AB) fermentation using *Clostridium pasteurianum*. Biocompatibility tests using 3% (vol.) surfactants showed that

L62, L62LF, and L61 did not show inhibition to AB production in 72-h fermentation using *C. pasteurianum*, while L64 reduced the AB yield and Triton X 114 inhibited the AB production. The results showed that L62 is a good extractant for isolating butanol from the fermentation broth, with the partition coefficient of 3–4 for butanol, and it significantly enhanced the butanol production with a butanol yield of 225% higher than the control using 6% (vol.) L62 (Dhamole *et al.* 2012).

However, extractive fermentation with *in situ* product removal, most previously researched of this kind, may not be suitable for large-scale production due to slow mass transfer into solvent phase, formation of emulsions through agitation, cell inhibition by solvent (interface toxicity), loss of cells at interfaces, difficult process control, and so forth (Kraemer *et al.* 2011). For this reason, fermentation integrated with external product removal in an extraction column with a recycle of product-lean broth was proposed (Figure 1.8), and mesitylene was identified as novel solvent with excellent properties for ABE extraction from the fermentation broth by using the computer-aided molecular design (Kraemer *et al.* 2011). This hybrid process allows using the toxic-to-cells solvents having a very low solubility in water and high extraction performance.

Membrane-assisted solvent extraction (membrane extraction, or perstraction), can also be utilized for recovery and separation of organic acids (Schlosser, Kertész, and Marták 2005), biofuels, and other chemicals. Coupling membrane-assisted extraction with fermentation is an efficient process-intensification approach. Non-dispersive membrane extraction has been used to detoxify corn stover biomass hydrolysates pretreated using dilute acetic acid. Ethanol yields for hydrolysates detoxified using an organic phase consisting of 15% Alamine 336 in oleyl alcohol are about 10% higher than hydrolysates detoxified using ammonium hydroxide treatment. The results of this study indicate that membrane extraction could be a feasible unit operation for detoxification of biomass hydrolysates. Unlike many current detoxification processes, acetic acid is also removed. Since membrane based processes are modular, scale-up is straightforward. The commercial viability of membrane extraction will depend on selection of an optimal organic phase diluent (Grzenia, Schell, and Wickramasinghe 2012). Besides, membrane-assisted extractive (MAE) fermentation of acetone–butanol–ethanol (ABE) by *Clostridium saccharoperbutylacetonicum* N1-4 using a polytetrafluoroethylene (PTFE) membrane and 1-dodecanol was studied. The membrane separates the aqueous phase from the organic phase allowing the use of the toxic extractant 1-dodecanol with high distribution coefficients in extraction of butanol without contact and toxifying the microorganism. Compared to conventional batch fermentation, MAE–ABE fermentation with 1-dodecanol as a extractant decreased butanol inhibition and increased glucose consumption from 59.4 to 86.0 g/L, and total butanol production increased from 16.0 to 20.1 g/L. The maximum butanol production rate increased from 0.817 to 0.979 g/L/h. The butanol productivity per membrane area was remarkably high with this system—78.6 g/L/h/m^2 (Tanaka *et al.* 2012).

Chapter 15 by Yang *et al.* provides additional details on extractive fermentation and its applications in biorefineries.

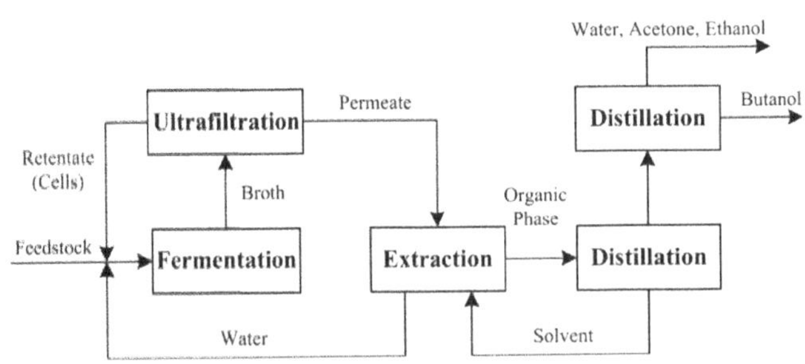

Figure 1.8 *Simplified block diagram of hybrid fermentation-external LLE process*

1.5.5.3 Reactive distillation

Reactive distillation, a combination of reaction and distillation in one unit, is a process intensification technique that can be applied successfully to produce biodiesel (Kiss, Dimian, and Rothenberg 2008), succinate ester (Orjuela *et al.* 2011), and upgrade the flash pyrolosys oil (Mahfud *et al.* 2007). By combining reaction and separation into a single unit, one can shift the reaction equilibrium toward the key product (ester) by continuous removal of byproduct (water), instead of using an excess of reactant. Rigorous process simulations show that combining metal oxide catalysts with reactive distillation technology is a feasible and advantageous solution for biodiesel production (Kiss, Dimian, and Rothenberg 2008). Orjuela *et al.* investigated the esterification of mixtures of succinic acid and acetic acid from fermentation of biomass carbohydrates with ethanol in a continuous reactive distillation system. The experimental results show that the conversions of both succinic acid and acetic acid were close to 100%; succinate ester (diethyl succinate) was separated as bottom products with 98% purity, and ethyl acetate was recovered in the distillate (Orjuela *et al.* 2011). Mahfud *et al.* studied the upgrading of flash pyrolysis oil in a reactive distillation using a high boiling alcohol such as n-butanol and a solid acid catalyst at 323–353 K under reduced pressure (<10 kPa). Results demonstrate that the water content of the pyrolysis oil reduced significantly. Using n-butanol and the solid acid Nafion SAC13, the product properties of the upgraded pyrolysis oils, particularly the heating value and the acidity were considerably improved (Mahfud, Melian-Cabrera, and Manurung 2007). Besides, reactive distillation can also be used for glycerol esterification with acetic acid for production of triacetin, which can be used as a biofuel additive (Hasabnis and Mahajani 2010).

Chapter 16 by Miller *et al.* provides additional details on reactive distillation and its applications in biorefineries.

1.5.5.4 Reactive absorption

Reactive absorption (RB), the combination of reaction and absorption in one unit, is another technology for process integration and intensification. RB can also be applied in the biodiesel production. An innovative technology based on RB using solid acid catalysts has been recently studied for biodiesel production. It was found that RB has many advantages over reactive distillation such as lower capital investment and operating costs due, higher conversion and selectivity, and no thermal degradation of products (Kiss and Bildea 2011). Also, this process eliminates all conventional catalyst-related operations, simplifying the production process (Kiss 2009). Details on the RB technology will be presented in a separate chapter of this book.

Chapter 17 by Kiss *et al.* provides additional details on reactive absorption and its applications in biorefineries.

1.6 Summary

This chapter attempted to provide an introduction and overview of the important biomass conversion processes including biochemical and thermochemical conversions and the potential separation and purification technologies in biorefineries. A number of representative value-added chemical building blocks and different biorefinery scenarios were introduced. Separation and purification technologies in current and future biorefineries were then reviewed. These included equilibrium-based processes such as absorption, distillation, liquid-liquid extraction, and supercritical fluid extraction; affinity-based separation such as adsorption, ion exchange, and simulated moving bed; membrane separation, solid–liquid extraction, as well as hybrid reaction-separation systems.

Separation and purification processes are one of the most important components of biorefineries. Generally, the costs of separation and purification processes account for 20–50% of the total costs of the biorefineries. Many biorefineries, especially the biochemical and biological approaches, have tremendous challenges in separation and purification due to number of factors including low feed concentration, product inhibition, and low product yield. There are number of significant challenges and opportunities in separation and purification in biorefineries including separation of phytochemicals from biomass, separation of biomass components including cellulose, hemicellulose, lignin and extractives, separation of fermentation inhibitors in hydrolyzates, separation and purification of different chemical species in the feed streams after initial pretreatment and hydrolysis, concentrating each of the species for varying end products and applications, separation of lignin and chemicals in spent pulping liquor, simultaneous removal of products which are also inhibitors during fermentation, integration of separation and purification technologies with bioprocessing, as well as downstream product separation and purification. These are just a few examples of challenges and opportunities that need to be addressed and solutions need to be developed and implemented for successful commercialization of biroefineries. They offer tremendous opportunities for research and development and it is imperative that both government and private industry continue to support research in this important area.

There are also significant opportunities for developing totally new approaches to separation and purification, especially suitable for biorefineries and their full integration in the overall biorefineries. Here are some examples of exciting potential approaches and opportunities. Ion exchange is the preferred approach for detoxification and will be still used in the future biorefinery because of its high detoxification efficiency, easy (continuous) operation and flexible combination of different anion and cation exchangers. Adsorption with a molecular sieve is efficient in breaking the ethanol–water or butanol-water azeotrope for biofuel dehydration (Huang *et al*. 2008). Membrane separation, especially ultrafiltration and nanofiltration represents a promising separation procedure for recovery of hemicelluloses from hydrolyzates and lignin from spent liquor. Hybrid separation systems such as extractive-fermentation and fermentation-membrane pervaporation are promising in removal of product inhibition, and hence are able to increase process performance. Fermentation, bipolar membrane electrodialysis, reactive distillation, and reactive absorption are suitable for separation of products obtained by esterification such as biodiesel production. Integrated bioprocessing, consolidated bioprocessing integrating pre-treatment, bioprocessing and separation and purification offer tremendously exciting new opportunities in future biorefineries.

References

Abdel-Rahman, M. A., Y. Tashiro, and K. Sonomoto. 2011. Lactic acid production from lignocellulose-derived sugars, using lactic acid bacteria: Overview and limits. *Journal of Biotechnology* 156 (4), 286–301.

Abels, C., C. Redepenning, A. Moll, T. Melin, and M. Wessling. 2012. Simple purification of ionic liquid solvents by nanofiltration in biorefining of lignocellulosic substrates, *Journal of Membrane Science* 405–406, 1–10.

Aden, A., M. Ruth, K. Ibsen, J. Jechura, K. Neeves, J. Sheehan, and B. Wallace. 2002. Lignocellulosic biomass to ethanol process design and economics utilizing co-current dilute acid prehydrolysis and enzymatic hydrolysis for corn stover. National Renewable Energy Laboratory report no. NREL/TP-510-32438.

Adsul, M. G., M. S. Singhvi, S. A. Gaikaiwari, and D. V. Gokhale. 2011. Development of biocatalysts for production of commodity chemicals from lignocellulosic biomass. *Bioresource Technology* 102, 4304–4312.

Akhtar, J. and N. A. S. Amin. 2011. A review on process conditions for optimum bio-oil yield in hydrothermal liquefaction of biomass. *Renewable and Sustainable Energy Reviews* 15, 1615–1624.

Al-Dajani, W. W. and U. W. Tschirner. 2008. Pre-extraction of hemicelluloses and subsequent kraft pulping Part 1: alkaline extraction. *Tappi Journal* 3–8.

Ali, O. F., J. T. Cenicola, J. Li, and J. D. Taylor, 2005. Process for producing alkaline treated cellulosic fibers, United States Patent 6,896,810.

Alriols, G. M, A. García, R. Llano-ponte, and J. Labidi. 2010. Combined organosolv and ultrafiltration lignocellulosic biorefinery process. *Chemical Engineering Journal* 157, 113–120.

Anastasakis, K. and A. B. Ross. 2011. Hydrothermal liquefaction of the brown macro-alga Laminaria Saccharina: Effect of reaction conditions on product distribution and composition. *Bioresource Technology* 102, 4876–4883.

Aşçi, Y. S. and İ. İnci. 2012. A novel approach for itaconic acid extraction: Mixture of trioctylamine and tridodecylamine in different diluents. *Journal of Industrial and Engineering Chemistry*, http://dx.doi.org/10.1016/j.jiec.2012.03.010 (accessed August 5, 2012).

Atadashi, I., M. K. Aroua, and A. A. Aziz. 2011. Biodiesel separation and purification: A review, *Renewable Energy* 36, 437–443.

Bacovsky, D., W. Körbitz, M. Mittelbach, M. Wörgetter. 2007. Biodiesel Production: Technologies and European Providers. IEA Task 39 Report T39-B6, 104 pp.

Bae, J. W., H. S. Potdar, S. H. Kang, and K. W. 2008. Coproduction of methanol and dimethyl ether from biomass-derived syngas on a Cu–ZnO–Al2O3/γ-Al2O3 hybrid catalyst. *Energy and Fuels* 22, 223–230. doi:10.1021/ef700461j.

Bauen, A., Berndes, G., Junginger, M., Londo, M., Vuille, F., Ball, R., Bole, T., Chudziak, C., Faaij, A., Mozaffarian, H. 2009. Bioenergy—a sustainable and reliable energy source—a review of status and prospects. IEA Bioenergy: ExCo: 2009:06IEA. 107pp., http://www.ieabioenergy.com/LibItem.aspx?id=6479 (accessed August 5, 2012).

Behrendt, F., Y. Neubauer, M. Oevermann, B. Wilmes, and N. Zobel. 2008. Direct liquefaction of biomass e review. *Chemical Engineering Technology* 31, 667–677.

Berchtold, K. A., R. P. Singh, J. S. Young, and K. W. Dudeck. 2012. Polybenzimidazole composite membranes for high temperature synthesis gas separations. *Journal of Membrane Science* 415–416, 265–270.

Berrios, M., M. A. Martín, A. F. Chica, and A. Martín. 2011. Purification of biodiesel from used cooking oils. *Applied Energy* 88, 3625–3631.

Berrios, M. and R. L. Skelton. 2008. Comparison of purification methods for biodiesel. *Chemical Engineering Journal* 144 (3), 459–465.

Biller, P. and A. B. Ross. 2011. Potential yields and properties of oil from the hydrothermal liquefaction of microalgae with different biochemical content. *Bioresource Technology* 102, 215–225.

Bond, J. Q., D. M. Alonso, D. Wang, R. M. West, and J. A. 2010. Dumesic, Integrated Catalytic Conversion of g-Valerolactone to Liquid Alkenes for Transportation Fuels. *Science* 327, 1110.

Bozell, J. J. 2010. Connecting Biomass and Petroleum Processing with a Chemical Bridge. *Science* 329, 522–523.

Bressler, E. and S. Braun. 1999. Separation mechanisms of citric and itaconic acids by water-immiscible amines. *Journal of Chemical Technology and Biotechnology* 74 (9), 891–896.

Bridgwater, A. V. 2012. Review of fast pyrolysis of biomass and product upgrading. *Biomass and Bioenergy* 38, 68–94.

Bulushev, D. A. and J. R. H. Ross. 2011. Catalysis for conversion of biomass to fuels via pyrolysis and gasification: A review. *Catalysis Today* 171, 1–13.

Caro, J. and Noack, M. 2010. Zeolite membranes—status and prospective. In: *Advances in Nanoporous Materials*, edited by Stefan Ernst. Elsevier B.V., Volume 1, pp. 1–96.

Chandel, A. K., R. K. Kapoor, A. Singh, and R. C. Kuhad. 2007. Detoxification of sugarcane bagasse hydrolyzate improves ethanol production by Candida shehatae NCIM 3501. *Bioresource Technology* 98 (10), 1947–1950.

Chang, C. D. 1992. The New Zealand Gas-to-Gasoline Plant: An Engineering Tour de force. *Catalysis Today* 13, 103–111.

Chapeaux, A., L. D. Simoni, T. Ronan, M. A. Stadtherr, and J. F. Brennecke. 2008. Extraction of alcohols from water with 1-hexyl-3-methylimidazolium bis(trifluoromethylsulfonyl)imide. *Green Chemistry* 10, 1301–1306.

Cheng, K.-K., X.-B. Zhao, J. Zeng, and J.-A. Zhang. 2012. Biotechnological production of succinic acid: current state and perspectives. *Biofuels, Bioproducts and Biorefining* 6, 302–318; DOI: 10.1002/bbb.

Cheng, L-H, Y.-F. Cheng, S.-Y. Yen, and J. Chen. 2009. Ultrafiltration of triglyceride from biodiesel using the phase diagram of oil-FAME-MeOH. *Journal of Membrane Science* 330, 156–165.

Cukalovic, A. and C. V. Stevens. 2008. Feasibility of production methods for succinic acid derivatives: a marriage of renewable resources and chemical technology. *Biofuels, Bioproducts and Biorefining* 2 (6), 505–529.

Czernik, S., A. V. Bridgewater. 2004. Overview of applications of biomass fast pyrolysis oil. *Energy and Fuels* 18, 590–598.

Dhamole, P. B., Z. Wang, Y. Liu, B. Wang, and H. Feng. 2012. Extractive fermentation with non-ionic surfactants to enhance butanol production. *Biomass and Bioenergy* 40, 112–119.

Dien, B. S., N. N. Nichols, and R. J. Bothast. 2001. Recombinant Escherichia coli engineered for production of L-lactic acid from hexose and pentose sugars. *Journal of Industrial Microbiology and Biotechnology* 27 (4), 259–264.

Drożdżyńska, A., K. Leja, and K. Czaczyk. 2011. Biotechnological production of 1,3-propanediol from crude glycerol. *Journal of Biotechnology, Computational Biology and Bionanotechnology* 92 (1), 92–100.

Dube, M. A., A. Y. Tremblay, and J. Liu. 2007. Biodiesel production using a membrane reactor. *Bioresource Technology* 98, 639–647.

Ecker, J., T. Raab, and M. Harasek. 2012. Nanofiltration as key technology for the separation of LA and AA. *Journal of Membrane Science* 389, 389–398.

Fang, K., D. Li, M. Lin, W. Wei, and Y. Sun. 2009. A short review of heterogeneous catalytic process for mixed alcohols synthesis via syngas. *Catalysis Today* 147, 133–138.

Gao, C., C. Ma, and P. Xu. 2011. Biotechnological routes based on lactic acid production from biomass. *Biotechnology Advances* 29, 930–939.

Gonzalez M. I., S. Alvarez, F. A. Riera, and R. Alvarez, 2008. Lactic acid recovery from whey ultrafiltrate fermentation broths and artificial solutions by nanofiltration. *Desalination* 228 (1–3), 84–96.

González-Pajuelo, M., I. Meynial-Salles, F. Mendes, J. C. Andrade, I. Vasconcelos, P. Soucaille. 2005. Metabolic engineering of Clostridium acetobutylicum for the industrial production of 1,3-propanediol from glycerol. *Metabolic Engineering* 7, 329–336. doi:10.1016/j.ymben.2005.06.001

Goudriaan, F. and J. E. Naber. 2008. HTU® Diesel from Wet Waste Streams. Symposium New Biofuels, Berlin, 2008.

Green, E. M. 2011. Fermentative production of butanol—the industrial perspective. *Current Opinion of Biotechnology* 22, 1–7.

Groot, W. J. 2011. Process for manufacturing succinic acid, PCT Int. Appl., WO 2011098598 A1 20110818.

Grzenia, D. L., X. Qian, S. S. da Silva, X. Wang, S. R. Wickramasinghe. 2011. Membrane Extraction for Biofuel Production. In: *Membrane Science and Technology*, edited by S. T. Oyama and S.M. Stagg-Williams. Elsevier B.V., Vol. 14, pp. 213–233.

Grzenia, D. L., D. J. Schell, and S. R. Wickramasinghe. 2012. Membrane extraction for detoxification of biomass hydrolysates. *Bioresource Technology* 111, 248–254.

Guo, Z., S. Wang, Y. Gu, G. Xu, X. Li, and Z. Luo. 2010. Separation characteristics of biomass pyrolysis oil in molecular distillation. *Separation and Purification Technology* 76, 52–57.

Guo, Z.-G., S.-R. Wang, Y.-Y. Zhu, Z.-Y. Luo, K. Cen. 2009. Separation of acid compounds for refining biomass pyrolysis oil. *Journal of Fuel Chemistry and Technology* 37 (1), 49–52.

Haelssig, J. B., A. Y. Tremblay, and J. Thibault. 2012. A new hybrid membrane separation process for enhanced ethanol recovery: Process description and numerical studies. *Chemical Engineering Science* 68 (1), 492–505.

Haelssig, J. B., A. Y. Tremblay, J. Thibault, and X.-M. Huang. 2011. Membrane dephlegmation: A hybrid membrane separation process for efficient ethanol recovery. *Journal of Membrane Science* 381 (1–2), 226–236.

Halim, R., B. Gladman, M. K. Danquah, and P. A. Webley. 2011. Oil extraction from microalgae for biodiesel production. *Bioresoruce Technology* 102 (1), 178–185.

Hasabnis, A. and S. Mahajani. 2010. Entrainer-based reactive distillation for esterification of glycerol with acetic acid. *Industrial and Engineering Chemistry Research* 49, 9058–9067.

He, Y., D. M. Bagley, K. T. Leung, S. N. Liss, and B.-Q. Liao. 2012. Recent advances in membrane technologies for biorefining and bioenergy production. *Biotechnology Advances* 30 (4), 817–858.

Hecke, W. V., P. V. Andezande, S. Claes, S. Vangeel, H. Beckers, L. Diels, and H. D. Wever. 2012. Integrated bioprocess for long-term continuous cultivation of Clostridium acetobutylicum coupled to pervaporation with PDMS composite membranes. *Bioresource Technology* 111, 368–377.

Holladay, J. E., J. J. Bozell, J. F. White, and D. Johnson. 2007. Top Value-Added Chemicals from Biomass Volume II—Results of Screening for Potential Candidates from Biorefinery Lignin. U.S. Department of Energy under Contract DE-AC05-76RL01830.

Huang, C., T. Xu, Y. Zhang, Y. Xue, and G. Chen. 2007. Application of electrodialysis to the production of organic acids: State-of-the-art and recent developments. *Journal of Membrane Science* 288, 1–12.

Huang, H.-J. and S. Ramaswamy. 2009. Modeling Biomass Gasification Using Thermodynamic Equilibrium Approach. *Applied Biochemistry and Biotechnology* 154, 93–204.

Huang, H.-J. and S. Ramaswamy. 2011. Thermodynamic analysis of black liquor steam gasification. *BioResources* 6 (3), 3210–3230.

Huang, H.-J. and S. Ramaswamy. 2012. Separation and purification of phytochemicals as co-products in biorefineries. In: *Biorefinery Co-Products: Phytochemicals, Primary Metabolites and Value-Added Biomass Processing*, edited by C. Bergeron, D. Julie Carrier, and Shri Ramaswamy. John Wiley & Sons, Ltd.

Huang, H.-J., S. Ramaswamy, W. Al-Dajani, and U. Tschirner. 2010. Process modeling and analysis of pulp mill-based integrated biorefinery with hemicellulose pre-extraction for ethanol production: A comparative study. *Bioresource Technology* 101, 624–631.

Huang, H.-J., S. Ramaswamy, U.W. Tschirner , B.V. Ramarao. 2008. A review of separation technologies in current and future biorefineries. *Separation and Purification Technology* 62, 1–21.

Huang, Y., Z. Li, K. Shimizu, and Q. Ye. 2012. Simultaneous production of 3-hydroxypropionic acid and 1,3-propanediol from glycerol by a recombinant strain of Klebsiella pneumoniae. *Bioresource Technology* 103 (1), 351–359.

Huber, G. W. and A. Corma. 2007. Synergies between bio- and oil refineries for the production of fuels from biomass. *Angewandte Chemie International Edition* 46 (38), 7184–7201.

Huh, Y. S., Y.-S. Jun, Y. K. Hong, H. Song, S. Y. Lee, and W. H. Hong. 2006. Effective purification of succinic acid from fermentation broth produced by Mannheimia succiniciproducens. *Process Biochemistry* 41 (6), 1461–1465.

Ji, P., W. Feng, and B. Chen. 2009. Production of ultrapure hydrogen from biomass gasification with air. *Chemical Engineering Science* 64, 582–592.

Jiang, L. Y., Y. Wang, T.-S. Chung, X.Y. Qiao, J.-Y. Lai. 2009. Polyimides membranes for pervaporation and biofuels separation. *Progress in Polymer Science* 34 (11), 1135–1160.

Jones, S. B. and Y. Zhu. 2009. Techno-economic Analysis for the Conversion of Lignocellulosic Biomass to Gasoline via the Methanol-to-Gasoline (MTG) Process, http://www.pnl.gov/main/publications/external/technical_reports/PNNL-18481.pdf (accessed 5 August 2012).

Jönsson, A.-S., A.-K. Nordin, and O. Wallberg. 2008. Concentration and purification of lignin in hardwood kraft pulping liquor by ultrafiltration and nanofiltration. *Chemical Engineering Research and Design* 86, 1271–1280.

Jönsson, A.-S. and O. Wallberg. 2009. Cost estimates of kraft lignin recovery by ultrafiltration. *Desalination* 237, 254–267.

Khodakov, A. Y., W. Chu, and P. Fongarland. 2007. Advances in the development of novel cobalt Fischer–Tropsch catalysts for synthesis of long-chain hydrocarbons and clean fuels. *Chemical Reviews* 107, 1692–1744. doi:10.1021/cr050972v.

Kim, S. D., S. C. Baek, Y. H. Lee, K. W. Jun, M. J. Kim, and I. S. Yoo. 2006. Effect of γ-alumina content on catalytic performance of modified ZSM-5 for dehydration of crude methanol to dimethyl ether. *Applied Catalysis A* 309, 139–143.

Kiss, A. A. 2009. Novel process for biodiesel by reactive absorption. *Separation and Purification Technology* 69, 280–287.

Kiss, A. A. and C. S. Bildea. 2011. Integrated reactive absorption process for synthesis of fatty esters. *Bioresource Technology* 102, 490–498.

Kiss, A. A., A. C. Dimian, and G. Rothenberg. 2008. Biodiesel by catalytic reactive distillation powered by metal oxides. *Energy Fuels* 22 (1), 598–604.

Kohl, A, and R. Nielsen. 1997. *Gas Purification* (fifth edition). Gulf Publishing Company.

Korens, N., D. R. Simbeck, and D. J. Wilhelm. 2002. *Process Screening Analysis of Alternative Gas Treating and Sulfur Removal for Gasification. Revised Final Report*, December, SFA Pacific, Inc. Mountain View, CA.

Kovacs, K. L., G. Maroti, and G. Rakhely, 2006. *International Journal of Hydrogen Energy* 31, 1460–1468.

Kraemer, K., A. Harwardt, R. Bronneberg, and W. Marquardt. 2011. Separation of butanol from acetone–butanol–ethanol fermentation by a hybrid extraction–distillation process. *Computers and Chemical Engineering* 35 (5), 949–963.

Kumar, S. and B. V. Babu. Process Intensification for Separation of Carboxylic Acids from Fermentation Broths using Reactive Extraction, http://discovery.bits-pilani.ac.in/~bvbabu/Susheel_Babu_JFET_Carboxylic%20acids_Review.pdf (accessed August 5, 2012).

Lancrenon, X. and J. Fedders. *2008*. An innovation in glycerin purification. *Biodiesel Magazine*, 2008.

Lange, J. P., R. Price, P. M. Ayoub, J. Louis, L. Petrus, L. Clarke, and H. Gosselink. 2010. Valeric biofuels: a platform of cellulosic transportation fuels. *Angewandte Chemie International Edition* 49, 4479–4483.

Laopaiboon, P., A. Thani, V. Leelavatcharamas, and L. Laopaiboon. 2010. Acid hydrolysis of sugarcane bagasse for lactic acid production. *Bioresource Technology* 101, 1036–1043.

Lee, J. W., H. U. Kim, S. Choi, J. Yi, and S. Y. Lee. 2011. Microbial production of building block chemicals and polymers. *Current Opinion in Biotechnology* 22 (6), 758–767.

Lee, H. J., Y. Xie, Y. M. Koo, and N.-H. L. Wang. 2004. Separation of lactic acid from acetic acid using a four-zone SMB. *Biotechnology Progress* 20, 179–192.

Leoneti, A. B., V. Aragão-Leoneti, S. V. W. B. de Oliveira. 2012. Glycerol as a by-product of biodiesel production in Brazil: Alternatives for the use of unrefined glycerol. *Renewable Energy* 45, 138–145.

Li, Q., D. Wang, and Y. Wu, W. Li, Y. Zhang, J. Xing, and Z. Su. 2010. One step recovery of succinic acid from fermentation broths by crystallization. *Separation and Purification Technology* 72 (3), 294–300.

Liu, Z., P. Fatehi, M. S. Jahan, and Y. Ni. 2011a. Separation of lignocellulosic materials by combined processes of prehydrolysis and ethanol extraction: effect of prehydrolysis step. *Bioresource Technology* 102, 1264–1269.

Liu, Z., P. Fatehi, S. Sadeghi, and Y. Ni. 2011b. Application of hemicelluloses precipitated via ethanol treatment of pre-hydrolysis liquor in high-yield pulp. *Bioresource Technology* 102, 9613–9618.

Lopes, F. V. S., C. A. Grande, and A. E. Rodrigues. 2011. Activated carbon for hydrogen purification by pressure swing adsorption: Multicomponent breakthrough curves and PSA performance. *Chemical Engineering Science* 66 (3), 303–317.

Luo, L. H., J.-W. Seo, D.-H. Kim, B.-R. Oh, S. K. Rhee, and C. H. Kim. 2010a. A novel microbial system for efficient production of 3-hydroxypropionic acid from glycerol using Klebsiella pneumoniae. *Journal of Biotechnology*, 150, Supplement, p.78.

Luo, L., E. van der Voet, and G. Huppes, 2010b. Biorefining of lignocellulosic feedstock—technical, economic and environmental considerations. *Bioresource Technology* 101, 5023–5032.

Maas, R. H., R. R. Bakker, M. L. Jansen, D. Visser, E. de Jong, G. Eggink, and R. A. Weusthuis. 2008. Lactic acid production from lime-treated wheat straw by Bacillus coagulans: neutralization of acid by fedbatch addition of alkaline substrate. *Applied Microbiology and Biotechnology* 78, 751–8.

Mahfud, F. H., I. Melian-Cabrera, and R. Manurung. 2007. Biomass to fuels—upgrading of flash pyrolysis oil by reactive distillation using a high boiling alcohol and acid catalysts. *Process Safety and Environmental Protection* 85, 466–472.

Mai, N. L., N. T. Nguyen, J.-I. Kim, H.-M. Park, S.-K. Lee, and Y.-M. Koo. 2012. Recovery of ionic liquid and sugars from hydrolyzed biomass using ion exclusion simulated moving bed chromatography. *Journal of Chromatography A* 1227 (2), 67–72.

Majlan, E. H., W. R. W. Daud, S. E. Iyuke, A. B. Mohamad, A. A. H. Kadhum, A. W. Mohammad, M. S. Takriff, and N. Bahaman. 2009. Hydrogen purification using compact pressure swing adsorption system for fuel cell. *International Journal of Hydrogen Energy* 34 (6), 2771–2777.

Manuale. D. L., V. M. Mazzieri, G. Torres, C. R. Vera, and J. C. Yori. 2011. Non-catalytic biodiesel process with adsorption-based refining. *Fuel* 90 (3), 1188–1196.

Mao, H., J. M. Genco, S. H. Yoon, A. Van Heiningen, and H. Oendse. 2008. Technical economic evaluation of a hardwood biorefinery using the "near-neutral" hemicellulose pre-extraction process. *Journal of Biobased Materials and Bioenergy* 2, 177–185.

Mendes, F. S., M. González-Pajuelo, H. Cordier, J. M. François, and I. Vasconcelos. 2011. 1,3-Propanediol production in a two-step process fermentation from renewable feedstock, http://repositorio.ucp.pt/bitstream/10400.14/7509/1/Filipa%20Mendes%20Appl%20Microbiol%20Biotechnol%202011.1.pdf (accessed August 5, 2012).

Minu, K., K. Kurian Jiby, and V. V. N. Kishore. 2012. Isolation and purification of lignin and silica from the black liquor generated during the production of bioethanol from rice straw. *Biomass and Bioenergy* 39, 210–217.

National Academy of Science, 2004. *The Hydrogen Economy: Opportunities, Costs, Barriers, and R&D Needs*. National Academies Press.

Ogawa, T., N. Inoue, T. Shikada, and Y. Ohno. 2003. Direct dimethyl ether synthesis. *Journal of Natural Gas Chemistry* 12, 219–227.

Öhman, F., H. Wallmo and H. Theliander. 2007a. Precipitation and filtration of lignin from black liquor of different origin. *Nordic Pulp and Paper Research Journal* 22 (2), 188–193.

Öhman, F., H. Wallmo, and H. Theliander. 2007b. An improved method for washing lignin precipitated from kraft black liquor—the key to a new biofuel. *Filtration* 7 (4), 309–315.

Öhman, F., H. Wallmo, and H. Theliander, 2007c. A novel method for washing lignin precipitated from kraft black liquor—laboratory trials. *Nordic Pulp and Paper Research Journal* 22 (1), 9–16.

Ohsaki, S., M. Tamura, and T. Yamaura. 2003. Method for purification of saccharide. U.S. Patent US 6,548,662.

Olajire, A. A. 2010. CO_2 capture and separation technologies for end-of-pipe applications—a review. *Energy* 35 (6), 2610–2628.

Oliveira , F. S., J. M. M. Araújo, R. Ferreira, L. P. N. Rebelo, I. M. Marrucho. 2012. Extraction of l-lactic, l-malic, and succinic acids using phosphonium-based ionic liquids. *Separation and Purification Technology*, 85 (2), 137–146.

Orjuela, A., A. Kolah, C. T. Lira, and D. J. Miller. 2011. Mixed succinic acid/acetic acid esterification with ethanol by reactive distillation. *Industrial and Engineering Chemistry Research* 50, 9209–9220.

Pal, P., J. Sikder, S. Roy, and L. Giorno. 2009. Process intensification in lactic acid production: a review of membrane based processes, *Chemical Engineering and Processing: Process Intensification* 48 (11–12), 1549–1559.

Park, H. B., S. H. Han, C. H. Jung, Y. M. Lee, and A. J. Hill. 2010. Thermally rearranged (TR) polymer membranes for CO2 separation. *Journal of Membrane Science* 259, 11–24.

Persson, T. and A.-S. Jönsson. 2010. Isolation of hemicelluloses by ultrafiltration of thermomechanical pulp mill process water—Influence of operating conditions. *Chemical Engineering Research and Design* 88, 1548–1554.

Persson, T., A.-S. Jönsson, and G. Zacchi. 2005. Fractionation of hemicelluloses by membrane filtration. Fourteenth European Biomass Conference. Paris, France.

Pflügl, S., H. Marx, D. Mattanovich, and M.Sauer. 2012. 1,3-propranediol production from glycerol with Lactobacillus diolivorans. *Bioresource Technology*, 119, 113–140.

Posada, J. A., L. E. Rincón, and C. A. Cardona. 2012. Design and analysis of biorefineries based on raw glycerol: Addressing the glycerol problem. *Bioresource Technology* 111, 282–293.

Raj S. M., C. Rathnasingh, J.-E. Jo, and S. Park. 2008. Production of 3-hydroxypropionicacid from glycerol by a novel recombinant Escherichia coli BL21 strain. *Process Biochemistry* 43 (12), 1440–1446.

Rasrendra, C. B., B. Girisuta, H. H. van de Bovenkamp, J. G. M. Winkelman, E. J. Leijenhorst, R. H. Venderbosch, M. Windt, D. Meier, and H. J. Heeres. 2011. Recovery of acetic acid from an aqueous pyrolysis oil phase by reactive extraction using tri-n-octylamine. *Chemical Engineering Journal* 176–177, 244–252.

Rathnasingh, C., S. M. Raj, Y.-J. Lee, C. Catherine, S. Ashok, and S. Park. 2010. Production of 3-hydroxypropionic acid through malonyl CoA pathway from glucose by recombinant *Escherichia coli* BL21. *Journal of Biotechnology*, 150, Supplement, p. 22.

Rezkallah, A. 2010. Method for purification of glycerol. US Patent 7,667,081.

Ribeiro, A. M., J. C. Santos, and A. E. Rodrigues. 2010. PSA design for stoichiometric adjustment of bio-syngas for methanol production and co-capture of carbon dioxide, *Chemical Engineering Journal* 163 (3), 355–363.

Ríos, S. D., J. Salvadó, X. Farriol, and C. Torras. 2012. Antifouling microfiltration strategies to harvest microalgae for biofuel. *Bioresource Technology* 119, 406–418.

Rossi, N., P. Jaouen, P. Legentilhomme, and I. Petit, 2004. Harvesting of Cyanobacterium Arthrospira platensis using organic filtration membranes, *Transactions of the Institution of Chemical Engineers Part C* 82, 244–250.

Rossi, N. Rossi, I. Petit, P. Jaouen, P. Legentilhomme, and M. Derouiniot, 2005. Harvesting of Cyanobacterium Arthrospira platensis using inorganic filtration membranes. *Separation Science and Technology* 40, 3033–3050.

Rossignol, N., L. Vandanjon, P. Jaouen, and F. Quéméneur, 1999. Membrane technology for the continuous separation microalgae/culture medium: compared performances of cross-flow microfiltration and ultrafiltration. *Aquacultural Engineering* 20, 191–208.

Sánchez, J. M., M. M. Barreiro, and M. Maroño. 2011. Hydrogen enrichment and separation from synthesis gas by the use of a membrane reactor. *Biomass and Bioenergy 35*, Supplement, S132–S144.

Sandström, L., E. Sjöberg, and J. Hedlund. 2011. Very high flux MFI membrane for CO2 separation. *Journal of Membrane Science* 380, 232–240.

Saxena, R. K., P. Anand, S. Saran, and J. Isar. 2009. Microbial production of 1,3-propanediol: Recent developments and emerging opportunities. *Biotechnology Advances* 27, 895–913.

Schlesinger, R., G. Götzinger, H. Sixta, A. Friedl, and M. Harasek. 2006. Evaluation of alkali resistant nanofiltration membranes for the separation of hemicellulose from concentrated alkaline process liquors, *Desalination* 192, 303–314.

Schlosser, Š., R. Kertész, and J. Marták. 2005. Review: recovery and separation of organic acids by membrane based solvent extraction and pertraction—an overview with a case study on recovery of MPCA, *Separation and Purification Technology* 41, 237–266.

Simoni, L. D., A. Chapeaux, J. F. Brennecke, and M. A. Stadtherr. 2010. Extraction of biofuels and biofeedstocks from aqueous solutions using ionic liquids. *Computers and Chemical Engineering* 34, 1406–1412.

Soh, L. and J. Zimmerman. 2011. Biodiesel production: the potential of algal lipids extracted with supercritical carbon dioxide. *Green Chemistry* 13, 1422–1429.

Song, Q.-H., J.-Q. Nie, M.-G. Ren, and Q.-X. Guo. 2009. Effective phase separation of biomass pyrolysis oils by adding aqueous salt solutions. *Energy and Fuels* 23, 3307–3312.

Strathmann, H. 2010. Electrodialysis, a mature technology with a multitude of new applications. *Desalination* 264, 268–288.

Tanaka, S., Y. Tashiro, G. Kobayashi, T. Ikegami, H. Negishi, and K. Sakaki. 2012. Membrane-assisted extractive butanol fermentation by Clostridium saccharoperbutylacetonicum N1-4 with 1-dodecanol as the extractant. *Bioresource Technology* 116, 448–452.

Teella, A. V. P. R. 2011. Separation of Carboxylic Acids From Aqueous Fraction of Fast Pyrolysis Bio-Oils Using Nanofiltration and Reverse Osmosis Membranes. Open Access Dissertations. Paper 485, http://scholarworks .umass.edu/open_access_dissertations/485 (accessed August 5, 2012).

Theegala, C. S. and J. S. Midgett. 2012. Hydrothermal liquefaction of separated dairy manure for production of bio-oils with simultaneous waste treatment. *Bioresource Technology* 107, 456–463.

Tilman, D., J. Hill, and C. Lehman. 2006. Carbon-negative biofuels from low-input high-diversity grassland biomass. *Science* 314, 1598–1600.

Toledano, A., A. García, I. Mondragon, and J. Labidi. 2010a. Lignin separation and fractionation by ultrafiltration. *Separation and Purification Technology* 71, 38–43.

Toledano, A., L. Serrano, A. Garcia, I. Mondragon, and J. Labidi. 2010b. Comparative study of lignin fractionation by ultrafiltration and selective precipitation. *Chemical Engineering Journal* 157, 93–99.

Toor, S. S., L. Rosendahl, and A. Rudolf. 2011. Hydrothermal liquefaction of biomass: A review of subcritical water technologies. *Energy* 36, 2328–2342.

Umpuch, C., S. Galier, S. Kanchanatawee, and H.R. de Balmann. 2010. Nanofiltration as a purification step in production process of organic acids: selectivity improvement by addition of an inorganic salt. *Process Biochemistry* 45 (11), 1763–1768.

U.S. Department of Energy. 2004. Top Value Added Chemicals from Biomass. *Volume I—Results of Screening for Potential Candidates from Sugars and Synthesis Gas*.

U.S. Department of Energy. 2007. *Hydrogen*, Fuel Cells and Infrastructure Technologies Program, Multi-Year Research, Development and Demonstration Plan.

U.S. Department of Energy. 2011. *U.S. Billion-Ton Update: Biomass Supply for a Bioenergy and Bioproducts Industry*. R.D. Perlack and B.J. Stokes, ORNL/TM-2011/224. Oak Ridge National Laboratory, Oak Ridge, TN.

Vardon, D. R., B. K. Sharma, G. V. Blazina, K. Rajagopalan, and T. J. Strathmann. 2012. Thermochemical conversion of raw and defatted algal biomass via hydrothermal liquefaction and slow pyrolysis. *Bioresource Technology* 109, 178–187.

Vardon, D. R., B. K. Sharma, J. Scott, G.Yu, and Z. Wang, 2011. Lance Schideman, Yuanhui Zhang, Timothy J. Strathmann. Chemical properties of biocrude oil from the hydrothermal liquefaction of Spirulina algae, swine manure, and digested anaerobic sludge. *Bioresource Technology* 102, 8295–8303.

Vegas, R., J. L. Alonso, H. Domínguez, and J. C. Parajó. 2005. Manufacture and refining of oligosaccharides from industrial solid wastes. *Industrial Engineering and Chemical Research* 44, 614–620.

Vijayakumar, J., R. Aravindan, and T. Viruthagiri. 2008. Recent trends in the production, purification and application of lactic acid. *Chemical and Biochemical Engineering Quarterly* 22 (2), 245–264.

Vitasari, C. R., W. Meindersma, and A. B. de Haan. 2011. Water extraction of pyrolysis oil: The first step for the recovery of renewable chemicals. *Bioresource Technology* 102 (14), 7204–7210.

Wallmo, H., H. Theliander, A.-S. Jönsson, O. Wallberg, and K. Lindgren. 2009. The influence of hemicelluloses during the precipitation of lignin in kraft black liquor. *Nordic Pulp and Paper Research Journal* 24 (2), 165–171.

Wang, J. L., X. H. Wen, and D. Zhou. 2000. Production of citric acid from molasses integrated with in-situ product separation by ion-exchange resin adsorption. *Bioresource Technology* 75, 231–234.

Wang, S., Y. Gu, Q. Liu, Y. Yao, Z. Guo, Z. Luo, and K. Cen. 2009. Separation of bio-oil by molecular distillation, *Fuel Processing Technology* 90, 738–745.

Wang, Y., C. Huang, and T. Xu. 2011a. Which is more competitive for production of organic acids, ion-exchange or electrodialysis with bipolar membranes? *Journal of Membrane Science* 374, 150–156.

Wang, Y., J. Nie, M. Zhao, S. Ma, L. Kuang, X. Han, and S. Tang. 2010. Production of biodiesel from waste cooking oil via a two-step catalyzed process and molecular distillation. *Energy Fuels* 24, 2104–2108.

Wang, Y., N. Zhang, C. Huang, and T. Xu. 2011b. Production of monoprotic, diprotic, and triprotic organic acids by using electrodialysis with bipolar membranes: effect of cell configurations. *Journal of Membrane Science* 385–386, 226–233.

Wee, Y. J. and H. W. Ryu. 2009. Lactic acid production by *Lactobacillus sp*. RKY2 in a cell-recycle continuous fermentation using lignocellulosic hydrolyzates as inexpensive raw materials. *Bioresource Technology* 100, 4262–4270.

Wickramasinghe, S. R. and D. L. Grzenia. 2008. Adsorptive membranes and resins for acetic acid removal from biomass hydrolysates. *Desalination* 234, 144–151.

Wilkens, E., A. K. Ringel, D. Hortig, T. Willke, and K.-D. Vorlop. 2012. Highlevel production of 1,3-propanediol from crude glycerol by Clostridium butyricum AKR102a. *Applied Microbiology and Biotechnology* 93, 1057–1063. doi:10.1007/s00253-011-3595-6.

Willke, T. and K. D. Vorlop. 2001. Biotechnological production of itaconic acid. *Applied Microbiology and Biotechnology* 56, 289–295.

Wu, B. C.-P., C. A. DeLuca, and E. K. Payne. 2010. Systems and methods for hydrothermal conversion of algae into biofuel, USPTO, 20100050502.

Yang, S. T., H. Huang, A. Tay, W. Qin, L. De Guzman, and E.C. San Nicolas. 2006. Extractive fermentation for the production of carboxylic acids. In: *Bioprocessing for Value-Added Products from Renewable Resources*, edited by S.T. Yang, Elsevier Science, pp. 421–446.

Yin, S., R. Dolan, M. Harris, and Z. Tan. 2010. Subcritical hydrothermal liquefaction of cattle manure to bio-oil: Effects of conversion parameters on bio-oil yield and characterization of bio-oil. *Bioresource Technology* 101, 3657–3664.

Yori, J. C., S. A. D'Ippolito, C. L. Pieck, and C. R. Vera, 2007. Deglycerolization of biodiesel streams by adsorption over silica beds. *Energy Fuels* 21 (1), 347–353.

Zeikus, G. J., M. K. Jain, and P. Elankovan, 1999. Biotechnology of succinic acid production and markets for derived industrial products. *Applied Microbiology and Biotechnology* 51, 545–552.

Zhang, Y., M. A. Dube, and D. D. McLean. 2003. Biodiesel production from waste cooking oil: 1 Process design and technological assessment. *Bioresource Technology* 89, 1–16.

Zhang, Z. Y., B. Jin, J. M. Kelly. 2007. Production of lactic acid from renewable materials by Rhizopus fungi. *Biochemical Engineering Journal* 35, 251–263.

Zhang, Y., J. Xiao, and L. Shen. 2009. Simulation of Methanol Production from Biomass Gasification in Interconnected Fluidized Beds. *Industrial Engineering and Chemical Research* 48, 5351–5359.

Zhao, L., E. Riensche, R. Menzer, L. Blum, and D. Stolten. 2008. A parametric study of CO2/N2 gas separation membrane processes for post-combustion capture. *Journal of Membrane Science* 325, 284–294.

Žilnik, L. F. and A. Jazbinšek. 2011. Recovery of renewable phenolic fraction from pyrolysis oil. http://www.biocoup.com/fileadmin/user/december/Update_December_2011/109_NIC_May11.pdf (accessed August 5, 2012).

Zou, S., Y. Wu, M. Yang, I. Kaleem, C. Li, and J. Tong. 2010. Production and characterization of bio-oil from hydrothermal liquefaction of microalgae Dunaliella tertiolecta cake. *Energy* 35, 5406–5411.

Part II

Equilibrium-Based Separation Technologies

2

Distillation

Zhigang Lei and Biaohua Chen

State Key Laboratory of Chemical Resource Engineering, Beijing University of Chemical Technology, China

2.1 Introduction

Distillation, with its unique operation and control advantages, has become a very powerful separation tool in the laboratory and in industry. Although many promising separation methods are constantly put forwarded by engineers and scientists, most of them cannot become alternatives to distillation on a large scale. Undoubtedly, minor improvements in distillation could bring significant economical and social benefits, so research on distillation (especially distillation in special applications such as biorefineries) should be increased, although it is always regarded as a mature separation technology.

In recent years, the distillation has been extended from its original field of chemical engineering to fields such as biotechnology, bioengineering, environmental engineering, biofuel and bioenergy engineering. This means that distillation technology will face new issues and challenges. We present some special distillation technologies associated with modern biorefineries.

Two prerequisites must be satisfied simultaneously to ensure that distillation can take place: the relative volatility as a driving force and the separation equipment. Where the relative volatility is close to unity, a third component, namely a separating agent, solvent or entrainer in this book, should be added into the mixture to enhance the relative volatility. Where the relative volatility is far from unity, no separating agent is needed. The latter involves ordinary distillation and molecular distillation, while the former involves azeotropic distillation and extractive distillation. We first introduce ordinary distillation in Section 2.2, due to the simplicity of its separation mechanism, and recent distillation equipment is discussed with respect to biorefineries. Then azeotropic distillation and extractive distillation are introduced in Sections 2.3 and 2.4 respectively, but the description of extractive distillation is given in more detail because it seems to be more promising than azeotropic distillation. Molecular distillation and its application is discussed in Section 2.5. These distillation processes are compared in Section 2.6 from the viewpoint of energy

Separation and Purification Technologies in Biorefineries, First Edition.
Edited by Shri Ramaswamy, Hua-Jiang Huang, and Bandaru V. Ramarao.

consumption, production scale, investment cost, and operation complexity. Finally, concluding remarks are given in Section 2.7.

2.2 Ordinary distillation

2.2.1 Thermodynamic fundamental

Fundamental knowledge of phase equilibrium is very important for understanding various separation processes. In particular, vapor–liquid equilibrium (VLE) is commonly encountered in distillation, and its calculable form plays a major role in establishing equilibrium stage (EQ) and nonequilibrium stage (NEQ) mathematical models of distillation columns [1]. Equilibrium is defined as a state that will be returned to its initial state after any short, small mechanical disturbance of external conditions. When equilibrium is reached for a p-phase, i-component system, the following criteria should be satisfied:

$$\mu_i^{(1)} = \mu_i^{(2)} = \mu_i^{(3)} = \ldots = \mu_i^{(p)} \tag{2.1}$$

$$T^{(1)} = T^{(2)} = T^{(3)} = \ldots = T^{(p)} \tag{2.2}$$

$$P^{(1)} = P^{(2)} = P^{(3)} = \ldots = P^{(p)} \tag{2.3}$$

which correspond to chemical potential μ_i, thermal (i.e. temperature T) and mechanical (i.e. total pressure P) equilibrium, respectively. For a vapor–liquid two phases system, Eq. (2.1) can be rewritten in the form of partial fugacity in vapor (V) and liquid (L) phases as

$$\bar{f}_i^{\mathrm{V}} = \bar{f}_i^{\mathrm{L}} \tag{2.4}$$

The partial fugacity can be expressed in different forms by introducing fugacity coefficient $\bar{\phi}_i$ in the mixture:

$$\bar{f}_i^{\mathrm{V}} = Py_i\bar{\phi}_i^{\mathrm{V}}, \bar{f}_i^{\mathrm{L}} = Px_i\bar{\phi}_i^{\mathrm{L}} \tag{2.5}$$

or activity coefficient γ_i:

$$\bar{f}_i^{\mathrm{V}} = y_i\gamma_{i\mathrm{V}}f_{i\mathrm{V}}^0, \ \bar{f}_i^{\mathrm{L}} = x_i\gamma_{i\mathrm{L}}f_{i\mathrm{L}}^0 \tag{2.6}$$

where the bar stands for a mixture fugacity, and the superscript "0" for pure component.

The following equations can be derived:

$$y_i\bar{\phi}_i^{\mathrm{V}} = x_i\bar{\phi}_i^{\mathrm{L}} \tag{2.7}$$

$$Py_i\bar{\phi}_i^{\mathrm{V}} = x_i\gamma_{i\mathrm{L}}f_{i\mathrm{L}}^0 \tag{2.8}$$

Equilibrium ratio (or phase equilibrium constant) K_i is defined as the ratio of mole fractions of a component in the vapor and liquid phases at equilibrium, i.e.

$$K_i = \frac{y_i}{x_i} = \frac{\bar{\phi}_i^{\mathrm{L}}}{\bar{\phi}_i^{\mathrm{V}}} = \frac{\gamma_{i\mathrm{L}}f_{i\mathrm{L}}^0}{\bar{\phi}_i^{\mathrm{V}}P} \tag{2.9}$$

where

$$\bar{\phi}_i^{\mathrm{V}} = \phi\left(T, P, y_1, y_2, \ldots, y_{n-1}\right) \tag{2.10}$$

$$\bar{\phi}_i^{\mathrm{L}} = \phi\left(T, P, x_1, x_2, \ldots, x_{n-1}\right) \tag{2.11}$$

$$\gamma_{iL} = \gamma\left(T, P, x_1, x_2, \ldots, x_{n-1}\right) \tag{2.12}$$

$$f_{iL}^{0} = f\,(T, P) \tag{2.13}$$

The ease of separation of a given mixture with the key components i and j is dependent on the relative volatility

$$\alpha_{ij} = \frac{y_i/x_i}{y_j/x_j} = \frac{K_i}{K_j} \tag{2.14}$$

While at low pressure the relative volatility can be expressed in a more simplified form:

$$\alpha_{ij} = \frac{K_i}{K_j} = \frac{\gamma_i P_i^0}{\gamma_j P_j^0} \tag{2.15}$$

where x and y represent molar fractions in the liquid and vapor phases, respectively, and P_i^0 is the saturated vapor pressure of the pure component i.

Where the relative volatility of the components to be separated is close to unity, a solvent is introduced to change the relative volatility as far away from unity as possible. Since the ratio of P_i^0/P_j^0 remains almost constant for a small temperature perturbation, the only way to modify the relative volatility is to change the ratio γ_i/γ_j. This ratio, in the presence of the solvent, is called selectivity S_{ij}:

$$S_{ij} = \left(\frac{\gamma_i}{\gamma_j}\right)_{\mathrm{S}} \tag{2.16}$$

Besides altering the relative volatility, the solvent should also be easily separated from the distillation products; that is, there should be a high boiling point difference between the solvent and the components or immiscibility with the components to be separated. Other criteria, such as corrosion, price, and source, should also be taken into consideration [2].

2.2.2 Distillation equipment

In most cases, either tray or packing (structured or random) columns are adopted in distillation. In general, vacuum applications are dominated by packing columns, while at the middle or even high pressure, it is better to choose tray columns. At the normal pressure, tray, structured or random packings can find suitable applications under different operating conditions. In order to decide which one is more suitable for biorefineries, some uniquely suitable special distillation equipment is introduced.

Where there is a high liquid-to-vapor flowrate ratio along the distillation column, the plate trays, especially double overflow trays, are generally used as internal fittings. Both double overflow valve trays and double overflow slant-hole trays have been adopted. However, it is reported that, if double overflow valve trays are replaced by double overflow slant-hole trays in a column, the amount of feed to be treated can be increased to over 50% with a tray efficiency similar to or higher than that of the valve trays and an energy saving of 10% by decreasing the drop in pressure to about one-third of the original [3].

The slant-hole tray, as an excellent and extensively applied tray, has an opposite stagger arrangement of the slant-holes, which causes rational flowing of vapor and liquid phases and level blowing of vapor. It permits a high vapor speed, the reduction of mutual interference between vapors, a steady liquid level and high efficiency of trays. After combining the columns with multiple downcomer (MD) trays, a new type of multi-overflow compound slant-hole tray was proposed, which adopts the downcomer similar to MD trays. The number of downcomers used is normally two. The downcomer has a simple structure, liquid flows for a longer distance, it has a higher column capacity, and high efficiency of trays. The configuration

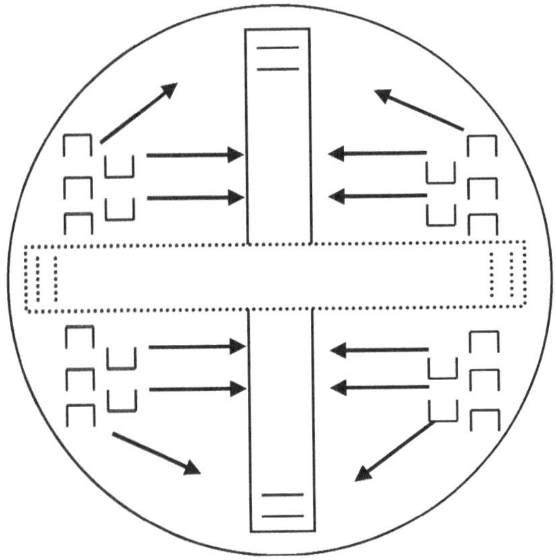

Figure 2.1 Configuration of multi-overflow (two flow) slant-hole tray. Reprinted from [4] © 2002, with permission from Elsevier

of the double overflow slant-hole trays is shown in Figure 2.1. In terms of the vapor and liquid load in the distillation column, the tray parameters are obtained by a computer program for tray design. The designed values should be within the range of normal operation conditions [4].

In the case of unclean feeding materials which are easy to polymerize or contain solid particles, the big-hole, flow-guided sieve trays are desirable because the holes with large diameters of 10–15 mm prevent the trays from blocking due to the formation of polymer or due to solid particles entering into the distillation column [5, 6]. Moreover, this type of tray is different from the ordinary sieve tray in that it has two modifications: one is that it opens a number of proper flow-guided sieves, which ensures that the gas and liquid flowing path is reasonable, and the other is the installation of bubble-promoting devices near the entrance to the liquid, which ensures that there is very small height difference in the liquid layer from the inlet to outlet on the tray, as shown in Figure 2.2.

This type of tray eliminates the gradient of liquid layer and causes the rational flowing of vapor and liquid phases on the tray, with the help of directed holes arranged for decreasing the radial mixing, and bubble promoter installed in the outlet of the downcomer.

Figure 2.3 (a) and (b) illustrate the gradient of liquid layer on the traditional tray and the improved gradient of liquid layer on the flow-guided sieve tray, respectively. Figure 2.4 shows the flowing direction of liquid phase on the flow-guided sieve tray, indicating that there is no liquid reflux on the tray.

In the case of clean feeding materials with high surface tension (e.g. dilute aqueous solutions and glycerol) or having a very strict separation requirements up to ppm level, new BeiHua (BH) structured packing has been proposed. In general, the corrugation angle in the conventional structured packing is designed to be 30° or 45° from the vertical position [7–9]. However, the BH packing has two kinds of transition, or wave-like, structures, which consist of central, upper and lower segments [10]. Each segment accounts for one-third of the total sheet height. The sheet corrugations vary in the order of 45°–30°–45° or 30°–45°–30° as shown in Figure 2.5, and the different segments connect smoothly. The purpose of this

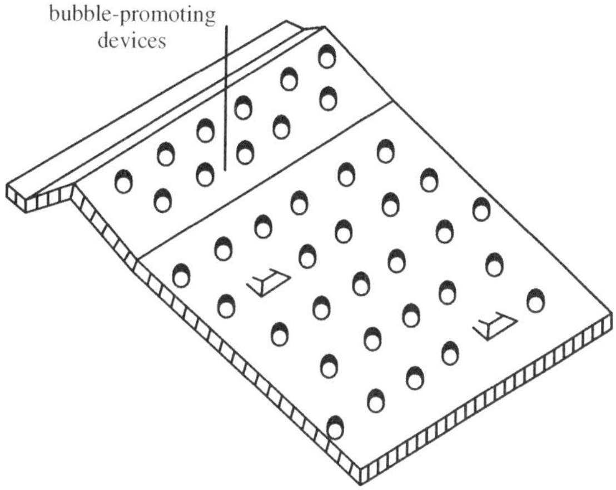

Figure 2.2 Structure of bubble-promoting device installed in the high-efficiency flow-guided sieve tray. Reprinted from [1] © 2005, with permission from Elsevier

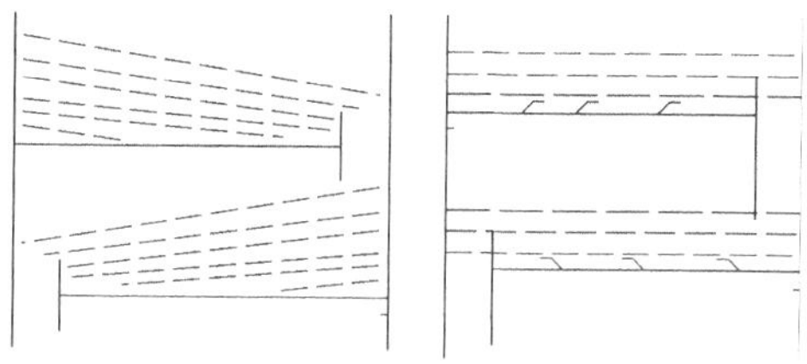

Figure 2.3 Gradient of liquid layers on the traditional tray (a) and on the flow-guided sieve tray (b). Reprinted from [1] © 2005, with permission from Elsevier

geometry design is to spread the liquid film on the sheet surface as completely as possible and enhance the local gas–liquid contact area.

It was found that the corrugation angles with two transition structures (i.e. 30°–45°–30° and 45°–30°–45°) is favorable when considering pressure drop and mass transfer coefficients together. As expected, a low ratio of packing height to diameter is favorable for increasing mass transfer coefficients, but leads to increasing pressure drop like common structured packings. It is necessary for us to identify the relationship between geometric configuration and performance of pressure drop and mass transfer coefficient so as to tailor the desirable packings.

2.2.3 Application in biorefineries

When conventional tray or packing columns cannot meet separation requirements in biorefineries, a process intensification approach to the existing distillation internals is needed. As the addition of extra streams and equipment to the original processes is not required, it is convenient to implement it in biorefinery plants. The special distillation internals, as mentioned earlier, are not complicated in geometric structure and can

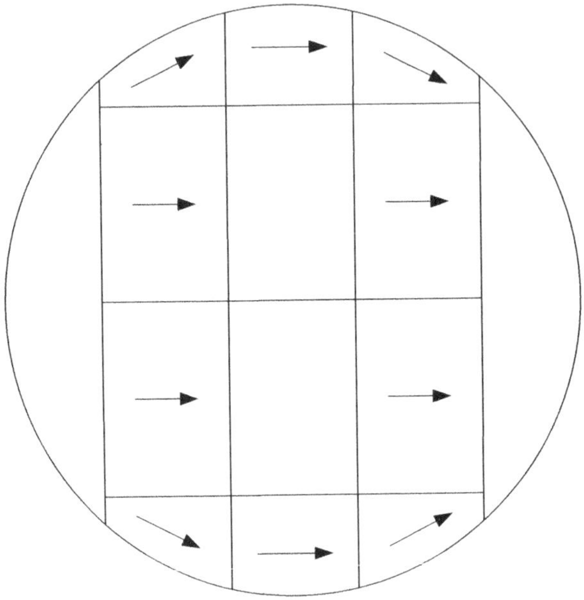

Figure 2.4 Flowing direction of liquid phase on the flow-guided sieve tray. Reprinted from [1] © 2005, with permission from Elsevier

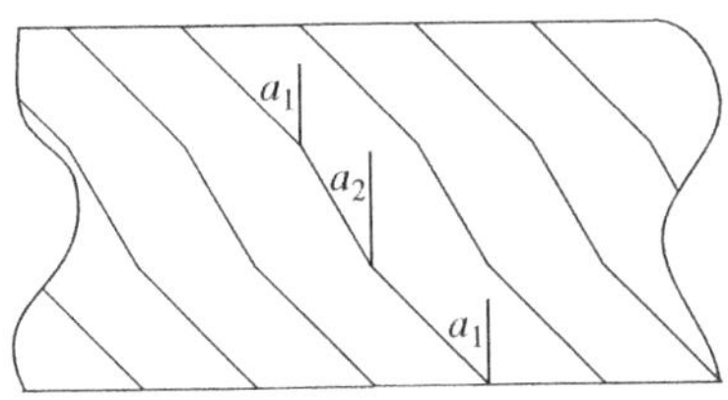

(a)

(b)

Figure 2.5 Transition structure of BH packing ($\alpha_1 = 45°$ and $\alpha_2 = 30°$; or $\alpha_1 = 30°$ and $\alpha_2 = 45°$). (a) Side view; (b) actual photography. Reprinted from [10] © 2009, with permission from Elsevier

be manufactured easily and cheaply. The investment in technology is therefore small and may be used for solving the following separation problems encountered in biorefineries:

- Separation of aqueous organic solutions with low concentration, such as in biomass-to-ethanol biorefineries, where the ethanol stream coming from the fermentor [11] is at a low concentration of about 5–10 wt%. In this case the ratio of liquid to vapor flowrates along the distillation column will be high and thus the multi-overflow (double flow) slant-hole tray is suitable for the separation to obtain approximately 92.5 wt% ethanol.

- Separation of unclean feeding materials containing solid particles from pulp mills, forest products and wood wastes. In this case the big-hole flow-guided sieve tray is more suitable than other trays and packings because the latter will bring about a column jam.
- Separation of clean feeding materials with high surface tension or very strict separation requirements. The high surface-tension materials include aqueous solutions (not containing undissolved fiber, germ and gluten), purification of biodiesel (fatty acid methyl ester) and byproduct glycerol [12, 13]. Materials with very strict separation requirements include various wastewaters from biodiesel and other biorefinery production plants. Before discharging into the environment, the concentration of volatile organic compounds (VOCs) should be very low, usually at ppm level. In this case, a very large number of theoretical stages (up to a few hundred) are needed in order to achieve an environmentally friendly separation process. Addressing environmental requirements should therefore be an ongoing challenge in separation and purification in biorefineries.

2.3 Azeotropic distillation

2.3.1 Introduction

In azeotropic distillation, a third volatile component (i.e. separating agent) is added to the components to be separated, which may have close boiling point or form an azeotrope. The added separating agent or entrainer must be sufficiently volatile so that it can be collected overhead with one of the two components at the top of azeotropic distillation column [14–17]. However, if the azeotrope formed by the components to be separated is heterogeneous, there is no need to add the separating agent, as shown in Figure 2.6. The separation process can be carried out because, in this case with the azeotrope at the top, where the liquid composition is x_0, it can be split into two phases automatically with the respective liquid compositions of x_1 and x_2 (see Figure 2.7). That is to say, in the x-y phase diagram, one can go across the intersection point of equilibrium and diagonal lines, and thus the required number of theoretical stages is not infinite.

For the homogeneous azeotrope formed by the components to be separated, a separating agent should be added. Figure 2.8 (called Process A) shows the azeotropic distillation process for the separation of ethanol and water. The product at the top of azeotropic distillation column may be either heterogeneous

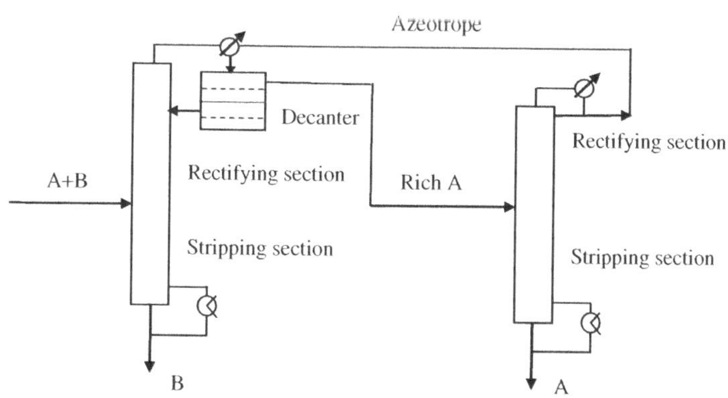

Figure 2.6 Two-column process for separating heterogeneous azeotrope

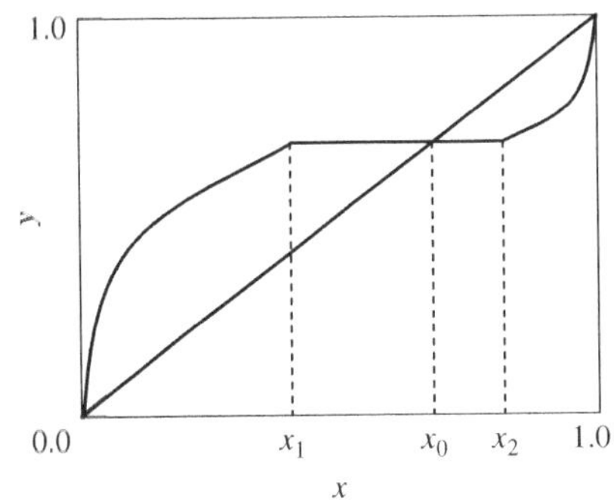

Figure 2.7 *x-y phase diagram for separating heterogeneous azeotrope*

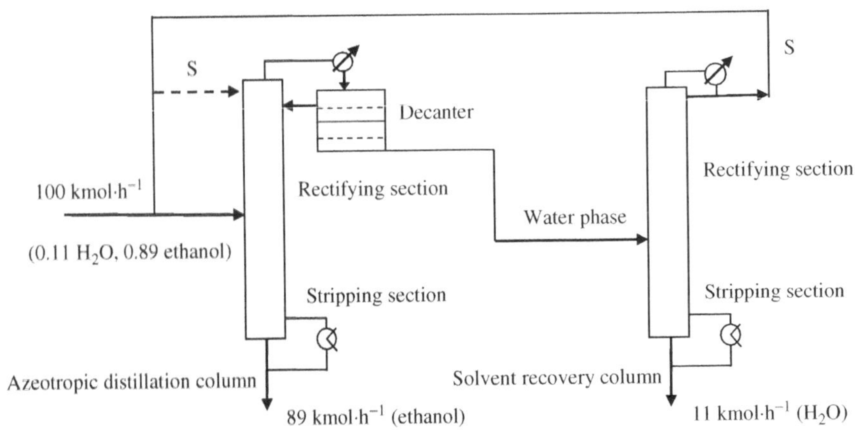

Figure 2.8 *Azeotropic distillation process for the separation of ethanol and water (S= Solvent) (Process A)*

(where a decanter should be set) or homogeneous, while the separating agent is recovered in a solvent recovery column.

2.3.2 Example in biorefineries

Another important separation method for the homogeneous azeotrope is pressure-swing distillation (PSD), of which the separation mechanism is based on the principle that a simple change of operating pressure can alter the relative volatility of the mixture with close boiling point or forming azeotrope (see Figure 2.9), and thus the separation process can be carried out without the addition of a third component. In this regard, PSD can be taken on as an environmentally friendly process.

A problem associated with PSD is how to arrange the column sequence at operating pressures P_1 and P_2. As shown in Figure 2.10, for the separation of ethanol and water in biorefineries, there are two kinds

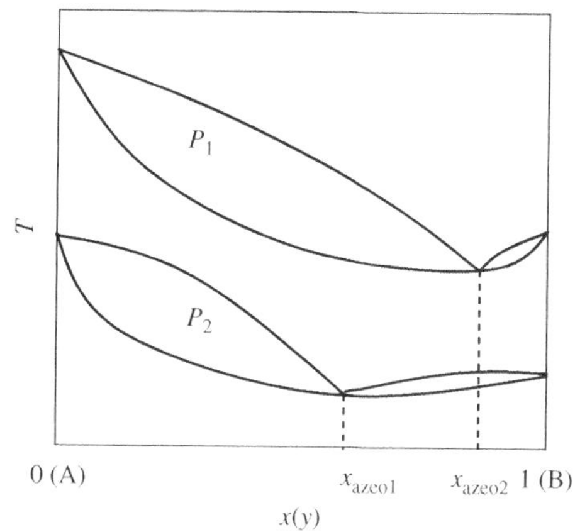

Figure 2.9 *T-x-y diagram at pressures P_1 and P_2 with different azeotropic compositions*

of column sequence in the order of P_1-P_2 (called Process B) and P_2-P_1 (called Process C). The question arises of which is more feasible. It is also interesting to compare the energy consumption between PSD and azeotropic distillation (Process A). Comparison of the operating performance among Processes A, B and C is given in Table 2.1. It can be seen that for the same separation requirements, i.e. ethanol product purity and production capacity, the heat duty on the condenser and the reboiler is Process A < Process B < Process C. Therefore, azeotropic distillation is better than PSD in saving energy because a third volatile component is added to increase the relative volatility of the components to be separated. However, if the feeding mixture is at a low ethanol concentration, it is advisable first to concentrate it above 85 wt% by ordinary distillation in order to decrease the amount of separating agent needed in azeotropic distillation and the cycled flowrate from the second distillation column to the first one in PSD afterwards.

Some applications of azeotropic distillation in biorefineries are summarized in Table 2.2. More details can be found in the cited references.

2.3.3 Industrial challenges

In the design and control of azeotropic distillation columns, we should pay attention to the sensitive plate, around which there is an abrupt change of temperature and/or composition. For instance, for the separation of ethanol and water with pentane as separating agent, stage no.3 (the theoretical plate is numbered from the top to the bottom) is a sensitive plate, as shown in Figure 2.11. In order to find the sensitive plate, it is necessary to establish the suitable mathematical models for azeotropic distillation column. A detailed review on mathematical models of azeotropic distillation can be found elsewhere [14].

As we know, in azeotropic distillation, the entrainer must be vaporized into the column top, and the amount of entrainer is usually large, which leads to large energy consumption compared to extractive distillation. For this reason, extractive distillation is used more often than azeotropic distillation, and azeotropic distillation is also less common than extractive distillation in industry. The future development of azeotropic distillation should therefore be addressed by exploring effective and efficient entrainers so as to decrease the amount of entrainers needed for separation.

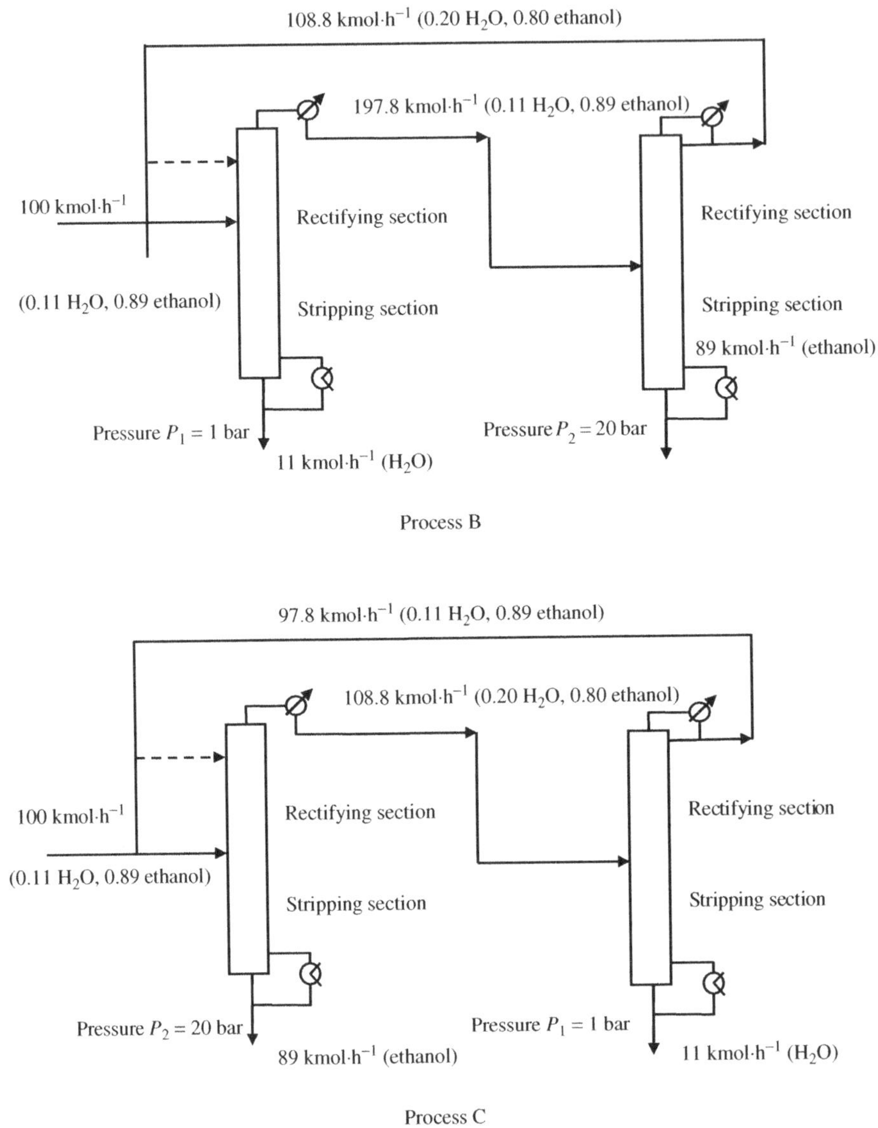

Figure 2.10 *Two kinds of column sequence of PSD for the separation of ethanol and water (Processes B and C)*

2.4 Extractive distillation

2.4.1 Introduction

Extractive distillation has become an important separation method in industry, and is suitable for the separation of hydrocarbons with close boiling points and mixtures forming azeotropes [2]. Like azeotropic distillation, it also requires the addition of a third component that modifies the relative volatility of the

Table 2.1 Comparison of the operating performance of azeotropic distillation and PSD under different pressures

Contents	Azeotropic distillation (Process A)	PSD (Process B)	PSD (Process C)
Ethanol product purity (mol %)	0.9960	0.9950	0.9950
Ethanol molar flowrate ($kmol \cdot h^{-1}$)	89.0	89.0	89.0
Heat duty on condenser (kW)	15843.4	18469.0	22195.4
Heat duty on reboiler (kW)	16069.6	19052.1	22904.5

Table 2.2 Examples of applications of azeotropic distillation in biorefineries

No.	Mixtures to be separated	Separating agents
1	Ethanol/water; isopropanol/water; tert-butanol/water	Benzene, pentane, toluene, hexane, cyclohexane, methanol, diethylether, methyl-ethyl-ketone[a–f]
2	Products from the synthesis of glycerol carbonate from glycerol and dimethyl carbonate	Benzene[g]
3	Removal of byproducts in lipase-catalyzed solid-phase synthesis of sugar fatty acid esters	Ethyl methylketone or acetone[h]
4	Combining fungal dehydration and lipid extraction	Chloroform, cyclohexane and hexane[i]

[a]A. Szanyi, P. Mizsey and Z. Fonyo, Novel hybrid separation processes for solvent recovery based on positioning the extractive heterogeneous-azeotropic distillation, *Chem. Eng. Process.*, **43**, 327–338 (2004).
[b]T.L. Junqueira, M.O.S. Dias, R.M. Filho, M.R.W. Maciel and C.E.V. Rossell, Simulation of the azeotropic distillation for anhydrous bioethanol production: Study on the formation of a second liquid phase, *Computer Aided Chem. Eng.*, **27**, 1143–1148 (2009).
[c]V.V. Hoof, L.V.D. Abeele, A. Buekenhoudt, C. Dotremont and R. Leysen, Economic comparison between azeotropic distillation and different hybrid systems combining distillation with pervaporation for the dehydration of isopropanol, *Sep. Purif. Technol.*, **37**, 33–49 (2004).
[d]P.A.M. Springer, S.V.D. Molen and R. Krishna, The need for using rigorous rate-based models for simulations of ternary azeotropic distillation, *Comput. Chem. Eng.*, **26**, 1265–1279 (2002).
[e]F.J.L. Castillo and G.P. Towler, Influence of multicomponent mass transfer on homogeneous azeotropic distillation, *Chem. Eng. Sci.*, **53**, 963–976 (1998).
[f]C.J. Wang, D.S.H. Wong, I.L. Chien, R.F. Shih, S.J. Wang and C.S. Tsai, Experimental investigation of multiple steady states and parametric sensitivity in azeotropic distillation, *Comput. Chem. Eng.*, **21**, S535-S540 (1997).
[g]J. Li and T. Wang, Coupling reaction and azeotropic distillation for the synthesis of glycerol carbonate from glycerol and dimethyl carbonate, *Chem. Eng. Process.*, **49**, 530–535 (2010).
[h]Y. Yan, U.T. Bornscheuer, L. Cao and R.D. Schmid, Lipase-catalyzed solid-phase synthesis of sugar fatty acid esters: Removal of byproducts by azeotropic distillation, *Enzyme Microb. Tech.*, **25**, 725–728 (1999).
[i]A.J. Tough, B.L. Isabella, J.E. Beattie and R.A. Herbert, Hetero-azeotropic distillation: Combining fungal dehydration and lipid extraction, *J. Biosci. Bioeng.*, **90**, 37–42 (2000).

components to be separated. Therefore, the separating agent is the key technology of extractive distillation in order to ensure an effective and economical separation process.

By far, there have been five kinds of separating agents used in extractive distillation, [18] i.e. liquid solvents, solid salts (or dissolved salts), the mixture of liquid solvent and solid salt, [3] ionic liquids [19–22] and hyperbranched polymers, [23–25] which are discussed as follows.

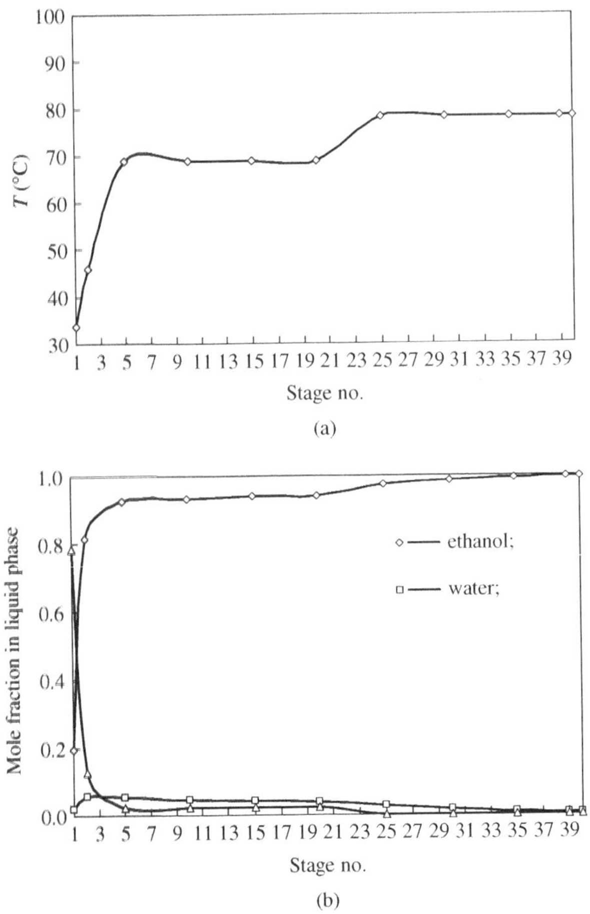

Figure 2.11 *Profiles of temperature (a) and liquid phase composition (b) along the azeotropic distillation column: the stages are numbered from the top to the bottom*

2.4.2 Extractive distillation with liquid solvents

The conventional extractive distillation process for a two-component separation is shown in Figure 2.12, which consists of an extractive distillation column including a solvent-recovery section, a rectifying section, a stripping section, and a solvent-recovery column. For instance, for the separation of ethanol and water by extractive distillation, ethanol and water respectively are obtained from the top of two columns, while the separating agent with high boiling point is recovered at the bottom of solvent recovery column. Therefore, when compared to azeotropic distillation, the separating agent is not needed to evaporate. The separating agents commonly used for ethanol dehydration are ethylene glycol and gasoline.

2.4.3 Extractive distillation with solid salts

For some systems, for example ethanol-water systems, it is feasible to use a solid salt dissolved into the liquid phase, when solubility permits, as the separating agent in extractive distillation. The relative

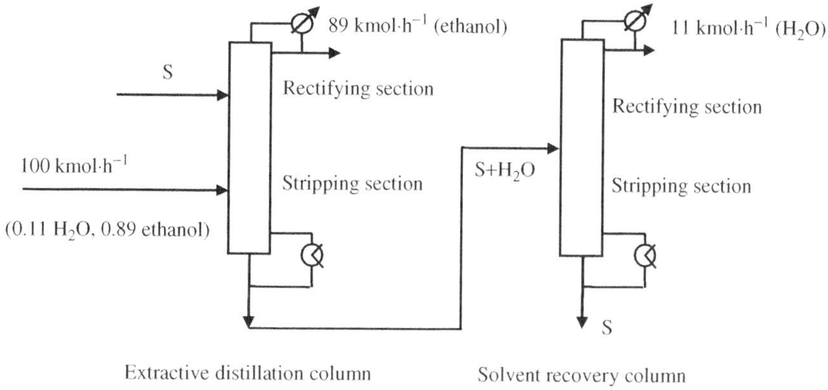

Figure 2.12 *Two column process for the separation of ethanol and water with extractive distillation with liquid solvent*

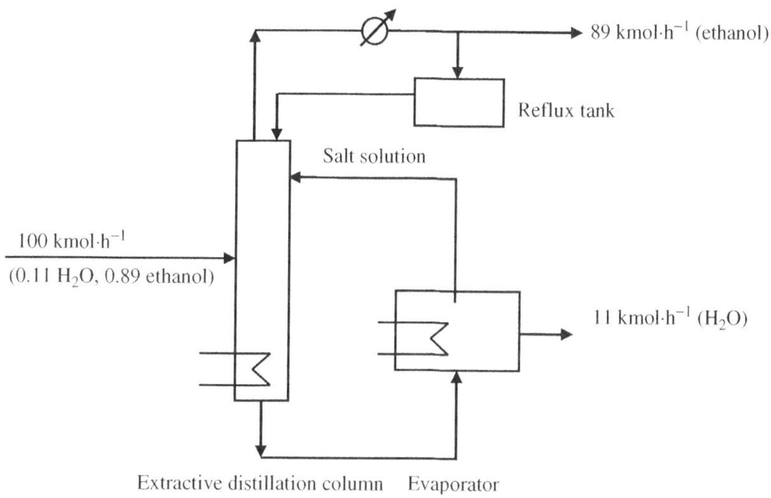

Figure 2.13 *Flowsheet for the separation of ethanol and water with extractive distillation with solid salt*

volatility of the components to be separated can be significantly changed due to the so-called salting effect. Figure 2.13 shows a typical flowsheet of extractive distillation using dissolved salt as separating agent, which is different from the flowsheet shown in Figure 2.12 in that the dissolved salt is recovered by evaporation rather than by distillation. It seems, therefore, that the energy consumption for recovering separating agent can be reduced. The separation of ethanol and water is the most common application of extractive distillation with solid salts. Moreover, the higher the valence of the metal ions, the stronger the salt effect. For instance, the salt effect is generally in the order of $Al^{3+} > Ca^{2+} > Na^{+}$ at a given anion and $Ac^{-} > Cl^{-} > NO_3^{-}$ for a given cation.

2.4.4 Extractive distillation with the mixture of liquid solvent and solid salt

Extractive distillation with the mixture of a liquid solvent and a solid salt as separating agent combines the advantages of a liquid solvent (easy operation) with those of a solid salt (high separation ability). It is therefore a very promising separation method, whether for the separation of polar or non-polar systems. We have measured the vapor–liquid equilibria (VLE) of three systems, i.e. ethanol-water, ethanol-water-ethylene glycol and ethanol-water-ethylene glycol-calcium chloride ($CaCl_2$) at finite concentration and normal pressure. It was found that ethylene glycol with added salt is more effective than a single ethylene glycol for the separation of ethanol and water by extractive distillation. For this reason, this technology has been widely used to produce anhydrous ethanol in industry in China. The flowsheet of extractive distillation with the mixture of liquid solvent and solid salt is the same as in Figure 2.12 because the mixture of liquid solvent and solid salt is homogeneous.

However, the amount of solid salt added into the liquid solvent is often limited. Moreover, liquid solvents are volatile and thus can pollute the top product of extractive distillation column. New environmentally friendly separating agents should therefore be further explored.

2.4.5 Extractive distillation with ionic liquids

In recent years, ionic liquids have been very popular for their potential as environment-friendly soft materials. What are ionic liquids? Typical ionic liquids are composed of a large organic cation and an inorganic polyatomic anion. In theory the number of combinations of cations and anions can be up to 10! [26] On the other hand, unlike solid salts, they are generally at liquid state at ambient temperature. That is to say, ionic liquids as the separating agents in extractive distillation have the advantages of both liquid solvents (easy operation) and solid salts (high separation ability). Moreover, ionic liquids do not pollute the top product of the extractive distillation column and thus can be applied to separation and purification in biorefineries, pharmaceutics and food processing, and so forth. In this regard, they are "green" solvents.

A large number of cations and anions can be combined, so it is necessary for us to identify the relation between the molecular structure of separating agents and separation performance in order to reduce the amount of experimental work. It is beyond our scope to discuss how to synthesize the specific ionic liquids. For the separation of ethanol and water, Figures 2.14 and 2.15 show the influence of alkyl chain length and group substitution on the selectivity of ethanol to water, respectively. It was found that the suitable ionic liquids are of small molecular volume, an unbranched group, and with no sterical shielding effect around the anion. This conclusion was obtained from the calculated results from the Conductor-like Screening for Real Solvents (COSMO-RS) model, which is a predictive molecular thermodynamic model especially suitable for systems containing ionic liquids. The calculated results are consistent with the experimental data.

The meaning of abbreviations for cations is: $[C_2MIM]^+$ (1-ethyl-3-methylimidazolium), $[C_2DMIM]^+$ (1-ethyl-2,3-dimethylimidazolium), $[C_4MIM]^+$ (1-butyl-3-methylimidazolium) and $[C_8MIM]^+$ (1-methyl-3-octylimidazolium), while for anions it is: $[OAc]^-$ (acetate), $[BBB]^-$(bis[1,2-benzenediolato(2-)-O,O']-borate), $[B(CN)_4]^-$ (tetracyanoborate), $[BF_4]^-$ (tetrafluoroborate), $[BMA]^-$ (bis(methylsulfonyl)amide), $[BMB]^-$ (bis(malonato(2-))borate), $[BOB]^-$(bis(oxalato(2-))borate), $[BSB]^-$ (bis(salicylato(2-))-borate), $[BTA]^-$ (bis(trifluoromethylsulfonyl)amide), $[CF_3SO_3]^-$ (trifluoromethylsulfonate), $[CH_3SO_3]^-$ (methylsulfonate), $[CH_3SO_4]^-$ (methylsulfate), $[C_2H_5SO_4]^-$ (ethylsulfate), $[C_8H_{17}SO_4]^-$ (octylsulfate), $[Cl]^-$ (chloride), $[DMPO_4]^-$ (dimethylphosphate), $[DEPO_4]^-$ (diethylphosphate), $[HSO_4]^-$ (hydrogensulfate), $[MAcA]^-$ (N-methylsulfonylacetamide), $[MDEGSO_4]^-$ (2-(2-methoxyethoxy)ethylsulfate), $[N(CN)_2]^-$ (dicyanamide), [BMA] (bis(methylsulfonyl)amide), $[PF_6]^-$ (hexafluorophosphate), $[SCN]^-$ (thiocyanate), $[TOS]^-$ (p-toluenesulfonate) and $[Sal]^-$ (salicylate).

The flowsheet of extractive distillation with ionic liquids is the same as in Figure 2.13 but the operation process would be stable and safe because we are not worried about the crystallization and concurrent jam that may be brought on by using solid salts as the separating agents.

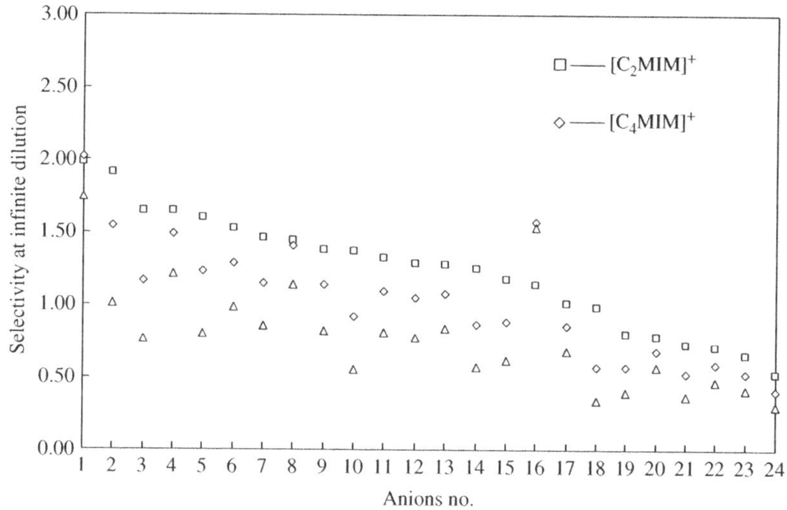

Figure 2.14 *Influence of alkyl chain length of the cations on the selectivity of ethanol to water at infinite dilution at 353.15 K. The corresponding no. of anions (1–24) is: 1—$[OAc]^-$; 2—$[HSO_4]^-$; 3—$[N(CN)_2]^-$; 4—$[DMPO_4]^-$; 5—$[SCN]^-$; 6—$[MAcA]^-$; 7—$[Sal]^-$; 8—$[CH_3SO_3]^-$; 9—$[CH_3SO_4]^-$; 10—$[BF_4]^-$; 11—$[BMA]^-$; 12—$[C_2H_5SO_4]^-$; 13—$[TOS]^-$; 14—$[CF_3SO_3]^-$; 15—$[BMB]^-$; 16—$[Cl]^-$; 17—$[MDEGSO_4]^-$; 18—$[PF_6]^-$; 19—$[BOB]^-$; 20—$[C_8H_{17}SO_4]^-$; 21—$[B(CN)_4]^-$; 22—$[BSB]^-$; 23—$[BBB]^-$; 24—$[BTA]^-$*

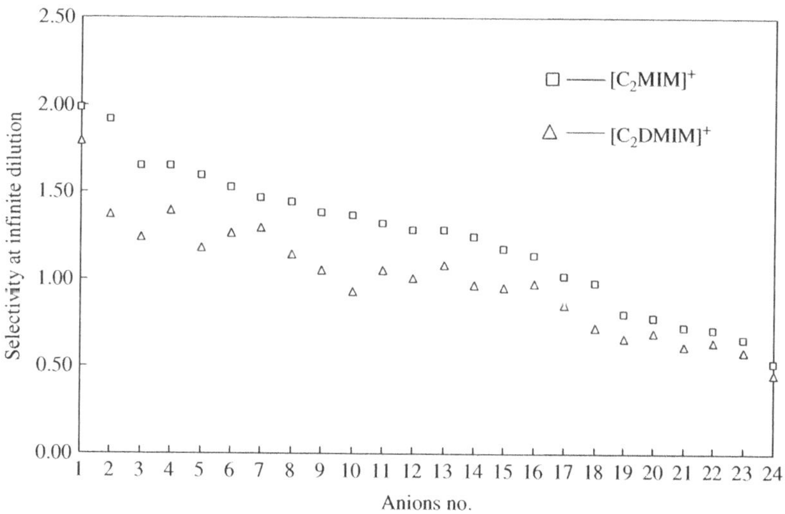

Figure 2.15 *Influence of group substitution between $[C_2MIM]^+$ and $[C_2DMIM]^+$ on the selectivity of ethanol to water at infinite dilution at 353.15 K. The corresponding number of anions (1–24) is the same as in Figure 2.14*

Hyperbranched polymers can also be used as separating agents in extractive distillation. However, research on them is very limited. This may be attributed to the difficulty in directly obtaining hyperbranched polymers from the chemical markets.

2.4.6 Examples in biorefineries

A typical application of extractive distillation in biorefineries is to produce anhydrous ethanol, so it is important for us to compare extractive distillation operating performance between liquid solvent, the mixture of liquid solvent and solid salt, and ionic liquids as separating agents, and to determine which is the most suitable. The results are given in Table 2.3, where commonly used separating agents, such as ethylene glycol, are compared, for example ethylene glycol, $[C_2MIM]^+[BF_4]^-$ and $[C_4MIM]^+[BF_4]^-$, and the physical properties of ionic liquids are taken from the references [27–29]. Under the same separation requirements—ethanol product purity and production capacity—the heat duty on the condenser and reboiler is in the order of Process E < Process D < Process F. For ionic liquids, the shorter the alkyl chain length, the smaller the energy consumption. In summary, extractive distillation with the mixture of liquid solvent and solid salt is better than the others. However, if the feeding mixture is at low ethanol concentration, it is suggested that it first be concentrated above 85 wt% by ordinary distillation in order to decrease the amount of separating agent in the following extractive distillation.

2.5 Molecular distillation

2.5.1 Introduction

Molecular distillation is generally accepted as the safest method to separate and purify thermally unstable compounds and substances having low volatility with high boiling points [30]. The process is distinguished by the following features: short residence time in the zone of the molecular evaporator exposed to heat; low operating temperature due to high vacuum in the space of distillation; and a characteristic mechanism of mass transfer in the gap between the evaporating and condensing surfaces. The separation principle of molecular distillation is based on the difference of molecular mean free path. The passage of the molecules through the distillation space should be collision free. Their mean free path, $\langle\lambda\rangle$, is defined by the following

Table 2.3 *Comparison of operating performance with various separation agents*

Contents	Liquid solvent (Process D)	Mixture of liquid solvent and solid salt (Process E)	Ionic liquids (Process F)	
Separating agent	ethylene glycol	ethylene glycol + $CaCl_2$	$[C_2MIM]^+$ $[BF_4]^-$	$[C_4MIM]^+$ $[BF_4]^-$
Ethanol product purity (mol %)	99.70	99.70	99.70	99.70
Ethanol molar flowrate (kmol·h^{-1})	89.2	89.2	89.2	89.2
Heat duty on condenser (kW)	1188.7	1122.2	1591.9	1640.1
Heat duty on reboiler (kW)	1882.4	1809.3	2213.0	2353.5

relation, derived from the theory of ideal gases:

$$\langle\lambda\rangle = \frac{1}{\sqrt{2}\pi d^3 n} = \frac{kT}{\sqrt{2}\pi d^3 P} = \frac{RT}{\sqrt{2}\pi d^2 N_A P} \tag{2.17}$$

where d is the molecular diameter (m), N_A is Avogadro constant ($=6.023 \times 10^{23}$ mol^{-1}), P is pressure (Pa) and T is temperature (K).

2.5.2 Examples in biorefineries

In theory, molecular distillation can also be used for separating mixtures with close boiling points or forming azeotrope because the constituents' molecular diameters are frequently not identical. The constituent with small molecular diameter should be more volatile than that with big molecular diameter according to Eq. (2.17). However, the motivation of molecular distillation arises not for this purpose, but for separating heat-sensitive compounds. By now, this technique is mostly applied in the fields of medicine and biology. Table 2.4 lists some applications of molecular distillation collected from SCI (Science Citation Index) database. The reason why molecular distillation is not widely applied as a special distillation process for separating the common mixture with close boiling point or forming azeotrope may be:

- limited theoretical stages, when compared with other special distillation columns;
- low production scale, but high energy consumption per amount of product;
- complicated equipment and high investment cost in order to achieve a high vacuum degree.

2.5.3 Mathematical models

The mathematical models for describing molecular distillation can be divided into two types: one is the phenomenological model, for example Response Surface Methodology (RSM) [31, 32], Artifical Neural Network (ANN) method [33], and the Takagi–Sugeno fuzzy model [34], which does not require knowledge specifically related to the separation process, but only the input and output data and the interior relationship between parameters; the other is mechanistic model, for example NEQ stage model where the mass and heat transfer rates are considered and the calculated results will be in quantitative agreement with the experimental data [35]. But the NEQ stage model is complicated and more design information on the configuration of molecular distillation apparatus must be specified so that mass transfer coefficients, interfacial areas, liquid hold-ups, and so forth, can be obtained. However, in modern chemical process simulation software such as ASPEN Plus and PRO/II, models of molecular distillation are not available [36]. To simplify, one can use a sequence of flash vessels to simulate the molecular distillation.

Molecular distillation is currently used in the production of high value-added products in fine chemicals, pharmaceutical, nutraceutical, petrochemical and biorefineries at a bench scale, as well as in scientific research on process organic chemicals with high molecular weight, high boiling point, and heat sensitivity. Due to the high vacuum degree at a pressure of about $10\text{–}10^{-1}$ Pa, molecular distillation on a large production scale is not possible compared to ordinary distillation and other special distillation processes.

2.6 Comparisons of different distillation processes

Four kinds of distillation processes are compared in Table 2.5, in which the numbers 1, 2 and 3 represent the low, middle and high degrees that they fit, respectively. All of them are currently used in industry and the choice of suitable distillation process depends on the specific separation task and economic considerations.

Table 2.4 *Applications of molecular distillation in biorefineries*

No.	Cases
1	High concentration of monoglycerides[a, b]
2	Fractionation of dimmers of fatty acids[c]
3	Examination of natural vitamin D[d]
4	Separation of crude meadowfoam fatty acids[e]
5	deacidification of crude low-calorie cocoa butter[f]
6	Removing terpenes from sweet orange oil[g]
7	Cell culture supports for slam-frozen[h]
8	Separation of triacylglycerol and free fatty acid mixtures[i]
9	Enriching tocopherols in soybean oil deodorizer distillate[j]
10	Polymerized drying oils[k]
11	Upgrading of bio-oil fraction in carboxylic acids and ketones[l]
12	Separation of tocopherols from soya sludge[m]
13	Production of biodiesel from waste cooking oil[n]
14	Recovering biodiesel and carotenoids from palm oil[o]
15	Recovering vitamin E from vegetal oils[p]
16	Concentration of monoglycerides[q]
17	Recovery of carotenoids from palm oil[r]
18	Separation of diacylglycerols from enzymatically hydrolyzed soybean oil[s]
19	Fructose monocaprylate and polycaprylate purification[t]
20	Enrichment of decanoic acid in cuphea fatty acids[u]
21	Grape seed oil deacidification[v]
22	Purification of crude octacosanol extract from rice bran wax[w, x]
23	Separation of bio-oil from biomass pyrolysis[y–aa]
24	Purification of 1,2-diacylglycerols from vegetable oils[bb, cc]
25	Extraction of tocotrienols from palm fatty acid distillates[dd]
26	Free fatty acid separation from vegetable oil deodorizer distillate[ee]
27	Recovering tocopherol and fatty acid methyl esters from rapeseed oil deodorizer distillate[ff, gg]
28	Concentration of tocopherols from natural sources[hh]
29	Vegetable oils refined by molecular distillation[ii]
30	Wheat germ oil refined by molecular distillation[jj]
31	Concentration of ethyl esters of eicosapentaenoic and docosahexaenoic acids[kk]
32	Hydroxytyrosol from *olea europaea* L. leaves[ll]

[a]H. Szelag and W. Zwierzykowski, The application of molecular distillation to obtain high concentration of monoglycerides, *Fett. Wiss. Technol.*, **85**, 443–446 (1983).

[b]L.V. Fregolente, P.B.L. Fregolente, A.M. Chicuta, C.B. Batistella, R.M. Filho and M.R. Wolf-Maciel, Effect of operating conditions on the concentration of monoglycerides using molecular distillation, *Chem. Eng. Res. Des.*, **85**, 1524–1528 (2007).

[c]H. Szelag and W. Zwierzykowski, Fractionation of dimers of fatty acids by molecular distillation, *Przem. Chem.*, **62**, 337–338 (1983).

[d]K.C.D. Hickman and E.L. Gray, Molecular distillation examination of natural vitamin D, *Ind. Eng. Chem.*, **30**, 796–802 (1938).

[e]C.C. Steven and A.I. Terry, Pilot-plant distillation of meadowfoam fatty acids http://www.sciencedirect.com/science/article/pii/S0926669001001054 - item1#item1, *Ind. Crop. Prod.*, **15**, 145–154 (2002).

[f]W.L. Wu, C. Wang and J.X. Zheng, Optimization of deacidification of low-calorie cocoa butter by molecular distillation, *LWT-Food Sci. Technol.*, **46**, 563–570 (2012).

[g]K. Liu, Y. Xu, K. and X. Wang, Microencapsulation of sweet orange oil terpeneless using the orifice method, *J. Food Eng.*, **110**, 390–394 (2012).

[h]S.L. White, D.A. Laska, P.S. Foxworthy, J.L. Gimple and D.M. Hoover, Cell culture supports for slam-frozen and molecular distillation dried procedures, *Microsc. Res. Tech.*, **26**, 184–185 (1993).

[i]G. Arzate-Martínez, A. Jimeénez-Gutiérrez, and H.S. Garcíahttp://pubs.acs.org/doi/abs/10.1021/ie200096q?prevSearch=%255BAbstract%253A%2Bmolecular%2Bdistillation%255D&searchHistoryKey=–cor1#cor1, Experimental analysis and modeling of the separation of triacylglycerol and free fatty acid mixtures using molecular distillation, *Ind. Eng. Chem. Res.*, **50**, 11237–11244 (2011).

[j]P.F. Martins, C.B. Batistella, R. Maciel-Filho and M.R. Wolf-Maciel, Comparison of two different strategies for tocopherols enrichment using a molecular distillation process, *Ind. Eng. Chem. Res.*, **45**, 753–758 (2006).

Table 2.4 *(continued)*

[k]R.S. Morse, Molecular distillation of polymerized drying oils, *Ind. Eng. Chem.*, **38**, 1039–1043 (1941).
[l]Z.G. Guo, S.R. Wang, G.H. Xu and Q.J. Cai, Upgrading of bio-oil molecular distillation fraction with solid acid catalyst, *Bioresources*, **6**, 2539–2550 (2011).
[m]E.B. de Moraes, P.F. Martins, C.B. Batistella, M.E. Alvarez, R. Maciel Filho and M.R. Maciel, Molecular distillation—A powerful technology for obtaining tocopherols from soya sludge, *Appl. Biochem. Biotechnol.*, **129–132**, 1066–1076 (2006).
[n]Y. Wang, J. Nie, M. Zhao, S. Ma, L. Kuang, X. Han and S. Tang, Production of biodiesel from waste cooking oil via a two-step catalyzed process and molecular distillation, *Energy Fuels*, **24**, 2104–2108 (2010).
[o]C.B. Batistella, E.B. Moraes, R. Maciel Filho and M.R. Wolf Maciel, Molecular distillation process for recovering biodiesel and carotenoids from palm oil, *Appl. Biochem. Biotechnol.*, **98**, 1149–1159 (2002).
[p]C.B. Batistella, E.B. Moraes, R. Maciel Filho and M.R. Wolf Maciel, Molecular distillation—Rigorous modeling and simulation for recovering vitamin E from vegetal oils, *Appl. Biochem. Biotechnol.*, **98**, 1187–1206 (2002).
[q]J. Kaplon, K. Minkowski and E. Kaplon, Distillation rate and separation efficiency of two-component mixture in wiped-film molecular evaporator, *Inz. Chem. Procesowa.*, **22**, 627–632 (2001).
[r]C.B. Batistella and M.R. Wolf Maciel, Recovery of carotenoids from palm oil by molecular distillation, *Comput. Chem. Eng.*, **22**, S53–S60 (1998).
[s]Y. Wang, M. Zhao, K. Song, L. Wang, X. Han, S. Tang and Y. Wang, Separation of diacylglycerols from enzymatically hydrolyzed soybean oil by molecular distillation, *Sep. Purif. Technol.*, **75**, 114–120 (2010).
[t]I. Redmann, D. Montet, F. Bonnot, M. Pina, J. Graille, Fructose mono- and polycaprylate purification by molecular distillation, *Biotechnol. Tech.*, **9**, 123–126 (1995).
[u]S.C. Cermak, A.L. John and R.L. Evangelista, Enrichment of decanoic acid in cuphea fatty acids by molecular distillation, *Ind. Crop. Prod.*, **26**, 93–99 (2007).
[v]M. Martinello, G. Hecker and M.D.C. Pramparo, Grape seed oil deacidification by molecular distillation: Analysis of operative variables influence using the response surface methodology, *J. Food Eng.*, **81**, 60–64 (2007).
[w]F. Chen, T. Cai, G. Zhao, X. Liao, L. Guo and X. Hu, Optimizing conditions for the purification of crude octacosanol extract from rice bran wax by molecular distillation analyzed using response surface methodology, *J. Food Eng.*, **70**, 47–53 (2005).
[x]F. Chen, Z. Wang, G. Zhao, X. Liao, T. Cai, L. Guo and X. Hu, Purification process of octacosanol extracts from rice bran wax by molecular distillation, *J. Food Eng.*, **79**, 63–68 (2007).
[y]S. Wang, Y. Gu, Q. Liu, Y. Yao, Z. Guo, Z. Luo and K. Cen, Separation of bio-oil by molecular distillation, *Fuel Process. Technol.*, **90**, 738–745 (2009).
[z]X. Guo, S. Wang, Z. Guo, Q. Liu, Z. Luo and K. Cen, Pyrolysis characteristics of bio-oil fractions separated by molecular distillation, *Appl. Energ.*, **87**, 2892–2898 (2010).
[aa]Z. Guo, S. Wang, Y. Gu, G. Xu, X. Li and Z. Luo, Separation characteristics of biomass pyrolysis oil in molecular distillation, *Sep. Purif. Technol.*, **76**, 52–57 (2010).
[bb]D.L. Compton, J. A. Laszlo, F. J. Eller and S. L. Taylor, Purification of 1,2-diacylglycerols from vegetable oils: Comparison of molecular distillation and liquid CO_2 extraction, *Ind. Crop. Prod.*, **28**, 113–121 (2008).
[cc]Y. Wang, M. Zhao, K. Song, L. Wang, X. Han, S. Tang and Y. Wang, Separation of diacylglycerols from enzymatically hydrolyzed soybean oil by molecular distillation, *Sep. Purif. Technol.*, **75**, 114–120 (2010).
[dd]L.R. Posada, J. Shi, Y. Kakuda and S.J. Xue, Extraction of tocotrienols from palm fatty acid distillates using molecular distillation, *Sep. Purif. Technol.*, **57**, 220–229 (2007).
[ee]P.F. Martins, V.M. Ito, C.B. Batistella and M.R.W. Maciel, Free fatty acid separation from vegetable oil deodorizer distillate using molecular distillation process, *Sep. Purif. Technol.*, **48**, 78–84 (2006).
[ff]S.T. Jiang, P. Shao, L.J. Pan and Y.Y. Zhao, Molecular distillation for recovering tocopherol and fatty acid methyl esters from rapeseed oil deodoriser distillate, *Biosyst. Eng.*, **93**, 383–391 (2006).
[gg]P. Shao, S.T. Jiang and Y. Li, Molecular distillation and enzymatic reaction for concentration of vitamin E from rapeseed oil deodorizer distillate, *Agr. Sci. (China)*, **39**, 2570–2576 (2006).
[hh]J. Green and P. R. Watt, The concentration of tocopherols from natural sources by molecular distillation, *J. Sci. Food Agr.*, **1**, 157–162 (1950).
[ii]P. Šimon and J. Cvengroš, Thermooxidative stability of vegetable oils refined by steam vacuum distillation and by molecular distillation, *Eur. J. Lipid Sci. Tech.*, **112**, 1236–1240 (2010).
[jj]M.A. Martinello, M. Villegas and M.D.C. Pramparo, Retaining maximum antioxidative potency of wheat germ oil refined by molecular distillation, *J. Sci. Food Agr.*, **87**, 1559–1563 (2007).
[kk]P.C. Rossi, M.D.C. Pramparo, M.C. Gaich, N. R. Grosso and V. Nepote, Optimization of molecular distillation to concentrate ethyl esters of eicosapentaenoic (20:5 ω-3) and docosahexaenoic acids (22:6 ω-3) using simplified phenomenological modeling, *J. Sci. Food Agr.*, **91**, 1452–1458 (2011).
[ll]M. Rada, Á. Guinda and J. Cayuela, Solid/liquid extraction and isolation by molecular distillation of hydroxytyrosol from Olea europaea L. leaves, *Eur. J. Lipid Sci. Tech.*, **109**, 1071–1076 (2007).

Table 2.5 *Comparison of four kinds of distillation processes*

Items	Ordinary distillation	Molecular distillation	Azeotropic distillation	Extractive distillation
Energy consumption	2	1	3	2
Production scale	3	1	3	3
Investment cost	1	3	2	2
Operation complexity	3	1	2	2

2.7 Conclusions and future trends

Ordinary distillation is the most mature of the separation processes, and thus there is very limited opportunity to improve internal distillation equipment. Molecular distillation is suitable for the production of high value-added products at a low scale due to the requirement for higher vacuums. Efforts should be made to increase the prediction accuracy of mathematical models of molecular distillation under extreme conditions to arrive at a "scientific design." Extractive distillation is used more in biorefineries than azeotropic distillation because the former saves energy. Since the separating agent is the key technology of extractive distillation, predictive thermodynamic models should be developed to screen the suitable separating agents rapidly in order to decrease the amount of experimental work. The COSMO-RS model can be used to make an a priori prediction of separation performance and the results are qualitatively consistent with the experimental data. But the UNIFAC Universal Quasichemical Functional-Group Activity Coefficients (UNIFAC) model can provide quantitative prediction even for systems containing ionic liquids. It is expected that more main groups and subgroups for ionic liquids will fill the gap in the current UNIFAC parameter matrix so that we can identify the relationship between the molecular structure of separating agents and separation performance.

Acknowledgement

This work was financially supported by the National Nature Science Foundation of China under Grants (Nos. 21121064 and 21076008), and the Projects in the National Science and Technology Pillar Program during the twelfth Five-Year Plan Period (No. 2011BAC06B04).

References

1. Z. Lei, B. Chen and Z. Ding, *Special Distillation Processes*, Elsevier, Amsterdam, 2005.
2. Z. Lei, C. Li and B. Chen, Extractive distillation: A review, *Sep. Purif. Rev.*, 32, 212–213 (2003).
3. Z. Lei, H. Wang, R. Zhou and Z. Duan, Influence of salt added to solvent on extractive distillation, *Chem. Eng. J.*, 87, 149–156 (2002).
4. Z. Lei, R. Zhou and Z. Duan, Process improvement on separating C4 by extractive distillation, *Chem. Eng. J.*, 85, 379–386 (2002).
5. H. Wu, Q. Li and Z. Li, Research of a new flow-guided sieve tray, *Chem. Ind. Eng. Prog. (China)*, 25, 21–24 (2006).

6. Q. Li, A. Wang, M. Mao and Z. Zhang, A theoretical research of high efficiency flow-guided sieve tray and its successful application in extractive distillation of isoprene, *Petrochem. Technol. (China)*, 33, 368–371 (2004).
7. Ž. Olujić, A.F. Seibert and J.R. Fair, Influence of corrugation geometry on the performance of structured packings: an experimental study, *Chem. Eng. Process.*, 39, 335–342 (2000).
8. S.Y. Wang, *Modern Packing Tower Technology*, China Petrochemical Press, Beijing, 1998.
9. J.R. Fair, A.F. Seibert, M. Behrens, P.P. Saraber and Ž. Olujić, Structured packing performances experimental evaluation of two predictive models, *Ind. Eng. Chem. Res.*, 39, 1788–1796 (2000).
10. Q. Li, Q. Chang, Y. Tian and H. Liu, Cold model test and industrial applications of high geometrical area packings for separation intensification, *Chem. Eng. Process.*, 48, 389–395 (2009).
11. H. Huang, S. Ramaswamy, U.W. Tschirner and B.V. Ramarao, A review of separation technologies in current and future biorefineries, *Sep. Purif. Technol.*, 62, 1–21 (2008).
12. N. Nguyen and Y. Demirel, Retrofit of distillation columns in biodiesel production plants, *Energy*, 35, 1625–1632 (2010).
13. I. Atadashi, M. K. Aroua and A. Abdul Aziz, Biodiesel separation and purification: A review, *Renew. Energ.*, 36, 437–443 (2011).
14. J. Li, Z. Lei, Z. Ding, C. Li and B. Chen, Azeotropic distillation: A review of mathematical models, *Sep. Purif. Rev.*, 34, 87–129 (2005).
15. S. Widagdo and W.D. Seider, Azeotropic distillation, *AIChE J.*, 42, 96–130 (1996).
16. R.F. Gould, *Extractive and Azeotropic Distillation*, American Chemical Society, Washington, DC, 1972.
17. E.J. Hoffman, *Azeotropic and Extractive Distillation*, John Wiley & Sons, Inc., New York (1964).
18. Z. Lei, B. Chen, C. Li and H. Liu, Predictive molecular thermodynamic models for liquids, solid Salts, polymers, and ionic liquids, *Chem. Rev.*, 108, 1419–1455 (2008).
19. Z. Lei, W. Arlt and P. Wasserscheid, Separation of 1-hexene and n-hexane with ionic liquids, *Fluid Phase Equilibr.*, 241, 290–299 (2006).
20. Z. Lei, W. Arlt and P. Wasserscheid, Selection of entrainers in the 1-hexene/n-hexane system with a limited solubility, *Fluid Phase Equilibr.*, 260, 29–35 (2007).
21. Z. Lei, J. Zhang, Q. Li and B. Chen, UNIFAC model for ionic liquids, *Ind. Eng. Chem. Res.*, 48, 2697–2704 (2009).
22. Z. Lei, B. Chen and C. Li, COSMO-RS modeling on the extraction of stimulant drugs from urine sample by the double actions of supercritical carbon dioxide and ion liquid, *Chem. Eng. Sci.*, 62, 3940–3950 (2007).
23. M. Seiler, C. Jork, A. Kavarnou, W. Arlt and R. Hirsch, Separation of azeotropic mixtures using hyperbranched polymers or ionic liquids, *AIChE J.*, 50, 2439–2454 (2004).
24. M. Seiler, D. Kohler and W. Arlt, Hyperbranched polymers: new selective solvents for extractive distillation and solvent extraction, *Sep. Purif. Technol.*, 29, 245–263 (2002).
25. M. Seiler, Hyperbranched polymers: Phase behavior and new applications in the field of chemical engineering, *Fluid Phase Equilib.*, 241, 155–174 (2006).
26. A. Szanyi, P. Mizsey and Z. Fonyo, Novel hybrid separation processes for solvent recovery based on positioning the extractive heterogeneous-azeotropic distillation, *Chem. Eng. Process.*, 43, 327–338 (2004).
27. J.O. Valderrama and K. Zarricueta, A simple and generalized model for predicting the density of ionic liquids, *Fluid Phase Equilib.*, **275**, —145–154 (2009).
28. Y. Ge, L. Zhang, X. Yuan, W. Geng and J. Ji, Selection of ionic liquids as entrainers for separation of (water + ethanol), *J. Chem. Thermodyn.*, 40, 1248–1252 (2008).
29. L.G. Remesh and J.A.P. Coutinho, A group contribution method for heat capacity estimation of ionic liquids, *Ind. Eng. Chem. Res.*, 47, 5751–5757 (2008).
30. J. Lutisan and J. Cvengros, Mean free path of molecules on molecular distillation, *Chem. Eng. J.*, 56, 39–50 (1995).
31. M. Martinello, G. Hecker and M.D.C. Pramparo, Grape seed oil deacidification by molecular distillation: Analysis of operative variables influence using the response surface methodology, *J. Food Eng.*, 81, 60–64 (2007).

32. F. Chen, T. Cai, G. Zhao, X. Liao, L. Guo and X. Hu, Optimizing conditions for the purification of crude octacosanol extract from rice bran wax by molecular distillation analyzed using response surface methodology, *J. Food Eng.*, 70, 47–53 (2005).
33. P. Shao, S.T. Jiang and Y.J. Ying, Optimization of molecular distillation for recovery of tocopherol from tapeseed oil deodorizer distillate using response surface and artificial neural network models, *Food Bioprod. Process.*, 85, 85–92 (2007).
34. N.M.N. Lima, L.Z. Liñan, F. Manenti, R.M. Filho, M.R.W. Maciel, M. Embiruçu and L.C. Medina, Fuzzy cognitive approach of a molecular distillation process, *Chem. Eng. Res. Des.*, 89, 471–479 (2011).
35. M.A. Durán, R.M. Filho and M.R.W. Maciel, Rate-based modeling approach and simulation for molecular distillation of green coffee oil, *Computer Aided Chem. Eng.*, 28, 259–264 (2010).
36. E.S. Mallmann, C.B.B. Costa, M.R.W. Maciel and R.M. Filho, A new computational tool for falling film molecular distillation performance prediction, *Computer Aided Chem. Eng.*, 27, 1905–1910 (2009).

3

Liquid-Liquid Extraction (LLE)

Jianguo Zhang and Bo Hu

Department of Bioproducts and Biosystems Engineering, University of Minnesota, USA

3.1 Introduction to LLE: Literature review and recent developments

Liquid-liquid extraction (LLE), also known as *solvent extraction*, is a typical ternary system in the chemical engineering field, which separates chemicals from one solution to another based on the different solubility of the solute chemical in two solvents. Three components are involved in the LLE process: *solute*, *diluent*, and *extractant*. *Solute*, which is dissolved in *diluent*, is extracted from the diluent and dissolved into another solvent, the *extractant*. A typical LLE process is as follows. Solute A represents impurities; solute B represents the component to be separated from diluent and/or solute A. Three steps are involved in the process. The first step is the addition of the extractant, which should be immiscible with diluent, where the solutes (A and B) are dissolved. The second step is to mix these two solvents. The interface between extractant and diluent can be significantly improved by mixing; most commonly either the extractant or diluent forms droplets due to the surface tension. Solutes (A and B) have the opportunity to choose the host solvent according to their different solubility in extractant and diluent. A well designed extraction system should have solute B mainly dissolved in the extractant, while solute A remains with the diluent. The final step is to layer the extractant and diluent. At this stage, it is easy to separate solute B from solute A simply by separating extractant from diluent. Solute B can be recovered from the extractant by other methods such as distillation or simple evaporation.

Multiple factors need to be considered when applying the LLE process. Determining how best to choose an extractant is usually the key to the successful design of LLE separation. Liquid-liquid equilibrium data for solvent systems with critical solvent condition (CST) are available in many comprehensive data collections; for instance, more than 6000 critical solution temperature observations are listed by Francis,

Separation and Purification Technologies in Biorefineries, First Edition.
Edited by Shri Ramaswamy, Hua-Jiang Huang, and Bandaru V. Ramarao.

and more than 300 of them involve water as one of the solvents [1]. The requirements for extractants are as follows:

- the extractant should have specific selectivity for the solute so that most of desirable solute can be dissolved in the extractant from the diluent, and vice versa for undesirable compounds;
- the solute should be easily separated from extractant after the LLE;
- a high partition ratio is another property allowing effective separation of a desirable compound from the diluent;
- the extractant should have high solubility for the target solute and low solubility for the diluent;
- the extractant should have a different density from diluents for easy layering;
- high interfacial tension should be present for rapid coalescence for extractant droplets after agitation;
- low toxicity for the extractant and diluents is needed for industrial use.

Ideal extractants that can meet all these requirements can be extremely difficult to find. Common issues related to LLE include emulsion formation and the use of large volumes of toxic organic solvents; in addition, LLE can be a time-consuming process [2]. Different miniaturization techniques have been developed by drastically reducing the volume of extractant in order to manage the toxic extractant solvents better in different micro-extraction systems, such as capillary liquid droplets [3], supported liquid films or droplets [4], and continuous forming and falling drop systems [5]. For example, Jeannot developed a well designed single-drop LLE system [6], and these types of the miniaturized LLE are widely used in analytical research [7, 8]. Modifiers are sometimes added to the LLE system to improve the properties of the extractant [9]. Nevertheless, permanent modifiers usually have disadvantages, such as poorly reproducible treatment technologies, impaired efficiency compared with the modifier addition to each sample aliquot, relatively short lifetimes, limitations imposed on temperature programs, and applicability to relatively simple matrices [10].

Mixing and layering are also important and necessary steps to increase the solute transfer from diluents to extractants and eventually to separate extractant from solute. The final productivity for the LLE ternary systems may depend on the interaction between diluent and extractant, as well as on the phase separation. Critical conditions are sometimes applied to create a homogenous diluent and extractant solution, which will maximize their interface for solute transfer, and no mechanical stir is needed in this process [11]. Some operational condition factors can be changed to induce and enhance the phase separation, such as temperature and composition [12–14].

Some LLE processes are slow due to the tiny density difference between two immiscible phases. To improve this process, an electrical driving force, such as a high-voltage electric field, is applied to the LLE process to control the shape, size, and dispersed drops, to increase the interface of two phases, and to increase the distribution rate of solute [15]. The application of an electric current can help mix the two phases efficiently and quickly through the droplet interaction. The droplets have high uniformity and widely controllable size range according to Sato *et al*. [16]. The disadvantage of electric LLE is that special equipment is needed. The equipment comprises three subgroups: perforated-plate and spray, mixed contactors, and liquid-film contactors.

3.2 Fundamental principles of LLE

The LLE process is based on the solubility of solute in different solvents. For a given system, the solute will distribute in different solvents at a certain partition ratio, and this ratio will not change as the system reaches equilibrium. The properties of the LLE system, and some important definitions, are as follows:

Equilibrium partition ratios. In the ideal equilibrium state, the solute concentration is low enough that every solute molecule is surrounded by solution and the two phases of the dissolved solute are in the same molecular state; therefore, the weight fraction of solute in extractant (y) divided by the weight fraction of solute in the diluent (x) at the given stable condition is constant, termed *partition ratio (K)*, as shown in Eq. (3.1):

$$K = \frac{y}{x} \tag{3.1}$$

Distribution coefficient. In a real situation, the solute will be at different states in different phases, and different states of solute will exchange depending on the solution. Therefore, the partition ratio is not a constant—it will change as the extractant changes. The distribution coefficient is introduced to describe the ratio of solute concentration in extractant to that in diluents Eq. (3.2):

$$A = \frac{S_e}{S_d} \tag{3.2}$$

S_e: solute concentration in extractant (mol/L);
S_d: solute concentration in diluent (mol/L).

In a single extraction stage (the extractant is mixed with the diluent once), y is same as A. In most cases, the LLE system is operated through several rounds of extraction. The distribution coefficient is affected by the concentration of solute and extractant, pH, and any other solute in the extractant, as well as diluents. With a high value of A, the solute is more easily extracted. For the diluted solutions of the solute, under steady-state conditions, it is generally assumed that K will reach a constant value.

Separation factor. The separation factor is a dimensionless factor that measures the relative enrichment of solute in extractant after the LLE process. For a given separation process, several solutes are in diluents. The desirable and unwanted solutes need to be separated by LLE, and the separation factor is a parameter to describe the selectivity of solvent, as in Eq. (3.3).

$$S_t = \frac{(S_e/R_e)}{(S_d/R_d)} \tag{3.3}$$

S_f: separation factor;
S_e: desirable solute in extractant;
S_d: desirable solute in diluent;
R_e: unwanted solute in extractant;
R_d: unwanted solute in diluent.

Extraction factor. The extraction factor is defined as the total amount of solute in the extractant phase divided by total amount solute in the raffinate phase. Thus the extraction factor is related to the volume of extractant used in the system. For many commercial processes, the extraction factor is in the range of 1.3 to 5. The extraction factor depends on the operation factors, such as temperature, solution volume, and concentration of salt in solution.

Phase diagrams (two phases). Triangle (equilateral) phase diagrams are commonly used in the LLE system in order to illustrate the ternary liquid-liquid equilibrium; these phase diagrams are easy to use in the solvent extraction. Other types of triangle diagrams are also used, and the phase diagram is not limited to triangle diagrams. Rectangular coordinates are also used for representation in a phase diagram [17].

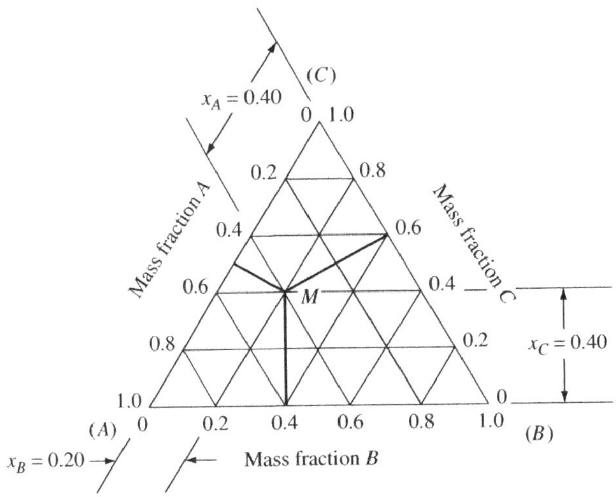

Figure 3.1 *The triangle phase diagram of the system with three components*

In Figure 3.1, an equilateral triangle phase diagram is chosen to explain how to use this diagram for the LLE calculation.

Each axis in the triangle can represent mass percentage, volume percentage, or molecular percentage of each component in the LLE process. For example, in Figure 3.1, the axis in the triangle represents the mass percentage. A represents the extractant, B represents the diluent, and C represents the solute. Three points (A, B, C) of the triangle represent the conditions of these three components with their own pure solution. The three side lines (AB, BC, AC) each represent a system with only two components in the mixture. The inside area of the triangle represents the mixture of three components in the LLE system. For example, the point M represents the system's three components (A, B, C). The concentration of each component at point M can be determined from the diagram in Figure 3.1. For the concentration of A, a parallel line of BC was draw through M, and the concentration of A (XA) is 0.4. Using the same method, concentration of B is 0.2, and concentration of C is 0.4. These types of calculation also can be determined with other kinds of phase diagram (right-angle triangle).

A typical phase diagram in the LLE process is shown in Figure 3.2. Solute A and solute B are partially miscible in each other and solute C is miscible in solute A or B. The curve apb (bimodal curve) in the triangle showed the miscible scope of A and B. The top part of triangle beyond curve apb is one phase (miscible). The remaining part of triangle is two phase. Only the ternary systems with miscibility gaps are suitable for extractions. Except for the plait point in the system, the points in the curve apb represent the critical condition. The lines that connect to a corresponding equilibrium point on the bimodal curve are termed *tie lines*. The equilibrium data must be experimentally obtained on a case-by-case basis. The points on the tie line are the two-phase system; the relative ratio of components A and B can be expressed by the length segment, such as M, the mass components of A/mass components of B = (length of MB)/(length of MA).

As shown in Figure 3.2, the bimodal curve is close to the line AB, which is a closed system. Several other types of phase diagrams are shown in Figure 3.3. Types I and III are open systems, in which the bimodal curve and line are not close. For example, in type I, B is not fully miscible in the whole range of A and C. In this condition, the capacity of mixture A and C is low, and the selectivity is usually high. Type II and IV are closed systems. The solute C is totally miscible with both A and B. The extraction is not possible if the concentration of C is higher than the point of C because there is a one-phase system in this condition. A is not fully miscible with the whole range mixture of B and C in type IV.

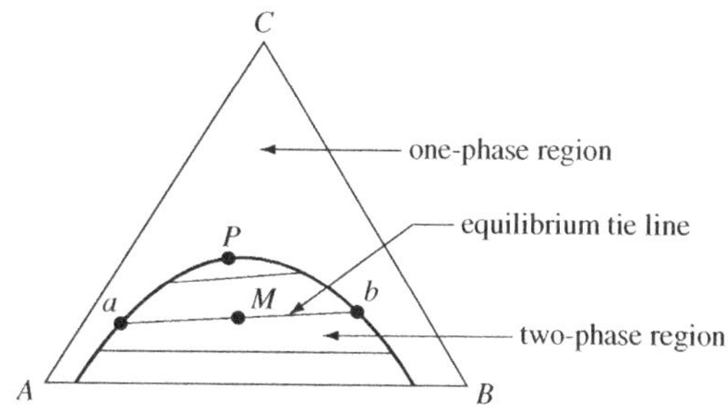

Figure 3.2 Typical triangle diagram of the liquid-liquid extraction (LLE) system

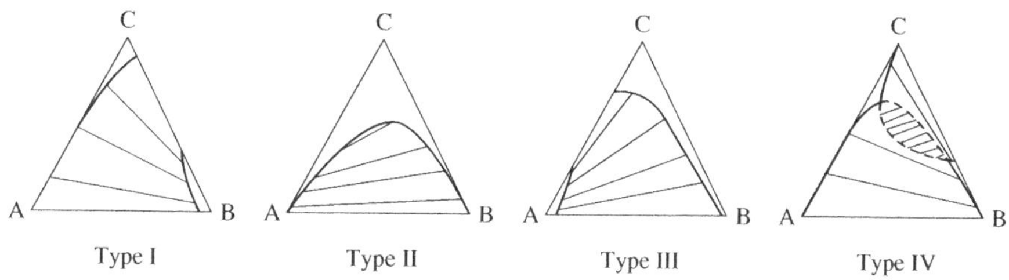

Figure 3.3 Four more types of triangle diagram in ternary liquid-liquid extraction (LLE) systems: Type I: Open system, with a miscibility gap between the key component of the solvent (B) and solvent (C). Type II: Closed system. C is completely miscible with the key component of the feed (A) and the key components of the extraction solvent (B). Type III: Open system, with a miscibility gap between A and C. Type IV: Closed system. C is completely miscible with the key component of the feed (A) and the key components of the extraction solvent (B). Also, A is not fully miscible in the mixture of B and C

3.3 Categories of LLE design

As an industrial operation unit, LLE technology is designed to fit into different processes for efficient extraction of different solutes. Several of the basic and most commonly used designs are described below.

Standard extraction. This extraction is the most basic and simple design of LLE. The extractant is mixed with diluents in a single stage, multiple stages, or continually at countercurrent flow direction (Figure 3.4). In most cases, extractant and diluent are fed into mixture vessel continually, and then the LLE is followed by evaporation to recycle the extractant. The continuous standard extraction process can extract solutes from diluents step by step, resulting in nearly complete extraction. Contaminants are a challenge for this simple process, as they transfer into extractant from diluents together with solute. To avoid unwanted chemicals, it is recommended that the ratio of Ki/Kj of solvents, should reach at least 20:1, depending on the solute and contaminant concentration in diluents.

Fraction extraction. Fraction extraction, known also as *dual solvent extraction*, is a version of standard extraction, modified by adding a washing step (Figure 3.5). The diluents enter in the middle of the solvent extraction process, and the washing solvent is fed into extractant in the countercurrent direction. Specifically, extractants will mix with two types of liquids sequentially. The extractant will first strip solute from diluents; then the extractant will be mixed with washing solvent to transfer unwanted chemicals into the washing solvent. The advantage of fraction extraction, compared with standard extraction, is the high

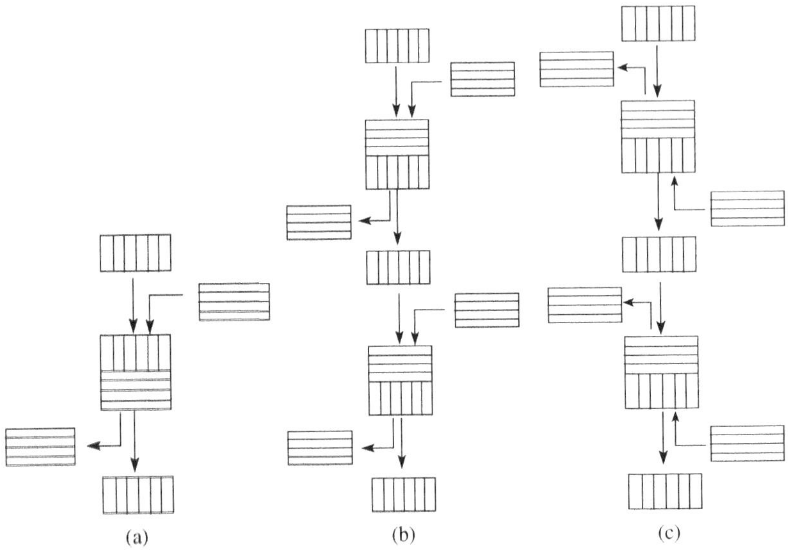

Figure 3.4 *Three types of standard extraction. Key: : diluent, : extractant, (a) single stage, (b) multiple stages, (c) countercurrent flow direction*

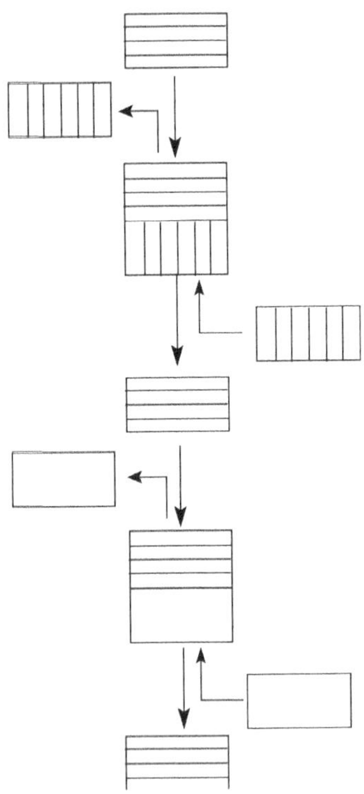

Figure 3.5 *Typical fraction extraction. Key: : diluent, : extractant, : washing solution*

recovery of solute and the low amount of contaminant transferred during the process. Fraction extraction can be applied to the system with the separation factor as low as 4. In addition, fraction extraction can be applied to several special cases, for example, recycling the washing solvent after solute is separated from extractant. Some components in the extractant after solute separation are recycled as the washing solvent; this is termed *single solvent fraction extraction*. Another approach is to recycle the extractant and washing solvent at the same time. The final stage is to reuse the extractant and the raffinate of diluents.

Dissociative extraction. Dissociative extraction is usually applied to the chemical solute with weak organic acids or bases in its solution. The solubility of solute with a neutral state in organic solvent is higher than that of one with a charged state condition. A charged solute has higher solubility in aqueous solution than in organic solution. The extraction process is based on the different solubility of the different states of solute in different solutions. The concept of dissociative extraction is explored in other chemical separations, such as the separation of antibiotics by pH-shift extraction, which means that a solute transfers from one solution to another by pH changes. Likewise, the temperature shift strategy is also used for solute extraction because the partition ratio changes with temperature.

Reaction-enhanced extraction. To change the solubility of a solute, some solute is used to react with an agent to obtain high solubility in an extractant. With the next step of LLE, further development of the reaction step is combined with separation to simplify the process, and this is termed *reaction extraction*. In most cases, the extractive reaction happens in the presence of the diluents and extractant. The modified solute is obtained as well as the solute that was dissolved into extractant. At the same time, low solute concentration in the diluents helps the extractive reaction to modify solute formation.

Hybrid extraction processes. These processes include any combination of extraction processes for extraction improvement, such as extraction-separation, reverse osmosis, and extraction. These combinations are based on specific products so it is difficult to describe all the combinations used industrially and in laboratories.

3.4 Equipment for the LLE process

3.4.1 Criteria

There are various requirements for designing and building an extractor. For best performance, extractors are expected to have the capabilities described below. In practice, however, meeting all the criteria is technically challenging. The criteria are listed here from the most important to least important for most cases. However, the sequence of importance may be reordered in specific conditions.

High volumetric efficiency is also known as the *product-specific extract ratio per unit area*. High volumetric efficiency is usually achieved with a high number of theoretical stages per unit and thus less volume required for the extractor column.

Avoiding surface-active impurities is important for columns with a long life. Surface-active impurities can reduce column capacity by more than 20% and efficiency by 60%. The efficiency of a rotating-disk contactor reduced faster than that of a small-diameter agitated column when trace surface-active contaminants were added [18].

Production capacity is determined according to the products. Some products need small extraction operations because of low throughput demand whereas others need large operations to make this process economically possible. It is necessary to understand the scale of operation that is required. A common strategy is to gain familiarity with the extractor on a small scale, and then extract the product on a large scale industrially.

Flexibility and the ease of scaling up refer to the ability to successfully follow the LLE process from small scale to large scale. In some cases, researchers can apply one extractor to several LLE processes; thus, this flexibility in terms of the extractor can create cost savings.

Tolerance for fouling, and ease of cleaning are important characteristics for any industrial equipment. The LLE extractor aspect of this process is the most serious concern because of the frequent handling of this chemical solution. Tolerance for fouling makes it possible to maintain a long operation cycle. Ease of cleaning saves break time in the process for maintenance.

Easy operation for the extractor can simplify the operation protocol and reduce the operator requirements. This type of operation can be more easily transferred to another plant or new plant and also facilitates troubleshooting during the years of operation.

The **availability of the extractor** is the last criterion listed here but it can still be a very important one. If the local extractor dealer can provide timely instructions, maintenance, and a troubleshooting service, this will show advantages during the long-term operation of the extractor.

For the industrial operation of LLE, better process control is needed for a large-scale system. Solvent loss during the process is an important problem. The reasons for solvent loss are solubility, vaporization, stable emulsions, entrainment of solvent, and sampling during the process. The equipment is another factor leading to solvent loss. Better operation with less solvent loss is more environmental friendly, especially for a large-scale operation.

3.4.2 Types of extractors

Several types of extractors were designed and are currently used in large-scale industrial production. These extractor types can be categorized into three groups: the static extraction column, the rotary or agitated column, and the centrifugal extractor.

The **static extraction column** is an extractor without any stirring mechanism to separate the extractant from diluents. Two types of liquid solvent are mixed and separated in the column after the solutes are transferred. Liquid solvents are moved into the static extraction column, and separated after the extraction, due to their different densities. As shown in Figure 3.6, three kinds of static extraction column have been designed, to give better performance. First, the spray column is the basic extractor without any facilities in the column. Next, the packed column is filled with metal, plastic, or ceramic materials to improve the separation efficiency. Last, the sieve-plate column disperses solvent into small drops using a plate with holes. The solvent dispersal enhances mass transfer and surface interaction.

Agitated columns were developed in order to control the efficiency of LLE. Three types of major agitated columns are illustrated in Figure 3.7: the rotary-impeller column, the reciprocation plate column, and the rotating-disk column.

The **centrifugal extractor** is an advanced design to separate extractant from diluents using a rapid centrifuge. Figure 3.8 shows the basic structure of a centrifugal extractor (Costner Industries—CINC). A hollow rotor is rotated inside the column. The light and heavy phases are separated by the centrifugal rotor. These phases run out as the mixed phase is fed in. With the development of LLE equipment, many commercial derivatives are possible [19].

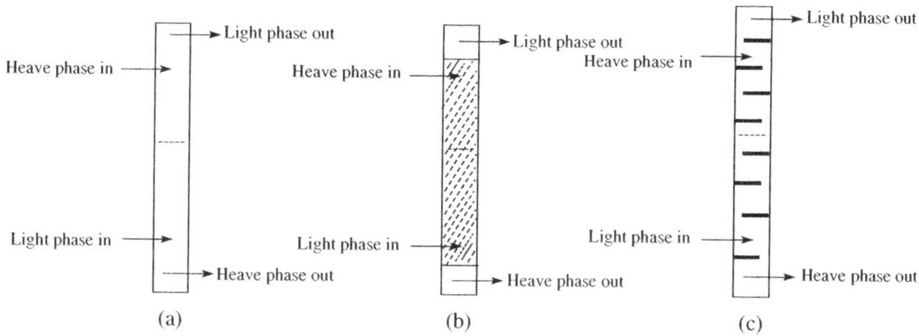

Figure 3.6 *Three kinds of static columns: (a) spray column; (b) packed column; (c) sieve plate column*

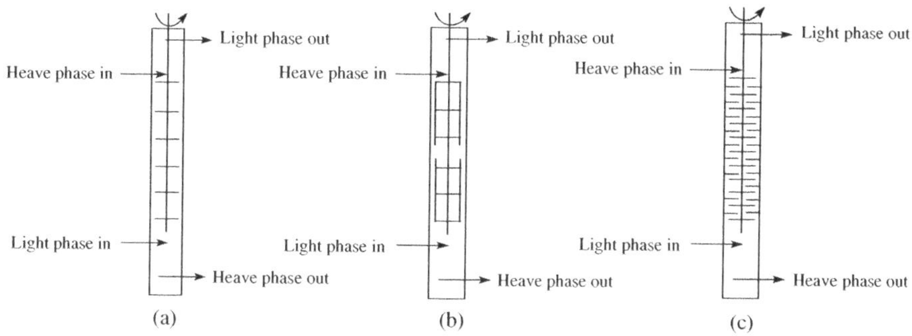

Figure 3.7 *Three kinds of rotary or agitated columns: (a) rotary-impeller column; (b) reciprocating-plate column; (c) rotating-disk column*

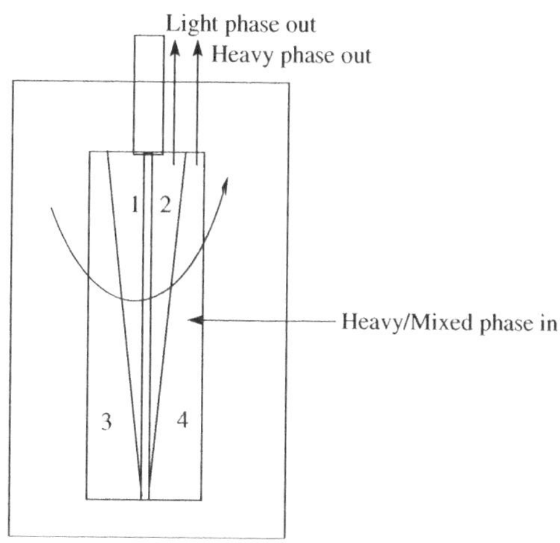

Figure 3.8 *Costner Industries centrifugal extractor. Area 1 and 2: light phase, Area 3 and 4: heavy phase*

3.4.3 Issues with current extractors

Some difficulties for LLE equipment to overcome are noted here. For static extraction columns, the process offers the advantages of a large diameter, simple construction and operation, only a single operating interface, and small footprint compared to mixer-settler equipment. However, some parameters still need to be improved, such as the interfacial area, drop size, and drop velocity. Another issue of concern is mass-transfer efficiency, which should be improved. An important development in static column extractor is redistribution in the column. The minimization of packing size and drop size increases the interface of solvent. An agitated extraction column is an improvement over the static column. Several kinds of agitation were reported recently, such as the SCHEIBEL column, the Kuhni column, and the KARR reciprocating column [19]. Centrifugal extractors also improve the LLE process because they reduce diffusion path lengths and increase the driving force for separation. The typical application of this design is the one-stage or multistage centrifugal extractor applied for penicillin extraction [20].

3.5 Applications in biorefineries

Liquid-liquid extraction has been widely applied in many industries, including petrochemical refining, pharmaceutical production, the food industry, metals, and acid production. A selected list of ternary systems for such applications was produced by Robbins [19]. Liquid-liquid extraction technology has also been applied in new areas, such as biotechnology, the pharmaceutical industry, agriculture, and wastewater treatment. The biorefinery domain is also an important area for LLE applications. Currently, many case studies are available on bioproduct separation by LLE, although no comprehensive review has been published on this topic to date. The LLE process features mild operational conditions and the process is easy to control, which is especially suitable for the biorefinery process with conversion using microorganisms. Some examples follow of LLE application in biorefineries.

3.5.1 Ethanol

Applications in the biorefinery industry are predominantly in lignocellulosic ethanol production. Ethanol is one of the major biorefinery products widely produced all over the world, especially in Brazil and the United States. Liquid-liquid extraction was also proposed for application in the pretreatment and hydrolysis step. Furfural, a common inhibition byproduct generated during the dilute acid pretreatment of lignocellulosic biomass, can be removed by the LLE process. In recent research, toluene was proven as the most effective solvent tried for the removal of furfural from aqueous solution after analyzing three systems (water-furfural-methyl isobutyl ketone, water-furfural-toluene, and water-furfurfal-isobutyl acetate) at 30 °C, although isobutyl acetate may be preferred because of its low toxicity [21]. Furfural, once purified, is also a raw material used to synthesize pharmaceutical precursors, nylons, lubricants, adhesives, and plastics, and it is also being investigated as a potential candidate for biofuel products. Furfural can be separated in LLE with very stable imidazolium-based ionic liquids for which equilibrium can be achieved within 30 minutes. A small amount of NaCl or Na_2SO_4 in the aqueous phase leads to a significant increase in the partition coefficients of furfural, plausibly due to hydrophobic interaction mechanisms [22].

Liquid-liquid extraction applications in lignocellulosic ethanol production are mainly focused on ethanol separation and purification, primarily due to the difficulty in separating ethanol from the ethanol-water mixture. Vane summarized the key technologies for ethanol separation from a fermentation broth [23]. Distillation is still the prevalent industrial practice, although it has the significant drawback of high energy and high temperature requirements. Liquid-liquid extraction is widely researched in the academia as an alternative solution; most studies on LLE for ethanol separation have focused on using ionic liquids as solvents.

Distribution coefficients and selectivities of a number of mixed solvent systems have been determined in order to assess their suitability for preferentially extracting ethanol from aqueous solution [24]. Considering that the composite of an ethanol water mixture is not consistent in industrial operation, the screening method for measuring equilibrium distribution coefficients to minimize the variation is necessary for the LLE application. The measures include entrainment, incomplete equilibration, impurities, and temperature. Several solvents were tested to obtain an efficient solvent [25]. The solvent was analyzed systematically by Offeman *et al*. [26] based on distribution coefficients. This study focused on varying chemical the structure of 57 alcohol solvents, including chain length, hydroxyl position for x-alcohols, and branch structure. For the unbranched alcohols, the separation factors increases as the hydroxyl group moved toward the middle of the chain, and with increasing molecular weight. For the branched alcohols with same molecular weight, the separation factor increases as the molecular weight increases. Factors such as temperature and solvent-to-feed ratio were optimized by Koullas [27]. Other solvents were also explored to extract ethanol from an aqueous solution. For instance, three ternary liquid-liquid systems containing water and ethanol were investigated, including amyl acetate, benzyl alcohol, and methyl isobutyl ketone, at 298.15 K and amyl acetate was found to be a better solvent than methyl isobutyl ketone and benzyl alcohol [28]. Another example includes three solvents considered as promising extractants: isoamyl acetate, isooctyl alcohol, and n-butyl acetate. In isoamyl acetate or isooctyl alcohol, the ethanol distribution coefficients were higher than 1 and the separation factors in Bancroft's coordinates of the order of 70 and 2000. At room temperature (30 °C), the solubility of ethanol in these systems increases with decreasing number of carbon atoms in the chain of solvents, giving higher values of the distribution coefficient and consequently lowering the separation factor. The distribution coefficients are greater than 1 and the separation factors are considerably greater than 2 for all the solvents studied (ethanol-water-I-butanol, ethanol-water-1-pentanol, and ethanol-water-1-hexanol ternary systems) [29].

Ionic liquids (ILs) are also applied to extract ethanol with much of the research focusing on imidazolium-based ILs as solvents. Recent studies have shown phosphonium-based ILs to have many advantages over imidazioum-based ILs for ethanol separation. Relative to imidazioum-based ILs, phosphonium-based ILs are less expensive, more thermodynamically stable, available on a multi-ton scale, and have already been used in industrial processes [12]. Phosphonium-based ILs also differ from most ILs because they have densities less than that of water. Since these ILs are less dense than water, conventional units used for the decantation of aqueous phases may be used. A 2011 study showed phosphonium-based ILs to be significantly more effective in ethanol extraction [12], with a reported maximum ethanol concentration range of 65% to approximately 90% using nine different phosphonium-based ILs [12].

To describe the effect of temperature on ethanol extraction by LLE, the bimodal curves were determined by the cloud-point method at 298.15, 308.15, and 318.15 K. The universal functional activity coefficient (UNIFAC) method proved to be more accurate than the non-random two-liquid (NRTL) and universal quasichemical (UNIQUAC) equations fitted to the experimental data. Under the experimental conditions used, ethanol extraction by 1,2-dichloroethane appears to be independent of the temperature [30].

Different chemicals or processes are added to the LLE process as needed in order to improve the efficiency of ethanol extraction. The process can be facilitated by technologies such as gas stripping, in which ethanol is dissolved into a solvent and then stripped by gas to improve the ethanol recovery ratio [31]. The addition of sodium chloride to the LLE process, using 2-ethylhexanol as the solvent, offers limited advantages because of the high cost [32]. However, ethanol solubility is significantly improved by adding a secondary solvent (capric acid, 1-hexanol and 2-ethyl hexanol) to the primary solvent (m-xylene); a small amount of 2-ethyl hexanol can enhance the distribution coefficient and maintain the separation selectivity constant [33].

A very important application of ethanol LLE process is to integrate ethanol fermentation in order to reduce product inhibition. The plug flow reactor is used with the LLE process to removal dodecanol.

By this new method, the ethanol productivity is multiplied by 5 and a solution of 407 g/L of glucose is totally fermented with *Saccharomyces cerevisiae*, which cannot normally work with more than 200 g/L glucose [34]. To maximize LLE for ethanol production, operation conditions, such as temperature, time, feed concentration, and phase ratio, are optimized. The cooling temperature and time have been investigated with the finding that the maximum extraction was obtained at 95 °C with 5 minutes of contact time [35]. Matsumura *et al*. [36] tested selectivity ratios of 25 types of solvents, including alcohol and ester. Tri-n-butyl phosphate, 2-ethyl-1-butanol, 3-phenyl-1-propanol, sec-octanol, and polypropylene glycol P1200 were considered to be good extractants, and 100 g/L ethanol was considered to be the right concentration for solvent extraction. Polypropylene glycol P-1200, 2-ethyl-1,3-hexanediol, and methyl crotonate had a relatively low toxicity to microbial growth. Modified chemicals are also used for better separation of ethanol. For example, 1-ethyl-3-methylimidazolium methanesulfonate leads to higher values of solute distribution ratios and selectivities than 1-ethyl-3-methylimidazolium trifluoromethanesulfonate, except for selectivity values at high solute concentrations [37].

The problem of LLE technology related to the fermentation of *S. cerevisiae* or other microorganisms is that cells will be attracted to the liquid-liquid interface and then a yeast layer will build at the interface, which creates a mass transfer barrier to reduce the rate of ethanol extraction [38].

On an industrial scale, a continuous pilot plant has been constructed for fermentative production of ethanol, using LLE to remove the product, with recycling of the fermented broth raffinate. The plant was operated for up to 18 days with feed glucose concentrations in the range 10.0%–45.8% (w/w). The solvent was *n*-dodecanol, and immobilized yeast was used to overcome the problem of emulsification without adverse effect on the ethanol production rate [39].

3.5.2 Biodiesel

Biodiesel production is another area where LLE can be used. Biodiesel, usually as fatty acid methyl ester (FAME), is generated from the esterification reaction of long-chain fatty acid or triglyceride with methanol. The methanol is not only the reactant, but the methanol can serve as a solvent to extract the oil from the feedstock for biodiesel production. This concept was studied on cottonseeds to generate cottonseed meal and biodiesel products. An extraction rate of 98.3% could be achieved for cottonseed oil, while the free fatty acid (FFA) and water content of cottonseed oil were reduced to 0.20% and 0.037%, respectively, meeting the requirement of alkali-catalyzed transesterification [40]. A similar concept was applied to rapeseed oil, with sodium hydroxide used as catalyst, and 98.2% triglyceride conversion ratio was reached at 9:1 methanol to oil [41].

Besides alcohol serving as the extractant, oil/fatty acid can also serve as the extractant for the alcohol, which is specifically applied in the ethanol recovery from aqueous fermentation broth. Different vegetable oils and their fatty alcohol and fatty ester derivatives were studied for the distribution coefficients and separation factors in the partitioning of ethanol and water from the aqueous mixtures. The results showed that castor oil, ricinoleyl alcohol, and methyl ricinoleate all had higher ethanol distribution coefficients and similar or reduced separation factors [42]. As ethanol can be utilized to generate fatty acid ethyl ester (FAEE), another form of biodiesel, the ethanol extraction process was integrated with biodiesel production from oleic acid by using lipase as the catalyst. This enzymatic esterification of ethanol and oleic acid resulted in higher than 50% conversion with simultaneous reduction of ethanol content in the broth [43].

Liquid-liquid extraction is also used to reduce the glycerol concentration in the biodiesel product in order to meet the ASTM D6751-02 standard [44]. It is possible to separate glycerol from methyl oleate by using different combinations of solvents (hexane, methanol, and water) [45].

Biodiesel was also proposed as the extractant for butanol production because butanol and biodiesel can dissolve each other. This strategy benefits the process in that biodiesel-based glycerol serves as a substitute for butanol production. The production of biodiesel/butanol blend will be an integrated process without any cost of butanol separation. The overall economic value of butanol production is therefore improved [46].

3.5.3 Carboxylic acids

Liquid-liquid extraction technology has also been used to remove organic acids, such as tartaric acid and lactic acid, from agroindustrial wastewater. The recovery of carboxylic acids from the byproducts of sugar-cane treatments by the LLE technique was investigated for pollution control and food safety. This kind of separation can be achieved by single contact and multiple contacts in a continuous countercurrent MORRIS extractor [47]. The mixture of tributylphosphate and dodecane was optimized for appropriate partition ratio, pH to separate aconitic acid with purity of 98% w/w [48], which is possible to apply on an industrial scale. A high concentration of acrylic acid was obtained from sugar through LLE using di-isopropyl ether as extractant after selection and modeling of several solvents [49]. The LLE system includes the alamine series and alkane, which can separate lactic acid without any toxicity to microbes (*Lactobacillus casei* subsp. rhamnosus (ATCC 11443)) [50].

However, LLE is not widely used in separating carboxylic acids from the fermentation broth, even though much research has been focused on this area. The issue is the lack of an appropriate extractant that has favorable distribution coefficients for carboxylic acid extraction. Amine extractant, due to high alkalinity, can often react with organic acids in order to increase the extraction yield and the selectivity. A simple mixer-settler setup can easily perform this type of reactive extraction, and amine solvent can be recycled after back extraction of the organic acid with trimethylamine or the pH-swing-, diluent-swing-, or temperature-swing regeneration methods [51]. Some examples of reactive extraction for carboxylic acids follow. Lactic acid was separated by alamine 336 solution (20%–40%) in toluene at 25 °C–60 °C [52]. Acetic acid was purified by LLE through equal volumes of alamine-336 and 2-ethyl-1-hexanol resulting in 85% extraction efficiency below pH 3.5. The esterification of ethyl and butyl alcohols was conducted at 3:1 molar ratio of alcohol to acetic acid. Water inhibited esterification to 5%–20%, although the theoretical value is 65%–75% [53].

Many researchers have been working on the succinic acid extraction from the fermentation broth, in which acetic acid is generated as a byproduct and strongly inhibits the fermentation process. When applying amine to extract succinic acid, applying emulsion to liquid membranes can help the separation ratio reach 98% [54]. Considering the potential environmental effects when applying organic solvent as extractants, ionic liquid was tested to extract organic acids, including L-lactic, L-malic, and succinic acids, with the result that phophonium-based ionic liquids are better extractants than the organic solvents traditionally used.

3.5.4 Other biorefinery processes

Many other biorefinery products have been investigated for extraction by LLE technology. Hydrocarbons can only be generated by a limited number of microbial species, and LLE technology can be integrated with cell cultivation for hydrocarbon production. Brief contact between concentrated algae *Botryococcus braunii* and hexane can extract hydrocarbon generated by this species effectively; moreover, the algae's cell growth and hydrocarbon production are not impaired during the repeated extractions by hexane [55]. Extraction of 1,3-propanediol from aqueous solution was fairly difficult due to the lack of a simple efficient extractant. A solvent screening study revealed that no solvent could yield satisfactory separation results [56]. The cosolvent of ethyl acetate and ethanol was recently proposed by Boonsongsawat [57] for 1,3-propanediol (1,3-PD) extraction. Glycerol supplement can also increase the distribution coefficient of 1,3-PD. When applied in fermentation broth, the result was an increase in the distribution coefficient from 0.14 to 0.2. Overall, the conventional LLE for 1,3-propanediol is not industrially feasible, considering that the extraction process requires the handling of large quantities of solvents and, in particular, the 1,3-PD extraction and separation efficiency is too low [58]. An alternative method is to convert 1,3-propanediol with aldehyde to form highly hydrophobic acetals (2-methyl-1,3-dioxane [2MD]) through a cyclic reaction, then extract them using an organic solvent such as o-xylene, toluene, or ethylbenzene, and finally hydrolyze the acetals to 1,3-PD. Liquid-liquid extraction is also used as an alternative enantioseparation for the crystallization approach. Bisnaphthyl phosphoric acid was used as extractant for phenylglycinol separation at a laboratory and industrial scale. After six extraction stages, the purity and yield were 70% and 36%, respectively [59]. The chiral n-protected alpha-amino acid derivatives transfer from an aqueous solution to the organic phase for enantioselective separation by lipophilic carbamylated quinine as chiral selector and phase carrier. The enantiomeric purity of N-(3,5-dinitrobenzoyl)-leucine exceeded 95% enantiomeric excess with 70% overall yield with a single extraction and back-extraction step [60].

3.6 The future development of LLE for the biorefinery setting

With the development of LLE technology, the environmental effect of solvent attracts more consideration than ever. Formaldehyde, dioxane, formic acid, and acetonitrile are toxic to the operator. More environmentally friendly and more efficient extractants are expected to arise with the further development of LLE technology. New applications have been developed recently, especially in the fields of aqueous two-phase systems, supercritical solvent extraction, and ionic liquid extraction.

Water is definitely a good extractant from the environmental point of view because it is an inexpensive, easily obtained, nontoxic, and recyclable liquid. At very high temperatures (above boiling point), the extensive hydrogen bonds break down, which reduces the polarity of water. High-temperature water can be used as modifier for an organic solvent, increasing the solubility. In recent years, water has been used to extract oil from shale [61] and oil from plants [62], which provided a good starting point for this new extractant. However, the drawbacks of high-temperature water are its low solubility, and the possibility of solute precipitation at low temperature. These disadvantages can be overcome by organic solvent collection [63] and filtration of solute by filter [64].

The two-phase aqueous extraction system uses mixtures of aqueous solutions containing polymers and inorganic salts, which will separate into two phases that are predominantly water. The development of such extractants is especially applicable in biotechnology and biorefinery contexts, in which biologically active compounds may lose activity with organic diluents. With two-phase aqueous system, it is possible to separate sensitive biological molecules, such as intracellular enzymes and other microbial proteins, without denaturation. Aqueous two-phase solvent (ATPS) contains polymer, such as polyethylene glycol (PEG), and salt (e.g., phosphate, sulfate) or two polymers and water. The PEG-dextran-water ATPS system to separate protein has been most widely investigated. The ATPS containing PEG has been applied to several protein separations with further analysis of its partition coefficients. In addition, the salt type and concentration are important factors for the LLE performance [65].

The process of supercritical fluid extraction uses the solvent so that its pressure and temperature are higher than its critical point, according to the phase diagram. There are several candidates for supercritical fluid preparation, such as hydrocarbon [66], fluorinated hydrocarbons [67], and nitrous oxide [68]. Carbon dioxide is the most common material for supercritical fluid based on its qualities of being nontoxic, nonflammable, and low cost. Supercritical carbon dioxide is good for lipophilic compounds, but poor for polar solute extraction. However, supercritical carbon dioxide can be combined with polar co-solvent to extract polar compounds [69]. Extractants in a supercritical condition show superior characteristics for LLE such as density similar to liquid and viscosity similar to gas. The drastically reducing viscosity enables the extractant to move freely and provides a large surface area. The supercritical liquid has the advantages of high solubility in organic solute. This technology can therefore be used for high value-added chemicals production. However, supercritical fluid extraction technology requires special machinery, which is the major barrier to its industrial application.

Ionic liquids are salts with melting points lower than 100 °C; some ionic salts even form at room temperature. Ionic liquids have many unique properties such as exceedingly low vapor pressure and superior stability at wide range of temperature. They can be dissolved into the aqueous phase of LLE for polar change for better separation. Ionic liquids may improve the volatile compound separation through high solubility, and they have been used widely in recent years such as for separation of metal [70] and other materials [71]. Ionics liquids have their own challenges in terms of contamination by ions in the extractant. Ionic liquids can sometimes decrease the volatility, which can complicate the extractant recovery. Another disadvantage is the high cost of ionic liquid systems.

The application of LLE technology to the biorefinery process presents several challenges. First, many biorefinery systems are involved with bioactive organisms or biological catalysis, which may limit the application of organic extractants. It is therefore necessary to find the right extractant, which has a tolerance for microbes. Second, the LLE process has been conducted either at high temperature or high pressure, which are mostly detrimental conditions for active organisms. The extraction process in a biorefinery setting must be conducted under mild temperature conditions. Future breakthroughs can occur in the development of new extractants, new processes that can handle living organisms, or new active organisms that can endure the harsh conditions of LLE. Finally, biological systems are usually much more complex because they contain many more components and chemicals than regular chemical extraction processes. A more selective solvent is therefore required. It is more complicated to scale up an LLE technology in the biorefinery process, and the interaction of components in biosystems will be more complex than that of a chemical process. With the development of these new types of extraction processes, the use of LLE in the biorefinery field will have a promising future.

References

1. Stephen, H.S.T., *Solubilities of Inorganic and Organic Compounds*, Pergamon Press, 1963.
2. Pena-Pereira, F., I. Lavilla, and C. Bendicho, Miniaturized preconcentration methods based on liquid-liquid extraction and their application in inorganic ultratrace analysis and speciation: A review. *Spectrochimica Acta Part B-Atomic Spectroscopy*, 2009. 64(1): 1–15.
3. Pereira, E.A. and P.K. Dasgupta, Measurement of atmospheric formaldehyde using a drop collector and in-situ colorimetry. *International Journal of Environmental Analytical Chemistry*, 1997. 66(3): 201–213.
4. Cardoso, A.A. and P.K. Dasgupta, Analytical-Chemistry in a Liquid-Film Droplet. *Analytical Chemistry*, 1995. 67(15): 2562–2566.
5. Borges, S.D. and B.F. Reis, An automatic falling drop system based on multicommutation process for photometric chlorine determination in bleach. *Analytica Chimica Acta*, 2007. 600(1–2): 66–71.
6. Jeannot, M.A. and F.F. Cantwell, Mass transfer characteristics of solvent extraction into a single drop at the tip of a syringe needle. *Analytical Chemistry*, 1997. 69(2): 235–239.
7. Jain, A. and K.K. Verma, Recent advances in applications of single-drop microextraction: A review. *Analytica Chimica Acta*, 2011. 706(1): 37–65.
8. Lindenburg, P.W., *et al.*, Online capillary liquid-liquid electroextraction of peptides as fast pre-concentration prior to LC-MS. *Electrophoresis*, 2010. 31(23–24): 3903–3912.
9. Bonnesen, P.V., *et al.*, Development of effective solvent modifiers for the solvent extraction of cesium from alkaline high-level tank waste. *Solvent Extraction and Ion Exchange*, 2003. 21(2): 141–170.
10. Tsalev, D.L., *et al.*, Permanent modification in electrothermal atomic absorption spectrometry—advances, anticipations and reality. *Spectrochimica Acta Part B-Atomic Spectroscopy*, 2000. 55(5): 473–490.
11. Brunner, G., *Supercritical Fluids as Solvents and Reaction Media*, Elsevier, 2004.

12. Alizadeh, N. and K. Ashtari, Coalescence extraction of silver(I) using the temperature-induced phase separation (TIPS) process. *Separation and Purification Technology*, 2005. 44(1): 79–84.
13. Ullmann, A., Z. Ludmer, and R. Shinnar, Phase-transition extraction using solvent mixtures with critical-point of miscibility. *Aiche Journal*, 1995. 41(3): 488–500.
14. Persson, J., *et al.*, Purification of recombinant proteins using thermoseparating aqueous two-phase system and polymer recycling. *Journal of Chemical Technology and Biotechnology*, 1999. 74(3): 238–243.
15. Bailes, P.J., Solvent-extraction in an electrostatic-field. *Industrial and Engineering Chemistry Process Design and Development*, 1981. 20(3): 564–570.
16. Sato, M., T. Hatori, and M. Saito, Experimental investigation of droplet formation mechanisms by electrostatic dispersion in a liquid-liquid system. *IEEE Transactions on Industry Applications*, 1997. 33(6): 1527–1534.
17. Müller, E., *et al.*, Liquid–liquid extraction, in *Ullmann's Encyclopedia of Industrial Chemistry*, Wiley-VCH Verlag GmbH & Co. KGaA, 2000.
18. Karr, A.E., T.L. Holmes, and R.W. Cusack, Effect of trace contaminants on performance of agitated extraction equipment. *Solvent Extraction and Ion Exchange*, 1990. 8(3): 515–528.
19. Timothy C., F.L.D.B.S. Holden, William D. Prince, A. Frank Seibert, Loren C., Wilson, Liquid-liquid extraction and other liquid-liquid opertions and equipment, in *Perry's Chemical Engineers' Handbook*, Don W.Green, Robert H. Perry, Editor, McGraw-Hill, 2008.
20. Likidis, Z. and K. Schugerl, Recovery of penicillin by reactive extraction in centrifugal extractors. *Biotechnology and Bioengineering*, 1987. 30(9): 1032–1040.
21. Croker, J.R. and R.G. Bowrey, Liquid extraction of furfural from aqueous-solution. *Industrial and Engineering Chemistry Fundamentals*, 1984. 23(4): 480–484.
22. Pei, Y.C., *et al.*, Recovery of furfural from aqueous solution by ionic liquid based liquid-liquid extraction. *Separation Science and Technology*, 2008. 43(8): 2090–2102.
23. Vane, L.M., Separation technologies for the recovery and dehydration of alcohols from fermentation broths. *Biofuels Bioproducts and Biorefining-Biofpr*, 2008. 2(6): 553–588.
24. Mitchell, R.J., A. Arrowsmith, and N. Ashton, Mixed-solvent systems for recovery of ethanol from dilute aqueous-solution by liquid-liquid-extraction. *Biotechnology and Bioengineering*, 1987. 30(3): 348–351.
25. Offeman, R.D., *et al.*, Solvent extraction of ethanol from aqueous solutions. I. Screening methodology for solvents. *Industrial and Engineering Chemistry Research*, 2005. 44(17): 6789–6796.
26. Offeman, R.D., *et al.*, Solvent extraction of ethanol from aqueous solutions. II. Linear, branched, and ring-containing alcohol solvents. *Industrial and Engineering Chemistry Research*, 2005. 44(17): 6797–6803.
27. Koullas, D.P., O.S. Umealu, and E.G. Koukios, Solvent selection for the extraction of ethanol from aqueous solutions. *Separation Science and Technology*, 1999. 34(11): 2153–2163.
28. Solimo, H.N., H.E. Martinez, and R. Riggio, Liquid liquid extraction of ethanol from aqueous-solutions with amyl acetate, benzyl alcohol, and methyl isobutyl ketone at 298.15-k. *Journal of Chemical and Engineering Data*, 1989. 34(2): 176–179.
29. Rahman, M.A., M.S. Rahman, and M.N. Nabi, Extraction of ethanol from aqueous solution by solvent extraction-liquid-liquid equilibrium of ethanol-water-1-butanol, ethanol-water-l-pentanol and ethanol-water-1-hexanol systems. *Indian Journal of Chemical Technology*, 2001. 8(5): 385–389.
30. Solimo, H.N. and J.L. Zurita, Influence of temperature on the liquid-to-liquid extraction of ethanol from (water plus ethanol plus 1,2-dichloroethane). *Canadian Journal of Chemistry-Revue Canadienne de Chimie*, 1992. 70(8): 2310–2313.
31. Egan, B.Z., D.D. Lee, and D.A. Mcwhirter, Solvent-extraction and recovery of ethanol from aqueous-solutions. *Industrial and Engineering Chemistry Research*, 1988. 27(7): 1330–1332.
32. Gomis, V., *et al.*, Salt effects in extraction of ethanol from aqueous solution: 2-ethylhexanol plus sodium chloride as the solvent. *Industrial and Engineering Chemistry Research*, 1998. 37(2): 599–603.
33. Habaki, H., *et al.*, Extraction equilibrium of ethanol for bioethanol production-solvent selection and liquid-liquid equilibrium measurement. *Journal of the Japan Petroleum Institute*, 2010. 53(3): 135–143.
34. Minier, M. and G. Goma, Production of ethanol by coupling fermentation and solvent-extraction. Biotechnology Letters, 1981. 3(8): 405–408.

35. Rahman, M.A., M.S. Rahman, and M. Asadullah, Production of fuel grade ethanol from dilute-solution by liquid-liquid-extraction using vegetable-oils as solvents. *Indian Journal of Chemical Technology*, 1995. 2(2): 90–92.
36. Matsumura, M. and H. Markl, Application of solvent-extraction to ethanol fermentation. *Applied Microbiology and Biotechnology*, 1984. 20(6): 371–377.
37. Arce, A., H. Rodriguez, and A. Soto, Effect of anion fluorination in 1-ethyl-3-methylimidazolium as solvent for the liquid extraction of ethanol from ethyl tert-butyl ether. *Fluid Phase Equilibria*, 2006. 242(2): 164–168.
38. Crabbe, P.G., C.W. Tse, and P.A. Munro, Effect of microorganisms on rate of liquid extraction of ethanol from fermentation broths. *Biotechnology and Bioengineering*, 1986. 28(7): 939–943.
39. Gyamerah, M. and J. Glover, *Production of ethanol by continuous fermentation and liquid-liquid extraction*. *Journal of Chemical Technology and Biotechnology*, 1996. 66(2): 145–152.
40. Qian, J.F., Z. Yun, and H.X. Shi, Cogeneration of biodiesel and nontoxic cottonseed meal from cottonseed processed by two-phase solvent extraction. *Energy Conversion and Management*, 2010. 51(12): 2750–2756.
41. Shi, H.X. and Z.H. Bao, Direct preparation of biodiesel from rapeseed oil leached by two-phase solvent extraction. *Bioresource Technology*, 2008. 99(18): 9025–9028.
42. Offeman, R.D., *et al*., Solvent extraction of ethanol from aqueous solutions using biobased oils, alcohols, and esters. *Journal of the American Oil Chemists Society*, 2006. 83(2): 153–157.
43. Csanyi, E., *et al*., Study on ethanol fermentation integrated with simultaneous solvent extraction and enzymatic reaction. *Acta Alimentaria*, 2004. 33(1): 63–70.
44. (ASTM), A.S.f.T.a.M., *ASTM D6751-02*, in *Standard Specification for Biodiesel Fuel (B100) Blend Stock for Distillate Fuels*.
45. Tizvar, R., *et al*., Optimal separation of glycerol and methyl oleate via liquid-liquid extraction. *Journal of Chemical and Engineering Data*, 2009. 54(5): 1541–1550.
46. Adhami, L., *et al*., Liquid-liquid extraction of butanol from dilute aqueous solutions using soybean-derived biodiesel. *Journal of the American Oil Chemists Society*, 2009. 86(11): 1123–1128.
47. Achour, D., B. Abdi, and J.R. Molinier, Recovery of lactic and tartaric acid from dilute aqueous effluents by liquid extraction. *Afinidad*, 2001. 58(492): 112–114.
48. Malmary, G.H., *et al*., Recovery of aconitic acid from simulated aqueous effluents of the sugarcane industry through liquid-liquid-extraction. *Bioresource Technology*, 1995. 52(1): 33–36.
49. Alvarez, M.E.T., *et al*., Evaluation of liquid-liquid extraction process for separating acrylic acid produced from renewable sugars. *Applied Biochemistry and Biotechnology*, 2007. 137: 451–461.
50. Demirci, A., A.L. Pometto, and K.R. Harkins, Rapid screening of solvents and carrier compounds for lactic acid recovery by emulsion liquid extraction and toxicity on *Lactobacillus casei (ATCC 11443)*. *Bioseparation*, 1999. 7(6): 297–308.
51. Kurzrock, T. and D. Weuster-Botz, Recovery of succinic acid from fermentation broth. *Biotechnology Letters*, 2010. 32(3): 331–339.
52. SanMartin, M., C. Pazos, and J. Coca, Liquid-liquid extraction of lactic acid with alamine 336. *Journal of Chemical Technology and Biotechnology*, 1996. 65(3): 281–285.
53. Katikaneni, S.P.R. and M. Cheryan, Purification of fermentation-derived acetic acid by liquid-liquid extraction and esterification. *Industrial and Engineering Chemistry Research*, 2002. 41(11): 2745–2752.
54. Lee, S.C., Extraction of succinic acid from simulated media by emulsion liquid membranes. *Journal of Membrane Science*, 2011. 381(1–2): 237–243.
55. Frenz, J., C. Largeau, and E. Casadevall, Hydrocarbon recovery by extraction with a biocompatible solvent from free and immobilized cultures of botryococcus-braunii. *Enzyme and Microbial Technology*, 1989. 11(11): 717–724.
56. Malinowski, J.J., Evaluation of liquid extraction potentials for downstream separation of 1,3-propanediol. *Biotechnology Techniques*, 1999. 13(2): 127–130.
57. Boonsongsawat, T., *et al*., Solvent extraction of biologically derived 1,3-propanediol with ethyl acetate and ethanol cosolvent. *Separation Science and Technology*, 2010. 45(4): 541–547.

58. Xiu, Z.L. and A.P. Zeng, Present state and perspective of downstream processing of biologically produced 1,3-propanediol and 2,3-butanediol. *Applied Microbiology and Biotechnology*, 2008. 78(6): 917–926.
59. Schuur, B., *et al.*, Enantioselective liquid-liquid extraction of (R,S)-phenylglycinol using a bisnaphthyl phosphoric acid derivative as chiral extractant. *Tetrahedron*, 2011. 67(2): 462–470.
60. Kellner, K.H., *et al.*, Enantioseparation of N-protected alpha-amino acid derivatives by liquid-liquid extraction technique employing stereoselective ion-pair formation with a carbamoylated quinine derivative. *Chirality*, 1997. 9(3): 268–273.
61. Ogunsola, O.M. and N. Berkowitz, Extraction of oil shales with sub-critical and near-critical water. *Fuel Processing Technology*, 1995. 45(2): 95–107.
62. Cassel, E., *et al.*, Steam distillation modeling for essential oil extraction process. *Industrial Crops and Products*, 2009. 29(1): 171–176.
63. Ayala, R.S. and M.D.L. de Castro, Continuous subcritical water extraction as a useful tool for isolation of edible essential oils. *Food Chemistry*, 2001. 75(1): 109–113.
64. Wennrich, L., B. Popp, and J. Breuste, Determination of organochlorine pesticides and chlorobenzenes in fruit and vegetables using subcritical water extraction combined with sorptive enrichment and CGC-MS. *Chromatographia*, 2001. 53: S380–S386.
65. Raghavarao, K.S.M.S., M.R. Guinn, and P. Todd, Recent developments in aqueous two-phase extraction in bioprocessing. *Separation and Purification Methods*, 1998. 27(1): 1–49.
66. Favati, F., J.W. King, and M. Mazzanti, Supercritical carbon-dioxide extraction of evening primrose oil. *Journal of the American Oil Chemists Society*, 1991. 68(6): 422–427.
67. Levy, J.M., E. Storozynsky, and R.M. Ravey, The use of alternative fluids in online supercritical fluid extraction—capillary gas-chromatography. *HRC—Journal of High Resolution Chromatography*, 1991. 14(10): 661–666.
68. Alexandrou, N., M.J. Lawrence, and J. Pawliszyn, Cleanup of complex organic mixtures using supercritical fluids and selective adsorbents. *Analytical Chemistry*, 1992. 64(3): 301–311.
69. Huang, K.J., *et al.*, Designed polar cosolvent-modified Supercritical CO2 removing caffeine from and retaining catechins in green tea powder using response surface methodology. *Journal of Agricultural and Food Chemistry*, 2007. 55(22): 9014–9020.
70. Visser, A.E., *et al.*, Liquid/liquid extraction of metal ions in room temperature ionic liquids. *Separation Science and Technology*, 2001. 36(5–6): 785–804.
71. Werner, S., M. Haumann, and P. Wasserscheid, Ionic liquids in chemical engineering. *Annual Review of Chemical and Biomolecular Engineering*, 2010. 1: 203–230.

4
Supercritical Fluid Extraction

Casimiro Mantell, Lourdes Casas, Miguel Rodríguez and Enrique Martínez de la Ossa

Chemical Engineering and Food Technology Department, Science Faculty, University of Cadiz, Spain

4.1 Introduction

The use of supercritical extraction to obtain natural products is an application of supercritical fluids that has undergone major developments in recent years. Numerous reports in the literature focus on the development of such processes (Reverchon and De Marco, 2006). The great interest in supercritical fluids (SCF) arises from their properties, which are intermediate between those of gases and liquids and make them suitable for use in extraction processes. Supercritical fluid extraction (SFE) is a mass transfer process under pressure and temperature conditions above the critical point of the solvent.

The biorefinery concept arose from the need to replace the use of non-renewable energy sources in the production of fuels and chemicals. In brief, in a biorefinery a variety of chemical products are obtained from renewable sources, using separation, isolation and chemical or biochemical transformation techniques. The possibility of using supercritical extraction at different stages in a biorefinery acquires greater significance considering the excellent characteristics of this technology (Temelli, 2009). From the point of view of green chemistry, extraction processes would need to use natural solvents, which include water, ethanol and carbon dioxide. Carbon dioxide in the supercritical state can also be used as an extraction solvent and this is considered as *Generally Recognized As Safe* (GRAS), which, within the concept of green chemistry, means that it is a non-polluting chemical that allows products to be obtained with an extremely high purity.

To compare alternative approaches, it is necessary to take into account the questions presented in Figure 4.1. However, one of the main issues to be considered in the optimization of a process is the removal of the solvent from the solute in order to obtain the product at the maximum level of purity and supercritical extraction is the best choice in a large number of cases.

Separation and Purification Technologies in Biorefineries, First Edition.
Edited by Shri Ramaswamy, Hua-Jiang Huang, and Bandaru V. Ramarao.

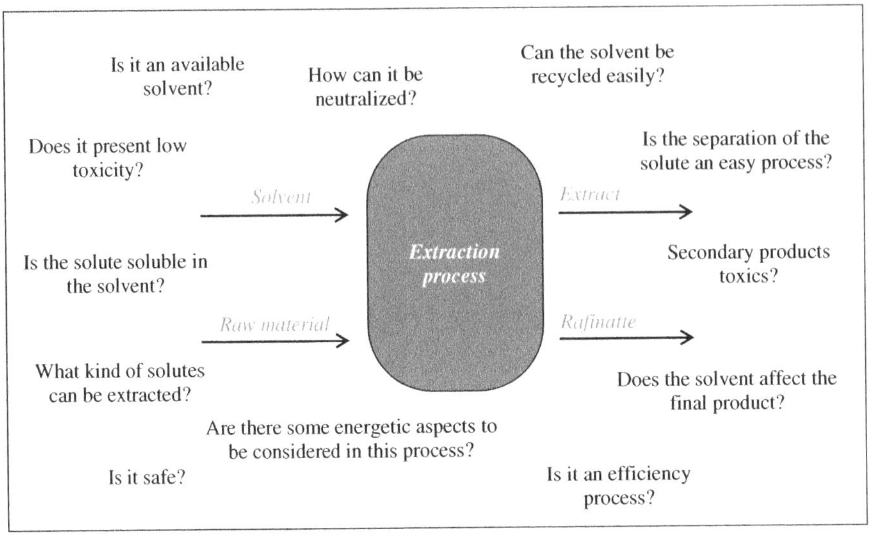

Figure 4.1 *Factors to be taken into account in the design of an extraction process*

The main raw materials that can be used in a biorefinery are described below along with different possibilities for the use of supercritical technology:

- ***Sugar/starch-rich crops***. These are the most common types of raw material used in biorefineries today. This resource stores large amounts of saccharose, which can easily be extracted from the plant material for subsequent fermentation to ethanol or bio-based chemicals. Sugar cane is currently the preferred feedstock from an economic and environmental perspective due to its relative ease of production. In this case, the solutes can be extracted easily using water as solvent and supercritical fluid extraction is not employed in this step (Prado *et al*., 2011). Nevertheless, there are numerous papers that concern the application of supercritical fluid extraction using a countercurrent column in the separation of ethanol from aqueous solutions (Pereyra *et al*., 1995; Di Giacomo *et al*., 1991; Gamse *et al*., 1999; Señorans *et al*., 2001; Ruiz-Rodríguez *et al*., 2010).
- ***Vegetable oil***. In this case, the raw material is normally used in the production of biodiesel by the transesterification process. The oil can originate from two sources—it can be obtained directly from natural materials or feedstock, like palm, soybean, rapeseed and sunflower seed, and it can be obtained as a waste product from a different process, for example cooking oil or animal fat. The sustainable and economic production of biodiesel from vegetable oils has proven to be a challenge. This is due to the significant land-use change and sustainability issues resulting from pure plant oil production and the high costs associated with the refinement of waste oil due to its unavoidable lack of purity. In this application, the use of supercritical fluid extraction has seen significant developments in recent years (Sahena *et al*., 2009).
- ***Lignocellulosic biomass***. This biomass feedstock is obtained from inedible plant material that is mainly composed of cellulose, hemicellulose and lignin. It is deemed likely that this type of second-generation feedstock will be used for the production of biofuels and bio-based chemicals in the future using different conversion technologies. There are numerous papers in the literature that focus on the application of supercritical fluid extraction in the separation of different secondary metabolites from agricultural wastes (Casas *et al*., 2010; El Marsni *et al*., 2011; Fernandez-Ponce *et al*., 2011; Daukšas *et al*., 2002).

In these cases, the use of this technology constitutes an initial step prior to the treatment of the biomass to decompose the cellulose and transform it into ethanol.

- ***Microalgae***. These are a large and diverse group of unicellular photo- and heterotrophic organisms that have attracted a great deal of attention in recent years due to the large amounts of interesting solutes that they can produce. These include carotenoids, polyunsaturated fatty acids, and antioxidants that can be used in the food industries, and the wide variety of lipids that can be used to synthesize biodiesel through transesterification. The remaining carbohydrate content can also be converted to bioethanol by fermentation. The application of supercritical fluid extraction is centered on the development of processes that are capable of separating these compounds from the raw material with a high efficiency.

4.2 Principles of supercritical fluids

The term "supercritical fluid" refers to any substance that is in an aggregated state at pressures and temperatures higher than the critical temperature and pressure. A typical temperature/pressure diagram for a pure substance is shown in Figure 4.2. Three regions can be clearly differentiated in this diagram and these correspond to three aggregation states: solid, liquid, and gas. Two other characteristic points can also be seen in the diagram: the triple point, at which three states (solid, liquid and gaseous) coexist, and the critical point, at the end of the vaporization curve, which is characterized by a critical pressure and a critical temperature. The critical point represents the pressure and temperature conditions under which phases such as liquid and gas cease to exist, and the supercritical fluid phase appears.

Changes in pressure or temperature, or both, also modify the density of the SCF. This fact is of great importance if we consider that the solvating power of a substance depends on its density as solvation results from intermolecular forces due to the packing of molecules in the solvent around the molecules of the solute, forces which in turn depend on the density. In this way, the density of SCF solvents can be as low as those of gases or as high as those of liquids, depending on the pressure and temperature. Furthermore, the solubility of a solute in the SC solvent can be modified on an ongoing basis from low to high values and vice versa.

The possibility of changing the density of the solvent in a process can be used to increase the selectivity of the solvent or for the fractionation of multiple solutes by a gradual reduction in the density of the solvent.

The pressure and temperature conditions that are most suitable to achieve these effects are those that are close to the critical point, because in this area there are major variations in the density of the SCF (and in its solvent power) with only minor changes in pressure and temperature.

In addition to the unique features outlined above for the control of density and solvent power, supercritical fluids have other specific physico-chemical properties related to mass transfer. Furthermore, although a supercritical fluid has a density and a solvent power similar to that of the liquid solvent, it has a viscosity and diffusivity of the same order of magnitude as gases. For this reason, this system improves the transfer characteristics of the liquid solvents. On the basis of these properties, one can consider that supercritical fluids move like gases and dissolve substrates in a similar way to a liquid.

The unique physico-chemical properties described above mean that supercritical fluids have major advantages over traditional liquid solvents. Efficiencies achieved in separations must be significantly higher and, at the same time, the solvent must easily and completely separate products simply by changing the pressure so that the supercritical fluid changes to the gaseous state.

Carbon dioxide is the main supercritical solvent used because the critical temperature and pressure (73.8 bar and 31.1 °C) are easily achieved. In addition, this solvent is completely miscible with hydrocarbons of low molecular weight and oxygenated organic compounds, so it is a good solvent for many organic compounds. The mutual solubility with water is small so carbon dioxide can be used as a solvent to extract

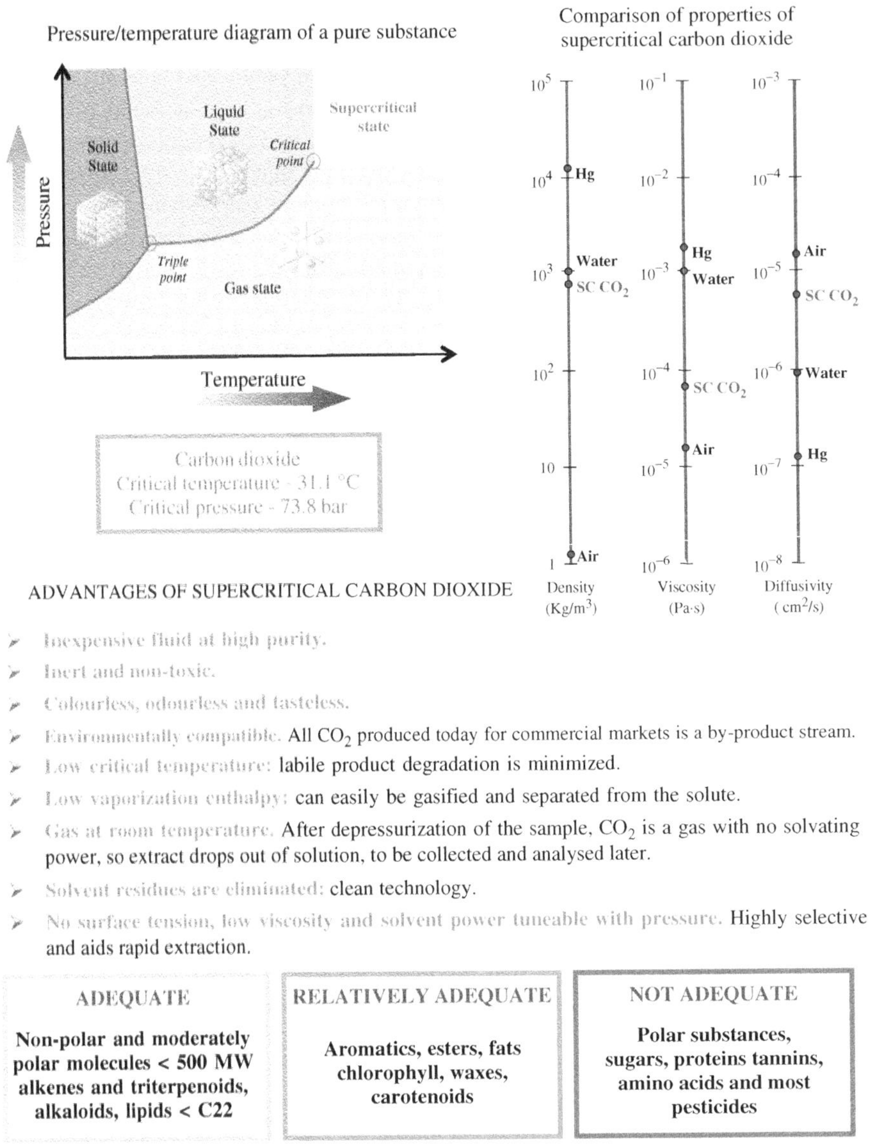

Figure 4.2 *Main characteristics of supercritical carbon dioxide*

organic products from aqueous solutions. Carbon dioxide is also non-toxic, non-flammable, non-corrosive, and non-polluting, it is cheap, abundant, and it can easily be obtained at different purity levels. Carbon dioxide also has favorable transport properties: low viscosity, high diffusion coefficients, appropriate heat conductivity and heat of vaporization, especially near the critical point, so the energy needs in many processes are low. Finally, it has the advantage that, once the extraction process has been completed and the decompression carried out, the solvent is gaseous under environmental conditions and escapes into the atmosphere (or it can be compressed and recycled for further use) and no longer remains in the extracted product.

4.3 Market and industrial needs

The main important applications of extraction using solvents under supercritical conditions are the extraction, refinement and fractionation of edible oils, fats, and waxes. The aim of the process is to separate certain solutes present in solid natural materials such as seeds, fruits and citrus peels. The term "refined" is applied to the separation of certain compounds, such as carotenoids, phospholipids and free fatty acids, that promote oxidation and therefore contribute to the rancidity of oils. Fractionation allows the selective separation of short-chain triglycerides and unsaturated vegetable and animal oils, as well as certain compounds present in natural products, generally of high value, such as vitamins, flavors and polyunsaturated fatty acids, among others.

The extraction of alkaloids from plant matrices is historically one of the first applications of supercritical fluid extraction technology (Palmer and Ting, 1995). A typical example of this is the decaffeination of coffee and tea. With the intention of minimizing the loss of aromas and flavors, in most decaffeination processes the extraction of the caffeine takes place on the green coffee, before roasting and milling. The decaffeination of coffee and tea with supercritical carbon dioxide currently represents 20% of world production.

The current trend in the use of supercritical fluid extraction of natural substances is related to the need to adapt research to the increasingly high demand of consumers for food additives with a natural origin and/or with beneficial health properties. In addition, the market increasingly demands extraction techniques that allow the product obtained "naturally," and, in turn, in an environmentally friendly way. In this sense, the application of supercritical technology has made an industrial breakthrough in recent years.

An example of this trend is the application of SFE in the production of food additives such as colorants, flavors and antioxidants from natural products. The possibility of obtaining these compounds from a natural source increases in importance when the extraction is carried out on agricultural by-products. A large number of studies has also now been carried out on the analysis of other properties that are beneficial in terms of health. The presence of anticancer, antimutagenic or anticonvulsive properties adds more interest to the products obtained.

Supercritical fluid extraction can be applied to systems on various scales, from the laboratory scale (a few grams) to the pilot plant scale (several hundred grams of sample), through to the industrial scale (tons of raw materials). Several examples of industrial applications are given in Table 4.1, which includes details of different industrial plants that use supercritical technology.

In the case of SFE on solid matrices using SC-CO_2, the basic scheme of operation is presented in Figure 4.3. The main characteristics of this process are as follows:

- The SC processes tends to work in a discontinuous way with respect to the solid and in a continuous way with respect to the SCF. In order to increase the efficiency of the process, it is often carried out with three extractors that operate independently. Thus, while one extractor is in operation, another is in the loading/pressurization step and the third extractor is in the depressurization/unloading stage.
- A process of progressive depressurization can fractionate the extracts. The current techniques in the SFE process make it possible to fractionate the products obtained in the extraction process by fractional precipitation using various separators in cascade. This mode of operation relies on the characteristics of the carbon dioxide to modify the solubility by varying the pressure/temperature conditions. After the extraction process, the extract passes through various separators that work under different pressure and temperature conditions, thus progressively decreasing the solvent density and therefore its solvent power. In this process the substance that is least soluble in the supercritical solvent will precipitate in the first separator and the more soluble solutes will be retained in solution, precipitating in turn through the system until the last separator is reached.

- Finally, one of the most economically important aspects is the recovery and recycling of the solvent. This process can usually be carried out in two tanks. One tank operates at low pressure and is used to receive the solvent after the separation process, and the other works at higher pressure where the solvent that is lost in the process can be added and the SCF is prepared to be re-introduced in the process.

The use of equipment with extraction columns operating counter currently is necessary for the extraction of liquid samples. In this way, the raw material is introduced into the equipment from the top of the column, while the supercritical solvent is introduced at the bottom. Once again, the extract can be separated in a cyclone separator (or several) in order to fractionate the products. A possible alternative to direct SFE is to extract the raw material by conventional methods to obtain oil and then concentrate it in the substances of interest using a fractionation countercurrent column with supercritical carbon dioxide. A plant working with fractionation countercurrent columns is shown in Figure 4.4.

4.4 Design and modeling of the process

The extraction process for solid substances primarily consists of two stages, the extraction and the separation of the product from the solvent. At the extraction stage, the solvent passes through a fixed bed of solid particles and dissolves the solute from the solid. The solvent moves from the extractor to the separator, where the solvent is recovered and the extracted product is obtained. The design and modeling of the process is centered on the extraction stage.

In order to analyze this stage in a generic manner, it can be considered that the extraction of solutes from a solid matrix occurs in five basic stages that can occur in series/parallel (Brunner, 1993):

- The solvent is placed in contact with the surface of the solid matrix.
- Intraparticular solvent diffusion occurs as the solvent accesses the interior of the matrix. The separation process begins in this way.
- The retained solutes are moved by simple drag or displacement from the active sites of the matrix due to the higher affinity and/or concentration of solvent molecules. Solubilization (solvation) occurs immediately in the solvent.
- Solutes are transported from inside the matrix to its surface, essentially by diffusion, which is the most important mechanism of transport at this stage.
- The dissolved solutes cross through the interstitial film fluid that surrounds the solid and they are transported to the bulk solvent and removed from the bed of solid particles.

Numerous parameters must be taken into account when modeling an extraction process and these include the solid matrix, the accessibility of the solute to the solvent in the bed, the chances of ascending and descending flow, and the flow distribution of the solute within the solid matrix. However, variation of the amount of extracted product with time is relatively simple.

The variation of the extraction yield with the extraction time for the extraction of carotenoids from marine microalgae is represented in Figure 4.5 (Macías-Sánchez *et al.*, 2009a). This figure shows the typical variation of the extraction yield for this kind of process.

It can be observed in Figure 4.5 that two stages are clearly differentiated. In the first stage, the variation in the quantity of extracted product with time fits a straight line, whereas in the second stage the yield follows a curve that tends to approach asymptotically a value that corresponds to the maximum amount of product that can be extracted. This variation indicates that the largest amount of solute is removed from the matrix during a short period of time at the beginning of the extraction. In the second stage the rate

Table 4.1 Some examples companies that employ supercritical fluid extractions at different scales

Organization	Location	System and application
Agrisana (www.agrisana.it)	Notaresco (Italy)	Extracts from plants for cosmetics, food industries
ALTEX (www.altex.es)	Valencia (Spain)	Three extractors of 1500 L for the extraction of flavors, material processing, spices, etc.
Applied separations (www.appliedseparations.com)	Allentown, PA (United States)	R&D supercritical fluids process
Eden Botanicals (www.edenbotanicals.com)	Hyampom, CA (United States)	Extracts of flavors and spices
Flavex (www.flavex.com)	Rehlingen (Germany)	Cosmetics, perfumery, food additives
Fuji Flavor Co. (www.fjf.co.jp)	Tokyo, (Japan)	High-quality extracts from natural materials, such as flowers, leaves and roots.
India Glycols Limited (www.indiaglycols.com)	New Delhi (India)	3 × 300 L extractors in spices and natural products
Indo-global Spices Ltd. (www.indoglobalspices.com)	Karnataka (India)	Three extractor of 300 L in spices and natural products
Industrial research limited IRL (www.irl.cri.nz)	Wellington (New Zealand)	R&D Supercritical Fluids Process
Kraft products (www.kraftfoodscompany.com)	Glattpark (Switzerland)	Decaffeination of coffee (HAG Coffee) and flavors
NATEX (www.natex.at)	Terniz (Austria)	R&D Supercritical Fluids Process
Organix South Inc. (organixsouth.com)	Bowling Green, FL (United States)	Production of Neem Bark Supercritical Extract
Phasex corporation (www.phasex4scf.com)	Lawrence, MA (United States)	R&D Supercritical Fluids Process
Philip Morris (www.pmi.com)	NY (United States)	Production of Tobacco without nicotine
Raps & Co. (www.raps.de)	Kulmbach (Germany)	Extracts of flavors and spices
Separex (www.separex.fr)	Champigneulles (France)	R&D supercritical fluids process
SKW/Trotsberg	Dusseldorf (Germany)	Different capacities—food technology applications
SMS Natural Products	Indore (India)	Two extractors of 1100 L for spices and natural products
SOLUTEX (www.solutex.es)	Zaragoza (Spain)	Two extractors of 3800 L for flavors and fragrances
Talent Natural Extract Co. Ltd (www.naturalcn.com)	Wuho (China)	Decaffeination of tea, production of flavors
TharProcess (www.thartech.com)	Pittsburg PA (United States)	R&D supercritical fluids process
The herbaria (www.theherbarie.com)	Prosperity, SC (United States)	Flavors, essential oils from plants
Xspray (www.xspray.com)	Stockholm (Sweden)	Various high-pressure systems in pharmaceutical application

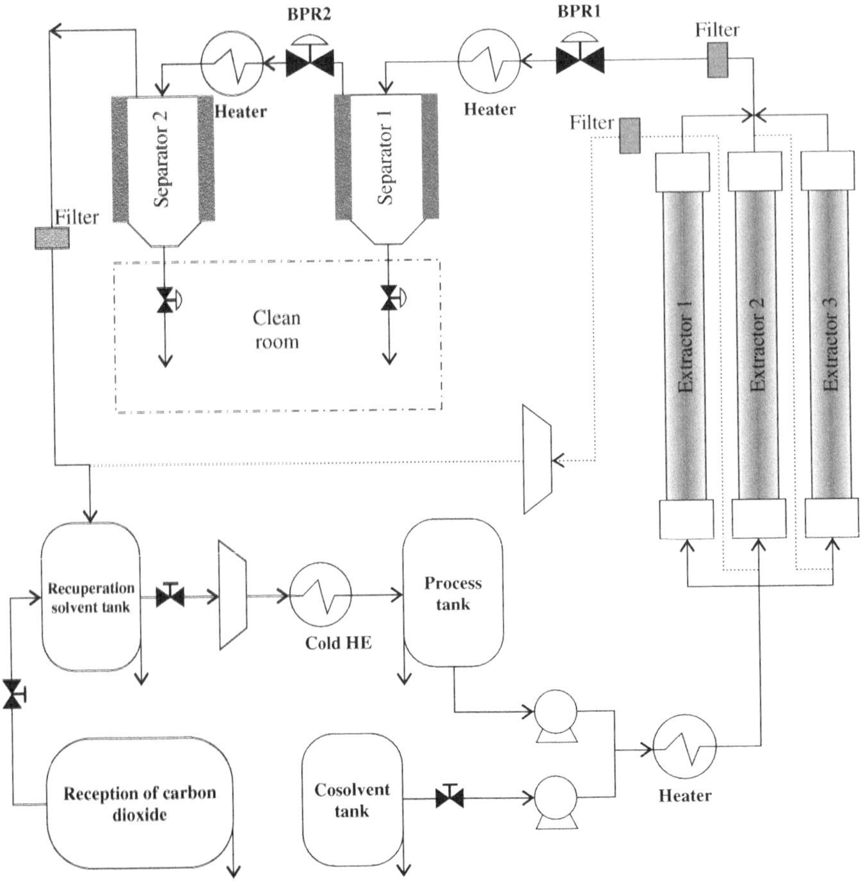

Figure 4.3 *Typical diagram of a supercritical extraction plant for solid samples*

of extraction decreases with time. This general behavior can be caused by different factors and these are discussed later. This phenomenon may present a problem in the design of a process and in the efficiency of an extraction processes on an industrial scale.

The slope of the first part of the graph can be defined by the solubility equilibrium, although this is not always the case as a straight line may appear merely due to the existence of a constant resistance to mass transfer. In fact, such constant resistance is common in the extraction process and therefore the solubility does not control the process, especially if the extraction is carried out in a dynamic way. In many cases, the solute is present in the matrix in small quantities and during the extraction process the concentration of the solute in the supercritical fluid is well below the limit of solubility.

The graph presented in Figure 4.5 is very limited for the comparison of extraction processes for different materials extracted using different equipment. There are numerous variables that can affect the overall process. However, the information provided by this curve is very useful for comparing the results of a series of extractions of a given substance using the same apparatus operating under different conditions.

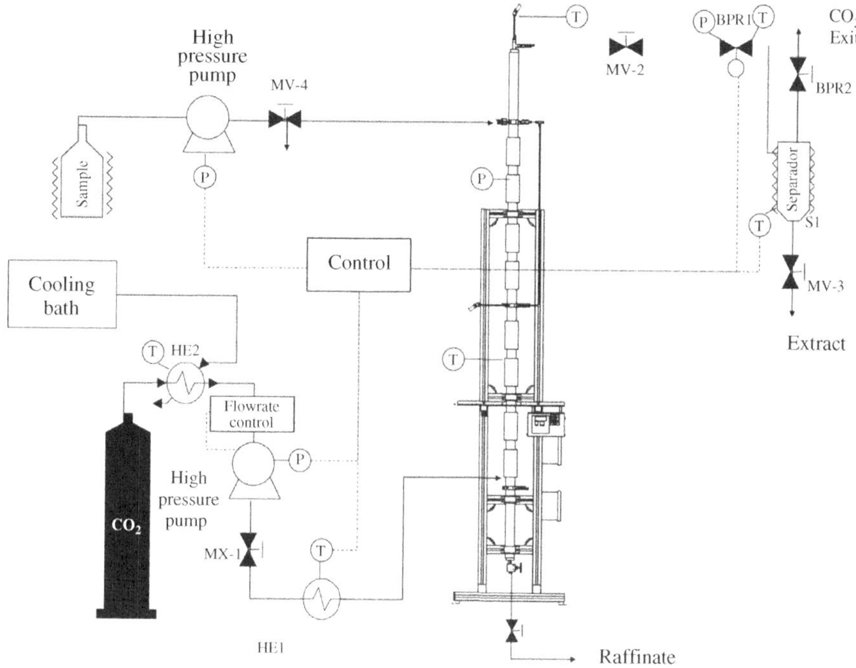

Figure 4.4 *Diagram of a supercritical extraction plant using a countercurrent column*

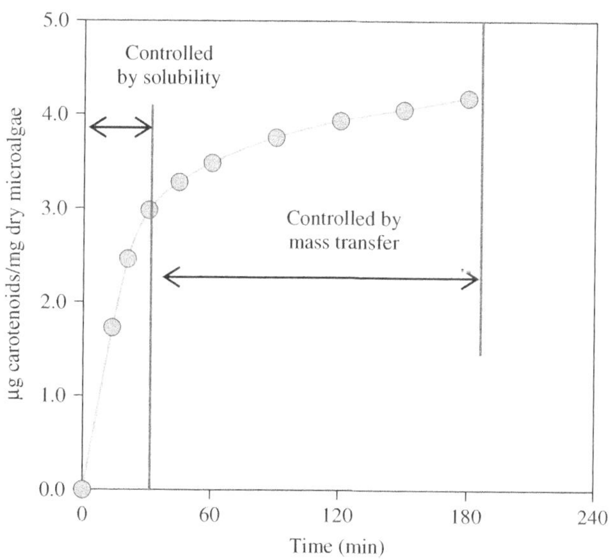

Figure 4.5 *Variation of the extraction yield with time for the SFE of carotenoids from a marine microalga Nanochlorposis gaditana*

A range of theories have been published in the bibliography to describe extraction processes (Al-Jabari, 2002). All of these approaches can be classified into two types of theory: film theory and penetration theory. The Biot number makes it possible to select the most appropriate mass transfer model:

$$Bi = \frac{\mathrm{Ke}2L}{\mathrm{Di}} \tag{4.1}$$

If the Biot number is greater than 10, then internal diffusion is the controlling stage of the extraction process (Perez Galindo *et al*. 2000). In these cases, the application of a penetration model is more appropriate. On the other hand, if $\mathrm{Bi} < 10$, the controlling stage is the mass transfer in the interstitial fluid and in this case the most appropriate model is the film model.

4.4.1 Film theory

A wide variety of approaches can be used to develop mass transfer equations. On the one hand, it must be borne in mind that the double-layer theory uses mass transfer coefficients and the driving forces of the process are expressed as units proportional to concentration (usually in molar fraction). The flow rate of a solute from the solid phase (which exists as a liquid) can be expressed in the following way:

$$\mathrm{Flux} = \mathrm{k_a A}\left(\mathrm{x}-\mathrm{x_i}\right) \tag{4.2}$$

On the other hand, the flow rate of the solute from the interface to the bulk of the solvent is given by the expression:

$$\mathrm{Flux} = \mathrm{k_b A}\left(\mathrm{y_i}-\mathrm{y}\right) \tag{4.3}$$

where $\mathrm{k_a}$ and $\mathrm{k_b}$ are transfer coefficients, $\mathrm{x_i}$ $\mathrm{y_i}$ are the concentrations of solute in the interface, x is the concentration of solute in the inert phase, y is the concentration of solute within the solvent, and A is the transfer area.

If the interface reaches the steady state, these fluxes are equivalent and the following equation can be obtained:

$$\frac{y_i - y}{x_i - x} = -\frac{Ak_a}{AK_b} = -\frac{K_a}{K_b} \tag{4.4}$$

The numerical value of this relationship is the slope of a line on an *x-y* diagram and depends on the conditions of the main stage and is called the resistance phase relationship. In the development of this equation it is assumed that the interface does not offer any resistance, which has proven to be true in most cases.

4.4.2 Penetration theory

These kinds of models are the most commonly used in systems to describe extraction processes using supercritical fluids. These models usually involve a series of initial considerations that can simplify the resolution of the system. These theories considerer a transport system that is composed of fluid elements with the following assumptions:

- the fluid is completely mixed and at isothermal conditions;
- the fluid is in contact with the other phase, maintaining a determinate concentration in the surface;
- there is no pressure drop along the bed of particles;
- the particles are packed with a constant porosity and apparent density along the whole bed;

- finally, if we suppose that the concentration of the solute in the supercritical phase is low, then the density of the fluids, the axial dispersion, and the flow-rate of the fluid are constant throughout the process.

The following equation represents the global balance for the fluid phase in the supercritical extraction of solids (Oliveira *et al*., 2011):

$$\varepsilon_c \frac{\partial C_i}{\partial t} = -u_G \frac{\partial C_i}{\partial Z} + \varepsilon_c D_{ax,i} \frac{\partial^2 C_i}{\partial Z^2} + (1 - \varepsilon_c) J_f a_p \tag{4.5}$$

where C_i is the concentration of component i in the supercritical phase, $D_{ax,i}$ is the coefficient of axial dispersion of component i, J_f is the flux density of material from the solid to the fluid phase, u_G is the interstitial fluid velocity and a_p is the specific surface area of the particle. Join to this equation, some initial and boundary conditions are required. This conditions and different assumption depend on the definition realized in the model.

The solution to the equation of balance is treated differently according to different authors. Oliveira *et al*. (2011) classified the models according to the following criteria:

- Linear driving force model. In this model it is assumed that the mass transfer flux is proportional to the difference between the mean concentration of the solute in the particle and the concentration of the solute in equilibrium with the fluid phase. In many cases, the particle is porous and the solute is present in the solid phase and in the fluid inside the pores—in this instance, two linear driving-force approximations are employed, one between the fluid phase and the fluid in the pores of the solid, and the other between the fluid phase in the pores and the solute in the solid phase.
- The shrinking core model: In this model it is assumed that there is a sharp boundary between the extracted and non-extracted parts of the particle. As the extraction proceeds, the boundary recedes until it reaches the center of the particle and all the solutes are extracted.
- The broken plus intact cell model. This model was initially proposed by Sovova (1994), Sovova *et al*. (1994) and Stastova *et al*. (1996) and it has the advantage of providing a reasonably simple analytical solution to the mass balance equation and a good physical description of the process. In this model a similar physical representation is proposed in which the particles are composed of cells that are broken up during grinding and cells that remain intact. The existence of two mass transfer resistances during SFE was hypothesized. The first resistance is located in the supercritical mixture and controls the extraction process until all the essential oil in the broken cells is exhausted. The second resistance is in the walls of the intact cells and controls the remaining part of the process. Reverchon and Marrone (2001) proposed a modification of this model, with the existence of a "parallel resistances" mechanism where both broken and intact cells transfer solute to the fluid with different kinetics.
- Finally, the same authors also worked with a combined model: broken plus intact cells and a shrinking core. A new model was proposed by Fiori *et al*. (2009) and this combined the concepts of the two previous models. It is assumed that, in a particle obtained from milled grape seed, there are N concentric layers. The cells can either be broken by the milling process or remain intact. It is assumed that the broken cells are located in the outer layer of the particle.

4.5 Specific examples in biorefineries

The potential for the application of supercritical technology in a biorefinery is a function of the different raw materials that may form part of the biorefinery. The best extraction method depends on the nature of the raw material and the characteristics of the process.

4.5.1 Sugar/starch as a raw material

After the fermentation of sugars it is necessary to separate ethanol from the aqueous solution. Several authors have proposed the use of supercritical fluids to separate ethanol from aqueous solutions with two objectives: firstly, to obtain a fermentation broth with low alcohol content and, secondly, to have an alternative production process for ethanol. According to data published by Pereyra *et al.* (1995) and Di Giacomo *et al.* (1991), one possible process for the production of ethanol from its fermentation broth (wine), in which the ethanol is at concentration about 6–10% by weight, is presented in Figure 4.6. The process consists of two stages: (i) concentration of the aqueous solutions in the organic compound and (ii) the removal of the organic compounds from the concentrated solutions using carbon dioxide. The extraction process with SC carbon dioxide in the recovery of ethanol below the azeotrope concentration from wine seems very promising.

In a similar context, Gamse *et al.* (1999) analyzed the application of an extraction column using supercritical carbon dioxide for use in the over-production of low-quality wine to isolate ethanol and aromatic components. The optimization of the separation process in a countercurrent system and the thermodynamic modeling of the process were presented in the bibliography (Señoráns *et al.*, 2001; Ruiz-Rodríguez *et al.*, 2010).

4.5.2 Supercritical extraction of vegetable oil

Several examples of processes involving the supercritical fluid extraction of lipids are given in Table 4.2. Supercritical fluid extraction has proven to be effective in the separation of essential oils and their derivatives for use in the food, cosmetics, pharmaceutical and other related industries (Sahena *et al.*, 2009). In general, the results obtained are comparable with traditional methods such as Soxhlet extraction, with similar extraction yields, but a higher quality extract can be obtained. In the production of biodiesel in a biorefinery, SFE can be an interesting technique for the extraction step before the transesterification process. Hernandez-Duran *et al.* (2010) proved that the extract obtained from the waste from the sugarcane

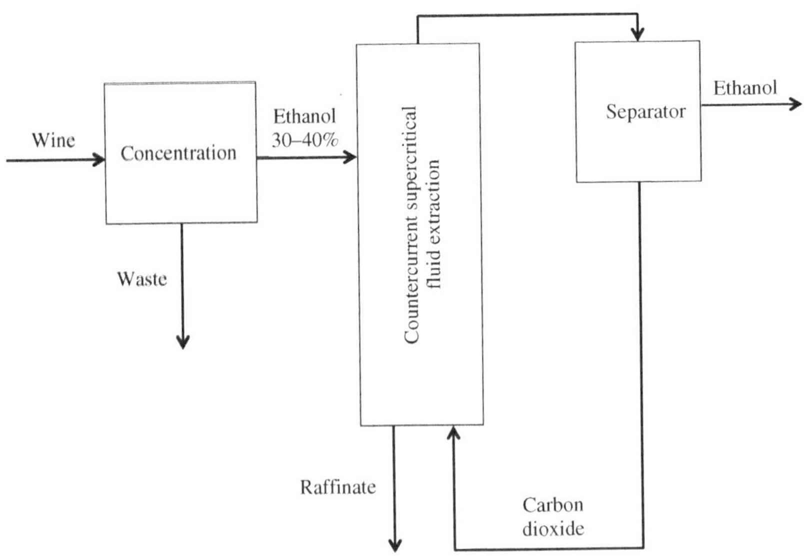

Figure 4.6 Process proposed for the supercritical separation of ethanol from wine

Table 4.2 *Examples of supercritical fluid extraction of lipids*

Raw material	Application	Process	Reference
Some genetically modified varieties of corn	Fatty acid profile	30 MPa and 333 K	Toribio *et al.*, 2011
Broccoli leaves	Fractionation of lipids	30 MPa and 333 K	Aenaiz *et al.*, 2011
Shrimp waste	Lipids, PUFA and astaxanthin	20–40 MPa and 313–333 K	Sanchez-Camargo *et al.*, 2011
Used frying oil	Triglycerides	Liquid and supercritical ethane at 15–25 MPa and 278–353 K	Rincón *et al.*, 2011
Hibiscus cannabinus seeds	Fatty acids, tocopherols and sterols	40–60 MPa and 313–353 K	Mariod *et al.*, 2011
Patinopecten yessoensis viscera	Fatty acid and sterols	Enzyme-assisted solvent and SFE methods	Zhou *et al.*, 2011
Prunus persica almond oil	Phenolic content	10–30 MPa and 303–323 K with cosolvent	Mezzomo *et al.*, 2010
Triticale bran	Alkylresorcinol	Two-step SFE first with CO_2 and then CO_2 + cosolvent	Athukorala *et al.*, 2010
Used frying oil	Fractionation the lipids	30–40 MPa and 298–353 K with different cosolvents	Rincon *et al.*, 2011
Spent coffee grounds	Fatty acid composition	15–30 MPa and 313–333 K	Couto *et al.*, 2009
Butter oil	Fractionation of fatty acid ethyl esters	9–17 MPa and 321–333 K	Torres *et al.*, 2009
Different examples	A review about lipid extraction		Sahena *et al.*, 2009
Different examples	A review about lipid extraction		Reverchon and De Marco 2006

industry with a SFE process gave better results in the transesterification process in comparison with the use of a traditional solvent. Data for the conversion in the base catalyzed transesterification process on oil obtained from sugar cane process waste are presented in Figure 4.7. In this case the transesterification process analyzed was the base-catalyzed process with different oil/methanol ratios. The results indicate that the oil obtained using the SFE process is more suitable for the transesterification reaction.

4.5.3 Supercritical extraction of lignocellulose biomass

Lignocellulose biomass can be classified into six main groups: agricultural residues (bagasse from sugarcane, wheat straw, rice straw, bagasse of corn, straw of barley, etc.), wood (poplar or black poplar), hard wood (pine, spruce, etc.), cellulose waste (such as newspaper), recycled paper sludge, herbaceous biomass (alfalfa, canary grass, lemongrass, etc.) and solid urban waste. The main components are cellulose (30–60% dry weight), hemicellulose (10–40% dry weight) and lignin, although these materials also contain other minor components (proteins, fats, waxes, terpenes, phenols, mineral salts, etc.), which are responsible for the color, smell or taste in each case (Chiaramonti, *et al.*, 2007).

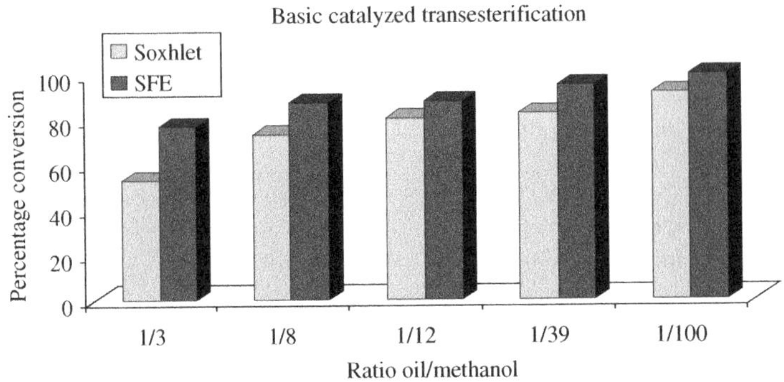

Figure 4.7 *Comparison of the conversion in the base catalyzed transesterification process on oil obtained from sugar cane process waste using Soxhlet-hexane and SFE techniques*

Sunflower seed is one of the main crops on a global scale with more than 35 00 000 MT of production per year. The residue of sunflower represents a significant volume of waste from the food industry. This residue basically consists of the stem and flower head without seeds and is normally burned or accumulated with the consequent environmental risk. There is great interest in the search for new alternatives for industrial uses. In fact, the stem and quinine have been used by various authors as raw materials for the production of bioethanol (Erzegin and Kükük, 1998; Sharma *et al*., 2002; Sharma *et al*., 2004; Ruiz *et al*., 2008). The remains of the seed after extraction of the oil are used to obtain a generic fermentation culture medium (Bautista *et al*., 1990) and for the production of bioethanol (Telli-Okur and Eken-Saracoglu, 2008), the husk and the comprehensive waste have been employed in the recovery of energy (Zabaniotou *et al*., 2008; Zabaniotou and Andreou, 2010) and the flower head without seeds has already been used for the extraction of pectin (Saharia *et al*., 2003) and, in a fermentation process, for the production of pectins (Patil and Dayanand, 2006). There are numerous studies in which supercritical technology has been used for the recovery and isolation of bioactive compounds from sunflower leaves for the production of an allelopathic extract. The application of this technology is a first stage prior to the lignocellulosic biomass decomposition and subsequent fermentation. This possibility can increase the efficiency of the process. The fact that carbon dioxide leaves the raw material unchanged facilitates the subsequent process of decomposition by means of physical-chemical or enzyme attack.

In the same way, there are several papers that describe the use of supercritical fluid extraction in order to recover high value-added compounds from various agricultural wastes (olive, mango, eucalyptus) or by-products of the food industry such as tomato processing or wine production (Table 4.3). Specifically, there are several works that concern the application of supercritical extraction for the recovery of polyphenols or resveratrol from grape pomace. These compounds have a high antioxidant capacity and are currently used in the pharmaceutical, cosmetics and food industries.

4.5.4 Supercritical extraction of microalgae

Marine microalgae offer great potential as a source for the extraction of substances with desirable properties and these compounds are receiving attention in many industries, including the food, pharmaceutical and cosmetics industries. These products are intended for direct human consumption and the mode of extraction is of paramount importance in terms of the technology applied.

Furthermore, microalgae could provide the raw material to produce renewable fuels such as biodiesel, methane, hydrogen and ethanol (Li *et al*., 2008). These biofuels are sulfur-free and perform the same

Table 4.3 *Examples of supercritical fluid extraction of agricultural wastes*

Raw material	Application	Process	Reference
Sunflower leaves	Bioactive compounds	10–50 MPa and 308–333 K and different cosolvents	Casas *et al.*, 2005, 2007, 2008, 2009, 2010
Sunflower leaves	Allelopathy	10–50 MPa and 308–333 K	El Marsni *et al.*, 2011
Grape pomace	Resveratrol	10–40 MPa and 308–333 K with ethanol as cosolvent	Casas *et al.*, 2010
Mango leaves	Phenolic compounds-Antioxidants	Carbon dioxide at 10–40 MPa and 313–333 K and subcritical water at 10 MPa and 353 K	Fernandez-Ponce *et al.*, 2011
Eucalyptus leaves	Antioxidants	20 MPA and 323 K	Fadel, 1999
Olive leaves	Tocopherols	25–45 MPa and 313–333 K	Daukšas *et al.*, 2002
Tomato paste waste	Pigments	20–30 MPa and 308–338 K	Baysal *et al.*, 2000
Red grape pomace	Anthocyanin	10–50 MPa and 313–333 K with methanol as cosolvent	Mantell *et al.*, 2003

functions as diesel oil while having the added benefit of reduced particulate, CO_2, and hydrocarbon emissions (Miao and Wu, 2006; Xu *et al.*, 2006; Chisti, 2007). The content in an algal oil can vary from 15 to 75% depending on the variety. The possibility of developing an extensive cultivation of the raw material means that production per hectare is 10 to 20 times higher than that of palm oil (the plant that represents the highest oil production per hectare). Likewise, microalgae grow extremely rapidly, doubling in biomass every 24 hours, and the oil produced by microalgae does not compromise food derived from seed production.

Supercritical fluid technology provides interesting alternatives for the extraction of substances from microalgae, as these are efficient and selective methods. The application of this extraction technique with supercritical carbon dioxide has been widely studied in recent years due to the clear advantages of carbon dioxide as a solvent—advantages that include low toxicity, low cost, and ease of separation of the extracted product (Mercer *et al*. 2011).

Numerous studies have been undertaken on the supercritical fluid extraction of wide varieties of compounds from different microalgae and these are summarized in Table 4.4. Hydrocarbons, carotenoids, lipids, fatty acids, and bioactive compounds can all be obtained with high extraction yields from microalgae using this technique. In many cases, it is necessary to add a small amount of modifier to the carbon dioxide to increase the extraction yields. Methanol is very efficient in removing large quantities of compounds in extraction processes but it is toxic to humans and, as a result, ethanol is usually selected as the cosolvent.

4.6 Economic importance and industrial challenges

The development of supercritical technology in the processing of natural products in industry and the increasing number of applications in other areas is mainly due to the factors listed below:

- Consumers tend to demand products made from clean technology, with liquid organic solvents replaced with solvents such as carbon dioxide.
- Products extracted with supercritical fluids are of higher quality than those obtained by extraction with organic solvents, mainly because of the absence of waste solvent residues in the products, but also

Table 4.4 *Examples of supercritical fluid extraction of microalga*

Varieties	Application	Process	Reference
Botryococcus braunii	Hydrocarbons	313 K and pressure up to 30 MPa	Palavra *et al.*, 2011
Chlorella vulgaris	Carotenoids	313 K and 35 MPa	Palavra *et al.*, 2011
Chlorella sp.	Biodiesel production	15–30 MPa and 313–333 K with hexane/methanol as cosolvent	Char *et al.*, 2011
Scenedesmus dimorphus	Lipids for biodiesel production	16–48 MPa and 323–373 K	Soh and Zimmerman, 2011
Schizochytrium limacinum	Lipid—DHA	35 MPa and 313 K with ethanol as cosolvent	Tang *et al.*, 2011
Chlorococcum sp.	Oil for biodiesel production	10–50 MPa and 333–353 K	Halim *et al.*, 2011
Chlorella vulgaris	Active compounds	30 MPa and 323 K and H_2O/ethanol as cosolvent	Wang *et al.*, 2010
Nannochloropsis oculata	Bioactive compounds	SFE and anti-solvent purification	Liau *et al.*, 2010
Scenedesmus almeriensis	Lutein	20–60 MPa and 305–333 K	Macías-Sánchez *et al.*, 2010
Crypthecoodinium cohnii	Lipids/PUFAs	20–30 MPa and 313–323 K	Couto *et al.*, 2010
Chlorcoccum littorale	Carotenoids	33 K and 30 MPa—ethanol as cosolvent	Ota *et al.* 2009
Cholrella vulgaris	Pigments	50 MPa and 353 K	Kitada *et al.*, 2009
Dunaniella salina	Carotenoids and chlorophyll	10–50 MPa and 313–333 K	Macías-Sánchez *et al.*, 2009b
Various microalgae and cyanobacterial species	Phenolic compounds	Combinations of solid-phase/supercritical-fluid extraction	Klejdus *et al.*, 2009
Haematococcus pluvialis	Astaxanthin	343 K and 40 MPa with vegetable oil as cosolvent	Krichnavaruk *et al.*, 2008
Various microalgae	Carotenoids and chlorophyll	20–50 MPa and 313–333K with ethanol as cosolvent	Macías-Sánchez *et al.*, 2008
Synechococcus sp.	Carotenoids and chlorophylls	10–50 MPa and 313–333 K	Macías-Sánchez *et al.*, 2008
Spirulina platensis	Vitamin E	8–36 MPa and 300–356 K with ethanol as cosolvent	Mendiola *et al.*, 2008
Spirulina platensis	Compounds with antioxidant and antimicrobial activities	22–32 MPa and 328 K with 10% of ethanol	Mendiola *et al.*, 2007
Chaetoceros muelleri	Compounds with antimicrobial activity	25 MPa and 333 K with ethanol as cosolvent	Mendiola *et al.*, 2007
Chlorella vulgaris	Carotenoids and fatty acids	30 MPa and 313 K	Gouveia *et al.*, 2007

Table 4.4 *(continued)*

Varieties	Application	Process	Reference
Schizochytrium sp.	Lipids—PUFAs	Experimental design and mathematical modeling	Zinnai *et al.*, 2006
Asthrospira maxima	Lipids—GLA	323–333 K and 25–35 MPa with ethanol as cosolvent	Mendes *et al.*, 2006
Haematococcus pluvialis	Astaxanthin and other carotenoids	20–30 MPa and 313–333 K with 10% ethanol as cosolvent	Nobre *et al.*, 2006
Spirulina platensis	Lipids—PUFAs	25–70 MPa and 313–328 K	Andrich *et al.*, 2006
Nannochloropsis sp.	Bioactive compounds—PUFAs	40–70 MPa and 313–328 K	Andrich *et al.*, 2005
Synechococcus sp.	Carotenoids	10–50 MPa and 313–333 K	Montero *et al.*, 2005
Nannochlorpsis gaditana	Carotenoids and chlorophylls	10–50 MPa and 313–333 K	Macías-Sanchez *et al.*, 2005
Spirulina platensis	Antioxidant compounds	22 MPa and 238 K with 10% ethanol as cosolvent	Mendiola *et al.*, 2005
Various microalga	Compounds with pharmaceutical importance	12.5–30 MPa and 313–333 K	Mendes *et al.*, 2003
Hematococcus pluvialis and *Spirulina maxima*	Astaxantine and phycocyanine	30 MPa and 333 K with ethanol as cosolvent	Valderrama *et al.*, 2003
Cholrella vulgaris	Carotenoids and other lipids	35 MPA and 313–328 K	Mendes *et al.*, 1995

because the materials are processed at moderate temperatures so that their properties are not altered. In the extraction of flavorings and fragrances, samples can undergo hydrolysis when subjected to distillation with steam, whereas the organoleptic properties are virtually unchanged when samples are processed by SFE. Similarly, Wagner and Eggers (1996) compared the refinement of oils by classical methods with those obtained by supercritical extraction and concluded that SFE allows the omission of several refinement steps after extraction with carbon dioxide, thus reducing the consumption of alkali and minimizing the loss of neutral lipids.

- There is no need for the separation of the solvent from the extract, a factor that reduces costs since the SFE is completely removed into the separator.
- Supercritical processes allow environmental problems to be solved, such as the reduction of emissions of volatile organic compounds and the replacement of conventional halogenated solvents used in wool, paints and metal, and textile drying and cleaning.
- Supercritical technology enables the production of various products from which it is difficult to remove traditional solvents (e.g. extracts of ginger, pepper and paprika) and other completely new products, such as so-called drug-distribution systems, a recent application in the pharmaceutical industry.
- The rapid rate of SFE processes in comparison to classical separations makes operation times far lower, a factor that significantly decreases staff costs. This is due to two factors: the SCF penetrates the solid

matrix more rapidly than liquid solvents and, secondly, a concentration stage is not required after the extraction.

In addition to the advantages outlined earlier, other— mainly economic—factors have impeded the rapid dissemination of supercritical technology. The systems operate at high pressures and this requires high investment costs for equipment. Currently, supercritical processes compete with traditional extraction processes when they are applied to high value-added products (polyunsaturated fatty acids, essential oils, vanillin extracts, etc.) or when large volumes of materials are processed, for example for coffee and tea, hops, the manufacture of paint, and the treatment of waste, among others. However, the increasingly strict regulations in relation to effects on the ozone layer, the discharge of volatile organic compounds, and waste concentrations in the final product for the protection of consumers and the environment, facilitate the development of supercritical extraction and fractionation processes, thus making them more competitive.

4.7 Conclusions and future trends

The use of supercritical technology is now a reality. The application of this approach to bio-based products including biofuels has great potential due to the excellent characteristics of this technology in relation to the preservation of the environment. Environmental regulations favor the trend for the use of natural solvents in the extraction process, and the use of supercritical carbon dioxide is the major option for the recovery of nonpolar compounds. The high selectivity of the extraction process and the reduction in the operation time allows the extraction of compounds with some added value. In the design of a biorefinery, the inclusion of processes with supercritical fluids is relevant in numerous stages. These include the separation of ethanol from aqueous solutions after fermentation, the extraction of oil from vegetable seeds for the transesterification process, the manufacture of biodiesel, and the SFE of secondary metabolites from agricultural waste as an initial stage prior to the degradation of lignocellulosic biomass feedstock. Future trends will focus on the development of processes for the purification and concentration of products that are of interest in supercritical technology. The use of step-by-step precipitation, fractionation in countercurrent columns, and supercritical chromatography will be future areas of research.

References

Aenáiz, E., Bernal, J., Martín, M. T., García-Viguera, C., Bernal, J. L., Toribio, L. 2011. Supercritical fluid extraction of lipids from broccoli leaves. *European Journal of Lipid Science and Technology*, 113, 479–486.

Al-Jabari, M. E. 2002. Kinetic models of supercritical fluid extraction. *Journal of Separation Science*, 25(8), 477–489.

Andrich, G., Nesti, U., Venturi, F., Zinnai, A., Fiorentini, R. 2005. Supercritical fluid extraction of bioactive lipids from the microalga *Nannochloropsis sp*. *European Journal of Lipid Science and Tecnology*, 107(6), 381–386.

Andrich, G., Zinnai, A., Nesti, U., Venturi, F., Fiorentini, R. 2006. Supercritical fluid extraction of oil from microalga *Spirulina (Arthrospira) platensis*. *Acta Alimentaria*, 35(2), 195–203.

Athukorala, Y., Hosseinian, F. S., Mazza, G. 2010. Extraction and fractionation of alkylresorcinols from triticale bran by two-step superctritical carbon dioxide. *LWT—Food Science and Technology*, 43, 660–665.

Bautista J., Parrado J., Machado A. 1990. Composition and fractionation of sunflower meal: use of the lignocellulosic fraction as substrate in solid-state fermentation. *Biological Wastes*, 32, 225–233.

Baysal, T, Ersus, S., Starmans, D. A. J. 2000. Supercritical CO_2 extraction of β-carotene and lycopene from tomato paste waste. *Journal of Agricultural and Food Chemistry*, 48, 5507–5511.

Brunner, G. 1993, *Gas Extraction*. Deutsche Bunsen-Gesellsacharft für Physikalische Chemie, Frankfurt.

Casas, L., Mantell, C, Rodríguez, M., Martinez de la Ossa, E., Roldán, A., De Ory, I., Caro, I., Blandino, A. 2010. Extraction of resveratrol from the pomace of Palomino fino grapes by supercritical carbon dioxide. *Journal of Food Engineering*, 96, 304–308.

Casas, L., Mantell, C., Rodríguez, M., Torres, A., Macias, F. A., Martinez de la Ossa, E. 2005. Effect of the pretreatment of the samples on the natural substances extraction from Helinathus annuus L. using supercritical carbon dioxide. *Talanta*, 67, 175–181.

Casas, L., Mantell, C., Rodríguez, M., Torres, A., Macias, F. A., Martinez de la Ossa, E. 2007. Effect of the addition of cosolvent on the supercritical fluid extraction of bioactive compounds from *Helianthus annuus* L. *Journal of Supercritical Fluids*, 41, 47–49.

Casas, L., Mantell, C., Rodríguez, M., Torres, A., Macias, F. A., Martinez de la Ossa, E. 2008. Supercritical fluid extraction of bioactive compounds from sunflower leaves with carbon dioxide and water on a pilot plant scale. *Journal of Supercritical Fluids*, 45, 37–42.

Casas, L., Mantell, C., Rodríguez, M., Torres, A., Macias, F. A., Martinez de la Ossa, E. 2009. Extraction of natural compounds with biological activity from sunflower leaves using supercritical carbon dioxide. *Chemical Engineering Journal*, 152, 301–306.

Char, JM., Wang, JK., Chow, TJ., Chien, QC. 2011. Biodiesel production from microalgae through carbon dioxide extraction. *Nihon Enerugi Gakkaishi/Journal of the Japan Institute of Energy*, 90(4) 369–373.

Chiaramonti, D., Oasmaa, A., Solantausta, Y. 2007. Power generation using fast pyrolysis liquids from biomass. *Renewable and Sustainable Energy Reviews*, 11, 1056–1086.

Chisti, Y. 2007. Biodiesel from microalgae. *Biotechnology Advances*, 25(3), 294–306.

Couto, R. M., Fernández, J., Gomes de Silva, M. D. R., Simoes, P. C. 2009. Supercritical extraction of lipids from spent coffee grounds. *Journal of Supercritical Fluids*, 51, 159–166.

Couto, R. M., Simoes, P.C., Reis, A., Da Silva, T. L., Martins, H. V., Sanchez-Vicente, Y. 2010. Supercritical fluid extraction of lipids from the heterotrophic microalgae Crypthecodinium cohnii. *Engineering in Life Sciences*, 10, 158–164.

Daukšas, E., Venskutonis, P. R., Sivik, B. 2002. Supercritical fluid extraction of tocopherol concentrates from olive tree leaves. *Journal of Supercritical Fluids*, 22(3), 221–228.

Daukšas, E., Venskutonis, P. R., Sivik, B., Nillson, T. 2001. Effect of fast CO_2 pressure changes on the yield of lovage (*Levisticum officinale* Koch.) and celery (*Apium graveolens* L.) extracts. *Journal of Supercritical Fluids*, 22, 201–210.

Di Giacomo, G., Brandani, V., Del Re, G., Martínez de la Ossa, E. 1991. Selectivity and loading behavior in liquid carbón dioxide extraction of etanol from dilute aqueous solutions. *Chemical and Biochemical Engineering Quarterly*, 5(3) 141–144.

El Marsni, Z., Casas, L., Mantell, C., Rodríguez, M., Torres, A., Macias, F. A., Martínez de la Ossa, E., Molinillo, J. R., Varela, R. 2011. Potential allelopathic of the fractions obtained from sunflower leaves using supercritical carbon dioxide. *Journal of Supercritical Fluids*, 60, 28–37.

Erzegin M., Kükük M.M. 1998. Liquefaction of sunflower stalk by using supercritical extraction. Energy Conversion and Management, 39(11), 1203–1206.

Fadel, J. G. 1999. Quantitative analyses of selected plant by-product feedstuffs, a global perspective. *Animal Feed Science and Technology*, 79, 255–268.

Fernández-Ponce, M. T., Mantell, C., Rodríguez, M., Casas, L., Martínez de la Ossa, E. 2011. Extraction of phenolic compounds from mango leaves (*Mangifera indica* L.) with supercritical fluids. Eleventh International Conference on Carbon Dioxide Utilization, 27–30 June, Dijon, France.

Fiori, L., Basso, D., Costa, P. 2009. Supercritical extraction kinetics of seed oil: A new model bridging the "broken and intact cells" and the "shrinking-core" models. *Journal of Supercritical Fluids*, 48, 31–138.

Gamse, T., Rogler, I, Marr, R. 1999. Supercritical CO_2 extraction for utilisation of excess wine of poor quality. *Journal of Supercritical Fluids*, 14(2), 123–128.

Gouveia, L., Batista, A. P., Miranda, A., Empis, J., Raymundo, A. 2007. Chlorella vulgaris biomass used as colouring source in traditional butter cookies. *Innovative Food Science and Emerging Technologies*, 8, 433–436.

Halim, R, Gladman, B., Danquah, M. K., Webley, P. A. 2011. Oil extraction from microalgae for biodiesel production. *Bioresoruce Tecnology*, 102(1), 178–185.

Hernández-Durán, Y., Casas, L., Mantell, C., Rodríguez, M., Casdelo, N., Martínez de la Ossa, E. 2010. Biodiesel from cachaza oil wax: transesterification with methanol. Third International Symposium on Energy from Biomass and Waste, 8–11 November, Venice, Italy.

Iglesias, M. T., Lozano, J. E. 2004. Extraction and characterization of sunflower pectin. *Journal of Food Engineering*, 62, 215–223.

Kitada, K., Machmudah, S., Sasaki, M., Goto, M., Nakashima, Y., Kumamoto, S., Hasegawa, T. 2009. Supercritical CO_2 extraction of pigment components with pharmaceutical importance from Chlorella vulgaris. *Journal of Chemical Technology and Biotechnology*, 84, 657–661.

Klejdus, B., Kopecký, J., Benešová, L., Vacek, J. 2009. Solid-phase/supercritical-fluid extraction for liquid chromatography of phenolic compounds in freshwater microalgae and selected cyanobacterial species. *Journal of Chromatography A*, 1216 (5), 763–771.

Krichnavaruk, S., Shotipruk, A. Goto, M., Pavasant, P. 2008 Supercritical carbon dioxide extraction of astaxanthin from *Haematococcus pluvialis* with vegetable oils as cosolvent. *Bioresoruce Technology* 99(13), 5556–5560.

Li, Y., Wang, B., Wu, N., Lan, C. Q., 2008. Effects of nitrogen sources on cell growth and lipid production of Neochloris oleoabundans. *Applied Microbiology and Biotechnology*, 81, 629–636.

Liau, B.C., Chun-Ting, S., Fong-Ping, L., Siang-En, H., Shih-Lan, H., Ting-Ting, J., Chieh-Ming, J. C. 2010. Supercritical fluids extraction and anti-solvent purification of carotenoids from microalgae and associated bioactivity. *Journal of Supercritical Fluids*, 55, 169–175.

Macías-Sánchez, M. D., Fernandez-Sevilla, J. M., Acién Fernández, F. G., Cerón García, M. C., Molina Grima, E. 2010. Supercritical fluid extraction of carotenoids from Scenedesmus almeriensis. *Food Chemistry*, 123, 928–935.

Macías-Sánchez, M. D., Mantell, C., Rodríguez, M., Martinez de la Ossa, E. 2009a. Kinetics of the supercritical fluid extraction of carotenoids from microalgae with CO_2 and ethanol as cosolvent. *Chemical Engineering Journal*, 150, 104–113.

Macías-Sánchez, M. D., Mantell, C., Rodríguez, M., Martinez de la Ossa, E., Lubián, L. M., Montero, O. 2005. Supercritical fluid extraction of carotenoids and chlorophyll *a* from *Nannochloropsis gaditana*. *Journal of Food Engineering*, 66, 245–251.

Macías-Sánchez, M. D., Mantell, C., Rodríguez, M., Martinez de la Ossa, E., Lubián, L. M., Montero, O. 2008. Supercritical fluid extraction of carotenoids and chlorophyll *a* from *Synechococcus* sp. *Journal of Supercritical Fluids*, 39, 323–329.

Macías-Sánchez, M. D., Mantell, C., Rodríguez, M., Martinez de la Ossa, E., Lubián, L. M., Montero, O. 2009b. Comparison of supercritical fluid and ultrasound-assisted extraction of carotenoids and chlorophyll a from Dunaliella salina. *Talanta*, 77, 948–952.

Mantell, C., Rodríguez, M., Martínez de la Ossa, E. 2003. A screening analysis of the high-pressure extraction of athocyanins from red grape pomace with carbon dioxide and cosolvent. *Chemical Engineering Technology*, 26(1), 38–42.

Mariod, AA., Matthäus, B., Ismail, M. 2011. Comparison of supercritical fluid and hexane extraction methods in extracting kenaf (*Hibiscus cannabinus*) seed oil lipids. *Journal of the American Oil Chemists' Society*, 88(7), 931–935.

Mendes, R. L., Fernández, H, L., Coelho, J. P., Reis, E. C., Cabral, J. M. S., Novais, J. M., Palavra, A. F. 1995. Supercritical CO_2 extraction of carotenoids and other lipids from Chlorella vulgaris. *Food Chemistry*, 53, 99–103.

Mendes, R. L., Nobre, B. P., Cardoso, M. T., Pereira, A. P., Palavra, A. F. 2003. Supercritical carbon dioxide extraction of compounds with pharmaceutical importance from microalgae *Inorganica Chimica Acta*, 356, 328–333.

Mendes, R. L., Reis, A. D., Palavra, A. F. 2006. Supercritical CO_2 extraction of γ-linolenic acid and other lipids from *Arthrospira* (*Spirulina*) maxima: Comparison with organic solvent extraction. *Food Chemistry*, 99, 57–56.

Mendiola, J. A., García-Martínez, D., Rupérez, F. J., Martín-Álvarez, P. J., Reglero, G., Cifuentes, A., Barbas, C., Ibañez, E., Señoráns, F. J. 2008. Enrichment of vitamin E from *Spirulina platensis* microalga by SFE. *Journal of Supercritical Fluids*, 43, 484–489.

Mendiola, J. A., Herrero, M., Cifuentes, A., Ibañez, E. 2007. Use of compressed fluids for sample preparation: Food applications. *Journal of Chromatography A*, 1152, 234–246.

Mendiola, J. A., Marín, F. R., Hernández, S. F., Arredondo, B. O., Señoráns, F. J., Ibañez, E., Reglero, G. 2005. Characterization via liquid chromatography coupled to diode array detector and tandem mass spectrometry of supercritical fluid antioxidant extracts of *Spirulina plantesis* microalga. *Journal of Separation Science*, 28(9–10), 1031–1038.

Mercer, P., Armenta, RE. 2011. Developments in oil extraction from microalgae. *European Journal of Lipid Science and Technology*, 113, 539–547.

Mezzomo, N., Mileo, B. R., Friedrich, M. T., Martínez, J., Ferreira, S. R. S. 2010. Supercritical fluid extraction of peach (*Prunus persica*) almond oil: process yield and extract composition. *Bioresource Technology*. 101, 5622–5632.

Miao X, Wu, Q. 2006. Biodiesel production from heterotrophic microalgal oil. *Bioresource Technology* 97(6), 841–846.

Montero, O., Macías-Sánchez, M. D., Lama, C. M., Lubián, L. M, Mantell, C., Rodríguez, M., Martínez de la Ossa, E. 2005. Supercritical CO_2 extraction of β-carotene from a marine strain of the cyanobacterium Synechococcus species. *Journal of Agriculture and Food Chemistry* 53, 9701–9707.

Nobre, B. Marcelo, F. Passos, R., Beirao, L. Palavra, A., Gouveia, L., Mendes, R. 2006. Supercritical carbono dioxide extraction of astaxanthin and other carotenoids from the microalga *Haematococcus pluvialis*. *European Food Research and Technology*, 223(6), 787–790.

Oliveira, E. L., Silvestre, A. J., Silva, C. M. 2011. Review of kinetic models for supercritical fluid extraction. *Chemical Engineering Research and Design*, 89, 1104–1117.

Ota, M., Watavabe, H., Kato, Y., Watanabe, M., Sato, Y., Smith Jr., R.L. Inomata, H. 2009. Carotenoid production from *Chlorococcum littorale* in photoautotrophic cultures with downstream supercritical fluid processing. *Journal of Separation Science*, 32(13), 2327–2335.

Palavra, A. M. F., Coelho, J. P., Barroso, J. G., Rauter, A. P., Fareleira, J. M. N., Mainar, A., Urieta, J. S., Nobre, B. P., Gouveia, L, Mendes, R. L., Cabral, J. M. S., Novais, J. M. XXXX. Supercritical carbon dioxide extraction of bioactive compounds from microalgae and volatile oils from aromatic plants. *Journal of Supercritical Fluids*, 60, 21–27.

Palmer, MV and Ting, S.S.T., 1995. Applications for supercritical fluid technology in food processing. *Food Chemistry*, 52(4), 345–352.

Patil, S. R., Dayanand, A. 2006. Optimization of process for the production of fungal pectinases from deseeded sunflower head in submerged and solid-state conditions. *Bioresource Technology*, 97, 2340–2344.

Pereyra, C., Molero, A., Martínez de la Ossa, E., 1995. Extracción supercrítica con dióxido de carbono. *Ingeniería Química Junio*, 181–184.

Pérez Galindo, J.A., López Miranda, J., Martín Domínguez, I.R. 2000, Geometric and Reynolds number effects on oregano (*Lippia Berlandieri Schauer*) essential oil extraction *Journal of Food Engineering*, 44, 127–133.

Prado, J. M., Prado, G. H. C., Meireles, M. A. A. 2011. Scale-up study of supercritical fluid extraction process for clove and sugarcane residue. *Journal of Supercritical Fluids*, 56(3), 231–237.

Reverchon, E., De Marco, I. 2006, Supercritical fluid extraction and fractionation of natural matter. *Journal of Supercritical Fluids*, 38, 146–166.

Reverchon, E., Marrone, C., 2001, Modelling and simulation of the supercritical CO_2 extraction of vegetable oils. *Journal of Supercritical Fluids*, 19, 161–171.

Rincón, J., Martínez , F., Rodríguez, L., Ancillo, V. 2011 Recovery of triglycerides from used frying oil by extraction with liquid and supercritical ethane. *Journal of Supercritical Fluids*, 56(1), 72–79.

Ruiz, E., Cara, C., Manzanares, P., Ballesteros, M., Castro, E. 2008. Evaluation of steam explosion pre-treatment for enzymatic hydrolysis of sunflower stalks. *Enzyme and Microbial Technology* 42, 160–166.

Ruiz-Rodríguez, A., Reglero, G., Ibáñez, E. 2010. Recent trends in the advanced analysis of bioactive fatty acids. *Journal of Pharmaceutical and Biomedical Análisis*, 51, 305–326.

Saharia, M.A., Akbarian, A. Hamedib, M. 2003. Effect of variety and acid washing method on extraction yield and quality of sunflower head pectin. *Food Chemistry*, 83, 43–47.

Sahena, F., Zaidul, I. S., Jinap, S., Karim, A. A., Abbas, K. A., Norulaini, N. A., Omar, A. K. 2009. Application of supercritical CO_2 in lipid extraction—A review. *Journal of Food Engineering*, 95, 240–253.

Sánchez-Camargo, A.P., Almeida Meireles, M.A., Lopes, B.L.F., Cabral, F.A. 2011. Proximate composition and extraction of carotenoids and lipids from Brazilian redspotted shrimp waste (*Farfantepenaeus paulensis*). *Journal of Food Engineering*, 102, 87–93.

Señoráns, F. J., Ruiz-Rodríguez, A., Ibáñez, E., Tabera, J., Reglero, G. 2001. Optimization of countercurrent supercritical fluid extraction conditions for spirits fractionation. *Journal of Supercritical Fluids*, 21, 41–49.

Sharma, S. K., Kalra, K. L., Grewal, H. 2002. Fermentation of enzymatically saccharified sunflower stalks for ethanol production and its scale up. *Bioresource Technology*, 85, 31–33.

Sharma, S. K., Kalra, K. L., Kocher, G. S. 2004. Fermentation of enzymatic hydrolysate of sunflower hulls for ethanol production and its scale-up. *Biomass and Bioenergy*, 27, 399–402.

Soh, L., Zimmerman, J. 2011. Biodiesel production: the potential of algal lipids extracted with supercritical carbon dioxide. *Green Chemistry*, 13, 1422–1429.

Sovova, H. 1994. Rate of the vegetable oil extraction with supercritical CO_2: I. Modelling of extraction curves. *Chemical Engineering Science*, 49, 409–414.

Stastova, J., Jez, J., Bartlova, M., Sovova, H. 1996. Rate of the vegetable oil extraction with supercritical CO_2 -III. *Extraction from sea buckthorn. Chemical Engineering Science*, 51, 4347–4352.

Tang, S. K., Qin, C. R., Wang, H. Q., Li, S. F., Tian, S. J. 2011. Study on supercritical extraction of lipids and enrichment of DHA from oil-rich microalgae. *Journal of Supercritical Fluids*, 57, 44–49.

Telli-Okur, M., Eken-Saracoglu, N. 2008. Fermentation of sunflower seed hull hydrolysate to ethanol by *Pichia stipitis*. *Bioresource Technology*, 99, 2162–2169.

Temelli, F. 2009. Perspectives on supercritical fluid processing of fats and oils. *Journal of Supercritical Fluids*, 47, 538–590.

Toribio, L., Del Nozal, M. J., Bernal, J. L., Bernal, J., Martín, M. T. 2011. Study of the enantiomeric separation of an acetamide intermediate by using supercritical fluid chromatography and several polysaccharide based chiral stationary phases. *Journal of Chromatography A*, 1218, 4886–4891.

Torres, C. F., Torrelo, G., Señoráns, F. J., Reglero, G. 2009. Supercritical fluid fractionation of fatty acid ethyl esters from butteroil. *Journal of Dairy Science*, 92, 1840–1845.

Valderrama, J. O., Perrut, M., Majewski, W. 2003. Extraction of astaxantine and phycocyanine form microalgae with supercritical carbon dioxide. *Journal of Chemical and Engineering Data* 48(4), 827–830.

Wagner, H., Eggers, R. 1996. Extraction of spray particles with supercritical fluids in a two-phase flow. *AIChE Journal* 42(7), 1901–1910.

Wang, H. M., Pan, J. L., Chen, C. Y., Chiu, C. C., Yang, H. M., Chang, H. W., Chang, J. S. 2010. Identification of anti-lung cancer extract form *Chlorella vulgaris* C-C by antioxidant property using supercritical carbon dioxide extraction. *Process Biochemistry*, 45(12), 1856–1872.

Xu, H., Miao, X., Wu, Q. 2006. High quality biodiesel production from a mocroalga Chlorella protethecoides by hetertrophic growth in fermenters. *Journal of Biotechnology*, 126, 499–507.

Zabaniotou, A. A., Andreou, K. 2010. Development of alternative energy sources for GHG emissions reduction in the textile industry by energy recovery from cotton ginning waste. *Journal of Cleaner Production*, 18, 784–790.

Zabaniotou, A. A., Kantarelis, E. K., Theodoropoulos, D. C. 2008. Sunflower shells utilization for energetic purposes in an integrated approach of energy crops: Laboratory study pyrolysis and kinetics. *Bioresource Technology*, 99, 3174–3181.

Zhou, D. Y., Zhu, B. W., Tong, L., Wu, H. T., Qin, L., Tan, H., Chi, Y. L., Qu, J. Y., Murata, Y. 2011. Extraction of lipid from scallop (*Patinopeten yessoensis*) viscera by enzyme-assisted solvent and supercritical carbon dioxide methods. *International Journal of Food Science and Technology*, 45(19), 1787–1793.

Zinnai, A., Nesti, U., Venturi, F., Andrich, G., Firoentini, R. 2006. Supercritical fluid extraction of oil from *Sxhizochytrium sp*. *Rivista italiana delle Sostanza Grasse*, 83(1), 1–5.

Part III

Affinity-Based Separation Technologies

5
Adsorption

Saravanan Venkatesan

Shell Global Solutions International B.V., Department of Innovation Biodomain, The Netherlands

5.1 Introduction

Growing energy demands, concerns over energy security, and environmental concerns present a major challenge to identify an alternative sustainable energy to replace fossil fuels. Many countries have set goals to partially replace liquid transport fuel with biomass-derived liquid fuels such as ethanol and biodiesel. Ligno-cellulosic materials are considered to be a promising source of feedstock for sustainable biofuels as well as industrial chemicals. These biomass materials contain mainly cellulose, hemicellulose and lignin, being present in a complex structural form. In biorefineries, these complex biomass feedstocks are broken down by enzymatic, chemical, mechanical, or thermo-chemical means. Cellulose and hemicellulose polymers are converted into glucose and xylose monomers, respectively, by enzymatic hydrolysis. These sugars can be further converted to transport fuels like ethanol, other oxygenates or hydrocarbons, by chemical or fermentation processes.

Separation and purification processes account for the major production and operating costs in biorefineries. It is therefore crucial to have robust separation technologies to make biorefineries economically viable. Purification of most of the products produced in biorefineries cannot be achieved by standalone conventional technologies and requires the use of combinations of separation and purification technology schemes discussed in other chapters of this book. For example, recovery of dilute alcohols produced by fermentation is very expensive and energy intensive using conventional distillation alone. An attractive alternative is adsorption, which can perform many selective separations that are not feasible by conventional techniques alone, such as distillation, extraction, absorption and even membrane-based systems.

In the adsorption process, molecules from bulk fluid (either liquid or gas) are preferentially bound to a solid surface. These surface phenomena can occur at any solid–fluid interface. The solid phase is called the adsorbent. The component which is in the adsorbed state on the adsorbent is called adsorbate and the component in the bulk fluid phase prior to adsorption is called adsorptive. Adsorption should not be

Separation and Purification Technologies in Biorefineries,
First Edition. Edited by Shri Ramaswamy, Hua-Jiang Huang, and Bandaru V. Ramarao.

confused with absorption, a process where a component penetrates or dissolves into the bulk of solid or liquid adsorbent. The term "sorption" refers to both adsorption and absorption.

The application of adsorption can be traced back as early as 1550 BC when Egyptians used charcoal for medicinal purposes [1]. In 1794, bone char was used as a decolorizing agent in the sugar industry in England. There was a revolution in adsorption technology when synthetic zeolites were produced commercially by Linde in 1956 [2]. Now, new adsorbents are constantly being produced with radically improved properties that can result in cost-effective separation process.

The most common current applications of adsorption are in the separation and purification of liquid and gas mixtures, bulk chemicals, isomers, and air; drying gases and liquids before loading them into industrial systems; removal of impurities from liquid and gas media; recovery of chemicals from industrial and vent gases; and water purification [1]. In bioprocess separation, liquid phase adsorption, particularly with aqueous media, is of most significance whereas gas-based adsorption plays a secondary role.

The overarching theme of this chapter is to highlight the role played by adsorption in the biorefinery separation processes. The chapter begins with a brief description of the underlying principles of adsorption, which are then invoked to describe adsorbent selection criteria. The properties of some of the current commercially available adsorbents, which are at different stages of implementation in biorefinery separations, are described, and some novel adsorbents, which are creating much excitement in this field, are also presented.

Regeneration of the adsorbents is a critical part of the adsorption process as it is necessary to complete the separation in a sustainable manner. The basic classification of adsorption separation processes is based on the regeneration step and it will be described below with specific emphasis given to temperature swing adsorption (TSA) and pressure swing adsorption (PSA). The scale-up of any adsorption process will critically hinge on the adsorption equilibrium and on breakthrough model predictions. Often the thermodynamic and rate models used are complex and require large amounts of existing data. In the young biorefinery context, this data might not be available and hence a simple model is presented here, which can easily be applied to suit different biorefining applications.

The role of adsorption in the biorefinery separation process will be brought alive by a broad review of existing applications followed by a deeper inspection via a particular case study: the recovery of 1-butanol from acetone-butanol-ethanol (ABE) fermentation broth using TSA technology. Despite the impressive progress in application and understanding of adsorption in biorefineries, many gaps remain and the chapter ends with some thoughts on focus areas for research in this exciting but challenging area of adsorption technology.

5.2 Essential principles of adsorption

Adsorption occurs due to the attractive forces existing between the adsorbent surface and adsorbate molecules. The basic types of attractive forces are van der Waals' forces (dispersion and repulsion), electrostatic forces, and chemical bonding. In the case of dense phase adsorption, solvent interactions with adsorbate and adsorbent also influence adsorption. Based on the type of attractive forces involved, adsorption is classified as physical adsorption or chemical adsorption. In physical adsorption or physisorption, the attractive force is relatively weak and is governed by weak van der Waals' and electrostatic forces. Hence, it is also called van der Waals' adsorption. It is generally an exothermic process and it can also be easily reversed by heating or decreasing the pressure of the adsorbate (as in the case of gases). In chemical adsorption or chemisorption, the attractive force is governed by chemical bonding. This type of adsorption tends to be irreversible. Separation and purification processes based on chemisorption, particularly involving π-complexation [3], have been explored recently, but otherwise, most of the industrial adsorption processes are based on the physisorption. As such, the main focus of the following discussions will be on physisorption.

5.2.1 Adsorption isotherms

Adsorption isotherm is the equilibrium relationship between the quantity of material adsorbed and its concentration or partial pressure in the bulk fluid phase. Since the equilibrium is a function of temperature, pressure, and composition, isotherm refers to equilibrium relation at constant temperature. Isotherms are ideally obtained experimentally and then correlated for use in design and modeling. There is no difference between adsorption from the liquid and gas phase, as thermodynamically the adsorbate concentration in equilibrium with a liquid must be the same as that which is in equilibrium with the saturated vapor. The simple isotherm models, developed to describe gas phase adsorption, predict well at low adsorbate concentrations but become unreliable close to saturation limit [4]. In liquid phase adsorption, the adsorbate concentration usually approaches saturation limit and hence these models have limited application. The most commonly used isotherm models for predicting adsorption equilibria of single and multicomponent adsorption are:

- Freundlich isotherm;
- Langmuir isotherm;
- BET isotherm;
- Ideal adsorbed solution theory (IAST).

5.2.1.1 *Freundlich isotherm*

Gas isotherm In 1909, Freundlich introduced a first empirical fit for an isotherm:

$$q = aP^{1/n} \tag{5.1}$$

where q is the mass of the adsorbate per mass of adsorbent; P is pressure; n is a constant greater than 1; a is also a constant.

At low pressure or high temperature n approaches unity. At high pressure or low temperature, $1/n$ approaches as low as 0.1.

Liquid isotherm The Freundlich isotherm for liquid phase adsorption is given by the correlation below:

$$q = aC^{1/n} \tag{5.2}$$

where C is the liquid phase concentration.

Recently, the Freundlich isotherm has been reported to best fit the adsorption of biofuel compounds (e.g. n-butanol and ethanol) from dilute aqueous solutions by mesoporous carbons and polymeric resins [5, 6]. The limitation of the Freundlich model is that it is generally applicable below saturation concentration/pressure.

5.2.1.2 *Langmuir isotherm*

Gas isotherm In 1918, Langmuir proposed a semi-empirical model for monolayer adsorption [7]. This model was originally developed for chemisorptions of gas molecules onto solids. The basic assumptions made in this model are:

- There is a fixed number of adsorption sites available on the solid surface;
- All adsorption sites are energetically equivalent;
- Each site can hold one adsorbate molecule and a monolayer adsorption is assumed;
- Adsorbed molecules do not interact with each other;
- Heat of adsorption is the same for all the molecules.

The Langmuir adsorption equation is given by:

$$\varphi = \frac{KP}{1 + KP} \tag{5.3}$$

where P is the equilibrium gas pressure; φ is the fraction of adsorption sites covered ($\varphi = \frac{V}{V_{mon}}$ where V = volume of gas adsorbed; V_{mon} = volume of adsorbed gas forming a monolayer; $K = \frac{k_a}{k_d}$ (k_a is the adsorption constant and k_d is the desorption constant).

At saturation pressure $P \to \infty$ and φ becomes unity showing monolayer adsorption. At low pressure, $\varphi = KP$, which is called Henry's law. The adsorption equilibrium constant should follow the Van 't Hoff equation:

$$K = K_o exp\left(\frac{-\Delta H}{RT}\right) \tag{5.4}$$

In the case of multicomponent adsorption, for j gas component, Eq. (5.3) can be modified as:

$$\varphi_j = \frac{K_j P_j}{1 + \sum_{i=1}^{i=n} K_i P_i} \tag{5.5}$$

Equation (5.5) is also called the extended Langmuir model.

Liquid isotherm The Langmuir isotherm is widely used to describe liquid-phase adsorption including multicomponent systems [8–10]. The Langmuir isotherm for a pure component adsorption is given by:

$$q = \frac{q_s kC}{1 + kC} \tag{5.6}$$

where q_s is the saturation loading of adsorbent; k is the Langmuir constant. For multicomponent mixtures containing (n) species, the extended Langmuir model for adsorption of component (j) is given by:

$$q_j = \frac{q_{m,j}\ k_j C_j}{1 + \sum_{i=1}^{i=n} k_i\ C_i} \tag{5.7}$$

The limitations of the Langmuir model are that it is applicable only to monolayer adsorption which may be possible only under low-pressure/concentration or high-temperature conditions. Under high-pressure/concentration and low-temperature conditions, multilayer adsorption is generally observed. The assumption of an energetically equivalent adsorption site may not be practical because real solid surfaces are heterogeneous. The assumption of no interaction between adsorbate molecules is also a simplification as weak force of attraction exists even between molecules of the same type. Despite the above limitations, the Langmuir model remains very useful for practical design purposes.

5.2.1.3 BET isotherm

Gas isotherm In 1938, Brunauer *et al.* [11] published a model known as the "BET model" to account for multilayer gas adsorption. This model provided a basis for the measurement of the surface area of microporous solid materials. The basic assumptions of the BET model are:

- gas molecules physically adsorb on a solid in layers infinitely;
- there is no interaction between each adsorption layer;
- the Langmuir theory can be applied to each layer.

The BET equation is expressed as:

$$\frac{1}{V\left[\frac{P_o}{P}-1\right]}=\frac{b-1}{V_{mon}b}\left(\frac{P}{P_o}\right)+\frac{1}{V_{mon}b} \tag{5.8}$$

where P_o is the saturation pressure of adsorbate at the temperature of adsorption; b is the BET constant expressed by $b=exp\left(\frac{h_1-h_L}{RT}\right)$ (h_1 is the heat of adsorption for the first layer and h_L is that for the second and higher layers and is equal to the heat of liquefaction).

From V_{mon} in Eq. (5.8), the total surface area (S_{tot}) of a solid can be calculated:

$S_{tot}=\frac{V_{mon}\ N_S}{\vartheta}$, where N is Avogadro's number; s is the area occupied by an adsorbate molecule; ϑ is the molar volume of adsorbate.

Liquid isotherm The liquid phase BET isotherm [12] is represented by

$$q=q_m\frac{b_S C_e}{(1-b_L C_e)\left[1-b_L C_e+b_S C_e\right]} \tag{5.9}$$

where q_m is the monolayer saturation capacity; C_e is the solute equilibrium concentration; b_s is the equilibrium constant of adsorption on the first layer; b_L is the equilibrium adsorption–desorption constant for upper layers of adsorbate.

5.2.1.4 Ideal adsorbed solution (IAS) theory

Liquid isotherm The IAS theory was originally proposed by Myers and Prausnitz [13] to predict the multicomponent gas adsorption isotherm using the pure component isotherm data for the same temperature and adsorbent. Radke and Prausnitz [14] modified this theory to describe liquid phase adsorption. Here, the IAS model for liquid phase adsorption is presented because of its more relevance to bioseparation processes. This theory assumes that the adsorbed mixture forms an ideal solution in equilibrium with liquid phase at a constant spreading pressure for each solute. The spreading pressure (π) is defined as the difference between the interfacial tension of the pure solvent-solid interface and that of the solution-solid interface [14]. The Gibbs relation for spreading pressure is:

$$\frac{\pi_i A}{RT}=\int_0^{C_i^o}\frac{q_i^o\left(C_i^o\right)}{C_i^o}dC_i^o \tag{5.10}$$

where the integrant $q_i^o\left(C_i^o\right)$ is the single solute adsorption isotherm; C_i^o is the single solute liquid phase concentration in equilibrium with q_i^o, the single solute adsorption capacity at the same temperature and spreading pressure of the multicomponent system; A is the specific adsorbent surface area; R is the universal gas constant; T is the absolute temperature.

The liquid–solid equilibrium, in analogy with Raoult's law, is represented by

$$C_T\ x_i = C_i^o\ z_i = C_i \tag{5.11}$$

where C_T is the total concentration of all solutes in the liquid phase; C_i is the concentration of solute i in the mixture; x_i is solvent-free liquid-phase mole fraction; z_i is the adsorbed phase mole fraction.

Since single-solute concentrations C_i^o are defined at the same spreading pressure as that of the mixture

$$\pi_i = \pi_{mixture} \tag{5.12}$$

Furthermore, sum of each mole fraction equals 1:

$$\sum_i x_i = \sum_i z_i = 1 \tag{5.13}$$

Assumption of constant adsorption area per mole of solute, for the mixture or the single solute, at the spreading pressure of mixture follows:

$$\frac{1}{q_t} = \sum_{i=1}^{i=n} \frac{z_i}{q_i^o} \tag{5.14}$$

where q_t is the total adsorption capacity which is related to the adsorption capacity of solute i in a mixture (q_i) by:

$$q_i = q_t z_i \tag{5.15}$$

If the single solute isotherm follows the Langmuir or Freundlich model, Eq. (5.10) simplifies to an algebraic equation; otherwise, it has to be solved numerically using experimental isotherm data. By specifying C_T and (n–1) independent liquid phase mole fractions (x_i) and simultaneously solving the Eqs. (5.10–5.13), the unknown variables (z_i, C_i^o, π_i and $\pi_{imixture}$) are calculated. Based on the single solute isotherm model that is applicable for a given solute i, C_i^o values are used to calculate q_i^o. Equations (5.14–5.15) are then solved to calculate q_t and q_i .

5.2.2 Types of adsorption isotherm

Brunauer *et al*. [15] classified physical adsorption isotherms into five types (Figure 5.1).

Type I isotherms correspond to a monolayer adsorption and can be explained by the Langmuir isotherm. It is characterized by the adsorbate concentration reaching a plateau. Microporous adsorbents exhibit a Type I isotherm. Type II isotherms describe adsorption in non-porous and macroporous adsorbents. At low pressure, they exhibit monolayer adsorption and, as pressure is increased, multi-layer adsorption is encountered followed by liquid condensation in the pores even below the saturation pressure (capillary condensation caused by the increase in capillary pressure). Type II isotherms can be described by the BET equation. Type III isotherms occur when the adsorbate-adsorbent interaction is weaker than the adsorbate-adsorbate interaction. Type IV isotherms are characterized by multi-layer formation and they are associated with capillary condensation taking place in mesoporous adsorbents. Type V isotherms are observed when inter molecular attraction effects are large. Types I, II and IV represent favorable equilibrium whereas Type III and V represent unfavorable ones. Among these, adsorbents that exhibit Type I behavior are generally preferred for cyclic operations.

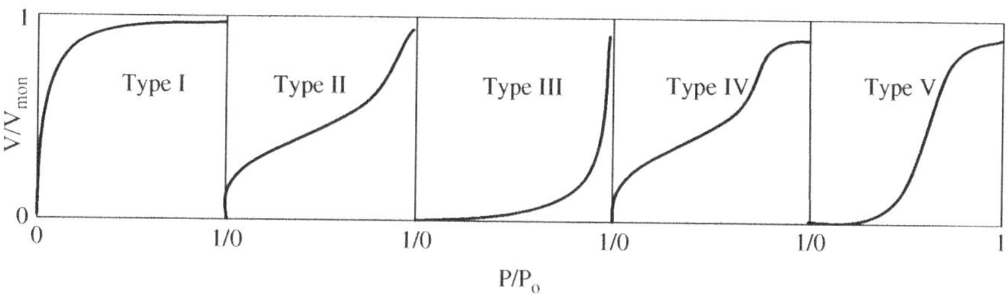

Figure 5.1 Adsorption isotherm classification. Reprinted with permission from [15] © 1940, American Chemical Society

5.2.3 Adsorption hysteresis

Adsorption hysteresis occurs when the adsorption and desorption isotherms deviate from one another. The specific causes of adsorption hysteresis are still an active area of research, but it is linked to differences in the adsorption and desorption mechanisms, usually as a result of capillary condensation inside mesopores. The IUPAC classified hysteresis loops into four types H1, H2, H3 and H4 [16]. These types are illustrated in Figure 5.2.

Types H1 and H4 are categorized as extreme types where branches are almost vertical (H1) and nearly parallel (H4) over an appreciable range of gas uptake. Types H2 and H3 are regarded as intermediate between these two extremes. The shapes of hysteresis loops are often identified with specific pore structures [16]. The Type H1 is characteristic of porous materials consisting of agglomerates or uniform spheres compacted approximately in a regular array. The Type H2 is observed for mesoporous materials

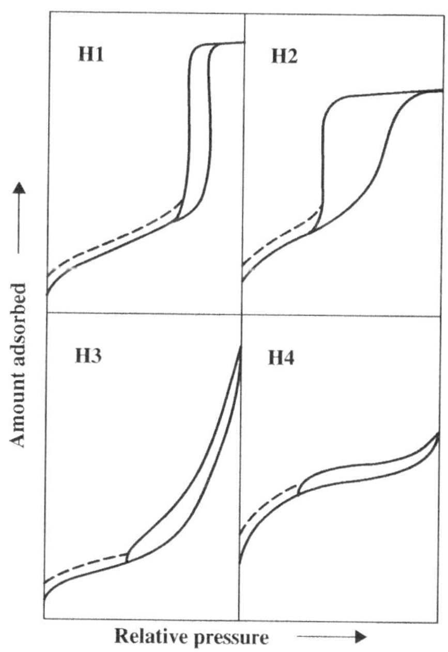

Figure 5.2 IUPAC classification of hysteresis loop. Reprinted with permission from [16] © 1982 International Union of Pure and Applied Chemistry

with undefined distribution of pore size and shape (e.g. silica gels). The Type H3 loop does not exhibit any adsorption limit at high relative pressure, which is associated with slit-shaped pores formed by aggregates of plate-like particles. Similarly, the Type H4 loop also appears to be associated with narrow slit-like pores. In Figure 5.2, the dashed lines indicates the low-pressure hysteresis where removal of the residual adsorbed material is possible only if the adsorbent is gassed out at higher temperatures.

There are many theories available such as density functional theory, self-consistent field theory, and lattice theory, and simulation methods to predict the adsorption hysteresis [17–20]. When an adsorption system exhibits hysteresis there may be an impact on the regenerability and kinetics, so it can become one of the important considerations while designing such an adsorption system on industrial scale.

5.2.4 Heat of adsorption

The measured heat of adsorption is the net enthalpy change upon the adsorbate binding to the adsorbent after accounting for solvent interaction. There are several definitions for heat of adsorption; isosteric, differential, integral, and equilibrium. The isosteric (i.e. at constant adsorbate loading) heat of adsorption for a component i is defined as [21]

$$h_i = RT^2 \left[\frac{\partial \ln (p_i)}{\partial T} \right]_{n_i} \tag{5.16}$$

where p_i and n_i are the partial pressure and adsorbent loading of ith component for a mixture. The isosteric heats of adsorption determine the extents of adsorbent temperature changes during the adsorption (exothermic) and desorption (endothermic) steps of the processes. The adsorbent temperature is a key variable in determining the local adsorption equilibria and kinetics on the adsorbent, which ultimately govern the separation performance of the processes [21]. Hence, it is a key thermodynamic parameter for designing an adsorption process. The plot of isosteric heat of adsorption and adsorbate loading can follow any one of the following trends: monotonously increasing, decreasing, or constant, as loading increases. The first case indicates that adsorption is weak at low concentration. Hence the regeneration is likely to be easy. The second indicates the reverse, strong adsorption at low concentration, probably due to the surface heterogeneity and regeneration may require more energy. The third case represents a neutral behavior.

5.3 Adsorbent selection criteria

Selection of a suitable adsorbent for a specific process is very important to make the separation process economically attractive. The most important criteria considered for selecting an adsorbent for any adsorptive separation process are: adsorbent loading, selectivity, regenerability, kinetics, compatibility, and cost. The following section briefly explains each criterion.

Adsorption capacity (or loading) is the most important characteristic of an adsorbent. It is the mass of material adsorbed per unit mass or volume of the adsorbent. It is clearly bound by equilibrium as described above in the isotherm section. Adsorption capacity dictates the amount of adsorbent required and thus the volume of the adsorber vessel. In cases where temperature swing (TSA) is used for regeneration, the adsorption capacity also has a direct effect on the energy requirements as the adsorbent needs to be heated as well. Maximum adsorption capacity data can be obtained from the adsorption isotherm. Other means to express adsorption capacity are isosteres, isobars, and various indices such as surface area, pore size distribution, iodine number and molasses number.

Selectivity is the ratio of adsorption capacity of one component to that of another at a given fluid concentration. Most of the adsorption processes depend on the equilibrium selectivity [4]. It can be defined as

$$\alpha_{ij} = \frac{z_i/z_j}{y_i/y_j} \tag{5.17}$$

where z_i and y_i are the mole fraction of component i in adsorbed and fluid phases at equilibrium. Equation 5.17 is analogous to the relative volatility, which measures the ease of separation by distillation. The selectivity varies with temperature and composition. For an ideal Langmuir system (both gas and liquid phase under low pressure/concentration), it is independent of composition and is equal to the ratio of the Henry's law constants of the two components:

$$\alpha_{ij} = \frac{K_i}{K_j} \tag{5.18}$$

where K_i and K_i are the Henry's constants for components i and j respectively. Thus, a preliminary selection of an adsorbent can be made from the Henry's constants of the components to be separated.

Regenerability is the ease with which the spent adsorbent can be brought back to its original state. The weaker the adsorption, such as physisorption, the easier would be the regeneration. The heat of adsorption is a measure of the energy required for the regeneration process. The working capacity of an adsorbent is the original capacity that is retained after cyclic regeneration. The working capacity gradually decreases with the ageing, poisoning of adsorbent, and other related causes. Thus the life of an adsorbent is dictated by its regenerability.

Adsorption kinetics are related to the intra-particle mass transfer resistance. It governs the cycle time of a fixed bed adsorption process. Fast kinetics results in a sharp breakthrough while slower kinetics results in a distended breakthrough curve. To overcome the distended curve effects of slow kinetics, the cycle time of the process should be long, which would require a huge adsorbent inventory. Hence, kinetics is an important criterion for adsorbent selection. Kinetics has also been exploited to separate different components using molecular sieve adsorbents [4]. This kinetic selectivity is measured by the ratio of the micropore diffusivities of the components. The most common example is the pressure swing adsorption process, which splits nitrogen from air using a carbon molecular sieve, which relies on the faster diffusion of oxygen compared to nitrogen.

Compatibility of an adsorbent covers both the effect of the process on the nature of the adsorbent and the effect of the adsorbent on the process components. The latter is more relevant for the *in situ* separation of alcohols from fermentation broths where the adsorbent can be toxic to the alcohol-producing microorganisms. The adsorbent should not irreversibly react with the adsorbed molecules.

Cost of the adsorbent may dominate over all the above criteria in many cases.

5.4 Commercial and new adsorbents and their properties

The advancement of adsorption technology primarily depends on the development of novel and robust adsorbents. These adsorbents take a broad range of chemical and geometrical structural forms. Table 5.1 shows their general classification.

The following section describes the important properties of some of the commercial and new adsorbents that may have potential application in biorefinery separation processes.

Table 5.1 Types of commercial adsorbents. Reprinted from [1] © 2001, with permission from Elsevier

Carbon adsorbents	Mineral adsorbents	Other adsorbents
Active carbons Activated carbon fibres Carbon molecular sieves Mesocarbon microbeads Fullerenes Heterofullerenes Carbonaceous nanomaterials	Silica gels Activated alumina Oxides of metals Hydroxides of metals Zeolites Clay minerals Pillared clays Porous clay hetero-structures (PCHs) Inorganic nanomaterials	Synthetic polymers Composite adsorbents: (complex mineral carbons, X-elutrilithe; X = Zn, Ca) Mixed sorbents Metal organic frameworks (MOF)

5.4.1 Activated carbon

Activated carbon is the most widely used adsorbent because of its large porous volumes and the resulting high surface area [3]. Activated carbon can be manufactured from any carbonaceous organic material. Commercial carbons are made from a range of materials, such as sawdust, wood, charcoal, peat, fruit nuts, lignite, petroleum coke, bituminous coal, and coconut shells. The activation steps that are commercially used are steam activation and chemical activation. The steam activation process consists of two steps: carbonization and activation. Carbonization is done by heating the material in the range of 400–500 °C in an oxygen-free atmosphere to remove the bulk of volatile matter. The carbonized particles are then "activated" by exposing them to an oxidizing agent, usually steam or carbon dioxide at 800–1000 °C [3]. This technique is used for activation of coal and coconut shell. This forms a porous, three-dimensional graphite lattice and a large surface area by removing the pore blocking pyrolysis materials created during the carbonization step.

Chemical activation is generally used for the activation of peat and wood-based raw materials. The raw material is impregnated with a strong dehydrating agent, typically phosphoric acid (H_3PO_4) or zinc chloride ($ZnCl_2$), mixed into a paste and then heated to temperatures of 500–800 °C to activate the carbon; the resulting activated carbon is washed, dried, and ground to powder. Activated carbons produced by chemical activation generally exhibit a very "open" pore structure, commonly referred to as "Macroporous," ideal for the adsorption of large molecules.

The important properties of an adsorbent, which will affect the adsorption characteristics, are: the pore size distribution, surface area, surface qualities (hydrophobic/hydrophilic nature and functional groups), and physical properties such as hardness, bulk density and particle-size distribution. These properties are generally controlled by fine tuning the different parameters of the manufacturing process. Some of these properties of activated carbon are briefly described below.

Activated carbon is found to have polymodal pore size distribution. The International Union of Pure and Applied Chemistry (IUPAC) defines the pore size distribution as:

- micropore radius <1 nm
- mesopore radius 1–25 nm
- macropore radius >25 nm

The macropores are used as the entrance to the activated carbon, the mesopores for transportation and the micropores for adsorption. The total pore volume of activated carbons can be up to 80%. Activated carbons used for gas-phase application are designed to have pore size ranging from 10 to 25 Å, whereas for liquid phase application, pore size will be larger than 30 Å in order to decrease the mass transfer resistance of large-size dissolved adsorbate molecules.

Surface area is the primary indicator of the activity level. Activated carbons have the largest surface area ranging from 300 to ~4000 $m^2\,g^{-1}$, as measured by the BET method [3]. The adsorption capacity is typically about 1 to 35 wt%. The surface of activated carbon is essentially nonpolar and a slight polarity may arise because of presence of surface oxide groups. Activated carbons have bulk density between 400 to 640 kg per cubic meter. They are broadly classified based on their physical characteristics as powdered activated carbon (PAC), granulated activated carbon (GAC), extruded activated carbon (EAC), and bead activated carbon (BAC). Careful consideration of particle size can provide significant operating benefits by balancing pressure drop and adsorption kinetics.

5.4.2 Silica gel

Silica gel is an amorphous inorganic adsorbent having mesoporous structure. It is well known for its desiccant property. Commercial silica gels are produced by polymerization of silicic acid. First, a sodium silicate solution is acidified using sulfuric or hydrochloric acid to produce silicic acid. Following this, silicic acid polymerizes into jelly precipitate, which is washed and dried to produce colorless silica gel. By varying the silica concentration, pH and temperature, the properties of silica gel such as pore volume, surface area and shape can be varied [22]. Two common types of silica gel are regular-density and low-density silica gels. The regular-density silica gel has a surface area of 750–850 $m^2\,g^{-1}$ and an average pore diameter of 22–26 Å, whereas the respective values of low-density gel are 300–350 $m^2\,g^{-1}$ and 100–150 Å [3]. The presence of hydroxyl groups makes its surface hydrophilic. Hence molecules such as water, alcohol, and phenol are adsorbed in preference to non-polar molecules [4]. Common forms are granules, extrudates (2 to 4 mm diameter) and beads (1 to 3 mm diameter).

5.4.3 Zeolites and molecular sieves

Zeolites are microporous crystalline aluminosilicate materials. The basic structural units of a zeolite framework are tetrahedra of silicon and aluminum, SiO_4 and AlO_4, which are cross-linked to each other by oxygen atoms. Clusters of these units form many secondary polyhedral building units, which are further linked to form entire three-dimensional frameworks. There are about 194 unique zeolite frameworks identified so far, and over 40 naturally occurring known zeolite frameworks. The structural formula of a zeolite unit cell can be represented by:

$$M_{x/n}\left[(AlO_2)_x\,(SiO_2)_y\right].zH_2O$$

where M is the cation such as Na^+, K^+, Ca^{2+}, Mg^{2+} and NH_4^+, x and y are integers and $y/x \geq 1$, n is the valence of the cation and z is the number of water molecules in each unit cell [3]. Each aluminum atom introduces a negative charge on the framework, which is compensated by an exchangeable cation. The location of cation on the framework plays a very important role in determining the adsorptive properties. Aluminum-rich zeolites have hydrophilic nature. The transition from hydrophilic to hydrophobic occurs at a Si/Al ratio of between 8 and 10 [4]. Thus zeolite with a specific adsorptive property can be prepared by appropriate choice of Si/Al ratio and cation type. Commercially significant zeolite types are A, X, Y, beta, ZSM-5, mordenite and silicalite. Type A and X zeolites can selectively adsorb (sieve) molecules depending on their relative sizes and the pore diameter of the adsorbent. Hence they are called molecular

Table 5.2 Commercial zeolites' characteristics

Zeolite type	Cation type	Nominal pore diameter (Å)	Number of tetrahedra in a ring	Si/Al ratio
3A	K	3	8	1
4A	Na	4	8	1
5A	Ca	5	8	1
10X	Ca	8	12	1.2
13X	Na	10	12	1.2
Y	K	8	12	2.4
Mordenite	H	7	12	5
ZSM-5	Na	6	10	31
Silicalite	–	6	10	∞

sieves. The zeolite framework has a very regular structure of cages, which are interconnected by windows in each cage. The window aperture size depends on the number of tetrahedra in a ring and also the type and number of cations present. Typical types of cation present in zeolites are alkali metal such as Na and K, alkaline earths such as Ca and Mg, transition metals such as Ti and V, and rare earths. Table 5.2 shows the characteristics of some of the commercial zeolite types.

Zeolites are usually manufactured by hydrothermal synthesis of sodium aluminosilicate from sodium hydroxide, sodium silicate and sodium aluminate. This is followed by ion exchange with cations and drying of the crystals, which can be pelletized with a binder to form macroporous pellets. By controlling the pH, temperature and concentration, different types of zeolites are produced. There are vast amount of literature reviews available on zeolites synthesis [23–26]. Recently, Chal *et al*. [27] presented a review on various synthesis strategies towards zeolites with mesopores.

5.4.4 Activated alumina

Activated alumina is produced from hydrated alumina ($Al_2O_3{\cdot}3H_2O$) by thermal dehydration and recrystallization. The effective surface area of activated alumina varies from 250 to 350 $m^2\,g^{-1}$ [3]. The surface is more polar than silica gel because of abundant Lewis acid sites (Al^{3+} sites). γ-alumina and η-alumina have very high acid sites (both Lewis and Brønsted) because of spinal defect forms. The pore structure and surface chemistry can be tailored by controlling the heat treatment conditions. Alumina tailored to have high Lewis acidity and low Brønsted acidity are found to be selective adsorbents for oxygenates such as alcohols, aldehydes, ketones, and carboxylic acids [3]. Recently, Candela *et al*. [28] patented a process in which activated alumina was found to be a very selective adsorbent for different oxygenate impurities present in tertiary butyl alcohol.

The adsorptive characteristics of zeolites, aluminas, and in some instances silicas are also used very effectively in catalysis to transport desired materials into catalytic sites that are configured deliberately into these materials to enable a reaction to take place. The actual adsorption and desorption steps for reagents are an integral part of the activity of a supported heterogeneous catalyst. Hence the opportunities for process intensification based on adsorptive behavior are exciting as they may apply to biorefineries

5.4.5 Polymeric resins

Polymeric resins are macroporous or macroreticular polymer beads. Most of the commercial resins are made from styrene/divinylbenzene (DVB) copolymers. Other than this, polymers of acrylates, methacrylates,

Table 5.3 Properties of typical commercial polymeric adsorbents (where available)

Trade name	Chemical name	Ionic functionalization	Nominal pore diameter (Å)	Specific surface area ($m^2 g^{-1}$)	Supplier
Dowex® Optipore L-493	Poly(styrene-co-DVB)	None	46	1100	Dow
Dowex® Optipore SD-2	Poly(styrene-co-DVB)	Tertiary amine (weak base)	50	800	Dow
Diaion HP-20	Poly(styrene-co-DVB)	None	260	500	Mitsubishi Chemicals
Diaion HP-2MG	Poly(methacrylate)	None	–	500	Mitsubishi Chemicals
Amberlite™ XAD-4	Poly(styrene-co-DVB)	None	100	750	Rohm and Haas
Amberlite™ XAD-16N	Poly(styrene-co-DVB)	None	150	800	Rohm and Haas
Ambersorb™ XE-563	Carbonaceous	–	38	550	Rohm and Haas
Purolite® PD206	Poly(styrene-co-DVB)	Sulfonic acid	–	–	Purolite

and vinylpyridine are also used as adsorbents. Sometimes, functional groups such as sulfonyl groups are attached to the benzene ring of these polymers and they are called ion exchange resins. Polymeric resins are usually available in the form of spherical beads and the size usually ranges from 0.3 to 1 mm in diameter. Each resin bead consists of large number of small "microbeads" joined together forming a macropore structure. These microbeads are made of microgel particles ranging in size between 0.01 μm to 15 μm [29]. The degree of cross-linking determines the micropore structure of these microbeads and also provides the high surface area and structural strength. The unfunctionalized polymeric resins are more hydrophobic than activated carbon because of presence of aromatic rings on the surface. The properties of some of the commercial resins are given in Table 5.3.

The distinct advantages of polymeric resin adsorbents are: greater phase stability (physically, chemically, and biologically), improved biocompatibility, complete immiscibility with the adsorbate medium, elimination of emulsification, and an increased potential for re-use [30]. There are a few drawbacks: they tend to shrink and swell on cyclic re-use and they are costlier than common available adsorbents. In some cases, better performance compensates for the resin cost.

5.4.6 Bio-based adsorbents

A wide range of agricultural materials are used as adsorbents. These bio-based adsorbents can be classified into starch-based and lignocellulosic adsorbents [31, 32]. Some of the starch-based adsorbents are corn grits, cornmeal, cooked corn, starch and other grains. Examples of lignocellulosic adsorbents include rice straw, bagasse, wheat straw, wood chips, and corn cob. Many of them have been reported to have potential application in biofuel downstream processes such as ethanol dehydration [33]. Bio-based adsorbents have many advantages over molecular sieves; for example, molecular sieves are highly selective, but water is very strongly adsorbed and high temperatures and/or low pressures are required to regenerate them [34], whereas, bio-based adsorbents have lower separation capacity than molecular sieves, but their regeneration temperature is much lower than molecular sieves. In addition, bio-based adsorbents are much cheaper than molecular sieves. The bio-based adsorbents can also be used as feedstock for upstream fermentation after saturation and thus avoiding pollution through disposal. Some of the disadvantages are the inherent variability of resources, supply fluctuation due to seasonal variation, large bulky nature, and thus constraint over transportation logistics.

5.4.7 Metal organic frameworks (MOF)

The potential of zeolitic imidazolate frameworks (ZIF), a MOF, to be a potential adsorbent for recovering alcohols from aqueous solutions has been recently explored [35]. The structure of ZIF is analogous to Zeolites, where tetrahedral Si (Al) and the bridging O are replaced with transition metal ion and Imidazolate link, respectively. Particularly ZIF-8 has been reported to have an exceptional thermal stability (up to 550 °C) and chemical stability in organic solvents like benzene, water and boiling alkaline water. Yaghi and co-workers synthesized this material by heating a solution of zinc nitrate and 2-methyl imidazole in dimethyl formamide [36]. The pore diameter, pore volume and surface area of ZIF-8 are 11.6 Å, $0.663\,m^3\,kg^{-1}$ and $1\,947\,000\,m^2\,kg^{-1}$ respectively.

5.5 Adsorption separation processes

Adsorption separation processes can be described in three different ways [37] with regard to:

- adsorbate concentration;
- modes of adsorber operation;
- adsorbent regeneration methods.

5.5.1 Adsorbate concentration

Based on adsorbate concentration, adsorption processes can be purification processes (where traces of contaminants are removed from the process stream) and bulk separation processes (where more than one component is recovered from a mixture at high concentration).

5.5.2 Modes of adsorber operation

Based on modes of adsorber operation, adsorption processes can be

- cyclic batch systems;
- continuous counter current configurations;
- chromatographic.

In cyclic batch, the adsorbent bed is alternatively saturated and regenerated in a cyclic manner. In a continuous system, adsorbent and feed are contacted in a counter-current manner. This was first proposed in the hypersorption process by Ruthven [4], but it tends to have difficulties with the solids circulation loop. This problem is avoided by simulating counter-current fluxes in a fixed bed of adsorbent (simulated moving bed). This is done by changing the injection and withdrawing points throughout the system at given time intervals [38, 39]. The time intervals can be made as short as possible thus providing a good approach to the real continuous counter-current process. In the chromatographic process, the feed stream is introduced as a pulse in a purge stream and the separation is accomplished according to the differing times the components remain in the absorption vessel after the pulse injection.

5.5.3 Adsorbent regeneration methods

Based on the regeneration method, adsorption process can be:

- Temperature swing adsorption (TSA). In TSA, the adsorbent is regenerated by heating the bed, usually by a hot gas stream, to a temperature at which the adsorbate is desorbed and removed from the bed with the gas stream.
- Pressure swing adsorption (PSA). In PSA, the system pressure is reduced at a constant temperature and the bed is purged at low pressure. It is generally applied in a gaseous system. Vacuum desorption is a special case of PSA.
- Inert purge gas stripping. In this method, desorption is done at constant temperature and pressure by passing an inert gas through the bed, which presents the loaded adsorbent with a low partial pressure bulk phase to affect the equilibrium. So it has elements of PSA and TSA except that preheating of the purge gas is not required. This method is applicable where adsorbate molecules are weakly bound to the adsorbent, otherwise, the quantity of purge gas required would be enormous. The desorbed gas is present at low concentration in the purge gas and hence this method is not used where desorbate has to be recovered [4]. Clearly this approach produces a regeneration stream that is a mixture of regenerant and adsorbate. So it may be more applicable for instances where purification of the original bulk fluid is desired.
- Displacement desorption. This system is similar to purge gas stripping but, instead of inert gas, a competitively adsorbing gas stream is used. Thus adsorbate is displaced by a competitive adsorbing species as in displacement chromatography mode.

5.5.3.1 Selection of regeneration method

The choice of regeneration method mainly depends on the nature of adsorption, economic factors and other technical considerations such as availability of cheap heat source. Table 5.4 provides some of the general considerations for selecting a regeneration method.

5.5.3.2 Temperature swing adsorption (TSA)

Principle of operation The TSA process normally operates in a cyclic batch mode where the adsorbent bed is saturated and regenerated alternatively. The regeneration is carried out by increasing the bed temperature usually by purging a hot inert gas at constant pressure. The principle of TSA operation is shown in Figure 5.3. It shows the effect of temperature on adsorbent equilibrium loading for a Type I isotherm at an adsorption pressure of P_{ads}. As the temperature is increased from adsorption temperature of T_{ads} to desorption temperature of T_{des}, the equilibrium loading is also reduced from q_{ads} to q_{des}. After the adsorbent is regenerated, it must be cooled down ready for a new adsorption step. The main disadvantage of TSA is that the number of cycles obtainable in any given time is limited by the relatively slow heating and cooling process steps. For this reason, TSA is limited to the removal of small quantities of strongly adsorbed impurities.

Table 5.4 *General considerations for selecting a regeneration method. Reprinted from [4] © 1984, with permission from John Wiley & Sons, Inc*

Method	Advantages	Disadvantages
TSA	Good for strongly adsorbed species; small change in temperatures results in large change in adsorbent loading	Thermal aging of adsorbent
	Desorbate may be recovered at high concentration	Heat loss mean inefficiency in energy usage
		Unsuitable for rapid cycling so adsorbent cannot be used with maximum efficiency
	Gases and liquids	In liquid systems high latent heat of interstitial liquid must be added
PSA	Good where weakly adsorbed species is required in high purity	Very low pressure may be required
		Mechanical energy is more expensive than heat
	Rapid cycling-efficient use of adsorbent	Desorbate recovered at low purity
Inert purge	Operation at constant temperature and pressure	Large purge volume required
Displacement desorption	Good for strongly held species	Product separation and recovery needed (choice of desorbent is crucial)
	Avoids risk of cracking reactions during regeneration	
	Avoids thermal aging of adsorbent	

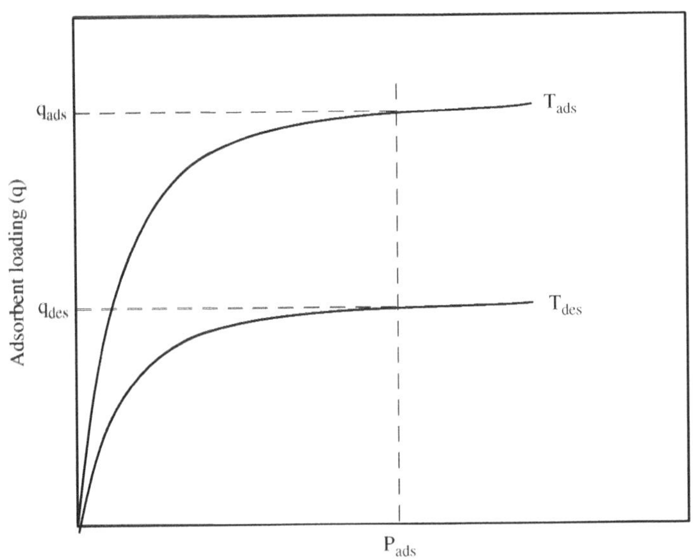

Figure 5.3 *Operating principle of a TSA system*

Two-bed TSA systems The simplest TSA system operates with two beds, one adsorbing and the other desorbing, in order to maintain continuous flow. The feed stream containing adsorbate is passed through the first bed at ambient temperature until the bed is saturated or breakthrough occurs. Then the bed is taken off-line and the feed is switched to the second bed. Simultaneously the first bed is regenerated by raising its temperature and purging with a hot inert gas normally in a flow direction opposite to that used for the adsorption step. The regenerated bed is then cooled to ambient temperature by using either cold feed or inert fluid. Although the desorption step can be accomplished in the absence of a purge by simply vaporizing the adsorbate at elevated temperature, re-adsorption of some solute vapor would occur upon cooling the bed.

When the adsorbate is valuable and easily condensed, the purge fluid might be a non-condensable gas. When the adsorbate is valuable but not easily condensed, and is essentially insoluble in water, steam may be used. Condensation of the steam allows the desorbed adsorbate to be separated. When the adsorbate is not valuable, fuel and/or air can be used as the purge fluid, followed by disposal, for example by incineration.

The heating and desorption steps must provide sufficient energy to perform the following functions:

- to raise the adsorbent, its associated adsorbate and the containment vessel to the desorption temperature;
- to provide the heat of desorption;
- to raise the adsorbent and vessel temperature to final regeneration temperature (if greater than that for desorption).

The adsorbent bed cannot normally be heated and cooled quickly and hence the cycle time of a typical TSA process may range from several hours for a bulk separation to several days for purification. Long cycle times inevitably mean large bed lengths resulting in high adsorbent inventories.

During the period when Bed-1 is adsorbing, Bed-2 is being desorbed, which includes the time required for heating and cooling. The two-bed TSA process requires that the time taken for desorbing gases from one bed matches the time allowed for adsorption in the other bed. Otherwise product flow would be discontinuous. If a longer period is required for desorption, then due to the time constraint, only a fraction of the adsorbate can be removed during the desorption step of the cycle. Bed capacity is consequently not fully utilized.

Because of the long cycle times required for TSA processes this mode of operation is used almost exclusively for the removal of low concentrations of adsorbable gases from feed streams.

Three-bed TSA systems For a fixed-bed system, the amount of adsorbent required to remove the contaminant from its inlet concentration to the desired level is termed the mass transfer zone (MTZ). As the MTZ progresses through the fixed beds, it reaches a point where the MTZ is longer than the remaining depth of the adsorbent in the vessel still capable of adsorption (not spent). At this point, the concentration of the contaminant begins to increase in the outlet of the adsorbent bed as the MTZ begins to exit the vessel; this is called breakthrough. Some processes may have a long length of unused bed (LUB, which is approximately one-half the mass transfer zone), which can result in huge adsorber size and inefficient usage of adsorbents. This problem can be overcome in a three-bed system where a guard bed is located between the primary adsorber bed and the bed undergoing regeneration [40]. Figure 5.4 is a schematic diagram of the operation.

In this operation, the feed first enters the adsorber bed in which adsorption occurs. When the concentration of primary adsorber effluent reaches nearly the feed concentration, the beds are switched. The guard bed becomes the primary adsorber, the regenerated bed becomes the guard bed, and the saturated bed goes for regeneration. This rotation is continued to keep the LUB section always in the guard bed and thus

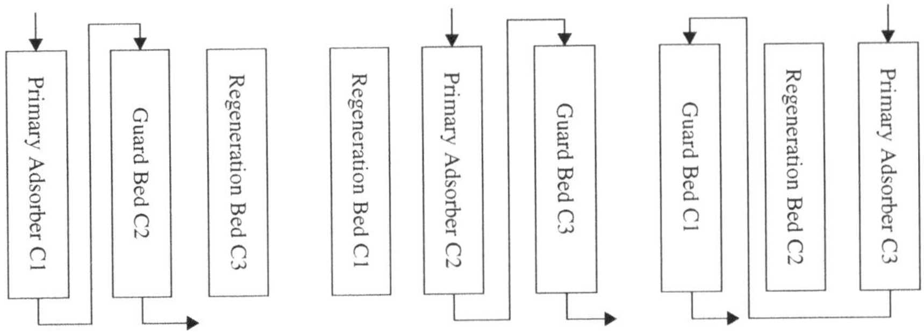

Figure 5.4 *Three-bed TSA system*

the primary adsorber is completely loaded when regeneration begins. Thus a complete bed utilization and economic regeneration are achieved.

Minimum purge temperature Efficient desorption is achieved above the minimum purge temperature T_o. It is based on the equilibrium theory proposed by Basmadjian [41]. According to this theory, T_o is the temperature at which the slope of the adsorption isotherm at the origin is equal to the ratio of the heat capacities of the solid phase (adsorbent plus adsorbate) and the inert carrier gas (C_{ps}/C_{pb}). For a Langmuir isotherm

$$V = \frac{V_{mon} K(T) P}{1 + K(T) P} \tag{5.19}$$

T_o is determined by

$$V_{mon} K(T_o) = \frac{C_{ps}}{C_{pb}} \tag{5.20}$$

As the temperature is increased beyond T_o, energy cost increases without a significant gain in desorption [3].

Common examples of TSA processes include solvent recovery with activated carbons, and drying of gases or liquids with type A zeolites, removal of water from VOCs with zeolites, gas sweetening, and so forth.

5.5.3.3 *Pressure swing adsorption (PSA)*

Principle of operation Pressure swing adsorption also operates as a cyclic batch. The regeneration of the adsorbent bed is achieved by reducing the total pressure and purging the bed at low pressure with a small fraction of the product stream. The operating principle of a PSA system is shown in Figure 5.5.

Figure 5.5 shows the effect of partial pressure on adsorbent equilibrium loading for a Type I isotherm at adsorption temperature of T_{ads}. As the partial pressure is reduced from P_{ads} to P_{des}, the equilibrium loading is also reduced from q_{ads} to q_{des}. As pressure changes can be effected much faster than temperature changes, the PSA process allows a much faster cycling. Thus it can remove large quantities of impurities. For strongly adsorbed species, PSA would require a very low pressure for desorption, which can increase the operating cost. Pressure swing adsorption processes are often operated at low adsorbent loadings because selectivity between gaseous components is often greatest in the Henry's law region. It is desirable to operate PSA processes close to ambient temperature to take advantage of the fact that, for a given partial pressure, the loading is increased as the temperature is decreased.

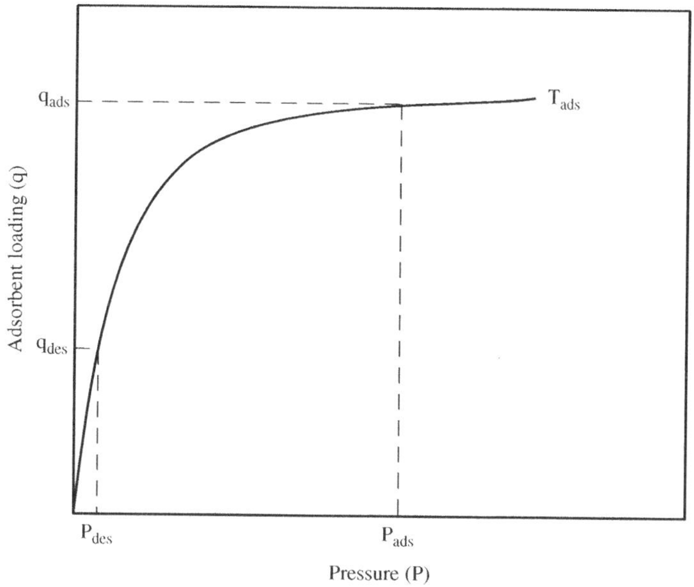

Figure 5.5 *Operating principle of a PSA system*

Skarstrom developed the first PSA system for air drying [42, 43]. Montgareuil and Daniel also invented a similar PSA cycle at about the same time [44]. Later, several improvements were made to increase the efficiency of the PSA system. The basic PSA process uses the two-bed system although multiple beds also can be operated in a staggered sequence. A typical PSA cycle consists of the following basic steps:

1. Adsorption.
2. Cocurrent depressurization.
3. Countercurrent depressurization.
4. Purge at low pressure.
5. Repressurization.

These steps are illustrated in Figure 5.6.

Adsorption (1→2) The gas mixture is fed into an adsorber bed under high pressure. The impurities are adsorbed and purified product is withdrawn. Flow is normally in the upward direction. When the adsorber reaches its adsorption capacity, it is taken off-line, and the feed is automatically switched to a fresh adsorber bed. This keeps the feed and product flows continuously.

Cocurrent depressurization (2→3) The gas mixture trapped in the void spaces of the adsorber is recovered by partly depressuring the bed from the product side in the same direction as the feed flow (cocurrent). This moves the impurity fronts migrating to the top of the adsorbent bed. Thus the cocurrent depressurization step can increase the concentration of the adsorbate recovered during the regeneration step.

Countercurrent depressurization (3→4) The saturated adsorber is then partly regenerated by depressurizing towards the feed end (counter current), and the desorbed impurities are rejected to the PSA offgas.

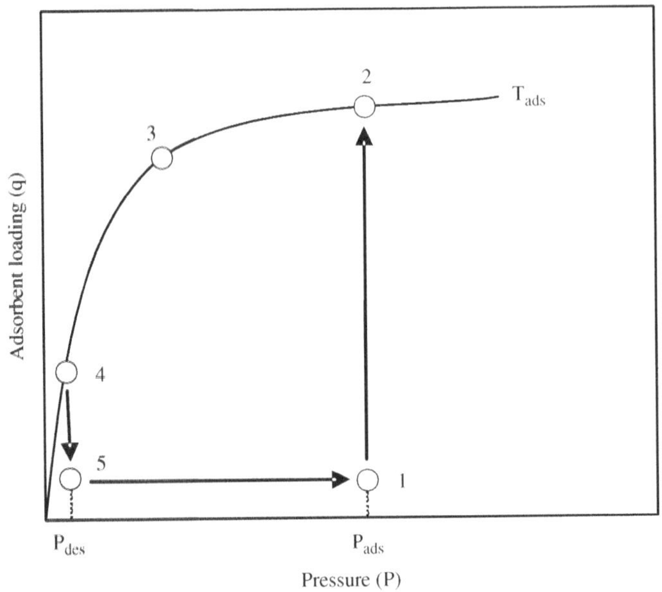

Figure 5.6 *PSA cycle steps*

Purge at low pressure (4→5) The adsorbent is then purged with purified product (taken from another adsorber during concurrent depressurization) at constant offgas pressure to further regenerate the bed.

Repressurization (5→1) The adsorber is then repressurized with product gas coming from the cocurrent depressurization, and with a slipstream from the product stream. When the adsorber has reached the adsorption pressure, the cycle has been completed, and the adsorber is ready for the next adsorption step.

Pressure equalizations The term "pressure equalization" (PE) refers to the action by which the pressure in two interconnected beds is equalized [3]. The main purpose of the PE step is to conserve the mechanical energy contained in the gas of a high-pressure bed. One of the ways to do this is to use the high-pressure gas removed during the cocurrent depressurization step to repressurize other adsorber by pressure equalizations. It is done by connecting the ends of two beds. In general, increasing the number of PEs also increases product recovery. However, the impurity of the gas stream coming from cocurrent depressurization increases with time. This can be overcome by having a minimum purge to reject the impurities.

Examples of PSA systems in biorefineries The PSA process is a well established technology for the dehydration of ethanol due to its low energy consumption and its capability of producing very dry ethanol [45–48]. The 3A molecular sieves are the most widely used adsorbents for this application. Jeong *et al*. [46] studied the dehydration of ethanol using 3A zeolite on a pilot scale PSA unit. They produced a 2 kL/day of dehydrated ethanol (99.5 wt%) from the feed ethanol concentration of 93.2 wt%.

In a biomass-to-liquid (BTL) process, the product gas from biomass gasification has to undergo a conditioning process prior to the Fischer-Tropsch conversion to remove CO_2 from the product gas. This is conventionally done by the Rectisol and Selexol processes. A PSA unit using activated carbon has been evaluated and found to be a promising alternative to these conventional processes [49].

5.6 Adsorber modeling*

The primary goal of adsorber modeling is to predict the breakthrough behavior of the adsorption bed. In this section, a simple model is presented for a liquid adsorber system. In a packed bed adsorber, the adsorbate concentration in liquid and solid phase changes as the feed moves along the bed. Hence, the process operates under un-steady state.

The following assumptions were made while developing the model:

- Mass transfer across the solid boundary is characterized by the inter-phase mass transfer coefficient ($k_p a$).
- The rate of mass transfer is assumed to be proportional to the adsorbate concentration difference between the bulk liquid and the solid phase.
- There is no radial adsorbate concentration gradient in the fluid phase.

The transport equations are developed using the mass balance for solid–liquid phases and the rate of adsorption [50, 51]. Considering a differential layer in a packed bed under un-steady state, the mass balance can be written as:

rate of adsorbate in − rate of adsorbate out − rate of adsorption = rate of accumulation

$$A\varepsilon\left(u_s C - D_L\frac{\partial C}{\partial z}\right)\bigg|_z - A\varepsilon\left(u_s C - D_L\frac{\partial C}{\partial z}\right)\bigg|_{z+\Delta z} - \frac{\partial q}{\partial t}\rho A(1-\varepsilon)\Delta z = \frac{\partial C}{\partial t}A\varepsilon\Delta z \tag{5.21}$$

where A is column cross-sectional area (m^2); ε is bed void fraction; ρ is bulk density of the adsorbent (kg m^{-3}); u_s is liquid interstitial velocity (m s^{-1}); q is adsorbate loading on adsorbent (kg adsorbate kg adsorbent^{-1}); C is aqueous phase adsorbate concentration (kg m^{-3}); z is bed axial distance (m); D_L is axial dispersion coefficient (m^2 s^{-1}); t is time (s).

By dividing Eq. (5.21) by $A\,\varepsilon\,\Delta z$

$$-u_s\frac{\partial C}{\partial z} + D_L\frac{\partial^2 C}{\partial z^2} - \rho\frac{(1-\varepsilon)}{\varepsilon}\frac{\partial q}{\partial t} = \frac{\partial C}{\partial t} \tag{5.22}$$

By re-writing Eq. (5.22) in the dimensionless form

$$\frac{\partial C}{\partial \theta} + \frac{\partial C}{\partial x} - \left(\frac{D_L}{u_s L}\right)\frac{\partial^2 C}{\partial x^2} + \rho\frac{(1-\varepsilon)}{\varepsilon}\frac{\partial q}{\partial \theta} = 0 \tag{5.23}$$

where $x = \frac{z}{L}$; $\theta = \frac{t}{\tau}$; $\left(\frac{D_L}{u_s L}\right)$ = Vessel dispersion number (VD); x is dimensionless distance; θ is dimensionless time; L is packed bed length (m); τ is liquid residence time (s).

The Vessel dispersion number is calculated using the following empirical correlation, which relates the axial dispersion with the particle Reynolds (Re) number in a packed bed column [52]. This correlation is based on numerous experiments over a broad range of Re number (10^{-3}–10^3).

$$\varepsilon\left(\frac{d_p u_s}{D_L}\right) = 0.2 + 0.011\left(\frac{\rho_l u_o d_p}{\mu}\right)^{0.48} \tag{5.24}$$

*Section 5.6 is from V. Saravanan, D.A. Waijers, M. Ziari and M.A. Noordermeer, Recovery of 1-butanol from aqueous solutions using zeolite ZSM5 with a high Si/Al ratio; suitability of a column process for industrial applications, *Biochem. Eng. J.*, **49**, 33–39 (2010). Reprinted with permission © 2010 Elsevier.

where d_p is average particle diameter (m); ρ_l is aqueous phase density (kg m^{-3}); μ is aqueous phase viscosity (kg m^{-1} s^{-1}); u_o is liquid superficial velocity (m s^{-1}).

$$\frac{\rho_l \, u_o \, d_p}{\mu} = \text{particle Reynolds number}$$

Thus from Eq. (5.24), the vessel dispersion number can be calculated as:

$$\text{VD} = \frac{D_L}{u_s L} = \frac{\left(\frac{d_p}{L}\right)\varepsilon}{\left(0.2 + 0.011\,Re^{0.48}\right)} \tag{5.25}$$

The adsorption rate is related to a linear driving force in terms of solid phase concentration as [51]:

$$\frac{\partial q}{\partial t} = k_p a \left(q^* - q\right) \tag{5.26}$$

where $k_p a$ is inter-phase mass transfer coefficient (s^{-1}); q^* is solid phase 1-butanol concentration which is in equilibrium with bulk liquid concentration (kg adsorbate kg adsorbent^{-1}).

Equation (5.23) can be written in finite difference form using the implicit method

$$\frac{C_i^{n+1} - C_i^n}{\Delta\theta} + \frac{C_{i+1}^{n+1} - C_{i-1}^{n+1}}{2\Delta x} - VD\,\frac{C_{i-1}^{n+1} - 2C_i^{n+1} + C_{i+1}^{n+1}}{\Delta x^2} + \rho\,\frac{(1-\varepsilon)}{\varepsilon}\,\frac{q_i^{n+1} - q_i^n}{\Delta\theta} = 0 \tag{5.27}$$

In the above equation, the temporal derivatives are written as forward difference and spatial derivatives as central difference. The scripts n and i refer to the time and space steps, respectively. The objective is to evaluate the concentration profile along the column length; Eq. (5.27) is further rearranged to solve the equation in $n+1$ time step as:

$$C_{i-1}^{n+1}\left(-0.5 - \frac{VD}{\Delta x}\right) + C_i^{n+1}\left(\frac{\Delta x}{\Delta\theta} + \frac{2\,VD}{\Delta x}\right) + C_{i+1}^{n+1}\left(0.5 - \frac{VD}{\Delta x}\right) = C_i^n\left(\frac{\Delta x}{\Delta\theta}\right) - \rho\Delta x\, k_p a\tau\left(\frac{1-\varepsilon}{\varepsilon}\right)\left(q_i^{n^*} - q_i^n\right) \tag{5.28}$$

Initial condition: At $\theta = 0$; $C = q = 0$ for all x

Boundary conditions: At $x = 0$; $C = Co$ for all θ; At $x = 1$; $\dfrac{\partial C}{\partial x} = 0$

The adsorption bed is divided into n nodes. From Equation 5.28 and the boundary conditions, correspondingly "n" simultaneous equations forming a tridiagonal matrix are generated. This matrix can be solved with a computer program using the Thomas algorithm [53]. Initially a value is assumed for the mass transfer coefficient ($k_p a$) and the equations are solved. This will generate liquid- and solid-phase concentration (C, q) profiles along the bed height at a given time. The same set of equations is solved at discrete time intervals to obtain the complete breakthrough curve. The curve obtained is compared with the experimental data. This procedure is repeated until the predicted data at 50% breakthrough nearly coincide with the experimental data at the same breakthrough point by iteratively changing the $k_p a$ value.

5.7 Application of adsorption in biorefineries

Ligno-cellulosic materials are considered to be a sustainable source of feedstock for biorefineries. The following are some of the major challenges in using them in commercial scale biorefineries:

1. The low energy content (energy density) of the biomass feedstocks.
2. Pre-treatment of cellulosic biomass produces degradation by-products which are toxic and inhibitory to enzymes [54] and fermenting microorganisms [55–57].
3. Recovery of products from dilute aqueous fermentation broth.
4. Product inhibition of fermenting species. In ethanol fermentation, when the ethanol concentration goes above $115\,g\,l^{-1}$, ethanol production is completely stopped [58]. In butanol fermentation, toxicity threshold of *Clostridia* sp is about 1.3% (w/v) of butanol [59].

The following section presents examples, available in the literature, for tackling challenges 2 through 4 using adsorption techniques.

5.7.1 Examples of adsorption systems for removal of fermentation inhibitors from lignocellulosic biomass hydrolysate

A major challenge faced in the commercial production of lignocellulosic bio-ethanol is the inhibitory compounds generated during the pretreatment of biomass. During the dilute acid and hot water pre-treatment, the major degradation by-products released are organic acids such as acetic acid (formed by hydrolysis of hemicellulose), furans such as 5-hydroxymethylfurfural (HMF), furfural (compounds derived from degradation of hexose and pentose sugars), and phenols such as vanillin, 4-hydroxybenzaldehyde (4-HB), and syringaldehyde (released by lignin degradation) [60]. These compounds are very toxic to fermenting species [55–57]. Each of these compounds may affect cell growth, sugar uptake, individual metabolic pathways, or all three [61]. A number of studies have been reported to reduce the concentration of these inhibitors using adsorption techniques such as adsorption on activated charcoal, adsorbent resins, ion-exchange resins, and zeolites. In this section, some of these studies are briefly described.

Acetic acid is produced from the hydrolysis of acetyl groups in the hemicellulose [62]. Acetic acid in protonated form can diffuse through the cytoplasmic membrane of cells and detrimentally affect cell metabolism [63]. Berson *et al.* [64] investigated adsorption of acetic acid from a dilute acid pretreated corn stover hydrolysate on activated carbon. The initial hydrolysate had acetic acid concentration of $16.5\,g\,l^{-1}$. The hydrolysate filtered from the corn stover hydrolysate slurry was contacted with activated carbon (Calgon BL) provided by Calgon in a shake flask. The operating conditions were: activated carbon concentration of $80\,g\,l^{-1}$; shaker speed 350 rpm; temperature 35 °C and contact time 10 minutes. At the end of experiment, carbon was removed from hydrolysate using centrifugation. To bring down the concentration of acetic acid from 16.5 to a level below $2\,g\,l^{-1}$, the hydrolysate was again brought in contact with activated carbon and the this step was repeated five times. Simultaneous adsorption of glucose and xylose was not measured during these stages. A US patent application [65] reported usage of a weak basic anion resin to adsorb inhibitors and mainly the acid molecules (such as acetic acid).

Wickramasinghe and Grzenia tested an adsorptive microporous membrane, Sartobind Q, and a weak base anion exchange resin, Amberlyst 21, to recover acetic acid from the hemicellulose hydrolysate on lab scale [66]. Sartobind Q is made of cross-linked regenerated cellulose membrane with a strong basic anion exchange group R-CH2-N^+-$(CH3)_3$ attached to their internal pores and Amberlyst 21 is made of Polystyrene macroreticular with styrene functional group. They showed that adsorptive membranes had many advantages over resin based system such as

- Reduced pore diffusion because adsorbate transport to the adsorbent site occurs by convection.
- Reduced processing time.
- Lower pressure drop compared to the resin packed bed.
- Scale-up for membrane system is easier than with the packed bed [67–69].

Lee *et al*. [70] reported a study on detoxification of inhibitors from prehydrolysate produced from auto-hydrolysis of mixed hardwood chips using activated carbon. The acid treated prehydrolysate had: xylose = 13.18 g l^{-1}; glucose = 2.42 g l^{-1}; acetic acid = 4.58 g l^{-1}; HMF = 0.08 g l^{-1}; furfural = 0.51 g l^{-1}; and formic acid = 4.04 g l^{-1}. The detoxification experiments were carried out in 250 ml flasks kept in a shaking water bath at 50 °C and 180 rpm speed. The concentrations of activated carbon (05-690A, 50–200 mesh, Fisher Scientific Co., Pittsburgh, PA, USA) used were 1.0, 2.5, 5.0, and 10.0 wt % with incubation time of 1 h. Figure 5.7 shows the percentage removal of toxic compounds against percentage activated carbon used.

From Figure 5.7, it can be seen that activated carbon at 2.5 wt% level in prehydrolysate could remove 96% of hydroxymethylfurfural (HMF), 93% of the furfural, 42% of the formic acid and 14% of the acetic acid. However, it also removed 8.9% of the xylose. The subsequent fermentation with *Thermoanaerobacterium saccharolyticum* strain MO1442 gave essentially a 100% yield. This study demonstrated that activated carbon is a good adsorbent to remove inhibitors like HMF and Furfural. Its adsorption capacity for acetic acid is limited probably due to the competitive inhibition of other compounds present in the complex hydrolysate. This is consistent with Berson *et al*.'s study [64], which showed that five adsorption cycles were required to remove acetic acid completely from corn stover hydrolysate.

Larsson *et al*. [57] used polystyrene divinylbenzene-based anion-exchange resin (AG 1-X8, Bio-Rad, Richmond, VA) to detoxify the hydrolysate of dilute acid pretreated Spruce. The adsorption treatment with anion resin at pH 10 showed the highest inhibitor removal compared with other treatments such as pH adjustment to 10 with NaOH or $Ca(OH)_2$, sulfite treatment, hydrolysate concentration by evaporation, enzymatic degradation using laccase, and microbial degradation using *Trichoderma reesei*. Table 5.5 shows reduction in hydrolysate inhibitor concentrations by adsorption with anion resin AG1-X8.

Carvalho *et al*. [71] studied detoxification of hemicellulosic hydrolysate produced from dilute sulfuric acid pre-treatment of Eucalyptus shavings using combination of activated carbon and adsorbent resins. Initially the hydrolysate was concentrated 5.8 times by vacuum evaporation, which helped to remove the

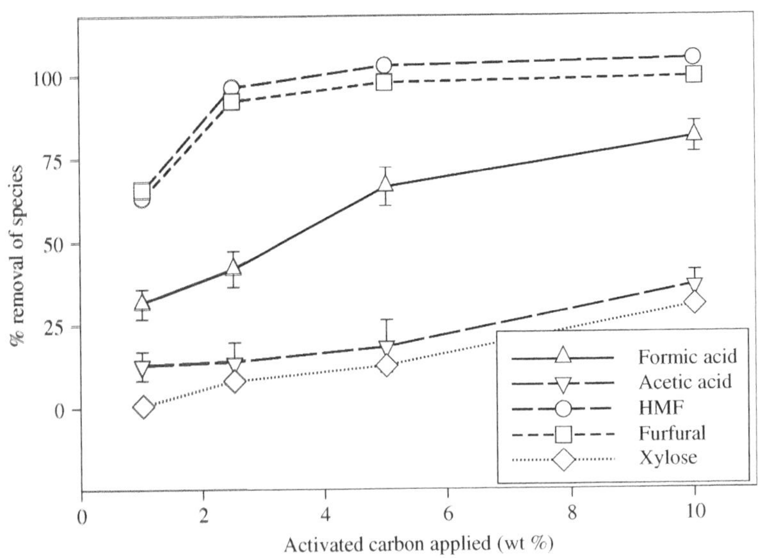

Figure 5.7 *Percentage removal of inhibitors versus activated carbon charge for 1 hour adsorption time. Reprinted from [70] © 2011, with permission from Elsevier*

Table 5.5 Detoxification of Spruce hydrolysate by adsorption using anion resin AG1-X8. Reference [57], with kind permission from Springer Science+Business Media © 1999

Treatment	glucose + mannose $g\,l^{-1}$	Levulinic acid $g\,l^{-1}$	Acetate $g\,l^{-1}$	Furfural $g\,l^{-1}$	5-HMF $g\,l^{-1}$	Total phenolic content $g\,l^{-1}$
Feed	32.2	2.6	2.4	1.0	5.9	0.48
Anion resin, pH 5.5 (0.45 g per g of hydrolysate)	29.6	0.36	0.26	0.69	4.37	0.15
Anion resin, pH 10 (0.49 g per g of hydrolysate)	23.8	0.18	0.10	0.27	1.77	0.043

more volatile furfural. The concentrated hydrolysate was treated with combination of different adsorbents as mentioned below:

1. Adsorption on activated charcoal and subsequently on the anionic resin Purolite A-860 S.
2. Adsorption on activated charcoal and subsequently on the adsorbent resin Purolite MN-150.
3. Adsorption on diatomaceous earths and subsequently on the anionic resin Purolite A-860 S.
4. Adsorption on diatomaceous earths and subsequently on the adsorbent resin Purolite MN-150.

The adsorption process on charcoal and diatomaceous earth was carried in a shake flask. The operating conditions were: adsorbent loading = 1.2 g adsorbent per 50 ml of hydrolysate; speed = 200 rpm; temperature = 30 °C; residence time = 34.5 min. The adsorption processes on resins were performed in a packed bed column with a bed volume of 200 cm^3 with a feed rate of 0.9 $cm^3\,min^{-1}$. Figure 5.8 shows the percent toxic compounds removed for each process.

From Figure 5.8, by comparing the four processes, the combination of charcoal and adsorbent resin MN-150 (Process B) was found to be the most effective for removal of acetic acid, HMF and phenolics with minimum sugar loss (5.9 and 1.3% for glucose and xylose respectively).

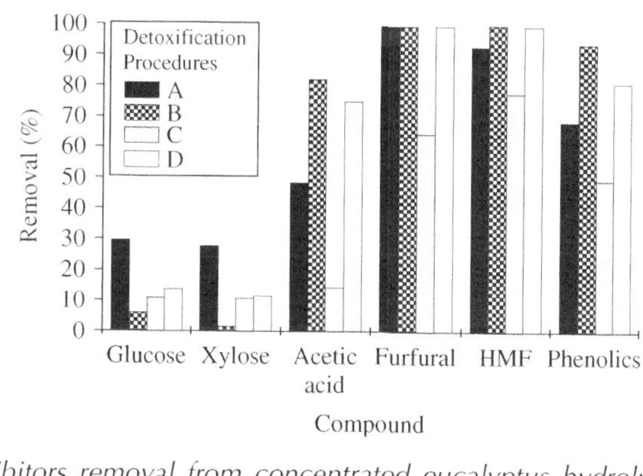

Figure 5.8 Sugar and inhibitors removal from concentrated eucalyptus hydrolysate for different adsorption processes. Reprinted from [71] © 2006, with permission from John Wiley & Sons

From the earlier examples it can be seen that anion exchange resins seem to be among the promising adsorbents for detoxifying the cellulosic hydrolysate. However it suffers from some disadvantages: (i) pH needs to be adjusted to 10 thus requiring significant quantities of acid and base chemicals; (ii) a significant amount of fermentable sugar loss up to 26%. Another promising adsorbent is activated charcoal, which does not require pH adjustment. Activated carbon also is found to adsorb sugar molecules to some extent. Adsorption process with activated charcoal can become expensive sometimes because the powdered form of activated charcoal cannot be regenerated and granular activated charcoal usually incurs a 10% loss during regeneration cycle [72]. Zeolite materials are found to overcome most of these disadvantages, especially the sugar loss.

Ranjan *et al.* [72] used different zeolite adsorbents (ZSM-5 (framework type MFI), beta, faujasite and ferrierite (framework type FER)) to recover 5-hydroxymethylfurfural (HMF), furfural and vanillin from lignocellulosic biomass hydrolysates. Hydrolysate was prepared from Aspen wood chips using dilute sulfuric acid pretreatment. The hydrolysate had total sugar content of $37.5\,g\,l^{-1}$, $1.45\,g\,l^{-1}$ of furfural, $0.16\,g\,l^{-1}$ of HMF and $0.05\,g\,l^{-1}$ of vanillin. The detoxification of hydrolysate was carried out by adding $10\,g\,l^{-1}$ β zeolite and the mixture was stirred overnight followed by zeolite removal by centrifugation. The treated hydrolysate had no detectable amount of inhibitors and the sugar loss was also negligible. This improved the ethanol yield of the fermentation process significantly (Figure 5.9a). Figure 5.9b shows the single component adsorption isotherm for furfural, vanillin and HMF on β zeolite with different Si/Al ratios. From Figure 5.9b it can be seen that high silica zeolite more efficiently removed furfural, vanillin and HMF.

Inhibitors such as HMF and furfural can be used as building blocks for the production of fine chemicals and plastics [73, 74]. Hence recovery of these inhibitors as pure compounds can be considered to have a promising future. Ranjan *et al.* [72] also reported a preliminary study on the selective adsorption of furfural over HMF on MFI/FER zeolite. They followed the screening process developed by Gaounaris *et al.* [75] to select a suitable adsorbent. Figure 5.10 shows the preferential adsorption of furfural over HMF on FER zeolite from the binary mixture. FER zeolite also showed negligible adsorption of sugars.

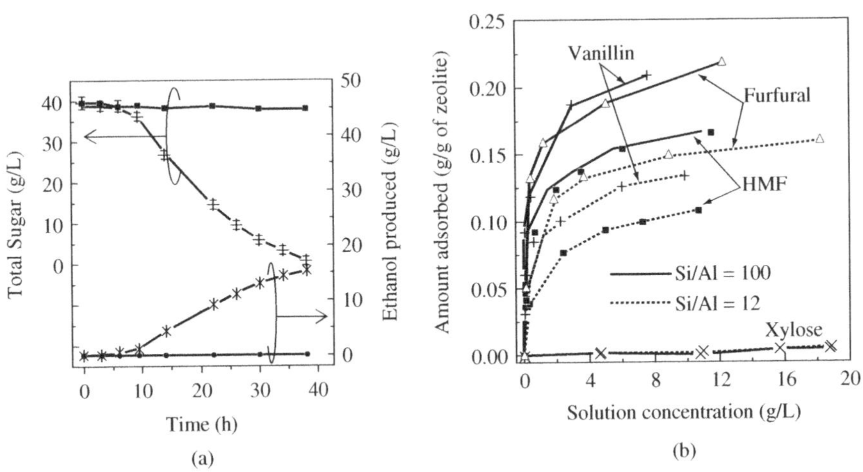

Figure 5.9 *(a) Effect of zeolite pretreatment on ethanol fermentation. (b) Single component adsorption of furfural, vanillin and HMF on β zeolite. Reprinted from [72] © 2009, with permission from Elsevier*

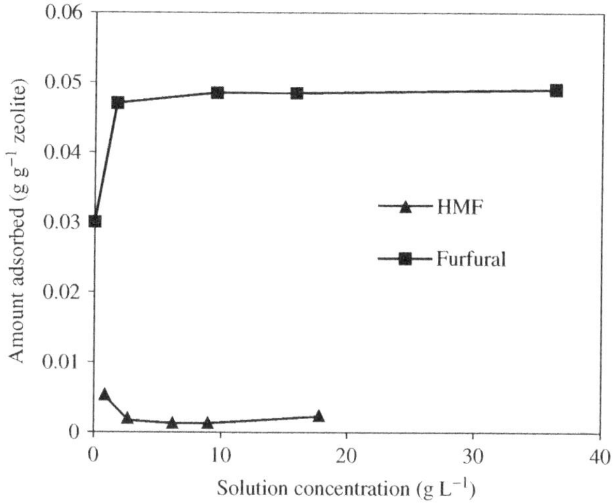

Figure 5.10 *Adsorption isotherm of HMF and Furfural binary mixture on FER zeolite. Reprinted from [72] © 2009, with permission from Elsevier*

5.7.2 Examples of adsorption systems for recovery of biofuels from dilute aqueous fermentation broth

5.7.2.1 In situ *recovery of 1-butanol*

1-Butanol is a relatively non-polar, long, high-energy compound. Hence, it is considered to be an interesting fuel-blending component. It has become more attractive because it can be produced from renewable resources by ABE fermentation using Clostridia bacteria (e.g. *C. acetobutylicum* or *C. beijerinckii*). However, when the 1-butanol concentration is above 1.3 (w/v)%, it becomes toxic to *Clostridia* sp [59]. Thus the product inhibition of fermenting bacteria leads to very dilute aqueous solutions (1–2 wt% 1-butanol). Due to low vapor pressure of butanol relative to water (0.0109 atm versus 0.0312 atm at 25 °C), recovery of butanol by distillation becomes energy intensive, requiring water to be vaporized first. Hence in order to remove the 1-butanol during fermentation, a number of alternative *in situ* or *ex situ* product isolation techniques such as adsorption, pervaporation, extraction, gas stripping, and ionic liquids have been investigated [76–82].

Oudshoorn *et al.* [83] presented a ranking of different recovery processes such as distillation, liquid mixing, freeze crystallization, pervaporation, supercritical extraction, gas stripping, extraction, and adsorption based on the selectivity of butanol over water. The selectivity estimate was based on a single-stage equilibrium operation. Among all the processes, adsorption-based process using non-polar adsorbents was found to be a promising option.

Qureshi *et al.* [84] presented a process scheme for butanol adsorption—desorption and concentration using silicalite (a lab-made Al-free zeolite analogue) from fermentation broth (Figure 5.11). An ultrafiltration (UF) membrane unit was placed between the fermentor and adsorption column to separate the cells from the broth otherwise it could foul the adsorbent column. The desorption was carried out by the thermal swing process using hot air. The material and energy balance was performed for the scheme considering the feed butanol concentration of 5 g l^{-1}. These calculations assumed that 50% of the adsorbed water was desorbed at 40 °C and butanol was desorbed at 150 °C. This sequential desorption resulted in the butanol

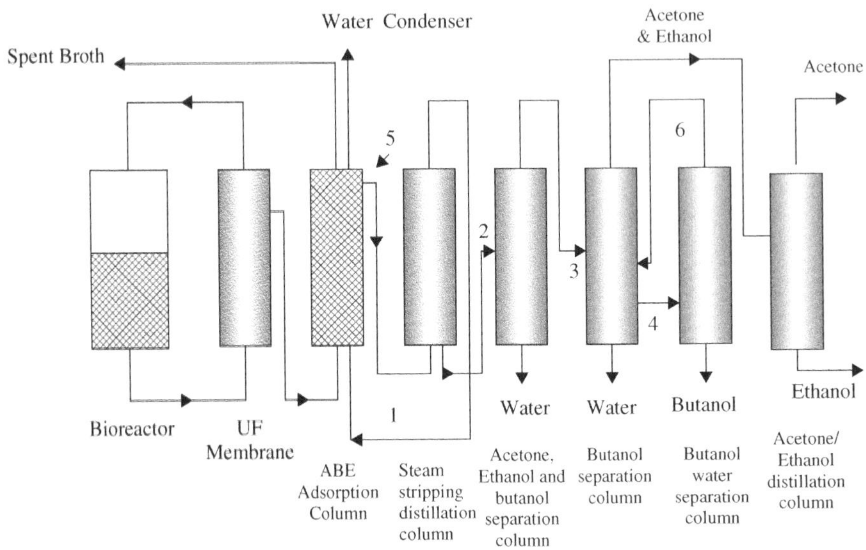

Figure 5.11 A schematic diagram of ABE separation and concentration from fermentation broth using adsorbent. 1 Hot air to remove water or ABE from silicalite column, 2 ABE and water stream, 3 ABE water stream, 4 Butanol water stream, 5 ABE water stream, 6 Butanol water stream. Reference [84], with kind permission from Springer Science+Business Media © 2005

concentration of 810 g l^{-1}. Based on this calculation, the energy requirement was calculated to be 1948 kcal per kg of 1-butanol recovered. A comparison of adsorption on to silicalite against other recovery processes such as steam stripping distillation, gas stripping, pervaporation, liquid–liquid extraction is presented in Figure 5.12. It shows adsorption on silicalite to be more attractive than other processes. In addition to a high adsorption capacity and 1-butanol selectivity, silicalite has small pores preventing adsorption of more complex molecules such as sugars, medium components and bacterial cells, which make them very suitable for *in situ* product removal.

Shao and Kumar used a ZSM-5 zeolite-filled polydimethylsiloxane membrane to separate 1-butanol from 1-butanol/2,3-butandiol mixture [85]. Milestone and Bibby also studied adsorption of alcohols to different forms of self-prepared ZSM-5 zeolites [86]. They found that the amount of alcohol adsorbed decreased as the ionic size of the cation increased (H, Na, K, and Cs). Furthermore, increasing percentages of Al_2O_3 (1, 2, and 4 %) led to lower adsorption capacities for 1-butanol and higher ones for ethanol. The presence of Al_2O_3 also catalyzed the dehydration of 1-butanol during desorption at higher temperatures from ZSM-5 containing 4% Al_2O_3. The (relative) amount of degraded ethanol and 1-butanol was not determined. Catalytic activity was not observed in Na-ZSM-5 Zeolites prepared by Falamaki *et al.* [87], and was ascribed to the different synthesis routes that were followed. A conceptual design for a 1-butanol adsorption/desorption process has been described by Holtzapple and Brown [88].

Nielsen and Prather studied a number of polymeric resins for *in situ* product recovery of 1-butanol and higher alcohols from fermentation broth [89]. The adsorption mechanism is considered to be governed by Van der Waals forces that exist between the resin surface and the alkyl chain of n-alcohols [90]. Resins derived from poly (styrene-co-divinyl-benzene) resins were found to have a very high 1-butanol adsorption potential (Figure 5.13). Among these potential resins, Dowex® Optipore SD-2 was found to enhance the butanol production the highest when tested under real fermentation conditions with *C. acetobutylicum* ATCC 824 (Figure 5.14). Diaion HP-20 and Dowex® M43 were found to be toxic to the culture. Dowex® Optipore SD-2 could be regenerated by heating the resin to 100 °C under vacuum. The average butanol

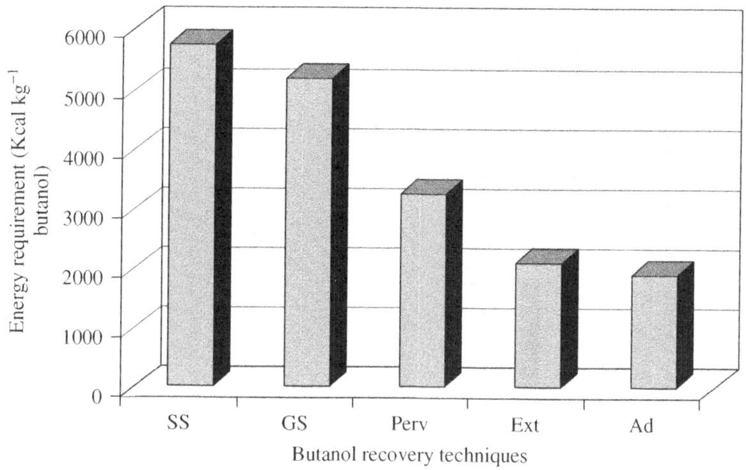

Figure 5.12 *A comparison of energy requirement to separate ABE from fermentation broth using various energy-efficient techniques. SS—steam stripping distillation; GS—gas stripping; Perv—pervaporation; Ext—liquid–liquid extraction; Ad—adsorption on to silicalite. Reference [84], with kind permission from Springer Science+Business Media © 2005*

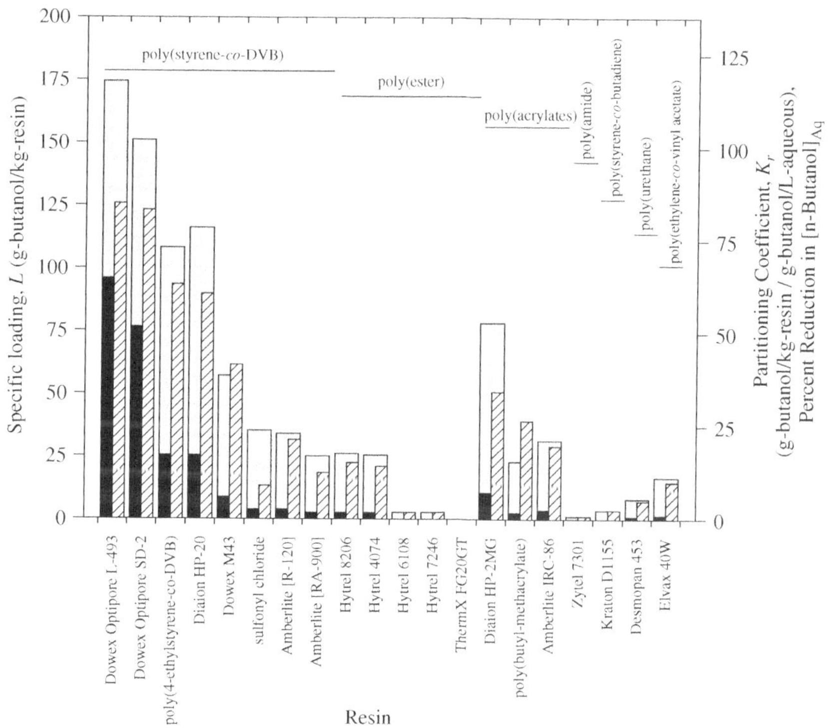

Figure 5.13 *Screening various commercial resins for n-butanol sorption affinity, as quantified by the specific loading of n-butanol (grey), equilibrium resin-aqueous partitioning coefficient (black), and percent reduction of [n-butanol]$_{Aq}$ (diagonally striped). For each test, initial n-butanol concentration = 20 g l^{-1} (2%, w/v) and resin concentration = 0.1 kg l^{-1}. Reprinted from [89] © 2009, with permission from John Wiley & Sons*

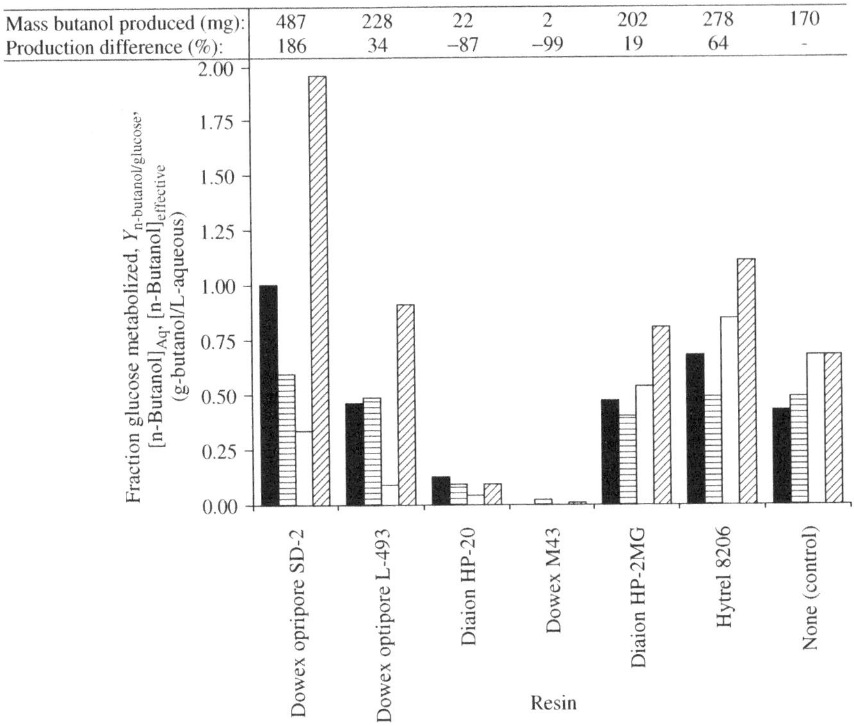

Figure 5.14 Evaluating the effect of various resins on n-butanol production by C. acetobutylicum ATCC824 in a 25 mL biphasic culture with 1xCRM, 8% (w/v) glucose and resin concentration = 0.1 kg l^{-1} by comparing the fraction of glucose metabolised (black), the molar butanol-to-glucose yield (horizontally striped), the final titre of n-butanol in the aqueous phase (white), and the effective final titre (diagonally striped). Inset table indicates the total mass of n-butanol in each system as well as the production difference relative to the control. Reprinted from [89] © 2009, with permission from John Wiley & Sons

recovery was 83 ±4% of the adsorbed butanol and the resin also could be reused after each cycle without any performance loss.

5.7.2.2 Recovery of other prospective biofuel compounds

Nielsen *et al*. [6] investigated the recovery and purification of many other emerging second-generation biofuel compounds such as iso-butanol, 2-methyl-1-butanol, 3-methyl-1-butanol, and n-pentanol using poly(styrene-co-divinyl-benzene) resins. Solute molecules were found to associate strongly with the π-π bonds within the phenyl side chain of the above resin matrix [91]. Van der Waals adsorption is generally enhanced by: high hydrophobicity of adsorbent material, specific surface area of adsorbent, and hydrophobicity of adsorbate molecule. Dowex® Optipore L-493 was found to show the highest adsorption capacity for all the alcohols series because of its higher hydrophobicity and specific surface area over other tested resins. Figure 5.15 shows the adsorption isotherm of the alcohol series on Dowex® Optipore L-493. From Figure 5.15, it is clear that as the alky chain length increases, the adsorption potential also increases as a result of increased adsorbate hydrophobicity. Thus 2-Methyl-1-butanol, 3-Methyl-1-butanol and n-pentanol are found to be potential molecules, which can be recovered by adsorption using hydrophobic resins.

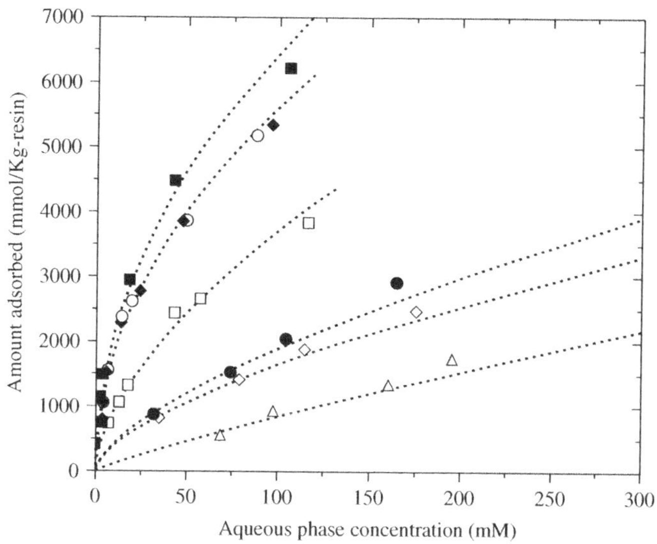

Figure 5.15 *Experimental and best-fit Freundlich adsorption isotherms of ethanol (△), iso-propanol (◇), n-propanol (●), iso-butanol (□), 2-methyl-1-butanol (◆), 3-methyl-1-butanol (○), and n-pentanol (■) using Dowex Optipore L-493. Reprinted from [6] © 2010, with permission from Elsevier*

5.7.2.3 *Ethanol dehydration*

Ethanol is the most promising biofuel produced from renewable resources. In a typical ethanol fermentation, broth containing dilute aqueous solution of about 5–12 wt% ethanol is produced. Separation of ethanol from this dilute solution accounts for a large fraction of the total production cost. Ethanol–water solution forms a minimum-boiling azeotrope at composition of 95.6 mol% ethanol at 78.15 °C and at standard atmospheric pressure. Distillation is found to be an effective separation process to concentrate the dilute solution up to 85 wt% [92]. To go above this concentration, distillation becomes expensive requiring high reflux ratios and additional equipment. Various techniques, such as adsorption, chemical dehydration, dehydration by vacuum distillation, azeotropic distillation, extractive distillation, membrane processes and diffusion distillation processes have been developed to break the azeotrope and produce anhydrous ethanol [92, 93]. Among these techniques, adsorption is particularly attractive because of its low energy consumption [33, 94, 95]. In this section, the drying of ethanol by adsorption is briefly reviewed.

There are two adsorption techniques used in ethanol-water separation: liquid phase adsorption of water from the fermentation broth and vapor-phase adsorption of water from the process stream coming out of distillation column [92] †.

Vapor phase adsorption of water The vapor phase adsorption consumes lower energy than distillation, because only a one-time vaporization is required [96]. The most potential adsorbents applied for vapor-phase adsorption of water from ethanol–water mixtures are molecular sieves [97], lithium chloride [98], silica gel [98], and activated alumina [99], and bio-based adsorbents such as corn grits [98, 100].

† The sections on liquid-phase adsorption and vapor-phase adsorption are from H.-J. Huang, S. Ramaswamy, U.W. Tschirner, and B.V. Ramarao, A review of separation technologies in current and future biorefineries, *Sep. Purif. Technol.*, **62**, 1–21 (2008). Reprinted with permission from © 2008 Elsevier.

Molecular sieve Molecular sieves selectively adsorb water on the basis of difference in molecular size between water and ethanol. The 3A zeolite molecular sieve, which has a nominal pore size of 3 Å, is most commonly used for dehydration of ethanol. Water molecules, with an approximate molecular diameter of 2.8 Å, can easily penetrate the pores of the molecular sieve adsorbent, while ethanol, with an approximate molecular diameter of 4.4 Å is retained [101]. Recently, Al-Asheh *et al*. [97] also studied ethanol–water separation using molecular sieves (3A, 4A and 5A). Molecular sieves are found to adsorb water up to 22% of their own weight. They are generally regenerated using temperature swing with hot carrier gas. Bed temperatures in the 175–260 °C range are usually employed for type-3A zeolites whereas for 4A, 5A and 13X sieves require temperatures in the 200–315 °C range. An alternative to this energy-intensive regeneration is to use desorbing agents such as methanol or acetone [93]. However, the latter method has not been reported to be practiced in large-scale operations.

Bio-based adsorbents The potential bio-based adsorbents include cornmeal, corn grits, starch, corn cobs, cassava, wheat straw, bagasse, cellulose, hemicellulose, wood chips, other grains, etc [102]. The mechanism of water adsorption is understood to involve hydrogen bonding with hydroxyl groups of the starch polymer chains. Ladisch and Dyck first investigated the biomass adsorption of water for ethanol dehydration and demonstrated that starchy and cellulosic biomass can be employed as an adsorbent to selectively adsorb water in the vapor mixture to obtain more than 99.5 wt% ethanol [33].

The adsorption of water from ethanol-water vapor mixture on a variety of starchy materials, such as cooked corn, corn grits and starch, which have different mean particle diameters and different relative amounts of amylose and amylopectin, has been experimentally measured at 90 °C. The results demonstrated that water selectivity over ethanol can be increased with the amylopectin/amylose ratio in starches [100]. Recently, the vapors of 92.4 wt% ethanol from distillation were passed over a fixed bed of corn grits, after which almost all the water was adsorbed on corn grits and anhydrous ethanol was obtained [98]. Chang *et al*. [103] investigated the cornmeal for ethanol dehydration on a pilot-scale fixed-bed adsorber at temperatures of 82–100 °C. Results showed that, for vapor containing 93.8 wt% ethanol, water selectivity over ethanol on the adsorbent at the breakthrough point is about 0.5–0.6 at a temperature of 91 °C.

As far as lignocellulosic adsorbents are concerned, bagasse, rice straw, and microcrystalline cellulose powder have been investigated for adsorption of water in the vapor mixture with 80–90% ethanol to produce anhydrous ethanol [104]. Al-Asheh *et al*. [97] studied corncobs, natural and activated palm stone and oak. Other lignocellulose-based adsorbents such as bleached wood pulp, oak sawdust, and kenaf core have also been explored for dehydrating the concentrated ethanol solution containing 90, 95, and 97 wt% ethanol in a thermal swing adsorption column. It was shown that water was selectively adsorbed and anhydrous ethanol was obtained [105]. Quintero and Cardona used corn (Zea mays), upright elephant ear (*Alocasia macrorrhiza*), cassava (*Manihot esculenta*), and sugar-cane bagasse (Saccharum) for their dehydration capacity [106]. When these materials were treated with α-amylase enzyme, the adsorption capacity increased because of better exposure of hydroxyl groups to the starch material. Corn starch was found to have the highest water adsorption capacity (19 g per 100 g of adsorbent).

Recently, compound starch-based adsorbents have been found still more efficient in anhydrous ethanol preparation. Wang *et al*. [107] tested a compound adsorbent ZSG-1, consisting of corn, sweet potatoes, and foaming agent, in a fixed bed adsorber to produce anhydrous ethanol. The adsorption capacity of ZSG-1 to water was found to be as high as that of molecular sieve and the cost was approximately twice that of the corn, and one-fifth of molecular sieve.

Liquid-phase adsorption of water In 1984, Type A zeolites were shown to have high capacity and selectivity in separating water from ethanol–water mixtures [4]. Recently, several combinations of starch-based and cellulosic materials, including white corn grits, α-amylase-modified yellow corn grits, polysaccharide-based synthesized adsorbent, and slightly gelled polysaccharide-based synthesized adsorbent, have also been tested and screened for liquid-phase adsorption of water. It was shown that starch-based adsorbents could remove liquid-phase water between 1 and 20 wt% from ethanol without the adsorbent being dissolved. Compared with silica gel and molecular sieves, these starch-based adsorbents have lower non-equilibrium adsorption capacity at water concentration below 10 wt%. At concentrations above 10 wt%, however, the starch-based adsorbents have similar non-equilibrium adsorption capacity to that of the inorganic adsorbents, under the same adsorption and regeneration conditions. The use of α-amylase to modify porosity and surface properties of starch resulted in materials with enhanced water sorption properties compared to the native material [108]. Among a variety of bio-based adsorbents, corn grits are reported as the only bio-based adsorbents that have been successfully applied in industry to produce 750 million gallons per year of anhydrous ethanol with purity of 99.8 wt%, although the other bio-based material such as cellulose and hemicellulose also have adsorptive properties [108].

5.7.2.4 *Biodiesel purification*

Biodiesel is an alternative fuel source to standard petrochemical diesel. It is derived from tricylglycerides of either vegetable or animal fat sources. Triglycerides are reacted with an alcohol (usually methanol, but sometimes ethanol or other alcohols) in the presence of a catalyst to produce biodiesel, or fatty acid esters (usually methyl esters). The reaction also produces glycerin, which must be separated from the biodiesel. After the separation process, biodiesel contains several contaminant materials such as soaps and traces of catalyst, along with ionic salts, which are detrimental to the quality of the fuel and thus must be eliminated from the product. The elimination of water-soluble portion of these materials is usually accomplished by water washing the biodiesel. However, with this method, the water-insoluble impurities remain in the biodiesel. There are also environmental concerns regarding the effluent water.

There are commercially available adsorbents that adsorb the contaminants without the water-washing step. Magnesol D60 (a synthetic magnesium silicate) is one such adsorbent introduced by the Dallas Group of America [109]. Amberlite™ BD10DRY™ is another resin supplied by Rohm & Haas Company, of the United States. These adsorbents remove both the water-soluble and water-insoluble contaminants such as soaps, free glycerin, free fatty acids, di-glycerides, monoglycerides, and sulfur. Table 5.6 shows the ASTM D6751 specification for biodiesel contaminants and the quality achieved by Magnesol D60 adsorbent [109].

Silica gel is also found to be an efficient adsorbent for removal of glycerol and free fatty acids (FFA) from biodiesel [110–112]. Silica gel had an adsorption capacity of 140 g per 100 g of FFA when tested with biodiesel prepared from chicken oil [110].

Table 5.6 *ASTM D6751 specification for biodiesel contaminants and the quality achieved by Magnesol D60 adsorbent [109]*

Parameter	Specification	Rapeseed methyl esters		Soybean methyl esters	
	ASTM D6751	*Initial sample*	*0.5% Magnesol D60*	*Initial sample*	*0.5% Magnesol D60*
Soap mg kg^{-1}	None	637	0	651	0
Free Glycerin %	0.02 max	0.053	0.005	0.033	0
Total Glycerin %	0.24 max	0.217	0.162	0.209	0.186
Water mg kg^{-1}	500 max	400	378	1000	300
Sulfated ash mass%	0.02 max	0.056	0	0.06	0
Methanol content %	None	0.19	0.009	0.15	0.011

5.8 A case study: Recovery of 1-butanol from ABE fermentation broth using TSA[‡]

5.8.1 Introduction

There are petrochemical processes and biochemical routes to produce 1-butanol. Conventionally 1-butanol was produced by fermentation of C_6 and C_5 sugars using *Clostridium* sp. The *Clostridum* sp have very low butanol tolerance, resulting in dilute aqueous solutions and thus requiring an energy-intensive process for the product recovery. The present case study describes recovery of 1-butanol from ABE fermentation broth by Temperature Swing Adsorption (TSA) using commercially available ZSM-5 zeolite (CBV28014) obtained from zeolyst. Zeolites in the form of extrudates were tested in a continuous column set-up to find its suitability for industrial application. The simulation of the breakthrough data obtained from the continuous column operation using the model described in the "adsorber modeling" section is also presented.

5.8.2 Adsorbent in extrudate form

CBV28014 has SiO_2/Al_2O_3 mole ratio of 280 and surface area of $400\,m^2\,g^{-1}$. It was supplied in powder form, particle size ranging from sub-nano to few microns. This can give a very high pressure drop for the liquid flow in a packed-bed column. The alternate solution would be to use zeolite extrudates of desired size having silica or alumina as binder. In this study, CBV28014 extrudates were prepared using 80% zeolite and 20% alumina binder. Zeolite powder, alumina powder and proportionate water were mixed in a Z-blade kneader to form an extrudable paste. Acetic acid was used as a peptizing agent. Additives like Superfloc N100 and Methocel K15 were dosed to improve the extrusion behavior. Extrusion was carried out over a 1 inch Bonnot single screw extruder, using a 0.8 mm trilobe die-plate. The extrudates were dried in a stationary ventilated oven at 120 °C for 3 hours; subsequently the temperature was raised to the calcination temperature of 550 °C and kept for 3 hours. The calcinated long extrudates (2.2–4.5 mm length) were broken and sieved to 30–80 mesh granules, 16–24 mesh granules and 12–24 mesh granules.

5.8.3 Adsorption kinetics

The influence of extrudate size on the adsorption kinetics was studied by contacting $64\,g\,l^{-1}$ of CBV28014 (both powder and extrudates) with 1 wt% butanol solution and measuring the aqueous phase butanol concentration with time. The experimental results are shown in Figure 5.16. From Figure 5.16, it can be seen that all the three forms of zeolite (zeolite powder, 30–80 mesh granules and long extrudate) have similar equilibrium loading, namely 0.12 g 1-butanol per g zeolite. However, it can be observed that CBV28014 powder has the highest adsorption rate, reaching 97.6% of the equilibrium loading within 2 minutes. Zeolite powder and 30–80 mesh extrudates reached equilibrium much faster than that of the long extrudates indicating the solid phase diffusion limitation.

5.8.4 Adsorption of 1-butanol by CBV28014 extrudates in a packed-bed column

The column adsorption experiments were carried out using CBV28014 extrudates of the following four different sizes:

(A) 30–80 mesh granules (0.18–0.60 mm);

[‡]Section 5.8 is from V. Saravanan, D.A. Waijers, M. Ziari and M.A. Noordermeer, Recovery of 1-butanol from aqueous solutions using zeolite ZSM5 with a high Si/Al ratio; suitability of a column process for industrial applications, *Biochem. Eng. J.*, **49**, 33–39 (2010). Reprinted with permission © 2010 Elsevier.

(B) 16–24 mesh granules and (0.71–1.0 mm);
(C) 12–24 mesh granules (0.71–1.4 mm);
(D) long extrudates.

For each run, 5 g of one of the above extrudate types was packed in a column of 1.08 cm diameter. A cell free ABE fermentation broth (gift from A. Lopez-Contreras, AFSG Wageningen) containing 1.28 wt% 1-butanol, 0.4 wt% acetone and 0.03 wt% ethanol or a model ABE solution containing the same composition of ABE was continuously fed through the column. When the liquid was fed from the top, the release of air from the extrudates disturbed the liquid flow. Hence, the column was fed from the bottom by an HPLC pump with a flow controller. Aliquots of 3 ml were collected and analyzed in GC.

The 30–80 mesh granules bed had a void fraction of 0.67 whereas it was 0.66 for the 16–24 mesh and 12–24 mesh granules bed. The liquid residence time was maintained at 7 minutes for the experiments with 30–80 mesh, 12–24 mesh and long extrudates. In the case of 16–24 mesh extrudates, the residence time was maintained at 9.3 minutes. Figure 5.17 shows the 1-butanol breakthrough curves for the four different extrudates loaded with a model ABE solution. Cell free ABE broth and model ABE solutions gave similar results, indicating that medium components like sugars and proteins do not influence 1-butanol adsorption. Hence the breakthrough for 1-butanol, acetone, and ethanol were alone measured.

The 1-butanol loadings at 10% breakthrough were 0.095, 0.085, 0.074, and 0.039 g 1-butanol per g zeolite of 30–80 mesh, 16–24 mesh, 12–24 mesh and long extrudates, respectively. Among these, the 30–80 mesh extrudates would be the best choice because of their highest loading at 10% breakthrough, but in a large-scale column at a higher velocity, they would offer a higher pressure drop compared to the rest. Hence, 16–24 mesh extrudates seemed to be the optimal choice considering the pressure drop and 1-butanol adsorption rate. The 1-butanol loading at 100% breakthrough was 0.11 g 1-butanol per g zeolite. Figure 5.17 also shows the acetone and ethanol breakthrough for 16–24 mesh extrudates. Initially acetone and ethanol removal was observed to be complete but in the later stage, acetone and ethanol sites were replaced by 1-butanol because of the high selectivity of extrudates for 1-butanol over acetone and ethanol. The replacement of acetone and ethanol sites by 1-butanol resulted in their concentration in the effluent

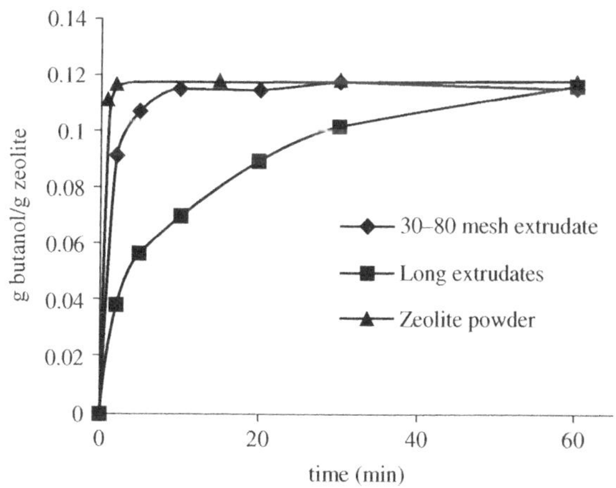

Figure 5.16 Adsorption kinetics of alumina based zeolite extrudates (CBV28014). Reprinted from [8] © 2010, with permission from Elsevier

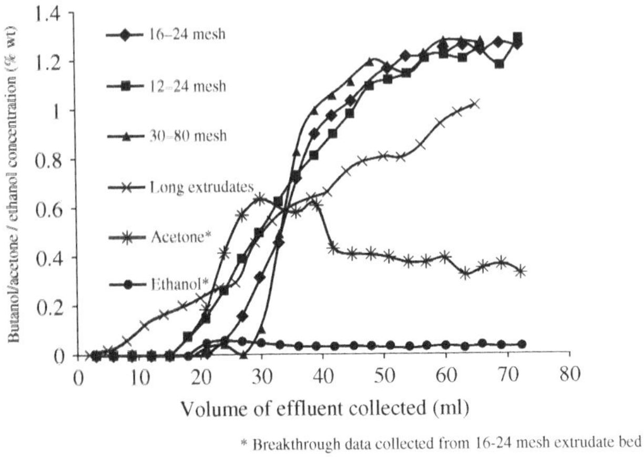

Figure 5.17 Butanol breakthrough curves for 30–80-mesh, 12–24 mesh, 16–24 mesh and long CBV28014 extrudates using the model ABE solution. Reprinted from [8] © 2010, with permission from Elsevier

being higher than that of their feed concentration. Total acetone loading at the end of the breakthrough was around 0.01 g acetone per g zeolite. Adsorption loading of ethanol was observed to be negligible.

5.8.5 Desorption

The desorption study was conducted in a glass column placed in an electrically heated oven. The column was filled with ABE loaded CBV28014 extrudates (16–24 mesh size). The column temperature was programmed using a PID controller. Argon gas was continuously passed through the column at 50 ml min^{-1} and the effluent was analyzed by online mass spectrometry. The m/e (mass/electron) signals of 18, 31 and 56 were used for the quantification of water, 1-butanol and 1-butene respectively. Initially the temperature of the column was maintained at 50 °C until the loosely bound water and traces of adsorbed ethanol were desorbed. The adsorbed acetone was also desorbed (its boiling point being 56 °C). The complete desorption of acetone and ethanol was ensured by observing their corresponding signals in the mass spectrometer returning to the base level. The temperature was then gradually raised at the rate of 5 °C per minute to 150 °C, and maintained at that level until the 1-butanol and water signal had returned to the base level. Following that the temperature was raised to 250 °C then to 350 °C and finally to 450 °C. Figure 5.18 shows the desorption profile of 1-butanol, water and 1-butene. The x-axis in Figure 5.18 denotes the data points collected in the mass spectrometer at a time interval of 14 s. By integrating the area under the different curves, the amount of 1-butanol, butene and water desorbed was calculated. Table 5.7 presents the composition of the 1-butanol, 1-butene and water desorbed from CBV28014 extrudate at different temperatures. The 1-butanol was found to be concentrated from 1.28 wt% to 84.3 wt%. No acetone and ethanol were present in the 1-butanol fraction. At 150 °C, approximately 80% of the adsorbed 1-butanol was desorbed, a further 18% could be desorbed and chemically converted to 1-butene at a temperature >250 °C. The same experiment was repeated by raising the temperature from 50 °C to 160 °C and similarly from 50 °C to 170 °C, which resulted in a recovery of 73% and 69% of adsorbed 1-butanol, whereas 27% and 31% could only be desorbed as 1-butene. Thus, desorption at 150 °C appeared to be optimal for 1-butanol recovery. When the column would be used in an industrial process with repeated adsorption and desorption at 150 °C, this would result in a 20% lower adsorption capacity after the first adsorption/desorption cycle but no further 1-butanol loss, as all stronger binding, adsorption sites will be blocked after the first

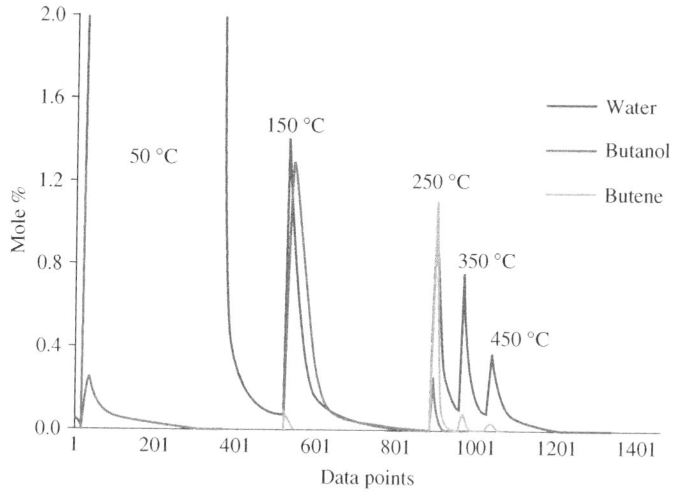

Figure 5.18 *Desorbed components from CBV28014 zeolite loaded with ABE solution (16–24 mesh extrudates). Reprinted from [8] © 2010, with permission from Elsevier*

Table 5.7 *Composition of 1-butanol, 1-butene and water desorbed from CBV28014 extrudate at different temperatures. Reprinted from [8] © 2010, with permission from Elsevier*

Temperature °C	Composition in g per g zeolite		
	1-butanol	1-butene	water
50	0.0241*	0*	0.9361*
150	0.0894	0	0.0167
250	0.0045	0.0137	0.0075
350	0.0005	0.0015	0.0059
450	0.0002	0.0005	0.0027

*Loosely bound molecules

cycle. The adsorption/desorption cycles were conducted for three cycles confirming the earlier mentioned adsorption capacities. The system can be used repeatedly and with confidence for a number of cycles.

5.8.6 Equilibrium isotherms

Different ratios of ABE model solution having 1.28 wt% of 1-butanol and CBV28014 extrudates (16–24 mesh size) were incubated in closed vessels and stirred on a roller overnight to allow adsorption to complete. The liquid phase was analyzed in GC. The equilibrium isotherm for 1-butanol on CBV28014 extrudate is shown in Figure 5.19.

The presence of acetone and ethanol could also lead to a multi-component adsorption problem. However, the weak adsorption of acetone and ethanol observed in desorption studies and column breakthrough justified to consider the overall net adsorption to be a single component one. The higher affinity of 1-butanol over acetone and ethanol on silicalite type materials have been reported elsewhere [76, 86].

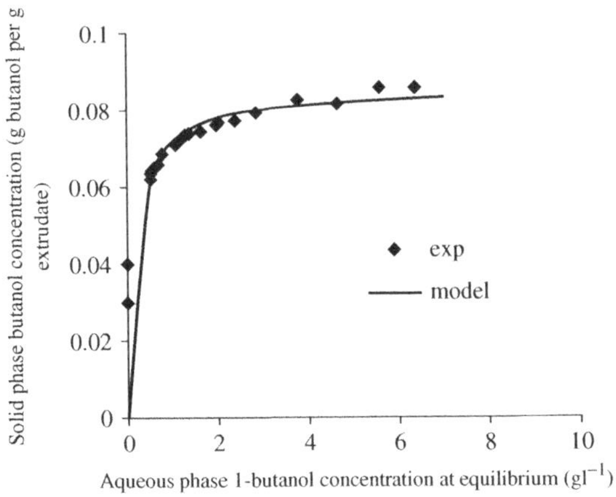

Figure 5.19 *Adsorption isotherm of CBV28014 extrudates with model fit. Reprinted from [8] © 2010, with permission from Elsevier*

Hence, the data were fitted to a simple Langmuir equation

$$q = \frac{q_s KC}{1 + KC} \tag{5.29}$$

where q_s is asymptotic maximum solid phase 1-butanol concentration (g 1-butanol per g extrudate); K is the Langmuir constant ($l\,g^{-1}$).

The data fitting was carried out with the Solver add-in function of Microsoft Excel, and q_o and K were calculated to be 0.0854 g 1-butanol per g extrudate and 5.2361 g^{-1} respectively.

The high SiO_2/Al_2O_3 ratio makes the zeolite more hydrophobic thus high affinity for 1-butanol and lower for water. The isotherms were generated in the presence of a swamping amount of solvent (water) and thus the equilibrium level measured accounts for the competition of the solvent with butanol for surface sites.

5.8.7 Simulation of breakthrough curves

The breakthrough data for 1-butanol shown in Figure 5.17 (except for the long extrudates) were simulated using the model described in the "adsorber modeling" section. The Langmuir isotherm described earlier was used to calculate the q^* value. Figures 5.20a to 5.20c show the graphical comparison of the model and experimental breakthrough data. The y-axis denotes the normalized 1-butanol concentration (C/C_o). The x-axis denotes the cumulative volume of effluent collected in terms of number of bed volumes. The mass transfer coefficients ($k_p a$) obtained from simulation were 0.10 min^{-1}, 0.052 min^{-1} and 0.042 min^{-1} for 30–80 mesh, 16–24 mesh and 12–24 mesh extrudates respectively. The differences in $k_p a$ value, 1-butanol adsorption kinetics (Figure 5.16) and its loading at 10% breakthrough for different particle sizes indicate that the lower the particle size the faster the adsorption rate. This leads to the conclusion that the solid phase diffusion controls the adsorption rate.

5.8.8 Summary from case study

This study indicated that alumina based extrudates of commercially available zeolites with a high Si/Al ratio are good adsorbents for 1-butanol recovery from ABE broth. The above described column set up is

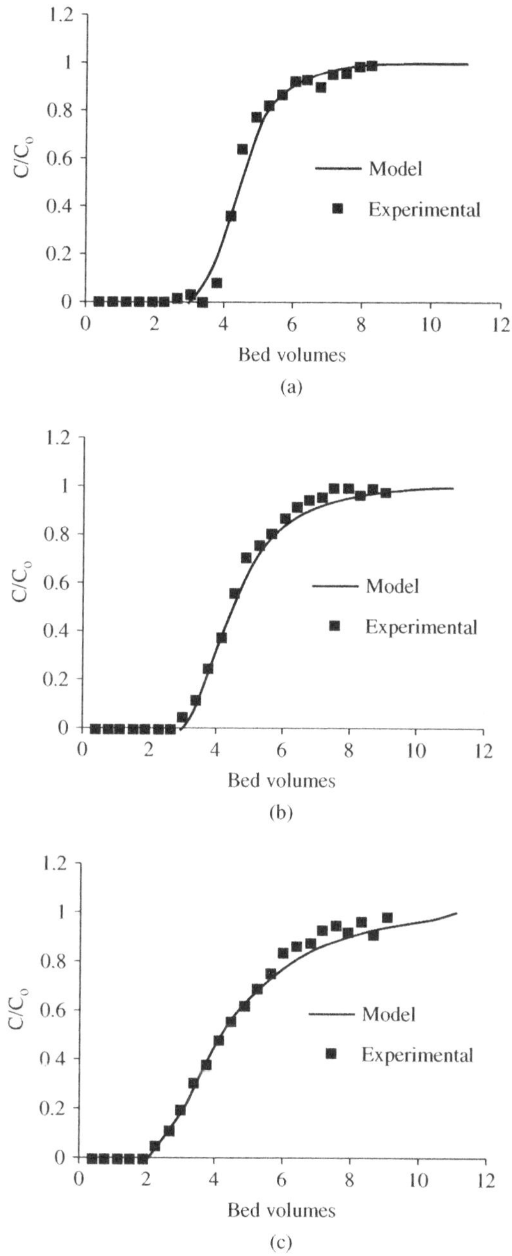

Figure 5.20 *Experimental and simulated breakthrough curves for an adsorption bed filled with (a) 8.0 ml of 30–80-mesh extrudate (b) 7.6 ml of 16–24 mesh extrudates (c) 8.0 ml of 12–24 mesh extrudates. Reprinted from [8] © 2010, with permission from Elsevier*

very suitable for a future industrial process, as 1-butanol could be concentrated from 1.28 wt% to 84.3 wt% in one step by TSA. Only 80% of the adsorbed 1-butanol could be recovered during desorption, but this is only the case in the first adsorption/desorption cycle and would not be a problem in an industrial process with repeated adsorption/desorption cycles. Comparison of simulated mass transfer coefficients for different sizes of extrudates clearly indicated that solid phase diffusion determines the adsorption rate. Thus, the calculated $k_p a$ can be a guiding parameter for estimating pilot-scale column performance.

5.9 Research needs and prospects

This chapter clearly indicates that adsorption can play a key role in the separation and purification processes of biorefineries. The number of examples presented in this chapter indicates that most of the studies are in the research phase requiring further improvements to make adsorption technology commercially feasible. Some of the areas where innovations are required are:

- *Developing new adsorbents*. The emerging processes for producing renewable fuels and chemicals require better adsorbents. Hence there is a continuous need to develop new adsorbents that will have (i) better selectivity for specific adsorbate, (ii) better stability under biorefinery process conditions, (iii) favorable geometries, and (iv) lower production cost.
- *Process improvements*. Along with new adsorbents development, process improvements are also necessary to make adsorption economically attractive—for example, using non-conventional ways to regenerate adsorbents such as sonic/ microwave energy can reduce the adsorption/desorption cycle time.
- *The need to demonstrate commercial feasibility*. While selecting a separation technology, the general tendency is to go for robust and conventional technologies even when newer technologies offer better economics. Hence, demonstrating the commercial feasibility of adsorption technology is absolutely necessary to change this trend.
- *The need for robust predictive model tools*. Predictive model tools play a key role in the adsorptive process design, which requires accurate data on adsorption equilibria, kinetics, and heats of adsorption. These data often may not be predicted by using today's models, particularly for complex systems such as cellulose hydrolysate, having adsorbates of different sizes and polarities, and the adsorbent being heterogeneous. There is a lack of "in-depth" understanding of the complex physicochemical phenomenon governing adsorption on practical heterogeneous adsorbents [113]. Hence the future need will be to develop a molecular modeling tool to predict the interaction of adsorbate and adsorbent. It should have the power to predict the best adsorbent for a given adsorbate molecule.
- *The need to generate and compile multicomponent adsorption data*. Considering adsorption use in biorefineries, such as in the recovery of products from a complex fermentation broth, there is a strong need to generate multicomponent adsorption database for better understanding of this complex phenomenon, for testing existing models, and for the development of new models.
- *Effective ways to dispose of spent adsorbents*. There is a serious environmental concern seen in disposing of environmentally unacceptable adsorbent materials from adsorption systems. Future research has to focus on effective ways of recycling spent adsorbents to make adsorption a truly green separation technology.

Innovation in adsorption technology to match the above needs can be aided by

- Advances in computational simulations.
- Development of molecular modeling methods for adsorption such as *ab initio* periodic density functional theory (DFT) [114].

- Structural techniques such as FTIR, variable temperature infrared (VTIR), X-ray crystallography, STM, AFM, XRD, and various kinds of NMR.
- Emergence of high-crystalline materials such as metal-organic frameworks, which are conducive to computational studies to complement experimental investigations of adsorption properties. Such well-characterized systems can aid the design of tailored adsorbents.

5.10 Conclusions

The success of biorefineries will be a strong function of how the industry manages its costs on a large scale. The separation and purification processes in the biorefinery currently account for much of the product and operating cost. It is critical that technical innovations in these processes are developed and implemented for the broader sustainable green industries to be a success.

This chapter has presented the role of adsorption technology as a candidate for solving some of these challenges. The state of this technology was described by presenting the underpinning science of adsorption and its application in developing engineering solutions. This was then illustrated with a detailed case study of the recovery of 1-butanol from an acetone-butanol-ethanol (ABE) fermentation broth using temperature-swing adsorption (TSA) technology. Academic innovation and advances in adsorption are ongoing but the challenge for adsorption in biorefineries still remains to make the process commercially viable. This challenge will be addressed by future research efforts and some of the potential options have been highlighted in the chapter. This chapter should allow readers to make a more knowledgeable decision on the applicability of adsorption in their specific bioseparation problems.

Acknowledgement

I would like to thank Dr. Jose Bravo, the chief scientist, Royal Dutch Shell, for his guidance and reviewing my work. I would like to thank Dr. Girish Rao, for his help in compiling this chapter. I am most grateful to my ever respectful well wisher Prof. P.V. Krishnan who taught me real character and how to use our intelligence for everyone's real welfare. Finally I would like to thank my wife, Malliga, for her support throughout this project.

References

1. A. Dabrowski, Adsorption from theory to practice, *Adv. Colloid Interface Sci.*, 93, 135–224 (2001).
2. D.W. Breck, W.G. Eversole, R.M. Milton, T.B. Read, and T.L. Thomas, Crystalline zeolites. I. The properties of a new synthetic zeolite Type A, *J. Am. Chem. Soc.*, 78, 5963–5971 (1956).
3. R.T. Yang, *Adsorbents: Fundamentals and Applications*, John Wiley & Sons, Inc., New York, 2003.
4. D.M. Ruthven, *Principles of Adsorption & Adsorption Processes*, John Wiley & Sons, Inc., New York, 1984.
5. T.J. Levario, M. Dai, W. Yuan, B.D. Vogt, and D.R. Nielsen, Rapid adsorption of alcohol biofuels by high surface area mesoporous carbons, *Microporous Mesoporous Matter*, 148, 107–114 (2012).
6. D.R. Nielsen, G.S. Amarasiriwardena and K.L.J. Prather, Predicting the adsorption of second generation biofuels by polymeric resins with applications for in situ product recovery (ISPR), *Bioresour. Technol.*, 101, 2762–2769 (2010).
7. I. Langmuir, The adsorption of gases on plane surfaces of glass, mica and platinum, *J. Am. Chem. Soc.*, 40, 1361–1403 (1918).
8. V. Saravanan, D.A. Waijers, M. Ziari and M.A. Noordermeer, Recovery of 1-butanol from aqueous solutions using zeolite ZSM5 with a high Si/Al ratio; suitability of a column process for industrial applications, *Biochem. Eng. J.*, 49, 33–39 (2010).

9. C. Efe, L.A.M. van der Wielen, and A.J.J. Straathof, High silica zeolites as an alternative to weak base adsorbents in succinic acid recovery, *Ind. Eng. Chem. Res.*, 49, 1837–1843 (2010).
10. T.C. Bowen and L.M. Vane, Ethanol, acetic acid, and water adsorption from binary and ternary liquid mixtures on high-silica zeolites, *Langmuir*, 22, 3721–3727 (2006).
11. S. Brunauer, P.H. Emmett and E. Teller, Adsorption of gases in multimolecular layers, *J. Am. Chem. Soc.*, 60, 309–319 (1938).
12. F. Gritti and G. Guiochon, New thermodynamically consistent competitive adsorption isotherm in RPLC. *J. Colloid Interface Sci.*, 264, 43–59 (2003).
13. A.L. Myers and J.M. Prausnitz, Thermodynamics of mixed-gas adsorption, *AIChE J.*, 11, 121–127 (1965).
14. C.J. Radke and J.M. Prausnitz, Thermodynamics of multisolute adsorption from dilute liquid solutions, *AIChE J.*, 18, 761–768 (1972).
15. S. Brunauer, L.S. Deming, W.E. Deming and E.J. Teller, On a theory of the van der Waals adsorption of gases, *J. Am. Chem. Soc.*, 62, 1723–1732 (1940).
16. K.S.W. Sing, Reporting physisorption data for gas/solid system, *Pure and Appl. Chem.*, 54, 2201–2218 (1982).
17. D.M. Razmus, and C.K. Hall, Prediction of gas adsorption in 5a zeolites using Monte Carlo simulation. *AIChE J.*, 37, 769–779 (1991).
18. J.P. Olivier, Modeling physical adsorption on porous and nonporous solids using density functional theory, *J. Porous Mater.*, 2, 9–17 (1995).
19. P.I. Ravikovitch, S.C.O. Domhnaill, A.V. Neimark, F. Schuth, and K.K. Unger, Capillary hysteresis in nanopores: theoretical and experimental studies of nitrogen adsorption on MCM-41, *Langmuir*, 11, 4765–4772 (1995).
20. M.D. Donohue and G.L. Aranovich, Adsorption hysteresis in porous solids, *J. Colloid Interface Sci.*, 205, 121–130 (1998).
21. S. Sircar, R. Mohr, C. Ristic, and M.B. Rao, Isosteric Heat of Adsorption: Theory and Experiment, *J. Phys. Chem. B*, 103, 6539–6546 (1999).
22. R. K. Iler, *The Chemistry of Silica*. John Wiley & Sons, Inc., New York, 1979.
23. D.W. Breck, *Zeolite Molecular Sieves*, John Wiley & Sons, Inc., New York, New York, 1974.
24. R. Szostak, *Molecular Sieves*, 2nd edn., Blackie Academic & Professional, New York, NY, 1998.
25. Mario L. Occelli and Harry E. Robson, *Zeolite Synthesis*, ACS symposium series, American Chemical Society, 1989.
26. M.E. Davis, R.F. Lobo, Zeolite and molecular sieve synthesis, *Chem. Mater.*, 4, 756–768 (1992).
27. R. Chal, C. Gérardin, M. Bulut, and S. van Donk, Overview and industrial assessment of synthesis strategies towards zeolites with mesopores, *Chem. Cat. Chem.*, 3, 67–81 (2011).
28. X. Li, L. Candela, Y. Han, and A.P. Kahn, Inventors; ARCO Chemical Technology, L.P., Assignee. Purification of tertiary butyl alcohol. US Patent 6770790, 2004 Aug 03.
29. R.L. Albright, Porous polymers as an anchor for catalysis, *React. Poly.*, 4, 155–174 (1986).
30. L. Rehmann, B. Sun, and A.J. Daugulis, Polymer selection for biphenyl degradation in a soli-liquid two-phase partitioning bioreactor, *Biotechnol. Progr.*, 23, 814–819 (2007).
31. H. Chang, X.G. Yuan, and H.T.A.W. Zeng, Experimental investigation and modeling of adsorption of water and ethanol on cornmeal in an ethanol–water binary vapor system, *Chem. Eng. Technol.*, 29, 454–461 (2006).
32. S.K. Rakshit, P. Ghosh, and V.S. Bisaria, Ethanol separation by selective adsorption of water, *Bioprocess Biosyst. Eng.*, 8, 279–282 (1993).
33. M.R. Ladisch and K. Dyck, Dehydration of ethanol: new approach gives positive energy balance, *Science*, 205, 898–900 (1979).
34. J.P. Crawshaw and J.H. Hills, Sorption of ethanol and water by starchy materials, *Ind. Eng. Chem. Res.*, 29, 307–309 (1990).
35. H.P.C.E. Kuipers, M.S. Rigutto, and H.A. Stil, Inventors; Shell Internationale Research Maatschappij B.V., Assignee. Process for producing alcohol. WIPO Patent Application WO/2010/012660. 2010 Feb 04.
36. K.S. Park, Z. Ni, A.P. Cote, J.Y. Choi, R. Huang, F.J. Uribe-Romo, H.K. Chae, M. O'Keeffe, and O.M. Yaghi, Exceptional chemical and thermal stability of zeolitic imidazolate frameworks, *PNAS*, 103, 10186–10191 (2006).

37. C.L. Cavalcante JR, Industrial adsorption separation processes: fundamentals, modelling and applications, *Lat. Am. Appl. Res.*, 30, 357–364 (2000).
38. R.W. Neuzil, Inventor; UOP Inc, Assignee. Process for separating para-xylene. US Patent 3997620, 1976 Dec. 14.
39. J. Johnson, Sorbex: Continuing innovation in liquid phase adsorption, in *Adsorption: Science and Technology*, A.E. Rodrigues, M.D. LeVan, and D. E. Tondeur (eds.), NATO ASI Series, Kluwer Academic Publishers, Netherlands, 1989.
40. C.W. Chi and W.P. Cummings, Adsorptive separation processes: gases, in *Kirk Othmer Encyclopedia of Chemical Technology*, 3rd edn., Vol. I. Wiley Interscience, New York, 1978.
41. D. Basmadjian, *The Little Adsorption Book*, CRC Press, Boca Raton, FL, 1997.
42. C.W. Skarstorm, Inventor; Exxon Research Engineering Co, Assignee. Method and apparatus for fractionating gaseous mixtures by adsorption. US Patent 2944627, 1960 Dec. 07.
43. C.W. Skarstorm, Heatless fractionation of gases over solid sorbents, in *Recent Developments in Separation Science*, (N.N.Li, ed), Vol. 2. CRC Press, Cleveland, 1972.
44. P.G.DE. Montgareuil and D. Daniel, Inventors; Air, Liquide, Assignee. Process for separating a binary gaseous mixture by adsorption. US Patent 3155468, 1964 Nov. 03.
45. M. Simo, S. Sivashanmugam, C.J. Brown and V. Hlavacek, Adsorption/desorption of water and ethanol on 3A zeolite in near-adiabatic fixed bed, *Ind. Eng. Chem. Res.*, 48, 9247–9260 (2009).
46. J.S. Jeong, B.U. Jang, Y.R. Kim, B.W. Chung, and G.W. Choi, Production of dehydrated fuel ethanol by pressure swing adsorption process in the pilot plant, *Korean J. Chem. Eng.*, 26, 1308–1312 (2009).
47. A.K. Frolkova and V.M. Raeva, Bioethanol dehydration: state of the art, *Theor. Found. Chem. Eng.*, 44, 545–556 (2010).
48. C. Boonfung and P. Rattanaphanee, Pressure swing adsorption with cassava adsorbent for dehydration of ethanol vapor, *Int. J. Chem. Biol. Eng.*, 3, 206–209 (2010).
49. A.M. Ribeiro, J.C. Santos, and A.E. Rodrigues, Pressure swing adsorption for CO2 capture in Fischer–Tropsch fuels production from biomass, *Adsorption*., DOI 10.1007/s10450-010-9280-9288 (2010).
50. G. Klein, Column design for sorption processes, *in Mass Transfer and Kinetics of Ion Exchange*, L. Liberti and F.G. Helfferich (eds.), 226–227, Martinus Nijhoff, Boston, 1982.
51. T.J. Tranter, R.S. Herbst and T.A. Todd, Determination of a solid phase mass transfer coefficient for modeling an adsorption bed system using ammonium molybdophosphate –polyacrylonitrile (AMP-PAN) as a sorbent for the removal of 137Cs from acidic nuclear waste solutions, *Adsorption*., 8, 291–299 (2002).
52. S.F. Chung and C.Y. Wen, Longitudinal dispersion of liquid flowing through fixed and fluidized beds, *AIChE. J.*, 14, 857–866 (1968).
53. S.D. Conte, C. de Boor, *Elementary Numerical Analysis*, McGraw-Hill, New York, 1972.
54. E. Ximenesa, Y. Kima, N. Mosiera, B. Diend, and M. Ladisch, Deactivation of cellulases by phenols, *Enzyme Microb. Technol.*, 48, 54–60 (2011).
55. J.P. Delgenes, R. Moletta, and J.M. Navarro, Effects of lignocellulose degradation products on ethanol fermentations of glucose and xylose by Saccharomyces cerevisiae, Zymomonas mobilis, Pichia stipitis, and Candida shehatae, *Enzyme Microb. Technol.*, 19, 220–225 (1996).
56. G.D. Gupta and A.M.D.K. Agarwal, Inhalation toxicity of furfural vapors: An assessment of biochemical response in rat lungs, *J. Appl. Toxicol.*, 11, 343–347 (1991).
57. S. Larsson, A. Reimann, N.O. Nilvebrant, and L.J. Jönsson, Comparison of different methods for the detoxification of lignocellulose hydrolyzates of spruce, *Appl. Biochem. Biotechnol*.,77, 91–103 (1999).
58. J.H.T. Luong, Kinetics of ethanol inhibition in alcohol fermentation, *Biotechnol. Bioeng.*, 27, 280–285 (1985).
59. D.T. Jones and D.R. Woods, Acetone-butanol fermentation revisited, *Microbiol. Rev.*, 50, 484–524 (1986).
60. H.B. Klinke, A.B. Thomsen, and B.K. Ahring, Inhibition of ethanol producing yeast and bacteria by degradation products produced during pretreatment of biomass, *Appl. Microbiol. Biotechnol.*, 66, 10–26 (2004).
61. E. Palmqvist and B. Hahn-Hägerdal, Fermentation of lignocellulosic hydrolysates. II: inhibitors and mechanisms of inhibition, *Biores. Technol*., 74, 25–33 (2000).
62. D. Nabarlatz, X. Farriol, and D. Montane, Kinetic modeling of the autohydrolysis of lignocellulosic biomass for the production of hemicellulose-derived oligosaccharides, *Ind. Eng. Chem.*, 43, 4124–4131 (2004).

63. E.M. Lohmeirer-Vogel, C.R. Sopher, and H. Lee, Intracellular acidification as a mechanism for the inhibition by acid hydrolysis-derived inhibitors of xylose fermentation by yeasts, *J. Ind. Microbiol. Biotechnol.*, 20, 75–81 (1998).
64. R.E. Berson, J.S. Young, S.N. Kamer, and T.R. Hanley, Detoxification of actual pretreated corn stover hydrolysate using activated carbon powder, *Appl. Biochem. Biotechnol.*, 121–124, 923–934 (2005).
65. H. Ren, H.Z. Huang, and J. Zheng, Inventors; Novozymes A/S and Cofco LTD, Assignee. Detoxifying and recylcing of washing solution used in pretreatment of lignocellulose-containing materials. US Patent Application 20090056889, 2009 Mar. 05.
66. S.R. Wickramasinghe and D.L. Grzenia. Adsorptive membranes and resins for acetic acid removal from biomass hydrolyzates, *Desalination*, 234, 144–151 (2008).
67. P.A. Belter, E.L. Cussler and W.S. Hu, *Bioseparations Downstream Processing for Biotechnology*, John Wiley & Sons, Inc., New York, 1988.
68. W.S.W. Ho and K.K. Sirkar, *Membrane handbook*, Van Nostrand Reinhold, New York, 1992.
69. W.K. Wang, *Membrane separations in biotechnology*, Marcel Dekker Inc, New York, 2001.
70. J.M. Lee, R.A. Venditti, H. Jameel, and W.R. Kenealy, Detoxification of woody hydrolyzates with activated carbon for bioconversion to ethanol by the thermophilic anaerobic bacterium *Thermoanaerobacterium saccharolyticum*, *Biomass and Bioenergy*, 35, 626–636 (2010).
71. G.B.M. Carvalho, S.I. Mussatto, E.J. Candido, and J.B. Almeida e Silva, Comparison of different procedures for the detoxification of eucalyptus hemicellulosic hydrolysate for use in fermentative processes, *J. Chem. Technol. Biotechnol.*, 81, 152–157 (2006).
72. R. Ranjan, S. Thust, C.E. Gounaris, M. Woo, C.A. Floudas, M.V. Keitz, K.J. Valentas, J. Wei, and M. Tsapatsis, Adsorption of fermentation inhibitors from lignocellulosic biomass hydrolyzates for improved ethanol yield and value-added product recovery, *Microporous Mesoporous Mater.*, 122, 143–148 (2009).
73. Y. Roman-Leshkov, J.N. Chheda, and J.A. Dumesic, Phase modifiers promote efficient production of hydroxymethylfurfural from fructose, *Science*, 312, 1933–1937 (2006).
74. J. Chheda, Y. Roman-Leshkov, and J. Dumesic, Production of 5-hydroxymethylfurfural and furfural by dehydration of biomass-derived mono- and poly-saccharides, *Green Chem.*, 9, 342–350 (2007).
75. C.E. Gounaris, C.A. Floudas, and J. Wei, Rational design of shape selective separation and catalysis—I: Concepts and analysis, *Chem. Eng. Sci.*, 61, 7933–7948 (2006).
76. I.S. Maddox, The acetone-1-butanol-ethanol fermentation: recent progress in technology, *Biotechnol. Genet. Eng. Rev.*, 7, 189–220 (1989).
77. W.J. Groot, R.G.J.M. van der Lans and K.Ch.A.M. Luyben, Technologies for 1-butanol recovery integrated with fermentations, *Process Biochem.*, 27, 61–75 (1992).
78. N. Qureshi, I.S. Maddox, and A. Friedl, Technologies for 1-butanol recovery integrated with fermentations, *Biotechnol. Prog.*, 8, 382–390 (1992).
79. L.M. Vane, A review of pervaporation for product recovery from biomass fermentation processes, *Biofuels, Bioprod. Biorefin.*, 2, 553–588 (2008).
80. T.C. Ezeji, N. Qureshi and H.P. Blaschek, Production of acetone 1-butanol (AB) from liquefied corn storch, a commercial substrate, using Clostridium beijerinckii coupled with product recovery by gas stripping, *J. Ind. Microbiol. Biotechnol.*, 34, 771–777 (2007).
81. K. Schugerl, Integrated processing for biotechnology products, *Biotechnol. Adv.*, 18, 581–599 (2000).
82. A.G. Fadeev and M.M. Meagher, Opportunities for ionic liquids in recovery of biofuels, *Chem. Comm.*, 3, 295–296 (2001).
83. A. Oudshoorn, L.A.M. van der Wielen, and A.J.J. Straathof, assessment of options for selective 1-butanol recovery from aqueous solution, *Ind. Eng. Chem. Res.*, 48, 7325–7336 (2009).
84. N. Qureshi, S. Hughes, I.S. Maddox, and M.A. Cotta, Energy-efficient recovery of butanol from model solutions and fermentation broth by adsorption, *Bioprocess Biosyst. Eng.*, 27, 215–222 (2005).
85. P. Shao and A. Kumar, Separation of 1-butanol/2,3-butanediol using ZSM-5 zeolite-filled polydimethylsiloxane membranes, *J. Membr. Sci.*, 339, 143–150 (2009).

86. N.B. Milestone and D.M. Bibby, Concentration of alcohols by adsorption on silicalite, *J. Chem. Technol. Biotechnol.*, 31, 732–736 (1981).
87. C. Falamaki, M. Sohrabi, G. Talebi, The kinetics and equilibrium of ethanol adsorption from aqueous phase using calcined (Na-1,6-hexanediol)-ZSM-5, *Chem. Eng. Technol.*, 24, 501–506 (2001).
88. M.T. Holtzapple and R.F. Brown, Conceptual design for a process to recover volatile solutes from aqueous solutions using silicalite, *Sep. Technol.*, 4, 213–229 (1994).
89. D.R. Nielsen and K.L.J. Prather, In situ product recovery of n-butanol using polymeric resins, *Biotechnol. Bioeng.*, 102, 811–821 (2009).
90. F.A. Carey and R.J. Sundberg, *Advanced Organic Chemistry*, Kluwer Academic/Plenum Publishers, New York, 2000.
91. N. Fontanals, R.M. Marce, and F. Borrull, New hydrophilic materials for solid-phase extraction, *Trac-Trend. Anal. Chem.*, 24, 394–406 (2005).
92. H.-J. Huang, S. Ramaswamy, U.W. Tschirner, and B.V. Ramarao, A review of separation technologies in current and future biorefineries, *Sep. Purif. Technol.*, 62, 1–21 (2008).
93. S. Kumar, N. Singh, and R. Prasad, Anhydrous ethanol: A renewable source of energy, *Renewable Sustainable Energy Rev.*, 14, 1830–1844 (2010).
94. A.A. Hassaballah and J.H. Hills, Drying of ethanol vapors by adsorption on corn meal, *Biotechnol. Bioeng.*, 35, 598–608 (1990).
95. Y. Wang, C. Gong, J. Sun, H. Gao, S. Zheng, and S. Xu, Separation of ethanol/water azeotrope using compound starch-based adsorbents, *Bioresour. Technol.*, 101, 6170–6176 (2010).
96. X. Hu and W. Xie, Fixed-Bed adsorption and fluidized-bed regeneration for breaking the azeotrope of ethanol and water, *Sep. Sci. Technol.*, 36, 125–136 (2001).
97. S. Al-Asheh, F. Ganat, and N. Al-Lagtah, Separation of ethanol–water mixtures using molecular sieves and biobased adsorbents, *Chem. Eng. Res. Des.*, 82, 855–864 (2004).
98. K.E. Beery and M.R. Ladisch, Adsorption of water from liquid-phase ethanol–water mixtures at room temperature using starch-based adsorbents, *Ind. Eng. Chem. Res.*, 40, 2112–2115 (2001).
99. M. Kondo, M. Komori, H. Kita, and K. Okamoto, Tubular type pervaporation module with zeolite NaA membrane, *J. Membr. Sci.*, 133, 133–141 (1997).
100. J.P. Crawshaw and J.H. Hills, Sorption of ethanol and water by starchy materials, *Ind. Eng. Chem. Res.*, 29, 307–309 (1990).
101. M.J. Carmo and J.C. Gubulin, Ethanol–water adsorption on commercial 3A zeolites: kinetic and thermodynamic data, *Braz. J. Chem. Eng.*, 14, 1–10 (1997).
102. M.R. Ladisch and G.T. Tsao, Inventors; Purdue Research Foundation, Assignee. Vapor phase dehydration of aqueous alcohol mixtures. US Patent 4345973, 1982 Aug. 08.
103. H. Chang, X-G. Yuan, H. Tian, and A-W. Zeng, Experimental study on the adsorption of water and ethanol by cornmeal for ethanol dehydration, *Ind. Eng. Chem. Res.*, 45, 3916–3921 (2006).
104. S.K. Rakshit, P. Ghosh and V.S. Bisaria, Ethanol separation by selective adsorption of water, *Bioprocess Biosyst. Eng.*, 8, 279–282 (1993).
105. T.J. Benson and C.E. George, Cellulose based adsorbent materials for the dehydration of ethanol using thermal swing adsorption, *Chem. Mater. Sci.*, 11, 697–701 (2005).
106. J.A. Quintero and C.A. Cardona, Ethanol dehydration by adsorption with starchy and cellulosic materials, *Ind. Eng. Chem. Res.*, 48, 6783–6788 (2009).
107. Y. Wang, C. Gong, J. Sun, H. Gao, S. Zheng, and S. Xu, Separation of ethanol/water azeotrope using compound starch-based adsorbents, *Bioresour. Technol.*, 101, 6170–6176 (2010).
108. K.E. Beery and M.R. Ladisch, Chemistry and properties of starch based desiccants, *Enzyme Microb. Technol.*, 28, 573–581 (2001).
109. B. Bertram, C. Abrams and B.S. Cooke, Inventors; The Dallas Group of America, Inc., Assignee. Purification of biodiesel with adsorbent materials. US Patent 7635398 B2, 2009 Dec. 22.
110. D.L. Manuale, V.M. Mazzieri, G. Torres, C.R. Vera, and J.C. Yori, Non-catalytic biodiesel process with adsorption-based refining, *Fuel*, doi:10.1016/j.fuel.2010.10.047 (2010).

111. V.M. Mazzieri, C.R. Vera and J.C. Yori, Adsorptive properties of silica gel for biodiesel refining, *Energy Fuels*, 22, 4281–4284 (2008).
112. J.C. Yori, S.A. D'Ippolito, C.L. Pieck, and C.R. Vera, Deglycerolization of biodiesel streams by adsorption over silica beds, *Energy Fuels*, 21, 347–353 (2007).
113. S. Sircar, Basic research needs for design of adsorptive gas separation processes, *Ind. Eng. Chem. Res.*, 45, 5435–5448 (2006).
114. L. Valenzano, B. Civalleri, S. Chavan, G. T. Palomino, C.O. Areán and S. Bordiga, Computational and experimental studies on the adsorption of CO, N2, and CO2 on Mg-MOF-74, *J. Phys. Chem. C*, 114, 11185–1119 (2010).

6
Ion Exchange

M. Berrios, J. A. Siles, M. A. Martín and A. Martín

Departamento de Química Inorgánica e Ingeniería Química, Universidad de Córdoba, Spain

6.1 Introduction

The term "sorption" is used to describe the capture of a substance from the external surface of solids, liquids, or mesomorphs as well as from the internal surface of porous solids or liquids (Skoulikides, 1989). Ion exchange is classified as electrostatic adsorption. This is a term reserved for Coulomb-attractive forces between ions and charged functional groups, and is commonly classified as ion exchange.

Ion exchangers are solid materials that are able to take up charged ions from a solution and release an equivalent amount of other ions into the solution. The structure of the materials determines their ability to exchange ions. The exchanger consists of what is known as a matrix, with a positive or negative excess charge. This excess charge is localized in specific sites in the solid structure or in functional groups. The charge of the matrix is compensated for by counter-ions, which can move within the free space of the matrix and can be replaced by other ions of equal charge (Helfferich, 1995). Counter-ions and solvent can be retained in the pores of the exchangers. This can cause swelling that depends on the nature of the counter-ions. Pores are defined as the open areas of variable size and shape that are in an exchanger. Hence, the exchangers exhibit a three-dimensional network of channels with irregular size. Some electrolytes can also penetrate into the exchanger along with the solvent. As a result, there are additional counter-ions, known as co-ions, which have the same charge as the fixed ions.

Sorption and ion exchange can be considered similar processes as a substance is captured by a solid in both processes. However, ion exchange is a stoichiometric process, in contrast to sorption (Helfferich, 1995). This means that for every ion that is removed in the ion-exchange process, another ion of the same charge is released into the solution. Ion exchange can be seen as a reversible reaction involving chemically equivalent quantities (Treybal, 1980; Perry and Green, 1999), but it is not exactly a chemical process. Ion exchange is, in principle, a redistribution of ions between two phases by diffusion, and chemical factors

Separation and Purification Technologies in Biorefineries, First Edition.
Edited by Shri Ramaswamy, Hua-Jiang Huang, and Bandaru V. Ramarao.

are less significant or even absent. Only when an ion exchange is accompanied or followed by a reaction such as neutralization can the whole phenomenon be characterized as a purely "chemical" process.

Ion exchange was first cited in Aristotle's *Problematica*. Ion exchange was used in sand filters to purify sea water and impure drinking water until the nineteenth century. However, two agricultural chemists, Thomson and Way, are credited with identifying the ion exchange phenomenon when they observed the discoloration and deodorization of urine during the filtration of liquid manure through a bed of an ordinary loamy soil. In 1848, Thomson and Way recognized the ion exchange phenomenon and its basic characteristics (Lucy, 2003). The first ion exchangers to be discovered were natural minerals, which are microporous, aluminosilicate minerals ($AlSiO_7H_6{}^-H^+$) commonly used as commercial adsorbents. Later on, zeolites were complemented with organic resins, which are more commonly used nowadays due to their high ion exchange capability (Gil-Rodríguez, 2005). After soil and clays, natural and synthetic aluminum silicates and synthetic zeolites were tested as ion-exchange materials. In 1905 Gans used zeolites to soften hard water; the first practical application of ion exchange. Adams and Holmes carried out the first synthesis of organic resins in 1935 (Helfferich, 1995). Much progress was made during World War II in the field of ion exchange, but the results were not published for some years due to reasons of confidentiality. Afterwards, ion-exchange materials and methods developed rapidly (Lucy, 2003).

6.1.1 Ion exchangers: Operational conditions—sorbent selection

The selection of sorbents is determined by their use in regenerative or non-regenerative systems. If a non-regenerative system is selected, the characteristics of the sorbent should be high capacity, a strongly favorable isotherm (if the sorbent is used for purification) and high selectivity (if the sorbent is used for separation). In contrast, if a regenerative system is employed, the most important factor is cost-effective regeneration, although high capacity and selectivity are required. Other factors to take into account in sorbent selection are mechanical and chemical stability, mass transfer characteristics, and cost. The classification of ion exchangers depends on their functionality and the physical properties of the support matrix. Depending on their ability to exchange positively or negatively charged species, the exchangers are classified as cation or anion exchangers, respectively. Strongly acidic and strongly basic ion exchangers are ionized and thus are effective at nearly all pH values (pH 0–14). Weakly acidic exchangers are typically effective in the range of pH 5–14. Weakly basic resins are effective in the range of pH 0–9. Weakly acidic and weakly basic exchangers are often easier to regenerate, but leakage due to incomplete exchange may occur. Achievable ion-exchange capacity depends on the concentration of ionogenic groups and their availability as an exchange site; the latter being a function of the support matrix. The degree of cross-linking can depend on the extent of swelling, exchange capacity, intra-particle diffusivity, ease of regeneration, and physic-chemical stability under operating conditions. In the case of resins with a lower degree of cross-linking, the change of one form to another is dramatic due to the size of the particles exchanged during the ion-exchange process. The concentration of ionogenic groups determines the capacity of the resin. Although the capacity per unit mass of dry resin is insensitive to the degree of cross-linking, the exchange capacity per unit volume of swollen resin increases significantly with the degree of cross-linking. The degree of cross-linking also affects the rate of ion exchange. Intra-particle diffusivity decreases nearly exponentially with the mesh size of the matrix. As a result, resins with a lower degree of cross-linking are normally required for the exchange of bulky species, such as organic ions with a molecular weight in excess of 100. The regeneration efficiency is typically greater for resins with a lower degree of cross-linking. Finally, the degree of cross-linking also affects the long-term stability of the resin (Perry and Green, 1999).

6.2 Essential principles

The change reaction between two non-miscible phases (in general solid phase with ionic aqueous solution) is called ion exchange. In this reaction, an ion in solution is replaced by another ion from the solid. It is important to note that the physical structure of the solid remains intact during the reaction. In practice, ion exchange is a dynamic phenomenon and its efficiency depends on the contact time between the solution and the resin (Savidan, 1967).

McCabe *et al.* (2002) reported some advantages and disadvantages to ion exchange. Among the advantages they found:

- in principle, all ions or ionizable species can be removed from aqueous liquids;
- recovery of valuable species is possible;
- high efficiency;
- a large variety of specific resins are available.

The disadvantages included:

- prefiltration is required (suspended particles in the feed should be less than about 50 mg/L to prevent plugging);
- interference of competing cations in the wastewater;
- low-temperature resistance of organic (resin) ion exchangers.

6.2.1 Properties of ion exchangers

1. Degree of cross-linking and porosity. Resins with a higher degree of cross-linking show more resistance to oxidizing conditions that tend to de-crosslink the polymer. Activation becomes difficult because access to the interior of the bead is hindered by the high density of the matrix. The rate of exchange increases in proportion to the mobility of the ions inside the exchanger bead. If the structure is too dense, ionic motion is slowed down, thus reducing the operating capacity of the resin. The greater the ionic mobility in the resin the poorer is the differentiation between the adsorption of ionic species with the same charge. Consequently, the degree of cross-linking in the resin must be increased when greater differences in ionic affinity are required.
2. Exchange capacity. The total exchange capacity of a resin, expressed in equivalents per unit weight (or per unit volume), represents the number of active sites that are available. Operating capacity is defined as the proportion of total capacity used during the exchange process. This can amount to a large or small proportion of the total capacity and depends on a number of process variables including concentration and type of ions to be adsorbed; rate of percolation; temperature; depth of resin bed; and type, concentration, and quantity of regenerant.
3. Stability and service life. Because ion-exchange resins are expected to have several years of service, their stability over long periods of life is of prime importance. The chemical stability of the matrix, thermal stability of active groups, mechanical stability, osmotic stability, and resistance to drying are aspects that must be evaluated for the use of ion exchangers.
4. Density. Resin density is an important property because it determines the hydrodynamic behavior in counterflow systems. Resin density normally lies within the following ranges:

- strongly acidic cation exchangers: 1.18–1.38 kg/m^3;
- weakly acidic cation exchangers: 1.13–1.20 kg/m^3;
- strongly basic anion exchangers: 1.07–1.12 kg/m^3;
- weakly basic anion exchangers: 1.02–1.10 kg/m^3.

   By choosing suitable particle sizes, several different types of resin can be used in the same column.

5. Particle size. For industrial use, particle size is a compromise between the speed of the exchange reaction (which is greater with small beads) and high flow rates (which require coarse particles to minimize the head loss). Standard resins contain particles with diameters measuring 0.3 to 1.2 mm, but coarser or finer grades are available.
6. Moisture content. Ion-exchange resins carry both fixed and mobile ions, which are always surrounded by water molecules located in the interior of the resin beads. The water retention capacity governs the kinetics, exchange capacity, and mechanical strength of ion-exchange resins (Elvers *et al*. 1989).

In practice, adsorption may be carried out as batch, semi-continuous or continuous processes. In a typical batch process with a liquid solution, a batch of the solution is mixed with a portion of the adsorbent for a determined length of time, after which the two phases are separated. In a semi-continuous process, the gas or the solution is continuously passed through a static bed of adsorbent in a column. In a continuous process, both the adsorbent and the gas (or liquid solution) are continuously fed into a system (usually in a counter-current fashion) where contact between the two phases occurs. Truly continuous adsorption processes are not commonly applied in the food industry.

Three main types of continuous counter-flow equipment are used in industry. Packed beds of resin moving intermittently are widely used; fluidized beds of resin, usually in series, are used to process dirty liquors and may also have cost advantages over moving packed-bed systems for clarified feeds; a series of stirred tanks offers simplicity, particularly for treating de-sanded ore leach pulps, and is suitable for large-scale use, but the system has a size disadvantage due to low resin hold-up as a consequence of low resin/solution flow ratios (Slater, 1979).

Ion exchangers are usually employed in cyclic processes, except on a very small scale. These cyclic operations involve sorption and desorption steps (Perry and Green, 1999). Water-treatment applications (for example for water softening) use a typical ion-exchange cycle with four steps (Figure 6.1):

1. Backwash: removal of accumulated solids obtained by an upflow of water to expand and fluidize the exchanger system.
2. Regeneration: in order to restore the original ionic form of the ion exchanger, a reagent called regenerant is passed slowly through the exchanger.
3. Rinse: regenerant is removed by passing water through the exchanger.
4. Loading: the contaminated solution is then passed through the exchanger until leakage is detected.

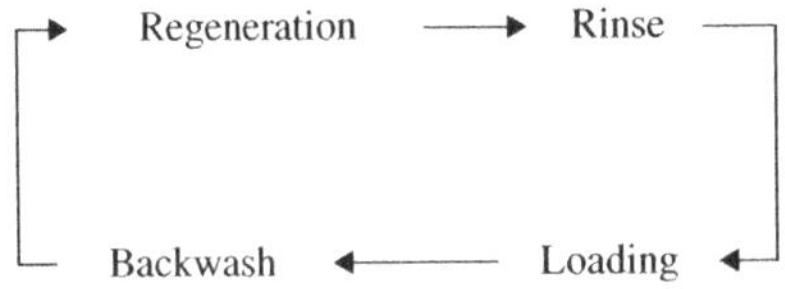

Figure 6.1 *Typical ion-exchange cycle for wastewater treatment*

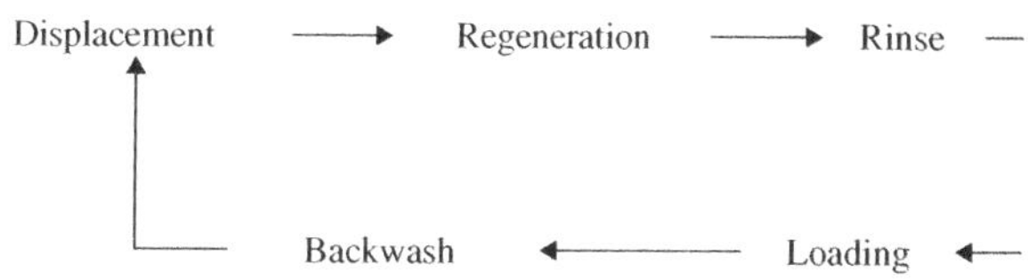

Figure 6.2 *Modified ion-exchange cycle to recover interesting compound*

Many ion-exchange columns operate in down-flow mode and are regenerated in the same direction. However, in order to obtain a better regeneration and lower leakage during loading, the regenerant should be passed counter-current to the loading flow.

In some cases, the retained solute can be interesting from any point of view. For example, the retained solute can be a useful compound in the chemical industry. The cycle is then modified to include a displacement step (Figure 6.2). Many of these compounds are amphoteric, their charge being dependent on solution pH.

6.3 Ion-exchange market and industrial needs

Due to the wide range of applications of ion-exchange resins worldwide, demand is expected to surpass \$535 million by 2015. This is mainly the result of the increasing need for pure water in a range of end use sectors, a rise in the population and the corresponding global increase in urban living, pollution, and need for resources.

Growth in the global population and the resulting urbanization, pollution, and resource shortages are the prime factors driving demand for ion-exchange resins worldwide. Demographic changes and the shift in production bases have altered market sizes across the globe. Life expectancy nearly doubled in the twentieth century as a result of improved health care. Presently, urban areas encompass about 50% of the world's population. Most of these urban areas are in developing regions. Out of 19 mega-cities in the world only three are in developed countries. New cities continue to emerge in developing countries to accommodate the growing population. These demographic trends are expected to increase the demand for food, water, power, and other resources, which in turn drive the demand for ion-exchange resins.

According to Global Industry Analysts Inc., irrespective of the geographic location, all industrialized countries are significant consumers of ion-exchange resins (Global Industry Analysts, Inc., 2010). The US, the UK, Japan, Russia, Germany, France and Italy are the major producers of ion-exchange resins. Major ion-exchange resin production facilities also exist in Canada, South Korea, China, India, Brazil, Mexico and Eastern Europe. In developing nations, tariff regulations frequently encourage the local functionalization of imported copolymers to provide ion-exchange resins (Global Industry Analysts, Inc., 2010).

Worldwide production of synthetic ion-exchange resins probably exceeds 1.5×10^5 m^3. In deionization and water-softening applications, ion exchangers can be considered commodity chemicals because of their characteristic large-scale production, limited market growth, and intense competition. However, resins used in smaller markets (e.g. chemical processing, sugar refining, pharmaceuticals, hydrometallurgy, catalysis and wastewater treatment) are classified more as specialty chemicals.

It is extremely difficult to find reliable quantitative estimates of the ion-exchange resin market. This is probably due to the factors mentioned above and the fact that market figures for ion-exchange resins are hidden in the overall commodity production data released by the few transnational corporations working in this sector. Correspondingly, the most diligent search yielded only cursory information (BV Sorbex, Inc., http://www.bvsorbex.net).

6.4 Commercial ion-exchange resins

There are different types of resins that may be classified as strong acid cation resins, weak acid cation resins, strong base anion resins and weak base anion resins (DeSilva, 1999). Commercial examples of these types of resins are shown in Table 6.1.

Ion-exchange resins are made by the combined polymerization of styrene and divinylbenzene, which leads to the generation of "gel"-type resins with limited porosity to the molecular dimensions.

Ion-exchange resins are most widely applied to decalcify water and obtain deionized water, which may be carried out under partial conditions (using strong acid cation resins in series with a weak base anion resin) or total demineralization (using strong anionic and cationic resins) (Gil-Rodríguez, 2005).

6.4.1 Strong acid cation resins

Sulfonic acid groups provide the functionality of strongly acidic cation resins. These strong acid exchangers operate at any pH, separate all salts, and require substantial amounts of regenerant. This is the resin of choice for almost all softening applications.

Strong acid cation resins contain sulfonic radicals (represented by $R\text{-}SO_3^-H^+$, sulfonated polystyrene) whose hydrogen ion can be exchanged with other cations. The structure of the resin is shown in Figure 6.3.

6.4.2 Weak acid cation resins

In the case of weak acid cation resins, carboxylic groups occupy the exchange sites. This type of resin is highly efficient, but is subject to reduced capacity from increasing flow rate, low temperatures, and a hardness-to-alkalinity ratio below 1.

Weak acid cation resins contain caboxilic radicals (represented by $R\text{-}CO_2^-H^+$) whose hydrogen ion can be exchanged with cations that are linked to weaker anions than acetic acid such as carbonates and bicarbonates. However, these weak acid resins are incapable of exchanging cations linked to chlorine, sulfate or nitrate. The acidic group of these resins shows a similar strength to acetic acid.

Table 6.1 Commercial examples

Commercial examples	Functionality	Applications
Amberlite™ FPA40 Cl	Strong base	Aminoglycoside purification
Amberlite™ FPA51	Weak base	Decolorization
Amberlite™ FPC22 H	Strong acid	Decolorization
Amberlite™ Cobalamion	Weak acid	Vitamin B12 purification
Dowex™ Marathon™ 650C (H)	Strong acid	Industrial and power applications
Dowex™ DR-G8	Strong acid	Biodiesel purification
Purolite® A400S	Strong base	Sugar applications
Purolite® A444DL	Strong base	Removal of organic matter from industrial & domestic water supplies—layered bed
Purolite® OL100	Strong acid	Removal of oil from condensate
Lewatit® S 7968	Strong acid	Biodiesel purification
Lewatit® MP 62	Weak base	Purification of chemicals / Organic matter removal from surface water
Lewatit® S 1567	Strong acid	Softening of drinking water
Lewatit® TP 207	Weak acid	Removal of heavy metals from contaminated matrixes

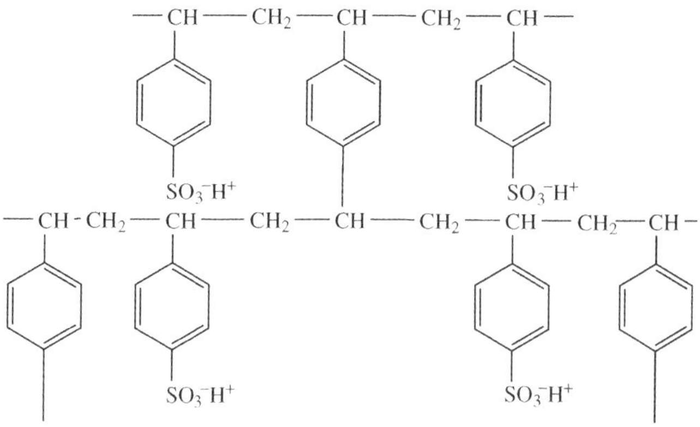

Figure 6.3 *Structure of strong acid cation resin (sulfonated polystyrene resin)*

6.4.3 Strong base anion resins

Strong base anion resins derive their functionality from quaternary ammonium exchange sites. Depending on the type of amine used during the chemical activation process, strong base anion resins can be divided into two main groups: type 1 (exchange sites have three methyl groups) and type 2 (the ethanol group replaces one of the methyl groups). Type-1 resins are more resistant to high temperatures than type 2 resins and should be used on highly alkaline and high silica waters.

Strong base anion resins contain quaternary ammonia radicals (represented by $R{-}N(CH_3)_3{}^+OH^-$) whose OH^- is exchangeable with other anions. The structure of the resin is shown in Figure 6.4.

6.4.4 Weak base anion resins

Weak base anion resins contain radicals of secondary or tertiary amines, which only exchange their OH^- with the anions of quite weak acids, such as carbonic or silicic acids.

The polyamine functional group is present in weak base anion resins. This functional group acts as an acid adsorber, removing strong acids from the cation effluent stream. This type of resin should be used on waters with high levels of sulfates and chlorides.

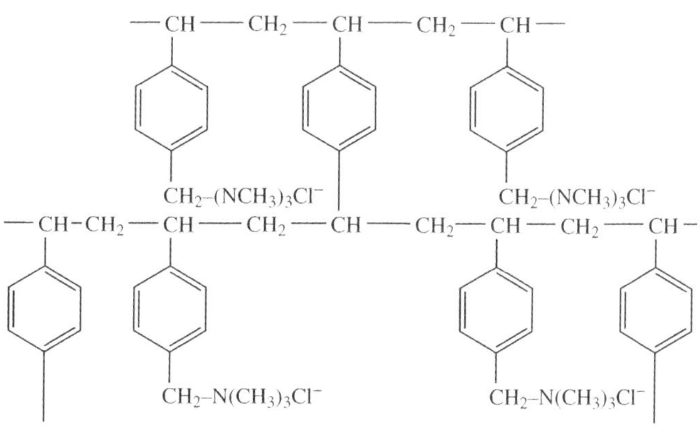

Figure 6.4 *Structure of stong base anion resin (chloro-ammonia resin)*

6.5 Specific examples in biorefineries

Ion exchange has a wide range of applications, and these will undoubtedly increase as awareness of the technology continues to grow.

Ion exchange is mostly used in the treatment of drinking water, for commercial and industrial use, and wastewater treatment. Ion exchangers can soften and deionize water, and they can even be used in desalination. The process is also used in industrial sectors where pure water is crucial to both product development and yield, as in the manufacturing of semiconductors. Recent developments and refinements in resin technologies make ion exchange one of the best and most complete forms of wastewater treatment available today. Ion exchange can also aid in the preparation of various acids, bases, salts and solutions, while the recovery of valuable metals is also possible using resins. The use of ion exchange and adsorbent technology for the processing of food streams in the nutrition market is well established, with a history stretching back decades. The food industry uses the process in a variety of ways, ranging from wine-making to sugar manufacture (Rochette, 2006).

6.5.1 Water softening

By far the most important practical application of ion exchange is in water softening. Hard water, which contains calcium and magnesium salts, is passed through a bed of a material containing exchangeable sodium ions. The calcium and magnesium ions are taken up by the exchanger and replaced by sodium. When the bed has lost so much sodium that the effluent from the bed is no longer soft, concentrated brine is passed; the adsorbed calcium and magnesium are displaced again by sodium, and the bed is ready to soften more water. The softening is discontinued as soon as the effluent contains appreciable quantities of calcium or magnesium.

The operating capacity depends on several factors besides the maximum exchange capacity of the bed, namely the rate of reaction between the exchanger and the solution and the equilibrium distribution of ions between the exchanger and the solution. Some of the industrial processes requiring softened water are:

- preparation of feed water for steam boilers to prevent scaling;
- treatment of water used for cooling containers after thermal processing to prevent unsightly "spots" left after the water drops dry out;
- preparation of softened process water for the production of beverages, for cooking legumes, and so forth.

The "hardness" of water is due to the presence of calcium and magnesium cations. Two kinds of "hardness" may be distinguished:

- Hardness due to calcium and magnesium salts in all forms. This kind of hardness affects the reaction of hard water with proteins (in legumes), certain anions, and particularly with fatty acid anions (soaps).
- Hardness due to calcium and magnesium salts in bicarbonate form. This kind of hardness causes the precipitation of insoluble carbonates (scaling) when the water is heated according to the following equation:

$$Ca\left(HCO_3\right)_2\,(\text{soluble}) \rightarrow CaCO_3\,(\text{insoluble}) + CO_2 + H_2O$$

In water softening by ion exchange, calcium and magnesium cations are exchanged with Na^+ or H^+ cations. In certain applications, the hardness cations are exchanged with Na^+ and the resin is regenerated with a concentrated solution of NaCl according to the following equation:

$$Ca^{2+}\ (soln.) + 2\ NaR \rightarrow CaR_2 + 2\ Na^+$$

In the softening stage, the medium (hard water) is a dilute solution and the resin is therefore highly selective for the bivalent calcium and magnesium ions. During the regeneration stage, the medium (concentrated brine) is a concentrated solution and the resin is therefore selective for the monovalent sodium ion. Practical water softening processes by ion exchange are based on this shift in selectivity (Berk, 2009).

6.5.2 Total removal of electrolytes from water

An alternative process is the total demineralization of water using a double anion-and-cation exchange process. When the cations of hard water can only be replaced by other metallic cations such as sodium ion, the total electrolyte content of water cannot be reduced by cation exchange. Carbonaceous exchangers, however, make it possible to replace metallic cations by hydrogen ions, and thus open the way for the complete removal of electrolytes from water.

The cation exchanger is in the H^+ form and the anion exchanger in the OH^- form. The cation exchanger adsorbs the cations in the feed water and releases H^+ ions. The anion exchanger exchanges the anions in the water with OH^- anions that are neutralized by the H^+ ions:

$$Ca^{2+} + 2HR \rightarrow CaR_2 + 2H^+$$

$$2Cl^- + 2R'OH \rightarrow 2R'Cl + 2\ OH^-$$

$$2H^+ + 2OH^- \rightarrow 2H_2O$$

The cation exchanger is regenerated with HCl and the anion exchanger with NaOH.

If natural water containing only bicarbonates is passed over an exchanger saturated with hydrogen ions, the effluent contains only carbonic acid in solution, which can be removed by aeration, leaving pure water. If the raw water also contains chlorides or sulfates, these ions will, of course, remain behind as hydrochloric or sulfuric acid (Berk, 2009).

6.5.3 Removal of nitrates in water

A nitrate removal process that drastically reduces salt consumption and waste discharge has been developed on a bench scale. Nitrate is removed by chloride ion-exchange, and the strong-base anion resin is completely regenerated under mild reaction conditions (i.e. ambient temperature, atmospheric pressure) in a closed circuit containing a single-flow fixed-bed reactor packed with a Pd-Cu/γ-Al_2O_3 catalyst. This combined treatment system avoids direct contact between the denitrification reactor and the water to be treated, while minimizing operational problems associated with each separate technique. The dissolution of the Pd and Cu metallic phases was not observed under the given operating conditions (Pintar *et al*., 2001).

6.5.4 Applications in the food industry

Some of the engineering applications of adsorption as a separation process in the food industry are:

- decolorization of edible oils with "bleaching earths" (activated clays);
- decolorization of sugar syrup with activated carbon in the manufacture of sugar;

- removal of bitter substances from fruit juices by adsorption on polyamides;
- odor abatement by passing gaseous emanations through activated carbon;
- removal of chlorine from drinking water by adsorption on carbon;
- various applications of ion exchange.

With regard to the reduction of excess acidity in fruit juices, when ion exchange is applied to citrus juices this treatment has been found to remove some of the bitterness of the product. The following equation illustrates the reduction of acidity due to citric acid, using an anion exchanger in OH^- form. The three-basic citric acid is the main source of acidity in citrus fruit juices:

$$H_3Citrate + 3ROH \rightarrow R_3Cit + 3H_2O$$

The citrate ion is relatively large. The resin used for this application is, therefore, a macro-reticular polymer, providing the internal porosity required for the accommodation of large counter-ions. Other carboxylic acids (malic, fumaric and lactic acids) are adsorbed in the same way. On the other hand, the purification of sugar juices was one of the first applications of synthetic cation exchangers. The juices were passed over a calcium exchanger, replacing the potassium and sodium ions for calcium. The resulting solution crystallized more readily and completely than the untreated juice to give a sugar of lower ash content. Cation exchangers can be used to remove lead from maple syrup, or heavy metals from sugar juices (Berk, 2009).

6.5.5 Applications in chromatography

Ion exchangers and adsorbents can also be employed for separating valuable substances which are present in solutions in comparable concentrations. Based on the knowledge gained from resin applications in chromatography, ion exchange is now being used at an industrial scale for the recovery of valuable substances.

Macromolecules with ionic groups may be separated by ion exchange chromatography. The macromolecule is absorbed by the carrier and is eluted with a solution of a defined ionic strength. Depending on the nature of the ionic strength of the eluent solution, the separation may be quite specific. Ion exchange chromatography has a wide range of uses for the purification of antibiotics from fermentation medium. Moreover, it is frequently used for the purification of proteins at full scale. The most frequently used materials for cationic exchange are Dowex HCR and OCR, Amberlite IR and IRC and Lewatit S, while those used for anionic exchanges are Dowex SAR and MSA, Amberlite IRA and Lewatit M.

Ion-exchange chromatography can be subdivided into three types as follows:

1. *Separation of non-ionic products*. Such applications utilize not only the general adsorption by resins of the substances being separated but also their differing diffusion rates and solubilities in the water of the resin matrix. Among other processes, the separation of polyalcohols such as sorbitol and mannitol by applying low-cross linked cation exchangers have been described (Martinola, 1986). Wide use is made at industrial scale of the separation of fructose from its mixture with glucose generated during the hydrolysis of saccharose. Fructose possesses a dietetic value and can be obtained in this way in a very pure state. Resin beds of up to 100 000 L are in operation.
2. *Separation of ionic and non-ionic compounds*. The ion exclusion process is the most important application. The first tests were carried out on mixtures of glycol or glycerine and common salt but it was soon discovered that sugar solutions could also be purified in this way. In the meantime, several industrial units have been built in which the sugar contained in the molasses is separated from other

substances. The process operates in plants containing up to 60 000 L ion-exchange resin. Water is the only desorption agent required in the process.

3. *Separation of ionic compounds*. Several hundred cubic meters of ion-exchange resin are employed in large-scale plants for the separation of mixtures of rare earths. Buffer solutions are used for developing the chromatogram.

Essential amino acids for baby foods and special diets can be isolated from effluents from the sugar industry or from animal wastes. The mixtures are passed through ion-exchange resins of various strengths and separated in the column with ammonia. Significantly large quantities of resin are used in these applications (Martinola, 1986).

6.5.6 Special applications in water treatment

It is sometimes desirable to remove small quantities of specific impurities from water even though complete electrolyte removal or softening may not be necessary. One example of this is the removal of traces of fluoride by means of anion exchange with basic tricalcium phosphate. A similar example is the use of cation exchange to remove small amounts of heavy metals from drinking water. Most heavy metals such as copper and lead are absorbed strongly by an exchanger, even when the latter is saturated with calcium from hard water. Traces of iron and manganese can be removed in the same way, but are removed more efficiently by oxidation with an activated oxide of manganese supported on a cation exchanger as a carrier.

6.5.7 Metal recovery

Ion exchangers may similarly be used for the recovery or concentration of valuable substances present as ions in solution in small amounts. Copper, for example, can be recovered from rayon-spinning waste liquors that contain copper ammonia complex ions, while metals can be recovered in a similar manner from electroplating wastes. In such applications, carbonaceous exchangers have clear advantages over siliceous exchangers as they can be regenerated with acid and used in low pH solutions. Organic bases such as alkaloids and even amino-acids can be recovered from solutions in the same way.

Ion exchangers were first used in metallurgy to extract uranium as it is not easy to concentrate this mineral by precipitation methods. Poor ores must be treated by lixiviation, acidic or alkaline lixiviation, although the first is the most common. Under oxidative conditions, Uranium is dissolved in diluted sulfuric acid leading to the generation of uranyl ions (UO_2^{2+}), which become more or less complex through the following equilibrium:

$$UO_2^{2+} + nSO_4^{2-} \leftrightarrows UO_2\left(SO_4\right)_n^{2-2n}$$

The complex shows a negative charge for $n \geq 2$. If this anion comes in contact with an ion exchanger, the following interchange reaction will take place:

$$4R^+X^- + UO_2\left(SO_4\right)_3^{4-} \leftrightarrows \left(R^+\right)_4 UO_2\left(SO_4\right)_3^{4-} + 4X^-$$

where R is the resin cation and X^- the anion (Cl^- or NO_3^-). As uranium is one of the few metals that may form anions in sulfuric acid solutions, the interchange is quite selective, leading to the retention of the other cationic metals (Ca, Fe, etc.) in the aqueous phase. It is important to point out that the resin capacity may reach a value in the range of 3–4 miliequivalents per gram of dry resin, which corresponds to 50–100 g U_3O_8 per liter of granular resin, reaching a maximum at pH values between 1.5 and 2.0 (Gutierrez-Miravete, 1987).

6.5.8 Separation of isotopes or ions

An unusual application of cation exchangers is the separation of isotopes or to separate anions from cations, which may then be determined separately. Moreover, carbonaceous exchangers containing exchangeable hydrogen ions can be used to promote acid-catalyzed reactions such as the hydrolysis of starch to glucose. This process has an advantage over sulfuric acid (which is normally used to catalyze such reactions) in that the resulting glucose solution can be concentrated and crystallized more easily. Ion exchange may be used as a method for preparing salts and has actually been employed for making sodium nitrate from calcium nitrate and sea water. The exchanger is saturated with sodium by passing sea water, and then calcium nitrate solution is passed (Walton, 1941).

6.5.9 Applications of zeolites in ion-exchange processes

There are several applications in which the ion-exchange properties of zeolites are exploited directly. These include water softening (or 'building') in detergents, wastewater treatment (including municipal, industrial, agricultural and radioactive wastes) and animal food supplementation (e.g., to regulate ammonia or ammonium levels in the stomach). Some of these direct applications of zeolite based ion exchange will be discussed below.

1. *Detergent building*. Until concerns arose in the 1970s about its environmental impact, sodium tripolyphosphate was water softener of choice for detergent formulations. Sodium tripolyphosphate forms strong complexes with Ca^{2+} and Mg^{2+} ions in solution, preventing their precipitation with the surfactant or as carbonates (reducing deposition of solids onto clothing and avoiding loss of detergent surfactant). Zeolites possessing selectivity for calcium and/or magnesium over sodium are obvious candidates to replace phosphates; indeed, legislative and environmental pressures favor the use of zeolites over phosphates in detergents around the world.

   While most zeolite types, both naturally occurring and synthetic ones, have been tested for their water softening capabilities, zeolite A has been chiefly used (high calcium selectivity and kinetics); sometimes in conjunction with zeolite X (which removes magnesium more effectively than zeolite A). It is estimated that approximately 800 000 tonnes of zeolite A are currently used in laundry detergents per annum. Recently, a P-type zeolite possessing a Si:Al ratio of 1:1 (Maximum Aluminum P, or MAP) has been developed and introduced into detergent formulations.
2. *Radionuclide separation*. Zeolites, as well as other materials with ion exchange properties, are commonly used to remove certain radio-nuclei from low- and medium-level nuclear waste. Natural zeolites are of great interest due to their (sometimes) high abundance and low cost. The principal radioactive components of nuclear waste are typically $^{90}Sr^{2+}$ and $^{137}Cs^{+}$. The removal of these nuclides from effluents which may contain significant quantities of competing ions such as Ca^{2+}, Mg^{2+}, Na^{+} and a range of anions may be carried out by ion exchange.
3. *Wastewater treatment*. Zeolites are commonly used to remove ammonia and ammonium ions from municipal and agricultural wastewater. Where deposits of natural zeolites are abundant, especially clinoptilolite, chabazite, mordenite and phillipsite, they have been used extensively for this purpose.
4. *Other applications*. The ion-exchange technique is frequently used to alter the properties of zeolites; for example to prepare acidic or basic catalysts, to tailor the pore size for specific adsorption processes, and to introduce specific adsorption sites. With regard to the preparation of catalysts, the most common application of ion exchange is the preparation of ammonium forms of zeolites *en route* to the generation of acidic catalysts (Townsend and Coker, 2001).

   Ion exchange is mainly used in wastewater treatment but it is also employed to reduce gas emissions, particularly the removal of hydrogen sulphide and ammonia by utilizing carboxylic acid resins and ammonium anion-exchange resins. Noble and Terry (2004) provided a summary of some characteristic environmental applications of ion exchange:

- treatment of mine drainage water: removal of metal cations and anions using silico-titanates and layered titanates;
- removal of nitrates and ammonia from groundwater;
- treatment of nuclear waste solutions;
- the plating industry.

6.5.10 Applications of ion exchange in catalytic processes

Blagova *et al*. (2006) studied the use of sulfonated resins in the esterification of *n*-butanol. They analyzed the kinetics of side reactions of the formation of *n*-butyl acetate in the heterogeneously catalyzed esterification of *n*-butanol with acetic acid in an isothermal fixed-bed flow reactor at temperatures between 100 and 120 °C. The observed side reaction products were isomers of butene, di-*n*-butyl ether, *sec*.-butyl-*n*-butyl ether as well as *sec*.-butanol and *sec*.-butyl acetate. Three ion-exchange resin catalysts with a similar matrix but different sulfonation were compared: Purolite™ CT 269, which is mono-sulphonated; Amberlyst ™ 46, which is surface-sulphonated; and Amberlyst ™ 48, which is bi-sulphonated. Purolite™ CT 269 and Amberlyst™ 48 are fully sulfonated in the gel phase, whereas Amberlyst™ 46 is only surface-sulphonated. The ion-exchange capacities of Purolite™ CT 269 and Amberlyst™ 48 were found to be similar, while the capacity of Amberlyst™ 46 was observed to be lower by a factor of 5. Despite this, all three catalysts showed only minor differences in terms of their esterification activity. Regarding the formation of side products, Purolite™ CT 269 and Amberlyst™ 48 showed similar results: side reactions proceed to a significant extent. For Amberlyst™ 46, however, side reactions were found to be almost negligible. The authors concluded that esterification occurs mainly on or near the external surface of catalyst particles, whereas side reactions occur mainly in the pores. This work shows that surface-sulphonated catalysts like Amberlyst™ 46 are very attractive for the production of esters by reactive distillation.

Umar *et al*. (2009) applied ion-exchange resins to ethyl *tert*-butyl ether (ETBE) synthesis. They studied ethyl *tert*-butyl ether (ETBE) synthesis from ethanol (EtOH) and *tert*-butyl alcohol (TBA) using different macroporous and gelular ion-exchange resin catalysts such as Purolite® (CT-124, CT-145 H, CT-151, CT-175 and CT-275) and Amberlyst™ (15 and 35) ion-exchange resins. The effect of various parameters such as catalyst type, temperature, reactants feed molar ratio, and catalyst loading were studied for the optimization of the reaction condition. Among the catalysts studied, Purolite CT-124 gave the best results for TBA conversion and selectivity towards ETBE. Kinetic modeling was performed with the catalyst and activation energy and water inhibition coefficient determined. Heterogeneous kinetic models, such as Eley–Rideal (ER) and Langmuir–Hinshelwood–Hougen–Watson (LHHW), were unable to predict the behavior of this etherification reaction, whereas the quasi-homogeneous (QH) model represented the system very well over a wide range of reaction conditions.

Petrus *et al*. (1984) studied the kinetics and equilibrium of the direct hydration of propene over a strong acid ion-exchange resin (C8P ex AKZO). The experiments were carried out with only one water-rich liquid phase present in the reactor apart from the solid resin catalyst. The authors found that besides 2-propanol, a certain amount of diisopropylether was formed, although other side products were not observed. The kinetics of the 2-propanol and diisopropylether formation reactions can be well represented by a scheme in which the isopropylcarbenium ion is the intermediate. The numerical values of the rate constants and equilibrium constants involved were determined. The presence of 2-propanol was found to lead to a larger reduction in reaction rate than expected by the approach of chemical equilibrium; a phenomenon that is ascribed to the solvent effect of 2-propanol. The authors found that intra-particle diffusion limitations only occur at temperatures above 130 °C when ion-exchange resin particles of normal commercial size (approximately 0.8 mm diameter) are used.

Goud *et al*. (2007) studied the catalysis of epoxidation reaction with ion-exchange resins. They analyzed the kinetics of epoxidation of jatropha oil by peroxyacetic/peroxyformic acid formed *in situ* by the reaction of aqueous hydrogen peroxide and acetic/formic acid in the presence of an acidic ion-exchange resin as catalyst in or without toluene. The presence of an inert solvent in the reaction mixture appeared to stabilize the epoxidation product and minimize side reactions such as the opening of the oxirane ring. The effect of several reaction parameters, such as stirring speed, hydrogen peroxide-to-ethylenic unsaturation molar ratio, acetic/formic acid-to-ethylenic unsaturation molar ratio, temperature, and catalyst loading on the epoxidation rate as well as on the oxirane ring stability and iodine value of the epoxidized jatropha oil were examined. The multiphase process consisted of a consecutive reaction, acidic ion-exchange resin catalyzed peroxyacid formation followed by epoxidation. The catalytic reaction of peroxyacetic/peroxyformic acid formation was found to be characterized by adsorption of only acetic (or formic) acid and peroxyacetic/peroxyformic acid on the active catalyst sites, and the irreversible surface reaction was the overall rate determining step. The proposed kinetic model considered two side reactions, namely, epoxy ring opening involving the formation of hydroxy acetate and hydroxyl groups and the reaction of the peroxyacid and epoxy group. The kinetic and adsorption constants of the rate equations were estimated by the best fit using a nonlinear regression method. A good fit between the experimental and predicted data validated the proposed kinetic model. From the estimated kinetic constants, the apparent activation energy for epoxidation reaction was found to be 53.6 kJ/mol. This value compares well with those reported by other investigators for the same reaction over similar catalysts.

6.5.11 Recent applications of ion exchange in lignocellulosic bioefineries

According to Huang *et al*. (2008), the ion-exchange resin (IER) method is currently the preferred choice for the detoxification of fermentation hydrolyzates in the conversion of cellulosic biomass to fuel ethanol and will continue to be used in biorefinery in the future due to its high detoxification efficiency, easy (continuous) operation, and flexible combination of anion and cation exchangers, while the enzymatic treatment will grow in the future.

Another application of ion exchange in lignocellulosic biorefineries is the purification of succinic acid where the ion exchangers make the acidification with a simultaneous crystallization process (Luo *et al*., 2010).

6.5.12 Recent applications of ion exchange in biodiesel bioefineries

Ion exchange is mainly used in biodiesel production during the purification process. Ion exchange has permitted the reduction of the water supply and subsequent wastewater treatment when biodiesel purification is carried out by water washing.

These adsorbents consist of acidic and basic adsorption (binding) sites and have strong affinity for polar compounds such as methanol, glycerin, glycerides, metals and soap. This technique is followed with the use of a filter to enable the process to be more effective and efficient as shown in Figure 6.5. According to Van Gerpen (2008), dry washing is usually carried out at a temperature of 65 °C and the process is mostly completed within 20–30 min. This permits the amount of glycerides and total glycerol in crude biodiesel to be lowered to a reasonable level during the washing process. Advantages in the use of ion exchange in dry washing are the high affinity for polar compounds present in biodiesel, it is a waterless process that is quicker than water-washing processes, it is easy to incorporate in industrial plants, it is easy to install, no wastewater is generated and the solid waste can be used (Atadashi *et al*., 2011).

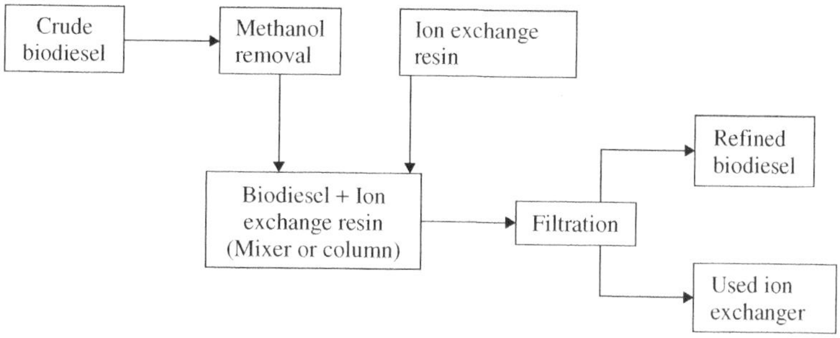

Figure 6.5 *Biodiesel purification with ion-exchange resin*

The application of ion-exchange resins as a dry washing agent is being widely used by resin manufacturers, particularly Purolite (PD206) and Rohm & Haas (BD10 Dry). Purolite (PD206) is a dry-polishing medium specifically formulated to remove by-products remaining after the production of biodiesel. Although they are sold as ion-exchange materials, none of the suppliers advocates its regeneration as they act as adsorbents (Atadashi *et al*., 2011).

Berrios and Skelton (2008) studied the effects of ion-exchange resins on the purification of crude biodiesel from used cooking oils and rapeseed oil. In their experimental setup, feed was passed through a column of resin in a glass tube. A metered pump was used to control the flow, and restricted outlets were employed to ensure a liquid head above the resins. They noted that the initial loading and flows of the resins were in accordance with the recommendations of Rohm & Haas company trade literature. The authors analyzed the samples at two-hour intervals for methanol and glycerol, demonstrating that ion-exchange resin has the capability to reduce glycerol to a value of 0.01 %wt and considerably remove soap, but can not successfully remove methanol. Figure 6.6 shows the evolution of glycerol content versus L biodiesel/kg resin in their work.

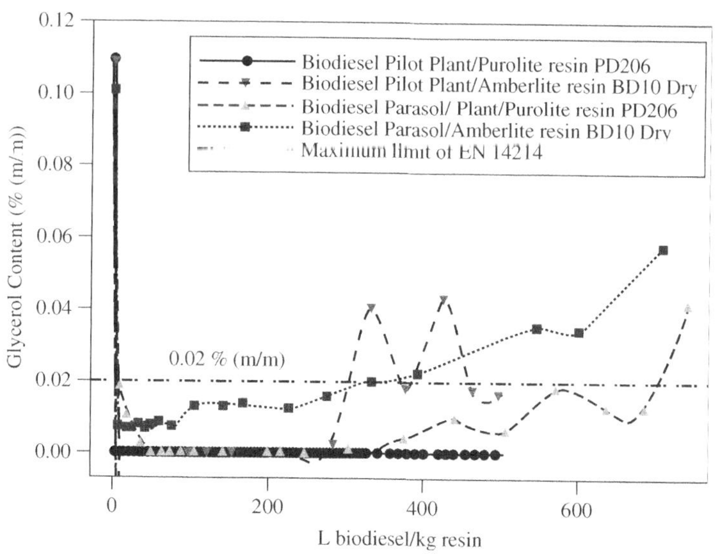

Figure 6.6 *Evolution of glycerol content versus biodiesel loading (L biodiesel/kg resin). Reprinted from Berrios, M* et al., *© 2008, with permission from Elsevier*

Berrios *et al*. (2011) used ion-exchange resin (Lewatit® GF202) in the purification of used cooking oil biodiesel. The tested resin showed no effect on density, kinematic viscosity or glyceride, FAME, FFA and water content. Soap, methanol and glycerol removal were 52.2%, 98.8% and 20.2%, respectively. An advantage to this resin is that it can be regenerated and reused, whereas other resins can be used only once. The authors proposed regeneration as an interesting step in spite of the high volume of methanol required for the process. They explained that the methanol used can be reused in the transesterification process.

6.6 Conclusions and future trends

This chapter reviewed some interesting aspects of the use of ion-exchange methodologies. Ion-exchange resins are useful in industry for research aimed at improving the quality of several biorefinery processes. The chapter described different types of resins and gave commercial examples. It also provided an extensive review of applications ranging from softening water to biodiesel purification. In our opinion, the development of ion-exchange methodologies should be directed toward a wider range of ion exchangers, higher efficiency, and the use of renewable materials or cost reduction.

In recent years, ion-exchange resins have been used in membrane processes. Thus new approaches to the use of ion-exchange resins should focus on the production of more efficient resins and a wider range of applications in order to be more competitive with membrane systems.

References

Atadashi, I. M., Aroua, M. K., Abdul Aziz, A. R., Sulaiman, N. M. N., Refining technologies for the purification of crude biodiesel. *Applied Energy*, doi:10.1016/j.apenergy.2011.05.029, 2011.

Berk, Z., *Food Process Engineering and Technology*. Elsevier Inc., 292–293, 2009.

Berrios, M., Martín, M. A., Chica, A. F., Martín, A., Purification of biodiesel from used cooking oils, *Applied Energy*, 88, 3625–3631, 2011.

Berrios, M., Skelton, R. L., Comparison of purification methods for biodiesel. *Chemical Engineering Journal*, 144, 459–465, 2008.

Blagov, S., Parada, S., Bailer, O., Moritz, P., Lam, D., Weinand, R., Hasse, H., Influence of ion-exchange resin catalysts on side reactions of the esterification of n-Butanol with acetic acid. *Chemical Engineering Science*, 61, 753–765, 2006.

DeSilva, F. J., *Essentials of Ion Exchange*. 25th Annual WQA Conference March 17, 1999.

Elvers, B., Hawkins, S., Ravenscroft, M., Schulz, G., *Ullmann's Encyclopedia of Industrial Chemistry*, 5th edn., Volume A14, VCH Publishers, Weinheim,1989.

Gil-Rodríguez, M., *Procesos de descontaminación de aguas. Cálculos avanzados informatizados*, International Thompson Editores, Spain, Paraninfo S.A. 2005.

Global Industry Analysts, Inc., *Ion-Exchange Resins—A Global Strategic Business Report*, Global Industry Analysts, Inc., 1 February, 2010. (Obtainable from www.companiesandmarkets.com.)

Goud, V. V., Patwardhan, A. V., Dinda, S., Pradhan, N. C., Kinetics of epoxidation of jatropha oil with peroxyacetic and peroxyformic acid catalysed by acidic ion-exchange resin. *Chemical Engineering Science*, 62, 4065–4076, 2007.

Gutierrez-Miravete, E., *Fundamentos de metalurgia extractive*, Editorial Limusa S.A. México, D.F., pp. 471–472, 1987.

Helfferich, F., *Ion Exchange*, Dover Publications, 1995.

Huang, H.-J., Ramaswamy, S., Tschirner, U. W., Ramarao, B. W., A review of separation technologies in current and future biorefineries. *Separation and Purification Technology*, 62, 1–21, 2008.

Lucy, C. A., The evolution of ion exchange: from Moses to the Manhattan Project to modern times. *Journal of Chromatography A*, 1000, 711, 2003.

Luo, L. Van, der Voet, E., Huppes, G., Biorefining of lignocellulosic feedstock—technical, economic and environmental considerations. *Bioresource Technology*, 101, 5023–5032, 2010.

Martinola, F., Ion exchangers and adsorbents—versatile aids for the materials processing industry. *Materials and Design*, 7, 1, 1986.

McCabe, J., Smith, J. C. and Harriot, P., *Operaciones Unitarias en Ingeniería Química*, McGraw-Hill Editorial, Mexico, 2002.

Noble, R. D., Terry, P. A., *Principles of Chemical Separations with Environmental Applications*, Cambridge University Press, 2004.

Perry, R. H. and Green, D. W., Perry's *Chemical Engineers' Handbook*, 7th edn., McGraw-Hill, 1999.

Petrus, L., De Roo, R. W., Stamhuis, E. J., Joosten, G. E. H., Kinetics and equilibria of the hydration of propene over a strong acid ion exchange resin as catalyst. *Chemical Engineering Science*, 39, 433–446, 1984.

Pintar, A., Batista, J., Levec, J., Integrated ion exchange/catalytic process for efficient removal of nitrates from drinking water. *Chemical Engineering Science*, 56, 1551–1559, 2001.

Rochette, F., New technology: Fresh approach to countercurrent ion exchange. *Filtration and Separation*, 43, 7, 18–19, 2006.

Savidan, L., *Resinas Intercambiadoras de Iones*, Editorial Alhambra S.A., Madrid, 1967.

Skoulikides, T. N., *Physical Chemistry I*, Symetria Editions, 1989.

Slater, M. J., Continuous ion exchange plant design methods and problems. *Hydrometallurgy*, 4, 299–316, 1979.

Townsend, R. P. and Coker, E. N., *Ion exchange in zeolites* In. *Studies in Surface Science and Catalysis*, H. Van Bekkum, E. M. Flanigen, P. A. Jacobs and J. C. Jansen (editors), Elsevier Science B.V., 2001, pp. 137, 467–524.

Treybal, R. P., *Mass Transfer Operations*, 3rd edn., McGraw-Hill, 1980.

Umar, M., Patel, D., Saha, B., Kinetic studies of liquid phase ethyl tert-butyl ether (ETBE) synthesis using macroporous and gelular ion exchange resin catalysts. *Chemical Engineering Science*, 64, 4424–4432, 2009.

Van Gerpen, J. H., Oilssed and Biodiesel Workshop Billings, Montana, January 9, 2008.

Walton, H. F., Ion exchange between solids and solutions. *Journal of the Franklin Institute Devoted to Science and the Mechanic Arts*, 232, 305–337, 1941.

7

Simulated Moving-Bed Technology for Biorefinery Applications

Chim Yong Chin[1] and Nien-Hwa Linda Wang[2]

[1]*PureVision Technology, Inc., Ft. Lupton, USA*
[2]*School of Chemical Engineering, Purdue University, USA*

7.1 Introduction

Simulated moving-bed (SMB) technology is a continuous separation technique that improves upon traditional batch chromatography. It continuously separates liquid products or purifies feed streams using very much the same chromatography mechanisms familiar to most practicing chromatographers, including adsorption, ion exchange, size exclusion, hydrogen bonds, complexation, or a combination of these mechanisms. The first recorded SMB chromatographic technique was developed in the petrochemical industry in the late 1940s [1] and has been widely used industrially in petrochemical refineries since the 1970s, in high-fructose corn-syrup processing since the 1980s, and enantiomer separations in the pharmaceutical industries since the 1990s. This chapter provides an introduction with an overview of the principles of SMB, its essential design tools, and several examples of potential SMB applications in biorefineries.

7.1.1 Principles of separations in batch chromatography and SMB

In conventional batch preparative chromatography, a feed mixture is first loaded into a column (or a series of columns) filled with a sorbent or stationary phase (such as activated carbon), and the feed pulse is then eluted with a desorbent (the mobile phase), isocratically (no change in the desorbent composition) or otherwise. A solute in the feed, which has a high affinity for the sorbent, has a high partition coefficient. This means that a larger fraction of the solute exists in the sorbent phase than in the mobile phase. Since only solutes in the mobile phase migrate downstream in the column, the average migration velocity of a solute is proportional to its fraction in the mobile phase. For this reason, a higher affinity solute migrates more slowly than a low-affinity solute, resulting in separation of the various feed components in the column.

Separation and Purification Technologies in Biorefineries, First Edition.
Edited by Shri Ramaswamy, Hua-Jiang Huang, and Bandaru V. Ramarao.

The exit stream from the column is either collected as waste or diverted as product. After product collection is completed, the column is regenerated, reequilibrated, and reused. Batch chromatography is most useful as a polishing step, where either the solute of interest (product) or a small amount of impurity is retained in a high-capacity, highly selective stationary phase. However, in most industrial separations, some impurity components in the feed have a slightly higher affinity and some have a slightly lower affinity than the product. Consequently, a high degree of product separation from both groups of impurities is needed to achieve high purity. In these instances, batch chromatography suffers from low yield, high solvent consumption, high product dilution, and poor column utilization. It is generally difficult to achieve high product purity and high yield simultaneously in batch chromatography.

Simulated moving-bed chromatography overcomes many of the drawbacks of conventional batch chromatography. It employs a series of linked columns, often in an unbroken loop, with inlet and outlet ports between the columns, as shown in Figure 7.1 for a standard system with eight columns. The loop of columns is divided by four inlet or outlet ports into four zones, and each zone has a different mobile phase velocity or flow rate. The ports move periodically and synchronously along the loop to follow the migrating

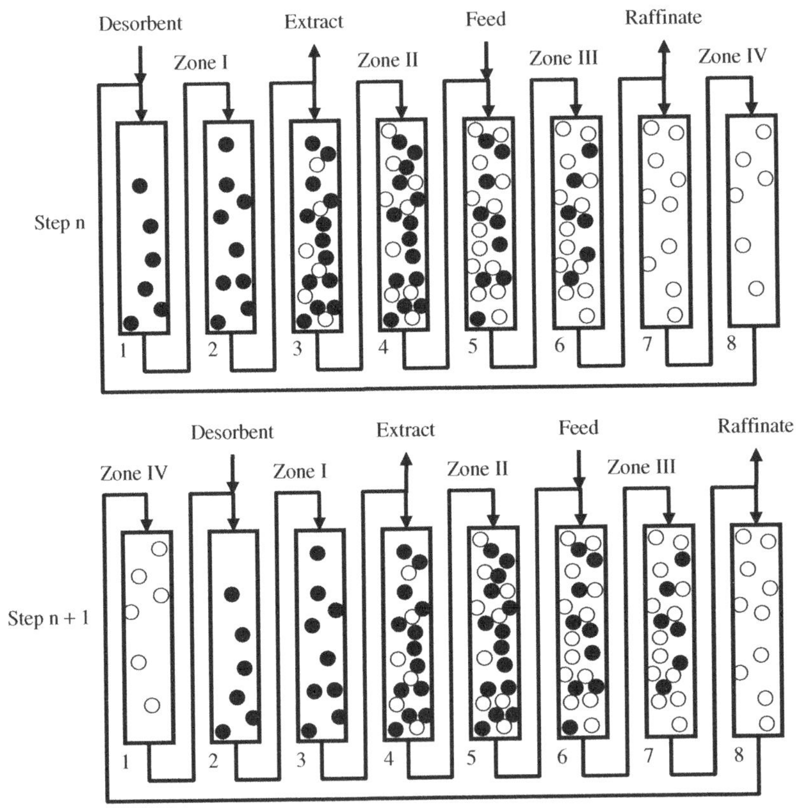

Figure 7.1 *Principle of binary separation in a conventional four-zone SMB with two columns in each zone. The black-filled circles represent the high-affinity (slow-moving) solutes, and the white circles represent the low-affinity (fast-moving) solutes. In step n + 1, the desorbent, extract, feed, and raffinate ports all move one column downstream from step n*

solute bands. In step n, for example, desorbent and feed, which has both the high-affinity (slow-moving) and the low-affinity (fast-moving) solutes, are continuously introduced into the loop through the desorbent port (inlet of Column 1) and the feed port (inlet of Column 5), respectively; some fast-migrating solute is continuously removed from the raffinate port (end of Column 6), which is located downstream from the feed port; and some slow-migrating solute is continuously removed from the extract port (end of Column 2), which is located upstream from the feed port. In step n + 1, the desorbent and feed ports move one column downstream to the inlets of columns 2 and 6, respectively. Similarly, the extract and raffinate ports move one column downstream to the end of columns 3 and 7, respectively.

The average port velocity and the zone flow rates are designed so that feed is added continuously into the mixed region in the loop, while the two products are removed continuously from the pure component regions on both ends of the mixed region, thus achieving high purity and high yield for both products. In essence the ports, on average, travel faster than the adsorption front of the slow-moving solute (black-filled circles) in zone III, which is the region between the feed and raffinate ports. As a result, the slow solute never reaches the raffinate port, allowing the raffinate to recover only the fast-moving solute. Similarly, the ports on average move slower than the desorption tail of the fast-moving solute in zone II (between the extract and feed ports). The extract port never reaches the tail of the fast-moving solute (white circles). The extract thus collects only the slow-moving solute (black circles). Zone IV (between the raffinate and desorbent ports) allows for solute-free desorbent to travel back into zone I. The slow-moving solute is eluted from the first column of zone I before the column is moved to zone IV in the next step. The periodic port movement achieves a simulated countercurrent movement of the stationary phase (or bed) relative to the fluid phase and maintains high purity and high yield for both products. If a great many very small columns are connected, the ports will appear to move continuously in the loop. This limiting example is called a continuous moving-bed (CMB) or a true moving-bed (TMB).

7.1.2 The advantages of SMB

Simulated moving-bed technology provides significant advantages over conventional batch chromatography because (i) it requires only partial band separation; (ii) the two products are continuously withdrawn from the pure component ends of the moving solute band; and (iii) the solvent is automatically recycled. Simulated moving-bed technology can thus achieve both high product purity and high yield while improving stationary phase productivity and desorbent utilization. These benefits result in significant cost savings over batch chromatography [2–6]. Furthermore, the lowered desorbent consumption reduces the environmental impact of the process. A recent comparison of different chromatographic techniques for separation of a racemic mixture found that the SMB technique has the lowest solvent consumption and the best column utilization at production scale [7]. Simulated moving-bed technology also reduces floor space, equipment size, and manpower, reportedly as much as four times for an industrial chiral separation process [8].

7.1.3 A brief history of SMB and its applications

Simulated moving-bed technology, the use of fixed beds and moving ports to simulate countercurrent movement of solids and liquids, was first used in the 1840s in England as the Shank's system for leaching [9]. Eagle and Scott of California Research Corp. (Richmond, California, U.S.) in 1949 reported SMB as the Cyclic Adsorption process used for the commercial recovery of aromatics and olefins from petroleum [1, 10, 11]. UOP LLC (Des Plaines, Illinois, U.S.), now part of Honeywell (Morristown, New Jersey, U.S.), developed a family of SMB processes for the petrochemical industry. UOP's Sorbex family of SMB processes includes Molex for the recovery of linear paraffins, Parex for the recovery of para-xylene, Ebex for the recovery of ethyl benzene from mixed C8 aromatic isomers, and Olex for the recovery of olefins from paraffins. UOP has received more than 300 U.S. patents for SMB-related operations, equipment and applications in petrochemical derivatives, carbohydrates, fatty acids, biochemicals, and pharmaceuticals [2, 3, 12–14].

Most, if not all, of the industrially relevant SMB processes were in the petrochemical industry until the late 1970s. The next industrial adoption occurred in the sugar industry, predominantly in the corn-derived sugar industry, in the 1980s. The Sarex process was developed by UOP for sugar purification [2, 14]. Many other similar SMB sugar-purification processes were patented and adopted industrially: see U.S. Patent No. 4 157 267 assigned to Toray Industries, Inc. (Tokyo, Japan) for SMB recovery of fructose on zeolite from a fructose-glucose feed mixture [15]; No. 4 182 633 assigned to Mitsubishi Chemical Industries Ltd. (Tokyo, Japan) for operational improvements on fructose-glucose SMB separation on acidic cation exchange resin [16]; and No. 4 412 866 assigned to the Amalgamated Sugar Co. (Ogden, Utah, U.S.) for backwashing the bed in a fructose-glucose SMB separation [17]; among many others. Examples of SMB sugar separations in the literature have also been reported [2, 3, 18–22]. We will later illustrate in detail the development of an SMB process for the purification of sugars from biomass hydrolysate.

The 1990s saw the development of selective but high-priced chiral stationary phases for enantiomer separation. Simulated moving-bed technology, with its significant improvement in stationary phase utilization, became the technique of choice for enantiomeric separation. Miller and colleagues repeatedly showed that SMB is superior to other process-scale enantiomer separations [5, 7, 8]. Lists of published SMB enantiomer separations and further discussion on the adoption of SMB in the pharmaceutical industry can be found in the literature [6, 23, 24].

The literature provides a long list of potential SMB applications, including waste removal, purifications of fine chemicals, organic acids [25–28], and pharmaceutics [4], which include enzymes [29], monoclonal antibodies [30], paclitaxel (a chemotherapeutic drug) [31–33], ascomycin derivative (an anti-inflammatory drug) [34], clarithromycin (an antibiotic) [35], cyclosporin (an immunosuppressive drug) [36], escitalopram (an antidepressant) [37], biosynthetic human insulin [38–45], and many others.

All of the above commercial applications were binary separations (i.e., a feed stream is split into two product streams), although some of the feeds contain more than two components, such as those in the UOP's petrochemical applications. This is also true with the lab-scale SMB processes reported in the literature, except for a few examples where ternary or higher order separations were required for feed streams with three or more components. For example, in the biosynthetic human insulin studies conducted by the chapter authors [38–45], the insulin product was the middle component with a fast-moving impurity (high-molecular-weight proteins) and a slow-moving impurity (zinc chloride). The authors designed a tandem SMB process using the knowledge-driven design process, explained later, that increases yield by 10%, reduces solvent consumption by two-thirds, and improves bed throughput threefold over the existing commercial batch process.

As an active research field, SMB has yet to generate a single comprehensive book written for practicing engineers. Much of the knowledge is buried in chapters of highly technical books of chromatography, spread out in technical journals, and the often intentionally obtuse published patents and patent applications, as obvious from the list of references herein. It has, however, garnered sufficient attention to be published in recent editions of the oft-cited *Perry's Chemical Engineering Handbook*, which now includes a brief account of SMB. Recent technical reviews can be found in Chin and Wang (2004), who discussed a variety of SMB process configurations and associated equipment [46]; Seidel-Morgenstern *et al*. (2008), who provided a review of new general developments in SMB [47]; and Rajendran *et al*. (2009) who provided a review of the SMB design method (called the triangle design) and details on new variations of SMB operations and chiral SMB separations [24].

7.1.4 Barriers to SMB applications

An SMB process, however, requires substantially more complex design than batch chromatography. For the standard four-zone SMB illustrated in Figure 7.1, for example, given the desorbent and stationary phase, four zone lengths, four zone flow rates, and the average port movement velocity (or switching time), a total of nine design parameters, must be specified. Trial and error in the nine-parameter space is challenging and time consuming. Even though the technique has been applied industrially for more than half a century, it is still considered a research-level process design and consequently has a much smaller trained expert base than traditional chromatography.

The equipment and operation of SMB are significantly more complex than those of batch chromatography. Although considerable work has been reported in the past few decades, process optimization, operation error identification and correction, and process robustness are still being actively researched.

The continuous recycle operation in SMB, for example, results in much longer residence times for the solutes [39]. By contrast, the solute residence times in batch chromatography are easily found, and they are always shorter than the batch-cycle times. The long solute residence times in SMB can be a critical issue for biologically sensitive products.

Furthermore, the feed port in SMB is located between the two product ports. For this reason, the residence times of fast-moving solutes are different from those of slow-moving solutes [39]. The residence time of a given component also depends on when it enters the SMB during a feed-injection step. A product collected at a given time consists of molecules from different feed injections. All of these characteristics of SMB result in complex residence time distribution for each component. An innovative feed strategy is needed to control the integrity of a feed batch [44], which is critical in pharmaceutical and other regulated industries.

More important, almost all feeds from biological sources have three or more major components. Very few examples of successful separations of multicomponent mixtures have been reported in the literature. The design method and splitting strategies for these mixtures are not widely known, and the SMB equipment that can split and obtain a relatively pure product from complex feed mixtures is not readily available commercially.

The rest of this chapter is focused on (i) the design methods to achieve high product purity and high yield for multicomponent separations, (ii) simulation tools to reduce the actual number of experiments for process development, (iii) SMB equipment designs for multicomponent separations, (iv) a knowledge-driven design method, which is based on intrinsic adsorption and mass-transfer parameters, and (v) examples of SMB biorefinery applications.

7.2 Essential SMB design principles and tools

The design of an SMB unit operation is an involved exercise even for a simple system involving binary separation. The next few pages will attempt to provide a working basic knowledge of SMB design principles. They are by no means comprehensive and are intended only to give readers broad guidance for further studies.

Simulated moving-bed technology can be applied to various chromatography modes: liquid, gas, and supercritical fluid [48–50]. Aside from the effects of pressure, which can be substantial in gas and supercritical applications, the principles of SMB design and operation are similar for the different fluid phases. We will briefly delve into SMB with reactions, and other more atypical variants later in this chapter.

Generally, SMB operations can simply be classified by its need to achieve a binary, ternary, or higher order separation. Industrial applications to date are often binary separations: recovering a particular class of aromatics from a petrochemical stream, separating fructose from glucose for the production of high-fructose corn syrup, and separating an enantiomer from a racemic mixture. However, some processes require a ternary or pseudo-ternary split. This is often so when more than two products need to be recovered separately from the feed stream, or when a single desired component is buried in the middle of a chromatographic sequence of undesirable components, as with insulin purification. Insulin is first separated from fast-moving impurities (high-molecular-weight proteins) in a first SMB; then it is separated from a slow-moving impurity (zinc chloride) in a second SMB. A comprehensive study of different strategies for splitting a multicomponent mixture has been reported [51].

7.2.1 Knowledge-driven design

Wang *et al*. have synthesized a knowledge-driven SMB design method that takes advantage of the intrinsic engineering parameters, and the design and simulation tools that have been developed at Purdue University. We believe the method is also applicable with tools designed by other research groups or commercial entities.

The knowledge-driven design approach, as shown in Figure 7.2, starts with the screening and selection of suitable pairs of desorbent and stationary phases. What drives the initial selection of viable desorbents here is often high feed solubility, product compatibility, desorbent cost, and desorbent recovery cost. A common desorbent is often the same carrier as in the feed. The selection of desorbent and stationary phase pairings is complex, and many of the same rules of thumb that govern method development in batch chromatography are applicable here. Simplistically, the best pairing preferably should be safe, stable, and low-cost, and have high separation selectivity (the components are easily separated in the column) and high capacity for the impurities or for the products. Many screening attempts rely on experimental trials and guidance from proprietary manufacturer databases.

Once a stationary phase with a corresponding desorbent is selected from screening tests, intrinsic parameters are then estimated using batch equilibrium tests or small column tests. The intrinsic parameters include

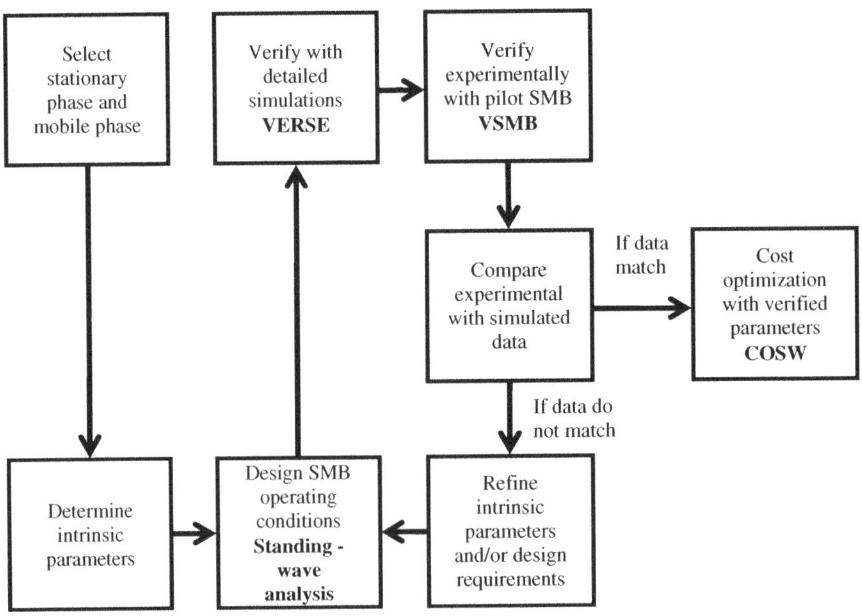

Figure 7.2 *Knowledge-driven SMB design using intrinsic parameters and the platform technologies; Standing-Wave Analysis, VERSE, VSMB and COSW*

adsorption isotherms, mass-transfer parameters, and stationary phase parameters (particle radius, bed-void fraction, and particle porosity). The intrinsic parameters along with the SMB system parameters (column length, column diameter, zone lengths, and extra-column dead volumes) are used in a systematic design tool, called the standing-wave analysis, to find the operating zone flow rates and switching time of the optimal SMB process. The standing-wave analysis has been validated for linear and Langmuir-type isotherms for binary and multicomponent separations, and can easily be extended to several other isotherms. It accounts for extra-column dead volume, pressure limits, and mass-transfer limits. It also allows an adjustable margin to counter known fluctuations or inaccuracies in the operating flow rates or switching time. Standing-wave analysis is easily extended to complex systems involving multiple components and complex SMB schemes. For multicomponent systems, product dilution and solvent usage depend on the separation sequence for product recovery (or splitting strategy). Standing-wave analysis is often used to identify the most favorable splitting strategy. The operating design from the standing-wave analysis can then be verified with a detailed rate-model simulation package such as VERSE. These simulations can be used to verify the product purity and yield that are specified in the design, enabling a deeper understanding of the complex dynamics of separation in an SMB system.

In the SMB development phase, the equipment size and corresponding system size is often limited to the available pilot unit on hand. The authors have developed a patented versatile SMB unit, called the VSMB, which allows for great experimental flexibility, a vital need in the development phase. The resulting experimental data—component profiles, effluent histories, product yields, and product purities—are then compared against the simulation to learn whether refinements of the system and intrinsic parameters are needed. This frequently happens because SMB amplifies subtle differences, which often are not discernible in the earlier parameter estimation efforts. The cycle consists of refining parameters, designing operating conditions with standing-wave analysis, verifying the operating design with VERSE simulations, generating experimental results on the VSMB, and comparing the experimental results with the simulations. This cycle eventually results in a verifiable set of intrinsic parameters, which can be used for design and simulation of SMB processes at any scale. Simulations are also used to explore many variations of advanced SMB operations, some which we will discuss later.

The verified intrinsic parameters determined at the pilot scale are then used in a general cost optimization program based on the standing-wave analysis to design a large-scale process. The optimization program can account for a wider variety of design variables, including different particle sizes, column configurations, cost functions, and such. At process scale-up, the same cyclic sequence of comparing design, simulation, and experimental results leading to a new optimal design is used to refine the process at production scale.

These platform technologies for design, optimization, simulation, and pilot testing equipment are broadly applicable to many different separations. They also enable fast, systematic, knowledge-driven design and scale-up of SMB operations. The next section covers each of these platform technologies in greater detail.

7.2.2 Design and optimization for multicomponent separation

7.2.2.1 Standing-wave analysis (SWA)

The design of the operating conditions of an SMB process requires specifying the zone flow rates and the switching time. A zone is the region between two successive ports, and each zone has a different mobile phase velocity or flow rate. A solute band can be considered as a region between an adsorption-concentration wave and a desorption-concentration wave (Figure 7.3).

Ma and Wang (1997) first proposed the standing-wave analysis for the design of an SMB for binary separation for linear isotherm systems [52]. It is a powerful concept based on a relatively simple idea. For a continuous moving-bed (CMB) at steady state, we would have fixed-column profiles for two components in a four-zone system for a linear, ideal system (without any wave spreading resulting from diffusion or other

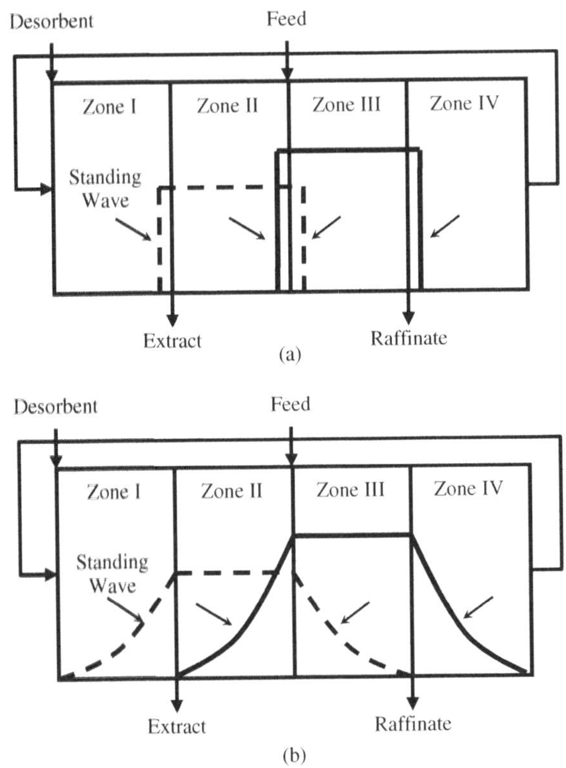

Figure 7.3 *The four-zone SMB systems shown in CMB diagrams. The continuous line denotes the low-affinity (fast) component 1, and the dashed line denotes the high-affinity (slow) component 2. The vertical scale represents concentration in the liquid phase and the horizontal scale represents bed length from the desorbent port. The arrows indicate standing waves. (a) Both components have linear isotherm and no significant mass transfer effects (ideal). (b) Linear isotherm system with significant mass transfer effects (nonideal)*

dispersion effects), as shown in Figure 7.3a. Component *1* is the fast-moving solute, and component *2* is the slow-moving solute. Components *1* and *2* are collected at the raffinate and extract ports, respectively. The ports would be moving at a port velocity v. For a linear, ideal system in CMB, if the port velocity is matched with the migration velocity of the desorption wave of component *2* in zone *I*, the desorption wave of component *1* in zone *II*, the adsorption wave of component *2* in zone *III*, and the adsorption wave of component *1* in zone *IV*, the wave in each zone will appear to be "standing" to an observer moving with the ports. This is illustrated in Figure 7.3a. Under such standing-wave conditions, the product purity and yield for both components are 100% for an ideal system.

A nonideal system has significant wave spreading resulting from diffusion or other dispersion effects in CMB. The concentration profiles for a linear isotherm system are shown in Figure 7.3b. To prevent the waves from spreading to other zones, the key wave velocities in zones *I* and *II* mentioned earlier are set to be faster than the port velocity, whereas the key wave velocities in zones *III* and *IV* are set to be slower than the port velocity. The velocity differences between the wave velocities and the port velocity tend to "focus" the waves and thus maintain high product yield and purity in a nonideal system.

SWA for linear, ideal systems To maintain high purities at the product ports and to prevent wraparound of the components, the key wave in each desorption zone must be equal to or faster than the port velocity ν, but the key wave in each adsorption zone must be equal to or slower than the port velocity. For the typical four-zone system illustrated in Figure 7.3a, this gives:

$$u_2^I \geq \nu \tag{7.1}$$

$$u_1^{II} \geq \nu \tag{7.2}$$

$$u_2^{III} \leq \nu \tag{7.3}$$

$$u_1^{IV} \leq \nu \tag{7.4}$$

where u_i^j is the ith solute velocity in zone j.

At the limits of Eqs. (7.1) to (7.4), when the key wave velocities are matched with the port velocity, the system is at its minimum solvent consumption and maximum throughput for a given feed flow rate. Throughput is defined as the volume of feed processed per unit time and per unit bed volume. Thus the standing-wave principle sets the key wave velocities u_i^j equal to that of the port velocity ν.

For a linear isotherm system, u_i^j, the ith wave velocity in zone j, can be related to u_o^j, the interstitial velocity in zone j, as follows [53]:

$$u_2^I = \frac{u_O^I}{1+P\delta_2} = \nu \tag{7.5}$$

$$u_1^{II} = \frac{u_O^{II}}{1+P\delta_1} = \nu \tag{7.6}$$

$$u_2^{III} = \frac{u_O^{III}}{1+P\delta_2} = \nu \tag{7.7}$$

$$u_1^{IV} = \frac{u_O^{IV}}{1+P\delta_1} = \nu \tag{7.8}$$

where P is the phase ratio, defined as $P = (1-\varepsilon_b)/\varepsilon$; ε_b is the interparticle void fraction; and, δ_i is the effective retention factor of solute i. For a linear system, δ_i is defined as [52]:

$$\delta_i = Ke_i\varepsilon_p + \left(1 - Ke_i\varepsilon_p\right)a_i + \frac{DV}{PL_C S\varepsilon_b} \tag{7.9}$$

Ke_i is the size-exclusion factor of solute i (the fraction of pores in the particles particle that is accessible by solute i); ε_p is the particle porosity; a_i is the partition coefficient of solute i; DV is the extra-column dead volume per column; S is the column cross-sectional area; and L_C is the single column length. The last term in Eq. (7.9) is meant to consider the additional retention due to extra-column dead volume, which can be important for a nonideal system.

From the mass balance at the feed port, the feed flow rate F^{feed} is related to the interstitial velocities of zones II and III:

$$\frac{F^{feed}}{\varepsilon_b S} = u_O^{III} - u_O^{II} \tag{7.10}$$

Equations (7.5)–(7.8) and (7.10) represent the set of five equations for the linear, ideal standing-wave design for the conventional four-zone SMB. Given the feed flow rate and system parameters, the five unknowns to be solved are the four zone flow rates (or their corresponding interstitial velocities) and the port velocity. Notice, however, that any feed velocity will produce pure products, since the concentration waves are square waves (Figure 7.3a). The resulting values will also show that $u_O^I = u_O^{III}$ and $u_O^{II} = u_O^{IV}$ for the linear ideal system. The maximum feed flow rate is then limited only by the pressure limit or the port velocity.

The standing-wave analysis for CMB can provide an approximate solution to a simulated moving-bed system by defining an average port velocity as

$$\nu = L_C / t_S \tag{7.11}$$

where t_S is the switching time or step time. The standing-wave solution for CMB is generally valid for SMB when every zone has two or more columns.

SWA for linear, nonideal systems In systems with significant mass-transfer effects (nonideal systems), the concentration waves spread out, as illustrated in Figure 7.3b. To prevent the waves from spreading beyond the respective zones and resulting in significant yield loss, the trailing edge of the desorption waves of solute *2* in zone *I* (u_2^I) and those of solute *1* in zone *II* (u_1^{II}) should migrate faster than the port velocity ν, whereas the leading edge of the adsorptive waves of solute 2 in zone *III* (u_2^{III}) and those of solute *1* in zone *IV* (u_1^{IV}) should migrate more slowly than the port velocity (Eq. (7.1) to (7.4)). To achieve such results, Ma and Wang (1997) derived a mass-transfer correction factor Δ for the key wave velocity in each zone (Eq. (7.12) to (7.15)) [52]. The wave velocity of solute *2* in zone *I* and that of solute *1* in zone *II* are higher than the port velocity by Δ_2^I and Δ_1^{II}, respectively, whereas the wave velocity of solute *2* in zone *III* and that of solute *1* in zone *IV* are lower than the port velocity by Δ_2^{III} and Δ_1^{IV}, respectively.

$$u_2^I = \frac{u_O^I}{1 + P\delta_2} = \nu + \Delta_2^I = \nu + \frac{\beta_2^I}{(1 + P\delta_2) L_C N_C^I}\left(E_{b2}^I + \frac{P\nu^2\delta_2^2}{k_2^I}\right) \tag{7.12}$$

$$u_1^{II} = \frac{u_O^{II}}{1 + P\delta_1} = \nu + \Delta_1^{II} = \nu + \frac{\beta_1^{II}}{(1 + P\delta_1) L_C N_C^{II}}\left(E_{b1}^{II} + \frac{P\nu^2\delta_1^2}{k_1^{II}}\right) \tag{7.13}$$

$$u_2^{III} = \frac{u_O^{III}}{1 + P\delta_2} = \nu - \Delta_2^{III} = \nu - \frac{\beta_2^{III}}{(1 + P\delta_2) L_C N_C^{III}}\left(E_{b2}^{III} + \frac{P\nu^2\delta_2^2}{k_2^{III}}\right) \tag{7.14}$$

$$u_1^{IV} = \frac{u_O^{IV}}{1 + P\delta_1} = \nu - \Delta_1^{IV} = \nu - \frac{\beta_1^{IV}}{(1 + P\delta_1) L_C N_C^{IV}}\left(E_{b1}^{IV} + \frac{P\nu^2\delta_1^2}{k_1^{IV}}\right) \tag{7.15}$$

where Δ_i^j is the mass-transfer correction term for solute i in zone j; β_i^j is an index of product purity and yield and is defined as the logarithm of the ratio of the highest concentration to the lowest concentration of the standing wave i in zone j; N_C^j is the number of columns in zone j; E_{bi}^j is the axial dispersion coefficient of solute i in zone j; k_i^j is the lumped mass-transfer parameter for solute i in zone j and can be estimated for a linear isotherm system according to the equation below:

$$\frac{1}{k_i^j} = \frac{R^2}{15Ke_i\varepsilon_p D_{pi}} + \frac{R}{3k_{fi}^j} \tag{7.16}$$

where R is the particle radius; D_{pi} is the intraparticle diffusivity of solute i; and k_{fi} is the film mass-transfer coefficient of solute i.

The five equations, (7.10) and (7.12)–(7.15), can be solved for the operating velocities and the port velocity if we know the feed flow rate, the system parameters (particle radius, column length and diameter, number of columns in each zone, and extra-column dead volume per column), the intrinsic parameters (particle porosity, bed void fraction, isotherms, size exclusion parameters, axial dispersion coefficient, and the lumped mass-transfer parameters), and the index for product purity and yield β_i^j. Component mass-balance equations have been used to derive an explicit function of the β_i^j values as a function of the specified yield of each component for a linear isotherm system [51] and for a Langmuir system [54]. The operating conditions solved in the standing-wave equations automatically give the minimum solvent consumption for the specified product purities (or yield) and the feed flow rate.

Without any pressure-drop limit, the maximum allowable feed flow rate for a given SMB system can be calculated from the earlier set of equations:

$$F_{\max}^{feed} = \varepsilon_b S \left(\frac{P\,(\delta_2 - \delta_1)^2}{4\left(\dfrac{\beta_2^{III}\delta_2^2}{k_2^{III} L_C N_C^{III}} + \dfrac{\beta_1^{II}\delta_1^2}{k_1^{II} L_C N_C^{II}} \right)} - \left(\frac{\beta_2^{III} E_{b2}^{III}}{L_C N_C^{III}} + \frac{\beta_1^{II} E_{b1}^{II}}{L_C N_C^{II}} \right) \right) \tag{7.17}$$

The intraparticle diffusivity, $D_{pi,}$ is usually estimated from pulse or frontal data. The axial dispersion coefficient of solute i in zone j, E_{bi}^j, can be determined experimentally or estimated from the Chung and Wen correlation [55]. In the author's reported work with SMB insulin purification, an apparent E_b value, which is 40 times of that estimated from the Chung and Wen correlation, was used to account for fronting of the insulin wave as a result of aggregation [38]. The film mass-transfer coefficient for zone j, k_i^j, can be estimated from the Wilson and Geankoplis correlation [56].

The standing-wave analysis explicitly relates product purities and yields to the zone flow rates, the port velocity, zone lengths, voids, isotherms, and mass-transfer parameters. The maximum feed flow rate for a mass-transfer limited system can be determined from Eq. (7.17). The maximum feed flow rate for a pressure-limited system can be determined from the Ergun equation [54]. The design can also account for zone-specific (and solute-specific) mass-transfer effects, extra-column dead volume, different isotherms, and size-exclusion factors. Standing-wave analysis gives the highest throughput and the lowest solvent consumption for the specified configuration, feed flow rate and the yield of each component.

The linear standing-wave design with significant mass-transfer effects has been successfully used to experimentally separate amino acids in a four-zone SMB [25, 26], to purify insulin from a ternary feed mixture in a tandem SMB (two SMBs in series) [38], and to purify paclitaxel [31]. For each separation, the design was first verified with simulations (Section 7.2.3) before experimental verification. Simulated concentration waves agree closely with the experimental data in these systems based on the standing-wave design.

The standing-wave analysis has since been extended to binary separations in Langmuir, anti-Langmuir [19] and modified Langmuir isotherms [28]. Further refinements to improve the solution algorithm used in these complex systems have also been reported [27, 54]. The nonlinear nonideal standing-wave design has been validated for organic acid [27, 28] and enantiomers separations [54, 57, 58]. The standing-wave design for a three-zone ion-exchange SMB-like process to remove zinc ions was recently published [59].

The standing-wave analysis gives the lowest desorbent consumption and the highest bed throughput for a desired feed-flow rate for a given system size. The design, however, has no room for operational discrepancies, such as deviations of zone flow rates or port velocity from the set point values. A more robust method for designing SMB operating conditions, pinched-wave analysis, was recently introduced to account for expected operational discrepancies [43, 60]. A larger difference in the port velocity and the key wave velocity in each zone is used to counter the operational discrepancies. In essence, the center of a key wave is focused closer to a port to provide a safety margin for its trailing edge (of a desorption wave) or its leading edge (of an adsorption wave). Pinched wave refers to the wave appearing to be more focused (or pinched) compared to the corresponding standing wave. The pinched-wave analysis assumes known discrepancies in the pumps' flow rates, switching time, bed voidage, size-exclusion factors, and extra-column dead volume. It then adjusts the zone flow rates and switching time accordingly; thus at the limits of the combined discrepancies, the key concentration waves become standing waves.

Another extension of the standing-wave design, the pseudolinear standing-wave concept, has been demonstrated [61]. An effective pseudolinear isotherm "a" value for a nonlinear component was estimated from an experimental frontal curve of a typical feed stream. An average flow rate based on estimated zone flow rates was used in the experiment. The standing-wave analysis was successfully used to isolate six sugars from a mixture of ten components [61]. The example is summarized in further detail later.

The pseudolinear approach based on experimental frontal curves allows for the fast design of SMB operating conditions if isotherm types are unknown or inaccurate. The pinched-wave with pseudolinear approach can be used effectively for systems with significant mass-transfer effects. This approach is fast, efficient, and does not need tedious and sometimes difficult, or expensive, estimation of the single component isotherms. This approach is especially useful for biorefinery streams, including hydrolysates and fermentation broths with a great many known and unknown components. It is also useful for proof-of-concepts and developmental work and for processing feeds with changing composition profiles.

7.2.2.2 Splitting strategies for multicomponent SMB systems

Hritzko *et al*. (2002) extended the linear, nonideal standing-wave design for a binary system to a general multicomponent system [51]. If the product is either the first or the last component in the elution sequence, only one SMB is needed to recover it. The impurities can be treated as a group or a pseudospecies in the standing-wave design; the adsorption wave of this group can be represented by that of the fastest moving impurity, whereas the desorption wave of this group can be represented by that of the slowest impurity. The remaining impurities will thus be bound by the two waves.

However, for a product buried within the elution sequence, two SMBs in series (or a tandem SMB) are needed to recover the product. Either the faster or the slower group of impurities is removed in the first SMB (strategies 1 and 2 in Figure 7.4). The remaining impurities are subsequently removed in the second SMB. The easier separation (defined by the larger difference in δ) should be performed first to reduce product dilution and desorbent consumption. If only one product is required, some impurities in the first ring should be allowed to distribute (or wraparound) throughout the first ring to reduce the separation load in the second ring (strategies 3 and 4 in Figure 7.4). For example, in strategy 3, the first ring is designed to remove the fast-moving impurity completely; the slow-moving impurity is distributed and collected at both product ports. The first ring raffinate will have the fast-moving impurity and some of the slow-moving impurity (which wraps around from zone I into zones IV and III). The first ring extract will have the middle-eluting product and reduced concentration of the slow-moving impurity, thereby reducing the separation load of the second ring. This strategy, which recovers only a pure product of the middle-eluting component, results in a lower desorbent consumption and a higher product concentration than in strategy 1, which recovers all three components with high purity and high yield. Strategy 3 was applied in the tandem SMB separation of insulin [38] and in the isolation of sugars from corn stover hydrolysate [61].

7.2.2.3 Comprehensive optimization with standing-wave (COSW)

The standing-wave analysis solves for the SMB operating parameters (zone flow rates and switching time) given the intrinsic parameters and system parameters (column size and zone configuration). In the process development stage, the lab-scale SMB system usually has a fixed column size or zone configuration. However, scale-up to pilot or plant-scale SMB requires comprehensive optimization of the system parameters

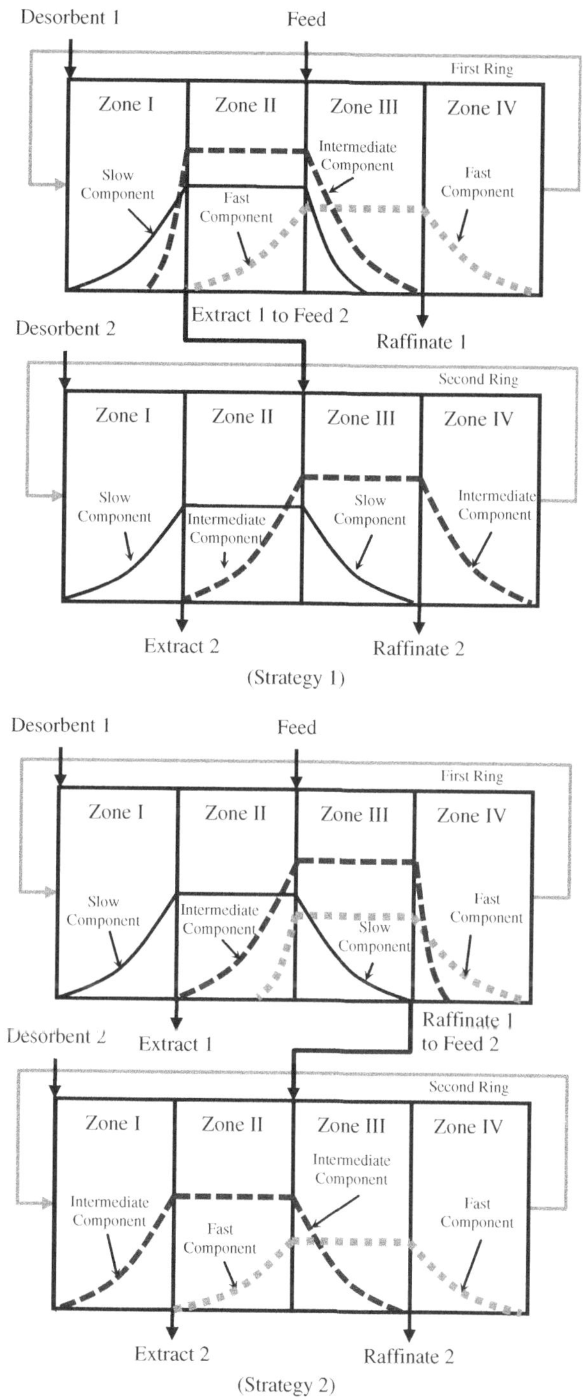

Figure 7.4 *Splitting strategies 1 to 4 for multicomponent tandem SMB system*

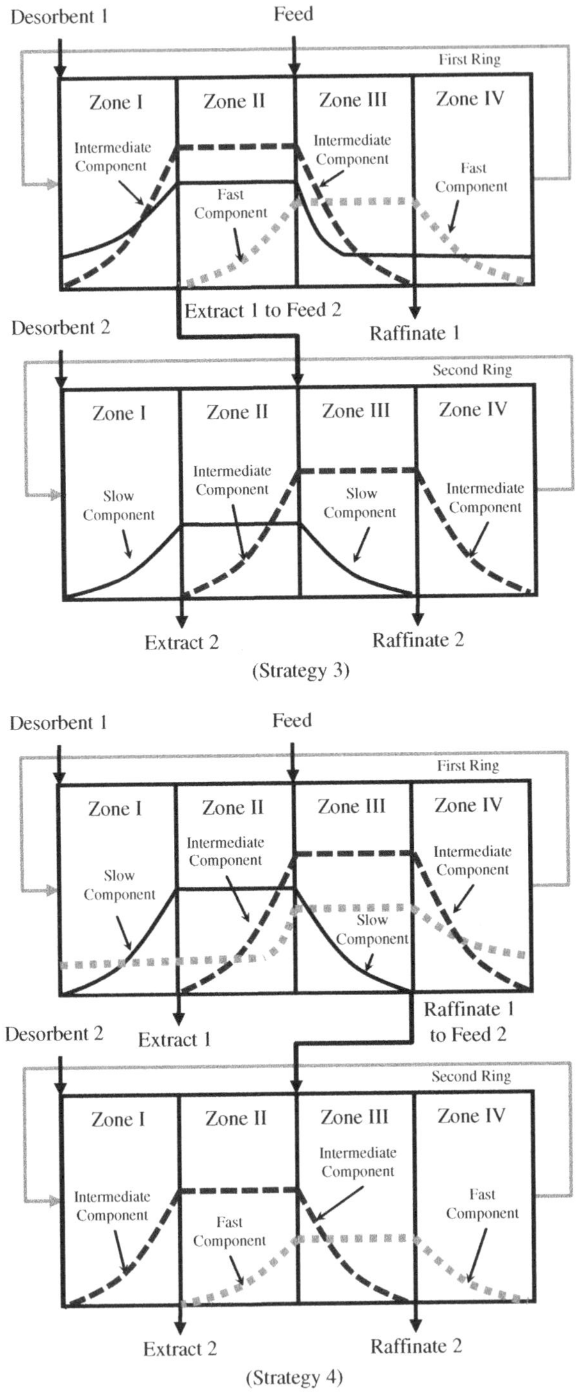

Figure 7.4 *(continued)*

and other variables, which include feed concentration, pump placement, particle size, splitting strategies for multicomponent systems, and integration with other unit operations.

Objectives for SMB optimization may include the lowest purification cost, the highest bed throughput (especially for expensive stationary phases), the lowest product dilution, or the lowest solvent consumption. These objectives are not always mutually exclusive and are highly dependent on the yield and purity requirements [54]. Cost functions for the energy, equipment, solvent, stationary phase, and manpower are needed for these general optimizations [42, 54].

Three optimization algorithms with standing-wave analysis at its core have been developed: a general grid search, genetic algorithm, and simulated annealing algorithms. The grid-search strategy uses an intensive multilevel search of the whole response surface at spaced intervals [42]. The first-level search uses wide intervals with subsequently refined searches at smaller intervals. Genetic algorithm is an optimization strategy that imitates natural evolution. The genetic algorithm uses the nondominated sorting genetic algorithm with a jumping genes operator for multiobjective optimization [62].

Simulated annealing is a generalization of a Monte Carlo method that resembles the way a metal cools after heating and eventually freezes into a minimum energy state. A simulated annealing algorithm for binary four-zone SMB separation of a linear system has been developed for multiobjective optimization [63]. The work has been further extended to a nonlinear system and now uses a hybrid simulated annealing and genetic algorithm (SAGA) to further improve the optimization [64]. These advanced algorithms generally are an order of magnitude more efficient than the grid search.

7.2.2.4 *Other design methodologies*

Various other methodologies for designing SMB operating conditions besides the standing-wave analysis have also been presented in the literature, including the triangle theory [65–67], the empirical safety margin factor [2, 68, 69], and the separation volume analysis [22, 70]. The triangle theory specifies a triangular region for an ideal system (without mass transfer effects) in a plot of the net flow ratios of mobile phase to solid phase in the zones immediately upstream and downstream from the feed port. The enclosed region represents the feasible operating conditions for high product purity and yield. For nonideal systems (with significant mass transfer effects), numerical simulations are used to refine the triangular region [50]. Interested readers are directed to Rajendran *et al*. (2009) for a more thorough description and review of the triangle theory [24]. The safety margin method provides an empirical correction to the zone flow rates for an ideal system to counter the mass transfer in a nonideal system. The separation volume analysis is an extension of the two previous methods where the operating conditions are now contained within a pyramidical region of a 3-D domain.

7.2.3 SMB chromatographic simulation

The propagation of concentration waves in SMB for a nonlinear, nonideal system is affected by competitive adsorption in the sorbent phase, convection and dispersion in the different zones, input and output flow rates, and port movement velocity. The detailed dynamic concentration profiles (the solute concentrations in the loop as a function of both position and time) and the effluent histories from the product streams cannot be obtained from analytical solutions. They can only be found from numerical solutions that satisfy the differential mass balance equations for all the components, the competitive adsorption rate equations or the equilibrium adsorption isotherm equations, the initial conditions, and the boundary conditions, which are related to the operating conditions [51, 52]. The following input parameters are required for the numerical solutions (or computer simulations): the intrinsic parameters (particle size, bed and particle void fractions, adsorption isotherms or rate equations, and mass transfer parameters), the system parameters (column diameter, column length, and zone lengths), the initial column concentration profiles, and the operating conditions (feed composition, solvent composition, zone flow rates, and the step time). The purity and yield of each product can be calculated readily from the simulated effluent histories.

The standing-wave designs can be tested experimentally or with computer simulations. The latter are more efficient and cost effective but require accurate intrinsic parameters. The product purities or yields targeted in the design can be verified with simulations before any SMB experiment. Furthermore, simulated dynamic column profiles and effluent histories can also be compared with experimental results for further verification of the design and the simulation parameters.

Several chromatographic simulation packages exist with very different capabilities and costs, including Aspen Chromatography (ASPEN Tech, Massachusetts, U.S.) and several research-level simulation packages [22, 65, 68]. This section focuses on the VErsatile Reaction and SEparation chromatographic simulator, or VERSE [71], developed in Wang's group (Figure 7.5). VERSE is an expanded version of an earlier rate model for batch chromatography, which was based on axial dispersion, film mass transfer, intraparticle pore diffusion, and equilibrium competitive adsorption and ion exchange [72]. To model complex protein adsorption and desorption phenomena in chromatography [71], Berninger, Whitley, and Wang further expanded the rate model by Yu and Wang (1989) to include nonequilibrium (or slow) adsorption and desorption [73], aggregation reactions in the mobile phase [74–76], and denaturation reactions in the stationary phase [77].

The pore-diffusion model, however, was found to be inaccurate for systems with high-affinity solutes at high loading, where surface diffusion results in asymmetric breakthrough curves with a sharp rise first and then a slow approach to saturation. As a solute concentration in the adsorbed phase approaches saturation, the driving force (surface concentration gradient) for surface diffusion diminishes, resulting in the slow approach to saturation. In contrast, the driving force for pore diffusion does not have this limitation, usually resulting in symmetric breakthrough curves, except for very low mass transfer efficiency systems. For this reason, Ma *et al*. (1996) added surface diffusion and parallel pore and surface diffusion features to the rate model [78].

Koh *et al*. (1995 and 1998) expanded the parallel pore and surface diffusion model in VERSE to fluidized and expanded beds [79, 80]. Ernest *et al*. (1997) further expanded the VERSE model for single-column batch chromatography to carousel systems [81], and Hritzko *et al*. (2000 and 2002) expanded and experimentally verified the VERSE model for SMB systems [51, 82].

VERSE has been validated with experimental data in 39 papers reported in the literature, including the separations of ions and small organic compounds [78, 81, 82], amino acids [25, 27, 79, 80, 83, 84], lactic acid [28], sugars [61, 85], chiral compounds [54, 57, 58, 62], antibiotics [35, 86], paclitaxel [31–33, 87, 88], insulin [38–40, 44, 45, 59, 60, 89–92], and other proteins [73–77].

In summary, VERSE can consider a host of adsorption and ion exchange, transport, and reaction phenomena (Figure 7.5) in fixed beds, expanded beds, fluidized beds, carousel, and SMB chromatography modes. Adsorption mechanisms include size exclusion and multicomponent competitive adsorption or ion exchange. Both equilibrium and nonequilibrium adsorption can be considered. Mass transfer mechanisms include axial dispersion, film mass transfer, pore diffusion, surface diffusion, and parallel pore and surface diffusion. Reactions include aggregation in the solution phase and reactions on the surface of the stationary phase. Mixing in the extra-column dead volume is simulated as a continuous stirred tank reactor (CSTR) attached to a column's inlet and outlet. VERSE also allows for frontal chromatography, isocratic or gradient elution chromatography, elution with periodic total recycle, displacement chromatography, and other cyclic operations, involving concentration changes in a column inlet.

Recently a user-friendly graphical interface has been developed for VERSE to facilitate data input and results output. Figure 7.6 shows a screenshot of part of the VERSE input screen. The interface has seven

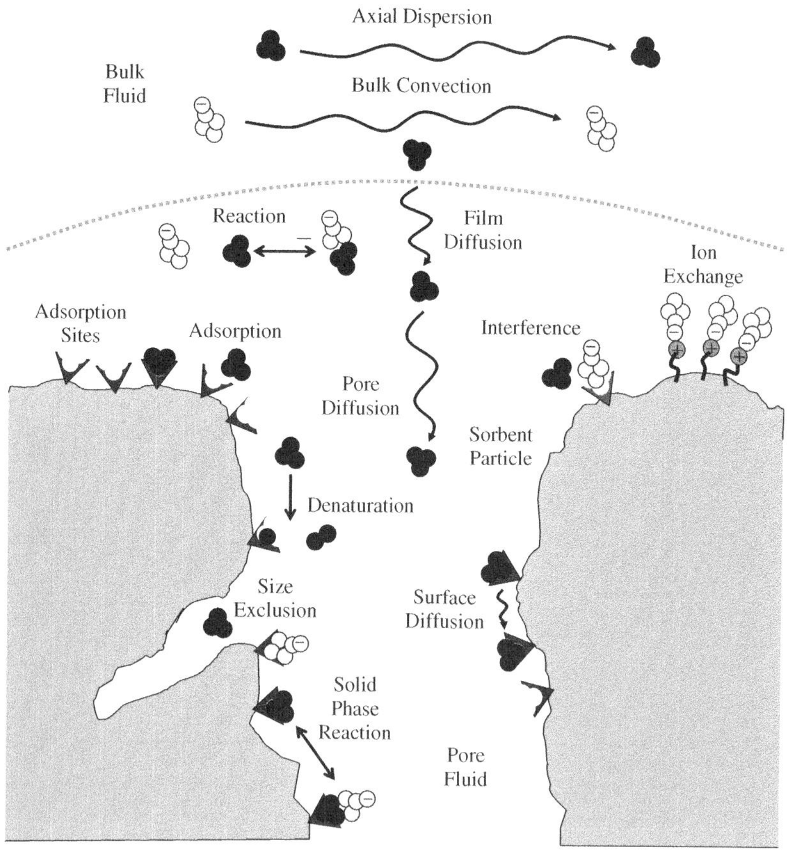

Figure 7.5 Competitive adsorption, transport, and reaction mechanisms used in SMB chromatography and considered in VErsatile Reaction and SEparation (VERSE) chromatographic simulator

sections: (i) system parameters, (ii) intrinsic parameters (isotherms, kinetics, mass transfer, reactions), (iii) initial conditions, (iv) batch or SMB operating conditions, (v) numerical parameters, (vi) output options, and (vii) simulation results (column profiles as a function of time or effluent histories). One can also export the results to spreadsheets for further data processing.

Figure 7.7a shows the VERSE simulation results of an experimentally verified four-zone SMB system with two-columns-per-zone (Figure 7.1). Two amino acids, phenylalanine and tryptophan are separated in SMB with water as the desorbent and poly(4-vinylpyridine) (PVP) as the stationary phase [27]. The figure shows the column profile at 8.5 times the switching time. The significant undulations on the tryptophan plateau are due to the interacting nature of the Langmuirian isotherms and the movement of the feed port downstream. Upon mixing with the feed, the concentration downstream from the feed port increases abruptly in the beginning of zone III. As the feed port moves one column downstream, the abrupt peak is moved to zone II and spread out by dispersion effects, resulting in a spread peak. The three spread peaks in zone II result from the three prior port switches.

Notice how the component waves in the output streams, extract in Figure 7.7b and raffinate in Figure 7.7c, follow a sawtooth pattern corresponding to movement of the ports downstream. In the extract port, at the start of the port switch, the slow-moving tryptophan component is first recovered at a high concentration

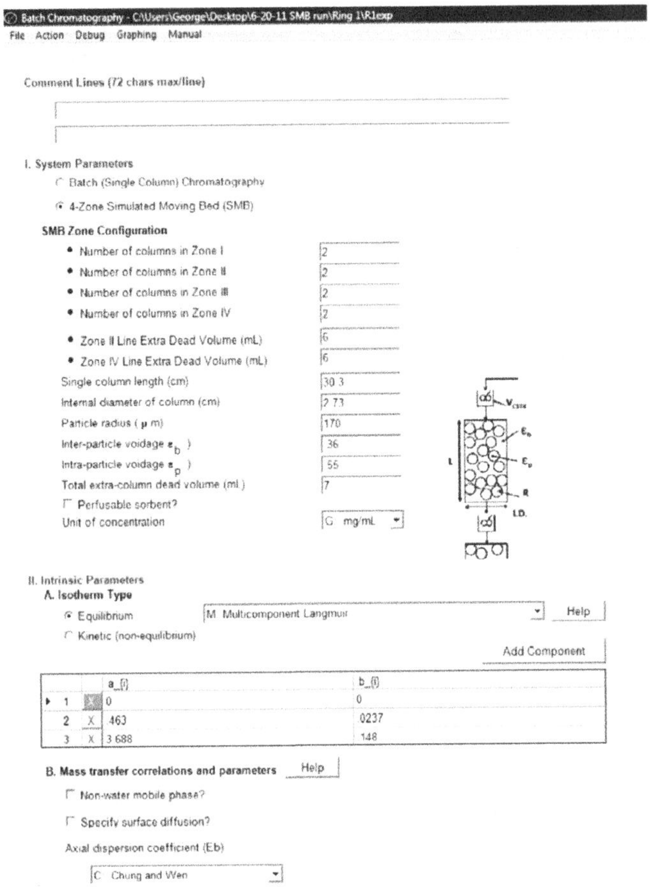

Figure 7.6 Part of VERSE input screen for a 4-zone SMB

that tails off as the desorption curve of the tryptophan moves through zone I into zone II. On the other hand, the raffinate port commences with low concentration of the fast-moving phenylalanine component at the start of each port switch that increases as the adsorption front of phenylalanine moves from zone III into zone IV. Sawtooth patterns in the output streams are common in well designed SMB systems. An instructive discussion on SMB wave dynamics of a linear, nonideal system can be found in the insulin purification literature [45].

7.2.4 SMB equipment

A simulated moving-bed unit essentially consists of pumps, valves, columns, stationary phase, and a control system. Each major piece of equipment includes a multitude of options and technologies, much of it proprietary to the equipment or stationary phase manufacturer, and is beyond the scope of this chapter. Simulated moving-bed technology primarily differs from other chromatographic techniques by its extensive use of valving to switch the location of the inlet and outlet ports, and thus we will restrict our focus to just SMB valves. For process development, the unit must be versatile and have a small nonseparation volume, known as the dead volume. System versatility refers to the ability to add extra columns or zones and to

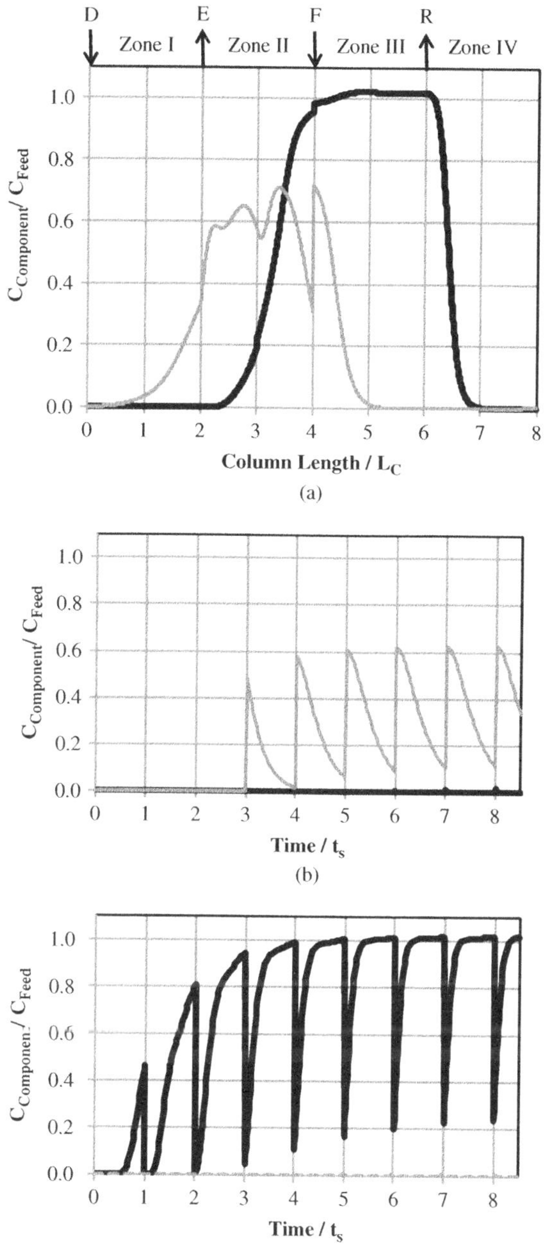

Figure 7.7 *Example of results from a VERSE simulation. (a) Column profile of phenylalanine-tryptophan four-zone SMB separation at 8.5 switching time (t_s), (b) Extract history, (c) Raffinate history. Phenylalanine is the fast-moving component (thick line), and tryptophan is the slow-moving component (thin line). D = desorbent, E = extract, F = feed, and R = raffinate*

change flow paths between the different ports; it also refers to asynchronous switching and other operational variations. In the next section, we will briefly touch on the many different inventions introduced in the past few decades that demand even more from SMB equipment, especially a versatile SMB valving system. However, many of these versatility requirements disappear at production scale, where SMB systems are often built new for a specific product, and reliability and equipment costs are then much more crucial.

Simulated moving-bed valving systems can be broadly categorized into central and distributed valve designs. Central valve designs have complex single valves that do it all. These designs are limited to a single switching time, suffer from scale-up issues, and once built have a limited number of configurations. Some of the designs require a carousel to move the columns, which at a large scale becomes increasingly difficult and unreliable. The designs include a variety of UOP single-valve designs, the ISEP/CSEP valve, and many others among the published patents.

Distributed valve designs generally use simpler two-way valves or rotary-type valves. Two-way (and the slightly more complex three-way) valves are common and come in a wide variety of types (such as knife, butterfly, and needle) and attributes (wetted surface and cleaning-in-place, among others). These versatile two-way valves can be thought of as the building blocks of any SMB system. Their designs require a great many valves, which increase the system's complexity. They tend to suffer from cross-contamination and a large dead volume, which are critical issues for small lab-scale SMBs used in process development. These drawbacks diminish for preparative or industrial-scale SMBs, and two-way valves are then preferred because of low cost, wide availability, and improved reliability over the more complex valves.

Rotary valves are made of a rotating core, called the rotor, and a static surface, called the stator. Each shift of the rotor results in a different liquid flow path through the valve. An effective designation the authors preferred is that by Valco Instruments Co. (Houston, Texas, U.S.). The Select-Dead-end (SD) is the simplest design, where a single port leads to one of multiple ports. Two SMB designs based on the SD valves are most common: (i) an SD valve per column, where the valve selects the desired port for each column, and (ii) an SD valve per stream, where the valve selects the column for each stream. The SD valves were used in some SMBs from UOP and Daicel (Osaka, Japan), and in the ADSEP SMB system of United States Filter (Palm Desert, California, U.S.).

The authors patented a new SMB valve design called the VSMB, which uses a select trapping (ST) valve to interrupt the flow between sequential columns [93]. The ST valves act like a pair of synchronized SD valves, where both SDs simultaneously select a pair of junction ports. This allows flexible exploitation of the flow between columns. The flow can be simply directed to the next column, can be partially withdrawn (such as at the extract or raffinate ports), or can have a new stream added, such as at the desorbent or feed ports. The stream can also be redirected to another part of the SMB in its entirety and a completely new stream will enter the next column. See Figure 7.8 for examples of the four basic port types.

The VSMB has a wide range of configurations for different SMB schemes. Either tandem SMB or parallel SMB can be configured. The columns do not rotate, and columns can be added to a zone easily. The valve design minimizes line sharing and reduces cross-contamination to ensure high product purity and high yield. The design also allows a strong regenerant and periodic cleaning-in-place (without physical removal of the columns). It allows for independent port switching or asynchronous switching; the step time in the regeneration zone can be different from the step time in the separation zones. Such features save labor, increase sorbent productivity, and allow savings of regenerant and desorbent. VSMB enables most, if not all, of the new process schemes invented in the SMB field in the past few decades. The authors believe that the VSMB is the most versatile SMB valve configuration for SMB developmental and pilot work. It will prove extremely useful for operation requiring frequent changes to its operating SMB schemes, such as in contract operations.

The VSMB has been used successfully to develop and test separations of organic acids and enantiomers in the traditional 4-zone configurations illustrated earlier. It has also been used to purify insulin in a tandem

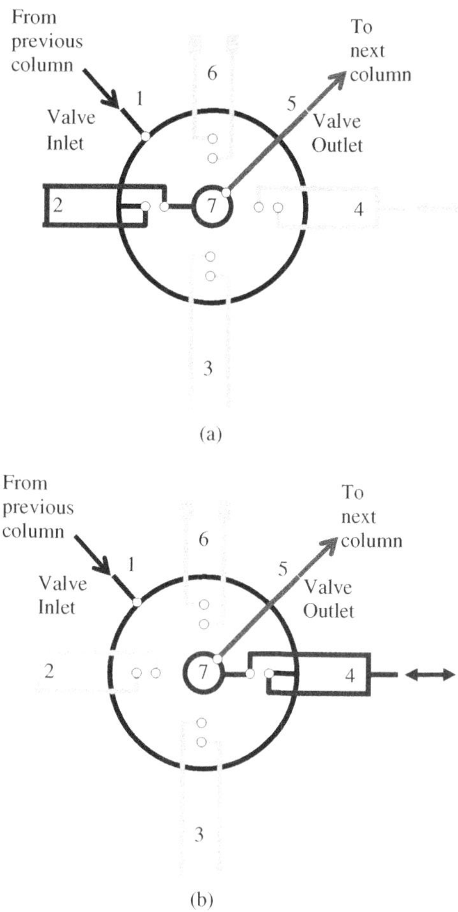

Figure 7.8 *VSMB's one-ST-valve–between-columns arrangement. Port 2 is "closed," port 4 is "teed," port 6 is "dead-ended," and port 3 is "opened." Figure 7.8a shows that the inlet stream from the preceding column is directed into inlet port 1 to the outer ring, then to "closed" port 2, and then through inner ring 7 and out through outlet port 5 to the next column. Figure 7.8b shows that the stream between the column is interrupted by the "teed" port 4, allowing for the addition or partial removal of the stream*

SMB (Figure 7.9a) and in a three-zone carousel (Figure 7.9b). It has also been used to recover sugar from biomass hydrolysate in a more complex five-zone configuration, detailed later. Readers interested in SMB equipment are directed to the authors' published review [46].

Online control and optimization of SMB have been actively studied since the early 2000s. Highly optimized SMB operating conditions are often prone to small perturbations of the operating conditions in actual production. Aside from incorporating these known perturbations into the design, such as through the pinched-wave design mentioned earlier, an online feedback and control could provide corrective actions. Unlike most other unit operations, controlling SMB operations is challenging because of the periodic, pseudo-steady-state nature, delayed cause-and-effect, and the continuous aging and fouling of the stationary phase. The methods most often used are based on model predictive control, where a dynamic model of the system, often simplified and linearized, is used to drive process control. Interested readers are directed to Rajendran *et al*. (2009) for references and additional details [24].

(a)

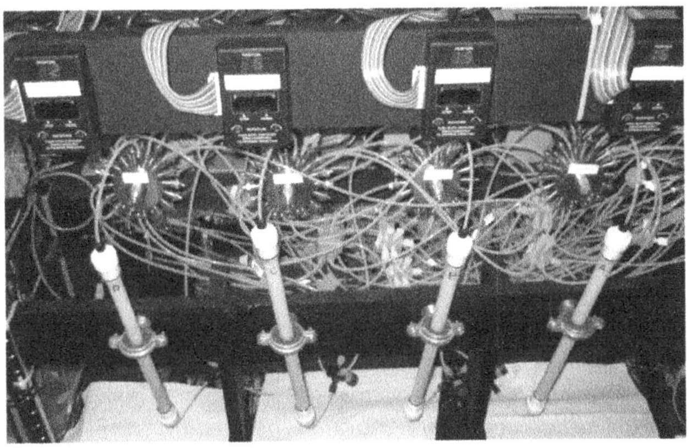

(b)

Figure 7.9 (a) Tandem VSMB system for size-exclusion purification of biosynthetic human insulin, using Sephadex G50 columns. (b) VSMB system for removal of zinc ions from a biosynthetic human insulin feed stream, using Chelex 100 resin. Reprinted with permission from [59] © 2006, American Chemical Society

7.2.5 Advanced SMB operations

Simulated moving-bed schemes refer to the arrangements of flows and zones or corresponding ports and columns. In general, SMB schemes can be broadly classified by the number of components to be separated and by the number of zones. Most SMB schemes are for binary separations; a few are for ternary separations, with higher order separations most often being a theoretical study. Typical zone numbers are 2 to 12, more typically 3 to 5. Chemical or thermal regeneration and reequilibration normally take up the additional zones.

Complex schemes beyond the traditional four-zone scheme of Figure 7.1 for binary separation arise from attempts to increase bed throughput, reduce the amount of desorbent or expensive stationary phase, and to overcome practical limitations such as pressure limits or stationary phase aging and fouling.

Tandem SMB or SMB rings in series are the easiest schemes for the separations of more than two components, as noted earlier. Tandem SMB allows for more effective independent optimization of both the first and second separations, especially solvent-stationary phase choices, column sizes, and switching times. Parallel schemes, where the rings in series are combined into a single ring, have reduced performance as a result of a single switching time and the same stationary phase and column dimensions. The authors of the this chapter had designed and experimentally verified a nine-zone parallel SMB, illustrated conceptually in Figure 7.10, to purify glucose and xylose from a hardwood yellow poplar hydrolysate [20]. The sugars were the intermediate group of components. The sulfuric acid used in the dilute acid pretreatment was the fast-moving impurity; acetic acid, the primary degradation product of the pretreatment step, was the slow-moving impurity. Sulfuric acid was completely removed in the first SMB segment. The acetic acid was partially removed as the extract of the first SMB segment (zones I to V) to reduce the load on the second segment (zones VI to IX). A solution containing the sugars and reduced amount of acetic acid was removed between the first extract and the feed port and used as feed into the second SMB segment, where the sugars were separated from the remaining acetic acid.

One of the advanced SMB operations, called Varicol SMB, uses asynchronous port switching. Varicol was successfully commercialized by Novasep (Pompey, France) for chiral separations. Other advance operations that have allowed prescribed changes to the constant operation of the traditional SMB scheme include intermittent feeding and withdrawal schemes, cyclic flow rate changes (Powerfeed), and modulation of the feed concentration (Modicon). The JO process uses intermittent feeding and withdrawal to achieve ternary split. In an innovative scheme, a different unit operation that concentrates the liquid stream, such as an evaporator, is integrated between zones. In a nonlinear system, the increased concentration can lead to improved performance. Changing the mobile phase composition, a common operation in batch chromatography, is far more complex in an SMB and has been studied extensively in the past decade. Likewise, changing stationary phase (such as mixing and alternate phases) in some zones is more complex and generally not regarded as economically applicable.

In our earlier studies of insulin purification with a tandem size exclusion SMB, it became apparent that the identity of a feed batch in SMB must be maintained to track each batch throughout the production process [44]. Solutes from a certain batch of feed can be mixed in SMB with solutes from adjacent feed batches. The mixing is due to the locations of the two product ports relative to the feed port and due to the

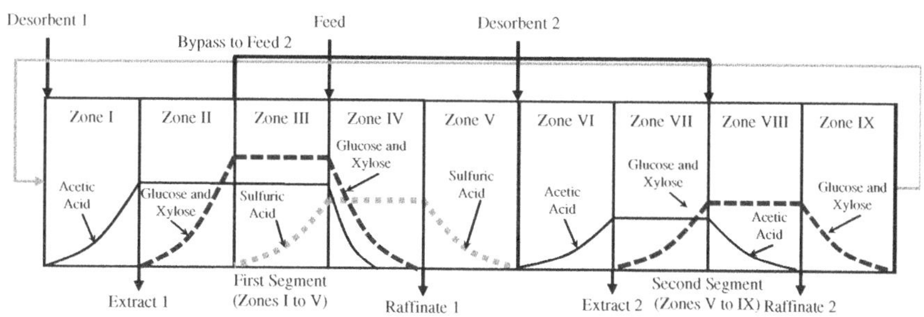

Figure 7.10 *Nine-zone parallel SMB separation of glucose and xylose from fast-moving sulfuric acid impurity and slow-moving acetic acid impurity. The feed hydrolysate was pretreated with 0.3% wt. concentration sulfuric acid at 195°C for 4.5 min in a Sunds reactor at the National Renewable Energy Laboratory, at Golden, Colorado. The SMB uses water as the desorbent and Dowex 99 as the stationary phase*

periodic port switching, recycle, and dispersion effects. The degree of mixing or the size of the overlap region in a product stream is proportional to the solute residence time [39]. If the product is a fast-moving solute (raffinate product), we can reduce its residence time by injecting feed only during the first half of the switching period. If it is a slow-moving solute (extract product), we can reduce its residence time by injecting feed only during the second half of the switching period [39]. We can eliminate the overlap by injecting desorbent instead of feed or by closing the feed port for certain steps [44]. The control of the feed batch identity is achieved here by trading off stationary and mobile phase productivities.

The study on insulin purification eventually led to the development of fast start-up and shutdown procedures that improved productivity and reduced transient times [40] and solute residence times [44]. The same studies also led to the development of a decoupled regeneration scheme to allow column cleaning-in-place without physical removal of the column from the loop; a column was periodically regenerated for a period much longer than the switching time [93]. Traditional methods required substantial columns in the regeneration zone (in multiples of the switching time) or a manual replacement of the column. Interested readers are directed to Chin and Wang (2004) [46], Seidel-Morgenstern *et al.* (2008) [47], and Rajendran *et al.* (2009) [24] for references and further details.

7.2.5.1 Simulated moving-bed reactors

Simulated moving-bed reactors, or SMBRs, were introduced as early as 1977 by Stine and Ward, both from UOP [94]. They involve simultaneous chemical reaction and separation. The reaction can be catalyzed by the stationary phase. The continuous removal of desired product(s) from the reaction zone minimizes the reverse reaction and improves product yield. Simulated moving-bed reactors have been researched for production of β-phenethyl acetate, methyl acetate ester, enzymatic inversion of sucrose to glucose and fructose, lipase-catalyzed diol esterification, and others. Several of these novel SMBR studies are of particular interest to the emerging biorefinery industry. We will briefly highlight two such recent reports.

Archer-Daniels-Midland Co. (Decatur, Illinois, U.S.) recently reported the use of SMBR to produce fatty acid methyl esters (biodiesel) through esterification of fatty acids with methanol [95]. These reactions commonly have low yields and suffer from difficult separations. In its work, the team designed an SMBR process that continuously produced high-purity biodiesel at near theoretical yield [95].

The Rodrigues research group, based in University of Porto, Portugal, is one of the leading SMBR research groups and has published several studies about it. Recently, the group reported preliminary studies on the synthesis of acetals for use as oxygenated additives in liquid fuel [96]. The batch syntheses of acetaldehyde dibutylacetal from butanol and acetaldehyde on Amberlyst-15 resin are encouraging. We expect the team to report success in applying their work to an SMBR process.

7.2.6 SMB commercial manufacturers

Commercial SMB design and construction companies are few. Most of the traditional engineering, procurement, and construction (EPC) businesses have limited abilities to design complex SMB separations, leaving it to specialty design companies. These SMB specialists include Honeywell UOP (Des Plaines, Illinois, U.S.), a large global operation offering process design through optimization of a large multitude of chemical operations, including SMB; and Novasep (Pompey, France) a midsized specialty design company providing designs and equipment for of purification processes and custom manufacturing for the life science industries. UOP, as noted earlier, has extensive SMB industrial experience, especially in the petrochemical industry. UOP is currently building a pilot biorefinery plant with Ensyn Technologies Inc.'s pyrolysis technology (Ottawa, Ontario, Canada) and UOP's hydroconversion technology. Novasep has built itself a sterling reputation in the pharmaceutical industry and a growing portfolio of advanced SMB technologies.

It partnered with Rohm and Haas Co. (Philadelphia, Pennsylvania, U.S.) to develop Rohm and Haas's Ambersep™ BD50 resin in an SMB process to purify crude glycerol from biodiesel production. SepTor Technologies (Utrecht, Netherlands), part of the Outotec Group (Espoo, Finland), reportedly provides SMB process design service and SMB equipment, with a pharmaceutical and fine-chemical focus.

Another SMB design company with a pharmaceutical focus, AMPAC Fine Chemicals (Rancho Cordova, California, U.S.), formerly Aerojet Fine Chemicals, has SMB design and manufacturing capabilities. The SMB design specialty company Amalgamated Research, LLC (Twin Falls, Idaho, U.S.) has an exclusive partnership with BlueFire Renewables (Irvine, California, U.S.) for the use of its SMB technology for acid recovery from biomass hydrolysate. Chiral Technologies, Inc. (West Chester, Pennsylvania, U.S.) offers SMB design and custom manufacturing primarily for enantiomers. Similarly, Sigma-Aldrich Corp. (St. Louis, Missouri, U.S.), through its SAFC division, provides SMB design and custom manufacturing, primarily for enantiomer purification.

Several companies provide commercial off-the-shelf bench-scale SMBs. The SMB of Knauer (Berlin, Germany) uses a central valve, the CSEP valve of Calgon Carbon (Pittsburg, Pennsylvania, U.S.). The Octave SMB of Semba Biosciences (Madison, Wisconsin, U.S.) uses a set of proprietary valve blocks, each with 36 valves, which claims minimum dead volume.

7.3 Simulated moving-bed technology in biorefineries

The nascent biorefinery industry is attempting to compete against the much more mature petrochemical industry in supplying both fuels and chemicals from a sustainable and renewable feed source. Lignocellulosic biorefineries generate a great many intermediate streams that often require product recovery or separation. Examples include syngas from gasification, bio-oil from pyrolysis, sugar hydrolysates from biomass hydrolysis, and fermentation broths. These biorefinery streams frequently contain very many useful products and a significant amount of impurities or reagents, which must be removed or recovered. Simulated moving-bed technology, which has proven to offer substantial cost savings over conventional chromatography, is expected to be a critical core technology for biorefineries in the future.

Simulated moving-bed (SMB) technology can be used to extract high-value components from biomass, such as paclitaxel recovery [31–33], to purify sugars, such as in high-fructose corn syrup processing, or to recover sugars from biomass hydrolysate [20, 61, 85]. It is usable for reagent recovery, such as the separation of concentrated sulfuric acid from sugars after acid hydrolysis of lignocellulosic biomass [97]. Further downstream, it will recover fermentation broth products, such as lactic acid [28]. It is also good for end-product or by-product cleanup.

Several challenges exist in fully utilizing SMB technology in the emerging biorefinery industry, including the limited spread of design and scale-up knowledge and the demanding characteristics of the feed streams. Many of the reported works in SMB applications in biorefineries are still at the research level, often with small pilot units with synthetic feed and relatively short run times. However, the long successful history of SMB operations in the petrochemical and sugar industries provides significant experience to draw on in designing, building, and operating biorefinery SMBs.

Simulated moving-bed feed streams in biorefineries, unlike most existing SMB applications, are often biologically sensitive and contain a large amount of both desirable and undesirable components. Their compositions often change with the season, site, and harvest. The undesirable nonhomogenous components, including tar and phenolics, are often present in significant amounts, and these impurities can quickly foul up the stationary phase. The interactions of minor and often unknown components with one another and with the stationary phase are often difficult to distinguish, making the iterative knowledge-driven SMB design process challenging and time-consuming. The ongoing intense international race to build and operate an economically viable biorefinery demands a fast design of an SMB process that is flexible to significant process changes—most of the major upstream or downstream processes are still being worked out in pilot and demonstration facilities. The knowledge-driven SMB design of Figure 7.2, which relies on detailed intrinsic parameters of all components, is often time consuming. A speedy variant is being developed and tested by the authors. Some of the efficiencies of a well-optimized SMB process may be traded for a fast-working SMB process. The fast-working SMB, within the limits of its initial equipment selection, will evolve over time toward a more efficient process design with an online optimization algorithm.

Simulated moving-bed biorefinery products such as sugars and fermentation products are often low-valued commodities compared to pharmaceuticals and fine chemicals. The lower product values require a low production cost to be commercially viable. On the other hand, very high product purity and yield are not usually required, and the environmental effects of these processes are often significantly better than existing industrial processes. The multicomponent nature of biorefinery feed streams and the need to reduce production costs often require complex higher order SMB schemes, such as the tandem SMB (Figure 7.4), parallel SMB (Figure 7.10), and others.

Although challenges still exist to effectively and efficiently design SMB processes for multicomponent separation, the knowledge-driven design of Figure 7.2 and its attendant standing-wave analysis, the related comprehensive optimization (COSW), the detailed VERSE simulator, and the versatile VSMB equipment, address many of these challenges. The following four processes were chosen as examples of SMB applications in biorefineries.

7.3.1 SMB separation of sugar hydrolysate and concentrated sulfuric acid

BlueFire Renewables is currently commercializing a biorefinery process based on concentrated acid hydrolysis. In this process, biomass is first decrystallized in an extruder reactor by adding 75% sulfuric acid to biomass at 85 °C to form a delignified amorphous gel. Water is then added to dilute the acid to 20%–30%. Further heating in a plug flow reactor results in a near-complete hydrolysis to monomeric sugars. Lignin is then removed via a filter press.

In the late 1980s, researchers at Tennessee Valley Authority (Knoxville, Tennessee, U.S.) and the University of Southern Mississippi (Hattiesburg, Mississippi, U.S.) had improved the process economics substantially compared to the older "Peoria Process" developed by the USDA during World War II. An ion exclusion chromatography step was introduced to achieve high yield in separating the sugars from the acid [97]. This improvement substantially reduces both the amount of acid consumed and the production of waste gypsum from neutralization of the acid with lime. They reportedly achieved 95% sugars and 98% acid recovery in a small pilot-scale 18-column, four-zone SMB system using water as the desorbent and a synthetic feed [98]. The researchers had several patents on the use of ion exclusion chromatography for separation of acid from sugars in SMB [99].

In the BlueFire process, the filtered acidic hydrolysate is separated into acid and sugar syrup streams in an SMB. Sugar at a concentration of about 18 wt. % is recovered in the raffinate, and sulfuric acid at a concentration of 18%–20% is recovered in the extract. Desorbent uses water recovered from the post-SMB acid concentration step. In its Fulton Project, BlueFire claims 98% sulfuric acid yield and 99.5% sugar yield [100]. In its recent program review with the U.S. Department of Energy (DOE), BlueFire further reported achieving less than 0.2% loss of sugars sugar in its SMB step [101]. The slightly acidic sugar extract stream is neutralized with lime to form gypsum, which is removed via a filter press. The pH-adjusted sugar syrup is then fermented to ethanol. Amalgamated Research helped develop the SMB process in an exclusive license with BlueFire.

7.3.2 Five-zone SMB for sugar isolation from dilute-acid hydrolysate

Figure 7.11 shows a dilute-acid hydrolysis scheme to produce ethanol from biomass. Dilute acid is first used to hydrolyze (or pretreat) the hemicellulose in biomass. The more recalcitrant cellulose residue is then separated from the hydrolysate. The primary hydrolysate components are six sugars (glucose, xylose, cellobiose, galactose, arabinose, and mannose) and four impurities (sulfuric acid, acetic acid, hydroxymethyl furfural, and furfural). These impurities inhibit downstream fermentation and must be removed in a purification step. After purification, the purified hydrolysate is added back to the solid residue for simultaneous enzymatic hydrolysis of the cellulose and fermentation of the sugars to ethanol.

The knowledge-driven SMB design, Figure 7.2, was used to develop SMB processes for the purification of the hydrolysate [61]. The six sugars were recovered from corn stover hydrolysate produced by the National Renewable Energy Laboratory (Golden, Colorado, U.S.), and fermented to ethanol [61, 85]. Figure 7.11 shows the central role of the SMB in this biorefinery scheme.

Two different resins, Dowex99 and poly(4-vinylpyridine) (PVP), were tested in this study, which was funded by the U.S. Department of Energy. The sugars were the "center cut" in the Dowex99 column. Thus a tandem SMB with two four-zone SMBs as illustrated in Figure 7.12a was considered for the Dowex99 resin. Sulfuric acid as the fastest moving impurity was first removed at the raffinate port in the first SMB. Its adsorption wave was the designated standing wave in zone IV, and its desorption wave was standing in zone II. The adsorption wave of glucose, the fastest moving component of the sugar group, was the

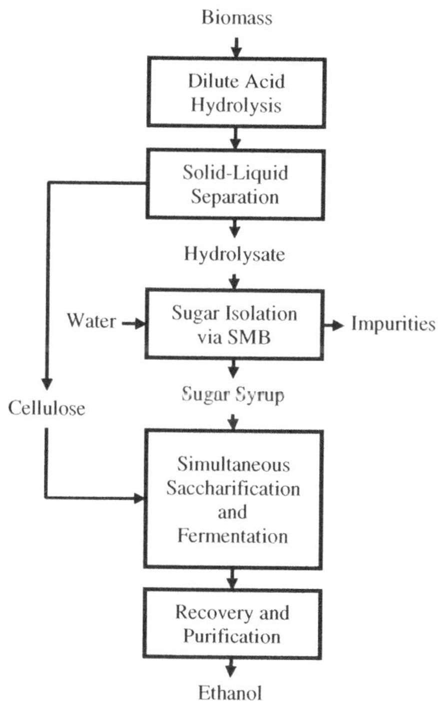

Figure 7.11 *Schematic diagram of SMB operation in the simplified bioethanol production from biomass*

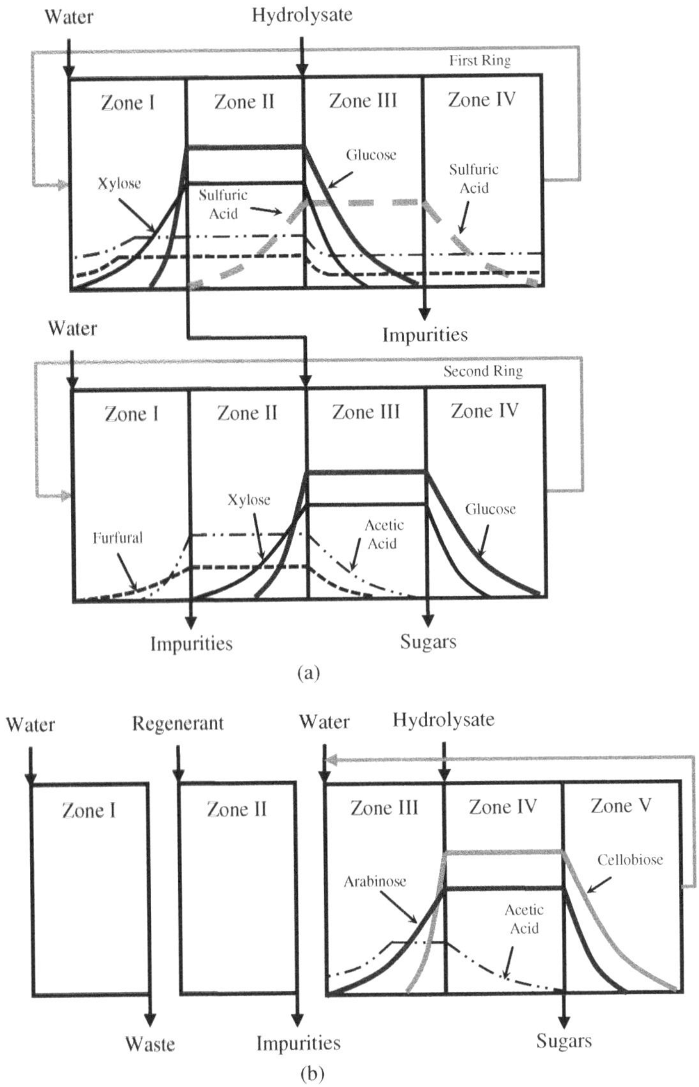

Figure 7.12 *Simulated moving-bed isolation of sugars from impurities. (a) Standing-wave design for tandem Dowex99 SMB; (b) Standing-wave design for five-zone PVP SMB. The component indicated in each zone is the standing component*

designated standing wave in zone III, and the desorption wave of xylose, the slowest moving component in the sugar group, was the designated standing wave in zone I. The remaining impurities (acetic acid, hydroxymethyl furfural, and furfural), which were slower moving than the sugar group, were allowed to distribute throughout the first SMB. These standing wave designations allowed for the complete removal of sulfuric acid and the partial removal of the remaining impurities (thus reducing the load for the second SMB).

In the second SMB, the desorption wave of furfural, the slowest impurity, was designated as the standing wave in zone I; the adsorption wave of acetic acid, the fastest impurity (now that sulfuric acid is removed), was designated as the standing wave in zone III. As such, acetic acid was prevented from reaching the raffinate port for sugars. Furfural, with its desorption wave standing in zone I and adsorption wave standing in zone III, was prevented from reaching the raffinate port. Glucose and xylose were respectively standing in zones II and IV; thus all the sugars were recovered at the raffinate port, resulting in high sugar yield and little loss of sugars to the impurity port.

Sugars, as a group, were the fast-moving components in the PVP column, and only a single five-zone SMB (Figure 7.12b) was needed to recover sugars from the hydrolysate. The impurities had high affinities for the PVP sorbent, and caustic regeneration was needed to remove the impurities efficiently. The feed (hydrolysate) was introduced between zones III and IV. The sugar port was between zones IV and V. The adsorption wave of the fastest moving component in the PVP resin, cellobiose, was designated as the standing wave in zone V. Conversely, the desorption wave of arabinose, the slowest moving sugar, was designated as the standing component in zone III. The adsorption wave of acetic acid, the fastest moving impurity, was designated as the standing wave in zone IV, thereby preventing the impurities from reaching the sugar port. The use of zone V allowed water to be recycled back into zone III. Zone II was the regeneration zone, where the impurities were removed, and zone I was the wash zone, where the regenerant was replaced with water.

The standing-wave analysis was used in the two different SMB designs and verified with VERSE. Additional cost analysis and fermentation tests with genetically modified pentose-fermentable yeast (LNH-ST 424) indicated that the PVP SMB was more cost-effective than the Dowex99 SMB. The PVP design was experimentally verified on a VSMB. Two different feed compositions, two different feed flow rates, and two different regenerants were tested in three PVP SMB experimental runs. All SMB runs had 99+% experimental yields. The sugar purities ranged from 93% to 95% because some unanticipated sulfate and acetate salts coeluted with the sugars. Discounting the salts resulted in >99% purities. General optimization to pilot-scale 50 gal feed/min (or 18.6 mt_{Sugars}/d) operation improved product concentration and reduced purification costs further to 4.7 cents per lb for sugars using the PVP 5-zone SMB. Water and regenerant contributed a major 68.3% to total cost and stationary phase 28.2%.

We also performed a preliminary batch-only investigation with a new stationary phase, activated carbon-type CPGLF from Calgon Carbon (Pittsburg, Pennsylvania, U.S.). This groundwork showed improved hydrolysate decolorization and impurities removal (especially beyond the four-tracked impurities), resulting in experimental fermentability comparable to pure sugars and estimated total cost from 1.2 to 2.4 cents per lb. The estimates were based on the feed concentration of 68 $g_{Sugar}/L_{Hydrolysate}$. A higher feed concentration was expected to improve bed throughput and desorbent efficiency. Further cost improvement to under 1 cent per lb sugars can be attained by increasing feed hydrolysate concentration, scale-up to commercial scale (a 1000 mt/d biorefinery would generate ~10X more hydrolysate), and additional system-wide integration.

This work showed that the knowledge-driven design and its tools could be used to systematically design economical SMB processes for feeds with a great many desired components and impurities. To the best of our knowledge, this was the first attempt in SMB process design to isolate six sugars from a mixture of ten identified components. The work discussed above should easily translate to other sugar hydrolysate streams with a need to remove dilute acids and inhibitors.

7.3.3 Simulated moving-bed purification of lactic acid in fermentation broth

Lactic acid is an important food ingredient and a major starting chemical block to many other valuable products, including ethyl lactate, acrylic acid, propylene glycol, and biodegradable polylactic acid polymer. Traditional lactic acid fermentation in industry starts with carbohydrates, typically glucose or sucrose, and lime or chalk (for pH control) to produce a crude calcium lactate fermentation broth. After gypsum removal, the crude lactic acid undergoes purification and concentration. Several research efforts have identified and constructed fermentation microorganisms capable of using both pentoses and hexoses from biomass hydrolysates [102–105].

Simulated moving-bed technology can play an important role, as illustrated in the earlier two examples, in purifying lignocellulosic biomass hydrolysates for effective fermentation, and in recovering reagents and other valuable by-products of biomass hydrolysis. The resulting lactic acid fermentation broth, like other fermentation products (such as citric acid fermentation), can also be recovered and purified efficiently using SMB technology [106, 107].

Although the use of chromatographic techniques to recover lactic acid has been reported before, UOP first patented the use of basic anionic exchange resin in SMB for purification of lactic acid from fermentation broth [108]. The desorbent of choice is water or dilute inorganic acid such as sulfuric acid. Vogelbusch (Vienna, Austria) later patented the use of a two-step chromatographic process to recover lactic acid at higher purity [109]. Lactic acid mash from *Lactobacillus delbrueckii* fermentation with ammonia water as the neutralizing agent is first separated from the solids, decolorized and concentrated to 30–50 wt% crude ammonium lactate. The salt lactate solution is first converted to free acid by ion exchange with a weakly acidic cation exchanger in the H+ form. Water is used as the desorbent and 1–2N sulfuric acid as the regenerant. In the second step, lactic acid is separated from the impurities, primarily ethanoic acid, in a strongly acidic cation exchanger in the H+ form. Deionized water is used as the desorbent. Although both UOP and Vogelbusch reported only batch data in their patents and no actual SMB design or experimental data, both noted that SMB is the much-preferred mode of operation for the recovery of lactic acid with substantial improvement in resin and desorbent productivity.

In an earlier work, the authors of this chapter designed and tested a four-zone SMB for the recovery of lactic acid from a synthetic broth of lactic acid and acetic acid. The latter was the major competing impurity in the fermentation broth of *Lactobacillus rhamnosus*. The SMB process was developed using the knowledge-driven design of Figure 7.2. The design, based on the Langmuir isotherms, achieved 93% yield and 99.9% purity [28]. The yield loss was due to inaccuracy of the Langmuir isotherm equations in correlating the equilibrium data at low concentrations. The equilibrium data of both components were better correlated with modified Langmuir isotherm equations. VERSE-simulated column profiles and effluent histories based on the modified Langmuir isotherms agreed closely with the experimental data. The standing-wave design method for the modified Langmuir isotherms was developed and tested. VERSE simulations showed that the operating conditions based on the modified standing-wave design could achieve higher lactic acid purity (>99.9%) and yield (>99.9%).

7.3.4 SMB purification of glycerol by-product from biodiesel processing

Transesterification of triglycerides (such as vegetable oils, animal fats, and waste cooking oil) with alcohol, typically ethanol or methanol, results in biodiesel and glycerol as a by-product. This crude glycerol by-product stream is typically contaminated with excess alcohol, water, inorganic salts, and salts of fatty acids, typically 1–5 wt.% sodium or potassium salts. Rohm and Haas and Novasep recently introduced a commercial SMB process for continuous purification of crude glycerol by-product from biodiesel production [110]. The process uses the Ambersep BD50 resin from Rohm and Haas. Glycerol with a purity of 99.5% is recovered in the extract; the fast-moving salts and other colored impurities are recovered in the raffinate. Water is used as the desorbent and recycled from the downstream glycerol product concentration step [111].

7.4 Conclusions and future trends

Simulated moving-bed technology has been proven for product recovery or purification. The basic binary-separation technology is well established in the petrochemical, sugar, and pharmaceutical industries, where SMB has provided significant cost savings over traditional batch chromatography, and new advances promise further cost reductions. SMB technology is expected to have important applications in the emerging biorefining industry. However, serious challenges still exist because of the complex characteristics of biorefinery feed streams, the multicomponent separation requirements, and the demand for low separation costs.

The fundamental understanding of multicomponent separation in SMB has been growing over the past decade, and the design and optimization tools developed can help meet these challenges. The standing-wave analysis (SWA) and its comprehensive optimization algorithms (COSW) can achieve high-purity, high-yield, and efficient separations of multicomponent feed streams. The detailed comprehensive VERSE simulation package allows for fast and accurate *in silico* exploration and can reduce the number of experiments required for process development. The highly versatile VSMB pilot unit allows experimental testing of complex SMB schemes required for multicomponent separations for biorefinery applications.

SMB is being applied in at least one commercial biorefinery demonstration. The authors have seen significant interest from other biorefinery developers and expect SMB technology to play an important role in future biorefineries. Advances in the future are expected in the application of supercritical fluid chromatography in SMB, especially supercritical carbon dioxide, and in the integration of SMB with other separation techniques. These advances will include crystallization and distillation operations between SMB zones. The authors also expect SMB reactors to see significant development in the future and the use of SMB principles in other unit operations.

References

1. S. Eagle and J.W. Scott. Refining by adsorption: New cyclic adsorption process separates high purity aromatics and olefins from petroleum. *Petrol. Process.*, Aug., 881–884 (1949).
2. D.M. Ruthven and C.B. Ching. Countercurrent and simulated countercurrent adsorption separation processes. *Chem. Eng. Sci.*, 44, 1011–1038 (1989).
3. G. Ganetsos and P.E. Barker. *Preparative and production scale chromatography*. Marcel Dekker, New York, 1993.
4. D.W. Guest. Evaluation of simulated moving bed chromatography for pharmaceutical process development. *J. Chromatogr. A*, 760, 159–162 (1997).
5. L. Miller, C. Orihuela, R, Fronek, D. Honda and O. Dapremont. Chromatographic resolution of the enantiomers of a pharmaceutical intermediate from the milligram to the kilogram scale. *J. Chromatogr. A*, 849, 309–317 (1999).
6. M. Juza. Development of a high performance liquid chromatographic simulated moving bed separation from an industrial perspective. *J. Chromatogr. A*, 865, 35–49 (1999).
7. C.M. Grill, L. Miller and T.Q. Yan. Resolution of a racemic pharmaceutical intermediate: A comparison of preparative HPLC, steady state recycling and simulated moving bed. *J. Chromatogr. A*, 1026, 101–108 (2004).
8. L. Miller, C. Grill, T. Yan, O. Dapremont and M. Juza. Batch and simulated moving bed chromatographic resolution of a pharmaceutical racemate. *J. Chromatogr. A*, 1006, 267–280 (2003).
9. P. Wankat. *Large-scale Adsorption and Chromatography*, CRC Press, Boca Raton, FL, 1986.
10. S. Eagle and C.E. Rudy. Separation and desulphurization of cracked naphtha. *Ind. Eng. Chem.*, 42(7), 1294–1299 (1950).
11. R.L. Humpreys. Cyclic adsorption refining. *Petrol. Refiner.*, 32(9), 99–101 (1953).

12. D.B. Broughton. Molex: Case history of a process. *Chem. Eng. Prog.*, 64, 60–65 (1968).
13. G. Maher. *UOP experience with the Sorbex process*, AIChE Chicago section symposium: Discussion of batch and simulated moving bed chromatography design and applications, Chicago, IL, 2001.
14. A.J. Rosset, R.W. Neuzil and D.B. Broughton. Industrial application of preparative chromatography, *in Percolation processes: Theory and application*, A.E. Rodrigues and D. Tonder (Eds.); Sijthoff & Noordhoff International, Rockville, MD, 1981.
15. H. Odawara, M. Ohno, T. Yamazaki and M. Kanaoka. *Continuous separation of fructose from a mixture of sugars*. US Patent 4 157 267 (1979).
16. H. Ishikawa, H. Tanabe and K. Usui. *Process of the operation of a simulated moving bed*. US Patent 4 182 633 (1980).
17. K.W.R. Schoenrock, M.M. Kearney and D.E. Rearick. *Method and apparatus for the sorption and separation of dissolved constituents*. US Patent 4 412 866 (1983).
18. C.B. Ching, C. Ho and D.M. Ruthven. An improved adsorption process for the production of high-fructose syrup. *AIChE J.*, 32(11), 1876–1880 (1986).
19. T. Mallmann, B.D. Burris, Z. Ma and N.-H.L. Wang. Standing wave design of nonlinear SMB systems for fructose purification. *AIChE J.*, 44(12), 2628–2646 (1998).
20. R. Wooley, Z. Ma and N.-H.L. Wang. A nine-zone simulated moving bed for the recovery of glucose and xylose from biomass hydrolyzate. *Ind. Eng. Chem. Res.*, 37, 3699–3709 (1998).
21. D.C.S. Azevedo and A.E. Rodrigues. Obtainment of high-fructose solutions from cashew (*Anacardium occidentale*) apple juice by simulated moving-bed chromatography. *Sep. Sci. Technol.*, 35(16), 2561–2581 (2000).
22. D.C.S. Azevedo and A.E. Rodrigues. Fructose-glucose separation in a SMB pilot unit: modeling, simulation, design and operation. *AIChE J.*, 47, 2042–2051 (2001).
23. M. Schulte and J. Strube. Preparative enantioseparation by simulated moving bed chromatography. *J. Chromatogr. A*, 906, 399–416 (2001).
24. A. Rajendran, G. Paredes and M. Mazzotti. Simulated moving bed chromatography for the separation of enantiomers. *J. Chromatogr. A*, 1216, 709–738 (2009).
25. D.J. Wu, Y. Xie, Z. Ma and N.-H.L. Wang. Design of simulated moving bed chromatography for amino acid separations. *Ind. Eng. Chem. Res.*, 37, 4023–4035 (1998).
26. Y. Xie, D. Wu, Z. Ma and N.-H.-L. Wang. Extended standing wave design method for simulated moving bed chromatography: Linear systems. *Ind. Eng. Chem. Res.*, 39, 1993–2005 (2000).
27. Y. Xie, C.A. Farrenburg, C.Y. Chin and N.-H.L. Wang. Design of SMB for a nonlinear amino acid system with mass transfer effects. *AIChE J.* 49, 2850–2863 (2003).
28. H.J. Lee, Y. Xie, Y.M. Koo and N.-H.L. Wang. Separation of lactic acid from acetic acid using a four-zone SMB. *Biotechnol. Prog.*, 20, 179–192 (2004).
29. N. Gottschlich, S. Weidgen and V. Kasche. Continuous biospecific affinity purification of enzymes by simulated moving bed chromatography: Theoretical description and experimental results. *J. Chromatogr. A*, 719, 267–274 (1996).
30. N. Gottschlich and V. Kasche. Purification of monoclonal antibodies by simulated moving bed chromatography. *J. Chromatogr. A*, 765, 201–206 (1997).
31. D.J. Wu, Z. Ma and N.-H.L. Wang. Optimization of throughput and desorbent consumption in simulated moving bed chromatography for paclitaxel purification. *J. Chromatogr. A*, 855, 71–89 (1999).
32. M.A. Cremasco, B.J. Hritzko and N.-H.L. Wang. Experimental purification of paclitaxel from a complex mixture of taxanes using a simulated moving bed. *Braz. J. Chem. Eng.*, 26(1), 207–218 (2009).
33. S.Y. Mun and N.-H.L. Wang. Optimization of productivity in solvent gradient simulated moving bed for paclitaxel purification. *Process Biochem.*, 43(12), 1407–1418 (2008).
34. E. Küsters, C. Heuer and D. Wieckhusen. Purification of an ascomycin derivative with simulated moving bed chromatography: A case study. *J. Chromatogr. A*, 874, 155–165 (2000).
35. C.A. Farrenburg, N.-H.L. Wang, Y. Xie and B.J. Hritzko. *Method and apparatus for separating a component from a mixture*. US Patent 6 843 854 (2005).
36. U. Voight, J. Kinkel, R. Hempel and R.M. Nicoud. *Chromatographic process for obtaining highly purified Cyclosporin A and related Cylcosporins*. US Patent 6 306 306 (2001).

37. A. Paulon, O. DeD. Lucchi, A. Castellin, F. Fabris, F. Sbrogio, E. Ceron, H. Petersen and R. Dancer. *Method for the preparation of escitalopram*. US Patent Application 2009/0247772 (2009).
38. Y. Xie, S.Y. Mun, J.H. Kim and N.-H.L. Wang. Standing wave design and experimental validation of a tandem simulated moving bed process for insulin purification. *Biotech. Prog.*, 18, 1332–1344 (2002).
39. S.Y. Mun, Y. Xie and N.-H.L. Wang. Residence time distribution in a tandem simulated moving bed process for insulin purification. *AIChE J.*, 49(8), 2039–2058 (2003).
40. Y. Xie, S.Y. Mun and N.-H.L. Wang. Startup and shutdown strategies of simulated moving bed for insulin purification. *Ind. Eng. Chem. Res.*, 42, 1414–1425 (2003).
41. Y. Xie, S.Y. Mun, C.Y. Chin and N.-H.L. Wang. Simulated moving bed technologies for producing high purity biochemicals and pharmaceuticals, *in New Frontiers in Biomedical Engineering*, N.H.C. Hwang and S.L.Y. Woo (Eds.), Kluwer Academic Publishers, New York, NY, 2003.
42. S.Y. Mun, Y. Xie, J.H. Kim and N.-H.L. Wang. Optimal design of a size-exclusion simulated moving bed for insulin purification. *Ind. Eng. Chem. Res.*, 42, 1977–1993 (2003).
43. S.Y. Mun, Y. Xie and N.-H.L. Wang. Robust pinched wave design of a size-exclusion simulated moving bed process for insulin purification. *Ind. Eng. Chem. Res.*, 42, 3129–3143 (2003).
44. S.Y. Mun, Y. Xie, and N.-H.L. Wang. Strategies to control batch integrity in size exclusion simulated moving bed chromatography. *Ind. Eng. Chem. Res.*, 44(9), 3268–3283 (2005).
45. S.Y. Mun and N.-H.L. Wang. Insulin wave dynamics in size-exclusion simulated moving bed with residence time control. *Ind. Eng. Chem. Res.*, 45, 1454–1465 (2006).
46. C.Y. Chin and N.-H.L. Wang. Simulated moving bed equipment designs. *Separ. Purif. Rev.*, 33(2), 77–155 (2004).
47. A. Seidel-Morgenstern, L.C. Keβler and M. Kaspereit. New developments in simulated moving bed chromatography. *Chem. Eng. Technol.*, 31(6), 826–837 (2008).
48. H. Ikeda, M. Negawa and F. Shoji. *Simulated moving bed chromatographic separation process*. US Patent 5 770 088 (1998).
49. A. Depta, T. Giese, M. Johannsen and G. Brunner. Separation of stereoisomers in a simulated moving bed-supercritical fluid chromatography plant. *J. Chromatogr. A*, 865, 175–186 (1999).
50. G. Biressi, F. Quattrini, M. Juza, M. Mazzotti, V. Schurig and M. Morbidelli. Gas chromatographic simulated moving bed separation of the enantiomers of the inhalation anesthetic enflurane. *Chem. Eng. Sci.*, 55, 4537–4547 (2000).
51. B.J. Hritzko, Y. Xie, R. Wooley and, N.-H.L. Wang. Standing wave design of tandem SMB for linear multi-component systems. *AIChE J.*, 48(12), 2769–2787 (2002).
52. Z. Ma and N.-H.L. Wang. Standing wave analysis of chromatography system. Linear systems. *AIChE J.*, 43, 2488–2508 (1997).
53. P.C. Wankat. *Rate-controlled separations*. Blackie Academic and Professional, Glasgow, London, 1994.
54. K.B. Lee, C.Y. Chin, Y. Xie, G.B. Cox and N.-H.L. Wang. Standing wave design of simulated moving bed under a pressure limit for enantioseparation of phenylpropanolamine. *Ind. Eng. Chem. Res.*, 44, 3249–3267 (2005).
55. S.F. Chung and C.Y. Wen. Longitudinal dispersion of liquid flowing through fixed and fluidized beds. *AIChE J.*, 14(6), 857–866 (1968).
56. E.J. Wilson and C.J. Geankoplis. Liquid mass transfer at very low Reynolds numbers in packed beds. *Ind. Eng. Chem. Fund.*, 5, 9–14 (1966).
57. Y. Xie, B. Hritzko, C.Y. Chin and N.-H.L. Wang. Separation of FTC-ester enantiomers using a simulated moving bed. *Ind. Eng. Chem. Res.*, 42, 4055–4067 (2003).
58. K.B. Lee, S.Y. Mun, F. Cauley, G.B. Cox and N.-H. L. Wang. Optimal standing wave design of nonlinear simulated moving bed systems for enantioseparation. *Ind. Eng. Chem. Res.*, 44, 3249–3267 (2006).
59. S. Mun, C. Chin, Y. Xie and N.-H.L. Wang. Standing wave design of carousel ion-exchange processes for the removal of zinc ions from a protein mixture. *Ind. Eng. Chem. Res.*, 45(1), 316–329 (2006).
60. S.Y. Mun, N.-H.L. Wang, Y.M. Koo and S.C. Yi. Pinched wave design of a four-zone simulated moving bed for linear adsorption system with significant mass-transfer effects. *Ind. Eng. Chem. Res.*, 45, 7241–7250 (2006).

61. Y. Xie, C.Y. Chin, D. Phelps, C.H. Lee, K.B. Lee, S. Mun and N.-H.L. Wang. A five-zone simulated moving bed for the isolation of six sugars from biomass hydrolyzate. *Ind. Eng. Chem. Res.*, 44, 9904–9920 (2005).
62. K.B. Lee, R. Kasat, G.B. Cox and N.-H.L. Wang. Simulated moving bed multiobjective optimization using standing wave design and genetic algorithm. *AIChE J.*, 54(11), 2852–2871 (2008).
63. F.G. Cauley, Y. Xie and N.-H.L. Wang. Optimization of SMB systems with linear adsorption isotherms by standing wave annealing technique. *Ind. Eng. Chem. Res.*, 43, 7588–7599 (2004).
64. F.G. Cauley, S.F. Cauley and N.-H.L. Wang. Standing wave optimization of SMB using a hybrid simulated annealing and genetic algorithm (SAGA). *Adsorption*, 14(4–5), 665–678 (2008).
65. G. Storti, M. Mazzotti, S. Carra and M. Morbidelli. Optimal design of multicomponent counter-current adsorption separation processes involving nonlinear equilibria. *Chem. Eng. Sci.*, 44, 1329–1345 (1989).
66. G. Storti, M. Mazzotti, M. Morbidelli and S. Carra. Robust design of binary countercurrent adsorption separation processes. *AIChE J.*, 39, 471–492 (1993).
67. M. Mazzotti, G. Storti and M. Morbidelli. Optimal operation of simulated moving bed units for nonlinear chromatographic separations. *J. Chromatogr. A*, 769, 3–24 (1997).
68. G. Zhong and G. Guiochon. Simulated moving bed chromatography. Effects of axial dispersion and mass transfer under linear conditions. *Chem. Eng. Sci.*, 52, 3117–3132 (1997).
69. C.B. Ching, D.M. Ruthven and K. Hidajat. Experimental study of a simulated counter-current adsorption system. III- Sorbex operation. *Chem. Eng. Sci.*, 40, 1411–1417 (1985).
70. D.C.S Azevedo and A.E. Rodrigues. Design of a simulated moving bed in the presence of mass-transfer resistances. *AIChE J.*, 45, 956–966 (1999).
71. J.A. Berninger, R.D. Whitley, X. Zhang and N.-H.L. Wang. The VERSE model: Simulation of reaction and non-equilibrium dynamics in multicomponent fixed-bed adsorption processes. *Comput. Chem. Eng.*, 15, 749–768 (1991).
72. Q. Yu and N.-H.L. Wang. Computer simulations of the dynamics of multicomponent ion exchange and adsorption in fixed beds—gradient-directed moving finite element method. *Comput. Chem. Eng.*, 13(8), 915–926 (1989).
73. R.D. Whitley, K.E. Van Cott and N.-H.L. Wang. Analysis of nonequilibrium adsorption/desorption kinetics and implications for analytical and preparative chromatography. *Ind. Eng. Chem. Res.*, 32(1), 149–159 (1993).
74. R.D. Whitley, K.E. Van Cott, J.A. Berninger and N.-H.L. Wang. Effects of protein aggregation in isocratic nonlinear chromatography. *AIChE J.*, 37(4), 555–568 (1991).
75. R.D. Whitley, J.A. Berninger, N. Rouhana and N.-H.L. Wang. Nonlinear gradient isotherm parameter estimation for proteins with consideration of salt competition and multiple forms. *Biotechnol. Prog.*, 7(6), 544–553 (1991).
76. K.E. Van Cott, R.D. Whitley and N.-H.L. Wang. Effects of temperature and flow rate on frontal and elution chromatography of aggregating systems. *Separ. Technol.*, 1(3), 142–152 (1991).
77. R.D. Whitley, X. Zhang and N.-H.L. Wang. Protein denaturation in nonlinear isocratic and gradient elution chromatography. *AIChE J.*, 40(6), 1067–1081 (1994).
78. Z. Ma, R.D. Whitley and N.-H.L. Wang. Pore and surface diffusion in multicomponent adsorption and liquid chromatography systems. *AIChE J.*, 42(5), 1244–1262 (1996).
79. J.H. Koh, N.-H.L. Wang and P.C. Wankat. Ion exchange of phenylalanine in fluidized/expanded beds. *Ind. Eng. Chem. Res.*, 34(8), 2700–2711 (1995).
80. J.H. Koh, P.C. Wankat and N.-H.L. Wang. Pore and surface diffusion and bulk-phase mass transfer in packed and fluidized beds. *Ind. Eng. Chem. Res.*, 37(1), 228–239 (1998).
81. M.V. Ernest Jr., J.P. Bibler, R.D. Whitley and N.-H.L. Wang. Development of a carousel ion-exchange process for removal of cesium-137 from alkaline nuclear waste. *Ind. Eng. Chem. Res.*, 36(7), 2775–2788 (1997).
82. B.J. Hritzko, D.D. Walker and N.-H.L. Wang. Design of a carousel process for cesium removal using crystalline silicotitanate. *AIChE J.*, 46(3), 552–564 (2000).
83. S.U. Kim, J.A. Berninger, Q. Yu and N.-H.L. Wang. Peak compression in stepwise pH elution with flow reversal in ion exchange chromatography. *Ind. Eng. Chem. Res.*, 31(7), 1717–1730 (1992).
84. M.A. Cremasco, A. Starquit and N.-H.L. Wang. Separation of L-tryptophan present in an aromatic amino acids mixture in a four-column simulated moving bed: experimental and simulation studies. *Braz. J. Chem. Eng.*, 26(3), 611–618 (2009).

85. Y. Xie, D. Phelps, C.H. Lee, M. Sedlak, N. Ho, and N.-H.L. Wang. Comparison of two adsorbents for sugars recovery from biomass hydrolyzate. *Ind. Eng. Chem. Res.*, 44, 6816–6823 (2005).
86. Y. Xie, E. Van de Sandt, T. de Weerd and N.-H.L. Wang. Purification of adipoyl-7-amino-3-deacetoxycephalosporanic acid from fermentation broth using stepwise elution with a synergistically adsorbed modulator. *J. Chromatogr. A*, 908(1–2), 273–291 (2001).
87. Z. Ma, D. Tanzil, B.W. Au and N.-H.L. Wang. Estimation of solvent-modulated linear adsorption parameters of taxanes from dilute plant tissue culture broth. *Biotechnol. Prog.*, 12(6), 810–821 (1996).
88. D.J. Wu, Z. Ma, B.W. Au and N.-H.L. Wang. Recovery and purification of paclitaxel from using low pressure liquid chromatography. *AIChE J.*, 43(1), 232–242 (1997).
89. C.Y. Chin, Y. Xie, J.S. Alford and N.-H.L. Wang. Analysis of zone and pump configurations in simulated moving bed purification of insulin. *AIChE J.*, 52(7), 2447–2460 (2006).
90. C.-M. Yu, S. Mun and N.-H.L. Wang. Theoretical analysis of the effects of reversible dimerization in size exclusion chromatography. *J. Chromatogr. A*, 1132(1–2), 99–108 (2006).
91. C.-M. Yu, S. Mun and N.-H.L. Wang. Phenomena of insulin peak fronting in size exclusion chromatography and strategies to reduce fronting. *J. Chromatogr. A*, 1192(1), 121–129 (2008).
92. P.-L. Chung, J.G. Bugayong, C.Y. Chin and N.-H.L. Wang. A parallel pore and surface diffusion model for predicting the adsorption and elution profiles of lispro insulin and two impurities in gradient-elution reversed phase chromatography. *J. Chromatogr. A*, 1217(52), 8103–8120 (2010).
93. N.H.-L. Wang and C.Y. Chin. *Versatile simulated moving bed systems*. US Patent 7 141 172 (2006).
94. L.O. Stine and D.J. Ward. *Simulated moving bed reaction process*. US Patent 4 028 430 (1977).
95. D.F. Geier, A.K. Hilaly and J.G. Soper. *Method of preparing fatty acid alkyl esters from waste or recycled fatty acid stock*. US Patent Application 2005/0 245 405 (2005).
96. N.S. Graça, L.S. Pais, V.M.T.M. Silva and A.E. Rodrigues. Oxygenated biofuels from butanol for diesel blends: synthesis of the acetal 1,1-dibutoxyethane catalyzed by Amberlyst-15 ion-exchange resin. *Ind. Eng. Chem. Res.*, 49, 6763–6771 (2010).
97. D.R. Nanguneri and R.D. Hester. Acid/sugar separation using ion exclusion resins: A process analysis and design. *Sep. Sci. Technol.*, 25(13–15), 1829–1842 (1990).
98. R.M. Springfield and R.D. Hester. Continuous ion exclusion chromatography system for acid/sugar separation. *Sep. Sci. Technol.*, 34(6), 1217–1241 (1990).
99. R.D. Hester, G.E. Farina and S. Nanguneri. *Process for separating acid-sugar mixtures using ion exclusion chromatography*. US Patent 5 407 580 (1995).
100. AECOM. *Final environmental assessment and notice of wetland involvement. Construction and operation of a proposed cellulosic biorefinery*, BlueFire Fulton Renewable Energy, LLC, Fulton, Mississippi. Prepared by AECOM (Minneapolis) for US Department of Energy, 2010.
101. N. Sumait and J Cuzens. *BlueFire Fulton renewable energy project*. DOE Biomass Program Peer Review Integrated Biorefineries & Infrastructure Presentation, 2011.
102. B.S. Dien, N.N. Nichols and R.J. Bothast. Recombinant *Escherichia coli* engineered for production of L-lactic acid from hexose and pentose sugars. *J. Ind. Microbiol. Biotech.*, 27(4), 259–264 (2001).
103. A. Garde, G. Johnsson, A.S. Schmidt and B.K. Ahring. Lactic acid production from wheat straw hemicellulose hydrolysate by *Lactobacillus pentosus* and *Lactobacillus brevis*. *Biores. Technol.*, 81(3), 217–223 (2002).
104. M.A. Patel, M.S. Ou, R. Harbrucker, H.C. Aldrich, M.L. Buszko, L.O. Ingram and K.T. Shanmugam. Isolation and characterization of acid-tolerant, thermophilic bacteria for effective fermentation of biomass-derived sugars to lactic acid. *Appl. Environ. Microbiol.*, 72(5), 3228–3235 (2006).
105. M.A. Abdel-Rahman, Y. Tashiro and K. Sonomoto. Lactic acid production from lignocellulose-derived sugars using lactic acid bacteria: Overview and limits. *J. Biotechnol*. doi:10.1016/j.jbiotec.2011.06.017 (2011).
106. S. Kulprathipanja and A. Oroskar. *Separation of lactic acid from fermentation broth with an anionic polymeric absorbent*. US Patent 5 068 418 (1991).
107. J. Wu, Q. Peng, W. Arlt and M. Minceva. Model-based design of a pilot-scale simulated moving bed for purification of citric acid from fermentation broth. *J. Chromatogr. A*, 1216, 8793–8805 (2009).

108. S. Kulprathipanja and A.R. Oroskar. *Separation of an organic acid from a fermentation broth with an anionic polymeric absorbent*. US Patent 5 068 419 (1991).
109. S. Sarhaddar, A Scheibl, E. Berghofer and A. Cramer. *Lactic acid extraction and purification process*. US Patent 5 641 406 (1997).
110. A. Rezkallah. *Method for purification of glycerol*. US Patent 7 667 081 (2010).
111. X. Lancrenon and J. Fedders. An innovation in glycerin purification. *Biodiesel Mag*. June 2008, www.biodieselmagazine.com/articles/2388/an-innovation-in-glycerin-purification (accessed August 26, 2012).

Part IV

Membrane Separation

8

Microfiltration, Ultrafiltration and Diafiltration

Ann-Sofi Jönsson

Department of Chemical Engineering, Lund University, Sweden

8.1 Introduction

Process streams in biorefineries are complex mixtures of various kinds of substances that have to be separated and concentrated. The separation processes used in processing fossil raw materials are generally state-of-the-art, whereas there is still a considerable need for research and development of separation processes for use in plants based on renewable resources [1]. Membrane processes are expected to be key components in biorefineries because of their low energy requirement and low chemical consumption. Fractionation is controlled by the membrane pore size and the operating parameters, and process streams can be treated without adjustment of the temperature or pH.

The main components in biorefinery process streams are cellulose, hemicelluloses, lignin and extractives. Because of the complex nature of these streams, fractionation often has to be performed in a number of successive stages, using a cascade configuration. A schematic illustration of a membrane cascade configuration is shown in Figure 8.1.

Solids that could block the flow channels of membrane modules in the subsequent stages are removed, for example, by filtration or centrifugation (F&C) of the process stream. Suspended solids and colloidal matter (e.g. extractives) are separated by microfiltration (MF) [2–5]. Macromolecules, for example hemicelluloses and protein, are concentrated by ultrafiltration (UF) [4, 6–9]. Monosaccharides, multivalent inorganic ions and low-molar-mass lignin are concentrated by nanofiltration (NF) [4] and salts are removed by reverse osmosis (RO).

Depending on the water quality required, permeate from various stages of the cascade can be reused. When treating an alkaline hydrolysate, NaOH can be recovered by recycling the UF permeate to the process [10]. Nanofiltration can be necessary to obtain a solution pure enough to be reused in the plant

Separation and Purification Technologies in Biorefineries, First Edition.
Edited by Shri Ramaswamy, Hua-Jiang Huang, and Bandaru V. Ramarao.

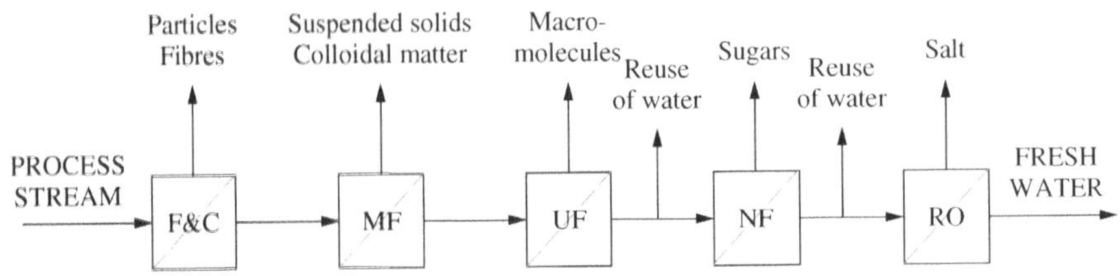

Figure 8.1 *A schematic of a process for the fractionation of substances in biorefinery process streams*

if the solution contains low-molar-mass lignin [4]. Multivalent metal ions are retained by nanofiltration membranes and also to a large extent by UF membranes, when treating pulp mill process streams [11–15]. In order to remove monovalent ions, RO may be needed to produce high-quality water.

Retained solutes can be purified by diafiltration (DF). Diafiltration refers to the mode of operation in which the retentate is diluted with additional solvent and further filtered for the selective removal of low-molar-mass components [16]. Purification of lignosulfonates is usually achieved by DF, for example [17].

Basic principles, process design and optimization of MF, UF and DF are reviewed in this chapter. The requirements for membrane plants in pulp and paper mills are very similar to the industrial challenges in lignocellulosic biorefineries. This survey is therefore to a large extent based on experience from MF and UF applications in the pulp and paper industry.

The distinction between the three pressure-driven membrane processes MF, UF and NF, is somewhat arbitrary. The following definitions are used in this chapter: MF membranes retain suspended particles in the range 0.1–10 µm and UF membranes retain macromolecules in the range 1–20 nm [16]. Nanofiltration membranes have smaller pores than UF, but larger than RO membranes. The classification of MF membranes is based on nominal pore size and UF membranes on nominal molar-mass cut-off in the range 1–1000 kDa. The driving force is usually 2–10 bar during UF and less than 2 bar during MF.

8.1.1 Applications of microfiltration

Microfiltration is used for clarification in the beverage and brewing industries [18–21], the separation of casein micelles and milk proteins [22], the removal of bacteria from milk [23–25], the separation of cells and protein in fermentation processes [26], and the treatment of municipal and domestic wastewater for bacterial control and reuse [27–31].

MF is a feasible technique for the removal of colloidal suspended matter in pulp and paper mills as it removes a substantial part of wood extractives, but not lignin, from integrated kraft pulp and paper mill wastewater [32]. High retention of extractives, but only low retention of lignin, has also been found during MF of thermomechanical pulp-mill process water [4]. Pitch retention has been found to be above 99.9% during MF of hardwood kraft black liquor [2], and terpene-like oils have been removed before anaerobic digestion of evaporator condensate [33]. Removal of toxins and inhibitors prior to anaerobic digestion of stillage [34] and the MF of thin stillage as an alternative to evaporation [35] have been evaluated.

8.1.2 Applications of ultrafiltration

Ultrafiltration has been in industrial use since the 1960s, and is used in a variety of applications [36]. Common applications are the treatment of oily wastewater [37–40], the treatment of electrodeposition paint [41–43], the separation of whey proteins [22, 44, 45] and the treatment of process streams in pulp and paper mills. The majority of the full-scale UF plants in pulp and paper mills are used for the fractionation of spent sulfite liquor and the treatment of bleach plant effluent.

In the mid-1970s, interest grew in the fractionation and concentration of spent sulfite liquors. The aim was to purify lignosulfonate in spent sulfite liquor in order to use it as a chemical product [46–50]. The first commercial, full-scale plant treating spent sulfite liquor was installed at the Borregaard Sarpsborg pulp mill in Norway in 1978 [51, 52]. Tembec, Canada [53] and Domsjö Fabriker AB, Sweden [54] are other companies using UF to isolate lignosulfonates from spent sulfite liquor.

An UF plant for pitch removal was installed at Domsjö sulfite mill, Sweden, in 1985. The plant was equipped with tubular polyvinylidene fluoride membranes with a cut-off of 100 kDa. The membrane area was 728 m^2 [52].

An UF plant treating 400 m^3 bleach plant effluent per hour has been in operation at the Stora Enso Nymölla sulfite pulp and paper mill, Sweden, since 1995 [55, 56]. The plant is the largest of its kind in the world, with a total membrane area of 4800 m^2 divided between two separate lines, the softwood line (2900 m^2) and the hardwood line (1900 m^2). The UF plant removes non-biodegradable, high-molar-mass organic matter from the effluent, reducing the total chemical oxygen demand of the effluent by 50% [56].

The main interest in most investigations on UF of kraft black liquor has been directed towards the use of kraft lignin as an external biofuel [12]. However, a method based on UF of kraft black liquor was developed for producing high-molar-mass lignin compounds for use in plywood by the late 1970s [17, 57]. A process for producing vanillin from kraft lignin has been developed more recently [58]. Today, the focus is on upgrading lignin in kraft black liquor for further conversion, for example to carbon fibres, adhesives, and phenol-based polymers. Brodin *et al*. obtained lignin permeates virtually free of carbohydrates after UF of hardwood kraft black liquor [59].

In most investigations of membrane performance when treating kraft pulping liquors, black liquor is withdrawn before it enters the evaporation unit [12–14, 60–63]. However, Jönsson and Wallberg [64] concentrated kraft black liquor withdrawn after the third evaporation stage in the evaporation unit with a dry solids content of 30 wt% to a volume reduction factor of 3. Ultrafiltration of cooking liquor withdrawn from batch and continuous digesters has been studied at temperatures below 100 °C [15, 65] and at 150 °C [66]. In the latter case cooking liquor was taken directly from a continuous digester without cooling or adjustment of pH.

Isolation of hemicelluloses in pulp mill process streams has attracted interest recently. Galactoglucomannan has been isolated from thermomechanical pulp-mill process streams [4, 67–70], xylan from hardwood kraft black liquor [71] and oligosaccharides have been recovered after the fermentation of biomass [72].

Lipnizki [73] presents a positive outlook on membrane processes in bioethanol plants. Ultrafiltration of stillage in the bioethanol process has been evaluated as an alternative to evaporation [74] and for the recovery of nutrients [7]. A zero-discharge bioethanol process has been developed by recycling the permeate after UF of stillage [75].

8.2 Membrane plant design

The most important factors in the design of a membrane filtration plant, are whether it should be a single-stage or a multistage plant, be operated in batch or continuous mode, and the kind of module configuration that should be used for the membranes.

8.2.1 Single-stage membrane plants

Batch operation using a single-membrane module is the simplest membrane system. Batch operation is used when treating small feed volumes and is commonly employed in laboratory experiments. The solid lines in the single-stage plant shown in Figure 8.2, illustrate the operation of a system in batch mode. Pressure and cross-flow velocity are regulated by the feed pump and the retentate valve.

Internal **circulation** of part of the retentate in a circulation loop (the dashed line in Figure 8.2) can be used when operating at high cross-flow velocity and high pressure. In this operation mode, the feed pump provides the operating pressure and the circulation pump the cross-flow velocity.

Dead-end operation describes the operation when all the retentate is circulated in the internal circulation loop. Dead-end operation is used, for example, in applications where MF is used as a prefiltration stage, to remove small amounts of particles before nanofiltration and reverse osmosis.

Feed-and-bleed operation is a continuous process characterized by continuous addition of feed and partial removal of retentate (dotted lines in Figure 8.2). No retentate is removed during start-up, i.e., the plant is operated in dead-end mode until the desired concentration of the retentate is reached. When the desired concentration is reached, retentate is bled off. Feed is supplied at the same rate as the withdrawal of retentate and permeate.

8.2.2 Multistage membrane plants

Most continuous full-scale membrane plants are multistage plants. A multistage plant consists of a number of feed-and-bleed stages in series, as shown in Figure 8.3. The concentration is constant in each stage and increases along the process.

The flux is higher in a multistage plant than in a single-stage feed-and-bleed plant, which always operates at the final concentration. As the number of stages increases, the flux of the multistage plant approaches the flux of a batch plant. The cost of concentration varies with the number of stages in the membrane plant. Calculating the optimum number of stages in a multistage plant is an iterative process. Cross [76] presents methods of achieving optimum process design for various kinds of systems. A minimum of 3 stages is usually required, and 6–7 stages are quite common in commercial plants [56]. The optimal number of stages has been found to be 8 when concentrating hemicelluloses in a process stream from

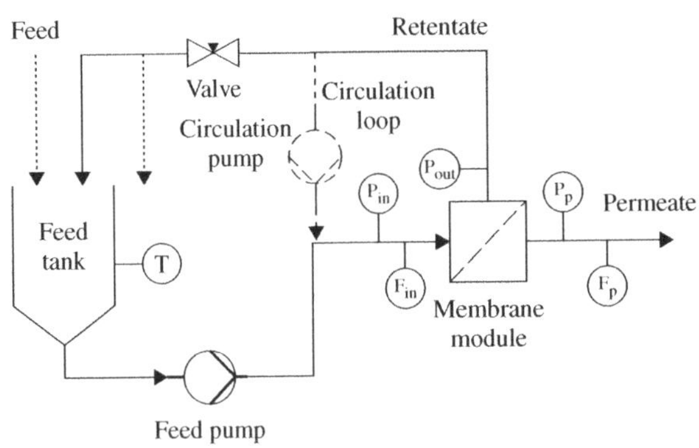

Figure 8.2 *Illustration of a single-stage membrane plant. Solid lines indicate batch operation, the dashed line internal circulation of part of the retentate, and dotted lines feed-and-bleed operation. T = temperature regulator, P_{in}, P_{out}, P_p = pressure transmitters, F_{in}, F_p = flow meters*

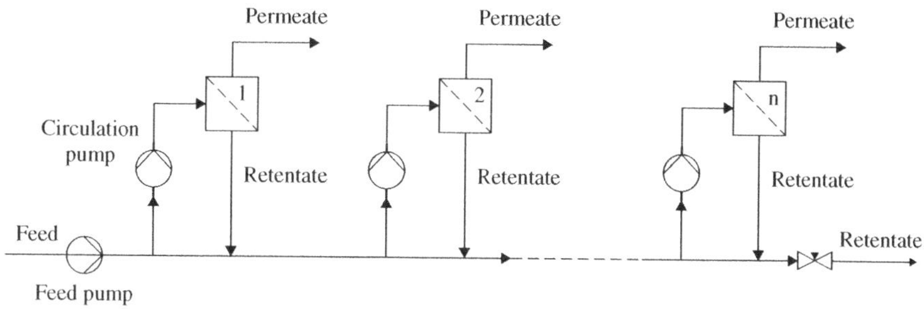

Figure 8.3 Schematic of a continuous multistage membrane plant

a thermomechanical pulp mill [67]. In larger plants one stage is often out of operation at any one time for cleaning.

8.2.3 Membranes

The separation characteristics and the mechanical and chemical stability of the membrane are important parameters when selecting a membrane for a specific application. The mechanical and chemical stability of the membrane are given in terms of the maximum pressure and temperature it can withstand. The chemical stability is given as a pH range and the resistance to solvents. Membranes are manufactured from a variety of materials, both polymeric and ceramic. Temperature and pH resistance are generally higher for ceramic membranes, whereas the maximum pressure is usually higher for polymeric membranes. Common polymeric materials are polysulfone, polyethersulfone, polyvinylidene fluoride and regenerated cellulose. Ceramic membranes are usually made of α-Al_2O_3 and TiO_2.

There is a variety of membranes with different separation characteristics within each group of materials. Microfiltration membranes are characterized by the average pore diameter and UF membranes by their cut-off. The retention of a solute is influenced not only by the size, but also by the shape of the molecule, and inter-molecular and membrane-molecular interactions. The cut-off and pore size of a membrane therefore give only an approximate indication of the separation ability of the membrane. The retention of a substance in a multicomponent solution can be quite different from the retention of a single solute of similar size. Furthermore, although the two applications appear to be similar, the optimal pore size for fractionation can differ due to the heterogeneity of the raw material, and differences in extraction process variables. Screening tests using a number of membranes are therefore advisable when designing a plant for a new application.

8.2.4 Membrane modules

Membranes are integrated into modules. Economic considerations and chemical engineering aspects are of prime importance in the choice of membrane modules. Five basic designs of modules are available on the market [16, 77, 78]:

- tubular modules;
- plate-and-frame modules;
- spiral-wound modules;
- hollow-fibre modules; and
- rotating and vibrating modules.

The principal differences between these module configurations are the price, footprint and risk of blockage of the feed flow channel. Spiral-wound modules are inexpensive and, together with hollow-fibre modules, have the smallest footprint. However, they are susceptible to blockage of the feed channel and extensive pretreatment of the feed is often necessary. Tubular modules have the largest foot-print and highest energy requirement, but the need for pretreatment is low. Plate-and-frame modules have the advantage of being well suited for treating viscous solutions, but the capital cost is high. Rotating and vibrating modules are used to increase shear forces, without increasing the cross-flow velocity. However, these modules are usually expensive.

Tubular modules usually contain one to 18 tubular membranes with tube diameters from 4 to 25 mm. The design of plate-and-frame modules has its origin in the conventional filter press concept. The flat membranes are attached to porous support discs, or a fine-meshed spacer, with a coarse spacer keeping the membranes apart forming the feed flow channels. A package of membranes, discs and spacers is clamped together between two end-plates. A spiral-wound module is, in principle, a plate-and-frame module that has been rolled up and inserted into a tubular pressure housing. The permeate is collected in a tube at the centre of the roll. A hollow-fiber module consists of a large number of membrane capillaries with diameters in the range of 0.5 to 1.5 mm.

8.2.5 Design and operation of membrane plants

When designing a membrane plant, attention is naturally directed towards membranes and modules. A great deal of information is available from membrane manufacturers and in scientific articles. However, advice on the design and operation of full-scale membrane plants is rare. Two exceptions are the article by Brouckaert *et al.* [79] and the *Membrane Filtration Handbook* by Wagner [80]. A number of points that should be borne in mind based on the author's own experience are given below.

- **Pumps** are of crucial importance. Follow the recommendations of the membrane manufacturer. Experience has shown that installing a cheap pump is often a false economy.
- **Construction materials** must be of good quality, preferably stainless steel for all parts and devices that come into contact with liquids. Rust will not only cause fouling, but can also increase the osmotic pressure.
- **Dead spaces** in the tubing system must be avoided as material will be trapped in these spaces and the system will never be thoroughly cleaned. These spaces are also the source of bacterial growth and associated biological fouling.
- **Rinse and clean** the system thoroughly before it is shut down. Never leave the process solution in a membrane plant! Even if it is possible to displace the remaining material afterwards (which is not always the case), extensive and time-consuming cleaning of the plant will be needed.
- **Start-up and shut-down** of a membrane plant are critical stages. The pressure should be increased gradually at start-up in order to avoid high initial concentrations at the membrane surface when the flow resistance of the clean membrane is low. Gradual changing of the temperature when rinsing the system during shut-down is recommended, in order to prevent the precipitation of solutes. Advice on how to start-up and shut-down a pilot plant treating kraft black liquor at temperatures above 100 °C has been presented by Wallberg and Jönsson [66].

8.3 Economic considerations

Separation accounts for 60 to 80% of the process cost in most mature chemical processes [81]. Design and optimization of the separation process are therefore vital in order to reduce costs. Minimizing the cost of membrane plants is generally a trade-off between high flux and a low energy requirement. Flux usually increases with increasing pressure and cross-flow velocity, but so does the energy required [64]. In most cases, the capital cost is the largest cost in a membrane plant. The main operating costs are: for electricity, replacement of membranes and cleaning.

8.3.1 Capital cost

The size of the plant, and hence, the investment cost, is related to the flux. The capital cost is given by:

$$\text{capital cost} = a \cdot Cost_{inv} \cdot A_{tot} = a \cdot Cost_{inv} \cdot \left(\frac{Q_{feed} \cdot VR}{J_{av}} + A_{clean} \right) \tag{8.1}$$

where a is the annuity (i.e. interest and pay-back time), $Cost_{inv}$ the investment cost per unit membrane area and A_{tot} the total membrane area in the plant. The membrane area required is determined by the feed flow, Q_{feed}, the volume reduction, VR, and the average flux, J_{av}, in the plant. A certain membrane area, A_{clean}, is required to allow for cleaning the membranes, which is dependent on the cleaning frequency.

The degree of concentration is often expressed as the volume reduction (VR) or volume reduction factor (VRF). Volume reduction defines how much of the initial feed volume, V_0, has been withdrawn as permeate, V_p, and VRF is the ratio between the initial feed volume and the remaining retentate volume, V_r.

$$VR = \frac{V_p}{V_0} \tag{8.2}$$

$$VRF = \frac{V_0}{V_r} = \frac{1}{(1 - VR)} \tag{8.3}$$

The limiting value of VR depends on the concentration and viscosity of the feed. The value of VR is often in the range 0.8–0.9. The solids content and purity of the retentate increase with increasing VR, and a high VR is therefore desirable. A UF plant treating bleach plant effluent with a VR of 0.98–0.99 shows that it is possible to operate at high VR in commercial plants [56].

8.3.2 Operating costs

The operating cost arises from costs associated with electricity, the replacement of membranes, cleaning, maintenance and labour.

Electricity is used mainly for pumping. In a multi-stage plant, a feed pump is needed to deliver the inlet pressure of the plant and circulation pumps to compensate for the frictional pressure losses and to maintain a certain circulation flow in the membrane modules (see Figure 8.3).

The energy required by the feed pump per m^3 can be expressed as:

$$W_{feed} = \left(\frac{\Delta p \cdot Q_{feed}}{\eta \cdot (VR \cdot Q_{feed})} \right) = \frac{\Delta p}{\eta \cdot VR} \tag{8.4}$$

and that required by each circulation pump:

$$W_{circ} = \frac{\Delta p_{f,mod} \cdot Q_{mod}}{\eta \cdot (J_{mod} \cdot A_{mod})} = \frac{\Delta p_{f,mod} \cdot u \cdot d_h}{4\ \eta \cdot J_{mod} \cdot L} \tag{8.5}$$

where Q_{mod} is the feed flow in the membrane module, A_{mod} is the membrane area in the module, η is the pump efficiency, $\Delta p_{f,mod}$ and J_{mod} are the average frictional pressure drop and the average flux in the module during concentration to VR, u is the cross-flow velocity, d_h is the hydraulic diameter and L the length of a flow channel in the membrane module.

During UF and MF the energy required by the circulation pumps is significantly higher than that of the feed pump because of the high cross-flow velocity and the relatively low feed pressure. In NF and RO plants, on the other hand, the feed pump often requires most energy.

As the concentration in the plant increases, the flux decreases and the frictional pressure drop increases, which leads to an increase in the amount of energy required. This is why the energy requirement is higher (usually much higher) in the last stage than in the first stage of a multistage plant [56].

The membrane lifetime is affected by the operating conditions (temperature, pH, fouling) in the specific application. High temperature, an extreme pH and aggressive particles shorten the lifetime. Frequent cleaning also shortens the lifetime, as the chemical conditions used during cleaning are usually very harsh. The lifetime of ceramic membranes is usually much longer than that of polymeric membranes. The lifetime of polymeric membranes in pulp and paper applications is usually about 18 months while the standard lifetime of ceramic membranes is six years [15].

The cost of cleaning depends on the fouling potential of the feed and the cleaning interval and includes the costs of cleaning chemicals and heating of the rinsing water and cleaning solution. Assuming that cleaning is carried out every two days, this leads to a cleaning cost of about €50 per m^2 per year in 2010 prices [82].

Maintenance and labour costs are commonly estimated to be 5% of the capital cost [64].

The parameters used to calculate the cost of a UF plant equipped with a tubular polymeric membrane and a ceramic membrane are listed in Table 8.1. The values in the table are based on data from membrane manufacturers and membrane plants in the pulp and paper industry.

Cost estimates of UF plants recovering lignin from kraft black liquor have been presented recently. The extraction of lignin fuel from softwood kraft black liquor by UF was estimated to cost about €20 per MWh of calorific value of the lignin fuel produced [15]. It has also been estimated that lignin can be recovered from a hardwood cooking liquor at a cost of about €60 per tonne of lignin and from hardwood black liquor withdrawn from the evaporation unit for about €33 per tonne [64].

The cost of the daily production of 4 tonnes of hemicelluloses has been calculated to be €670 per tonne [67]. The process was based on treatment of thermomechanical pulp process water by MF and UF. The product contained 30 g hemicelluloses per litre with a purity (defined as the ratio between the hemicelluloses and the total solids) of approximately 80%. The total operating cost of an MF plant treating thin stillage from a 150 million litre per year ethanol plant has been estimated to be $274 340 per year [35].

Table 8.1 *Parameters used in cost estimates of a UF plant treating bleach plant effluent (2010 prices). Reprinted from [82] © 2010, with permission from Elsevier*

Membrane material	Polymeric	Ceramic
Investment cost/€ per m^2 membrane area	2000	3300
Membrane cost (excluding housing)/€ per m^2	120	630
Membrane lifetime/years	1.5	6
Electricity price/€ per MWh	38	38
Pump efficiency	0.8	0.8
Cleaning cost/€ per m^2 per year	50	50
Labour and maintenance costs/% of capital cost per year	5	5
Annuity factor	0.1	0.1
Operating time/h per year	8000	8000

8.4 Process design

Optimizing the process design of membrane plants is generally a trade-off between high flux, high recovery and high purity. Unfortunately, high recovery and high purity are not compatible, as will be shown below. A high flux is desirable because the capital cost is inversely proportional to the flux (see Eq. (8.1)). A high recovery is desirable to reduce the cost of the raw material, and a high-purity product is often required.

8.4.1 Flux during concentration

The flux decreases as VR increases, as shown in Figure 8.4. The influence of VR on flux can be described by a polynomial relation:

$$J = a + b \cdot VR + c \cdot VR^2 + d \cdot VR^3 \tag{8.6}$$

where a, b, c and d are polynomial coefficients. The average flux during batch concentration is obtained by integrating Eq. (8.6), and is given by [14]:

$$J_{av} = \frac{\int_0^{VR} J dVR}{VR} = a + \frac{b}{2} \cdot VR + \frac{c}{3} \cdot VR^2 + \frac{d}{4} \cdot VR^3 \tag{8.7}$$

8.4.2 Retention

The separation properties of a membrane are usually specified in terms of the retention of the membrane. A membrane that retains all solutes has a retention of 100%, and the retention is 0 if no solutes are retained. It is important to be aware of the fact that the retention depends on the operating conditions. The retention in a specific application is therefore seldom the same as the retention given in the membrane manufacturer's data sheet. The purpose of the retention values presented by the manufacturer is to simplify the choice of membranes to be evaluated for new applications, and to simplify the comparison of membranes from different manufacturers.

The retention is defined as:

$$R = 1 - \frac{C_p}{C_r} \tag{8.8}$$

where C_p and C_r are the concentrations in the permeate and in the retentate.

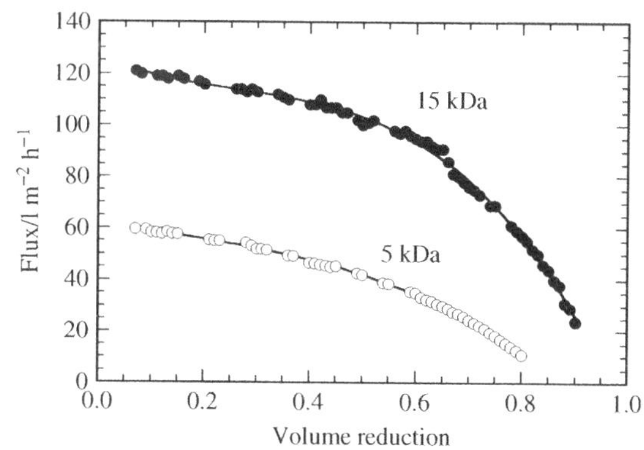

Figure 8.4 Flux during the concentration of kraft black liquor at 90°C with ceramic membranes with cut-offs of 5 and 15 kDa. The symbols are experimental values and the curves were obtained by regression analysis of the data using a third-order polynomial (Eq. (8.6)). Reprinted from [14] © 2003, with permission from Elsevier

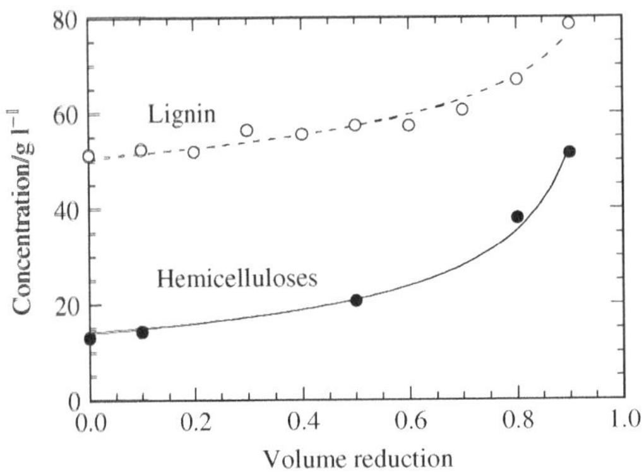

Figure 8.5 *Concentration of lignin and hemicelluloses in the retentate during UF of hardwood kraft black liquor*

If the retention is constant during concentration, the correlation between the retentate concentration and VR in a batch process follows the relationship [16]:

$$C_r = C_0 \cdot \left(\frac{1}{1 - VR}\right)^R \tag{8.9}$$

where C_0 is the initial concentration.

The retention of multidisperse compounds commonly increases during concentration. The mean retention during batch concentration can be obtained by regression analysis of the experimental data using Eq. (8.9). For example, the retention of lignin increased from about 10% at VR = 0 to almost 30% at VR = 0.9, and the retention of hemicelluloses from 83% to 90% during UF of kraft black liquor (see Figure 8.5). The mean retention of lignin and hemicelluloses was 18% and 57%.

It is important to bear in mind that the retention of a substance in a multicomponent solution is often higher than the retention of a single model substance of similar size. For example, the retention of polyethylene glycol with molar mass of 4 and 40 kDa was 0.5% and 6%, whereas the retention of galactoglucomannan (GGM) with a mean molar mass of 9 kDa was >50% during MF of process water from a thermomechanical pulp mill [83]. The higher retention of GGM may be due to the formation of a filter cake at the membrane surface, consisting of suspended solids and colloids retaining molecules that would pass through a clean membrane, and to the aggregation of GGM in a layer of increased concentration at the membrane surface.

8.4.3 Recovery and purity

The VR and the retention govern the recovery and purity of the product.

The **recovery** is the fraction of a component in the feed that is collected as a useful product. If the retention is constant during concentration and the retentate is the product, recovery is given by:

$$\text{recovery} = \frac{m_r}{m_0} = (1 - VR)^{1-R} \tag{8.10}$$

where m_r and m_0 are the amounts of substance in the retentate and feed. If R < 100%, part of the product is lost with the permeate during concentration, which reduces the recovery, as shown in Figure 8.6. The retention values used in the figure are representative of the retention of hemicelluloses (95%) and lignin (50%) in thermomechanical pulp process water.

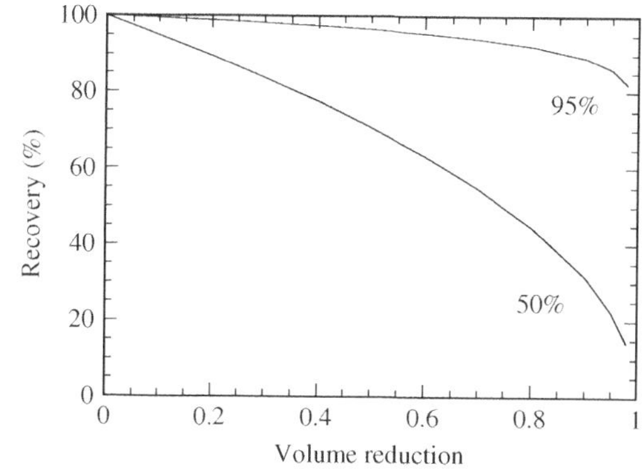

Figure 8.6 *Influence of volume reduction on recovery during UF of two compounds with retentions of 95% and 50%. Recovery was calculated using Eq. (8.10)*

If the retention is zero, the concentration in the permeate is similar to that in the retentate, and the amount removed depends only on the VR, according to Eq. (8.10).

The purity of the product depends on the concentration of all compounds in the product stream. The purity of the product in the retentate is:

$$\text{purity} = \frac{C_r}{\sum C_i} \tag{8.11}$$

where C_r and ΣC_i are the concentrations of the product and of all compounds in the retentate.

The purity depends on the difference in retention of the compounds in the feed. Concentration and purity both increase with increasing VR, as shown in Figure 8.7, but at the expense of recovery (see Figure 8.6).

The purity of the high-molar-mass product increases with increasing VR as more of the low molar-mass components is removed. A steep increase in purity is often obtained at high VR. However, the increase in purity diminishes when the VR is so high that most of the remaining compounds have similar, or higher, retention than the product [67].

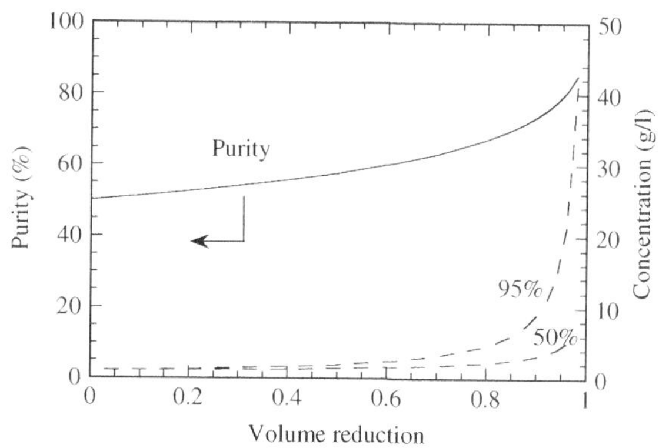

Figure 8.7 *Influence of volume reduction on concentration (dashed lines) of two compounds with retentions of 95% and 50%. Purity (solid line) was calculated using Eq. (8.11). The initial concentration of each substance was 1 g l*$^{-1}$

8.5 Operating parameters

Costs and membrane performance are affected by a number of operating parameters, that must be optimized in each specific application. Operating parameters commonly studied are:

- the transmembrane pressure (TMP);
- the cross-flow velocity;
- the temperature; and
- the pH.

Parameter studies are often performed at a number of concentrations in order to simulate the conditions in different stages of a multistage plant. The importance of each operating parameter depends on the characteristics of the specific membrane and the feed in question. The TMP and the cross-flow velocity are the most important operating parameters and are therefore often optimized in pilot-scale investigations.

Some general trends in UF and MF will be illustrated below.

- When the operating **pressure** is increased the flux first increases almost linearly. It then levels off as the pressure is raised further, and finally may even decrease at elevated pressures.
- An increase in the **cross-flow velocity** causes an increase in the flux. The effect is reduced as the velocity is increased further.
- Flux increases with **temperature**.
- Flux decreases as the solute **concentration** is increased.

Typical flux curves are shown in Figure 8.8. Pure water flux (PWF) is defined as the flux of deionized water, and the limiting flux is the highest flux that can be obtained when increasing the TMP within a given set of operating conditions. There are many definitions of the concept of critical flux, all excellently reviewed by Bacchin *et al*. [84]. A common definition of critical flux is the flux at which the operation shifts from reversible to irreversible during the increase in pressure—i.e., the point below which the flux remains constant with time. The most important differences between UF and MF are that the PWF is generally much higher, and that the limiting flux is reached at lower pressure, for MF membranes.

The PWF is defined as:

$$PWF = \frac{\Delta p}{\mu \cdot R_m} \tag{8.12}$$

where Δp is the TMP, μ is the viscosity of water and R_m is the hydraulic resistance of the membrane. The PWF is used to calculate the membrane resistance and to control the efficiency of cleaning. The PWF is a

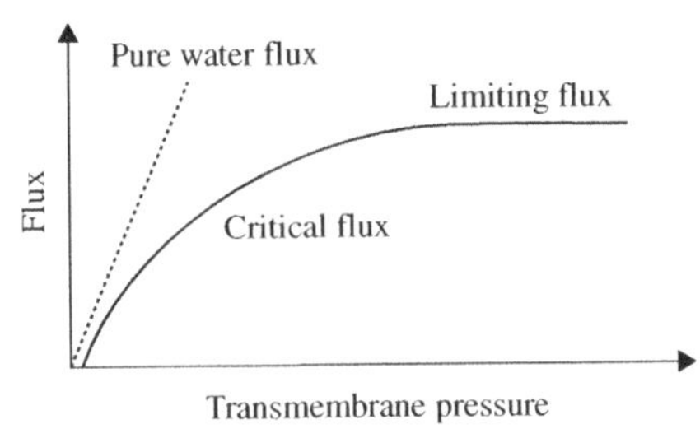

Figure 8.8 Influence of transmembrane pressure on pure water flux (dashed line) and on the flux of a solution (solid line)

linear function of pressure until the maximum operating pressure of the membrane is reached. At this point compressibility of the membrane limits the flux increase. UF and MF membranes present a wide variety of PWF values depending on the pore size and porosity of the membrane. The PWF of UF membranes at 1 bar and 25 °C is usually in the interval 10–500 l/m^2 h, and that of MF membranes 500–2000 l/m^2 h.

There are two basic flux models. In the osmotic pressure model, the driving force is reduced by the osmotic pressure across the membrane. In the cake filtration model, the flux is reduced due to the additional flow resistance offered by material retained by the membrane at the membrane surface.

In the **osmotic pressure model**, the flux is given by:

$$J = \frac{\Delta p - \Delta \pi}{\mu_p \cdot R_m} \tag{8.13}$$

where $\Delta \pi$ is the osmotic pressure difference across the membrane and μ_p is the viscosity of the permeate. The osmotic pressure of suspended solids and colloids retained by MF membranes is negligible and the osmotic pressure in the bulk solution of macromolecules retained by UF membranes is usually insignificant. However, when there is a permeate flow through the membrane, solutes are transported together with the solvent to the membrane surface. This means that retained compounds will accumulate near the membrane, and the concentration at the membrane surface will be higher than the concentration in the bulk solution. This phenomenon occurs to varying degrees in all membrane processes, and is commonly referred to as concentration polarization [16, 77, 78]. The osmotic pressure of macromolecules at the concentrations prevailing at the membrane surface can be significant, markedly reducing the flux [85–87].

In the **cake filtration model**, it is assumed that a layer of concentrated solute, a cake or a gel, is formed at the membrane surface. The flux is then expressed as:

$$J = \frac{\Delta p}{\mu_p \cdot (R_m + R_c)} \tag{8.14}$$

where R_c is the hydraulic resistance of the filter cake. In Eq. (8.14), the effect of the osmotic pressure on the driving force is neglected since the additional resistance of the filter cake is large in comparison.

The osmotic pressure model applies fairly well when treating solutions containing small, non-interacting solutes, whereas when treating solutes with a molar mass greater than 100 kDa, the cake filtration model gives a better description of the concentration profile. The cake filtration model is thus often used during MF and the osmotic pressure model during UF.

Flux decline with time (fouling) is not included in Eqs (8.13) and (8.14). The **resistance-in-series model** is often used to interpret and quantify flux decline behaviour. In this model, which is an extension of the cake filtration model, the resistance to flow is accounted for by several resistances in series with the membrane:

$$J = \frac{\Delta p}{\mu_p \cdot (R_m + R_{cp} + R_f)} \tag{8.15}$$

where R_{cp} is the reversible resistance to flow due to concentration polarization and R_f is the fouling resistance. The fouling resistance is divided into different components, presented in more detail in the later section on fouling.

8.5.1 Pressure

The pressure in a membrane plant is measured at the inlet, p_{in}, and outlet, p_{out}, of the membrane module and on the permeate side, p_p, as shown in Figure 8.2.

The **transmembrane pressure**, is the driving force in UF and MF. The TMP used in Eqs (8.12)–(8.15) is generally the average pressure difference across the membrane:

$$\Delta p = \frac{p_{in} + p_{out}}{2} - p_p \tag{8.16}$$

The TMP is usually regulated by the retentate valve. However, when using ceramic membranes the TMP is often regulated by a valve on the permeate side of the module. If there is no permeate valve, the permeate pressure equals atmospheric pressure.

The frictional pressure drop is the pressure difference between the inlet and outlet of the module.

$$\Delta p_f = p_{in} - p_{out} \tag{8.17}$$

The pressure drop is often negligible in small bench-scale membrane modules, but can be significant in full-scale modules. If the inlet pressure is high, as during UF, the difference between the pressure at the inlet and the outlet is usually small enough to be considered negligible. However, when treating viscous fluids at high cross-flow velocities, the pressure drop can be so high that the flux is zero, or there may even be a reverse flow of permeate, in the last part of the feed flow channel [56, 88].

In MF applications, relatively low pressures and high cross-flow velocities are used. This means that the frictional pressure drop is often of the same magnitude as the inlet pressure. As a result of this, a filter cake may be formed due to the high flux at the entrance of the module, while the flux towards the end of the module is zero. One method employed to obtain a uniform TMP throughout the entire membrane module is the Bactocatch process. Here a forced flow on the permeate side of the membrane is adjusted so that a pressure drop corresponding to the pressure drop on the feed side is obtained [89]. There are also ceramic membranes with longitudinal permeability gradients that allow a constant TMP along the whole length of the module [90]. Ways of achieving a uniform TMP in the feed channel have been reviewed by Vadi and Rizvi [91].

The frictional pressure drop increases during concentration due to the increase in the viscosity. Higher frictional pressure drop leads to an increase in the energy required during concentration (see Eq. (8.5)). The energy required in the last stage is 12 times higher than in the first stage in a six-stage UF plant treating bleach-plant effluent, for example [56].

8.5.2 Cross-flow velocity

Cross-flow operation is employed to reduce concentration polarization and increase mass transfer [92–94]. In the film theory, the fluid flow in the boundary layer adjacent to the membrane surface is assumed to be laminar, whereas the fluid flow outside this layer is turbulent with complete mixing of the solute. At steady state, the convective solute transport in the boundary layer is equal to the permeate flow, and the diffusive back transport of solute into the bulk solution. The correlation between flux and concentration at the membrane surface given by the film theory model is:

$$J = k \cdot \ln\left(\frac{C_m - C_p}{C_b - C_p}\right) \tag{8.18}$$

where $k = D/\delta$ is the mass transfer coefficient and C_m, C_b and C_p are the concentrations at the membrane surface, in the bulk solution and in the permeate. D is the diffusion coefficient and δ is the thickness of the boundary layer. The thickness of the boundary layer depends on the rheological properties of the solution and the shear forces.

Traditionally, the mass transfer coefficient is estimated from the Sherwood number [95]:

$$Sh = \frac{k \cdot d_h}{D} = A \cdot \mathrm{Re}^a \cdot Sc^b \tag{8.19}$$

where Re is the Reynolds number, Sc is the Schmidt number and A, a and b are empirically determined constants. In order to increase the flux, the mass transfer coefficient is normally increased by decreasing the boundary layer thickness. In most modules the shear rate is enhanced by increasing the cross-flow velocity. The dependence of the mass transfer coefficient on cross-flow velocity is given by [96]:

$$k \propto u^{0.33} \text{ (laminar flow)} \tag{8.20}$$

$$k \propto u^{0.69-0.8} \text{ (turbulent flow)} \tag{8.21}$$

where u is the cross-flow velocity. The flow in the feed channel is usually turbulent.

The mass transfer coefficient is also influenced by the viscosity, and hence by the concentration of the feed solution. The influence of bulk viscosity on the mass transfer coefficient under turbulent flow conditions is given by [97]:

$$k \propto \mu_b^{-0.33} \tag{8.22}$$

The influence of cross-flow velocity on flux during UF of black liquor is shown in Figure 8.9. As can be seen, the cross-flow velocity has a significant effect, on both the increase in flux with TMP and the value of the limiting flux. However, not only the flux, but also the frictional pressure drop, increases with increasing velocity. The frictional pressure drop in the experiment shown in the figure was 40 kPa at 2 m s^{-1}, 90 kPa at 4 m s^{-1} and 160 kPa at 6 m s^{-1} [98]. It is therefore necessary to take into account both the increased flux and the increased energy requirement when optimizing the cross-flow velocity.

It is important to realize that the filtration characteristics of liquors can change with concentration. It is thus not sufficient to perform a parameter study only at the initial concentration of the feed solution.

8.5.3 Temperature

Increasing the feed solution temperature leads to three flux enhancing effects: (i) The viscosity of the permeate decreases which results in a higher flux, according to Eqs (8.12)–(8.15). (ii) The lower viscosity

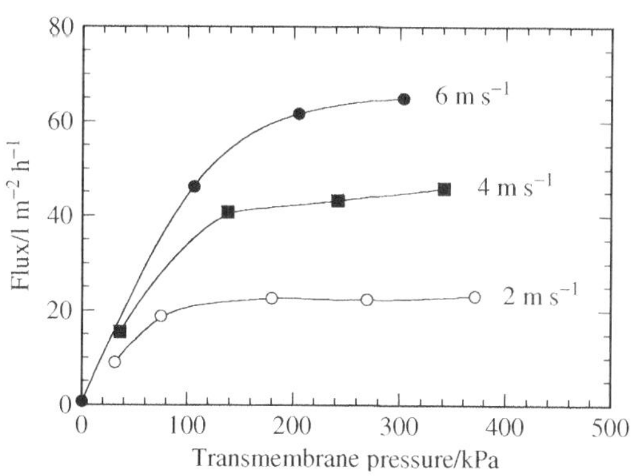

Figure 8.9 *Influence of transmembrane pressure and cross-flow velocity on flux during UF of kraft black liquor at 90°C with a 15 kDa ceramic membrane. The dry solids content was 31%. Reprinted from [98] © 2008, with permission from Elsevier*

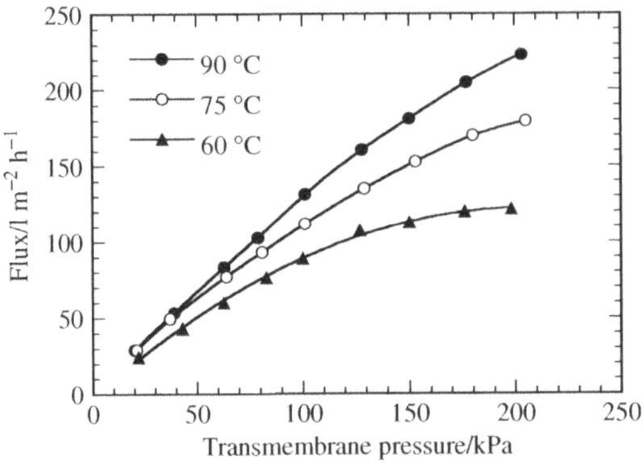

Figure 8.10 *The influence of transmembrane pressure on the flux of kraft black liquor during UF with a ceramic membrane at different temperatures. Reprinted from [12] © 2003, with permission from Elsevier*

of the solution on the feed side of the membrane improves the mass transfer coefficient, as shown by Eq. (8.22). (iii) The lower bulk solution viscosity increases the Reynolds number, and hence, lowers the frictional pressure drop, which also has a positive effect on flux. An example of the influence of temperature on flux during UF of kraft black liquor is shown in Figure 8.10. A similar increase in flux with increasing temperature was found during UF of black liquor above 100 °C [66].

8.5.4 Concentration

Membrane processes are limited in the degree of concentration of the feed solution. No membrane process can concentrate solutes to dryness. In RO and NF processes, it is frequently the osmotic pressure of the concentrated solute that limits the process. In MF and UF processes, it is rarely the osmotic pressure, but rather the low mass transfer rate of the high-molar-mass substances retained by these membranes, and the high viscosity, that make pumping of the retentate difficult, limiting the final concentration.

Flux decreases as the concentration increases, as shown in Figure 8.4. According to Eq. (8.18) the flux varies proportionally with the logarithm of the bulk concentration, i.e. when J is plotted against ln C_b, the intercept on the ln C axis is at a concentration corresponding to the limiting concentration. However, Eq. (8.18) only applies when the flux is independent of the pressure, i.e. at limiting flux. A straight line could be drawn between the flux and the logarithmic concentration only at a TMP of 6 bar, but not at 2 or 4 bar, during concentration of pulp-mill process water, as shown in Figure 8.11. Thus, according to Eq. (8.18), limiting flux was only reached at 6 bar. Extrapolation of the straight line in the figure indicates a limiting concentration of about 110 g l^{-1}.

The limiting concentration of a 5 kDa membrane when concentrating kraft black liquor withdrawn at two different positions at the pulp mill was 240 and 250 g l^{-1}, as can be seen in Figure 8.12.

8.5.5 Influence of concentration polarization and critical flux on retention

A high flux reduces the capital cost, but is not entirely positive. The concentration at the membrane surface increases as the flux increases, according to the film theory model.

$$C_m = C_p + (C_b - C_p) \cdot \exp\left(\frac{J}{k}\right) \tag{8.23}$$

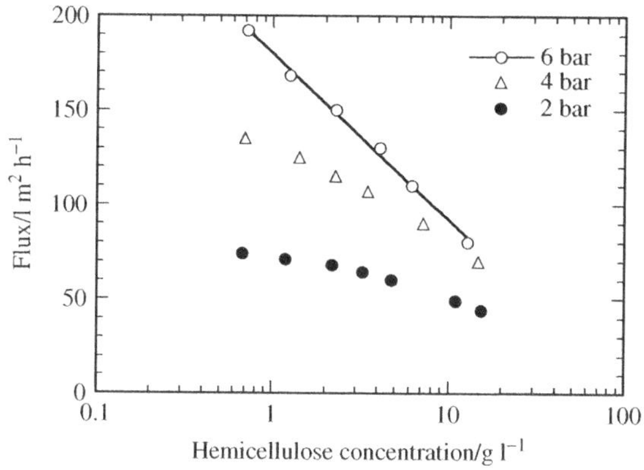

Figure 8.11 *Influence of hemicellulose concentration on flux at various transmembrane pressures during UF of thermomechanical pulp mill process water using a membrane with a cut-off of 5 kDa. Reprinted from [70] © 2010, with permission from Elsevier*

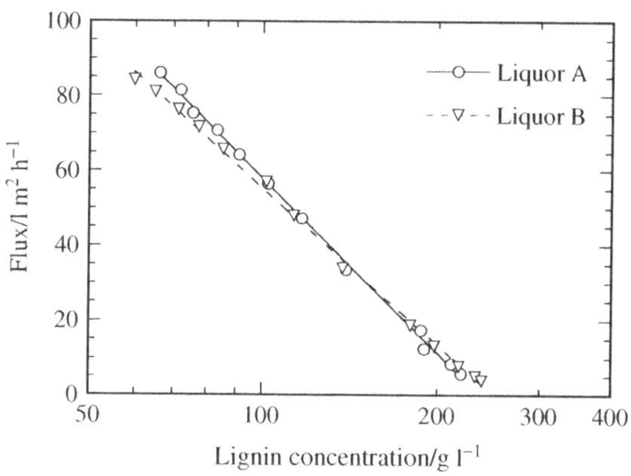

Figure 8.12 *Influence of lignin concentration on flux during UF of kraft cooking liquor. Symbols are experimental data and the lines show the values calculated with Eq. (8.18). Reprinted from [65] © 2005, with permission from Elsevier*

The increased concentration at the membrane may decrease retention and increase fouling. An example of this can be seen in Figure 8.13, which shows the result of concentrating hemicelluloses in thermomechanical pulp mill process water by UF. A marked reduction in retention with increasing flux was noted.

The hemicelluloses in the process water are polydisperse. The increase in retention with increasing hemicellulose concentration observed in Figure 8.13 is probably due to an increase in the mean molar mass of the hemicelluloses during concentration of the process water.

Retention decreases with increasing flux below the critical flux, but increases when a cake or a gel is formed on the membrane [99]. The critical flux was not reached in the experiment shown in Figure 8.13. When using a more open membrane (with lower R_m), the critical flux was reached, and the retention was found to increase above the critical flux, as shown in Figure 8.14. The increase in retention above the critical flux is due to compression of the material retained at the membrane surface when the flux (i.e. the TMP) is increased.

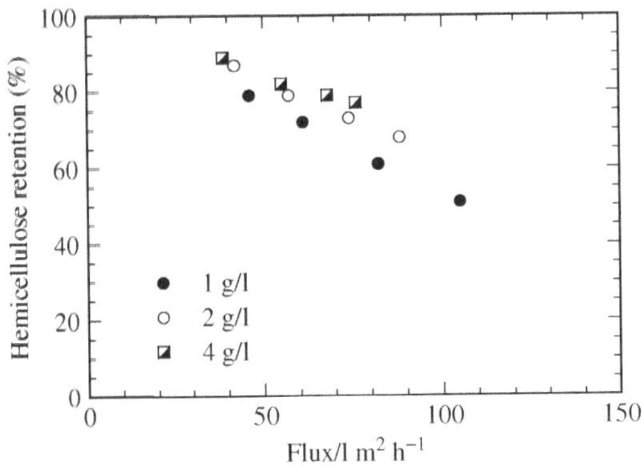

Figure 8.13 Influence of flux on the retention of hemicelluloses at various hemicellulose concentrations during UF with a polymeric membrane with a cut-off of 1 kDa. Reprinted from [70] © 2010, with permission from Elsevier

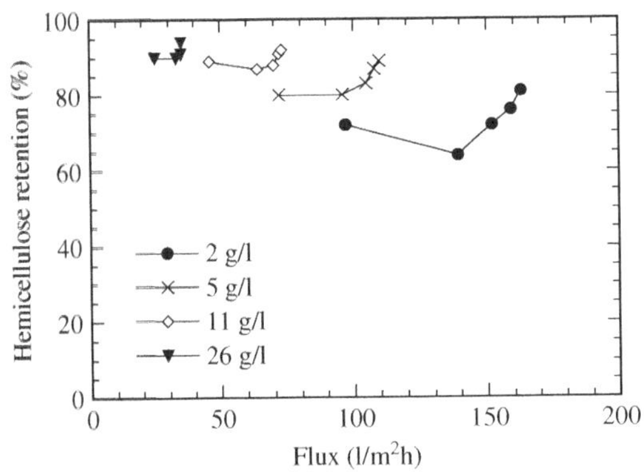

Figure 8.14 Influence of flux on the retention of hemicelluloses at various hemicellulose concentrations during UF with a polymeric membrane with a cut-off of 10 kDa. Reprinted from [70] © 2010, with permission from Elsevier

8.6 Diafiltration

In diafiltration (DF) the product is purified by removing permeable solutes by dilution with water. Diafiltration is used to purify the product when the purity is not sufficient after UF. It is the state-of-the art method of purification in many industries, for example during fractionation of spent sulfite liquor. Haagensen [100] reported in an early study that 82% of the dry matter in the retentate was made up of lignosulfonate after UF. A final lignin purity of 95% was achieved after DF of the retentate.

Diafiltration can be carried out in two ways: discontinuous or continuous. In discontinuous DF, freely permeable solutes are removed from the retentate by volume reduction using UF (preconcentration), followed by dilution with a solvent (usually water) and concentration of the diluted solution by UF (final concentration). Dilution and UF are repeated until the desired purity is achieved. A schematic flow scheme of a discontinuous DF process is shown in Figure 8.15.

- **Preconcentration**—The concentration is increased until the flux becomes too low. The VR during this phase is normally 0.8–0.95.

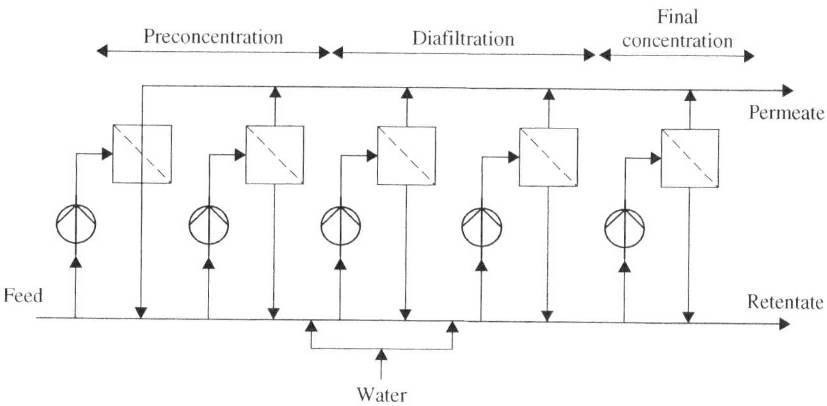

Figure 8.15 *Schematic flow scheme of a discontinuous diafiltration process*

- **Diafiltration**—Water is added to one or more stages to wash out undesired low-molar-mass substance. The purity of the final product depends to a large extent on the amount of diafiltration water added.
- **Final concentration**—The concentration is increased in the last stage to the desired level. Final dry solids contents of 25–30% from kraft black liquor [65], about 10% from thermomechanical pulp process water [4] and about 30% from bleach plant effluent [82] have been reported.

In continuous DF water is added at the same rate as the permeate flux, thus keeping feed volume constant. The basic principles of DF are presented in detail by Cheryan [16].

It must be remembered that purity increases at the expense of recovery. When increasing the purity of hemicelluloses in pulp mill process water from 57 to 77% by DF, the recovery simultaneously decreased by 16% [101].

Whether or not DF should be used depends on the circumstances. Continuous DF is especially applicable when the concentration of the retained solute is so high that the flux will fall to uneconomical values. However, more diafiltration liquor is needed to obtain the same purity of the retentate during continuous DF. This results in a larger volume of permeate requiring subsequent processing. The smallest amount of dilution liquid is needed during discontinuous DF with a high VR during preconcentration.

If DF is started early (at a low VR), the permeate flux will be high, but a large amount of fresh water must be added to obtain a given purity. The membrane area required will thus be large. On the other hand, if DF is started late (at a high VR), only a small amount of fresh water is required to obtain the same purity. However, as the flux is low in this case, a large membrane area is still required. Consequently, there is an optimum value of the conversion ratio at which DF should be started in order to minimize the membrane area required.

Foley [102, 103] has investigated the possibility of finding the optimum macrosolute concentration for the start of DF, with the aim of minimizing the total process time. Classic concentration polarization analysis of UF has been used [102], as well as the theory of UF based on osmotic pressure and the viscosity dependency of the mass transfer coefficient [103]. When using polarization analysis it is always possible to find an optimum for macrosolutes with a retention of 100% when operating at the limiting flux. However, when the retention is less than 100% and the variation in macrosolute concentration is taken into account, an optimum can only be found for macrosolutes with a retention higher than 85%. When the retention is lower, Foley [102] suggests that the entire volume reduction should be completed before DF is performed.

A new concept of counter-current DF has been introduced. In this concept, part of the permeate is recycled as diafiltration water, which reduces both the amount of diafiltration water required and the amount of permeate produced. Batch, continuous and counter-current DF have been compared in a case study performed by Lipnizki *et al.* [104]. The separation performance of all three concepts was found to be very similar although it was shown that the conventional concepts, batch and continuous process, had several advantages and disadvantages. This leads to case-specific concept selection. However, a disadvantage of both batch and continuous DF is their high diafiltration water consumption. The new concept of counter-current DF reduces the diafiltration water consumption and thus the cost of pre-treatment of diafiltration liquid and post-treatment of the permeate. A disadvantage of counter-current DF is the need for a larger membrane area, which is directly linked to an increase in capital cost [104].

During the design of a membrane plant the number of stages, the membrane area per stage, diafiltration factors and pressures can be varied in different ways to minimize capital and operating costs. Morison *et al.* [105] have suggested a method for the design and optimization of continuous multistage plants.

8.7 Fouling and cleaning

Minimization of fouling and successful cleaning are of great importance if a membrane process is to be implemented on an industrial scale. Bench-scale tests are commonly used to study the fouling characteristics of different membranes, and pilot plant investigations to develop an appropriate cleaning schedule.

8.7.1 Fouling

The flux of a solution is always lower than the PWF. This is due to two phenomena – concentration polarization and fouling. These two phenomena are distinguished by the reversibility of the flux decline. Concentration polarization is the part of the flux decline that is reversible by simply changing the operating conditions, whereas when the flux decline is due to fouling membrane cleaning is required to restore the flux. Flux decline caused by concentration polarization and fouling is illustrated in Figure 8.16.

There are three main fouling mechanisms: cake layer formation, pore blocking and adsorption [106]. Cake layer formation and pore blocking can occur if the solute molecules are large enough to be retained by the membrane. If, on the other hand, the solute molecules are small enough to enter the pores they may

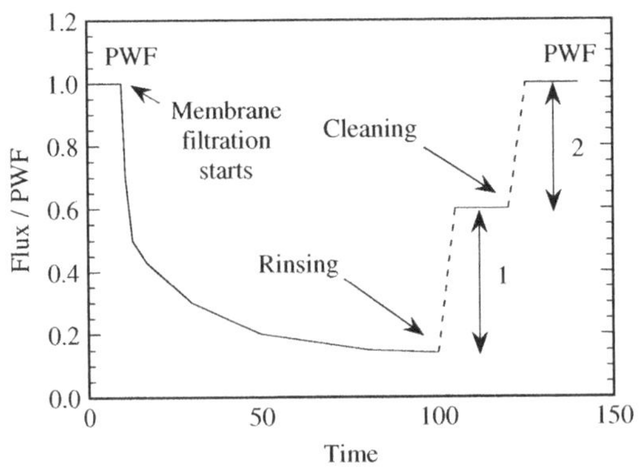

Figure 8.16 *Flux decline caused by concentration polarization (1) and fouling (2)*

be adsorbed onto the pore walls of the membrane. The adsorbed molecules can then reduce the effective pore diameter resulting in an increase in the membrane resistance.

One should be cautious in making general statements about the influence of different parameters on fouling. The many exceptions to the general rules reflect the complexity of the fouling phenomenon. However, there are some general trends.

- Hydrophobic membranes have a greater fouling tendency than hydrophilic membranes.
- Hydrophobic solutes are known to be more readily adsorbed onto membrane surfaces than hydrophilic solutes.
- The performance of a membrane is very dependent on the history of the membrane. For example, a membrane that is cleaned before use shows quite a different performance from that of one that has not been cleaned.

The measures appropriate to reduce fouling depend on the nature of the actual fouling phenomenon. Cake formation is reduced by operating below the critical flux [107, 108] and using backpulsing [2], pore blocking by using membranes with smaller pores [9, 83], and adsorption is commonly reduced by using hydrophilic membranes [9, 109, 110] and adjusting the pH [111]. The resistance-in-series model (Eq. (8.15)) is often used to distinguish between the influence of different flux deterioration phenomena [112–119]. A review of fouling and fouling reducing methods has been presented by Hilal *et al*. [120].

8.7.2 Pretreatment

The fouling tendency of a solution may be reduced by suitable pretreatment of the solution, be it mechanical, thermal or chemical. Common modifications to the feed solution include adjustment of the pH and the removal of fibres and particles. As a rule of thumb, the maximum particle size in the feed should be one-tenth of the smallest dimension of the feed channel [16].

8.7.3 Cleaning

Unfortunately, there are few theoretical guidelines to cleaning, and development of cleaning procedures is usually a matter of trial-and-error. However, a cleaning cycle generally includes the following stages:

- removal of product from the system;
- rinsing the system with water or permeate;
- cleaning in one or several steps;
- rinsing the system with water; and
- disinfection of the system (if required).

Cleaning begins with displacement of the retentate and thorough rinsing of the system. If it is possible, it can be advantageous to collect the permeate and use it for the first rinse. When treating kraft black liquor it was found that cleaning was much easier when permeate was used instead of deionized water to rinse the membrane because lignin and other substances that fouled the membrane were dissolved in the permeate which had the same pH and solubility properties as the original solution [65].

It is important that the temperature of the rinsing liquid is the same as that of the solution treated in order to avoid precipitation of solutes during rinsing. Rinsing should continue until both the retentate and permeate streams are totally clear and neutral. Normally quite large rinsing volumes are needed.

A large number of cleaning recommendations are available, but the temperature, time, concentration and type of cleaning agent must be chosen to suit each application. As a rule, mineral deposits are removed by acids, and organic matter by alkaline solutions. Alkaline cleaning solutions usually contain sodium hydroxide, phosphate, sequestering agents and surface active agents. In some applications, it is necessary to use different chemicals in succession to obtain satisfactory cleaning. In order to obtain a good mechanical cleaning effect, the circulation flow rate is often higher and the pressure lower during cleaning than during normal operation.

The PWF is usually measured before and after treatment of the process stream in order to establish the extent of fouling. The PWF can also be measured to check whether cleaning has been successful.

8.8 Conclusions and future trends

Ultrafiltration and MF have the necessary prerequisites to be key components in biorefineries as they are cost-efficient and environmentally friendly separation processes. Separation can be performed without adjusting the pH or temperature of the feed, and fractionation can be fine-tuned by the choice of appropriate membrane and operating conditions.

In the future, all the compounds in biomass will be recovered and used to a much higher degree than today. In the sustainable society, biomass will be used as food and fuel, and to produce value-added products. To achieve this, MF and UF will be used to separate colloids and macromolecular matter, and to fractionate polydisperse compounds (lignin and hemicelluloses), according to molecular mass.

References

1. B. Kamm and M. Kamm, Biorefinery – systems, *Chem. Biochem. Eng. Q.*, 18, 1–6 (2004).
2. S. Cortinas, S. Luque, J.R. Alvarez, J. Canaval and J. Romero, Microfiltration of kraft black liquor for the removal of colloidal suspended matter (pitch), *Desalination*, 147, 49–54 (2002).
3. P. Gullon, M.J. Gonzalez-Munoz, H. Dominguez, and J.C. Parajo, Membrane processing of liquors from Eucalyptus globulus autohydrolysis, *J. Food Eng.*, 87(2), 257–265 (2008).
4. T. Persson, H. Krawzcyk, A.-K. Nordin and A.-S. Jönsson, Fractionation of process water in thermomechanical pulp mills, *Bioresource Technol.*, 101, 3884–3892 (2010).
5. A. Arora, B.S. Dien, R.L. Belyea, V. Singh, M.E. Tumbleson and K.D. Rausch, Heat transfer fouling characteristics of microfiltered thin stillage from the dry grind process, *Bioresource Technol.*, 101(16), 6521–6527 (2010).
6. G. Schild, H. Sixta and L. Testova, Multifunctional alkaline pulping, delignification and hemicellulose extraction, *Cell. Chem. Technol.*, 44, 1–45 (2010).
7. A. Arora, B.S. Dien, R.L. Belyea, V. Singh, M.E. Tumbleson and K.D. Rausch, Nutrient recovery from the dry grind process using sequential micro and ultrafiltration of thin stillage, *Bioresource Technol.*, 101(11), 3859–3863 (2010).
8. R. Zeitoun, P.Y. Pontalier, P. Marechal and L. Rigal, Twin-screw extrusion for hemicellulose recovery: Influence on extract purity and purification performance, *Bioresource Technol.*, 101(23), 9348–9354 (2010).
9. J. Leberknight, B. Wielenga, A. Lee-Jewett and T.J. Menkhaus, Recovery of high value protein from a corn ethanol process by ultrafiltration and an exploration of the associated membrane fouling, *J. Membrane Science*, 366(1–2), 405–412 (2011).
10. K.R. Colyar, J. Pellegrino and K. Kadam, Fractionation of pre-hydrolysis products from lignocellulosic biomass by an ultrafiltration ceramic tubular membrane, *Separ. Sci. Technol.*, 43(3), 447–476 (2008).
11. S.A. Kirbawy and M.K. Hill, Multivalent ion removal from kraft black liquor by ultrafiltration, *Ind. Eng. Chem. Res.*, 26, 1851–1854 (1987).
12. O. Wallberg, A.-S. Jönsson and R. Wimmerstedt, Fractionation and concentration of kraft black liquor with ultrafiltration, *Desalination*, 154(2), 187–199 (2003).

13. O. Wallberg, A.-S. Jönsson and R. Wimmerstedt, Ultrafiltration of kraft black liquor with a ceramic membrane, *Desalination*, 156, 145–153 (2003).
14. O. Wallberg and A.-S. Jönsson, Influence of the membrane cut-off during ultrafiltration of kraft black liquor with ceramic membranes, *Chem. Eng. Res. Des.*, 81(10), 1379–1384 (2003).
15. A. Holmqvist, O. Wallberg and A.-S. Jönsson, Ultrafiltration of kraft black liquor from two Swedish pulp mills, *Chem. Eng. Res. Des.*, 83(A8), 994–999 (2005).
16. M. Cheryan, *Ultrafiltration and microfiltration handbook*, Technomic Publishing Co, Inc., Lancaster, Pennsylvania, USA, 1998.
17. O. Olsen, Membrane technology in the pulp and paper industry, *Desalination*, 35, 291–302 (1980).
18. A. Ludemann, Wine clarification with a crossflow microfiltration system, *Am. J. Enol. Viticult.*, 38(3), 228–235 (1987).
19. H.C. van der Horst and J.H. Hanemaaijer, Cross-flow microfiltration in the food industry. *State of the art, Desalination*, 77, 235–258 (1990).
20. Q. Gan, J. A. Howell, R. W. Field, R. England, M. R. Bird, C. L. O'Shaughnessy and M. T. MeKechinie, Beer clarification by microfiltration – product quality control and fractionation of particles and macromolecules, *J. Membrane Sci.*, 194, 185–196 (2001).
21. Q. Gan, Beer clarification by cross-flow microfiltration – effect of surface hydrodynamics and reversed membrane morphology, *Chem. Eng. Process.*, 40(5), 413–419 (2001).
22. V. Espina, M.Y. Jaffrin, L. Ding and B. Cancino, Fractionation of pasteurized skim milk proteins by dynamic filtration, *Food Res. Int.*, 43(5), 1335–1346 (2010).
23. U. Merin and G. Daufin, Crossflow microfiltration in the dairy industry: state-of-the-art, *Lait*, 70, 281–291 (1990).
24. I. Pafylias, M. Cheryan, M.A. Mehaia and N. Saglam, Microfiltration of milk with ceramic membranes, *Food Res. Int.*, 29(2), 141–146 (1996).
25. V.V. Mistry and J.-L. Maubois, Application of membrane separation technology to cheese production, in *Cheese: Chemistry, Physics and Microbiology, Vol*. 1, 3rd edn, (eds. P.F. Fox, P.L.H. McSweeney, T.M. Cogan and T.P. Guinee), Elsevier Ltd, pp. 261–285 (2004).
26. U. Frenander and A.-S. Jönsson, Cell harvesting by cross-flow microfiltration using a shear-enhanced module, *Biotechnol. Bioeng.*, 52, 397–403 (1996).
27. K.-H. Ahn and K.-G. Song, Treatment of domestic wastewater using microfiltration for reuse of wastewater, *Desalination*, 126(1–3), 7–14 (1999).
28. S.K. Karode, A new unsteady-state model for macromolecular ultrafiltration, *Chem. Eng. Sci*, 55(10), 1769–1773 (2000).
29. P. Côté and D. Thompson, Wastewater treatment using membranes: the North American experience, *Water Sci. Technol.* 41(10–11), 209–215 (2000).
30. S. Rosenberger, R. Witzig, W. Manz, U. Szewzyk and M. Kraume, Operation of different membrane bioreactors: experimental results and physiological state of the micro-organisms, *Water Sci. Technol*. 41(10–11), 269–277 (2000).
31. R. Bai and H.F. Leow, Microfiltration of activated sludge wastewater – the effect of system operation parameters, *Sep. Purif. Technol.*, 29(2), 189–198 (2002).
32. T. Leiviskä, J. Rämö, H. Nurmesniemi, R. Pöykiö and T. Kuokkanen, Size fractionation of wood extractives, lignin and trace elements in pulp and paper mill wastewater before and after biological treatment, *Water Res.*, 43(13), 3199–3206 (2009).
33. K. Minami, K. Okamura, S. Ogawa and T. Naritomi, Continuous anaerobic treatment of wastewater from a kraft pulp mill, *J. Ferment. Bioeng.*, 71(4), 270–274 (1991).
34. I.-S. Chang, K.-H. Choo, C.-H. Lee, U.-H. Pek, U.-C. Koh, S.-W. Kim and J.-H. Koh, Application of ceramic membrane as a pretreatment in anaerobic digestion of alcohol-distillery wastes, *J. Membrane Science*, 90(1–2), 131–139 (1994).
35. A. Arora, A. Seth, B.S. Dien, R.L. Belyea, V. Singh, M.E. Tumbleson and K.D. Rausch, Microfiltration of thin stillage: Process simulation and economic analyses. *Biomass Bioenerg.*, 35(1), 113–120 (2011).

36. A.-S. Jönsson and G. Trägårdh, Ultrafiltration Applications, *Desalination*, 77, 135–179 (1990).
37. J.M. Burke, Waste treatment of metalworking fluids, a comparison of three common methods, *Lubr. Eng.*, 47(4), 238–246 (1991).
38. M. Bodzek and K. Konieczny, The use of ultrafiltration membranes made of various polymers in the treatment of oil-emulsion wastewater, *Waste Manage.*, 12, 75–84 (1992).
39. R.A. Dangel, D. Astraukis and J. Palmateer, Fatty acid separation from hydrolyzer wastewater by ultrafiltration, *Environ. Prog.*, 14(1), 65–68 (1995).
40. M. Belkacem, H. Matamoros, C. Cabassud, Y. Aurelle, J. Cotteret, New results in metal working wastewater treatment using membrane technology, *J. Membrane Sci.*, 106, 195–205 (1995).
41. R. R. Koch and I. T. Selldorff, What ultrafiltration does for electrocoaters, *Products Finishing*, (Mar.), 54–60 (1972).
42. J. Zahka and L. Mir, Ultrafiltration of cathodic electrodeposition paints – seven years of field experience, *Plat. Surf. Finish.*, (Nov.), 34–39 (1979).
43. J. L. Short, Selection, applications and optimisation of hollow fibre UF membranes, *Filtr. Separat.*, (Sept./Oct.), 410–414 (1982).
44. V. Espina, M.Y. Jaffrin, P. Paullier, and L. Ding, Comparison of permeate flux and whey protein transmission during successive microfiltration and ultrafiltration of UHT and pasteurized milks, *Desalination*, 264(1–2), 151–159 (2010).
45. C. Marella, K. Muthukumarappan and L. E. Metzger, Evaluation of commercially available, wide-pore ultrafiltration membranes for production of alpha-lactalbumin-enriched whey protein concentrate, *J. Dairy Sci.*, 94(3), 1165–1175 (2011).
46. I.K. Bansal and A.J. Wiley, Fractionation of spent sulfite liquors using ultrafiltration cellulose acetate membranes, *Environ. Sci. Technol.*, 8(13), 1085–1090 (1974).
47. I.K. Bansal and A.J. Wiley, Membrane processes for fractionation and concentration of spent sulfite liquors, *Tappi J.*, 58(1), 125–130 (1975).
48. K. Forss, R. Kokkonen, H. Sirelius and P.E. Sagfors, How to improve spent sulphite liquor use, *Pulp Pap.-Canada*, 80(12), 411–415 (1979).
49. P. Eriksson, Ultrafiltration for recovery of lignosulfonates from spent sulfite liquor, *AIChE Symp.Series*, 76(197), 316–320 (1980).
50. K. Kovasin and H.V. Norden, Determination of lignosulfonate rejection from test results in the ultrafiltration of spent sulfite liquor, *Sven. Papperstidn.*, 87(6), 44–47 (1984).
51. *PCI Ultrafiltration at Borregaard industries, A case history dealing with the treatment of calcium sulphite lye for Borregaard industries*, TPRO 46.1 (1981), PCI Membrane Systems Ltd.
52. A.-S. Jönsson and R. Wimmerstedt, The application of membrane technology in the pulp and paper industry, *Desalination*, 53, 181–196 (1985).
53. L. Magdzinski, Tembec Temiscaming integrated biorefinery, *Pulp Pap.-Canada*, 107(6), 44–46 (2006).
54. Domsjö Fabriker AB, Örnsköldsvik, Sweden, http://www.domsjoe.com (accessed 21 August 2012).
55. P. Wickström, P., 1997, Membrane filtration in the closed-cycle bleach plant, Proc. 1997 *TAPPI Minimum Effluent Mills Symp*, October, San Francisco, CA, 1997, 79–83.
56. A.-K. Nordin and A.-S. Jönsson, Case study of an ultrafiltration plant treating bleach plant effluent from a pulp and paper mill, *Desalination*, 201, 277–289 (2006).
57. U.H. Haagensen, *Case-Rauma Repola Finland – Ultrafiltration of spent sulphite liquor*, 1923-GB-0482-50 (1982) DDS Filtration, (1982).
58. M. Zabkova, E A.B. da Silva and A.E. Rodrigues, Recovery of vanillin from lignin/vanillin mixture by using tubular ceramic ultrafiltration membranes, *J. Membrane Science*, 301(1–2), 221–237 (2007).
59. I. Brodin, E. Sjöholm and G. Gellerstedt, Kraft lignin as feedstock for chemical products: the effects of membrane filtration, *Holzforschung*, 63, 290–297 (2009).
60. I. Tanistra and M. Bodzek, Preparation of high-purity lignin from spent black liquor using ultrafiltration and diafiltration processes, *Desalination*, 115, 111–120 (1998).
61. O. Wallberg and A.-S. Jönsson, Lignin extraction from kraft black liquor by ultrafiltration, *Proc. 9th World Filtration Congress*, Session 232, April 18–22, 2004, New Orleans, Louisiana, USA, (2004).

62. A. Keyoumu, R. Sjödahl, G. Henriksson, M. Ek, G. Gellerstedt and M.E. Lindström, Continuous nano- and ultra-filtration of kraft pulping black liquor with ceramic filters: A method for lowering the load on the recovery boiler while generating valuable side-products, *Ind. Crop. Prod.*, 20(2), 143–150 (2004).
63. A. Dafinov, J. Font and R. Garcia-Valls, Processing of black liquors by UF/NF ceramic membranes, *Desalination*, 173(1), 83–90 (2005).
64. A.-S. Jönsson and O. Wallberg, Cost estimates of kraft lignin recovery by ultrafiltration, *Desalination*, 237, 254–267 (2009).
65. O. Wallberg, A. Holmqvist and A.-S. Jönsson, Ultrafiltration of kraft cooking liquors from a continuous cooking process, *Desalination*, 180, 109–118 (2005).
66. O. Wallberg and A.-S. Jönsson, Separation of lignin in kraft cooking liquor from a continuous digester by ultrafiltration at temperatures above 100 °C, *Desalination*, 195, 187–200 (2006).
67. T. Persson, A.-K. Nordin, G. Zacchi and A.-S. Jönsson, Economic evaluation of isolation of hemicelluloses from process streams from thermo-mechanical pulping of spruce, *Appl. Biochem. Biotech.*, 136–140, 741–752 (2007).
68. S. Willför, K. Sundberg, M. Tenkanen and B. Holmbom, Spruce-derived mannans – a potential raw material for hydrocolloids and novel advance natural materials, *Carbohyd. Polym.*, 72, 197–210 (2008).
69. C. Xu, S. Willför, P. Holmlund and B. Holmbom, Rheological properties of water-soluble spruce O-acetyl galactoglucomannans, *Carbohyd. Polym.*, 75, 498–504 (2009).
70. T. Persson and A.-S. Jönsson, Isolation of hemicelluloses by ultrafiltration of thermomechanical pulp mill process water – Influence of operating conditions, *Chem. Eng. Res. Des.*, 88, 1548–1554 (2010).
71. E. Ribe, M. Söderqvist Lindblad, O. Dahlman and H. Theliander, Xylan sorption kinetics at industrial conditions Part 1. Experimental results, *Nord. Pulp Paper Res. J.*, 25(2), 138–149 (2010).
72. J. Li and H.A. Chase, Applications of membrane techniques for purification of natural products, *Biotechnol. Lett.*, 32, 601–608 (2010).
73. F. Lipnizki, Membrane process opportunities and challenges in the bioethanol industry, *Desalination*, 250, 1067–1069 (2010).
74. R. Vegas, A. Moure, H. Dominguez, J.C. Parajo, J.R.Alvarez and S. Luque, Evaluation of ultra- and nanofiltration for refining soluble products from rice husk xylan, *Bioresource Technol.*, 99(13), 5341–5351 (2008).
75. J.-S. Kim, B.-G. Kim, C.-H. Lee, S.-W. Kim, H.-S Jee, J.-H. Koh, and A.G. Fane, Development of clean technology in alcohol fermentation industry, *J. Clean. Prod.*, 5(4), 263–267 (1997).
76. R.A. Cross, Optimum process designs for ultrafiltration and crossflow microfiltration systems, *Desalination*, 145(1–3), 159–163 (2002).
77. M. Mulder, *Basic Principles of Membrane Technology*, Kluwer Academic Publishers (1991).
78. R.W. Baker, *Membrane Technology and Applications*, John Wiley & Sons, Ltd (2004).
79. C.J. Brouckaert, C.A. Buckley, E.P. Jacobs, Common pitfalls in the design and operation of membrane plants – or how I should have done it!, *Water Sci. Technol.* 39(10–11), 107–114 (1999).
80. J. Wagner, Membrane Filtration Handbook, http://www.wagnerdk.dk/DownLoads.htm (accessed 21 August 2012).
81. A.J. Ragauskas, C.K. Williams, B.H. Davison, G. Britovsek, J. Cairney, C.A. Eckert, W.J. Frederick Jr.,, J.P. Hallett, D.J. Leak, C.L. Liotta, J.R. Mielenz, R. Murphy, R. Templer and T. Tschaplinski, The path forward for biofuels and biomaterials, *Science*, 311, 484–489 (2006).
82. A.-K. Nordin and A.-S. Jönsson, Influence of module configuration on total economics during ultrafiltration at high concentration, *Chem. Eng. Res. Des.*, 88, 1555–1562 (2010).
83. H. Krawzcyk and A.-S. Jönsson, Separation of dispersed substances and galactoglucomannan in thermomechanical pulp waste water by microfiltration, *Sep. Purif. Technol.*, 79, 43–49 (2011).
84. P. Bacchin, P. Aimar and R. W. Field, Critical and sustainable fluxes: theory, experiments and applications, *J. Membrane Sci.*, 281(1–2), 42–69 (2006).
85. G. Jonsson, Boundary layer phenomena during ultrafiltration of dextran and whey protein solutions, *Desalination*, 51, 61–77 (1984).
86. J.G. Wijmans, S. Nakao and C.A. Smolders, Flux limitation in ultrafiltration: osmotic pressure model and gel layer model, *J. Membrane Sci.*, 20, 115–124 (1984).

87. V.L. Vilker, C.K. Colton, K.A. Smith, and D.L. Green, The osmotic pressure of concentrated protein and lipoprotein solutions and its significance to ultrafiltration, *J. Membrane Sci.*, 20, 63–77 (1984).
88. M. Defossez, L. Ding, M. Jaffrin and M. Fauchet, Scintigraphic study of local flux and osmotic pressure distributions in ultrafiltration of blood plasma, *J. Colloid Interf. Sci.*, 177, 179–191 (1996).
89. L.V. Saboya and J.L. Maubois, Current developments of microfiltration technology in the dairy industry, *Lait*, 80, 541–553 (2000).
90. M. Skrzypek and M. Burger, Isoflux ceramic membranes – practical experiences in dairy industry, *Desalination*, 250(3), 1095–1100 (2010).
91. P.K. Vadi and S.S H. Rizvi, Experimental evaluation of a uniform transmembrane pressure crossflow microfiltration unit for the concentration of micellar casein from skim milk, *J. Membrane Sci.*, 189, 69–82 (2001).
92. V. Gekas and B. Hallström, Mass transfer in the membrane concentration polarization layer under turbulent cross flow 1. Critical literature review and adaption of existing Sherwood correlations to membrane operations, *J. Membrane Sci.*, 30(2), 153–170 (1987).
93. G.B. van den Berg, I.G. Racz and C.A. Smolders, Mass transfer coefficients in cross-flow ultrafiltration, *J. Membrane Sci.*, 47(1–2), 25–51 (1989).
94. J. Lipnizki and G. Jonsson, Flow dynamics and concentration polarisation in spacer-filled channels, *Desalination*, 146(1–3), 213–217 (2002).
95. R.B. Bird, W.E. Stewart, E.L. Lightfoot, *Transport Phenomena*, John Wiley & Sons, Inc. (2002).
96. W.F. Blatt, A. Dravid, A.S. Michaels and L. Nelsen, Solute polarization and cake formation in membrane ultrafiltration: causes, consequences, and control techniques, in *Membrane Science and Technology*, J.E. Flinn (ed.), Plenum Press (1970), pp. 47–97.
97. M. Pritchard, J.A. Howell and R.W. Field, The ultrafiltration of viscous fluids, *J. Membrane Sci.*, 102, 223–235 (1995).
98. A.-S. Jönsson, A.-K. Nordin and O. Wallberg, Concentration and purification of lignin in hardwood kraft pulping liquor by ultrafiltration and nanofiltration, *Chem. Eng. Res. Des.*, 86, 1271–1280 (2008).
99. O. F. von Meien and R. Nobrega, Ultrafiltration model for partial solute rejection in the limiting flux region, *J. Membrane Sci.*, 95, 277–287 (1994).
100. U.H. Haagensen and J. Wagner, *Recovery and purification of lignsulphonates and alkali-lignin from pulping waste liquor*, 1801-GB-0680-50 (1980) DDS Filtration, (1980)
101. A. Andersson, T. Persson, G. Zacchi, H. Stålbrand and A.-S. Jönsson, Comparison of diafiltration and size-exclusion chromatography to recover hemicelluloses from process water from thermo-mechanical pulping of spruce, *Appl. Biochem. Biotechnol.*, 136–140, 971–983 (2007).
102. G. Foley, Minimisation of process time in ultrafiltration and continuous diafiltration: the effect of incomplete macrosolute rejection, *J. Membrane Science*, 163, 349–355 (1999).
103. G. Foley and J. Garcia, Ultrafiltration flux theory based on viscosity and osmotic effects: application to diafiltration optimisation, *J. Membrane Science*, 176, 55–61 (2000).
104. F. Lipnizki, J. Boelsmand and R.F. Madsen, Concepts of industrial-scale diafiltration systems, *Desalination*, 144, 179–184 (2002).
105. K.R. Morison and X. She, Optimisation and graphical representation of multi-stage membrane plants, *J. Membrane Science*, 211, 59–70 (2003).
106. W.R. Bowen, J.I. Calvo and A. Hernandez, Steps of membrane blocking in flux decline during protein microfiltration, *J. Membrane Sci.*, 101, 153–165 (1995).
107. H. Carrère, F. Blaszkow and H. Roux de Balmann, Modelling the clarification of lactic acid fermentation broths by cross-flow microfiltration, *J. Membrane Sci.*, 186, 219–230 (2001).
108. P. Bacchin and P. Aimar, Critical fouling conditions induced by colloidal surface interaction: from causes to consequences, *Desalination*, 175, 21–27 (2005).
109. A.-S. Jönsson, Fouling during ultrafiltration of a low-molecular weight hydrophobic solute, *Separ. Sci. Technol.*, 33(4), 503–516 (1998).
110. L. Puro, M. Kallioinen, M. Mänttäri, G. Natarajan, C. Cameron and M. Nyström, Performance of RC and PES ultrafiltration membranes in filtration of pulp mill process waters, *Desalination*, 264(3), 249–255 (2010).

111. J. Brinck, A.-S. Jönsson, B. Jönsson and J. Lindau, Influence of pH on the adsorptive fouling of ultrafiltration membranes by fatty acid, *J. Membrane Sci.*, 164, 187–194 (2000).
112. M. Ousman and M. Bennasar, Determination of various hydraulic resistances during cross-flow filtration of a starch grain suspension through inorganic membranes, *J. Membrane Sci.*, 105(1–2), 1–21 (1995).
113. G. Liu, Y. Liu, J. Ni, H. Shi and Y.Qian, Treatibility of kraft spent liquor by microfiltration and ultrafiltration, *Desalination*, 160, 131–141 (2004).
114. H. Choi, K. Zhang, D.D. Dionysiou, D.B. Oerther and G.A. Sorial, Influence of cross-flow velocity on membrane performance during filtration of biological suspension, *J. Membrane Sci.*, 248(1–2), 189–199 (2005).
115. P. Rai, C. Rai, G.C. Majumdar, S. Das Gupta and S. De, Resistance in series model for ultrafiltration of mosambi (*Citrus sinensis* (*L.*) *Osbeck*) juice in a stirred continuous mode, *J. Membrane Science*, **283**, 116–122 (2006).
116. R.-S. Juang, H.-L. Chen and Y.-S. Chen, Resistance-in-series analysis in cross-flow ultrafiltration of fermentation broths of *Bacillus subtilis* culture, *J. Membrane Science*, 323, 193–200 (2008).
117. A. Arora, B.S. Dien, R.L. Belyea, P. Wang, V. Singh, M.E. Tumbleson and K.D. Rausch, Thin stillage fractionation using ultrafiltration: resistance in series model, *Bioproc. Biosyst. Eng.*, 32, 225–233 (2009).
118. E. Kujundzic, A.R. Greenberg, R. Fong, B. Moore, D. Kujundzic and M. Hernandez, Biofouling potential of industrial fermentation broth components during microfiltration, *J. Membrane Science*, 349, 44–55 (2010).
119. X. Zhang, Q. Hua, M. Sommerfeld, E. Puruhito and Y. Chen, Harvesting algal biomass for biofuels using ultrafiltration membranes, *Bioresource Technol.*, 101, 5297–5304 (2010).
120. N. Hilal, O.O. Ogunbiyi, N.J. Miles and R. Nigmatullin, Methods employed for control of fouling in MF and UF membranes: A comprehensive review, *Separ. Sci. Technol.*, 40(10), 1957–2005 (2005).

9

Nanofiltration

Mika Mänttäri[1], Bart Van der Bruggen[2] and Marianne Nyström[1]

[1]*Lappeenranta University of Technology, Department of Chemical Technology, Laboratory of Membrane Technology and Technical Polymer Chemistry, Finland*
[2]*K.U.Leuven, Department of Chemical Engineering, Laboratory for Applied Physical Chemistry and Environmental Technology, Belgium*

9.1 Introduction

Nanofiltration (NF) is one of the newest of the pressure driven membrane filtration methods for separation of molecules from liquids. The name "nanofiltration" came into use during the 1980s but the principle had been used under the names "tight ultrafiltration" or "loose reverse osmosis." Today it is applied, by itself or as a hybrid, for separations at the "nano" scale.

There are many ways to define nanofiltration (Figure 9.1):

- *According to cut-off*. Molecules with molar masses less than around 300 g/mol (Dalton) are separated from larger molecules. The membranes sold as NF membranes are from 200 to 1000 g/mol (molar mass).
- *According to size*. Based on the largest diameter of the projection of the molecule. Hence, molecular stiffness can play a role in separation. The general cut-off diameter is about 1 nm.
- *According to salt retention*. Based on the principle of charge and size it retains over 98% of magnesium sulfate and less than 50% of sodium chloride and thus it can separate multivalent ions from monovalent ions.

The first book on nanofiltration, *Nanofiltration—Principles and Applications* (Schäfer *et al*., 2005) appeared in 2005. It contains chapters on different applications (see the list on next page) using NF, as well as theoretical consideration on nanofiltration membranes. In practice, NF can be used to separate multivalent salts in the first stage of desalination or to separate small organic molecules from water into the retentate or from bigger organic solutes into permeate. In water treatment, NF is in principle sufficient on

Separation and Purification Technologies in Biorefineries, First Edition.
Edited by Shri Ramaswamy, Hua-Jiang Huang, and Bandaru V. Ramarao.

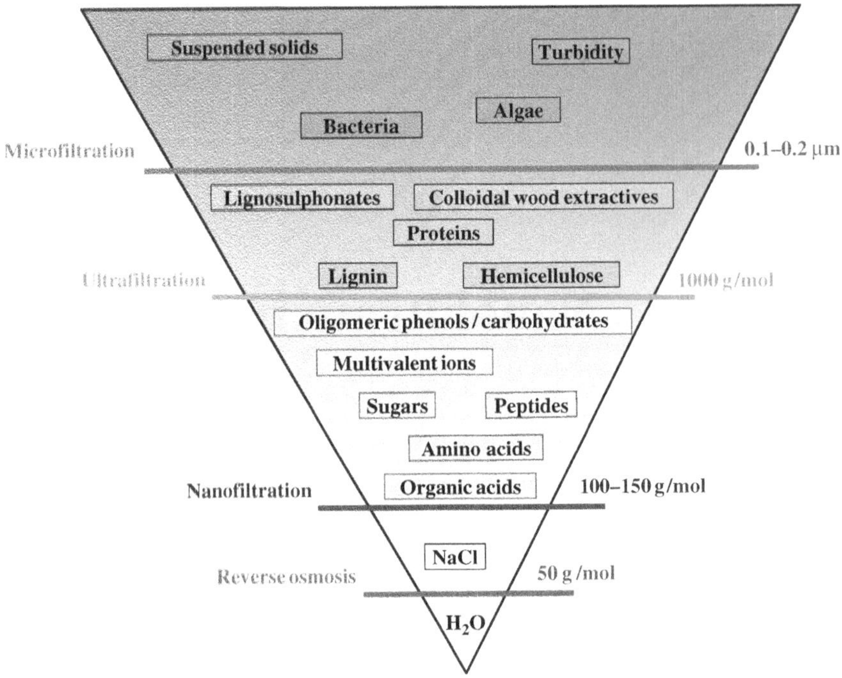

Figure 9.1 *Definition of NF and separation triangle*

its own for the production of drinking water, but normally either microfiltration (MF) or ultrafiltration (UF) are used as prefiltration methods before NF. Nanofiltration can also be used to remove special hazardous ions (e.g. arsenic) or organic pollutants (e.g. endocrine disrupters) from otherwise clean drinking water. Nanofiltration can also be used to remove radioactive ions, which was one of the first uses when membranes were developed altogether.

Nanofiltration can be used in the separation of solvents, but because NF membranes are still seldom made out of ceramics or metals, solvents might destroy the polymeric membrane structures. Some solvent-resistant polymeric NF membranes exist, but many of them are fairly thick and therefore do not have good enough fluxes.

In the metal and mining industry NF can be used to clean/separate some ions from others, for example sulfates from nitrates or copper ions from hydrogen ions (Tanninen *et al.*, 2006). Nanofiltration can also be used to recover nutrients, such as phosphorus and nitrogen compounds, from various waste waters.

In the pharmaceutical industry NF can be used to concentrate small organic molecules or to fractionate them from bigger molecules.

In the pulp and paper industry NF membranes can be used to concentrate dilute solutions containing carbohydrates or other organic molecules. From some of the process waters NF can fractionate small organic molecules from each other, for example, xylose from glucose, even at high solute concentration (Sjöman *et al.*, 2007).

In the dairy industry NF can be used to concentrate whey using a lactose and salt-penetrating membrane and then concentrate lactose from salt using a tighter membrane or RO can be used in the second stage. By tuning the membrane cut-off, salt and lactose can go to either the retentate fraction or to the permeate fraction.

Some possible applications of NF in various industries

Water production

Hardness removal
Removal of natural organic matter (NOM)
Removal of pesticides and endocrine disrupters
Removal of heavy metals

Food

Demineralization of whey or sugar solutions
Separation of sunflower oil from solvent
Recovery of solvents used in the extraction of edible oil processing
Fractionation of bioactive peptides from commercial whey isolates
Fractionation of monosaccharides
Purification of glucose syrap

Textile and leather industry

Removal of dyes from wastewaters
Removal and reuse of chromates
Recovery of water and salts from wastewater

Pulp and paper

Extraction of xylan before Kraft cooking
Fractionation of xylose from spent sulphite liquor
Recovery of caustic solution in viscose production

Metal and mining

Separation of heavy metals from acid solutions
Removal of metals from wastewaters
Recovery of copper from ore extraction streams

9.2 Nanofiltration market and industrial needs

Nanofiltration has come on the market during the last few decades, in part because there has been a trend over the world to be aware of nano-sized particles and their usefulness. Nanofiltration can fractionate small molecules from each other and can be used instead of chromatography, so the market for it has increased, for example in the manufacture of pharmaceuticals and biologically active substances. It is also used in the treatment of different effluents and in the production of drinking water. A major application of NF is in water treatment. In drinking water some salts need to remain in the water, but contaminants like pesticides and endocrine disrupters should be removed from the surface waters. NF has succeeded very well, for example, with pesticides at an installation in Méry-sur-Oise, France, removing most of the atrazine present in the Seine river (Wittmann and Thorsen, 2005).

The existing market for nanofiltration is approximately \$300 million and it is growing fast. Today polymeric membranes are still dominating the market but the demand for nonpolymeric materials, including ceramic, metal and composites, is increasing more than 10% annually due to their better performance in extreme temperatures and greater pH ranges. In biorefinery applications nanofiltration membranes will encounter various acidic or alkaline streams at high temperature and, therefore, a membrane having better resistance at extreme conditions is much needed.

9.3 Fundamental principles

9.3.1 Pressure and flux

Nanofiltration has similarities with reverse osmosis and ultrafiltration. Tight NF works much like RO and is considered to depend on solution diffusion. It also needs a high pressure and the working pressure is calculated taking into account the osmotic pressure gradient over the membrane. In more open NF, and working with organic molecules (having less osmotic pressure than salts), NF is more similar to ultrafiltration. Interestingly, with NF one can take into account both possibilities of separation transport (diffusion and convection) to achieve separation in the most successful way.

The working pressures for nanofiltration used to be higher than 10 bar (1 MPa), but today, if the osmotic pressure of the solution is not very high, pressures between 4–8 bar (400–800 kPa) can be used. The fluxes in NF depend naturally on the fouling propensity of the membranes. Normally fluxes not much higher than 20 L/(m^{2}h) are designed in constant flux mode, but in non-fouling conditions, like when using high flow velocity, fluxes as high as 100 L/(m^{2}h) can be achieved. If non-fouling conditions are desired for long-term use, without cleaning, the best is to work under the critical flux, which normally is lower than the maximum flux. Fouling or gel-layer formation prevents the permeation of target small molecules (Jönsson *et al.*, 2008). Non-fouling conditions are therefore quite important when specific membrane properties are utilized in fractionation.

9.3.2 Retention and fractionation

When considering fractionation with NF, many different possibilities can be used. As in UF, size can be the criterion for separation. Nanofiltration membranes retain 90% of the uncharged molecules above the cut-off molar mass, so a specific small molecule can be fractionated into permeate from a solution containing different molecules. Typical examples could be the fractionation of salt or lactose from milk/whey or monosaccharides from polysaccharides. In addition to the size of dissolved molecules compared to the size of the membrane pores, other physicochemical effects such as hydrophobic interactions (dispersion forces) and polar interactions (dipole-dipole, dipole-induced dipole forces) sometimes determine the retention achieved. For instance the retention of pesticides has been shown to depend on the dipole moment and the size of molecules. A high dipole moment (dielectric constant) enhances transmission. The molecule (dipole) turns itself against the membrane and can more easily pass through the membrane. This is important with membranes where the main pore size is bigger than the size of the molecule (Van der Bruggen *et al.*, 2001). For same-size molecules, retention of polar molecules is typically smaller, for example organic acid and sugar.

The feasibility of fractionation can also depend on charge. More charged ions are retained better than monovalent ions. An example of this could be the fractionation of copper sulfate from sulfuric acid, which is based on the higher positive charge of the copper ion compared to the hydrogen ion and on the larger size of the copper ion (Tanninen *et al.*, 2006). In this type of fractionation, based on the so-called Donnan effect, negative retention of the acid can be achieved. (Negative retention means that there is a higher concentration of a component in the permeate than in the feed.) This effect can be used particularly well when the NF membrane is also charged. A similar type of charge effect has also been seen with organic molecules (Mänttäri *et al.*, 2009).

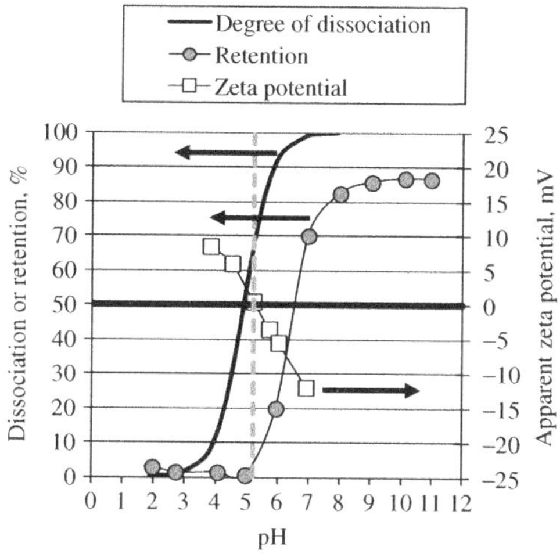

Figure 9.2 *Retention of octanoid acid as a function of pH*

The pH of the solution to be filtered can change both the charge of the membrane and the charge of some of the molecules to be filtered. Typically, for instance, weak organic acids and proteins have an isoelectric point (IEP) where the molecules are uncharged, and above and below this point they are negatively and positively charged respectively. In the same way membranes made of, for example, amine-containing groups can have an IEP. Some molecules can thus, having the same type of charge, repel each other or the membrane, while countercharged molecules attract each other or the membrane. A typical fractionation according to this principle could be the fractionation of an amino acid at its IEP from one being negatively charged similar to the membrane, even if they are similar in molecular size (Timmer *et al*., 1998). Figure 9.2 demonstrates the effect of charge repulsion on the retention of organic acids. The retention is zero until both the membrane and the acid become negatively charged, which occurs above the isoelectric point (pH 5.2) of the NF270 membrane. The NF270 membrane is a typical NF membrane showing an amphoteric behavior as a function of pH. Polyamide based NF membranes have frequently isoelectric points in the pH range 3.5–6.

9.3.3 Influence of filtration parameters

Filtration parameters have a remarkable effect on retention in nanofiltration. Due to the solution-diffusion mechanism of the tight NF membranes, retention typically decreases with constant flux conditions, when the concentration of solutes increases or when the temperature of the solvent increases. This is because diffusion increases with concentration and temperature. The increase of pressure increases the solvent flux but the solute flux is dependent on its concentration in the membrane, according to the solution-diffusion model, and as a result retention increases. The solution pH has an effect on the charged species and also on the membrane functional groups. At a certain pH when both have the same sign of charge electrostatic repulsion increases retention. A small change in pH can improve retention enormously (Figure 9.2). Generally, a minimum retention is seen at the isoelectric point. In addition, high temperature and/or pH might change the membrane polymeric structure and decrease retention. There are enormous amounts of molecules in the nano-size range, and, due to the different separation mechanisms existing in NF, the filtration conditions can also be used to tailor the separation within certain limits. This proves the great possibilities for NF in various purification and fractionation applications.

9.4 Design and simulation

For the design of nanofiltration and the simulation of the performance of any given application, two factors are to be considered: water permeation through the membrane, expressed by the water flux, and the retention of solutes. In addition, related parameters are to be taken into account, such as the stability of flux (flux decline and membrane fouling), and water yield (or recovery). These, however, are very application-dependent and difficult to generalize.

9.4.1 Water permeation

Nanofiltration membranes have an (often negative, at neutral pH) surface charge, which acts as a barrier to multivalent ions. Monovalent ions thus permeate through the membrane, which implies that they do not add to the osmotic pressure difference. Because the transmembrane pressure has to overcome the osmotic pressure, the applied pressures in nanofiltration can be much lower than, for example, in reverse osmosis.

The water flux is given by the Hagen–Poiseuille equation, Eq. (9.1):

$$J_v = \frac{\varepsilon r^2}{8\eta\tau}\frac{\Delta P}{\Delta x} \tag{9.1}$$

where

J_v = water flux through membrane ($m^3/(m^2s)$);
ε = porosity of membrane (−);
τ = tortuosity (−);
η = viscosity (Pas);
r = pore radius (m);
ΔP = pressure difference over the membrane (Pa);
Δx = membrane thickness (m).

This equation reflects Darcy's law for capillary flow ($J_v = K\Delta P$), where K is the Darcy permeability coefficient ($m^3/(m^2s$ Pa). It can be seen that membrane characteristics (porosity, pore size, tortuosity) as well as characteristics of the medium (temperature, viscosity) determine water transport through the membrane. In addition, however, it must be noted that when the solvent is not water, flux modeling becomes much more complex (Darvishmanesh *et al*., 2010).

9.4.2 Solute retention

The retention of any given component can theoretically be described with the Spiegler and Kedem (1966) equations for the transport of a dissolved component through the membrane. Two driving forces are important: transport is a combination of diffusion (first term) and convection (second term).

$$J_s = -P\Delta x\frac{dc}{dx} + (1-\sigma)J_v \tag{9.2}$$

where

J_s = flux of components through the membrane ($g/(m^2s)$);
P = permeability of a component (m/s);
c = concentration (g/m^3);
σ = reflection coefficient (−).

For every component, two parameters are needed: the permeability P and the reflection coefficient σ. The permeability is a measure of the importance of diffusion in the transport; the reflection coefficient is the maximal retention at infinite flux. The retention as a function of pressure can be calculated as:

$$R = \frac{\sigma(1-F)}{1-\sigma F} \tag{9.3}$$

$$F = \exp\left(-\frac{1-\sigma}{P}J_v\right) \tag{9.4}$$

P and σ can be determined experimentally, but in general the information about these parameters is not available for a given component and a given membrane. Models should therefore be developed to calculate the retention of a component. A difference can be made between inorganic molecules, where charge interactions are important, and organic molecules, where retention is based on effects of size and polarity.

9.4.2.1 *Retention of organic components*

Organic molecules are sieved from the feed solution because they are larger than the pores, or due to other interaction mechanisms such as polarity or charge, when they are charged. The pore "size" in nanofiltration is about 1 nm, which corresponds to a molar mass of 200 to 300 g/mol. Membranes do not have one single pore size, but rather a distribution around an average pore size, so the retention of uncharged components will increase gradually as the molar mass increases. Retention of uncharged molecules can thus be described by a "retention curve," where the retention for a given membrane is represented as a function of molar mass, with constant conditions of pressure and temperature, or the "reflection curve," where the maximal retention at large pressure is given as a function of molecular size. For every membrane, a reflection curve should be calculated. With the reflection curve for a given membrane, the retention can be estimated for every component at a given pressure; the equations above can be used to calculate retentions at other pressures.

The size of the molecules must be represented by an appropriate size parameter. The most simple size parameter is molar mass, but this is not a physical measure of size. Alternatives are the Stokes diameter, derived from Stokes' law for diffusion, or the effective diameter (Van der Bruggen *et al*., 1999), which is the statistical average of the projection of the molecule on the membrane surface.

Another requirement for the use of retention curves is the availability of a mathematical expression for the retention curve of a given membrane as a function of membrane-related parameters. For example, a (slightly moderated) log-normal distribution for the pore sizes can be assumed; in that case the retention curve is described by the following equation (Van der Bruggen *et al*., 2000):

$$\sigma(MM^*) = \int_0^{MM} \frac{1}{S_{MM}\sqrt{2\pi}}\frac{1}{MM}\exp\left(-\frac{(\ln(MM) - \ln(\overline{\overline{MM}}) + 0.56S_{MM})^2}{2S_{MM}^2}\right)dMM \tag{9.5}$$

MM^* is the molar mass of the molecule, S_{MM} is the standard deviation of the log-normal distribution and $\overline{\overline{MM}}$ is the molar mass cut-off, which is the molar mass of a component that is retained by 90%. An example of reflection curves obtained with this equation is given in Figure 9.3 for a typical nanofiltration membrane, i.e. NTR 7450 (Nitto-Denko, Japan).

Because the log-normal model is based on a sieving mechanism without assuming interactions between the component and the membrane, deviations from the model may occur. This is particularly noticeable for components causing fouling, or for polar components. Molecules with a high dipole moment have

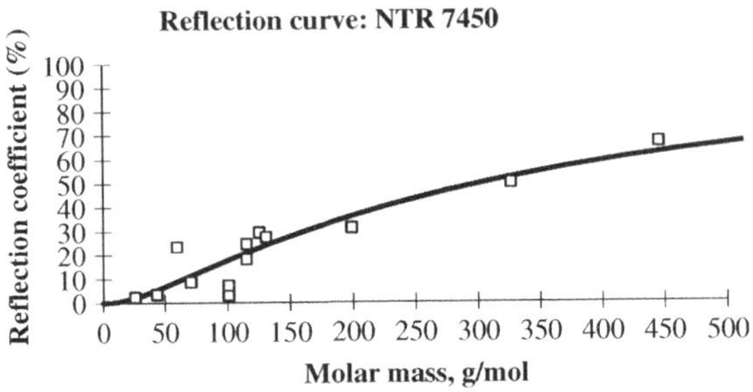

Figure 9.3 *Reflection curve for a typical nanofiltration membrane (NTR 7450, Nitto-Denko)*

lower retentions compared to non-polar molecules. This effect, which is independent of the sign of the membrane charge (positive or negative), can be explained by electrostatic interaction directing the dipole towards the membrane. Other more complex interactions between solutes and membrane may interfere with the effect of the dipole moment, and have a combined effect of flux decline and a shift in the retention of the component.

9.4.2.2 *Retention of inorganic components*

Several transport models are available to describe transport and retention of inorganic compounds (Zhu *et al*., 2011). The basis for simulation of ion retentions is the extended Nernst-Planck equation:

$$J_i = -D_{i,p}A_k\frac{dc_i}{dx} - \frac{z_i c_i A_k D_{i,p}}{RT}F\frac{d\psi}{dx} + K_{i,c}c_i V \tag{9.6}$$

where

J_i = flux of ion i (mol/(m^{2}h))
$D_{i,p}$ = diffusion coefficient in the membrane phase (m^2/s)
c_i = concentration of the ion (mol/m^3)
z_i = ion valence (−)
A_k = membrane porosity (−)
R = ideal gas constant (J/(mol K)
T = temperature (K)
F = Faraday's constant (C/mol)
ψ = electrical potential(V)
$K_{i,c}$ = hindrance coefficient for convection (−).

The terms on the right-hand side represent transport due to diffusion (concentration gradient), electromigration (electrical potential gradient) and convection (caused by the pressure gradient across the membrane) respectively. The superscript m denotes a membrane parameter. Two models developed on the basis of the extended Nernst–Planck equation are the hybrid model (HM) and the Donnan–Steric partitioning pore model (DSPM) (Bowen and Mukhtar, 1996).

The hindrance coefficients can be described by empirical equations. The conditions for electroneutrality and zero current inside the membrane are as follows:

$$\sum_{i=1}^{n} z_i c_i = 0 \quad \sum_{i=1}^{n} z_i c_i^m = -X \tag{9.7}$$

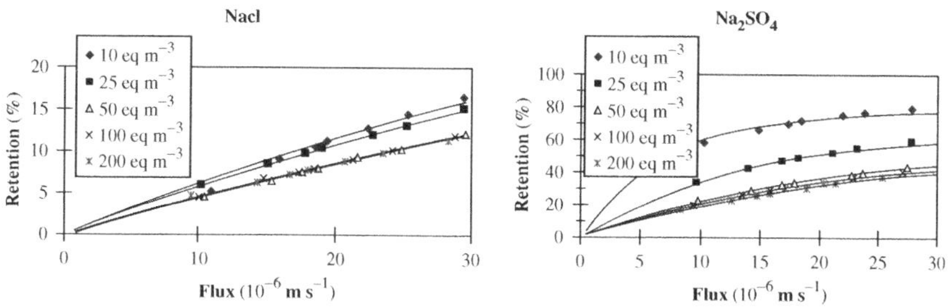

Figure 9.4 Simulation of NaCl and Na_2SO_4 retentions using the extended Nernst–Planck equation for the CA 30 membrane (1E-6 m/s is equal to 3.6 L/(m²h))

In which c_i is the bulk concentration of ion i, c_i^m is the concentration of ion i inside the membrane and X is the effective volumetric membrane charge density. X is assumed to be constant at all points in the active part of the membrane. Thus, by deriving this equation and introducing

$$\sum_{i=1}^{n} z_i \frac{dc_i^m}{dx} = 0 \tag{9.8}$$

into the extended Nernst–Planck equation, the following expression for the electrical potential gradient is obtained:

$$\frac{d\psi}{dx} = \frac{\sum_{i=1}^{n} \frac{z_i J_v}{D_{i,p}} \left(K_{i,c} c_i - C_{i,p}\right)}{\frac{F}{RT} \sum_{i=1}^{n} \left(z_i^2 c_i\right)} \tag{9.9}$$

The zero current condition inside the membrane is expressed as:

$$I_c = \sum_{i=1}^{n} F\left(z_i j_i\right) = 0 \tag{9.10}$$

By assuming the correct boundary conditions and by using the Poisson–Boltzmann equation, which expresses that the electrical potential gradient is a constant:

$$\frac{d^2 \psi^m}{dx^2} = -\frac{p^m}{\varepsilon^m} \tag{9.11}$$

(with ρ^m the space charge and ε^m the dielectric constant in the membrane) a set of equations is obtained from which the retention of an ion can be numerically obtained. Figure 9.4 shows simulation results of NaCl and Na_2SO_4 retention using the CA 30 membrane (Nadir, Germany).

9.5 Membrane materials and properties

Nanofiltration membranes are usually made of many layers of material—maybe as many as four different layers. The support layer is giving the strength to the membrane so that it can be handled without giving extra resistance to the final membrane. The support layer is often made of a fiber network. The second membrane layer also often has a supporting function, but can be a microfiltration or an ultrafiltration membrane.

The nanofiltration separation layer can be made on site out of interacting chemicals on top of the support, or it can be modified out of the top layer. The top NF layer gives the membrane its functional properties like cut-off, hydrophilicity, charge and possible special properties. The most used polymers in the selective layer are polyamides, polyamines (PA), cellulosic esters, for example cellulose acetate (CA), polyethersulfones (PES), polysulfones (PS), and polyvinylalcohol (PVA). In addition, polyimide (PI) and polydimethylsiloxane (PDMS) types of membranes are typically used in solvent nanofiltration.

9.5.1 Structure of NF membranes

Nanofiltration membranes are mostly made of organic polymers, but also ceramics are tried out as nanofiltration materials. Typically, a composite NF membrane consists of, for example a polysulfone sublayer and a top layer that is made of polyamide. Today, the cut-offs are still difficult to make out of ceramics or metals, but in the future, when it will be possible, these materials can stand more heat and solvents and could, therefore, be used more in for instance separations of oils and fuels as well as different extraction liquors in biorefineries. Also polymeric NF membranes have been made for solvent fractionation, but there are not many companies yet who make them and some of the membranes have fairly low fluxes (see Table 9.1). Anyway, solvent NF seems to belong to the most promising new membrane applications in the future.

In Figure 9.5 cross-sections of virgin and used NF membrane are presented. The membrane was used to purify xylose from pulping waste liquor at a temperature of 70 °C and a pressure of 30 bar. These extreme conditions have caused a significant change in membrane structure as can be seen in the figure.

9.5.2 Hydrophilic and hydrophobic characteristics

The hydrophilicity or hydrophobicity of a membrane is very important. Hydrophilic membranes are less fouling when dealing with water solutions, while hydrophobic membranes are important in the separation of oils and solvents. Figure 9.6 shows the contact angles of some common NF membranes (the lower the contact angle—the more hydrophilic the membrane). The most hydrophilic membranes are those made of regenerated cellulose but few such NF membranes are made. Polyvinyl alcohol is also a very hydrophilic material. Most often the membranes are modified with hydroxylated substances to make them more hydrophilic. Polyamides are generally more hydrophilic than other materials and most NF membranes are made of cross-linked polyamides. The cross-linking is needed to give a low cut-off and at the same time give more stability to the membrane. Hydrophilic membranes (contact angle <40) are preferred in water treatment but for solvent separations membranes with very high contact angles are used.

9.5.3 Charge characteristics

The charge of the membrane determines the repulsive conditions between the filtered molecules and the membrane, which is often of importance in the filtration of salts or other ionic molecules. Some of the NF materials, like amine- or imine-containing materials, are naturally charged. Some other materials often become charged by adsorption of ions from solution. Ions can also be implanted in the membrane material and, in this more controlled way, the spaces for flux in nanofiltration membranes keep fixed during casting. Typically implanted ions are zirconia and titania, and today many of the ions from transition metals in order to give the membrane a specific character. Gold can be implanted in membranes to give a non-reactive surface. The above characteristics are the most important for nanofiltration. Many other characteristics can also be measured when going to very subtle fractionations (Kallioinen and Nyström, 2008).

Table 9.1 *Typical commercial NF membranes (information collected from membrane manufacturers' homepages)*

Membrane	Manufacturer	Material	pH	T_{max} °C	P_{max} bar	Permeability L/(m^2h bar)	Retention %	Conditions °C; g/L; bar
Membranes for water treatment								
Nano-Pro	BPT*		0–14		120		>98.5 $MgSO_4$	
MPS-32	Koch SelRo	—	0–14	80				
NP030	Microdyn-Nadir	PES	0–14	95			25–35 NaCl/ 85–95 Na_2SO_4	20;–; 40
NP010	Microdyn-Nadir	PES	0–14	95			5–15 NaCl/ 30–60 Na_2SO_4	20; –; 40
Desal KH	GE Osmonics	—	0–11.5	80				
MPS-36	Koch	—	1–13	70	35	6.7	10 NaCl	30; 5; 30
NTR-7450 HG	Nitto Denko	Sulfonated PES	1–13	90	50	9	50 NaCl	25; 2; 10
NTR 7430 HG	Nitto Denko	PVA	2–11	90	30		30 NaCl	25; 2; 5
NTR 70 HG	Nitto Denko	PVA	2–8	60	50		93 NaCl	25; 1.5; 10
Desal-5 DK	GE Osmonics	PA	4–11	90	68	5.4	50 NaCl	25:01:07
AFC80	PCI membranes	PA	1.5–10.5	70	60		80 NaCl	
AFC40	PCI membranes	PA	1.5–9.5	60	60		60 $CaCl_2$	
NF270	Dow Filmtec	PA	3–10	40	41	13 / 10	40–60 $CaCl_2$/ >97 $MgSO_4$	25; 0.5; 4.8
NF90	Dow Filmtec	PA	3–10	45	41	6.7 / 8.5	85–95 NaCl/ >97 $MgSO_4$	25; 2; 4.8
NF200	Dow Filmtec	PA	3–10	45	41	7.1/6.0	35–50 $CaCl_2$/ >97 $MgSO_4$	25; 2; 4.8
TS80	Trisep	PA	4–11	45	41	5.5	99 $MgSO_4$	25; 2; 7.6
XN45	Trisep	PA-urea	3–11	45	41		95 $MgSO_4$	25; 2; 6.8
NF1	Sepro	PA	3–10	50	83	11	80 NaCl/90 $MgSO_4$	25; 2: 10
NF2	Sepro	PA	3–10	50	83	13	55 NaCl/97 $MgSO_4$	25; 2: 10
NF3	Sepro	PA	3–10	50	83	4	50 NaCl/98 $MgSO_4$	25; 2: 10
NF4	Sepro	PA	3–10	50	83	11	35 NaCl/98 $MgSO_4$	25; 2: 10
ESNA1	Hydranautics	PA composite	2–10	45	41	7.1	86 $CaCl_2$	25; 2; 5.2

(continued overleaf)

Table 9.2 (continued)

Membrane	Manufacturer	Material	pH	T_{max} °C	P_{max} bar	Permeability L/(m^2h bar)	Retention %	Conditions °C; g/L; bar
NE70	Woongjin Chemical	PA	2–11	45	41	5.7	40–70 NaCl / 97 $MgSO_4$	25; 2; 5.2
NE90	Woongjin Chemical	PA	2–11	45	41	6.1	85–95 NaCl / 97 $MgSO_4$	25; 2; 5.2
PVD-1	Hydranautics	PVA derivative	2–8	40		3.5	80	
Solvent-resistant membranes								
010206	SolSep			120	20	1 (acetone)	300 Da, fatty acid in acetone, 95	
030306	SolSep			150	40	2 (acetone)	500 Da, colorant in acetone, 99	
030705	SolSep	Silocon type		150	20	2–5 (ethanol)	1000 Da, oil in ethanol, 70	
MPF-50	Koch	PMDS					700 Da, Sudan IV in ethyl acetate	
MPF-44	Koch	PMDS		40		1.3	98 (5% sucrose)	
Starmem 122	W.R. Grace and Comp	PI		60			220 Da, n-alkanes in toluene	
Starmem 240	W.R. Grace and Comp	PI		60			400 Da, n-alkanes in toluene	
Duramem	MET	PI				1–8 (DMF)	180–1200 Da	
Ceramic membranes								
	TAMI Industries	TiO_2	2–14	<300			1 kDa	
1 nm	Inopor membranes	SiO_2					600 Da	
1 nm	Inopor membranes	TiO_2					750 Da	
0.9 nm	Inopor membranes	TiO_2					450 Da	

*Can be used in 20% sulfuric acid or NaOH

Polyethersulphone (PES), polyvinyl alcohol (PVA), polydimethylsiloxane (PDMS), plyimide (PI)

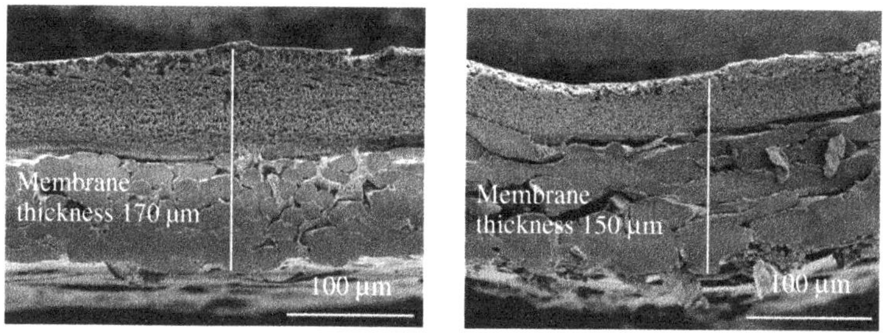

Figure 9.5 Electron microscope pictures of cross-sections of virgin and used NF membrane. The membrane was used in a biorefinery application at 70 °C and 30 bars for several months. Pictures courtesy of M. Kallioinen, 2007

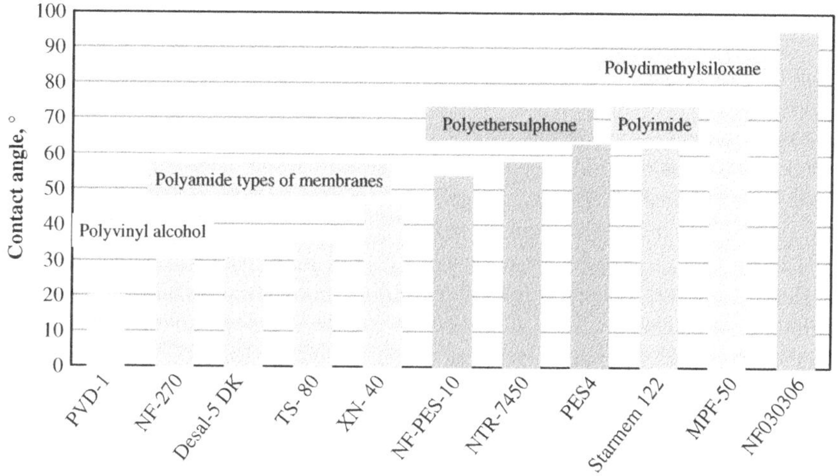

Figure 9.6 Sessile drop contact angles of NF membranes made from different materials

9.6 Commercial nanofiltration membranes

Typical NF membranes are made of polymers and consist of several layers, the selective layer being 40–250 nm thin. The most used polymers are polyamides, polyamines (PA), cellulosic esters, for example cellulose acetate (CA), polyethersulfones (PES), polysulfones (PS), and polyvinylalcohol (PVA). In addition, polyimide (PI) and polydimethylsiloxane (PDMS) types of membranes are typically used in solvent nanofiltration—compare Figure 9.6 and Table 9.1.

The membranes used in biorefinery applications should be compatible with "green" hydrophilic material and broadly resistant against fouling—they should be hydrophilic, like cellulose or PVA membranes. Unfortunately, hydrophilic membranes are not the most resistant against degradation at high temperature or in an extreme pH range. In addition, they might be susceptible to microbes, which might even use the membrane as a source of their nutrition. If more resistant materials are needed, hydrophobic membrane materials are modified to become more hydrophilic. Looking at pH resistance, the first-generation membranes were made from CA (pH range 4–8), the second from modified PA (pH range 2–11) and the third from modified PS (pH range 0–14). The very resistant membranes are, for instance, needed in filtration of very hot black liquor at pH close to 14. Membranes made out of ceramics or metal would be excellent to use but today there are not many that could have the small pore sizes needed in NF. Their use might come in the future.

Marketed NF membranes generally retain divalent ions more than 97%, see Table 9.1. Their sodium retention can vary a lot depending on the purpose for which they are produced. Today special membranes are manufactured for specific purposes, if the application has a high enough market value. An example is the NF200B membrane, which was developed to retain pesticides but partly permeate ions causing hardness in the Méry-sur-Oise plant in France (Wittmann and Thorsen, 2005).

9.7 Nanofiltration examples in biorefineries

The concept of biorefinery includes various types of raw materials and processes to produce value added products and energy. In the following sections some biorefineries, where nanofiltration is used or investigated, are reviewed. Generally, nanofiltration membranes have been studied for concentration and purification purposes. Separation is mostly based on molecular size exclusion. However, the special properties of nanofiltration membranes (for example, Donnan exclusion) discussed in Sections 9.5 and 9.6 can also be used in biorefinery applications especially when acidic compounds are recovered or fractionated. Donnan exclusion arises from the repulsion of charged species from similarly charged membrane. It makes it possible to efficiently separate charged and uncharged compounds from each other as well as to fractionate mono- and multivalent ions. Therefore, separation in nanofiltration can be tailored by adjustment of pH when the acid dissociation constants (k_a) of the separated compounds are different.

Membranes offer several advantages in the processing of biomaterials. They can operate at room temperature and without a phase change. They are relatively easy to scale up and they operate with low energy consumption. These properties are particularly important for the purification of protein-type compounds, which are easily deactivated or denatured under extreme conditions. On the other hand, novel nanofiltration membranes are also available for the treatment of alkaline or acidic streams, even at high temperature, and in the filtration of organic solvents. Therefore, NF can be used to fractionate compounds in various extraction liquors. Organic solvent nanofiltration (OSN) is a strongly increasing sector in membrane separation and several new applications are installed, for instance to purify solvents or products in the food and the pharmaceutical industry. The largest installation of OSN membrane (polyimide) is located at Exxon Mobil's Beaumont refinery (Texas, USA) where solvent, about 11 000 m^3/day, is recovered in a lubricant refining process (Peeva *et al.*, 2010). OSN could also be a possible technology for fractionation of dissolved compounds in a multicomponent biorefinery based on Organosolv pulping (Pan *et al.*, 2005; Alriols *et al.*, 2010).

9.7.1 Recovery and purification of monomeric acids

Green biorefineries are using naturally wet green biomass for the manufacture of industrial products. Raw materials such as grass land biomasses are often only partly utilized. The green biorefinery uses fresh materials or fermented materials. In many cases fermented raw materials contain more valuable compounds than fresh grass materials. In addition, a short growing season limits the availability of fresh raw materials. Fermentation products are typically in the nanofiltration size range. Therefore, NF is a potential concentration and purification process for many fermentation products. For instance, nanofiltration has been used to purify and concentrate antibiotics from fermentation broths (Tessier *et al.*, 2005; Kamm *et al.*, 2006).

9.7.1.1 *Separation of lactic acid and amino acids in fermentation plants*

Nanofiltration has been studied in the purification of amino and lactic acids. During the ensiling process of grass silage, significant amounts of amino acids and lactic acids are generated. The dry matter of silage press juice contains about 30% lactic acid and 30% crude proteins. Proteins exist mostly in the form of free amino acids (about 85–100%). These acids can be recovered and purified using membrane technology (Koschuh *et al*., 2005; Novalin and Zweckmair, 2009).

A demonstration plant was constructed in Austria to produce lactic acid and various amino acids from pressed silage juice. After sedimentation of the juice the first step is ultrafiltration, in which disturbing macromolecules and proteins are separated from the product permeate stream. The UF permeate is then treated by nanofiltration to produce phases enriched in amino acids (concentrate) and lactic acids (permeate).

Both streams are further purified by ion-exchange resins or electrodialysis. The product streams can be used for different applications such as in the animal food industry or in the chemical industry as a precursor of ethyl lactate (lactic acid) and in the food industry or in the cosmetics and pharmaceutical industry (amino acids). All by-products such as solid residues, UF, and ED concentrates are used in the adjacent biogas plant. About 0.1 ha of field is needed to produce 2220 kg silage (1000 kg total solids) and after fractionation as final products about 60–120 kg amino acids and 120–160 kg lactic acids. Ecker *et al*. (2011) optimized the separation of lactic and amino acids by controlling the feed pH and the filtration pressure. For instance, the ratio of lactic acid/amino acids mass flux increased from 7 to 37 by decreasing the feed original pH 3.9 to pH 2.5 (Steimüller, 2007; Mandl and Steinmüller, 2009; Ecker and Harasek, 2010; Ecker *et al*., 2011).

9.7.1.2 *Separation of lactic acid from cheese whey fermentation broth*

Lactic acid can also be produced from lactose by fermentation of lactose-rich raw materials like cheese whey. Li and Shahbazi (2006) showed that nanofiltration separates lactic acid from lactose and cells in the cheese whey fermentation broth. Cheese whey was previously a serious pollutant but today is an important by-product from the cheese manufacturing process. Ultrafiltration combined with diafiltration is used to concentrate and purify whey proteins from lactose, salt, and other components in the whey. Whey proteins are commonly used as food additives or as protein supplement. Lactose sugar can be used to produce value-added products such as lactic acid and ethyl alcohols by fermentation processes. Li *et al*. (2008) used the free living cell of *Bifidobacterium longum* to produce lactic acid from lactose. Nanofiltration retained about 97% of the lactose and 40% of the lactic acid from the fermentation broth. The permeated lactic acid was concentrated at 55 bar pressure using reverse–osmosis (RO) membrane. The RO membrane completely retained lactic acid.

9.7.2 Biorefineries connected to pulping processes

Lignocellulosic biomass is an abundant organic source on Earth. The annual production has been estimated to be about $170*10^9$ metric tons (Amidon and Liu, 2009). Forest covers over 40% of the annual biomass production. Biomass pretreatment has a crucial role in biorefineries based on lignocellulose feedstock. Biomass pretreatment technologies (chemical, physical and biological) open and break the structure of biomass to improve the hydrolysis rate and increase the yields of fermentable sugars or other chemicals from lignocellulosic biomass. The pretreatment is usually a costly process and, therefore, biorefineries integrated to existing biomass converting plants (for example, pulp mills) are more attractive than green-field investments.

Nanofiltration can be used to separate monomers from higher molar mass compounds or even fractionate small molar mass compounds. Separation of sugars from acids or other impurities can be done by NF and a simultaneous purification and concentration is possible to achieve. Fractionation of different organic acids might also be a future application for nanofiltration membranes in biorefineries. Purification of monosaccharides from lignin can partly be made by ultrafiltration or NF. However, there are very few industrial examples on the use of nanofiltration in forest-based biorefineries. One of them purifies sodium hydroxide from the viscose process (Schlesinger *et al.*, 2006). Some research articles on potential applications are reviewed in the next sections.

9.7.2.1 Valorization of black liquor compounds

The pulping industry produces about 190 million tons of pulp annually in the world. This is less than 1% of the annual growth of the forest. About 63% of the pulp is produced by the Kraft pulping process. Roughly about half of the wood material is converted to chemical pulp and the rest of it is dissolved during cooking and is usually burned in a chemicals and energy recovery boiler (about 120 Mt annually). The spent Kraft cooking liquor, referred to as black liquor, contains lignin-type compounds, organic acids, carbohydrates, extractives, and inorganic cooking chemicals, mostly NaOH and Na_2S. Many mills separate wood extractives (terpenoids, resin and fatty acids, sterols, and so forth) prior to the recovery boiler. The main by-products from Kraft pulping are tall oil (annual global production about 1.5 Mt) and turpentine (0.1 Mt). Furthermore, less than 0.1 Mt of Kraft lignin is isolated from black liquor for chemical applications (Pye, 2008). The existing sulfate (Kraft) pulp mills produce fibers, energy, and various chemicals, and can thus be called biorefineries although the biomass utilization is not currently completely optimized.

It can be said that spent Kraft cooking liquor is still an underutilized resource that has potential utility for production of bio-based chemicals or biofuels and polymeric materials. Today the black liquor, after removal of by-products, is evaporated and combusted in a recovery boiler to recycle and regenerate the cooking chemicals and to recover the energy content of the dissolved wood material as steam and electricity. Some compounds in the black liquor (such as organic acids formed mostly by degradation of hemicelluloses) have a relatively low heating value and, therefore, their separation could be possible without affecting significantly on the amount of the energy produced. The fraction of carbohydrate degradation products is mainly composed of hydroxy monocarboxylic and volatile acids (formic and acetic acids) along with lesser amounts of various dicarboxylic acids. The amount of organic acids in black liquor is almost 30% of the dry solids (DS) including volatile acids (formic and acetic acids) roughly 10% of the DS. The main hydroxyl carboxylic acids in black liquors are presented in Figure 9.7. The annual global generation of hydroxyl acids in Kraft pulping is about 30 Mt. The main hydroxy carboxylic acids (such as lactic and glycolic acids) have well-established applications. The properties and uses of especially isosaccharinic acids would still require further studies (Niemelä *et al.*, 2007). They may potentially be used to make biodegradable barriers.

Membrane filtration, especially ultrafiltration, has been studied since the 1970s to fractionate black liquors. Mainly the purpose has been the separation of lignin. For instance Jönsson *et al.* (2008) used a hybrid ultrafiltration/nanofiltration process to purify and concentrate lignin from black liquor (Figure 9.8). Ceramic UF membrane was first used to remove the hemicelluloses. Ultrafiltration was made at high cross-flow conditions (5 m/s) and at low pressure (1 bar) to reduce concentration polarization and improve the separation of lignin and hemicelluloses. The separation of lignin from hemicelluloses is needed when lignin is used for carbon fiber production. The original black liquor contained about 7% hemicelluloses in total dissolved solids (TDS). The UF permeate had less than 2% hemicelluloses in TDS. At a volume reduction of 90% the hemicellulose concentration in the UF retentate was about 50 g/L. After NF the lignin concentration was 165 g/L in the NF concentrate. The purity of the lignin in the NF concentrate was

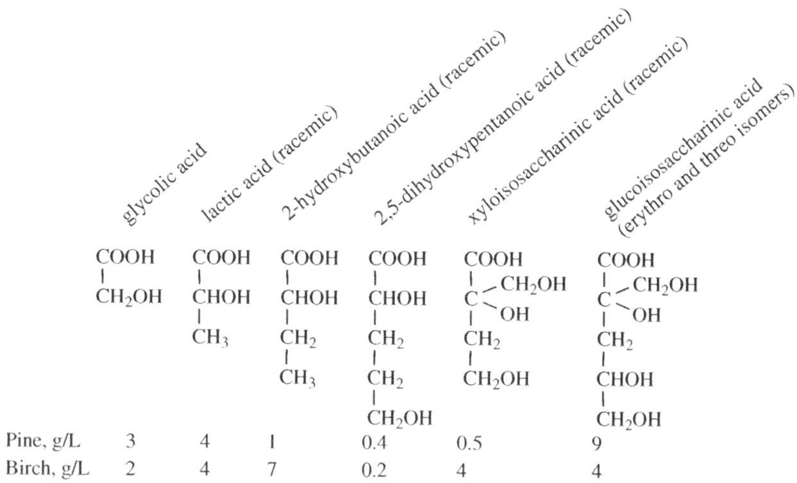

Figure 9.7 Main hydroxyl acids and their amount in black liquor

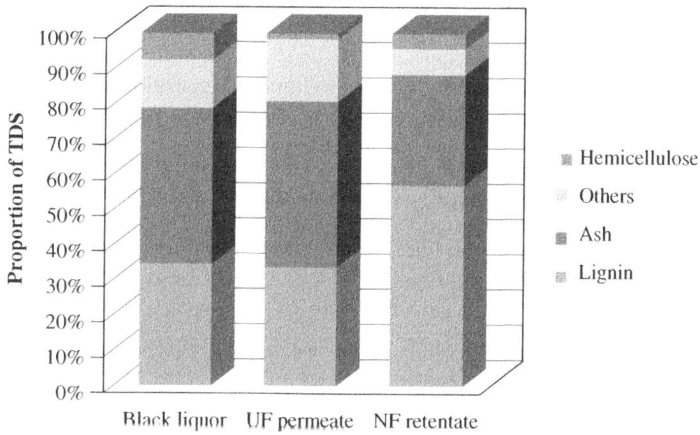

Figure 9.8 Proportion of various compounds in black liquor, UF permeate and NF concentrate calculated from the total dissolved solids (TDS). Based on data published by Jönsson et al. (2008)

about 57% compared to 35% in the black liquor. Further purification could probably have been done by diafiltration. Their cost estimate indicated a production cost of €33 per tonne of purified lignin (Jönsson *et al*., 2008).

Recently more interest has been directed to the recovery of hydroxyl acids. By ultrafiltration over 80% of the lignin can be removed without a significant loss of organic acids. Ultrafiltration permeate contains organic acids, inorganic compounds and low molar-mass lignin. The residual lignin can partly be precipitated by decreasing the pH of the black liquor. Nanofiltration has been studied to purify the residual liquor after lignin precipitation and salt crystallization (Niemi *et al*., 2011). The fractionation of different acids is partly possible by controlling the pH of the liquor (Figure 9.9). A negative retention of formic acid was seen when nanofiltration was used to fractionate hydroxyl acids from pretreated black liquor.

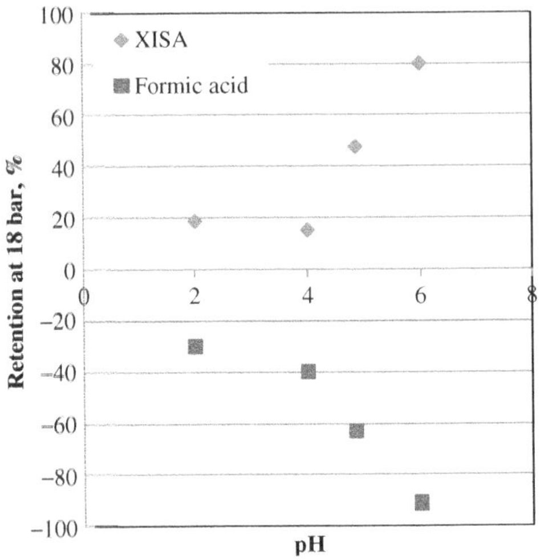

Figure 9.9 *Retention of xyloisosaccharinic acid and formic acid as a function of pH at a pressure of 18 bar when purified black liquor was filtered using a Desal-5 DK membrane (the feed was the mother liquor after lignin removal by UF, acid precipitation and salt removal by cooling crystallization made by adding 10% non-solvent (ethanol) at 2 °C)*

Electrodialysis (Rowe and Gregor, 1986) and chromatographic separation are also possible methods to purify organic acids originating from black liquor.

9.7.2.2 *Purification of pre-extraction liquors and hydrolysates*

In Kraft cooking of wood chips most of the alkaline is consumed by the saccharinic acids formed in the degradation of hemicelluloses. If part of the hemicelluloses can be removed prior to the cooking process, the alkali consumption would decrease in the cooking. Therefore, extracting the hemicellulose from wood chips prior to pulping could increase the pulp yield and the production, and/or produce higher value chemicals and polymers. Different pre-treatment processes such as pressurized hot-water extraction, steam explosion or alkaline/acid treatment have been widely studied to dissolve hemicelluloses from wood chips. Several studies have also been published on organosolv pretreatment of biomass (Carvalheiro *et al*., 2008; Zhao *et al*., 2009).

Depending on the final target different amounts of consequential process steps might be needed. High molar mass hemicelluloses can be separated from the pre-hydrolysate by ultrafiltration, and NF membranes can be applied to concentrate and purify low molar mass carbohydrates. The high molar mass compounds such as hemicelluloses are going to be used as barrier materials or they can be hydrolyzed to oligomeric or monomeric compounds. If the target is to recover carbohydrates to be further refined, for example to ethanol or other products by fermentation, several process steps are needed. Generally polymeric compounds need to be converted to monomers prior to fermentation. This conversion is typically done by acids or enzymes. As shown in Figure 9.10, nanofiltration can be applied in many stages to purify, recover and concentrate hydrolysates and products.

The pretreatment processes mentioned above are not selective only for carbohydrates. The pre-hydrolysates therefore contain a mixture of chemicals including hemicellulose-derived sugars, lignin-derived aromatic compounds, organic acids, methanol and furan compounds, and so forth. Some of

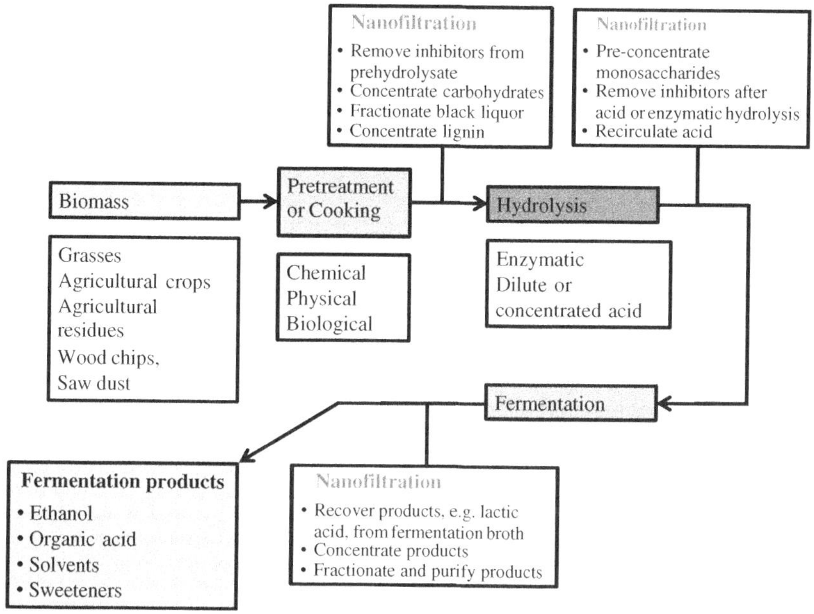

Figure 9.10 *Nanofiltration in biorefineries based on optional pretreatment, hydrolysis and fermentation of sugars to various end-products*

them are toxic for enzymes and inhibit fermentation by yeast and bacteria. The toxic compounds include different degradation products from wood such as acetate from the deacetylation of xylan, furfural, and hydroxyl-methylfurfural from furan dehydration, formic and levunic acids from sugars and various phenolic compounds formed in the degradation of lignin. In addition, Sun *et al*. (2011) reported that also particles play an important role in the inhibition of hydrolysate fermentation. Particles can be formed by condensation reactions of aromatic compounds such as furfural and phenols, or from precipitated lignin. They showed that, when particles were removed from the fermentation broth, the ethanol production increased tenfold.

Several authors have used NF to simultaneously concentrate carbohydrates and to remove inhibitors from hydrolysates (Liu *et al*., 2008; Amidon and Liu, 2009; Huang *et al*., 2010; Weng *et al*., 2010; Qi *et al*., 2011). Amidon and Liu (2009) extracted hemicelluloses from sugar maple wood chips prior to chemical pulping and fractionated the hydrolyzed extraction liquor by nanofiltration. Hot water extraction at 160 °C for 2 hours dissolved 23% of the wooden material. The main compound in the extract was xylan, which can be converted into ethanol or biodegradable plastics. The pre-extraction released over 70% of the xylan from the wood chips. In the Kraft cooking of hardwood, acetyl groups from xylan are liberated forming acetic acids, which consume a significant amount of the alkaline cooking chemicals, which is a reason for removing the xylan. Therefore, having a reduced amount of xylan in the wood chips less alkali is consumed in Kraft cooking. Furthermore, the heating value of wood carbohydrates is approximately half of that of lignin. The recoverable heating value of black liquor is therefore higher than without pre-extraction.

9.7.2.3 Examples of monosaccharides purification

Several authors have shown that a tight NF membrane retains about 99% of monosaccharides, for example xylose, and permeates many other monomers such as fermentation impurities like acetic acid and furan

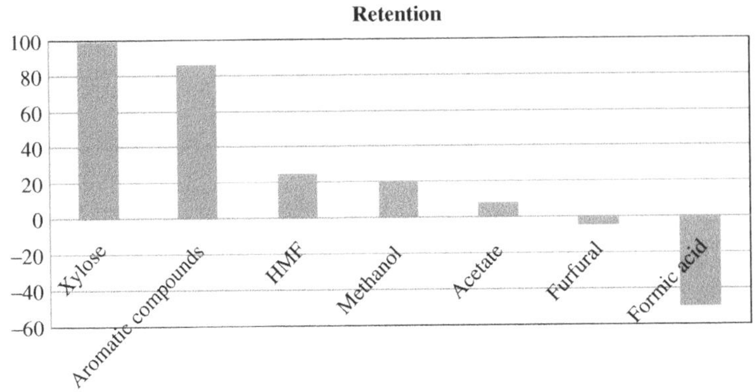

Figure 9.11 Retention of various monomers in nanofiltration of hot-water wood extract. Based on data published by Amidon and Liu (2009)

compounds. Amidon and Liu (2009) reported that the retention of formic acid was even negative meaning that its concentration was higher in permeate than in the feed solution (Figure 9.11). Liu *et al.* (2008) showed that a tight NF membrane had the capability to concentrate monosaccharides and permeate furfural, hydroxymethylfurfural, acetic and formic acid, and methanol, which are all known to be toxic for enzymes. Weng *et al.* (2010) separated furan and carboxylic acids from carbohydrates in dilute acid rice straw hydrolysates by nanofiltration (Desal-5 DK). The separation factors of acetic acid over xylose and arabinose were about 50. Qi *et al.* (2011) improved the separation of furfural from monosaccharides by diafiltration. Huang *et al.* (2010) developed a process model to simulate an integrated forest biorefinery manufacturing pulp and other co-products. Their model included nanofiltration to concentrate pre-extraction hydrolysate sugars. The simulation results showed that pre-extraction is a potential method to recover hemicelluloses prior to Kraft cooking.

As discussed above, nanofiltration is a potential technique to concentrate monosaccharides and simultaneously remove inhibiting compounds from hydrolysates. The concentrated carbohydrates can be fermented to alcohol after enzymatic hydrolysis. However, many authors have focused mostly on the separation efficiency, and have not intensively discussed membrane fouling and flux, which are crucial factors in industrial-scale processes. Almost no studies have been made to solve the fouling problems related to membrane filtration of hydrolysates (Koivula *et al.*, 2011).

9.7.2.4 Nanofiltration to treat sulfite pulp mill liquors

Although most of the chemical pulp is produced by the Kraft process, there are still several sulfite mills in operation. During the sulfite cooking process sulfonated lignin is formed and hemicelluloses are degraded mostly to monosaccharides. Therefore, a distinct hydrolysis of hemicelluloses is not needed if the aim is to use the monosaccharides. The spent sulfite pulping liquors contain, as typical components, lignosulfonates, sulfite cooking chemicals, xylonic acid, oligomeric sugars, dimeric sugars and monosaccharides and carboxylic acids, such as acetic acid, and uronic acids. The main compounds are lignosulfonates (50–65%) and carbohydrates (15–30%). The lignosulfonates are valuable by-products, which are recovered from spent liquor by UF (for example, Borregaard, Norway) and refined to different grades of vanillin products (about 3 kg vanillin from 1 t wood). The carbohydrates can be used as such after purification, or fermented or/and chemically converted to various products. Glucose can be fermented to ethanol, organic

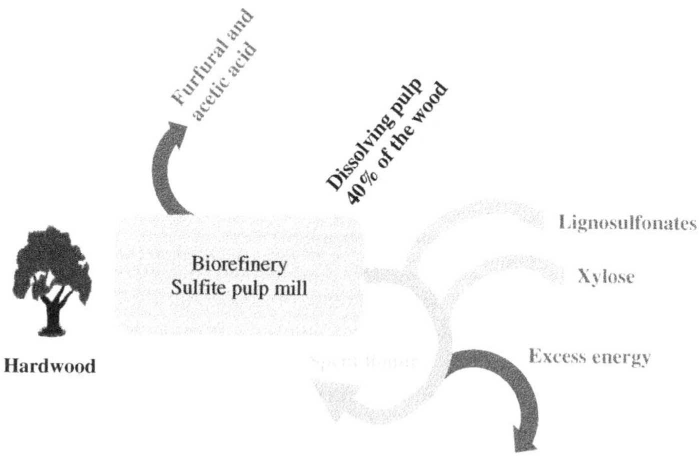

Figure 9.12 *Biorefinery based on the sulfite pulp mill*

acids (for example, lactic acid), solvents (acetone, butanol) and xylose to ethanol or xylitol. In addition, acids catalyze glucose dehydration to furfural (precursor of Nylons 6 and 6,6).

The Lenzing biorefinery (Acidic Mg-sulfite pulp mill) in Austria recovers acetic acid, furfural, magnesium lignin sulfonate and sodium sulfate. Furthermore, in close connection with the pulp mill Danisco Ltd. separates xylose from the spent cooking liquor. Figure 9.12 shows some of the products from the sulfite pulp mill. Nanofiltration is used on an industrial scale to purify NaOH for reuse in viscose fiber production (Section 9.7.4.1) and to recover and purify xylose from spent cooking liquor (Section 9.7.4.2).

9.7.3 Miscellaneous studies on extraction of natural raw materials

Several authors have extracted carbohydrates from wood materials, pulp or other biomasses using alkaline or acidic chemicals or enzymes. Svenson and Li (2010) presented a method to produce high-purity xylose from cellulosic fibers. A cold caustic extraction was used to extract xylan from lignocellulosic materials at a high purity (up to 95% purity). An alkali resistant membrane (MPS-34, Koch membranes) was used to concentrate and purify xylan solution, and to recover sodium hydroxide into permeate. The subsequent hydrolysis of the xylan results in a xylose product with a purity of greater than 80%. Therefore, additional purification steps such as chromatographic separation and crystallization are not needed and xylose can be hydrogenated to xylitol (Svenson and Li, 2010).

Murthy *et al*. (2005) compared nanofiltration and evaporation to concentrate xylose hydrolysate originated from acid hydrolysis of rice husk. Nanofiltration concentrated xylose from 2% to 10% at a reasonably high flux of 24 L/(m^{2}h) and a xylose rejection of >99% at 20 bar pressure. The NF permeate could be reused in acid hydrolysis. In addition, NF partially permeated unwanted monovalent salts. The operational cost of NF was only 4% of the evaporation costs.

Various oligosaccharides are possible products for biorefineries. Xylo-oligosaccharides are mixtures of xylose containing oligosaccharides having from two to seven molecules of xylose connected with β-1,4 linkages. They can be produced by hydrolysis of plant hemicelluloses. Typically xylan is first extracted from plant materials by acid, alkaline or steam treatment, and then hydrolyzed by enzymes. About 25% of birchwood and rice shell materials is xylan. In corncob the xylan content can even be 40%.

Yuan *et al*. (2004) extracted xylan by steaming treatment from corncob and enzymatically hydrolyzed it to oligomeric compounds, mainly xylobiose and xylotrioses. Nanofiltration concentrated and purified the xylo-oligosaccharides to a concentration of 15% and a purity of 74% at a pressure of 14 bar.

Gullón *et al*. (2011a) studied the separation of xylo-oligosaccharides from Eucaluptus wood autohydrolysates. The aqueous extraction was made in two stages. In the first stage (130 °C) the hemicellulose remained mostly unchanged but the extractives were partly removed. The second stage was made at 175 °C, and the hemicellulose (xylan) was extensively converted into soluble oligomers. This solution was purified by nanofiltration combined with diafiltration and resin adsorption.

Gullón *et al*. (2011b) also studied the simultaneous saccharification and fermentation of apple pomace to produce a medium containing mainly lactic acid and oligosaccharides. Lactic acid was removed by ion exchange and nanofiltration was used to concentrate the oligosaccharides. Combining diafiltration (diafiltration factor 5, amount of diafiltration water 5 × the feed volume) with concentration filtration a high yield (>90%) of oligosaccharides and approximately a 95% removal of impurities (NaCL, arabinose and lactic acid) was achieved. The concentration of the oligomeric compounds increased from 16.8 kg/m^3 up to 80.7 kg/m^3. Some prebiotic effects of the refined oligosaccharide concentrate were observed when its ability to support the growth of individual and mixed bacterial populations was measured.

9.7.4 Industrial examples of NF in biorefinery

9.7.4.1 *Recovery and purification of sodium hydroxide in viscose production*

The major part of the sulfite pulp is used as dissolving pulp in, for example, viscose production. In the production of regenerated cellulose highly purified celluloses are needed. The production of viscose-type cellulosic textile fibers therefore generates large volumes of sodium hydroxide waste liquor containing dissolved hemicelluloses. The liquor contains about 200 g/L sodium hydroxide and a significant amount of carbohydrates, which have a detrimental effect on the processability and quality of the viscose fibers. Therefore, at least part of the carbohydrates should be removed from the process with a simultaneous minimization of NaOH losses. Schlesinger *et al*. (2005, 2006) studied NF to purify NaOH for reuse in viscose fiber production. Although few NF membranes are resistant to extreme alkaline conditions their study showed that NF can be used to purify sodium hydroxide and to remove around 80–90% of the carbohydrates. Their study also showed that the NF membrane cut-off-values were somewhat higher than reported by the manufacturer due to the extremely alkaline conditions and high temperature. In Austria NF has been used since 2006 to separate hemicelluloses and alkaline liquor in viscose production (Schlesinger *et al*., 2006; http://www.lenzing.com/sites/nh08/english/html/1_4.htm).

9.7.4.2 *Xylose recovery and purification into permeate*

Because the pore sizes of nanofiltration membranes are in the same range as the sizes of monomeric carbohydrates the separation of carbohydrates depends significantly on the filtration conditions such as permeate flux, sugar concentration, and temperature. Nanofiltration membranes can therefore also be applied to increase the purity of permeating sugars, as will be discussed in the following paragraphs.

Xylose is a valuable raw material in the sweet manufacturing, aroma, and flavoring industries and particularly as a starting material in the production of xylitol. Xylose is formed in the hydrolysis of xylan-containing hemicellulose, for instance in sulfite pulping processes. The spent liquor originating from pulping of hardwood is an attractive resource for xylose, which can further be converted to low caloric value sweetener (xylitol). Traditionally chromatographic separation has been used to recover xylose on an industrial scale, and recently nanofiltration has shown to be an efficient recovery and purification method for xylose. The process efficiency depends on the filtration conditions, the feed solution composition and

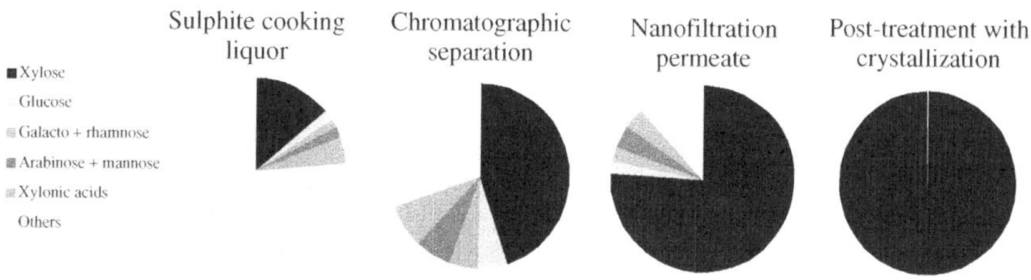

Figure 9.13 *Development of xylose purity (% on DS) in a three-stage purification process using chromatographic separation, nanofiltration and crystallization (Heikkilä et al. 2005)*

the type of membranes, and when everything is optimized over 70% purity of xylose can be achieved. Several purification processes are typically needed to achieve the highest xylose purity. Figure 9.13 shows the composition of the spent sulfite cooking liquor and how different purification processes improve the xylose purity on dry solids (DS). Chromatographic separation is used in this case as a pretreatment prior to nanofiltration and crystallization is the post-treatment. In this application xylose is recovered into NF permeate and this permeate is evaporated and crystallized to obtain a 99.8% xylose purity (Heikkilä *et al.*, 2005).

In biorefinery applications the feed concentrations are significantly higher than what is typically assumed in nanofiltration. A high feed concentration affects the retention and might facilitate a separation that is not possible at low concentration. As Figure 9.14 shows, the xylose purity in NF permeate depends strongly on the feed dry solids. In addition, an increase of the feed liquor pH from 3.4 to 5.9 increases the xylose purity in the nanofiltration permeate from 27% to 55% on DS (Heikkilä *et al.*, 2005).

9.7.4.3 *Purification of dextrose syrup*

Another example where a NF membrane purified the product into permeate is the purification of dextrose. Starch was first enzymatically degraded to dextrose syrup, which contains about 95% dextrose and about 5% di- and trisaccharide impurities. Then the used nanofiltration membrane, at about 30% dry solids

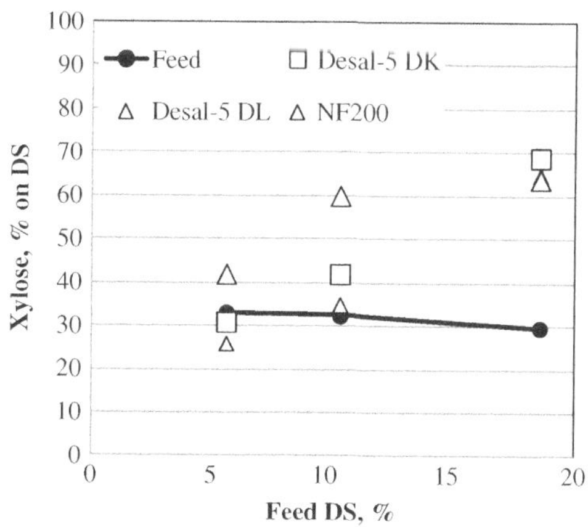

Figure 9.14 *Effect of feed dry solids concentration on the xylose purity in nanofiltration permeate when ultrafiltered and pH adjusted (pH 5.3) spent sulfite cooking liquor was nanofiltered (Heikkilä et al. 2005)*

concentration, permeated dextrose efficiently but retained di- and trisaccharides. As a result a dextrose purity over 99% was achieved in the NF permeate (Binder *et al.*, 1999).

9.8 Conclusions and challenges

In biorefinery applications, nanofiltration membranes are often severely fouled. This is a drawback and challenge for researchers. Fouling-resistant membranes are needed and efficient cleaning needs to be developed. In addition, novel ways to predict fouling are needed. Fouling caused by adsorption of organic compounds and biofouling are often the most difficult types of fouling. In addition, at high concentration a gel layer formation might surprise by dramatically decreasing flux. The post-treatment of existing membranes making them more hydrophilic, or the addition of catalytic particles might lower fouling of the membranes. Furthermore, more attention should be allocated to pre-treatments prior to membrane processes. It can be predicted that, in future biorefineries, NF will be used as a part of a hybrid process rather than alone.

Another drawback is the relatively wide pore size distribution of the commercial membranes. This will limit the separation efficiency in the fractionation of similar molar mass compounds. Therefore, membranes that have a narrow pore-size distribution are needed instead of integrated membrane cascades.

Despite these challenges, membrane processes dealing with biorefinery, including NF, are very much potential separation methods in future biorefineries. The most promising applications are those connected to already existing processes because they will be more economical, and will use existing knowledge on how to collect and handle raw materials. The recovery and purification of valuable solute compounds such as food additives or pharmaceutical products will improve the economy of the existing biomass refining plants like pulp mills. However, a big challenge will be how to recover or purify these products. Nanofiltration can hopefully be one of the future solutions.

References

Alriols M.G., Garcia A., Liano-ponte R., Labidi J., 2010. Combined organosolv and ultrafiltration lignocellulosic biorefinery process, *Chem. Eng. J.* 157(1), 113–120.

Amidon T.E., Liu S., 2009. Water-based woody biorefinery, *Biotech. Adv.* 27, 542–550.

Binder T.P., Hadden D.K., Sieverd L.J., 1999. Nanofiltration process for making dextrose, US patent 5 869 297, *February* 9, 1999.

Bowen W.R., Mukhtar H., 1996. Characterisation and prediction of separation performance of nanofiltration membranes, *J. Membr. Sci.* 112, 263–274.

Carvalheiro F., Duarte L.C., Gírio F.M., 2008. Hemicellulose biorefineries: a review on biomass pretreatments, *J. Sci. Ind. Res.* 67, 849–864.

Darvishmanesh S., Degrève J., Van der Bruggen, B., 2010. Mechanisms of solute rejection in solvent resistant nanofiltration: the effect of solvent on solute rejection. *Phys. Chem. Chem. Phys.* 12, 13333–13342.

Ecker J. Harasek M., 2010, Membrane based production of lactic acid and amino acids—green biorefinery Upper Austria, http://opensourceecology.org/wiki/Upper_Austria_Green_Biorefinery, referred 10.12.2011.

Ecker J, Raab, T., Harasek M., 2012. Nanofiltration as key technology for the separation of LA and AA, J. Membr. Sci. 389 389–398. Gullón P., González-Muñoz M. J., van Gool M. P., Schols H. A., Hirsch J., Ebringerová A., Parajó J. C., 2011, Structural features and properties of soluble products derived from *Eucalyptus globulus* hemicelluloses, *Food Chem.* 127(4), 1798–1807.

Gullón B., Gullón P., Sanz Y., Alonso J.L., Parajó J.C., 2011b. Prebiotic potential of a refined product containing pectic oligosaccharides, LWT—Food Sci.*Techn.*, 44, 1687–1696.

Heikkilä H., Mänttäri M., Lindroos M., Nyström M., 2005. Recovery of xylose, US patent 6,872,316 B2, March 29 2005.

Huang H.-J., Ramaswamy S., Al-Dajani W.W., Tschirner U., 2010. Process modeling and analysis of pulp mill-based integrated biorefinery with hemicelluloses pre-extraction for ethanol production: A comparative study, *Biores. Techn.* 101, 624–631.

Jönsson A.-S., Nordin A.-K., Wallberg O., 2008. Concentration and purification of lignin in hardwood Kraft pulping liquor by ultrafiltration and nanofiltration, *Chem. Eng. Res. Design* 8(11), 1271–1280.

Kallioinen, M. and Nyström, M., 2008. *Membrane Surface Characterization*. In: Advanced Membrane Technology and Applications, eds. Li, N.N., Winston Ho, W.S., Fane, A.G. and Matsuura, T., John Wiley & Sons, Inc., New Jersey, pp. 841–878.

Kamm B., Kamm M., Gruber P.R., Kromus S., 2006. Biorefinery systems—an overview. In Biorefineries—*Industrial Processes and Products, Status Quo and Future Directions*, Vol 1, eds. Kamm B., Gruber P.R., Kamm M., Wiley-VCH Verlag Gmbh & Co., Weinheim, pp. 3–40.

Koivula, E., Kallioinen, M., Preis, S., Testova, L., Sixta, H., Mänttäri, M., 2011. Evaluation of various pretreatment methods to manage fouling in ultrafiltration of wood hydrolysates, *Sep. Pur. Techn.* 85, 50–56.

Koschuh W., Thang V.H., Krasteva S., Novalin S., Kulbe K.D., 2005. Flux and retention behavior of nanofiltration and fine ultrafiltration membranes in filtrating juice from a green biorefinery: A membrane screening, *J. Membr. Sci.* 261, 121–128.

Li Y., Shahbazi A., 2006. Lactic acid recovery from cheese whey fermentation broth using combined ultrafiltration and nanofiltration membranes, *Appl. Biochem. Biotechn.* 129–132, 985–996.

Li Y., Shahbazi A., Williams K., Wan C., 2008. Separate and concentrate lactic acid using combination of nanofiltration and reverse osmosis membranes, *Appl. Biochem. Biotechn.* 147, 1–9.

Liu S., Amidon T.E., Wood C.D., 2008. Membrane filtration: Concentration and purification of hydrolyzates from biomass, J. *Biobased Material Bioenergy* 2(2), 121–134.

Mandl M., Steinmüller H., 2009, Green biorefinery in Europe. http://www.landbrugsinfo.dk/Planteavl/Afgroeder/Energiafgroeder/Filer/pl_09_016_11_Michael_Mandl.pdf (accessed September 28, 2012).

Mänttäri M., Sjöman E., Heikkilä H., Koivikko H., Lindell J., 2009. Separation process, US patent application 2009/0014386 A1, *January* 15, 2009.

Murthy G.S., Sridhar S., Sunder M.S., Shankaraiah B., Ramakrishna M., 2005. Concentration of xylose reaction liquor by nanofiltration for the production of xylitol sugar alcohol, *Sep. Pur. Techn.* 44, 221–228.

Niemelä K., Tamminen T., Ohra-aho YT., 2007. Black liquor components as potential raw materials, 14th International Symposium on Wood, Fibre and Pulping Chemistry, Durban, South Africa, 25–28 June, 2007.

Niemi, H., Lahti, J., Hatakka, H., Kärki, S., Rovio, S., Kallioinen, M., Mänttäri, M., Louhi-Kultanen, M., 2011. Fractionation of organic and inorganic compounds from black liquor by combining membrane separation and crystallization, *Chem. Eng. Techn.* 34(4), 593–598.

Novalin S., Zweckmair T., 2009. Renewable resources—green biorefinery: separation of valuable substances from fluid-fractions by means of membrane technology, Biofuels, *Bioprod. Bioref.* 3, 20–27.

Pan X., Arato C., Gilkes N., Gregg D., Mabee W., Pye K., Xiao Z., Zhang X., Saddler J., 2005. Biorefining of softwood using ethanol organosolv pulping: Preliminary evaluation of process streams for manufacture of fuel-grade ethanol and co-products, Biotechn. *Bioeng.* 90(4), 473–481.

Peeva L.G., Sairam M., Livingston A.G., 2010. *Nanofiltration operations in nonaqueous systems*. In: Comprehensive Membrane Science and Engineering, eds. Drioli E., Giorno L., Elsevier, Oxford, pp. 91–113.

Pye E.K., 2008. Industrial lignin production and applications. In: *Biorefineries—Industrial process and products, eds. Kamm B., Gruber P.R., Kamma, M., Wiley-VCH Verlag Gmbh & Co., Weinheim, Germany*, Vol. 2, pp.165–200.

Qi B., Luo J., Chen X., Hang X., Wan Y., 2011. Separation of furfural from monosaccharides by nanofiltration, *Bioresource Techn*. 102, 7111–7118.

Rowe J.W., Gregor H.W., 1986. Membrane processes for separation of organic acids from Kraft black liquors, US patent 4,584,057, *April* 22, 1986.

Schlesinger R., Götzinger G., Sixta H., Friedl A., Harasek M., 2006. Evaluation of alkali resistant nanofiltration membranes for the separation of hemicelluloses from concentrated alkaline process liquors, *Desalination* 192, 303–314.

Schlesinger R., Röder T., Götzinger G., Sixta H., Harasek M., Friedl A., 2005. Influence of hemicellulose aggregate and gel formation on flux and retention during nanofiltration of alkaline solutions, *Desalination* 175, 121–134.

Schäfer A.I. Fane A.G., Waite T.D., 2005. *Nanofiltration, Principles and Applications*, eds., Elsevier Advanced Technology, pp. 273–277.

Sjöman E., Mänttäri M., Nyström M., Koivikko H., Heikkilä H., 2007. Separation of xylose from glucose by nanofiltration from concentrated monosaccharide solutions, *J. Membr. Sci.* 292, 106–115.

Spiegler K.S., Kedem O., 1966. Thermodynamics of hyperfiltration (reverse osmosis): criteria for efficient membranes, *Desalination* 1(4), 311–326.

Steinmüller H., 2007, The Austrian green biorefinery, http://www.energytech.at/pdf/biorefinery_steinmueller.pdf.

Sun Z., Shupe A., Liu T., Hu R., Amidon T.E., Liu S., 2011. Particle properties of sugar maple hemicellulose hydrolysate and its influence on growth and metabolic behavior of Pichia stipitis, *Biores. Technol.* 102, 2133–2136.

Svenson D.R., Li J., 2010. Process for manufacturing high purity xylose. US Patent 7,812,153 B2, October 12, 2010.

Tanninen J., Mänttäri M., Nyström M., 2006. Nanofiltration of concentrated acidic copper sulfate solutions, *Desalination* 189(1), 92–96.

Tessier L., Bouchard P., Rahni M., 2005. Separation and purification of benzylpenicillin produced by fermentation using coupled ultrafiltration and nanofiltration technologies, *J. Biotechnol.* 116, 79–89.

Timmer J.M.K., Speelmans M.P.J., van der Horst H.C., 1998. Separation of amino acids by nanofiltration and ultrafiltration membranes, *Sep. Pur. Techn.* 14, 133–144.

Van der Bruggen B, Everaert K., Wilms D., Vandecasteele C., 2001. Application of nanofiltration for removal of pesticides, nitrate and hardness from ground water: rejection properties and economic evaluation, *J. Membr. Sci.* 193, 239–248.

Van der Bruggen, B., Schaep, J., Vandecasteele, C., Wilms, D. A., 2000. Comparison of models to describe the maximal retention of organic molecules, *Sep. Sci. Technol.* 35(2), 169–182.

Van der Bruggen, B., Schaep, J., Wilms, D., Vandecasteele, C., 1999. Influence of molecular size, polarity and charge on the retention of organic molecules by nanofiltration. *J. Membr. Sci.* 156, 29–41.

Weng Y.-H., Wei H.-J., Tsai T.-Y., Lin T.-H., Wei T.-Y., Guo G.-L., Huang C.-P., 2010. Separation of furans and carboxylic acids from sugars in dilute acid rice straw hydrolyzates by nanofiltration, *Bioresource Technol*. 101, 4889–4894.

Wittmann E., Thorsen T., 2005. *Water treatment*. In: Nanofiltration Principles and Applications, eds. A.I. Schäfer, A.G. Fane, T.D. Waite, Elsevier Advanced Technology, Bodmin, pp. 273–277.

Yuan Q.P., Zhang H., Qian Z.M. Yang X.J., 2004. Pilot-plant production of xylo-oligosaccharides from corncob by steaming, enzymatic hydrolysis and nanofiltration, *J. Chem. Technol. Biotechn.* 79, 1073–1079.

Zhao X., Cheng K., Liu D., 2009. Organosolv pretreatment of lignocellulosic biomass for enzymatic hydrolysis, Mini-review, *Appl. Microbiol. Biotechn.* 82, 815–827.

Zhu H.C., Szymczyk A., Balannec B., 2011. On the salt rejection properties of nanofiltration polyamide membranes formed by interfacial polymerization, *J. Membr. Sci.* 379, 215–223.

10

Membrane Pervaporation

Yan Wang, Natalia Widjojo, Panu Sukitpaneenit and Tai-Shung Chung

Department of Chemical and Biomolecular Engineering, National University of Singapore, Singapore

10.1 Introduction

Biofuels are gaining more importance in the energy supply chain for bolstering energy security, and sustainability, and lowering the intensity of greenhouse-gas emissions. Renewable liquid biofuel—bioalcohol—is a prominent biofuel, capable of replacing petroleum globally, and has received significant attention in both academia and industry. It is projected that the global production of liquid biofuels will increase significantly in the near future [1], especially after the successful development of second-generation biofuels from non-food crops and agricultural residues (for example, lignincellulosic residues such as corn stover and plant trimmings).

Bioalcohols are mainly produced from the digestion of lignocellulosic biomass by enzymes to release stored sugars, followed by yeast-based fermentation. Depending on the process conditions and microorganisms used in fermentation, the fermentation broth typically contains water, acetone, butanol, ethanol, and many other substances with various compositions, where alcohol content is in the range of 5–12% [2–5]. As a consequence, concentration and separation must be conducted to produce high-purity alcohol. The basic physicochemical properties of those main components in fermentation broths are summarized in Table 10.1 [6–10]. The success of biofuel development is not only dependent on the advances in genetic transformation of biomass into biofuel but also on the breakthroughs in "separation" techniques, which account for 60% to 80% of the biofuel production cost [11].

Conventional separation techniques for liquid mixtures have included distillation, low-temperature crystallization, adsorption, extraction, and chromatography. Distillation is still the dominant refinery process. However, due to their energy intensive nature, negative environmental impact, and complicated operation procedure, the above techniques are generally not economical and practical to stand alone for the entire bioalcohol separation process. For instance, a conventional distillation process is inadequate for small-scale

Separation and Purification Technologies in Biorefineries, First Edition.
Edited by Shri Ramaswamy, Hua-Jiang Huang, and Bandaru V. Ramarao.

Table 10.1 *Physicochemical properties of the main components in the fermentation broth*

Properties	ethanol	acetone	n-butanol	water
Chemical formula	C_2H_5OH	C_2H_6CO	C_4H_9OH	H_2O
Molecular weight (g/mol)	46.1	50.08	74.1	18.0
Molecular volume (Å^3) [a]	96.8	105.7	151.8	29.9
Kinetic diameter (nm) [7]	0.430	0.469	0.505	0.296
Radius of gyration (Å) [b]	2.259	2.746	3.251	0.615
Density (g/cm^3)	0.789	0.793	0.810	1.000
Boiling point (°C)	78.3	56.5	117.8	100.0
Vapor pressure 25 °C (kPa) [c]	7.795	30.810	0.915	3.169
Azeotrope with water (wt% water)	4.4	no azeotrope	44.5	–
Empirical polarity parameter E_T (30) (kcal/mol) [8]	51.9	42.2	49.7	63.1
Solubility parameter (J/cm^3)$^{1/2}$ [9]	26.5	19.9	23.1	47.8

[a] The molecular volume is calculated by the molecular weight divided by the density and the Avogadro number [10].
[b] The radii of gyration are calculated using AspenTech DISTIL (version 2004.1).
[c] The saturated vapor pressures at 25 °C are obtained using AspenTech DISTIL (version 2004.1).

operations, particularly in the presence of azeotropic mixtures. To improve the efficiency of current biofuel separation, new bio-refinery concepts and hybrid technologies must be developed.

The separation of bioalcohol (mainly ethanol and butanol) from fermentation broths using membrane-based pervaporation technology has been widely studied in recent years. Pervaporation is a membrane process technology combining membrane permeation and evaporation for molecular-scale liquid separation. This is a promising alternative to conventional technologies in the separation of liquid mixtures in the biorefinery, petrochemical, pharmaceutical industries, and so forth, as it is highly selective, economical, energy efficient, safe and ecofriendly. In pervaporation, the liquid feed mixture is in contact with one side of a non-porous polymeric membrane or molecularly porous inorganic membrane. The more permeable components are then sorbed into/onto the membrane, diffuse through the membrane, and evaporate as permeates because of the chemical potential differences across the membrane induced by vacuum or gas purge. The separation of different components is achieved based on the sorption and diffusion differences between the feed components. The partial vaporization of feed molecules and a phase change of the penetrants significantly determines the mechanism of separation process.

This chapter aims to provide an overview of the potential markets for and industrial needs of membrane pervaporation in biofuel separation. The fundamentals of the pervaporation process and performance evaluation are elucidated. The engineering principles for pervaporation membranes for different applications—dehydration or recovery, are discussed in detail, including material selection and morphology design. To understand the roles of pervaporation units in an integrated biorefinery process, different designs of hybrid processes involving pervaporation are compared and elaborated in the last section.

10.2 Membrane pervaporation market and industrial needs

The term "pervaporation" was introduced by Kober in 1917, in his report about the selective permeation of water from aqueous solutions of albumin and toluene through collodion (cellulose nitrate) films [12]. Between 1956 and 1962, Binning and co-workers established the principles of pervaporation and proposed the potential of pervaporation technology based on their research on the separation of hydrocarbon mixtures through a non-porous polyethylene film [13–15].

However, the low permeation flow rate through homogenous dense films was a critical problem hindering the large-scale industrial application of pervaporation, until the asymmetric membrane was fabricated using the phase inversion method developed by Loeb and Sourirajan [16]. The real breakthrough was achieved in 1980s by GFT (Gesellschaft für Trenntechnik, Hamburg, Germany) who developed a poly(vinyl alcohol) (PVA) and polyacrylonitrile (PAN) composite membrane for the dehydration of alcohol/water azeotropic mixtures [17]. From then until 1999, pervaporation has been progressively commercialized for large-scale processes, with more than 90 industrial pervaporation units installed worldwide [18]. Meanwhile, around 300 European and US patents about pervaporation were issued by then [18].

The major applications of pervaporation can be classified into three categories: (i) the dehydration of organic solutions; (ii) the removal or recovery of small amounts of organic volatiles from mixtures, and (iii) the separation of organic-organic mixtures. The last category will not be introduced in this chapter because it is not often involved in biofuel purification.

Of these three applications, organic dehydration is the most important in the industry, particularly for alcohol dehydration. In this application, low water content in the feed mixture is normally required to avoid extensive membrane swelling and to achieve a relatively stable and long-term separation performance. Due to a huge consumption of high-purity ethanol products in chemical, medical and biopharmaceutical industries (the ethanol easily forms an azeotropes with water at 95 wt%), more than 100 plants have been set up for the dehydration of ethanol since the first one was established by GFT in 1982. In addition to most of the early solvent dehydration systems aiming at ethanol purification, dehydration of other solvents such as isopropanol, glycols, acetone, and methylene chloride has received attention and several pervaporation dehydration technologies have been gradually developed.

Another application of pervaporation is the removal of organic solvents from solutions used in water purification, pollution control, and solvent recovery. This application is competing with distillation or solvent extraction for purifying solutions with less than 1–2 wt% of organic solvents. With recent high prices and fluctuations of crude oil, significant attention has been given to exploring the potential of pervaporation in the biorefinery and food industries [19, 20]. Most examples of this type of applications have been reported on a laboratory scale and very few for industrial applications. The first application of organic removal was developed by Membrane Technology and Research, Inc. (MTR) using PDMS-based membranes [21].

Nevertheless, so far, the number of pervaporation applications on an industrial scale remains rather small and the majority of industrial pervaporation units are still for the dehydration of a limited number of organic solvents. The major challenge to the industrialization of pervaporation is that the current permeation flux is too low to be economical. The high costs of membrane production and module fabrication, and problems associated with membrane reliability and resistance to harsh environments are other issues to be concerned. Here we mainly focus on the review of current pervaporation membranes and pervaporation-based hybrid processes.

10.3 Fundamental principles

10.3.1 Transport mechanisms

The separation characteristics of pervaporation are far more complex than gas and other liquid separations because a phase change (liquid to vapor) is involved in the process. This process therefore involves both mass and heat transfer. The membrane acts as a barrier layer between a liquid and a vapor phase, regulating the transport of the permeates. The driving force for the mass transport is the chemical potential gradient (partial pressure or/and concentration of each species) across the membrane, which can be created by applying either a vacuum pump or an inert purge on the permeate side to maintain the permeate vapor pressure lower than the partial pressure of the feed liquid. The ultimate separation performance of a pervaporation membrane towards the feed permeates is therefore caused by the differences in the permeability and fugacity between permeates. The differences in permeates are further determined by (i) the physicochemical properties of feed mixtures and their own interactions, (ii) the affinities of permeates toward the macromolecules of the membrane material, and (iii) the physical structure of the membrane.

There are many different views on the separation mechanism of pervaporation. Two of them are representative: the solution-diffusion mechanism theory [22] and the pore flow mechanism theory [23]. Both theories agree that the complicated chemical and physical interactions among feed components and membranes play important roles on determining the overall separation performance. Several models have been proposed by other researchers. Binning *et al*. [13] suggested that selectivity takes place in a boundary layer between the liquid regime and the gas regime in the membrane. Michaels *et al*. [24] interpreted the selectivity as a result of sieving by polymer crystals. Schrodt *et al*. [25] believed that the hydrogen bonding between the polymer and solvent components played an important role. Long *et al*. [26] considered that the diffusion and concentration gradients in different solvent components were the governing factors. Matsuura *et al*. [27, 28] regarded the pervaporation mechanism as a combination of reverse osmosis separation, followed by the evaporation and vapor transportation through capillary pores on the surface layer of a reverse osmosis membrane. Yoshikawa *et al*. [29–33] explained that the specific and selective separation of substances through a membrane might be realized by differences in the strength of hydrogen bonding interaction, leading to a selective separation through a membrane. Here we will describe only the two main models: the solution-diffusion model and the pore-flow model.

The solution diffusion model, a semi-empirical or phenomenological model, is generally used to describe the transport mechanism in non-porous membrane. It was originally developed by Graham [34] to describe gas permeations, and has been widely adopted by most researchers for the pervaporation process due to its good agreement with experiments [3, 6, 22, 35–43]. Referring to this mechanism, pervaporation consists of three consecutive steps (as shown in Figure 10.1): (i) sorption of the permeate from the feed liquid to the membrane, (ii) diffusion of the permeate through the membrane, and (iii) desorption of the permeate to the vapor phase on the downstream side of the membrane. It hypothesizes that the permeation rate and selectivity are governed by the solubility and diffusivity of the feed components permeating across the membrane. The solubility of a feed component in the membrane is determined primarily by the physicochemical nature of the membrane material and the permeating molecules, while the diffusivity is dependent on physical factors such as the size and shape of penetrate molecules as well as the micro-morphology of the membrane structure [44]. Generally, the sorption step prefers more condensable

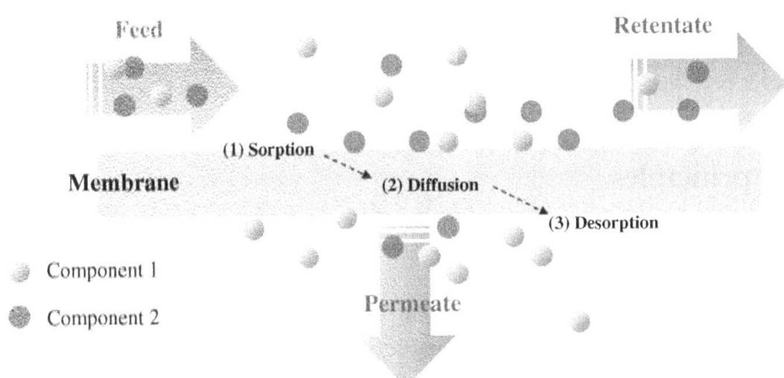

Figure 10.1 *Solution-diffusion mechanism for the permeants to transport through a dense membrane. Adapted from: Membranes for biofuel separation, Asia Pacific Biotech News (APBN), 16, 34–39, (2012). http://www.asiabiotech.com/articles/readmore/1605-article02.html*

molecules or molecules that have a special interaction and/or affinity with membrane materials [45]. The solubility of feed components in the membrane can be qualitatively described by their solubility parameters. The difference in the solubility parameter between the penetrants as well as their individual differences with the membrane may provide an important clue for the selectivity of the resultant membrane selectivity [23, 46, 47]. Besides the solubility parameter, polarity is another important index to judge the solubility of the permeating components in the membrane. On the other hand, the diffusion behavior of the permeants across the membrane depends greatly on (i) the penetrant size and shape, (ii) the mobility of polymer chains, (iii) the interstitial space between polymer chains, and (iv) the interactions between penetrants and between penetrant and membrane material [48].

The pore-flow model is an alternative for describing the transport mechanism in pervaporation membranes. The model was first proposed by Okada and Matsuura [49]. Figure 10.2 depicts the schematic diagram of the pervaporation transport on the basis of pore-flow concept. The model was established based on the following assumptions: (i) there is a bundle of straight cylindrical pores with an effective length δ penetrating across the selective layer of the membrane; (ii) all pores are operating under an isothermal condition. According to the model, the mass transport mechanism in pervaporation also consists of three successive steps: (i) the permeant transports through the liquid-filled portion of the pore with a distance δ_a, (ii) a liquid-to-vapor phase change takes place inside the pore, and (iii) the permeant transports through the vapor-filled portion of the pore with a distance δ_b. In other words, the pervaporation transport can be considered to be a combination of liquid-phase and vapor-phase transport in series. The predictability of the models has been elucidated through many studies using different membrane materials and various separating mixtures [50–54]. In the recent work by Sukitpaneenit *et al.* [53], the mass transport phenomenon in pervaporation of the ethanol/water system via poly(vinylidene fluoride) (PVDF)

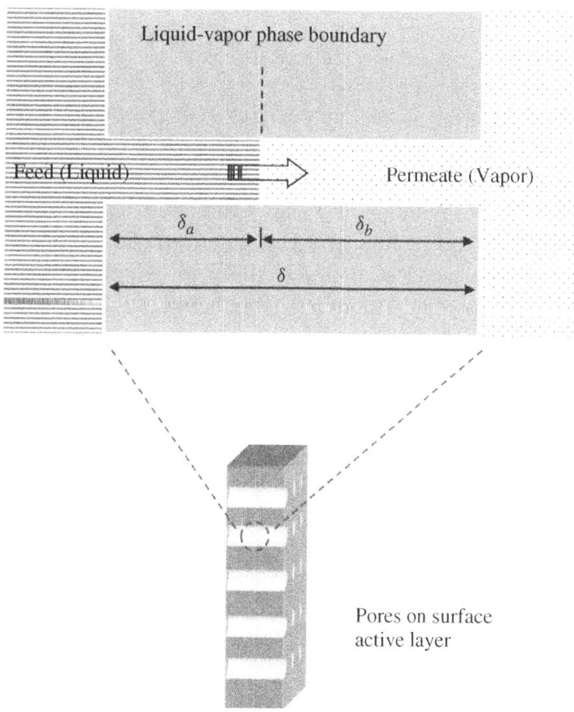

Figure 10.2 *Schematic representation of the pore-flow model in pervaporation transport*

asymmetric hollow fiber membranes has been demonstrated through the pore-flow model and a newly modified pore-flow model has been proposed. The newly modified pore-flow model differs from the original pore-flow model by factoring in the contribution of the Knudsen flow (in addition to the surface flow) to vapor transport.

10.3.2 Evaluation of pervaporation membrane performance

The performance evaluation of a pervaporation membrane is generally based on its capability to separate components from each other. There are two sets of interlinked parameters that have been widely used to describe this capability, namely: (i) the flux and separation factor, and (ii) permeability (or permeance) and selectivity.

Traditionally, the performance of a pervaporation membrane is characterized by the flux (J) and separation factor (α) as defined by the following equations:

$$J = \frac{Q}{A \cdot t} \tag{10.1}$$

$$\alpha_{i/j} = \frac{y_{w,i}/y_{w,j}}{x_{w,i}/x_{w,j}} \tag{10.2}$$

where Q is the total mass transferred over the operation time t, A is the membrane area, subscripts i and j refer to the two components to be separated in the feed mixture, y_w and x_w are the weight fractions of the components in the permeate and feed, respectively.

For flat-sheet dense membranes, the flux is also generally expressed in normalized flux (J_N), which is defined as the total flux J multiplied by the membrane thickness (l).

$$J_N = J \cdot l \tag{10.3}$$

In addition to the flux and separation factor, another set of performance parameters—permeability (or permeance) and selectivity, is more representative and accurate for the evaluation of intrinsic properties of a specific membrane-permeants system, since they significantly decouple the effects of process parameters on the performance evaluation. Wijmans [55] has elaborated on the importance and differences in using permeability (permeance) instead of flux to investigate intrinsic membrane properties. The detailed examples of using permeability (permeance) and selectivity in calculations and applications have been elucidated in our recent works [6, 36–43]. The relationship between permeability (or permeance) and flux can be expressed as below:

$$J_i = \left[\frac{P_i}{l}\right] \cdot \left(P_i^f - P_i^P\right) \tag{10.4}$$

where P_i is the membrane permeability of the component i, a product of diffusivity and solubility coefficients, and l is the thickness of the membrane selective layer. The term $[P_i/l]$ is also known as permeance. It is often employed for an anisotropic membrane with an unknown thickness of the dense selective layer. P_i^f is the partial vapor pressure of component i in a hypothetical vapor phase that is in equilibrium with the feed liquid, and P_i^P is the partial vapor pressure of component i in the membrane downstream. P_i^f and P_i^P can be expressed in following equations:

$$P_i^f = x_i \gamma_i P_i^{sat} \tag{10.5}$$

$$P_i^P = y_i P^P \tag{10.6}$$

where $x_{n,i}$ and $y_{n,i}$ are the mole fractions of the component i in the feed and permeate, respectively, γ_i is the activity coefficient, p_i^{sat} is the saturated vapor pressure, and p^P is the permeate pressure. p_i^{sat} and γ_i can be calculated by the Antoine and Wilson equations, respectively [36]. Therefore, the flux of the component i in Eq. (10.4) can be further expressed as below:

$$J_i = \left[\frac{P_i}{l}\right] \cdot \left(x_{n,i}\gamma_i P_i^{sat} - y_{n,i}P^P\right) \tag{10.7}$$

The difference in partial vapor pressure of the component i between the feed side and permeate side is its driving force to transport through the membrane, and can be written as follows.

$$\text{driving force} = \left(x_{n,i}\gamma_i P_i^{sat} - y_{n,i}P^P\right) \tag{10.8}$$

The permeability and permeance of the component i can be expressed by rearranging Eq. (10.4) as follows:

$$P_i = \frac{J_i}{x_{n,i}\gamma_i P_i^{sat} - y_{n,i}P^P} \cdot l \tag{10.9}$$

$$\overline{P}_i = \frac{P_i}{l} = \frac{J_i}{x_{n,i}\gamma_i P_i^{sat} - y_{n,i}P^P} \tag{10.10}$$

The total permeability (permeance) is therefore defined as the sum of permeabilities (permeances) of all individual components instead of the total flux divided by the total driving forces since each component has individual partial vapor pressures in both feed and permeate sides.

The ideal membrane selectivity β is defined as the ratio of permeability coefficients or permeance of two components as follows, indicating how efficient the two components can be separated:

$$\beta_{i/j} = \frac{P_i}{P_j} \quad or \quad \frac{\overline{P}_i}{\overline{P}_j} \tag{10.11}$$

The ideal selectivity can be weight-based ($\beta_{w,\, i/j}$) or mole-based selectivity ($\beta_{n,\, i/j}$), depending on the unit of the permeability coefficients or the permeance. Their relationship is as follows:

$$\beta_{n,i/j} = \beta_{w,i/j} \times \frac{M_j}{M_i} \tag{10.12}$$

where M_i and M_j are the molecular weights of the two components, respectively.

Based on their definitions, the discrepancy between separation factor and selectivity arises fundamentally from the fact that the latter has decoupled the activity coefficient and saturated vapor pressure. The detailed derivation of their relationships under different operation conditions has been presented elsewhere [41].

10.4 Design principles of the pervaporation membrane

One of the major hurdles for the expansion of pervaporation to the emerging markets is the lack of appropriate membranes. Based on numerous previous works on pervaporation membranes, two strategies have been adopted in parallel to design suitable synthetic membranes for biofuel separation: one is the molecular design of membrane materials with desired physicochemical properties (inherent solubility, diffusivity, and selectivity); the other is the macromolecular engineering of desired membrane morphology with ultrathin and defect-free dense selective layer. The former aims to tailor membranes' thermodynamic and/or kinetic responses to the penetrants, while the latter minimize defects and transport resistance. As a result, a synergistic combination among diffusion selectivity and sorption selectivity, high permeability and desired separation performance can be achieved. Therefore, in order to develop a high performance pervaporation membrane for biofuel separation, one must conduct material selection, membrane preparation, post-treatment and modification.

10.4.1 Membrane materials and selection

In principle, one should screen and choose membrane materials based on their separation factors and fluxes towards target components. As a rule of thumb, the selected membrane material must have a greater solubility or diffusivity towards preferred components than the non-preferred components. Furthermore, a high chemical resistance, thermal resistance, and stable mechanical strength to the target solution are also very important, particularly in harsh feed systems. Thus, when identifying an appropriate membrane material, extensive and tedious evaluations on material selection and tests cannot be evaded. In view of the merits of polymers relative to other materials (such as manufacturability and reasonable production costs), the emphasis of this chapter is on the design of polymeric-based membranes. A brief introduction to inorganic membranes for biofuel separation via pervaporation is also included.

Current commercially available polymers are limited for the development of pervaporation membranes in terms of balanced physicochemical properties, while synthesizing entirely new materials always involves risk, high cost, and considerable time; therefore modifying the existing membrane materials provides new prospects for overcoming current membrane limitations and creating new applications for better separation performance. Membrane modification includes chemical modifications and physical modifications. Common chemical modification methods including cross-linking, grafting, and plasma treatment may result in membranes with stable and customized physicochemical permeation properties. Cross-linking is a common strategy to suppress excessive membrane plasticization and swelling to enhance membrane stability. It may result in membranes with a higher selectivity but a lower permeability. Grafting is to graft some functional group to the polymer backbone in order to facilitate the membrane selectivity, while plasma treatment can optimize the surface hydrophilicity/hydrophobicity. Detailed examples are given in the following sections. Compared to chemical modifications, physical modifications such as thermal treatment and polymer blending focus on the optimization of membrane morphology and material matrix. Thermal treatment is a simple and efficient post-treatment method to improve the membrane performance because it can densify membrane structure, eliminate defects, and enhance selectivity. Blending is another powerful approach because it may synergistically combine the advantages of individual components and overcome their individual deficiencies with regard to separation applications. Generally, miscible polymer blends are desirable when fabricating the membrane, because homogeneous materials are essential to produce the membrane with uniform performance and stable thermal and mechanical properties.

Apart from polymeric membranes, inorganic membranes based on silica, alumina or zeolites have increasingly gained attention in recent years for biofuel separation. Since they are not subjected to any solvent-induced swelling and have a superior thermal and mechanical stability, they typically exhibit a greater selectivity and flux than most polymeric membranes. Moreover, they have uniform and molecular-sized pore structure, which allows molecular sieving and contributes to a significant difference in permeation rates of transporting molecules [56]. Inorganic membranes may therefore lead to a higher product quality and broaden the application range of pervaporation in industry, especially for some specific separations in harsh environments.

Despite so many advantages, the high cost and low processiblity of inorganic materials still limit their applications in membrane separation; they still cannot replace the dominant position of polymeric membranes at this moment. The combination of the polymeric membrane with inorganic membranes in different forms opens up new applications for membrane technology. Firstly, they can be combined in a composite membrane form where the polymeric layer is generally deposited on the ceramic or other inorganic membrane substrate, which is further discussed in the morphology part. Good separation performance is achieved compared to the conventional polymeric membranes because of the unique chemical, mechanical and thermal stability of the ceramic substrates. Another common form is the mixed matrix membranes (MMM), which involve embedded nano-size selective fillers (such as zeolite, silicalite and activated carbon) in the polymeric matrix. An improved separation performance can be yielded, due to the combined effects of molecular sieving, selective adsorption, and difference in diffusion rates of the permeants.

10.4.1.1 Polymeric pervaporation membranes for bioalcohol dehydration

Organic dehydration is the main application of membrane pervaporation in the industry. The development of membrane materials for alcohol dehydration is therefore more mature than alcohol recovery. A large amount of information on polymeric membranes for water/alcohol separation can be retrieved from the rich literature database. Among the polymers, PVA (Figure 10.3a), chitosan (CS) (Figure 10.3b) and sodium alginate (NaAlg) (Figure 10.3c) are the most common membrane materials for alcohol dehydration by pervaporation. This is because these macromolecules have plentiful polar functional groups (for example, hydroxyl, amino and acid groups) and thus display excellent hydrophilicity. Some other hydrophilic polymeric materials have also been extensively studied, including polyimide (PI), polyacrylic acid (PAA) (Figure 10.3d), PAN (Figure 10.3e), and cellulose sulfate, etc. Several reviews on polymeric membranes for alcohol dehydration have recently been documented [3, 35, 47, 57, 58].

Of all the membranes for pervaporation separation of aqueous organic mixtures, PVA-based membranes have been studied most intensively for alcohol dehydration. Poly(vinyl alcohol) films exhibit high-abrasion resistance, elongation, tensile strength, and flexibility, and have been used in a number of commercial membranes. In addition, PVA is a hydrophilic polymer whose hydroxyl groups have strong interactions with water through hydrogen bonding, thus it owns excellent water perm-selective properties. It is one of the very few high-molecular-weight water-soluble resins, and can be easily cross-linked chemically or thermally. So far, most researches are PVA-based membranes centered on chemical modifications of

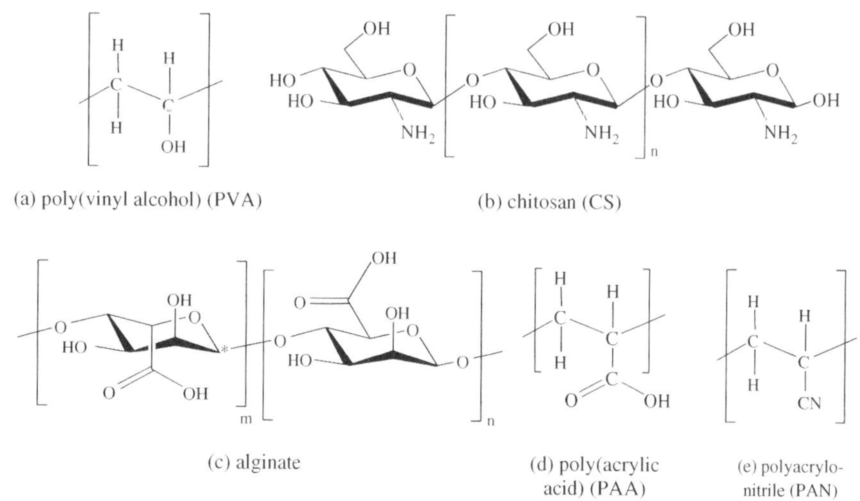

Figure 10.3 *Commonly used polymeric membrane materials for pervaporation dehydration*

Table 10.2 *Separation performance of various PVA-based membranes for ethanol and butanol dehydration*

Material	Feed composition (wt%)	T (°C)	Separation factor	Total flux (g/m^{2}hr)	Ref.
	ethanol/water				
PVA	95.6/4.4	60	~10	120	59
PVA cross-linked by amic acid	90/10	45	100	250	60
PVA cross-linked with dimethylolurea	90/10	60	~115	~120	61
PVA/NaAlg composite hollow fiber membrane with PSf support	90/10	45	384	384	62
PVA/PAA IPN dense membrane	95/5	75	4100	60	63
PVA/polyaniline nanocomposite dense membrane	90/10	30	564	69	64
Carboxymethyl CS /PVA blend membrane (blend ratio 8:2) cross-linked by GA with porous support	95/5	45	2959	140	65
polyelectrolyte and PEC membrane consisting of QPVA$^+$ and PPVA$^-$ (1:1 molar ratio)	95/5	75	973	403	66
	butanol/water				
PVA cross-linked by citric acid	90/10	30	171	82	67
PVA/polystyrene sulfonic acid composite membrane on PAN support, cured at 120 °C for 2h, followed by ionic cross-linking	95/5	60	18600	230	68
Ceramic supported PVA hollow fiber cross-linked by maleic anhydride	87/13	70	16	3900	69
	91/9	80	1600	2400	
PVA-NaAlg/PSf composite hollow fiber	90/10	45	606	585	62
Commercial Pervap®2510 multilayer membrane	85/15	60	60	1430	36

T: feed temperature
~Estimated values from graphs

PVA with improved permselectivity and stability. This approach has been used by GFT to develop its membranes for solvent dehydration from chemically cross-linked PVA. Table 10.2 summarizes separation performance of some PVA-based membranes for biofuel dehydration [36, 59–69].

Chitosan, a linear polymer comprising primarily of glucosamine, is another promising membrane material for bioalcohol dehydration with high water permselectivity and solvent stability. It is one of natural abundant polymers and is considered as a versatile material for various applications, due to the large variety, low cost, thermal stability, and excellent film-forming properties. Many chitosan membranes show superior separation performance to cross-linked PVA membranes on ethanol dehydration. Table 10.3 summarizes the separation performance of chitosan-based membranes for the biofuel dehydration [70–76]. Furthermore, with the aid of reactive hydroxyl and amino groups, chitosan can be further modified to suit specific pervaporation applications. For example, Lee and co-workers have studied chitosan membranes extensively with various modification methods for ethanol dehydration to enhance its stability and separation performance [76–82].

Alginate, one kind of hydrophilic polysaccharide type polymers, has gained special interest as a membrane material because it shows the highest flux and separation factor among the tested hydrophilic materials

Table 10.3 Separation performance of various chitosan-based membranes for ethanol dehydration

Material	Feed composition (ethanol wt%)	T (°C)	Separation factor	Total flux (g/m^2hr)	Ref.
Neat chitosan membrane	90	30	65.2	~52	70
CS/POSS hybrid membranes	90	30	373.3	~29	
CS/PAN composite membrane with carbopol as an bridging intermediate layer	90	80	256	1247	71
CS dense membrane cross-linked with H_2SO_4	90	60	1791	472	72
Two-ply CS/NaAlg dense composite membrane cross-linked with GA	90	60	~1000	~210	73
CS/PAA complex dense membrane	95	60	1008	132	74
CS/HEC blended composite membrane with cellulose acetate support	90	60	5469	424	75
Phosphorylated chitosan membrane with 56 mg/m^2 phosphorus content	90	70	541	182	78

for pervaporation dehydration. The effects of counter cations (Li^+, Na^+, K^+, Rb^+, and Cs^+) on permeation rate of alginic acid membranes were investigated around 1990s for alcohol dehydration by pervaporation [83, 84]. Among them, NaAlg membranes have received most attention and Table 10.4 summarizes some of their performance [85–90].

Despite the excellent separation performance of the hydrophilic materials mentioned earlier, they generally lack mechanical strength and stable membrane integrity in aqueous solutions. These membranes suffer severe swelling and flux decline with extended operation time. Therefore, further modifications via cross-linking, blending or hybrid with others, are needed to stabilize and improve their mechanical strength and long-term stability, as we mentioned above.

Recently polyimides have been gaining attention in the academia and industries for biofuel dehydration [3] because of their excellent thermal, chemical and mechanical stabilities. Commercial polyimides are extensively studied for pervapoartion applications, such as Torlon, Ultem, P84, Matrimid, etc. The selectivity of polyimide membranes towards water is attributed to the rigid chemical structure (high glass-transition temperature) and the preferential interaction between water molecules and the imide groups

Table 10.4 Separation performance of various NaAlg-based membranes for ethanol dehydration

Material	Feed composition (ethanol wt%)	T (°C)	Separation factor	Total flux (g/m^2hr)	Ref.
NaAlg dense membrane	90	60	10,000	290	85
NaAlg hybrid membranes containing 6 wt% of preyssler type heteropolyacid	96	30	59,976	43	86
NaAlg dense membrane cross-linked with phosphoric acid	95	30	2182	35	87
Alginate/CS two ply composite membranes supported by PVDF porous membrane	96	50	202	95	88
NaAlg membrane with PAN support	90	70	2750	942	89
Ca^{2+} cross-linked dense NaAlg membrane	90	50	300	230	90

Table 10.5 *Separation performance of various polyimide membranes for ethanol and butanol dehydration*

Material	Feed composition (wt%)	T (°C)	Separation factor	Total flux (g/m^2hr)	Ref.
	ethanol/water				
Aromatic polyimide PI-2080 asymmetric membrane	95/5	60	900	1000	92
Polyimide composite membrane with PSf porous support	90/10	40	240	1700	93
Matrimid® polyimide hollow fiber membrane, annealed at 260 °C for 5 hours	85/15	30	~195	~50	94
Torlon® polyimide hollow fiber membrane, annealed at 260 °C for 5 hours	85/15	30	128	~6.5	
6FDA-ODA-NDA/Ultem dual-layer hollow fiber	85/15	60	162	480	95
	butanol/water				
PAI/PEI dual-layer hollow fiber, heat treat at 75 °C for 2hr	85/15	60	1174	846	6
6FDA-ODA-NDA/Ultem dual-layer hollow fiber	85/15	60	0.38	2260	95
Ceramic-supported P84 polyimide membranes	95/5	150	360	1000–6000	96

through hydrogen bonding. In this regard, the selection of aromatic monomers (dianhydrides and diamines) for polyimide synthesis is critical to architecture its chemical structure and properties to achieve desirable membrane separation [91]. Since at least the early 2000s, an extensive body of experimental data on pervaporation for a wide array of polyimides (Table 10.5) has displayed encouraging potential for the dehydration of solvents [6, 92–96]. A specific review paper on this type of material for pervaporation application has been published recently [3].

Table 10.6 summarizes some other membrane materials and their pervaporation performance for biofuel dehydration, including PAA, PAN, cellulose sulfate, and some hydrophobic polymeric materials with improved hydrophilicity after modification [97–104]. In most cases, membranes show a trade-off relationship in separation performance; they possess either an improved separation factor with a significantly reduced flux or an enhanced flux with a lack of selectivity.

Inorganic pervaporation membranes, such as silica and zeolite, are reported for their unsurpassed separation performance. Silica membranes are generally prepared via sol–gel routes or chemical vapor deposition (CVD) methods on porous substrates for gas or pervaporation separation. Oxides such as ZrO_2 and TiO_2 are added in most cases in order to improve its stability in aqueous solution. Zeolites, are also widely studied as pervaporation membrane material, because of its unique pore structure, adsorption properties, as well as good mechanical, chemical resistance and thermal stabilities [105]. They are alumino-silicates with varied SiO_2/Al_2O_3 ratio and can form polycrystalline structure with well-defined nano-sized pores. They can be either hydrophobic at a high aluminium-to-silicon ratio, such as MFI membranes (silicalite-1 and ZSM-5) used for organic recovery, or hydrophilic at low aluminium to silicon ratio, such as zeolite A, mordenite and T-type zeolite membranes for dehydration application. Table 10.7 [106–122] gives a summary of some recently-reorted inorganic membranes for pervaporation dehydration of ethanol and butanol.

Organic–inorganic hybrid membranes are of great potential in pervaporation application. They combine the advantages of the exceptional high separation performance of inorganic elements, and

Table 10.6 Separation performance of other polymeric membranes for ethanol and butanol dehydration

Material	Feed composition (wt%)	T (°C)	Separation factor	Total flux (g/m^2 hr)	Ref.
	ethanol/water				
210 °C heat-treated PAN hollow fiber membranes	90/10	25	~1000	~180	97
Semi-IPN dense membrane of NR and cross-linked PAA	95/5	30	~305	~180	98
Polyaniline dense membrane	50/50	—	10000	1.3	99
Perfluorinated polymer on PAN support	98.7/1.3	50	387	1650	100
Asymmetric sulfonated PSf membranes	90/10	25	~00	~20	101
PSf/poly(ethylene glycol) blended dense membrane	90/10	25	325	600	102
Nylon 4 membrane grafted by n,n-dimethylaminoethyl methacrylate	90/10	25	28.3	439	103
	butanol/water				
Polyelectrolyte complex membrane based on cellulose sulfate (SYMPLEX) with PVDF support	90/10	25	300	1700	104

cost-effectiveness and ease of membrane fabrication of polymers. Besides multilayer composite membranes, organic and inorganic components can also be combined at a molecular level. For example, Lai and his co-workers [123–125] used inorganic cross-linker (nanosized silica particles with sulfonic acid groups (ST-GPE-S) or γ-glycidoxypropyltrimethoxysilane (GPTMS)) to modify chitosan membranes. The chitosan-silica hybrid membranes displayed enhanced hydrophicity and stability simultaneously and thereby improved both permeation flux and separation factor. Another successful example is the Hybsi® membrane developed by the Energy Research Centre at Netherlands (ECN). This organic–inorganic hybrid silica-based material has a unique molecule structure where each silicon atom is not only connected to oxygen atoms as in pure silica but also to an organic fragment, resulting in a truly organic–inorganic hybrid silica pore network. In the dehydration of n-butanol with 5% of water, the membrane shows a high separation factor of over 4000, and ultra-fast water transport at a rate of more than 20 kg/m^2h at 150 °C, far surpassing most reported polymeric membranes [126]. The pervaporation performance of some organic–inorganic hybrid membranes is listed in Table 10.8 [69, 86, 96, 125–135].

10.4.1.2 Pervaporation membranes for biofuel recovery

Together with bioalcohol dehydration, the potential of pervaporation membranes for bioalcohol recovery has also been recognized. However, the development of pervaporation membranes for bioalcohol recovery is still considered as an initial stage as compared with the membranes for pervaporation dehydration. A lack of desirable membrane materials that possess a sufficiently high selectivity/flux separation characteristic with a cost-effective fabrication has been a major hurdle hindering the advancement of this technology in the industrial practice. Since the 1980s, several membrane materials ranging from polymeric, inorganic and composite (mixed matrix) materials have been extensively investigated but no significant breakthrough has been made yet. Usually rubbery polymers are adopted as pervaporation recovery

Table 10.7 Separation performance of inorganic membranes for ethanol and butanol dehydration

Material	Feed composition	T (°C)	Separation factor	Total flux (Kg/m^2hr)	Ref.
	ethanol/water				
Zeolite NaA membrane with TiO_2 supporting tube	90/10	50	8500	0.8 ~ 1.0	106
Zeolite NaA membrane with α-alumina substrate	95/5	45	12500	0.23	107
Zeolite NaA membrane on tubular α-alumina monolayer substrate	90/10	75	>5000	5.6	108
Zeolite NaA membrane with α-alumina tubular supports	90/10	125	3600	3.8	109
Zeolite A membrane on top of tubular alumina supports	90.8/9.2	93	138	2.5	110
Hydrophobic DD3R zeolite membrane	82/18	100	1500	2	111
Microporous silica membrane	94/6	70	358	0.76	112
Silica molecular sieve membrane with α-alumina substrate	95/5	25	210	8	113
Silica membrane with γ-alumina intermediate layer and α-alumina substrate	90/10	80	800	1	114
Cobalt-doped silica membrane	90/10	75	2530	1.1	115
Silicalite on silica tube	97/3	60	95	0.58	116
Mordenite membrane with α-alumina substrate tube	90/10	80	139	0.16	117
Mordenite with ceramic tubular supports	85/15	90	60	0.06	118
Multilayer membranes with ZrO_2 top layer	95/5	70	>150	<0.1	119
	butanol/water				
Silica membrane with α- and γ- alumina support layers	95/5	75	600	4.5	120
Silica membranes with γ- alumina substrate tube	95/5	75	250	3	121
Silica hollow fiber on ceramic substrate	95/5	80	1200	2.9	122

membranes, such as poly(dimethylsiloxane) (PDMS) (Figure 10.4a), poly [1-(trimethylsilyl)-1-propyne] (PTMSP) (Figure 10.4b), PVDF (Figure 10.4c), polybutadiene, natural rubber, and polyether copolymers.

Poly(dimethylsiloxane), often referred to as "silicone rubber," is the most widely studied membrane material for bioalcohol recovery. Table 10.9 summarizes the separation performance of PDMS membranes reported in the literature for ethanol and butanol recovery [136–153]. From the table, PDMS membranes exhibit alcohol separation factors in the range of 4–15 and 27–50 for ethanol-water and butanol-water pairs, respectively. The variation of membrane selectivity and flux in PDMS membranes often arises from various factors including polymer starting materials (although they are called "PDMS" they are often different), the method of membrane casting, the degree of cross-linking, membrane module design and testing conditions. To achieve the flux enhancement, PDMS composite membranes comprising a PDMS thin film layer coated on a porous support have been developed. The supporting porous material also plays an important role determining both flux and membrane selectivity. Shi and co-workers reported that using polyamide (PA) as

Table 10.8 *Separation performance of organic–inorganic hybrid membranes for ethanol and butanol dehydration*

Material	Feed composition	T (°C)	Separation factor	Total flux (g/m^2hr)	Ref.
	ethanol/water				
Multilayer membrane with PVA/KA zeolite MMM top layer (20 wt% KA)	90/10	80	~3000	~21	127
NR/cross-linked PVA semi-IPN embedded with the zeolite 4A	90/10	30	1506	2800	128
CS/TiO_2 nanocomposite dense membrane with 6 wt% TiO_2	90/10	80	196	340	129
Alumino-phosphate (AlPO4-5/)-filled NaAlg membrane	96/4	30	980	104	130
Organic–inorganic hybrid membranes of PVA and 3% 1,2-bis(triethoxysilyl)ethane (BTEE)	85/15	50	60	244	131
Ceramic/polymeric membrane with free-radical graft polymerized PAA onto silylated ceramic support	95/5	30	~1440	540	132
Chitosan-silica hybrid membrane cross-linked by 3-aminopropyl-triethoxysilane (APTEOS)	85/15	50	597	887	133
NaAlg membranes incorporated with 20 wt% aluminum-containing mesoporous silica (Al-MCM-41), cross-linked with GA	90/10	30	1089	129	134
Chitosan-silica complex membranes cross-linked by ST-GPE-S	90/10	70	919	410	125
Cross-linked N-p-carboxy benzyl chitosan (NCBC)-silica nanocomposite membrane	90/10	30	3860	440	135
NaAlg hybrid membranes containing 6 wt% of preyssler type heteropolyacid	90/10	30	14,991	57	86
	butanol/water				
Cross-linked PVA membrane on hollow fibre alumina support	95/5	80	500 ~ 10000	800 ~ 2600	69
Organic–inorganic hybrid silica-based Hybsi® membrane	95/5	150	4000	20,000	126
Ceramic-supported P84 polyimide membranes	95/5	150	360	1000 ~ 6000	96

a supporting layer results in the PDMS composite membrane with a better separation and flux than those fabricated with polysulfone (PSf) substrates [143]. In work by Xiangli and co-workers [149], membranes consisting of a thin PDMS layer deposited on ZrO_2/Al_2O_3 porous ceramic supports displayed a remarkable total flux up to 19.5 kg/m^2h and a separation factor of 5.7 for 4.3 wt% ethanol feed solution at 70 °C.

Chemical modification of PDMS membranes to improve alcohol/water separation properties has been extensively studied and some studies with encouraging findings are listed in Table 10.10 [154–163]. Kashiwagi and co-workers [157] studied the modification of PDMS membranes using plasma-polymerization of silanes with different alkyl lengths. Silane monomers containing longer alkyl groups are better modification reagents and the resultant membranes exhibit better ethanol/water permselectivity. Chang and

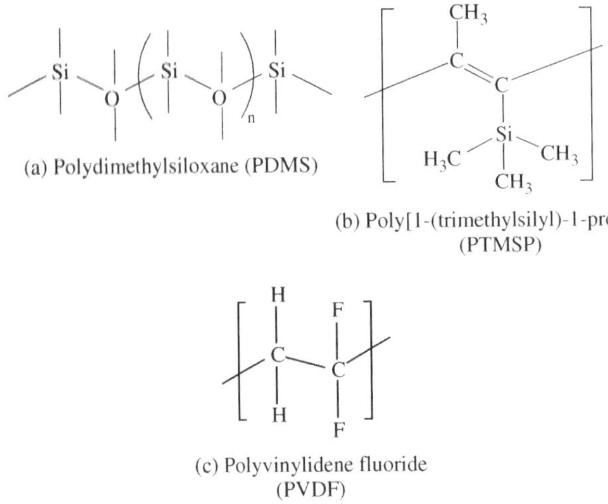

Figure 10.4 *Commonly used polymeric membrane materials for pervaporation recovery*

Chang fabricated plasma-induced grafted trimethoxyvinylsilane (TMVS)/PVDF membranes, coating by phosphate ester containing silicone copolymer [159]. The membranes displayed the separation factor up to 31 with the flux of 900 g/m^{2}h for a feed mixture containing 10 wt% ethanol at 30 °C. For the recovery of a feed solution containing 10 wt% ethanol at 60 °C, the separation factor of the PDMS-polystyrene interpenetrating polymer network (IPN) on polyethersulfone (PES) ultrafiltration support is 5.5 and the permeation flux is 160 g/m^{2}h [161]. Krea *et al*. synthesized polysiloxane-imide copolymers from α,ω-bis(3-aminopropyldimethyl) oligodimethylsiloxane (ODMS) and aromatic dianhydrides PMDA and 6FDA [160]. Compared to PDMS polymers, PDMS-imide comprising fluorinated imide blocks and siloxane blocks exhibit higher sorption affinities and selectivity for ethanol with enhanced mechanical properties.

Poly [1-(trimethylsilyl)-1-propyne], a glassy polymer with a large free volume, has been explored for alcohol recovery. The pervaporation performance of PTMSP membranes for ethanol/water and butanol/water separations is summarized in Table 10.11 [164–169]. The ethanol/water separation factor of PTMSP membranes falls in the range from 9 to 20, while the butanol/water separation factor can be as high as 115. Overall, PTMSP membranes exhibit greater membrane selectivity and flux relative to conventional PDMS membranes. The flux with PTMSP is about threefold higher than the corresponding flux obtained with conventional PDMS under similar operation conditions. Nevertheless, the separation performance of PTMSP membrane is not very stable and declines as a function of time probably due to the compaction of the polymer and/or the sorption of foulants within the membrane. The introduction of a cross-linked structure to PTMSP membranes could be a feasible approach to accomplish more stable performance and to strengthen the prospects of PTMSP membranes for alcohol recovery [170, 171].

Tremendous efforts have been expended on searching for other new polymeric materials with better separation characteristics than PDMS and PTMSP. These reported materials are very limited, and include styrene-fluoroalkyl acrylate graft copolymer [136], polyorganophosphazene [141], styrene-butadinene-styrene block copolymers [144], polyurethane [150], polyurethaneurea [155], poly(ether-b-amide) (PEBA) [42, 172], PVDF [54], fluorinated polyimides [173] and others. Some examples of their alcohol/water separation performance are listed in Table 10.12 [54, 136, 141, 144, 150, 155, 172]. So far, only a few specific cases have attained alcohol-water separation factors greater than that obtained by the common PDMS membranes. For instance, Kazuhiko and Kiyohide [136] reported that membranes fabricated using a styrene-fluoroalkyl acrylate graft copolymer on a cross-linked PDMS support displayed

Table 10.9 *Separation performance of PDMS membranes for ethanol and butanol recovery*

Membrane material	Feed composition (wt%)	*T* (°C)	Separation factor	Total flux (g/m^2hr)	Ref.
	ethanol/water				
PDMS flat sheet membrane	8/92	30	10.8	25.1	136
PDMS flat sheet membrane	10/90	30	5.0	20	137
PDMS flat sheet membrane (GE 615 membrane)	6/94	50	8.6	100	138
PDMS flat sheet membrane (Sulzer membrane)	6/94	35	6.0	34	139
PDMS flat sheet membrane on cellulose acetate support	5/95	40	8.5	1300	140
PDMS flat sheet membrane on Nylon support	10/90	40	5.0	160	141
PDMS flat sheet membrane on PTFE support	6/94	20	14.0	1530	142
PDMS flat sheet membrane on polyamide support	4/96	40	8.5	1400	143
PDMS flat sheet membrane on polysulfone support	4/96	40	4.5	1150	143
PDMS flat sheet membrane on polyimide support	3/97	41	4.6	120	144
PDMS flat sheet membrane on microporous support (MTR)	6/94	25	5.5	39	145
PDMS/polysulfone composite hollow fiber	8/92	50	6.6	576	146
PDMS/PVDF/non-woven-fiber, multi-layer flat sheet membrane	5/95	60	15	450	147
PDMS on ZrO_2/Al_2O_3 ceramic tubular support	4.2/95.8	60	7.93	4190	148
PDMS on ZrO_2/Al_2O_3 ceramic tubular support	4.3/95.7	70	5.7	19500	149
	butanol/water				
PDMS flat sheet membrane	1/99	50	37	70	150
PDMS flat sheet membrane (GFT 1060)	0.01/99.99	37	27	129	151
PDMS flat sheet membrane (MEM-100^{TM}, MemPro. Corp.)	2/98	62	50	250	152
PDMS flat sheet membrane (GE VTR615) on polyetherimide support	0.01/99.99 (10 g/L)	30	42	52.8	153
PDMS flat sheet membrane (GE VTR615) on polyetherimide support	0.01/99.99 (10 g/L)	50	49.6	145.3	153

an excellent ethanol/water separation factor of 46, which is significantly higher than the intrinsic PDMS separation factor of 11. Recently, Ghofar and Kokugan [174] investigated the pervaporation characteristics of microporous polytetrafluroethylene (PTFE) and polypropylene (PP) membranes for ethanol-water separation. They found that the resulting membranes are ethanol selective and the ethanol-water separation factor could reach as high as 75 at an optimal downstream pressure condition. Even though such membranes in both examples display a promising separation performance, further investigation and independent verification of the membrane's performance may be required.

Inorganic membranes based on silicalite-1 and hydrophobic zeolites for biofuel recovery typically display a greater separation factor and greater flux than polymer-based membranes. A comprehensive summary

Table 10.10 Separation performance of modified PDMS membranes for ethanol and butanol recovery

Membrane	Feed composition (wt%)	*T* (°C)	Separation factor	Total flux (g/m^2hr)	Ref.
	ethanol/water				
PPMS on cellulose acetate support	5/95	40	6.2	1432.6	154
PDMS-styrene graft copolymer on PTFE support	8.1/91.9	60	6.2	130	155
Plasma-induced grafted PDMS/PVDF composite	10/90	35	5.1	1650	156
PDMS (Plasma-polymerized tetramethoxysilane monomers)	4/96	25	4.6	380	157
PDMS (Plasma-polymerized hexamethyltrisiloxane monomers)	4/96	25	5.0	320	157
PDMS (Plasma-polymerized hexamethyltrisiloxane PDMS membranes and treated with octadecyldiethoxymethylsilane)	4/96	25	18.0	15	157
Plasma-induced grafted TMVS/PVDF, coating by phosphate ester containing silicone copolymer	10/90	30	4.6	2850	158
Plasma-induced grafted TMVS/PVDF, coating by phosphate ester containing silicone copolymer	10/90	30	31	900	159
PDMS-imide copolymers (synthesized from ODMS and PMDA)	10/90	40	10.6	560	160
PDMS-imide copolymers (synthesized from ODMS and 6FDA)	10/90	40	3.6	2120	160
PDMS-polystyrene interpenetrating polymer network on polyethersulfone ultrafiltration support	10/90	60	5.5	160	161
PDMS-polysulfone block copolymers	10/90	25	6.2	27	162
	butanol/water				
Polysiloxane (CMX-GF-010-D, CELFA AG, Switzerland)	0.2/99.8	40	39	330	163
PDMS containing dimethyl and methyl vinyl siloxane copolymers (PERTHESE®, France)	0.2/99.8	40	56	33	163

on zeolite materials and the fundamentals of using zeolites for pervaporation applications is provided in a review by Bowen *et al*. [175]. Table 10.13 shows the separation performance of silicalite-1 and other zeolite membranes investigated for ethanol and butanol recovery [176–190]. From the table, the average separation factors of silicalite-1 membranes are approximately 100 and 300 for ethanol/water and butanol/water separations, respectively. Recent studies revealed that silicalite-1 membranes coated with a thin PDMS layer can bring the separation factor up to 125 for ethanol/water pair [184] and to 465 for butanol/water pair [187].

The idea of incorporating zeolite/silicalite-1 into the polymer matrix has therefore been of great interest recently. Several research groups have investigated silicalite-1/PDMS mixed-matrix membranes for alcohol/water recovery. As summarized in Table 10.14 [144, 151, 153, 191–200], the separation factors of silicalite-1/PDMS membranes are in the range of 5–59 and 30–145 for ethanol/water and butanol/water pairs, respectively. Importantly, the wide range in separation performance of the mixed matrix membranes is attributed to the differences in silicalite-1 loading, particle size, source of silicalite-1, and membrane-casting

Table 10.11 Separation performance of PTMSP and modified PTMSP membranes for ethanol and butanol recovery

Membrane material	Feed composition (wt%)	*T* (°C)	Separation factor	Total flux (g/m^{2}hr)	Ref.
PTMSP polymers	**ethanol/water**				
PTMSP	6/94	30	19.9	325	164
PTMSP	6/94–7/93	50	10.3	480	165
PTMSP	6/94	75	9	700	166
PTMSP	10/90	50	14.5	210	167
Modified PTMSP polymers					
PTMSP/PDMS graft copolymer	7/93	30	28.3	62	168
Trimethylsilyl substituted PTMSP	6/94–7/93	50	17.6	590	165
n-Decane substituted PTMSP	6/94–7/93	50	17.8	430	165
PTMSP polymers	**butanol/water**				
PTMSP	2/98	53	115	1750	169
PTMSP	2/98	62	80	2600	169

Table 10.12 Separation performance of other hydrophobic membranes for ethanol and butanol recovery

Membrane material	Feed composition (wt%)	T (°C)	Separation factor	Total flux (g/m^{2}hr)	Ref.
	ethanol/water				
Flat sheet styrene-fluoroalkyl acrylate graft copolymer on PDMS support	8/92	30	16.3–45.9	5–14	136
Flat sheet polyorganophosphazene containing —OC_2H_5 pendant group on Nylon support	10/90	40	2.0	130	141
Flat sheet polyorganophosphazene containing —OCH_2CF_3 pendant group on Nylon support	10/90	40	6.1	260	141
Flat sheet polyorganophosphazene containting —$OCH_2CF_2CF_2CF_2CF_2H$ pendant group on Nylon support	10/90	40	4.0	96	141
Flat sheet polyurethaneurea containing PDMS	10/90	40	8.6	130	155
Flat sheet PEBA (polyether block amide) PEBAX® 2533, Antofina	5/95	23	2.5	117.5	172
Flat sheet SBS (styrene-butadiene-styrene) block copolymer, dense	3/97	41	5.5	125	144
Flat sheet SBS (styrene-butadiene-styrene) block copolymer, porous	3/97	41	3.5	1752	144
PVDF hollow fiber membrane	5/95	40	5–8	3500–8800	54
	butanol/water				
Flat sheet PEBA (polyether block amide)	1/99	50	20	278	150
Flat sheet PEBA (polyether block amide) PEBAX® 2533, Antofina	5/95	23	8.2	65.3	172
Flat sheet Polyurethane	1/99	50	9	88	150

Table 10.13 *Separation performance of silicalite-1 and hydrophobic zeolite membranes for ethanol and butanol recovery*

Membrane material	Configuration	Feed composition (wt%)	T (°C)	Separation factor	Total flux (g/m^2hr)	Ref.
Silicalite-1		**ethanol/water**				
Silicalite-1 on porous stainless steel support	Flat sheet	4/96	60	58	760	176
Silicalite-1 on porous stainless steel support	Flat sheet	5/95 (fermented ethanol)	30	21	104	177
Silicalite-1 on Al_2O_3 support	Flat sheet	9.7/90.3	32	11.5	100	178
Silicalite-1 on porous α-Al_2O_3 tube support	Tubular	5/95	60	39	1510	179
Silicalite-1 on Mullite tubular support	Tubular	5/95	60	106	930	180
Silicalite-1 on α-Al_2O_3 tubular support	Tubular	5/95	60	89	1800	181
Silicalite-1 on Mullite tubular support	Tubular	10/90	60	72	2550	182
Silicalite-1 coated with PDMS on Stainless steel support	Flat sheet	10/90 (fermented ethanol)	30	43	230	183
Silicalite-1 on stainless steel support	Flat sheet	4/96	30	26–51	150–330	184
Silicalite-1 coated with PDMS on stainless steel support	Flat sheet	4/96	30	125	140	184
Silicalite-1 on porous stainless steel disk support	Flat sheet	4.7/95.3	30	41	400	185
Silicalite-1 on porous stainless steel disk support	Flat sheet	4.6/95.4 (fermented ethanol)	30	88	500	185
Silicalite-1 treated with silane, $C_8H_{17}SiCi_3$, on porous stainless steel disk support	Flat sheet	4/96	50	44	650	186
Silicalite-1 treated with silane, $C_{18}H_{37}SiCi_3$, on porous stainless steel disk support	Flat sheet	4/96	50	45	133	186
Silicalite-1		**butanol/water**				
Silicalite-1 on porous α-Al_2O_3 tube support	Tubular	2/98	60	150	100	179
Silicalite-1 on tubular porous stainless-steel support	Tubular	1/99	45	96–325	18.7–55.0	187
Silicalite-1 coated with silicone rubber, on tubular porous stainless-steel support	Tubular	1/99	45	110–465	10.7–38.4	187
Other hydrophobic zeolite		**ethanol/water**				
Ge-ZSM-5, on stainless-steel support	Flat sheet	5/95	30	47	223	188
B-ZSM-5, on Al_2O_3-coated SiC multi-channel monolith support	Flat sheet	5/95	60	31	160	189
Ti-Silicalite, on α-Al_2O_3 capillary support	Tubular	5/95	45	16–62	700–2100	190

Table 10.14 Separation performance of silicalite-1/PDMS mixed-matrix membranes and other hybrid membranes for ethanol and butanol recovery

Membrane material	Particle size (μm)	Particle loading (%)	Feed composition (wt%)	T (°C)	Separation factor	Total flux (g/m^2hr)	Ref.
Silicalite-1/PDMS			**ethanol/water**				
Silicalite-1/PDMS	0.5–5	40	5–5.5/94.5–95	22.5	14.9	32.2	191
Silicalite-1/PDMS	0.5–5	60	5–5.5/94.5–95	22.5	16.5	44.9	191
Silicalite-1/PDMS	~1	60	5.3/94.7	50	21	105	192
Silicalite-1 filled PDMS on polyimide support	0.1–0.23	15	3/97	41	4.8	170	144
Silicalite-1 (UOP Inc.) /PDMS	1–10	70	7/93	22	29	52	193
Silicalite-1/PDMS	0.3–4	77	7/93	22	59	89	193
Silicalite-1/PDMS	~1	67	5.3/94.7	50	32	90	192
			butanol/water				
Silicalite/PDMS	3	60	0.01/99.99 (10g/L)	65	50	80	194
Silicalite/PDMS	0.1–0.2	60	0.01/99.99 (10g/L)	50	111.3	191	153
Silicalite/PDMS	0.1–0.2	60	0.01/99.99 (10g/L)	70	93	607	153
Silicalite-PDMS, PERVAP-1070 (Sulzer Chemtech, German)	–	60	0.01/99.99 (10g/L)	70	47.8	344.4	153
Silicalite/PDMS (PERVAP-1070, Sulzer Chemtech)	–	–	0.1/99.9	55	31	350	195
Silicalite filled PDMS (GFT 1070)			0.01/99.99 (10g/L)	37	36	90	151
Silicalite-1/PDMS	–	65	0.01/99.99 (10 g/L) (ABE solution)	50	145	60–250	196

(*continued overleaf*)

Table 10.14 (continued)

Membrane material	Particle size (μm)	Particle loading (%)	Feed composition (wt%)	T (°C)	Separation factor	Total flux (g/m^2hr)	Ref.
Hydrophobic zeolites/PDMS			**ethanol/water**				
ZSM-5/PDMS CBV-28014, Zeolyst International) (Si/Al = 137)	~2.4	50	5/95	50	37	70	197
ZSM-5/PDMS, on polyimide support, CBV 3002 (Si/Al = 240)	1–1.5	30	3/97	41	5.5	151	144
Hydrophobic zeolites/PDMS			**butanol/water**				
ZSM-5 zeolite/PDMS	Nil	50	1/99	40	45	1500	198
Other hybrid membranes			**ethanol/water**				
Polyphosphazene nanotubes/PDMS	50 μm in length, 0.040 in diameter	10	10/90	40	10	476	199
Carbon black/PDMS	0.051	10	6/94	35	9	51	200
PTMSP/nano-silica	0.0045–0.0065	1.5	10/90	50	15.3	400	167
Other hybrid membranes			**butanol/water**				
β-CD/PDMS	Nil	50	1/99	40	40	800	198

protocols. Usually, loadings of 60 wt% or higher are required to deliver a significant improvement in the alcohol/water separation factor. Besides mixed-matrix membranes comprising silicalite-1 or zeolite, other hybrid or composite membranes containing other different fillers, such as β-cyclodextrin (CD) [198], poly(hedral oligomeric silsesquioxane) (POSS) [42], polyphosphazene nanotubes [199], carbon black [200], and silica [167] have been studied, and their separation performance mostly falls in between those of PDMS and silicalite-1/PDMS membranes.

The alcohol/water separation factors of different membrane materials are typically ranked in the following order: PDMS < PTMSP < composite or mixed matrix membranes < inorganic membranes. Ongoing work on the development and exploration of membranes with better separation and physicochemical properties for alcohol recovery is of paramount importance and necessary. Without the development of an appropriate membrane, pervaporation is unlikely to be materialized as a viable alternative technique for ethanol recovery. Compared with works for ethanol recovery, pervaporation studies on butanol recovery are relatively limited.

10.4.2 Membrane morphology

The morphology of membranes can directly determine the physical properties of membranes, which closely relate to the ultimate separation efficiency of membranes. According to their physical structures, membranes can be classified into dense membranes, asymmetric membranes, and composite membranes, as schematically shown in Figure 10.5.

Dense membranes with a homogenous structure are usually prepared via slow evaporation of dilute polymer solution. These membranes are widely used in laboratories for the study of intrinsic material properties such as sorption ability and selectivity for a designated mixture. A relatively low flux is a major disadvantage of these membranes when they are used in pervaporation experiments because of their quite

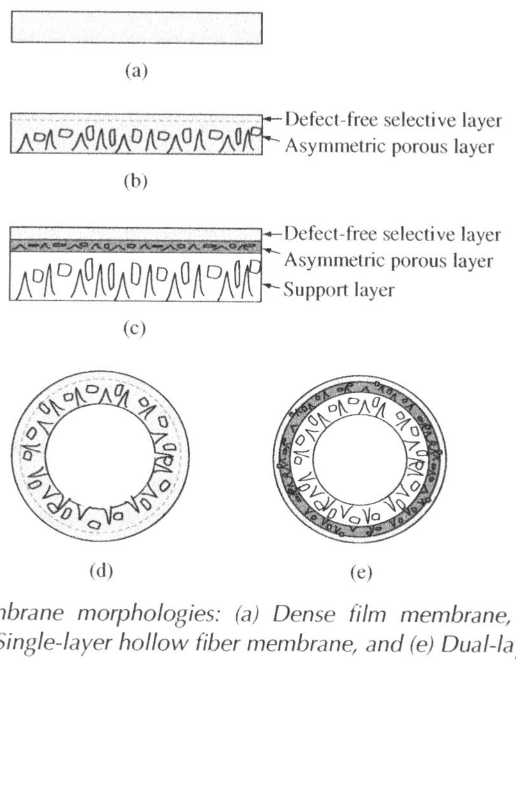

Figure 10.5 *Different membrane morphologies: (a) Dense film membrane, (b) Asymmetric membrane, (c) Composite membrane, (d) Single-layer hollow fiber membrane, and (e) Dual-layer hollow fiber membrane*

thick dense selective layer. Therefore, a membrane with a thinner selective layer is regarded as more valuable from the practical point of view.

Asymmetric membranes consist of a dense selective layer on the top and a microporous support layer at the bottom. For practical applications, the formation of anisotropic structure with an ultra-thin selective skin is the preferred choice for insuring high flux. With this in mind, membrane scientists are searching eagerly for better technologies to overcome problems such as skin-layer defects and structure integrity in order to fabricate asymmetric membranes with an ultrathin dense selective layer with a porous substrate layer. By doing so, it may have the same impact in terms of performance enhancement as a new material that can double or triple the permeability. These membranes are commonly prepared by the phase-inversion method, which was developed by Loeb and Sourirajan [16]. Kesting intensively investigated the association of final membrane structure with preparation parameters [201]. The resulting morphology and structure of asymmetric membranes are highly dependent on the various parameters of the membrane preparation process such as solution composition, coagulation composition, precipitate rate, temperature, and medium. As a result, a membrane with high flux as well as good selectivity can be obtained through the optimization of membrane preparation parameters.

For large-scale industrial applications, composite membranes are the most favored membrane structure, and have been reported extensively for pervaporation applications in recent years. One typical example is the ceramic-supported polymeric membrane, as mentioned at the beginning of this section. Many polymeric composite membranes have also been reported with impressive separation performance for bioalcohol separation. In composite membranes, the selective layer provides the overall separation function by employing the most suitable material, while the substrate may provide substantial mechanical strength, lower water sorption, or minimize material costs. The thickness of the selective layer can therefore be minimized without sacrificing the overall mechanical strength. In addition, its chemical structure can be further modified separately towards a desirable separation property. Since only a very small amount of expensive materials with superior performance is used as the ultra-thin selective layer, composite membranes are cost-effective. The typical example of commercial composite membrane is the GFT PVA composite membrane, which comprises three layers: a PVA selective layer, a PAN supporting layer, and a non-woven support layer. The most popular methods to prepare composite membrane include dip coating, photo-grafting, interface polymerization, and so forth.

On the other hand, compared to conventional flat-sheet asymmetric membranes or composite membranes, the development of asymmetric hollow fiber membranes for biofuel separation and pervaporation applications has gained much attention in recent years. From the application point of view, hollow fibers are more attractive because of the following advantages: (i) a larger membrane area per unit membrane module volume, resulting in a higher permeation flow per unit volume; (ii) self-supporting structure, allowing the membrane to be a self-contained vacuum channel, where the feed can be supplied from the shell side while vacuum is applied on the lumen side; and (iii) good flexibility and ease of handling during module fabrication and system operation. Like flat-sheet asymmetric membranes, hollow fiber membranes with an ultrathin selective skin and defect-free structure are desired to attain high flux and satisfactory selectivity.

For biofuel dehydration applications, asymmetric hollow fiber membranes prepared from various hydrophilic polymers, such as PVA [202], PVA/NaAlg [69], PP with grafted PAA [203], cellulose acetate (CA)/CS [204], PI [3 ,6, 39, 95, 96, 205, 206], and polyamide-imides (PAI) [6, 96], have been investigated. Among them, polyimide hollow fibers have shown a great potential because of the high separation factor and anti-swelling properties. In contrast to biofuel dehydration, relatively less effort has been expended on hollow-fiber membranes for biofuel recovery using hydrophobic materials. So far, most hollow-fiber membranes reported in the literature for biofuel recovery are composite membranes comprising a thin layer of PDMS coated on a strong porous support [146]. Recently, an asymmetric PVDF hollow-fiber membrane for ethanol recovery has been attempted by Sukitpaneenit *et al.* [53,

54]. The reported PVDF asymmetric hollow-fiber membrane shows an impressive performance with an extraordinary total flux and reasonable separation factor for ethanol-water separation.

Like flat-sheet composite membranes, asymmetric hollow-fiber membranes fabricated with the aid of dual-layer co-extrusion technology are of great interest for biofuel separation. Since single-layer hollow fibers may suffer severe swelling in the feed solution and lose their selectivity, the use of anti-swelling materials as the inner support layer can improve the overall separation pervaporation performance and long-term stability. Several studies on dual-layer asymmetric hollow fibers for pervaporation by Chung's group [6, 39, 41, 43, 95, 206, 207] recently have shown promising pervaporation performance. Without intensive thermal or chemical treatment, the dual-layer hollow fibers intrinsically demonstrate excellent synergistic separation performance for a series of biofuels and solvents purifications if inner- and outer-layer materials are properly selected. Recent works by Wang *et al.* [6] and Jiang *et al.* [39] are two most prominent examples. The former developed PAI/polyetherimide (PEI) hollow fiber membranes with synergized performance for dehydration of C_1–C_4 alcohols; while the latter demonstrated PSf/Matrimid dual-layer hollow fiber membranes with an outstanding separation performance that far surpasses those of most existing polymeric membranes and approaches those of ceramic membranes for *t*-butanol dehydration. In addition to the unique dehydration properties of the selective layer, the superior performance of dual-layer hollow fibers were also attributed to (i) the low water uptake and less swelling characteristics of the support layer; and (ii) the desirable membrane morphology consisting of a fully porous inner layer and an ultrathin dense-selective skin. Noticeably, the dual-layer fabrication technology may potentially open up new avenue for the development of next-generation pervaporation membranes for biofuel separation.

10.4.3 Commercial pervaporation membranes

Existing commercial membranes for pervaporation separation are still limited in the world, as summarized in Table 10.15 [105, 208–211]. Here, most of the information is from the web site of pervaporation.org [211].

For the current pervaporation dehydration market, PVA-based membranes have taken the lead while polyimide membranes have also attracted attention. However, ceramic membranes and organic–inorganic hybrid membranes start to show potential in recent years. Therefore, the dominant position by polymeric pervaporation membranes may be shifted if the cost of inorganic membranes can be further lowered by optimizing their synthesis routes or the combination of organic and inorganic materials can be materialized. In the case of organophilic membranes, so far only PDMS and silicalite-1/PDMS composite membranes have been commercialized. Alcohol-selective membranes produced by Sulzer Chemtech (Neunkirchen, Germany), SolSep BV (Apeldoorn, Netherlands), Pervatech BV (Enter, Netherlands) are the benchmarks [210].

10.5 Pervaporation in the current integrated biorefinery system

In general, five *in situ* alcohol recovery technologies are used in practice for biofuel separation from the acetone-ethanol-butanol (ABE) fermentation broth. These are as follows [212]: (i) stripping; (ii) adsorption; (iii) liquid-liquid extraction; (iv) pervaporation; and (v) membrane solvent extraction. Compared to the conventional process (Figure 10.6a), each integrated system consists of fermentor, separation and purification processes as shown in Figure 10.6b and 10.6c. The feed preparation for integrated systems can be simplified and more flexible in terms of fermentation types. In addition, the product recovery can be incorporated inside the fermentor (Figure 10.6b) or outside the fermentor in a closed loop (Figure 10.6c). The former can be applied only via liquid-liquid or supercritical extraction, while the latter is more flexible in terms of process such as stripping, pervaporation, reverse osmosis, and so forth.

Table 10.15 *Some commercial membranes fro biofuel dehydration and recovery*

Membrane	Company	Comments
NaA ceramic membranes	Jiangsu Jiuwu Hi-Tech, China	Pervaporation dehydration
NaA zeolite membrane	Misui, Japan	Demonstration plant for the dehydration of 100L/h of ethanol (feed 93% to product 99.5%)
NaA-zeolite membrane	Inocermic GmbH	Flux of 13 kg/m^{2}h for 85 wt% ethanol feed at 120 °C [105]
Siftek^{TM} polyimide hollow fiber	Vaperma, Canada	Pervaporation dehydration
ZeoSep A membrane	I3 Nanotec LLC, USA	Flux of 3.5-4.0 kg/m^{2}h and separation factor of 20,000–40,000 for 90 wt% ethanol feed at 75 °C [208]
Aromatic polyimide membrane	UBE America, USA	Pervaporation dehydration
PERVAP® cross-linked PVA membrane with PAN support	Sulzer Chemtech, Switzerland	Pervaporation dehydration
cross-linked PVA membrane	CM-Celfa AG, Switerland	Pervaporation dehydration
ceramic silica-based membrane	Pervatech BV, Netherlands	Flux of 2 kg/m^{2}h and a separation factor of 160 for 89 wt% ethanol feed at 70 °C [209]
HybSi® organic–inorganic hybrid silica-based membrane	ECN, Netherlands	Pervaporation dehydration with excellent hydrothermal stability at high temperarure
GKSS Simplex Complex polyelectrolytes/PAN	GKSS, Germany	Pervaporation dehydration
Ceramic membranes	IBMEM Industrial Biotech Membranes, Germany	ethanol dehydration and solvent recovery
cross-linked PDMS membrane with support	Membrane Technology and Research, USA	Pervaporation recovery
PERVAP® cross-linked PDMS membrane with PAN support	Sulzer Chemtech, Switzerland	Pervaporation recovery
PDMS-based membranes	GKSS, Germany	Pervaporation recovery
PDMS-based membranes	Solsep BV, Apeldoorn, Netherlands	Pervaporation recovery 210

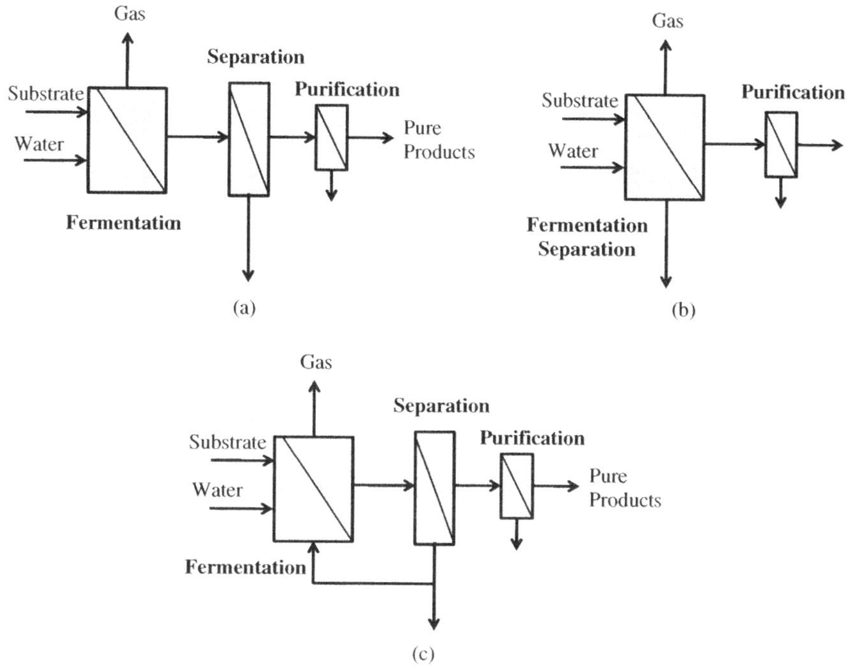

Figure 10.6 *Process designs of butanol production: (a) schematic of a conventional process; (b) schematic of integrated process with product separation inside the fermentor; (c) schematic of an integrated process with product separation outside the fermentor. Reprinted from [212] © 1992, with permission from Elsevier*

Qualitative comparison among five *in situ* technologies is shown in Table 10.16 [212] in terms of product capacity, selectivity, fouling tendency, and ease of operation. It can be seen that extraction and pervaporation have higher potential among all these technologies. The liquid-liquid extraction has its own attractive benefits including high selectivity and better flexibility to carry out the separation inside the fermentor. However, this type of separation is more prone to fouling as well as emulsion formation, which can lead to technical problems during operation. Meanwhile, although pervaporation cannot be incorporated inside the fermentor, it has the advantages of a low fouling tendency and ease of operation. In addition, pervaporation technology generally possesses good selectivity and high productivity if suitable membrane material is chosen. In terms of energy requirements, processes with a low selectivity, that is adsorption or stripping, require higher energy consumptions than pervaporation.

The integrated biorefinery system commonly consists of more than one separation unit. The pervaporation-hybrid system is the most popular combination and has been proven as an energy-efficient

Table 10.16 *Qualitative comparisons of in situ product recovery technologies. Reprinted from [212] © 1992, with permission from Elsevier*

	Stripping	Adsorption	Extraction	Pervaporation	Perstraction
Capacity	Moderate	Low	High	Moderate	Low
Selectivity	Low	Low	High	Moderate	High
Fouling	Low	High	Moderate	Low	Low
Ease of operation	High	Low	Low	High	High

process for achieving high-purity products [213]. By coupling pervaporation units with a distillation process, the number of trays in the distillation column is reduced, thereby reducing the complexity of the column design. In addition, the hybrid process lowers energy (thermal) consumption and brings down the operating cost, which can compensate for the high previous investment cost of distillation columns [214, 215].

Pervaporation-distillation hybrid systems can be of single- or multi-stages in different configurations. Depending on the layout of the pervaporation unit, the ethanol concentration of the final product can be in the range of 99.5 and 99.95 wt%. The placement of pervaporation units in the hybrid system itself is flexible and they can be divided into two categories: (i) recovery for low alcohol content in the feed (5–10 wt%); and (ii) dehydration for higher alcohol content of about 30 wt% and above. Matsumura *et al*. [216] have proposed an integrated system by the incorporation of pervaporation with a liquid membrane supported with a microporous polymeric film before the distillation process for bioalcohol recovery. Comparing distillation and a pervaporation-distillation hybrid system for the separation of 0.5 wt% butanol in the broth, it is claimed that the pervaporation-distillation system requires only one-tenth of the energy (7.4 MJ/kg of butanol production) as compared to the other system (79.5 MJ/kg of butanol production).

Alternatively, the distillation process can take place in the first stage, to concentrate ethanol from 8.8 to 80 wt% followed by two stages of pervaporation dehydration [217]. The first stage of pervaporation aims to split the azeotropic mixture with a high-flux but a low-selectivity dehydration membrane to obtain ethanol with 95 wt% purity. Meanwhile, the second pervaporation stage further purifies the ethanol up to 99.8 wt%, using a membrane with a low flux but a high selectivity. This patented technology may potentially replace four required distillation columns in the conventional process and significantly reduce energy consumption. Other works by Tusel and Bruschke [17], Cogat [218], and Fleming [219] also employed distillation-pervaporation hybrid processes, and demonstrated the capability to produce high-purity ethanol in the range of 99.5–99.95 wt%. Interestingly, Gooding and Bahouth [220] incorporated a pervaporation unit between two distillation columns, and convert ethanol of 4 mole% to 99.5 mole%. In this case, 81 mol% ethanol from the first distillation column was fed into the pervaporation system to break azeotrope, and the permeate from pervaporation was further purified by means of the second distillation column up to 99.5 mol% ethanol. Bruschke and Tusel [221] adopted a similar process and claimed that their approach could save up to 27% investment cost and 40% operating cost to concentrate ethanol from 94 wt% to 99.85 wt% when compared to the conventional process.

Besides distillation-pervaporation hybrid systems, biofuel separation via two stages of pervaporation units has also been proposed. As illustrated in Figure 10.7 [222], the first pervaporation unit concentrates ethanol up to 40 wt% in the ideal case with ethanol-selective membranes, while second pervaporation unit further dehydrates the permeate from the first stage of pervaporation with water selective membranes. Even though the permeate from the second stage may be 99 wt% ethanol, in reality, the purification of ethanol from 5 wt% to 95 wt% via two stages of pervaporation is not so attractive compared to that of pervaporation-distillation hybrid systems [221].

To improve the efficiency of biofuel purification, a dephlegmator unit is often employed together with pervaporation to enhance the efficiency of concentrating ethanol up to the azeotropic point. The so-called dephlegmator is actually a condenser to cool the vapor mixture from the pervaporation unit, which significantly improves the overall separation performance [223]. Vane *et al*. [19] suggested employing a dephlegmator between the recovery pervaporation and dehydration pervaporation. Since the feed for the dephlegmator is in the vapor phase from the first stage of recovery pervaporation, no reboiler is required in this process. In this stage, ethanol with purity more than 90 wt% can be produced and considered as the feed for the next dehydration pervaporation. Figure 10.8 represents the schematic diagram of alcohol purification via continuous hybrid system utilizing dual recovery-dehydration pervaporation with the aid of a dephlegmator [19]. This process design is considered more attractive than the conventional distillation process since a lower total energy is required for low ethanol content (less than 1 wt%) in the fermentation

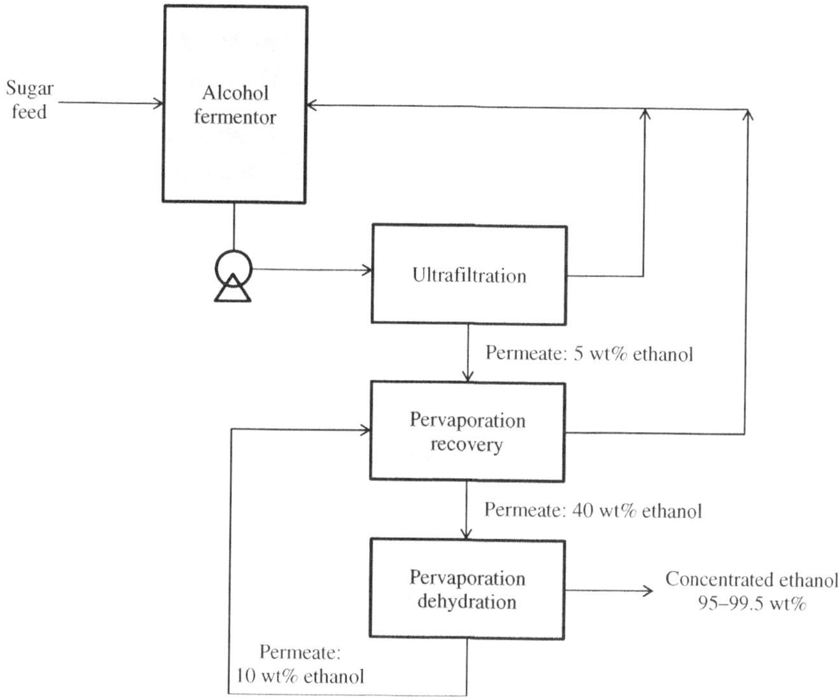

Figure 10.7 *Schematic presentation of membrane-controlled continuous fermentation process for the production of pure ethanol. Reprinted from [222] © 1983, with permission from Elsevier*

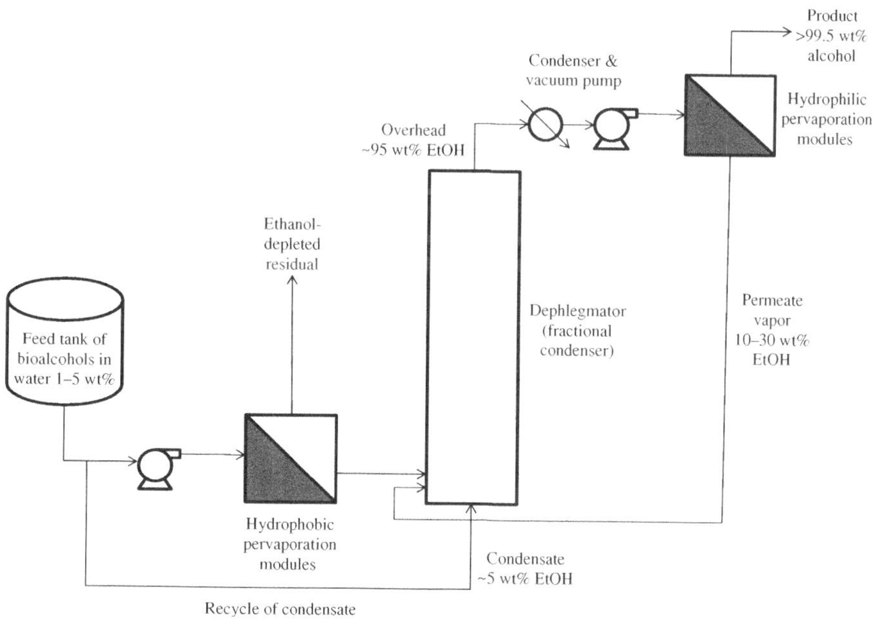

Figure 10.8 *Illustration of hybrid system for biorefinery via synergetic recovery-dehydration pervaporation processes. Adapted from [19] © 2005, with permission from John Wiley & Sons*

broth. Moreover, the hybrid process also overcomes the ethanol-water azeotropy problem (up to 95 wt% ethanol concentration) and complexity of distillation in the small-scale process.

Considering the carryover cells and the fouling tendency of pervaporation membranes, ultrafiltration or microfiltration is often placed between fermentor and pervaporation. Lee *et al*. [224] have proposed a conceptually continuous hybrid membrane system of ultrafiltration-pervaporation in line with an alcohol fermentor to produce alcohol with concentration up to 99.5 wt%. Figure 10.7 is a schematic diagram of hybrid membrane systems for ethanol production from fermentation process. The main advantages of an integrated ultrafiltration-pervaporation system in the fermentor are: (i) to recycle the cells back to the fermentor; (ii) to remove the ethanol itself, which acts as an inhibitor at high concentrations; (iii) to prevent fouling in the pervaporation membrane. Typically, the permeate of ultrafiltration consists of 5–10 wt% ethanol, while it can be much lower in the case of butanol fermentation [222].

Following the ultrafiltration-pervaporation concept, microfiltration and pervaporation hybrid systems have also been developed for ethanol production with enhanced ethanol productivity [212, 225]. Three different designs were studied: (i) direct coupling of pervaporation to the fermentor; (ii) direct coupling of pervaporation and microfiltration to the fermentor; (iii) direct coupling of microfiltration to the fermentor, and pervaporation of the cell-free broth. The first design significantly increases the substrate conversion (from glucose to ethanol) by approximately three times (from 118 kg/m^3 to 360 kg/m^3) as compared to the batch fermentor alone. For this purpose, the membrane materials must have high selectivity to minimize inhibitor compounds and decrease recovery product cost. By continuous product removal via pervaporation, the accumulation of inhibitors in fermentation broths can be minimized, increasing overall production. When both microfiltration and pervaporation incorporated into a fermentor (designs ii and iii), the ethanol productivity is higher by about three times than the system using only pervaporation (42 kg/m^{3}h versus 14 kg/m^{3}h).

10.6 Conclusions and future trends

It is well known that pervaporation technology may possess many benefits over conventional techniques for biofuel production and purification. However, current pervaporation technology has not been extensively utilized in the industry. This is due to the fact that most pervaporation membranes are tested under mild feed and bench-scale operating conditions. The separation performance of membranes for real industrial applications still remains unclear. Moreover, issues related to membrane materials and fabrication costs as well as the overall process design must be addressed.

The operating conditions in a separation process, such as operation temperature, feed/permeate pressure, feed composition, and operation duration, play significant roles in determining the separation efficiency of the membrane, since they not only manipulate the driving forces to transport permeants but also affect the physicochemical properties of the membrane itself. Thus, in addition to good selectivity and permeability, other critical factors to be considered for the membrane material selection include the thermal, mechanical, and chemical robustness of the membranes in the presence of aggressive feeds and harsh operating conditions.

Undoubtedly, choosing a membrane with good separation performance still outweighs other parameters but the focus of membrane material research needs to be prioritized according to real-life requirements. For alcohol dehydration, it is predicted that inorganic membranes and organic–inorganic hybrid membranes may share the industrial market with the cross-linked PVA membranes or take the dominant position. Polymeric membranes with higher flux and separation factors in harsh operating environment are in demand. This can be achieved via various modification routes to optimize physiochemical properties, stability, and integrity. Many novel promising polymeric materials have been reported recently, such as thermally rearranged (TR) polymers with well-tuned cavity size [226–232], PBI with dual-layer hollow fiber morphology [41, 43, 207] or loaded with nano-size zeolitic imidazolate frameworks (ZIF) [233], and polymers embedded with cyclodextrin via chemical bonding [40, 234]. These recently discovered polymers have shown superior separation performance, and will definitely shed useful light on nano-scale molecular separations, such as pervaporation, for the molecular design of suitable polymeric membranes in the near future.

For alcohol recovery, the development of novel hydrophobic materials is needed. Greater emphasis should be placed on the design and engineering of membranes that could deliver promising alcohol selectivity and flux with a sufficient stable performance. PVDF membranes, as reported recently, may be one of the promising alternatives to PDMS-based membranes since the former possesses relatively higher permeation flux and membrane stability than the latter [53, 54, 235]. More work should be done on mixed-matrix membranes, composite membranes, and hollow-fiber membranes with desirable membrane morphology.

In the industry, the pervaporation-distillation hybrid process is still the most popular configuration for biofuel separation. However, future works should focus on the science and hybrid engineering so that one can wisely select the best hybrid system for specific fermentation types, production scales, product purity requirements, and available investment.

Acknowledgements

The authors would like to thank the Singapore National Research Foundation (NRF) for support through the Competitive Research Program for the project titled, "New Biotechnology for Processing Metropolitan Organic Wastes into Value-Added Products" (grant number: R-279-000-311-281). The authors would like to acknowledge the valuable contributions from Prof. L.Y. Jiang in Central Southern University. Special thanks are given to Mr. X.K. Zheng for his help with checking the reference format and sequence.

References

1. Energy Information Administration (U.S.), *International Energy Outlook 2009*, U.S. Department of Energy, Washington D.C., 2009.
2. S.E. Koonin, Getting serious about biofuels, *Science*, 311, 435–435 (2006).
3. L.Y. Jiang, Y. Wang, T.S. Chung, X.Y. Qiao and J.Y. Lai, Polyimides membranes for pervaporation and biofuels separation, *Prog. Polym. Sci.*, 34, 1135–1160 (2009).
4. N. Qureshi, M.M. Meagher, J. Huang and R.W. Hutkins, Acetone butanol ethanol (ABE) recovery by pervaporation using silicalite-silicone composite membrane from fed-batch reactor of Clostridium acetobutylicum, *J. Membr. Sci.*, 187, 93–102 (2001).
5. T.C. Ezeji, N. Qureshi and H.P. Blaschek, Butanol fermentation research: Upstream and downstream manipulations, *Chem. Rec.*, 4, 305–314 (2004).
6. Y. Wang, S.H. Goh, T.S. Chung and N. Peng, Polyamide-imide/polyetherimide dual-layer hollow fiber membranes for pervaporation dehydration of C1-C4 alcohols, *J. Membr. Sci.*, 326, 222–233 (2009).
7. M.E. van Leeuwen, Derivation of Stockmayer potential parameters for polar fluids, *Fluid Phase Equlibr.*, 99, 1–18 (1994).
8. C. Reichardt, *Solvents and Solvent Effects in Organic Chemistry*, Wiley-VCH Verlag GmbH, Weinheim, 1988.
9. C.M. Hansen, *Hansen Solubility Parameters: A User's Handbook*, CRC Press, Boca Raton, 2007.
10. R.Y.M. Huang and N.R. Jarvis, Separation of liquid mixtures by using polymer membranes. 2. Permeation of aqueous alcohol solutions through cellophane and poly(vinyl alcohol), *J. Appl. Polym. Sci.*, 14, 2341–2356 (1970).
11. A.J. Ragauskas, C.K. Williams, B.H. Davison, G. Britovsek, J. Cairney, C.A. Eckert, W.J. Jr. Frederick, J.P. Hallett, D.J. Leak, C.L. Liotta, J.R. Mielenz, R. Murphy, R. Templer and T. Tschaplinski, The path forward for biofuels and biomaterials, *Science*, 311, 484–489 (2006).

12. P.A. Kober, Pervaporation, perstillation and percrystallization, *J. Amer. Chem. Soc.*, 39, 944–948 (1917).
13. R.C. Binning, R.J. Lee, I. Joseph, F. Jennings and E.C. Martin, Separation of liquid mixtures by permeation, *Ind. Eng. Chem.*, 53, 45–50 (1961).
14. R.C. Binning and F.E. James, Permeation. A new commercial separation tool, *Perot. Eng.*, 30, 6 (1958).
15. R.C. Binning, J.F. Jennings and E.C. Martin, US Patent 3035060, 1962.
16. S. Loeb and S. Sourirajan, The preparation of high-flow semi-permeable membranes for separation of water from saline solutions, US Patent 3133132, 1964.
17. G.F. Tusel and H.E.A. Bruschke, Use of pervaporation system in the chemical industry, *Desalination*, 53, 327–338 (1985).
18. A. Jonquieres, R. Clement, P. Lochon, J. Neel, M. Dresch and B. Chreticn, Industrial state-of-the-art of pervaporation and vapour permeation in the western countries, *J. Membr. Sci.*, 206, 87–117 (2002).
19. L.M. Vane, A review of pervaporation for product recovery from biomass fermentation processes, *J. Chem. Technol. Biotechnol.*, 80, 603–629 (2005).
20. F. Lipnizki, J. Olsson and G. Tragardh, Scale-up of pervaporation for the recovery of natural aroma compounds in the food industry. Part 2: Otimisation and integration, *J. Food Eng.*, 54, 197–205 (2002).
21. I. Blume, J.G. Wijmans and R.W. Baker, The separation of dissolved organics from water by pervaporation, *J. Memb. Sci.*, 49, 253–286 (1990).
22. J.G. Wijmans and R.W. Baker, The solution-diffusion model: a review, *J. Membr. Sci.*, 107, 1–21 (1995).
23. T. Matsuura, *Synthetic Membrane and Membrane Separation Process*, CRC Press, Boca Raton, 1994.
24. A.S. Michaels, R.F. Baddour, H.J. Bixler and C.Y. Choo, Conditioned polyethylene as a permselective membrane, *Ind. Eng. Chem. Process Des. Dev.*, 1, 14–25 (1962).
25. V.N. Schrodt, R.F. Sweeny and A. Rose, Division of Industrial and Engineering Chemistry, 144th Meeting, ACS, Los Angeles, 1963.
26. R.B. Long, Liquid permeation through plastic films, *Ind. Eng. Chem. Fundam.*, 4, 445–451 (1965).
27. S. Sourirajan, B. Shiyo and T. Matsuura, *in Proceedings of the Second International Conference on Pervaporation Processes in Chemical Industry*, Bakish Materials Corp., Englewood NJ, 1987.
28. S. Sourirajan and T. Matsuura, *Reverse Osmosis and Ultrafiltration/Process Principles*, National Research Council of Canada, Ottawa, 1985.
29. M. Yoshikawa, T. Yukoshi, K. Sanui and N. Ogata, Separation of water and ethanol by pervaporation through poly(acrylic acid-co-acrylonitrile) membrane, *J. Polym. Sci., Polym. Lett.*, 22, 473–475 (1984).
30. M. Yoshikawa, H. Yokoi, K. Sanui and N. Ogata, Pervaporation of water-ethanol mixture through poly(maleimide-co-acrylonitrile) membrane, *J. Polym. Sci. Polym. Lett.*, 22, 125–127 (1984).
31. M. Yoshikawa, H. Yokoi, K. Sanui and N. Ogata, Selective separation of water-alcohol binary mixture through poly(maleimide-co-acrylonitrile) membrane, *J. Polym. Sci.: Polym. Lett.*, 22, 2159–2168 (1984).
32. M. Yoshikawa, T. Yukoshi, K. Sanui and N. Ogata, Selective separation of water ethanol mixture through synthetic-polymer membranes having carboxylic-acid as a functional-group, *J. Polym. Sci.: Part A: Polym. Chem.*, 24, 1585–1597 (1986).
33. M. Yoshikawa, N. Ogata and T. Shimidzu, Polymer membrane as a reaction field. 3. Effect of membrane polarity on selective separation of a water ethanol binary mixture through synthetic-polymer membranes, *J. Membr. Sci.*, 26, 107–113 (1986).
34. T. Graham, On the law of diffusion of gases, *Philos. Mag.*, 32, 401–420 (1866).
35. X. Feng and R.Y.M. Huang, Liquid separation by membrane pervaporation: a review, *Ind. Eng. Chem. Res.*, 36, 1048–1066 (1997).
36. W.F. Guo, T.S. Chung and T. Matsuuraa, Pervaporation study on the dehydration of aqueous butanol solutions: a comparison of flux vs. permeance, separation factor vs. selectivity, *J. Membr. Sci.*, 245, 199–210 (2004).
37. X.Y. Qiao, T.S. Chung, W.F. Guo, T. Matsuura and M.M. Teoh, Dehydration of isopropanol and its comparison with dehydration of butanol isomers from thermodynamic and molecular aspects, *J. Membr. Sci.*, 252, 37–49 (2005).
38. Y. Wang, L.Y. Jiang, T. Matsuura, T.S. Chung and S.H. Goh, Investigation of the fundamental differences between polyamide-imide (PAI) and polyetherimide (PEI) membranes for isopropanol dehydration via pervaporation, *J. Membr. Sci.*, 318, 217–226 (2008).

39. L.Y. Jiang, H. Chen, Y.C. Jean and T.S. Chung, Ultra-thin polymeric interpenetration network with separation performance approaching ceramic membranes for biofuel, *AIChE J.*, 55, 75–86 (2009).
40. L.Y. Jiang, T.S. Chung, Homogeneous polyimide/cyclodextrin composite membranes for pervaporation dehydration of isopropanol, *J. Membr. Sci.*, 346, 45–58 (2010).
41. Y. Wang, T.S. Chung, B. Neo and M. Gruender, Processing and engineering of pervaporation dehydration of ethylene glycol via dual-layer polybenzimidazole (PBI) /polyetherimide (PEI) membranes, *J. Membr. Sci.*, 378, 339–350 (2011).
42. N.L. Le, Y. Wang, T.S. Chung, Poly(ether-block-amide)/POSS mixed matrix membranes for ethanol recovery from aqueous solutions, *J. Membr. Sci.*, 379, 174–183 (2011).
43. G.M. Shi , Y. Wang, and T.S. Chung, Dual-layer PBI/P84 hollow fibers for pervaporation dehydration of acetone, *AICHE J.*, 58, 1133–1145 (2012).
44. H. Eustache and G. Histi, Separation of aqueous organic mixtures by pervaporation and analysis by mass-spectrometry or a coupled gas chromatograph-mass spectrometer, *J. Membr. Sci.*, 8, 105–114 (1981).
45. J.G. Crespo and K.W. Böddeker, *Membrane Process in Separation and Purification*, Kluwer Academic Publishers, Boston, 1995.
46. B. Smitha, D. Suhanya, S. Sridhar and M. Ramakrishna, Separation of organic-organic mixtures by pervaporation—a review, *J. Membr. Sci.*, 241, 1–21 (2004).
47. P. Shao, R.Y.M. Huang, Polymeric membrane pervaporation, *J. Membr. Sci.* 287, 162–179 (2007).
48. M.H.V. Mulder, *Basic Principles of Membrane Technology*, Kluwer Academic Publishers, Boston, 1996.
49. T. Okada and T. Matsuura, A new transport model for pervaporation, *J. Membr. Sci.*, 59, 133–149 (1991).
50. T. Okada, M. Yoshikawa and T. Matsuura, A study on the pervaporation of ethanol/water mixtures on the basis of pore flow model, *J. Membr. Sci.*, 59, 151–168 (1991).
51. S. Deng, B. Shiyao, S. Sourirajan and T. Matsuura, A study of the pervaporation of isopropyl alcohol/water mixtures by cellulose acetate membranes, *J. Colloid Interface Sci.*, 136, 283–291 (1990).
52. T. Okada and T. Matsuura, Predictability of transport equations for pervaporation on the basis of pore-flow mechanism, *J. Membr. Sci.*, 70, 163–175 (1992).
53. P. Sukitpaneenit, T.S. Chung and L.Y. Jiang, Modified pore-flow model for pervaporation mass transport in PVDF hollow fiber membranes for ethanol–water separation, *J. Membr. Sci.*, 362, 393–406 (2010). (Erratum to this article: DOI: 10.1016/j.memsci.2010.06.062.)
54. P. Sukitpaneenit and T.S. Chung, Molecular design of the morphology and pore size of PVDF hollow fiber membranes for ethanol–water separation employing the modified pore-flow concept, *J. Membr. Sci.*, 374, 67–82 (2011).
55. J.G. Wijmans, Process performance = membrane properties plus operating conditions, *J. Membr. Sci.*, 220, 1–3 (2003).
56. S. Sommer and T. Melin, Influence of operation parameters on the separation of mixtures by pervaporation and vapor permeation with inorganic membranes. Part 1. Dehydration of solvents, *Chem. Eng. Sci.*, 60, 4509–4523 (2005).
57. S.I. Semenova, H. Ohya, K. Soontarapa, Hydrophilic membranes for pervaporation: an analytical review, *Desalination*, 110, 251–86 (1997).
58. P.D. Chapman, T. Oliveira, A.G. Livingston and K. Li, Membranes for the dehydration of solvents by pervaporation, *J. Membr. Sci.*, 318, 5–37 (2008).
59. M. Rafik, A. Mas, M.F. Guimon, C. Guimon, A. Elharfi and F. Schule, Plasma modified poly(vinyl alcohol) membranes for the dehydration of ethanol, *Polym. Int.*, 52, 1222–1229 (2003).
60. R.Y.M. Huang, *Pervaporation Membrane Separation Processes*, Elsevier, Amsterdam, 1991.
61. M.L. Gimenes, L. Liu and X.S. Feng, Sericin/poly(vinyl alcohol) blend membranes for pervaporation separation of ethanol/water mixtures, *J. Membr. Sci.*, 295, 71–79 (2007).
62. Y.Q. Dong, L. Zhang, J.N. Shen, M.Y. Song and H.L. Chen, Preparation of poly(vinyl alcohol)-sodium alginate hollow-fiber composite membranes and pervaporation dehydration characterization of aqueous alcohol mixtures, *Desalination*, 193, 202–210 (2006).
63. L. Liang and E. Ruckenstein, polyvinyl alcohol-polyacrylamide interpenetrating polymer network membranes and their pervaporation characteristics for ethanol-water mixtures, *J. Membr. Sci.*, 106, 167–182 (1995).

64. B.V.K. Naidua, M. Sairam, K.V.S.N. Raju and T.M. Aminabhavi, Pervaporation separation of water plus isopropanol mixtures using novel nanocomposite membranes of poly(vinyl alcohol) and polyaniline, *J. Membr. Sci.*, 260, 142–155 (2005).
65. J.N. Shen, H.M. Ruan and C.J. Ga, Preparation and characterization of CMCS/PVA blend membranes and its sorption and pervaporation performance (I), *J. Appl. Polym. Sci.*, 114, 3369–3378 (2009).
66. X.C. Liu, J.A. Chen, J.A. Zou, H.Z. Duan, J.D. Li, C.X. Chen and P.R. Meng, Preparation and membrane separation performances of quarternized ammonium cationic polyvinyl alcohol, *J. Appl. Polym. Sci.*, 119, 2584–2594 (2011).
67. M.C. Burshe, S.B. Sawant, J.B. Joshi and V.G. Pangarkar, Sorption and permeation of binary water-alcohol systems through PVA membranes crosslinked with multifunctional crosslinking agents, *Sep. Purif. Technol.*, 12, 145–156 (1997).
68. S. Takegami, H. Yamada and S. Tsujii, dehydration of water ethanol mixtures by pervaporation using modified poly(vinyl alcohol) membrane, *Polym. J.*, 24, 1239–1250 (1992).
69. T.A. Peters, C.H.S. Poeth, N.E. Benes, H.C.W.M. Buijs, F.F. Vercauteren and J.T.F. Keurentjes, Ceramic-supported thin PVA pervaporation membranes combining high flux and high selectivity; contradicting the flux-selectivity paradigm, *J. Membr. Sci.*, 276, 42–50 (2006).
70. D. Xu, L.S. Loo and K.A. Wang, Pervaporation Performance of novel chitosan-POSS hybrid membranes: Effects of POSS and operating conditions, *J. Polym. Sci., Part B: Polym. Phys.*, 48, 2185–2192 (2010).
71. J. Ma, M.H. Zhang, H. Wu, X. Yin, J. Chen and Z.Y. Jiang, Mussel-inspired fabrication of structurally stable chitosan/polyacrylonitrile composite membrane for pervaporation dehydration, *J. Membr. Sci.*, 348, 150–159 (2010).
72. J.J. Ge, Y.F. Cui, Y. Yan and W.Y. Jiang, The effect of structure on pervaporation of chitosan membrane, *J. Membr. Sci.*, 165, 75–81 (2000).
73. G.Y. Moon, R. Pal and R.Y.M. Huang, Novel two-ply composite membranes of chitosan and sodium alginate for the pervaporation dehydration of isopropanol and ethanol, *J. Membr. Sci.*, 156, 17–27 (1999).
74. J.J. Shieh and R.Y.M. Huang, Pervaporation with chitosan membranes. 2. Blend membranes of chitosan and polyacrylic acid and comparison of homogeneous and composite membrane based on polyelectrolyte complexes of chitosan and polyacrylic acid for the separation of ethanol-water mixtures, *J. Membr. Sci.*, 127, 185–202 (1997).
75. R. Jiraratananon, A. Chanachai, R.Y.M. Huang and D. Uttapap, Pervaporation dehydration of ethanol-water mixtures with chitosan/hydroxyethylcellulose (CS/HEC) composite membranes I. Effect of operating conditions, *J. Membr. Sci.*, 195, 143–151 (2002).
76. Y.M. Lee and E.M. Shin, Pervaporation separation of water ethanol through modified chitosan membranes .4. Phosphorylated chitosan membranes, *J. Membr. Sci.*, 64, 145–152 (1991).
77. S.Y. Nam and Y.M. Lee, Pervaporation and properties of chitosan poly(acrylic acid) complex membranes, *J. Membr. Sci.*, 135, 161–171 (1997).
78. Y.M. Lee, Modified chitosan membranes for pervaporation, *Desalination*, 90, 277–290 (1993).
79. Y.M. Lee, E.M. Shin and C.N. Chung, Pervaporation separation of water-ethanol through modified chitosan membranes. 3. Sulfonated chitosan membranes, *Polym. (Korea)*, 15, 497–500 (1991).
80. Y.M. Lee, E.M. Shin and S.T. Noh, Pervaporation separation of water-ethanol through modified chitosan membranes. 2. Carboxymethyl, carboxyethyl, cyanoethyl, and amidoxime chitosan membranes, *Angew. Makromol. Chem.*, 192, 169–181 (1991).
81. Y.M. Lee, E.M. Shin and K.S. Yang, Pervaporation separation of water-ethanol through modified chitosan membranes 1. Chitosan-acetic acid and -metal ion complex membranes, *Polym. (Korea)*, 15, 182–190 (1991).
82. Y.M. Lee, S.Y. Nam and D.J. Woo, Pervaporation of ionically surface crosslinked chitosan composite membranes for water-alcohol mixtures, *J. Membr. Sci.*, 133, 103–110 (1997).
83. T. Uragami and M. Saito, studies on syntheses and permeabilities of special polymer membranes .68. analysis of permeation and separation characteristics and new technique for separation of aqueous alcoholic solutions through alginic acid membranes, *Sep. Sci. Technol.*, 24, 541–554 (1989).

84. A. Mochizuki, S. Amiya, Y. Sato, H. Ogawara and S. Yamashita, Pervaporation separation of water ethanol mixtures through polysaccharide membranes. 4. The relationships between the permselectivity of alginic acid membrane and its solid-state structure, *J. Appl. Polym. Sci.*, 40, 385–400 (1990).
85. C.K. Yeom., J.G. Jegal and K.H. Lee, Characterization of relaxation phenomena and permeation behaviors in sodium alginate membrane during pervaporation separation of ethanol-water mixture, *J. Appl. Polym. Sci.*, 62, 1561–1576 (1996).
86. V.T. Magalad, A.R. Supale, S.P. Maradur, G.S. Gokavi and T.M. Aminabhavi, Preyssler type heteropolyacid-incorporated highly water-selective sodium alginate-based inorganic–organic hybrid membranes for pervaporation dehydration of ethanol, *Chem. Eng. J.*, 159, 75–83 (2010).
87. S. Kalyani, B. Smitha, S. Sridhar and A. Krishnaiah, Pervaporation separation of ethanol–water mixtures through sodium alginate membranes, *Desalination*, 229, 68–81 (2008).
88. R.Y.M. Huang, R. Pal and G.Y. Moon, Pervaporation dehydration of aqueous ethanol and isopropanol mixtures through alginate/chitosan two ply composite membranes supported by poly(vinylidene fluoride) porous membrane, *J. Membr. Sci.*, 167, 275–289 (2000).
89. Y.Q. Shi, X.W. Wang, G.W. Chen, G. Golemme, S.M. Zhang and E. Drioli, Preparation and characterization of high-performance dehydrating pervaporation alginate membranes, *J. Appl. Polym. Sci.*, 68, 959–968 (1998).
90. R.Y.M. Huang, R. Pal and G.Y. Moon, Characteristics of sodium alginate membranes for the pervaporation dehydration of ethanol-water and isopropanol-water mixtures, *J. Membr. Sci.*, 160, 101–113 (1999).
91. Y.X. Xu, C.X. Chen and J.D. Li, Experimental study on physical properties and pervaporation performances of polyimide membranes, *Chem. Eng. Sci.*, 62, 2466–2473 (2007).
92. H. Yanagishita, C. Maejima, D. Kitamoto and T. Nakane, Preparation of asymmetric polyimide membrane for water-ethanol separation in pervaporatin by the phase inversion process, *J. Membr. Sci.*, 86, 231–240 (1994).
93. J.H. Kim, K.H. Lee and S.Y. Kim, Pervaporation separation of water from ethanol through polyimide composite membranes, *J. Membr. Sci.*, 169, 81–93 (2000).
94. W.L. Qiu, M. Kosuri, F.B. Zhou and W.J. Koros, Dehydration of ethanol-water mixtures using asymmetric hollow fiber membranes from commercial polyimides, *J. Membr. Sci.*, 327, 96–103 (2009).
95. N. Widjojo and T.S. Chung, Pervaporation dehydration of C2-C4 alcohols by 6FDA-ODA-NDA/Ultem® dual-layer hollow fiber membranes with enhanced separation performance and swelling resistance, *Chem. Eng. J.*, 155, 736–743 (2009).
96. R. Kreiter, D.P. Wolfs, C.W.R. Engelen, H.M. van Veen and J.F. Vente, High-temperature pervaporation performance of ceramic-supported polyimide membranes in the dehydration of alcohols, *J. Membr. Sci.*, 319, 126–132 (2008).
97. H.A. Tsai, Y.L. Ye, K.R. Lee, S.H. Huang, M.C. Suen and J.Y. Lai, Characterization and pervaporation dehydration of heat-treatment PAN hollow fiber membranes, *J. Membr. Sci.*, 368, 254–263 (2011).
98. S. Amnuaypanich and N. Kongchana, Natural rubber/poly(acrylic acid) semi-interpenetrating polymer network membranes for the pervaporation of water–ethanol mixtures, *J. Appl. Polym. Sci.*, 114, 3501–3509 (2009).
99. I.J. Ball, S.C. Huang, K.J. Miller, R.A. Wolf, J.Y. Shimano and R.B. Kaner, The pervaporation of ethanol water feeds with polyaniline membranes and blends, *Synth. Met.*, 102, 1311–1312 (1999).
100. V. Smuleac, J. Wu, S. Nemser, S. Majumdar, D. Bhattacharyya, Novel perfluorinated polymer-based pervaporation membranes for the separation of solvent/water mixtures, *J. Membr. Sci.*, 352, 41–49 (2010).
101. S.H. Chen, R.M. Liou, Y.Y. Lin, C.L. Lai and J.Y. Lai, Preparation and characterizations of asymmetric sulfonated polysulfone membranes by wet phase inversion method, *Eur. Polym. J.*, 45, 1293–1301 (2009).
102. C.S. Hsu, R.M. Liou, S.H. Chen, M.Y. Hung, H.A. Tsia and J.Y. Lai, Pervaporation separation of a water–ethanol mixture by PSF-PEG membrane, *J. Appl. Polym. Sci.*, 87, 2158–2164 (2003).
103. Y.H. Wang, M.Y. Teng, K.R. Lee, D.M. Wang and J.Y. Lai, Application of pervaporation and vapor permeation processes to separate aqueous ethanol solution through chemically modified nylon 4 membranes, *Sep. Sci. Technol.*, 33, 1653–1665 (1998).
104. N. Scharnagl, K.V. Peinemann, A. Wenzlaff, H.H. Schwarz and R.D. Behling, Dehydration of organic compounds with SYMPLEX composite membranes, *J. Membr. Sci.*, 113, 1–5 (1996).
105. S.L. Wee, C.T. Tye and S. Bhatia, Membrane separation process—Pervaporation through zeolite membrane, *Sep. Purif. Technol.*, 63, 500–516 (2008).

106. M. Pera-Titus, M. Bausach, J. Llorens and F. Cunill, Preparation of inner-side tubular zeolite NaA membranes in a continuous flow system, *Sep. Purif. Technol.*, 59, 141–150 (2008).
107. J. Zah, H.M. Krieg and J.C. Breytenbach, Pervaporation and related properties of time-dependent growth layers of zeolite NaA on structured ceramic supports, *J. Membr. Sci.*, 284, 276–290 (2006).
108. K. Sato and T. Nakane, A high reproducible fabrication method for industrial production of high flux NaA zeolite membrane, *J. Membr. Sci.*, 301, 151–161 (2007).
109. M.P. Pina, M. Arruebo, M. Felipe, F. Fleta, M.P. Bernal, J. Coronas, M. Menendez and J. Santamaria, A semi-continuous method for the synthesis of NaA zeolite membranes, *J. Membr. Sci.*, 244, 141–150 (2004).
110. F. Tiscareno-Lechuga, C. Tellez, M. Menendez and J. Santamaria, A novel device for preparing zeolite-A membranes under a centrifugal force field. *J. Membr. Sci.*, 212, 135–146 (2003).
111. J. Kuhn, K. Yajima, T. Tomita, J. Gross and F. Kapteijn, Dehydration performance of a hydrophobic DD3R zeolite membrane, *J. Membr.Sci.*, 321, 344–349 (2008).
112. Y. Ma, J.H. Wang and T. Tsuru, Pervaporation of water/ethanol mixtures through microporous silica membranes, *Sep. Purif. Technol.*, 66, 479–485 (2009).
113. M.C. Duke, R. Campbell, X. Cheng, A. Leo and J.C.D. da Costa, Characterization and Pervaporation Study on Ethanol Separation Membranes, *Drying Technol.*, 27, 538–541 (2009).
114. J. Sekulic, J.E.T. Elshof and D.H.A. Blank, Separation mechanism in dehydration of water/organic binary liquids by pervaporation through microporous silica, *J. Membr. Sci.*, 254, 267–274 (2005).
115. J.H. Wang and T. Tsuru, Cobalt-doped silica membranes for pervaporation dehydration of ethanol/water solutions, *J. Membr. Sci.*, 369, 13–19 (2011).
116. H. Chen, C. Song and W. Yang, Effects of aging on the synthesis and performance of silicalite membranes on silica tubes without seeding, *Microporous Mesoporous Mater.*, 102, 249–257 (2007).
117. A. Navajas, R. Mallada, C. Tellez, J. Coronas, M. Menendez and J. Santamaria, Study on the reproducibility of mordenite tubular membranes used in the dehydration of ethanol, *J. Membr. Sci.*, 299, 166–173 (2007).
118. L. Casado, R. Mallada, C. Tellez, J. Coronas, M. Menendez and J. Santamaria, Preparation, characterization and pervaporation performance of mordenite membranes, *J. Membr. Sci.*, 216, 135–147 (2003).
119. T. Van Gestel, D. Sebold, H. Kruidhof and H.J.M. Bouwmeester, ZrO_2 and TiO_2 membranes for nanofiltration and pervaporation—Part 2. Development of ZrO2 and TiO_2 toplayers for pervaporation, *J. Membr. Sci.*, 318, 413–421 (2008).
120. H.M. Van Veen, Y.C. Van Delft, C.W.R. Engelen and P.P.A.C. Pex, Dewatering of organics by pervaporation with silica membranes, *Sep. Purif. Technol.*, 22–23, 361–366 (2001).
121. F. Petrus-Cuperus and R.W. Van Gemert, Dehydration using ceramic silica pervaporation membranes—the influence of hydrodynamic conditions, *Sep. Purif. Technol.*, 27, 225–229 (2002).
122. T.A. Peters, J. Fontalvo, M.A.G. Vorstman, N.E. Benes, R.A. van Dam, Z.A.E.P. Vroon, E.L.J. van Soest-Vercammen and J.T.F. Keurentjes, Hollow fibre microporous silica membranes for gas separation and pervaporation; synthesis, performance and stability, *J. Membr. Sci.*, 248, 73–80 (2005).
123. Y.L. Liu, C.Y. Hsu, Y.H. Su, J.Y. Lai, Chitosan-silica complex membranes from sulfonic acid functionalized silica nanoparticles for pervaporation dehydration of ethanol-water solutions, *Biomacromolecules*, 6, 368–373 (2005).
124. H.A. Tsai, W.H. Chen, C.Y. Kuo, K.R. Lee and J.Y. Lai, Study on the pervaporation performance and long-term stability of aqueous iso-propanol solution through chitosan/polyacrylonitrile hollow fiber membrane, *J. Membr. Sci.*, 309, 146–155 (2008).
125. Y.L. Liu, Y. H. Su, K. R. Lee and J.Y. Lai, Crosslinked organic–inorganic hybrid chitosan membranes for pervaporation dehydration of isopropanol-water mixtures with a long-term stability, *J. Membr. Sci.*, 251, 233–238 (2005).
126. H.L. Castricum, R. Kreiter, H.M. van Veen, D.H.A. Blank, J.F. Vente and J.E. ten Elshof, High-performance hybrid pervaporation membranes with superior hydrothermal and acid-stability, *J. Membr. Sci.*, 324, 111–118 (2008).
127. H.M. Guan, T.S. Chung, Z. Huang, M.L. Chng and S. Kulprathipanja, Poly(vinyl alcohol) multilayer mixed matrix membranes for the dehydration of ethanol-water mixture, *J. Membr. Sci.*, 268, 113–122 (2006).

128. S. Amnuaypanich, J. Patthana and P. Phinyocheep, Mixed matrix membranes prepared from natural rubber/poly(vinyl alcohol) semi-interpenetrating polymer network (NR/PVA semi-IPN) incorporating with zeolite 4A for the pervaporation dehydration of water-ethanol mixtures, *Chem. Eng. Sci.*, 64, 4908–4918 (2009).
129. D. Yang, J. Li, Z. Y. Jiang, L. Lu and X. Chen, Chitosan/TiO_2 nanocomposite pervaporation membranes for ethanol dehydration, *Chem. Eng. Sci.*, 2009, 64, 3130–3137.
130. S.D. Bhat, N.N. Mallikarjuna and T.M. Aminabhavi, Microporous alumino-phosphate (AlPO4-5) molecular sieve-loaded novel sodium alginate composite membranes for pervaporation dehydration of aqueous-organic mixtures near their azeotropic compositions, *J. Membr. Sci.*, 282, 473–483 (2006).
131. Q.G. Zhang, Q.L. Liu, A.M. Zhu, Y. Xiong and X.H. Zhang, Characterization and permeation performance of novel organic–inorganic hybrid membranes of poly(vinyl alcohol)/1,2-bis(triethoxysilyl)ethane, *J. Phys. Chem. B.*, 112, 16559–16565 (2008).
132. X.Z. Cao, T.Z. Zhang, Q.T. Nguyen, Y.Y. Zhang and Z.H. Ping, A novel hydrophilic polymer-ceramic composite membrane 1 acrylic acid grafting membrane, *J. Membr. Sci.*, 312, 15–22 (2008).
133. J.H. Chen, Q.L. Liu, X.H. Zhang and Q.G. Zhang, Pervaporation and characterization of chitosan membranes cross-linked by 3-aminopropyltriethoxysilane, *J. Membr. Sci.*, 292, 125–132 (2007).
134. M.B. Patil, R.S. Veerapur, S.A. Patil, C.D. Madhusoodana and T.M. Aminabhavi, Preparation and characterization of filled matrix membranes of sodium alginate incorporated with aluminum-containing mesoporous silica for pervaporation dehydration of alcohols, *Sep. Purif. Technol.*, 54, 34–43 (2007).
135. B. P. Tripathi; M. Kumar, A. Saxena and V.K. Shahi, Bifunctionalized organic–inorganic charged nanocomposite membrane for pervaporation dehydration of ethanol, *J. Colloid Interface Sci.*, 346, 54–60 (2010).
136. I. Kazuhiko and M. Kiyohide, Pervaporation of ethanol-water mixture through composite membranes composed of styrene-fluoroalkyl acrylate graft copolymers and crosslinked polydimethylsiloxane membrane, *J. Appl. Polym. Sci.*, 34, 437–440 (1987).
137. M.C. Conclaves, G.S.S Marquez and F. Galembeck, Pervaporation and dialysis of water-ethanol solution by using silicone rubber-membrane, *Sep. Sci. Technol.*, 18, 893–904. (1983).
138. B. Moermans, W.D. Beuckelaer, I.F.J. Vankelecom, R. Ravishankar, J.A. Martens and P.A. Jacobs, Incorporation of nano-sized zeolites in membranes, *Chem. Commun.*, 24, 2467–2468 (2000).
139. I.F.J. Vankelecom, D. Depré, S.D. Beukelaer and J.B. Uytterhoeven, Influence of zeolites in PDMS membranes: Pervaporation of water/alcohol mixtures. *J. Phys. Chem.*, 99, 13193–13197 (1995).
140. L. Li, Z. Xiao, S. Tan, L. Pu and Z. Zhang, Composite PDMS membrane with high flux for the separation of organics from water by pervaporation, *J. Membr. Sci.*, 243, 177–187 (2004).
141. Y. Huang, J. Fu, Y. Pan, X. Huang and X. Tang, Pervaporation of ethanol aqueous solution by polyphosphazene membranes: Effect of pendant groups, *Sep. Purif. Technol.*, 66, 504–509 (2009).
142. Y. Mori and T. Inaba, Ethanol production from starch in a pervaporation membrane bioreactor using clostridium thermohydrosulfuricum, *Biotechnol. Bioeng.*, 36, 849–853 (1990).
143. E. Shi, W. Huang, Z. Xiao, D. Li and M. Tang, Influence of binding interface between active and support layers in composite PDMS membranes on permeation performance, *J. Appl. Polym. Sci.*, 104, 2468–2477 (2007).
144. A. Dobrak, A. Figoli, S. Chovau, F. Galiano, S. Simone, I.F.J. Vankelecom, E. Drioli and B. Van der Bruggen, Performance of PDMS membranes in pervaporation: Effect of silicalite fillers and comparison with SBS membranes, *J. Colloid Interf. Sci.*, 346, 254–264 (2010).
145. S.L. Schmidt, M.D. Myers, S.S. Kelley, J.D. McMillan and N. Padukone, Evaluation of PTMSP membranes in achieving enhanced ethanol removal from fermentations by pervaporation, *Appl. Biochem. Biotech.*, 63–65, 469–482 (1997).
146. J. Guo, G. Zhang, W. Wu, S. Ji, Z. Qin and Z. Liu, Dynamically formed inner skin hollow fiber polydimethylsiloxane/polysulfone composite membrane for alcohol permselective pervaporation, *Chem. Eng. J.*, 158, 558–565, (2010).
147. X. Zhan, J. Li, J. Huang and C. Chen, Enhanced pervaporation performance of multi-layer PDMS/PVDF composite membrane for ethanol recovery from aqueous solution, *Appl. Biochem. Biotechnol.*, 160, 632–642 (2010).

148. F. Xiangli, W. Wei, Y. Chen, W. Jin and N. Xu, Optimization of preparation conditions for polydimethylsiloxane (PDMS)/ceramic composite pervaporation membranes using response surface methodology, *J. Membr. Sci.*, 311, 23–33 (2008).
149. F. Xiangli, Y. Chen, W. Jin and N. Xu, Polydimethylsiloxane (PDMS)/ceramic composite membrane with high flux for pervaporation of ethanol-water mixtures, *Ind. Eng. Chem. Res.*, 46, 2224–2230 (2007).
150. K.W. Böddeker, G. Bengtson, H. Pingel, Pervaporation of isomeric butanols, *J. Membr. Sci.*, 54, 1–12 (1990).
151. A. Jonquieres and A. Fane, Filled and unfilled composite GFT PDMS membranes for the recovery of butanols from dilute aqueous solutions: influence of alcohol polarity, *J. Membr. Sci.*, 125, 245–255 (1997).
152. A.G. Fadeev, S.S. Kelley, I.D. McMillan, Y.A. Selinskaya, V.S. Khotimsky and V.V. Volkov, Effect of yeast fermentation by-products on poly[1-(trimethylsilyl)-1-propyne] pervaporative performance, *J. Membr. Sci.*, 214, 229–238 (2003).
153. J. Huang and M.M. Meagher, Pervaporative recovery of n-butanol from aqueous solutions and ABE fermentation broth using thin-film silicalite-filled silicone composite membranes, *J. Membr. Sci.*, 195, 231–242 (2001).
154. Y. Luo, S. Tan, H. Wang, F. Wu, X. Liu, L. Li, and Z. Zhang, PPMS composite membranes for the concentration of organics from aqueous solutions by pervaporation, *Chem. Eng. J.*, 137, 496–502 (2008).
155. S. Takegemi, H. Yamada and S. Tusujii, Pervaporation of ethanol/water mixture using novel hydrophobic membrane containing polydimethylsiloxane, *J. Membr. Sci.*, 75, 93–105 (1992).
156. C.L. Chang and P.Y. Chang, Performance enhancement of silicone/PVDF composite membranes for pervaporation by reducing cross-linking density of the active silicone layer, *Desalination*, 192, 241–245 (2006).
157. T. Kashiwagi, K. Okabe and K. Okita, Separation of ethanol from ethanol/water mixtures by plasma-polymerized membranes from silicone compounds, *J. Membr. Sci.*, 36, 353–362 (1988).
158. C.L. Chang and M.S. Chang, Preparation of composite membranes of functionalised silicone polymers and PVDF for pervaporation of ethanol-water mixture, *Desalination*, 148, 39–42 (2002).
159. C.L. Chang and M.S. Chang, Preparation of multi-layer silicone/PVDF composite membranes for pervaporation of ethanol aqueous solutions, *J. Membr. Sci.*, 238, 117–122 (2004).
160. M. Krea, D. Roizard, N. Moulai-Mostefa and D. Sacco, New copolyimide membranes with high siloxane content designed to remove polar oranics from water by pervaporation, *J. Membr. Sci.*, 241, 55–64 (2004).
161. L. Liang and E. Ruckenstein, Pervaporation of ethanol-water mixtures through polydimethylsiloxane-polystyrene interpenetrating polymer network supported membranes, *J. Membr. Sci.*, 114, 227–234 (1996).
162. K. Okamoto, A. Butsuen, S. Tsuru, S. Nishioka, K. Tanaka, H. Kita and S. Asakawa, Pervaporation of water-ethanol mixtures through polydimetylsiloxane block-copolymer membranes, *Polym. J.*, 19, 747–756 (1987).
163. V. Garcia, E. Pongracz, E. Muurinen and R.L. Keiski, Recovery of n-butanol from salt containing solutions by pervaporation, *Desalination*, 241, 201–211 (2009).
164. A.G. Fadeev, S.S. Kelley, I.D. McMillan, Y.A. Selinskaya, V.S. Khotimsky and V.V. Volkov, Effect of yeast fermentation by-products on poly[1-(trimethylsilyl)-1-propyne] pervaporative performance, *J. Membr. Sci.*, 214, 229–238 (2003).
165. Y. Nagase, Y. Takamura and K. Matsui, Chemical modification of Poly(substituted-acetylene). V. Alkylsilylation of Poly(1-Trimethylsilyl-1-propyne) and improved liquid separating property at pervaporation, *J. Appl. Polym. Sci.*, 42, 185–190 (1991).
166. J.R. Gonzalez-Velasco, J.A. Gonzalez-Marcos and C. Lopez-Dehesa, Pervaporation of ethanol-water mixtures through poly(1-trimethylsilyl-1-propyne) (PTMSP) membranes, *Desalination*, 149, 61–65 (2002).
167. S. Claes, P. Vandezande, S. Mullens, R. Leysen, K.D. Sitter, A. Andersson, F.H.J. Maurer, H. Van den Rul, R. Peeters and M.K. Van Bael, High flux composite PTMSP-silica nanohybrid membranes for the pervaporation of ethanol/water mixtures, *J. Membr. Sci.*, 351, 160–167 (2010).
168. Y. Nagase, K. Ishihara and K. Matsui, Chemical modification of poly(substituted-acetylene): II. Pervaporation of ethanol/water mixture through poly(1-trimethylsilyl-1-propyne)/poly(dimethylsiloxane) graft copolymer membrane, *J. Poly. Sci. Part B: Polym. Phys.*, 28, 377–386 (1990).
169. A.G. Fadeev, M.M. Meagher, S.S. Kelley and V.V. Volkov, Fouling of poly[-1-(trimethylsilyl)-1-propyne] membranes in pervaporative recovery of butanol from aqueous solutions and ABE fermentation broth, *J. Membr. Sci.*, 173, 133–144 (2000).

170. C.J. Ruud, J. Jia and G.L. Baker, Synthesis and characterization of poly[(1-trimethylsilyl-1-propyne)-*co*-(1-(4-azidobutyldimethylsilyl)-1-propyne)] copolymers, *Macromolecules*, 33, 8184–8191 (2000).
171. J. Jia and G.L. Baker, Cross-linking of poly(1-trimethylsilyl-1-propyne) membranes using bis(aryl azides), *J. Appl. Polym. Sci. Part B: Polym. Phys.*, 36, 959–968 (1998).
172. F. Liu, L. Liu and X. Feng, Separation of acetone-butanol-ethanol (ABE) from dilute aqueous solutions by pervaporation, *Sep. Purif. Technol.*, 42, 273–282 (2005).
173. X. Jiang, J. Gu, Y. Shen, S. Wang and X. Tian, New fluorinated siloxane-imide block copolymer membranes for application in organophilic pervaporation, *Desalination*, 265, 74–80 (2011).
174. A. Ghofar and T. Kokugan, The pervaporation mechanism of dilute ethanol solution by hydrophobic porous membranes, *Biochem. Eng. J.*, 18, 235–238 (2004).
175. T.C. Bowen, R.D. Noble, and J.L. Falconer, Fundamentals and applications of pervaporation through zeolite membranes, *J. Membr. Sci.*, 245, 1–33 (2004).
176. T. Sano, H. Yanagishita, Y. Kiyozumi, F. Mizukami and K. Haraya, Separation of ethanol/water mixture by silicalite membrane on pervaporation, *J. Membr. Sci.*, 95, 221–228 (1994).
177. T. Ikegami, D. Kitamoto, H. Negishi, K. Haraya, H. Matsuda, Y. Nitanai, N. Koura, T. Sano and H. Yanagishita, Drastic improvement of bioethanol recovery using a pervaporation separation technique employing a silicone rubber-coated silicalite membrane, *J. Chem. Tech. Biotechnol.*, 78, 1006–1010 (2003).
178. Q. Liu, R.D. Noble, J.L. Falconer and H.H. Funke, Organics/water separation by pervaporation with a zeolite membrane, *J. Membr. Sci.*, 117, 163–174 (1996).
179. D. Shen, W. Xiao, J. Yang, N. Chu, J. Lu, D. Yin and J. Wang, Synthesis of silicalite-1 membrane with two silicon source by secondary growth method and its pervaporation performance, *Sep. Purif. Technol.*, 76, 308–315 (2011).
180. X. Lin, X. Chen, H. Kita and K. Okamoto, Synthesis of silicalite tubular membranes by in situ crystallization, *AIChE J.*, 49, 237–247 (2003).
181. X. Lin, H. Kita and K.I. Okamoto, Silicalite membrane preparation, characterization and separation performance, *Ind. Eng. Chem. Res.*, 40, 4069–4078 (2001).
182. X. Lin, H. Kita and K.I. Okamoto, A novel method for the synthesis of high performance silicalite membranes, *Chem. Comm.*, 1889–1890 (2000).
183. T. Ikegami, H. Yanagishita, D. Kitamoto, H. Negishi, K. Haraya and T. Sano, Concentration of fermented ethanol by pervaporation using silicalite membranes coated with silicone rubber, *Desalination*, 149, 49–54 (2002).
184. H. Matsuda, H. Yanagishita, H. Negishi, D. Kitamoto, T. Ikegami, K. Haraya, T. Nakane, Y. Idemoto, N. Koura and T. Sano, Improvement of ethanol selectivity of silicalite membrane in pervaporation by silicone rubber coating, *J. Membr. Sci.*, 210, 433–437 (2002).
185. M. Nomura, T. Bin and S.I. Nakao, Selective ethanol extraction from fermentation broth using a silicalite membrane, *Sep. Purif. Technol.*, 27, 59–66 (2002).
186. T. Sano, M. Hasegawa, S. Ejiri, Y. Kawakami and H. Yanagishita, Improvement of the pervaporation performance of silicalite membranes by modification with a silane coupling reagent, *Microporous Mater.*, 5, 179–184 (1995).
187. H. Negishi, K. Sakaki and T. Ikegami, Silicalite Pervaporation membrane exhibiting a separation factor of over 400 for butanol, *Chem. Lett.*, 39, 1312–1314 (2010).
188. S. Li, V.A. Tuan, J.L. Falconer and R.D. Noble, Properties and separation performance of Ge-ZSM-5 membranes, *Micropor. Mesopor. Mat.*, 58, 137–154 (2003).
189. T.C. Bowen, H. Kalipcilar, J.L. Falconer and R.D. Noble, Pervaporation of organic/water mixtures through B-ZSM-5 zeolite membranes on monolith supports, *J. Membr. Sci.*, 215, 235–247 (2003).
190. V. Sebastian, J. Motuzas, R.W.J. Dirrix, R.A. Terpstra, R. Mallada and A. Julbe, Synthesis of capillary titanosilicalite TS-1 ceramic membranes by MW-assisted hydrothermal heating for pervaporation application, *Sep. Purif. Technol.*, 75, 249–256 (2010).
191. H.J.C. te Hennepe, D. Bargeman, M.H.V. Mulder and C.A. Smolders, Zeolite-filled silicone rubber membranes Part 1. Membrane preparation and pervaporation results, *J. Membr. Sci.*, 35, 39–55 (1987).

192. S. Yi, Y. Su and Y. Wan, Preparation and characterization of vinyltriethoxysilane (VTES) modified silicalite-1/PDMS hybrid pervaporation performance and its application in ethanol separation from dilute aqueous solution, *J. Membr. Sci.*, 360, 341–351 (2010).
193. M.D. Jia, K.V. Peinemann and R.D. Behling, Preparation and characterization of thin-film zeolite-PDMS composite membranes, *J. Membr. Sci.*, 73, 119–128 (1992).
194. N. Qureshi, M.M. Meagher and R.W. Hutkins, Recovery of butanol from model solutions and fermentation broth using a silicalite/silicone membrane, *J. Membr. Sci.*, 158, 115–125 (1999).
195. E.A. Fouad and X. Feng, Pervaporative separation of n-butanol from dilute aqueous solutions using silicalite-filled poly(dimethyl siloxane) membranes, *J. Membr. Sci.*, 339, 120–125 (2009).
196. H. Zhou, Y. Su, X. Chen, S. Yi and Y. Wang, Modification of silicalite-1 by vinyltrimethoxysilane (VTMS) and preparation of silicalite-1 filled polydimethylsiloxane (PDMS) hybrid pervaporation membranes, *Sep. Purif. Technol.*, 75, 286–294 (2010).
197. L.M. Vane, V.V. Namboodiri and T.C. Bowen, Hydrophobic zeolite-silicone rubber mixed matrix membranes for ethanol-water separation: Effect of zeolite and silicone component selection on pervaporation performance, *J. Membr. Sci.*, 308, 230–241 (2008).
198. E. Favre, Q.T. Nguyen and S. Bruneau, Extraction of 1-butanol from aqueous solutions by pervaporation, *J. Chem. Tech. Biotechnol.*, 65, 221–228 (1996).
199. Y. Huang, P. Zhang, J. Fu, Y. Zhou, X. Huang and X. Tang, Pervaporation of ethanol aqueous solution by polydimethylsiloxane/polyphosphazene nanotube nanocomposite membranes, *J. Membr. Sci.*, 339, 85–92 (2009).
200. I.F.J. Vankelecom, J.D. Kinderen, B.M. Dewitte and J.B. Uytterhoeven, Incorporation of hydrophobic porous fillers in PDMS membranes for use in pervaporation, *J. Phys. Chem. B*, 101, 5182–5185 (1997).
201. R.E. Kesting, *Synthetic Polymeric Membranes: A Structural Perspective*, John Wiley & Sons, Inc., New York, 1985.
202. Y. M. Wei, Z.L. Xu, F.A. Qusay and K. Wu, Polyvinyl alcohol/polysulfone (PVA/PSF) hollow fiber composite membranes for pervaporation separation of ethanol/water solution. *J. Appl. Polym. Sci.*, 98, 247–254 (2005).
203. Z.K. Xu, Q.W. Dai, Z. M. Liu, R.Q. Kou and Y.Y. Xu, Microporous polypropylene hollow fiber membranes. Part II. Pervaporation separation of water/ethanol mixtures by the poly(acrylic acid) grafted membranes, *J. Membr. Sci.*, 214, 71–81 (2003).
204. H.A. Tsai, H.C. Chen, W.L. Chou, K.R. Lee, M.C. Yang and J.Y. Lai, Pervaporation of water/alcohol mixtures through chitosan/cellulose acetate composite hollow fiber membranes, *J. Appl. Polym. Sci.*, 94, 1562–1568 (2004).
205. L.Y. Jiang, T.S. Chung and R. Rajagopalan, Dehydration of alcohols by pervaporation through polyimide Matrimid® asymmetric hollow fibers with various modifications, *Chem. Eng. Sci.*, 63, 204–216 (2008).
206. R.X. Liu, X.Y. Qiao and T.S. Chung, Dual-layer P84/polyethersulfone hollow fiber for pervaporation dehydration of isopropanol, *J. Membr. Sci.*, 294, 103–114 (2007).
207. Y. Wang, M. Gruender and T.S. Chung, Pervaporation dehydration of ethylene glycol through polybenzimidazole (PBI)-based membranes. 1. Membrane fabrication, *J. Membr. Sci.*, 363, 149–159 (2010).
208. i3 Nanotec, ZeoSep A membrane applications, available from: http://www.i3nanotec.com/Products/AMemb.html (accessed August 28, 2012).
209. Pervatech, Applications and preferred combinations, available from: http://www.pervaporation-membranes.com/Applications-and-preferred-combinations.html (accessed August 28, 2012).
210. P. Peng, B. Shi and Y. Lan, A review of membrane materials for ethanol recovery by pervaporation, *Sep. Sci. Technol.*, 46, 234–246 (2011).
211. http://www.pervaporation.org/manufacturers/index.html (accessed August 28, 2012).
212. W.J. Grooth, R.G.J.M. van der Lans and K.Ch.A.M. Luyben, Technologies for butanol recovery integrated with fermentations, *Process Biochem.*, 27, 61–75 (1992).
213. A. Pucci, P. Mikitenko and L. Asselineau, Three-phase distillation. Simulation and application to the separation of fermentation products, *Chem. Eng. Sci.*, 41, 485–494 (1986).
214. F. Lipnizki, R.W. Field and P.K. Ten, Pervaporation-based hybrid process: a review of process design, applications and economics, *J. Memb. Sci.*, 153, 183–210 (1999).

215. M. Ishida and N. Nakagawa, Energy analysis of a pervaporation system and its combination with a distillation column based on an energy utilization diagram, *J. Membr. Sci.*, 24, 271–283 (1985).
216. M. Matsumura, H. Kataoka, M. Sueki and K. Araki, Energy saving effect of pervaporation using oleyl alcohol liquid membrane in butanol purification, *Bioprocess Eng.*, 3, 93–100 (1988).
217. G.F. Tusel and A. Ballweg, Method and apparatus for dehydrating mixtures of organic liquids and water, US Patent 4405409, 1983.
218. P.O. Cogat, Dehydration of ethanol pervaporation compared with azeotropic distillation, R. Bakish (Ed.), *in Proceedings of the Third International Conference on Pervaporation Processes in the Chemical Industry*, Bakish Material Corporation, Englewood, NJ, USA, 305–316 (1988).
219. H.L. Fleming, Membrane pervaporation: separation of organic/aqueous mixtures, *Sep. Sci. Technol.*, 25, 1239–1255 (1990).
220. C.H. Gooding and F.J. Bahouth, Membrane-aided distillation of azeotropic solutions, *Chem. Eng. Comm.*, 35, 267–279 (1985).
221. H.E.A. Bruschke and G.F. Tusel, Economics of industrial pervaporation processes, *in Proceedings of the Conference on Membranes and Membrane Processes*, 581–586 (1986).
222. M.H.V. Mulder, J.O. Hendrikman, H. Hegeman, C.A. Smolders, Ethanol-water separation by pervaporation, *J. Membr. Sci.*, 16, 269–284 (1983).
223. R.W. Baker, Membrane technology and applications, 2nd edition, John Wiley & Sons, Inc., New York (2004).
224. T.S. Lee, D. Omstead, N.H. Lu and H.P. Gregor, Membrane separations in alcohol productions, *Ann. N.Y. Acad. Sci.*, 369, 367–381 (1981).
225. H.J. Huang, S. Ramaswamy, U.W. Tschirner, B.V. Ramarao, A review of separation technologies in current and future biorefineries, *Sep. Purif. Technol.*, 62, 1–21 (2008).
226. H.B. Park, C.H. Jung, Y.M. Lee, A.J. Hill, S.J. Pas, S.T. Mudie, E. Van Wagner, B.D. Freeman and D.J. Cookson, Polymers with cavities tuned for fast selective transport of small molecules and ions, *Science*, 318, 254–258 (2007).
227. S.H. Han, N. Misdan, S. Kim, C.M. Doherty, A.J. Hill, and Y.M. Lee, Thermally rearranged (TR) polybenzoxazole: Effects of diverse imidization routes on physical properties and gas transport behaviors, *Macromolecules*, 43, 7657–7667 (2010).
228. H.B. Park, S.H. Han, C. H. Jung, Y. M. Lee and A. J. Hill, Thermally rearranged (TR) polymer membranes for CO_2 separation, *J. Membr. Sci.*, 359, 11–24 (2010).
229. M. Calle and Y.M. Lee, Thermally rearranged (TR) poly(ether-benzoxazole) membranes for gas separation, *Macromolecules*, 44, 1156–1165 (2011).
230. C. Ba, M. Leo, K. Czenkusch, C. Rebeiro, D. Paul, B. Freeman, G. Mensitieri, *Thermally rearranged (TR) polymers as membranes for ethanol dehydration*. Paper presented at the *Twenty-First Annual Meeting of the North American Membrane Society*, Las Vegas, USA, 2011.
231. H. Wang, S.L. Liu, T.S. Chung, H.M. Chen, Y.C. Jean and K.P. Pramoda, The evolution of poly(hydroxyamide amic acid) to poly(benzoxazole) via stepwise thermal cyclization: Structural changes and gas transport properties, *Polymer*, 52, 5127–5138 (2011).
232. Y.K. Ong, H. Wang and T.S. Chung, Thermally rearranged (TR) Membranes for dehydration of biofuels and solvents, *in 2011 AIChE Annual Meeting*, Minneapolis, USA, 2011.
233. T.X. Yang, Y.C. Xiao and T.S. Chung, Poly-/metal-benzimidazole nano-composite membranes for hydrogen purification, *Energy Environ. Sci.*, 4, 4171–4180 (2011).
234. Y. Wang, T.S. Chung, H. Wang, Polyamide-imide membranes with surface immobilized cyclodextrin for butanol isomer separation via pervaporation, *AICHE J.*, 57, 1470–1484 (2011).
235. P. Sukitpaneenit, T.S. Chung, PVDF/nano-silica dual-layer hollow fibers with enhanced selectivity and flux as novel membranes for ethanol recovery, *Ind. Eng. Chem. Res.*, 51, 978–993 (2012).

11
Membrane Distillation

M. A. Izquierdo-Gil

Department of Applied Physics I, Faculty of Physics, University Complutense of Madrid, Spain

11.1 Introduction

There has been increasing interest in the conversion of biomass to fuel-grade ethanol in recent years due to the need to minimize oil imports in a period of increasing global oil consumption, the need for a renewable source of fuels, the need to minimize greenhouse gas (GHG) emissions caused by the use of fossil fuel, and political pressures. Biorefineries for conversion of biomass to ethanol can be categorized into three types: corn-to-ethanol, basic lignocellulosic biomass-to-ethanol, and integrated lignocellulosic biomass-to-ethanol and other co-products including the concept of integrated forest biorefinery [1].

In order for future biorefineries to be successful, it is crucial to pre-extract as many value-added co-products as possible using highly efficient separation methods [1]. *Extractive distillation* with ionic liquids and hyperbranched polymers, *adsorption* with molecular sieve and bio-based adsorbents, and three specific *hybrid* methods are potentially significant separation methods especially suitable for future biorefineries (including pre-extraction of hemicelluloses and conversion to other value-added chemicals, ethanol product separation and dehydration, and detoxification of fermentation hydrolyzates). The hybrid methods couple separation and fermentation and include extractive-fermentation, a membrane pervaporation-bioreactor, and a vacuum membrane distillation (VMD)-bioreactor.

Extractive-fermentation, the membrane pervaporation-bioreactor hybrid, and the VMD-bioreactor hybrid are very promising processes for removing inhibitory compounds and increasing ethanol yield. Vacuum membrane distillation is one of the membrane distillation process configurations.

Membrane distillation (MD) [2, 3] is a relatively new process that is being investigated worldwide as a low-cost, energy-saving alternative to conventional separation processes such as distillation and reverse osmosis (RO). The driving force for the transport process is the vapour partial pressure gradient across the membrane. The liquid-vapour equilibrium determines the separation process. The membrane is not selective from the point of view of the separation process.

Separation and Purification Technologies in Biorefineries, First Edition.
Edited by Shri Ramaswamy, Hua-Jiang Huang, and Bandaru V. Ramarao.

The main characteristics of this process are:

- The membrane must be porous and hydrophobic (that is, the membrane must not be wetted by the process liquids).
- The membrane does not alter the vapour-liquid equilibrium of the different components in the process liquids.
- The driving force for each component is a partial pressure gradient in the vapour phase.
- At least one side of the membrane must be in direct contact with the process liquids.
- No capillary condensation must take place inside the pores of the membrane.

The main benefits of membrane distillation versus other conventional separation processes may be summarized as follows:

- High quality of distillate. Theoretically, 100% of ions, macromolecules, colloids, cells, and other non-volatile components are rejected.
- Lower operating temperatures.
- Low-grade waste and alternative energy sources such as solar and geothermal energy may be used.
- Lower operating pressures.
- Fewer demands about membrane mechanical properties.
- Chemical interaction between membrane and process liquids is very reduced. The membrane acts merely as a support for the vapour liquid interface.

The membrane is constituted by the matrix and the pores as shown in Figure 11.1.

There are four different configurations for MD processes [2], depending on the way in which the partial pressure gradient in the vapour phase is established: direct-contact membrane distillation (DCMD), air gap membrane distillation (AGMD), sweeping gas membrane distillation (SGMD), and vacuum membrane distillation (VMD).

11.1.1 Direct-contact membrane distillation (DCMD)

In this configuration the permeate side of the membrane is in direct contact with a cold aqueous solution. The main problem of this configuration is low energy efficiency. Direct-contact membrane distillation is

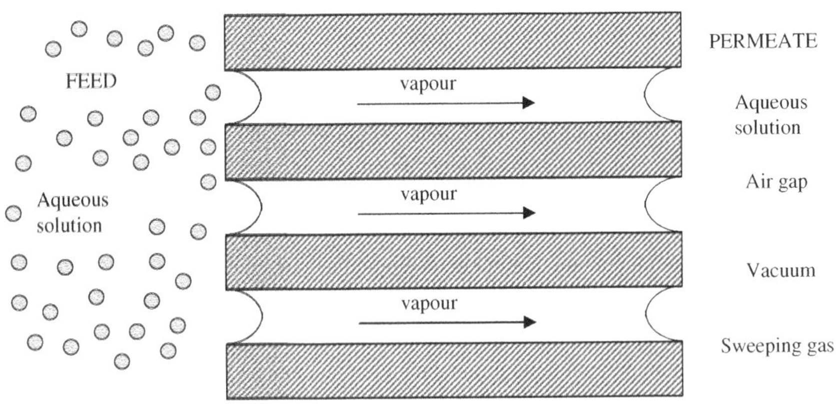

Figure 11.1 Schematic representation of membrane-solutions system

most suitable for applications in which the major permeating component is water, such as in desalination or the concentration of aqueous solutions. Indeed, DCMD desalination has been studied extensively and the performance of DCMD for concentrating fruit juices, blood and waste/process water has also been evaluated. Desalination was first performed with DCMD by Weyl in 1964 [4]. Weyl reported fluxes up to 1 kg/m^2h, which were in good agreement with theory, but these fluxes fell short of the 20 to 75 kg/m^2h fluxes commonly obtained in RO. With the advances made in membrane fabrication technology in the early 1980s came a renewed interest in MD. Fluxes as high as 75 kg/m^2h were measured in DCMD desalination [5, 6], which are competitive with the fluxes typically observed in RO. On the other hand, salt rejections of nearly 100% have been observed in DCMD desalination [5, 7], which cannot be accomplished with RO (at high fluxes).

Direct-contact membrane distillation has also been applied successfully to waste-water treatment, yielding a permeate that is less hazardous to the environment and a retentate that is concentrated in valuable chemicals. The process has been successfully applied to textile waste water contaminated with dyes [8], pharmaceutical waste water containing taurine [9], waste water contaminated with heavy metals [10], and sulfuric acid solutions rich in lanthane compounds [11]. It has also found success in areas where high-temperature applications lead to degradation of the process fluids.

Calabro *et al.* [12] and Kimura *et al.* [13] have shown DCMD to be effective in the concentration of fruit juices and Sakai *et al.* [14] have applied the process to the concentration of blood.

11.1.2 Air gap membrane distillation (AGMD)

One of the problems with DCMD is the relatively low efficiency of heat utilization. A large portion of the heat supplied to the feed solution is lost by conduction through the membrane. In this case, to improve energy efficiency, an air gap is set between the membrane and the cold surface. However, the mass transfer resistance is increased in this configuration, resulting in lower fluxes. Air gap membrane distillation is more versatile than DCMD because it is adaptable to more applications, due to the fact that permeate does not condense directly on the cold permeate solution, and for this reason there are fewer problems of membrane wetting. In DCMD the permeate must be dilute to prevent membrane wetting, while in AGMD the concentration of the condensed permeate is not a concern because it does not come into contact with the membrane. If DCMD were used to remove volatile components from an aqueous solution, an aqueous permeate solution would have to become contaminated, resulting in no net gain.

The AGMD process has been applied successfully to pure water production and concentration of various non-volatile solutes [13, 15, 16, 17]. The fluxes in AGMD water purification applications are typically as high as those obtained with DCMD (up to 75 kg/m^2h), decreasing linearly with the air gap thickness. Gostoli *et al.* [18, 19] have also examined the use of AGMD in ethanol-water separation. Ethanol preferentially vaporizes from the aqueous feed and is concentrated in the permeate. Permeates with ethanol concentrations up to double that of the feed were obtained with relatively low total fluxes (up to 4.7 kg/m^2h). A more recent investigation by Udriot *et al.* [20] has shown AGMD to be useful in breaking azeotropic mixtures of water and hydrochloric or propionic acids.

11.1.3 Sweeping gas membrane distillation (SGMD)

In this configuration an inert gas is blown over the membrane surface in the cold chamber and the permeate is condensed in an external condenser. Unfortunately, the condenser must do a lot more work in the SGMD configuration, because a tiny volume of permeate is vaporized in a large volume of sweep gas. As a result, very little work has been done with SGMD.

The sweeping gas configuration is the most difficult to describe, since in the downstream compartment, in contrast to other MD configurations, none of the variables remains constant along the module. All temperatures, concentrations, as well as heat and mass transfer rates change during the gas progression within the module. Additionally, in this configuration partial condensation inside the module should be taken into account [21].

In general, SGMD has received very little attention. Basini *et al.* [22] were the first to study this configuration for the desalination of water. They showed that the flux in SGMD is independent of the temperature of the sweep gas. Additionally, the SGMD flux increases through a maximum as the gas velocity increases, but then begins to decrease. Later SGMD was proposed for the concentration of sucrose solutions and the separation of ethanol–water mixtures.

The sweeping gas membrane distillation (SGMD) configuration holds great promise for the future, because as mentioned earlier it combines a relatively low conductive heat loss with a reduced mass transfer resistance. It provides much higher permeate fluxes than air-gap membrane distillation (AGMD) while maintaining high temperature polarization coefficient and evaporation efficiency. The advantages of SGMD over direct-contact membrane distillation (DCMD) could include better selectivity performance, smaller temperature polarization, and higher evaporation efficiency, albeit at a lower permeate flux.

11.1.4 Vacuum membrane distillation (VMD)

Applying vacuum to the permeate side of the membrane larger pressure gradients are achieved comparing with other MD configurations.

The VMD process is similar to and is often confused with pervaporation. The fundamental difference between VMD and pervaporation is the role that the membrane plays in the separation. Vacuum membrane distillation employs a microporous membrane that acts only as a support for a vapour-liquid interface. While the VMD membrane may impart some selectivity based on individual Knudsen diffusivities of diffusing species, the largest degree of the separation is determined by vapour-liquid equilibrium conditions at the membrane-solution interface. On the other hand, pervaporation uses a dense membrane, and the separation is based on the relative solubility and diffusivity of each component in the membrane material. Because of these differences, VMD typically achieves fluxes that are several orders of magnitude higher than pervaporation fluxes.

One of the benefits of VMD relative to the other MD configurations is that conductive heat loss through the membrane is negligible. As a separation tool, VMD is most often used to remove volatile components from dilute aqueous solutions [23–27]. Bandini *et al.* [25, 26] obtained permeates with ethanol concentrations up to ten times that of the feed, much better than for the AGMD separation of ethanol-water. Special care must be taken in VMD to prevent membrane wetting, because $\Delta P_{interface}$ (defined later in the "Basic principles of membrane distillation" section) is typically higher in VMD than in the other MD configurations.

The different MD configurations are shown in Figure 11.2.

11.2 Membrane distillation market and industrial needs

Industrial applications of MD may be classified into four main groups [28]:

- pure water production;
- waste water treatment;
- food industry;
- concentration of organic and biological solutions.

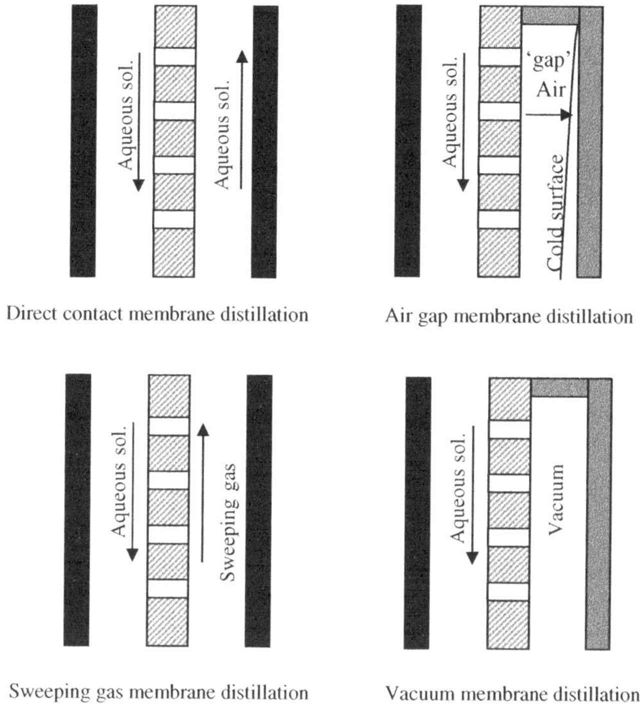

Figure 11.2 *Schematic representation of different MD configurations*

11.2.1 Pure water production

High-purity water production represents one of the main industrial applications of MD. This process has a very high rejection rate of 100% for non-volatile dissolved compounds, significantly higher than can be achieved with reverse osmosis (~90%).

In 1982, Gore proposed the use of two different MD membrane modules for desalting NaCl aqueous solutions: a flat membrane for AGMD (production rate: 7 l/m^{2}h, t_{feed} = 30 °C, $t_{distillate}$ = 20 °C), and a spiral-wound module (production rate: 3 l/m^{2}h, t_{feed} = 30 °C, $t_{distillate}$ = 20 °C) [29]. A few years later, literature related to the use of MD in desalination processes increased exponentially [30–33].

Direct-contact membrane distillation carried out by PTFE (polytetrafluoroethylene) microporous membranes was considered by Godino *et al.* [34] for obtaining pure water from NaCl brines. L. Banat and Simandl [35] used an AGMD module for carrying out desalination experiments on PVDF (polyvinylidene fluoride) membrane sheets. Very pure water with less than 5 ppm TDS was obtained in all experiments. The experimental analysis of Lawson and Lloyd [36] indicated that DCMD is a viable process for seawater desalination, with fluxes reaching up to 2 mol/m^{2}s working at feed temperature of 75 °C and distillate temperature of 20 °C; these fluxes are two times higher than commercial RO systems. In addition, concentration measurements carried out on the permeate stream revealed a quasi-total rejection of NaCl molecules. First assessments of the process economics gave indications that the use of PTFE membranes for desalting seawater raises the costs of MD to an excessively high level [31] mainly due to the elevated price of the commercial membrane modules; however, this trend is now reversing. The most interesting prospects for the development of membrane distillation technology are probably related to the possibility of combining them with other conventional pressure-driven membrane processes.

11.2.2 Waste water treatment

The possibility of removing heavy metals from waste water has been discussed by Zolotarev and colleagues [37]. In particular, a rejection coefficient close to unity was obtained by treating aqueous solutions of nickel sulphate in the range of 0.1–3.0 N. Membrane distillation has been applied in the recovery of HCl from acidic spent solutions generated by cleaning of electroplated surfaces. It has been used to concentrate sulfuric acid obtained after apatite phosphogypsum extraction used to recover lanthane compounds. The concentration process was protracted up to 40% of H_2SO_4; lanthane compounds were precipitated by cooling [11].

Membrane distillation has been investigated as treatment method for radioactive liquid wastes, generated from the nuclear industry, or by other end-uses of radioactive materials (hospitals, nuclear R&D centres and so forth) [38].

Membrane distillation operating under vacuum is an effective method for removing volatile organic components from dilute aqueous solutions [11, 26, 39] such as acetone and isopropanol, ethanol, methylterbutylether, ethylacetate, methylacetate, and benzene traces from contaminated water.

11.2.3 Concentration of agro-food solutions

Membrane distillation works at relatively low feed temperature: this makes it especially suitable for the food industry, where solutions are sensitive to high temperatures. With respect to standard concentration methods (generally a multistage vacuum evaporation) that involve a significant energy consumption and degradation of the organoleptic properties of juices, membrane distillation process represents a competitive alternative, able to increase the quality of concentrates.

Direct-contact membrane distillation has been successfully tested in the concentration of many juices [40–42]. Other authors [43] developed a model to predict the flux and outlet temperatures in the process for concentrating blackcurrant juice by DCMD. The model was based on heat and mass balances along a tubular membrane module. The mass balance was based on the 'dusty-gas model' (DGM) and the model included correction for heat and concentration polarization by the use of empirical correlations for the heat and mass transfer coefficients. The aim of setting up a model describing the flux of DCMD on black currant juice *a priori* was successfully accomplished.

The DCMD process has been used to investigate the concentration of apple juice [44]. Results showed that at a constant temperature of the juice in the hot cell, an increase in the permeate flux of DCMD resulted in reducing the temperature of cooling water in the cold cell. Increasing the temperature of juice in the hot cell reduced the influence of the cooling water temperature in the cold cell on the permeate flux of DCMD. The concentration of soluble substances in concentrate and hydrodynamic conditions in the experimental equipment was also studied. In the concentration of apple juice, 50% of solids content was obtained when the permeate flux reached about 9 l/m^2h. Further concentration of juice to 60–65% solids resulted in reduced productivity (3.8–3.0 l/m^2h) and therefore a decrease in the biological value of the concentrate.

In all cases, the concentration obtained (50–60 °Brix) is significantly higher than that achieved by pressure-driven membrane processes, such as RO. On the other hand, in the range of 10–20 °Brix, the MD fluxes at 25–30 °C were of the order of 1–3 l/m^2h, much lower than those measured for RO at the same temperature (10–15 l/m^2h). This behaviour may be observed in the Figure 11.3, where data corresponding to concentrating sucrose solutions have been added to the plot for comparison. They were measured by using DCMD, PTFE membranes, at $t_{permeate}$ = 22.5 °C and t_{feed} = 47.5 °C [45]. It may be observed that they are very similar to those obtained by Drioli *et al.* [40] with orange juice, given by a dotted line

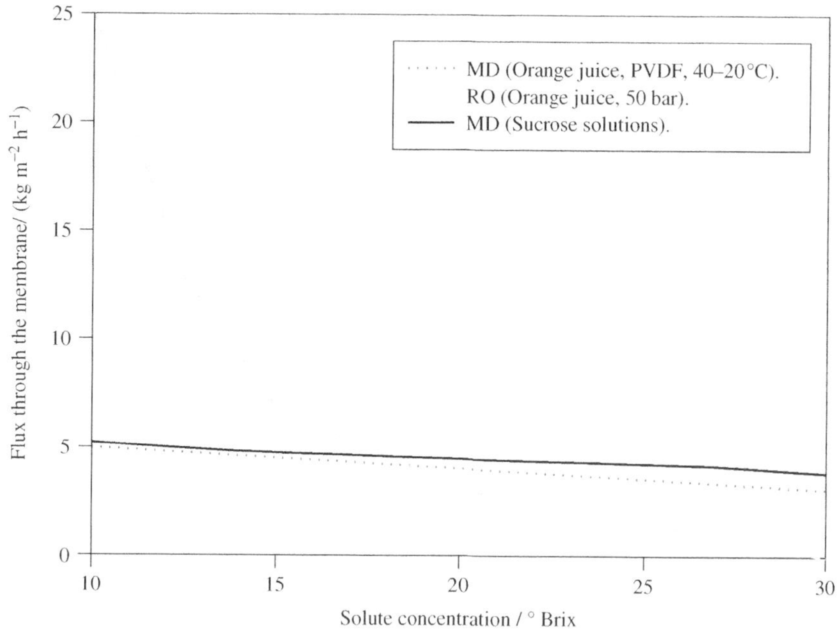

Figure 11.3 *Comparison between membrane distillation (MD) and reverse osmosis (RO) for orange juice concentration and membrane distillation for concentrating sucrose solutions*

in Figure 11.3. In general, a loss in taste and flavours of the concentrate juice was also observed, due to the evaporative nature of MD process.

11.2.4 Concentration of organic and biological solutions

Membrane distillation has been applied in the concentration of organic and biological solutions for selective extraction of volatile solutes and solvents. Blood and plasma have been treated by MD in order to promote a solute-free extraction of water from biomedical solutions without loss in quality [14, 46].

It is also used to recover volatile compounds such as ethanol. Vacuum membrane distillation is quite similar to pervaporation, the only difference being that the separation factor is established by vapour-liquid equilibrium of the feed solution, which is not affected by the membrane used [18]. Alcohol is produced by fermentation of biomass in batch fermentors. The ethanol excess in a fermentation broth inhibits the process, possibly leading to a zero rate of bioconversion. The integration of MD downstream of the fermentor improves the process. Due to the difference in volatility between water and ethanol, alcohol can be removed also using a non-selective microporous membrane [47–49]. Gryta and co-workers [50] observed that, in the case of fermentation combined with MD, an efficiency of 0.47–0.51 (g EtOH)/(g of sugar) and a production rate of 2.5–4 (g EtOH)/dm^3 h was achieved in relation to 0.35–0.45 (g EtOH)/(g of sugar) and 0.8–2 (g EtOH)/dm^3 h obtained in the classical batch fermentation. The ethanol flux measured in MD varied in the range of 1–4 (kg EtOH)/m^2 per day and was dependent on the temperature and the feed composition. They used a membrane distillation bioreactor where porous capillary polypropylene membranes were applied for separating volatile compounds, including ethanol and other inhibitors, from feed (broth), leading to an increase in the productivity and the sugar-to-ethanol conversion rate, as has been indicated earlier. Vacuum membrane distillation is commercially competitive because of its high selectivity of ethanol over water, large flux, high thermal efficiency and low energy cost [51].

Air-gap membrane distillation was tested by Banat and Simandl [52] for ethanol–water separation using PVDF membranes. The upper feed concentration tested was 10 wt.% ethanol. Within the feed temperature range of 40–70 °C, ethanol selectivity of 2–3.5 was achieved.

11.3 Basic principles of membrane distillation

The basic separation process principle of membrane distillation is a combined, simultaneous heat and mass transfer mechanisms.

11.3.1 Mass transfer

The "dusty-gas model" (DGM) is a general model for mass transport through porous media [53]. The porous medium is treated as one component of the gas mixture, consisting of giant molecules held fixed in space, and the highly developed kinetic theory of gases is applied to this supermixture. Initially the model was developed by James Clerk Maxwell and it may be applied to model other phenomena.

Different independent modes or mechanisms may be present in the gas transport through porous media, as follows:

- Free-molecule or Knudsen flow, in which the gas density is so low that collisions between molecules can be ignored compared to collisions of molecules with the walls of the porous medium or tube.
- Viscous flow, in which the gas acts as a continuum fluid driven by a pressure gradient, and molecule-molecule collisions dominate over molecule-wall collisions.
- Continuum diffusion, in which the different species of the mixture move relative to each other under the influence of concentration gradients, temperature gradients, or external forces. Here molecule-molecule collisions again dominate over molecule-wall collisions.
- Surface diffusion, in which molecules move along a solid surface in an adsorbed layer.

The different transport mechanisms combine in a similar way to the electric resistances in an electric circuit. The diffusive fluxes combine in series and the total diffusive flux combines in parallel with viscous and surface flows. This mass transfer resistances combination may be represented by an electric circuit like that shown in Figure 11.4.

The DGM, which was first described by Maxwell in 1860, does not present any new equations for diffusion in porous media; indeed, the same equations can be derived from momentum transfer arguments. It is simply a more theoretically sound way to regard diffusion through porous media. It is based on well developed kinetic theory rather than the heuristic arguments required by the momentum transfer method.

In MD, the molecule-membrane interaction is low and the area of surface diffusion is relatively small compared to the pore surface, so the surface diffusion is considered negligible in MD processes.

In its most general form, the DGM applicable to MD (neglecting surface diffusion) is given by equations:

Diffusive flux:

$$\frac{J_i^D}{D_{ie}^k} + \sum_{j=1\neq i}^{n} \frac{y_j J_i^D - y_i J_j^D}{D_{ije}^o} = -\frac{1}{RT}\nabla p_i \tag{11.1}$$

Viscous flux:

$$J_i^v = -\frac{B_o p_i}{RT\mu}\nabla P$$

where the effective diffusion coefficients (Knudsen and ordinary diffusion coefficients) are given by:

$$D_{ie}^{k} = K_0 \sqrt{\frac{8RT}{\pi M_i}} \quad ; \quad D_{ije}^{0} = K_1 D_{ij}^{0} \tag{11.2}$$

J_i^D is the diffusive molar flux given by the generalized Maxwell's equation. B_0, the viscous flux parameter, is a constant characteristic of the medium alone. K_0 is related in first approximation to a geometric constant characteristic of the dust particles and a quantity that depends on the angular scattering pattern with which the gas molecules rebound from the dust particles [53, 54]. It could be determined for a porous medium if the geometry and scattering law are known, but it is a complex mathematical task. The best option is to measure those parameters (K_0 and B_0) by means of gas permeation experiments. $D_{ij}{}^0$ is the ordinary diffusion coefficient, J_i^v is the viscous flux, P is the total pressure, p_i is the partial pressure of component i, μ is the fluid viscosity, T is the temperature, M_i is the molecular weight of component i, y_j the gas molar fraction of component j and R the gas's constant.

Assuming a structural model for the membrane with non-interconnected cylindrical pores, K_1, K_0, and B_0 depend on the geometry in the following way [2, 53, 54]:

$$B_0 = \frac{\varepsilon r^2}{8\chi}; \quad K_0 = \frac{2\varepsilon r}{3\chi}; \quad K_1 = \frac{\varepsilon}{\chi} \tag{11.3}$$

where r is the pore radius, ε is the porosity (void volume fraction) and χ is the tortuosity factor (effective pore length = $\chi \cdot$ pore length).

For each of the mass transport mechanisms, the associated resistance arises from collisions between diffusing molecules and either other molecules or the pore walls of the membrane. In a large volume of a pure gas, the mean free path of a molecule, λ is defined as the average distance the molecule travels between two successive collisions. The value of λ can be calculated from kinetic theory. The predominance, the coexistence or transition regime may be determined from Knudsen number, defined as:

$$\text{Kn} = \frac{\lambda}{d_{\text{poro}}} \tag{11.4}$$

When $\text{Kn} \ll 1$ the molecule-molecule collisions are predominant, whereas when $\text{Kn} \gg 1$ then molecule-pore wall collisions are predominant.

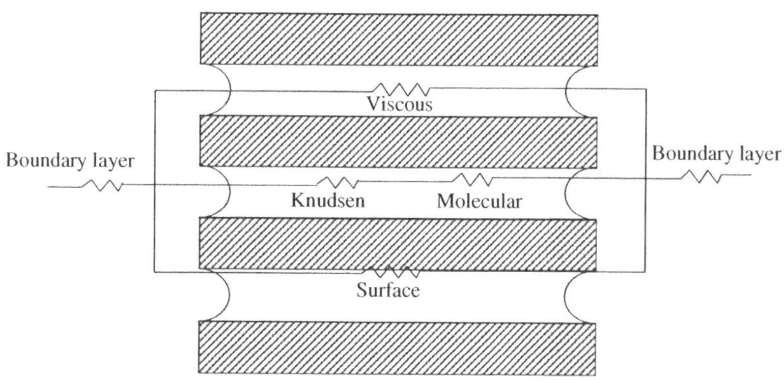

Figure 11.4 *Electrical circuit analogue for combining the different mass transfer mechanisms*

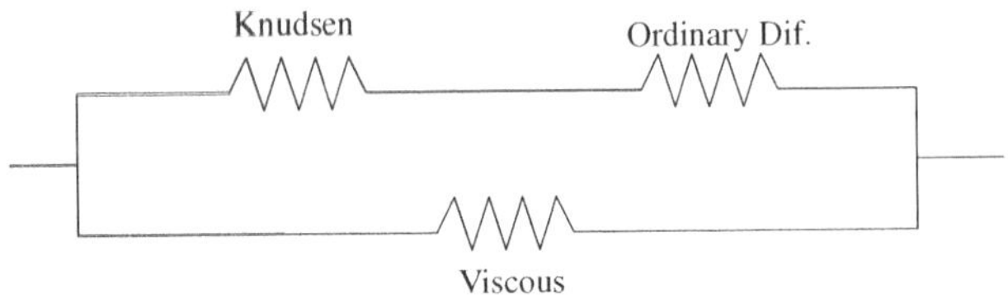

Figure 11.5 *Electrical circuit analogue neglecting the surface diffusion*

Table 11.1 *Simplified transport equations*

Configuration	Assumption	Transport equation	
VMD	$d_{pore} \ll \lambda$	$J_i = -\frac{D_{ie}^k}{RT} \nabla p_i$	(11.5)
AGMD	Air stagnant film, $\gamma = 2{,}334$, if only water diffuses	$J_i = -P D_{i-air,e}^0 \frac{T^{\gamma-1/\gamma}}{R \lvert p_{air} \rvert_{ln}} \nabla p_i$	(11.6)
DCMD (transition molecular-Knudsen)	$d_{pore} \sim \lambda$	$J_i = -\frac{1}{RT} \left[\frac{1}{D_{ie}^k} + \frac{y_{air}}{D_{ije}^0} \right]^{-1} \nabla p_i$	(11.7)
DCMD (without air, transition Knudsen-viscous and gas permeation)	Knudsen resistance dominates and there is viscous flux	$J_i = -\frac{1}{RT} \left[D_{ie}^k \nabla p_i + B_0 \frac{p_i}{\mu} \nabla P \right]$	(11.8)

Assuming the surface diffusion contribution to be negligible, the scheme shown in Figure 11.4 may be simplified as shown in Figure 11.5.

Depending on the MD configuration, the general equations for the transport (11.1) may be simplified, depending on which mechanisms are-to be considered. Table 11.1 shows the simplified transport equations for the different MD configurations and assumptions:

In the VMD configuration, water λ may reach very high values, because it is inversely proportional to P. In AGMD, air solubility in water is so low that air flow may be neglected, and the air may be treated as a stagnant film. For membranes with small pores filled with air, the molecule-pore wall collisions begin to be as frequent as molecule-molecule collisions, so transport mechanisms through the membrane may be considered to be in the Knudsen-molecule transition region. For DCMD experiments with previously deaerated systems and gas permeation experiments, the Knudsen-viscous transition describes the mass flows adequately. In SGMD the Stefan–Maxwell equations provide a correct description for the multicomponent diffusion.

11.3.2 Concentration polarization phenomena

Usually the bulk solution's concentration differs from the concentration close to the membrane surface. This phenomenon is known as concentration polarization. This effect may be more or less significant depending on the MD configuration and the operating conditions. In order to understand this phenomenon better, the concentration and temperature profiles for a DCMD system (where feed is considered to be water or an aqueous solution of non-volatile solute) are shown in Figure 11.6. The subscript *1* indicates the feed side and subscript *2* refers to permeate side. The subscripts *b* and *m*, indicate bulk solution and membrane interface, respectively. Q represents the heat flux and x denotes the corresponding mole fraction.

For practical purposes it is often convenient to perform MD with pure water, because boundary-layer resistance to mass transfer can be ignored. However, in a real MD separation process, concentration polarization must be given special attention because the boundary layer not only increases the overall resistance to mass transfer, but also the concentration of solute at the membrane surface can be too high causing spontaneous wetting of the membrane. The boundary layers can increase the overall resistance to mass transfer. Such mass transfer through the liquid phase can be adequately described by the film theory model [55, 56].

11.3.3 Heat transport

Similarly a temperature polarization phenomenon can take place in MD processes as shown in Figure 11.6. Temperatures measured in bulk solutions in contact with a membrane are not the same as those in the liquid-vapour interface, which are not measurable experimentally (that is, ΔT_{bulk} is not the same as $\Delta T_{membrane}$). This phenomenon may be represented in an electrical circuit analogue as shown

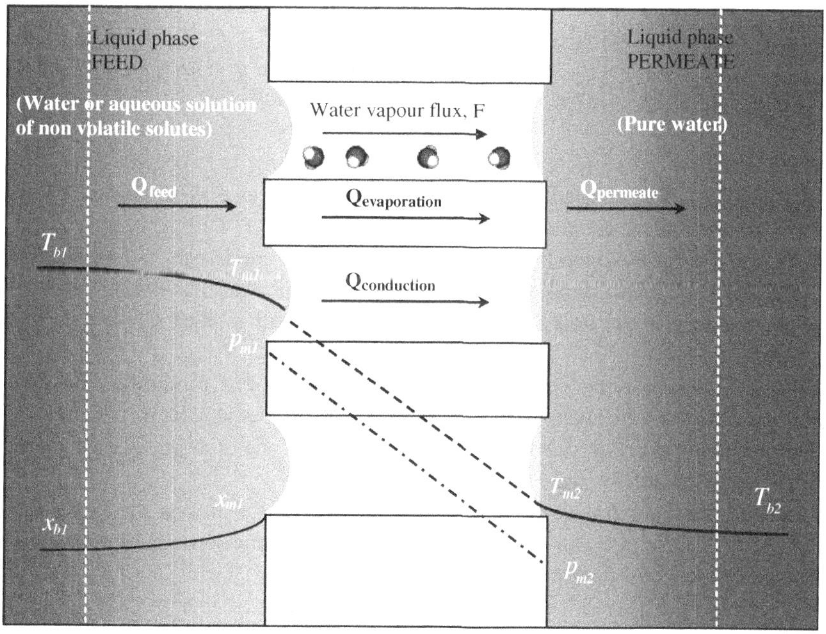

Figure 11.6 Concentration and temperature profiles for a DCMD system

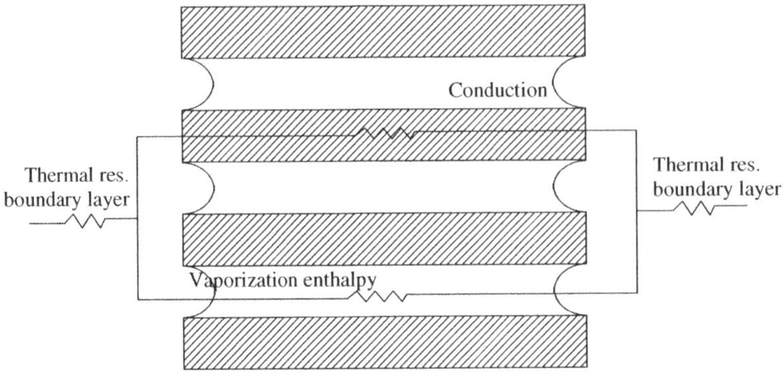

Figure 11.7 *Electrical circuit analogue for combining the different heat transfer mechanisms*

in Figure 11.7. The heat transfer in the boundary layers shown in the figure constitutes the main limiting factor in transport efficiency.

The total heat flux through the membrane is given by

$$Q = \left[\frac{1}{h_f} + \frac{1}{h_m + {}^{J\Delta H_v}/_{\Delta T_m}} + \frac{1}{h_p} \right]^{-1} \Delta T_{bulk} \tag{11.9}$$

where h_f and h_p are the heat-transfer coefficients at the feed and permeate solutions respectively, and ΔH_v is the evaporation enthalpy.

In order to quantify the effect of the temperature polarization phenomenon, the temperature polarization coefficient is defined by the following expression:

$$TPC = \frac{T_{m1} - T_{m2}}{T_{b1} - T_{b2}} = \frac{\Delta T_{membrane}}{\Delta T_{bulk}} \tag{11.10}$$

It must be as close to 1 as possible. Usually it is between 0.4 and 0.7. It represents the fraction of the total thermal driving force, ΔT_{bulk}, used to generate the mass transfer driving force, $\Delta T_{membrane}$.

The heat transfer coefficient is estimated from the Nusselt number. Different correlations may be used to determine the Nusselt number depending on the hydrodynamic conditions. Some of them are shown in Table 11.2 [57]. In Table 11.2, Nu denotes the Nusselt number, Re the Reynolds number, Pr the Prandlt number, d the tube diameter or the hydraulic diameter for noncircular tubes, L the tube length, and μ the fluid viscosity.

High recirculation fluxes may be achieved by using hollow fibre modules. The Reynolds number increases and the efficiency of the heat transfer is improved. By using channel spacers [58] membrane distillation performance is also improved. The flux is destabilized, creating eddies in the laminar flux favouring the momentum, heat and mass transfer.

As indicated earlier, the heat transfer in the boundary layers constitutes the main limiting factor in the transport efficiency. The temperature polarization coefficient may be used as an indirect index of the transport efficiency in membrane distillation.

11.3.4 Liquid entry pressure

As indicated in the section 11.1, the membrane must not be wetted by the process liquids. This is an important limitation in MD processes. The pressure difference at the interface must therefore not overcome

Table 11.2 Correlations for Nusselt number [57]

Equation	Conditions
$Nu = 3.66 + \frac{0.067\,(d/L)\,Re\,Pr}{1 + 0.04\left[(d/L)\,Re\,Pr\right]^{2/3}}$	Laminar flux
$Nu = 1.86\,(Re\,Pr\,(d/L))^{0.33}$	Laminar flux (Short tubes)
$Nu = 0.023 Re^{0.8} Pr^{0.33} (\mu/\mu_w)^{0.14}$	Turbulent flux Re>6000 y L/d>
$Nu = 0.036 Re^{0.8} Pr^{0.33} (d/L)^{0.055}$	Turbulent flux (Short tubes)

the liquid entry pressure. If $\Delta P_{interface}$ exceeds ΔP_{entry} the liquid can penetrate into and through the membrane pores. Once a pore has been penetrated it is said to be 'wetted' and the membrane must be completely dried and cleaned before the wetted pores can once again support a vapour-liquid interface [59]. Decreasing the hydrostatic pressure on the membrane will not restore the membrane to its un-wetted state.

Wetted pores allow liquid to pass directly through the membrane, which can result in several problems depending on the type of MD being used and the liquid pressures. For example, in DCMD and AGMD, the feed solution can flow directly across the membrane through the wetted pore, contaminating the permeate. If a large number of pores become wetted, permeate quality will deteriorate. This problem may be avoided in DCMD by keeping the hydrostatic pressure of the permeate higher than that of the feed. Then, if membrane pores become wetted, permeate will flow across the membrane to the feed reducing the overall flux but maintaining permeate quality. However, the goal of any MD system design should be to completely prevent pore wetting. Schneider *et al.* [60] recommend that a maximum pore radius of 0.5 to 0.6 μm be used in practice to ensure that process pressure and temperature fluctuations do not result in membrane wetting. In general, ΔP_{entry} may be determined experimentally [61]. Its dependence on the process parameters (type of membrane, temperature, type of alcohol, and alcohol concentration) was studied because these parameters are very important in membrane distillation processes. It was observed that the liquid entry pressure is strongly dependent on the alcohol concentration of the aqueous solution, on the type of alcohol used, and on the temperature. A linear relationship between the LEP value and its corresponding liquid surface tension was observed only within the range of alcohol concentration where the dispersion component of surface tension of the solution remained practically constant.

11.4 Design and simulation

A MD module design for a particular separation process is a trade-off of between a number of factors, such as cost, membrane performance—including thermal stability, selectivity, and thermal conductivity. There are number of different hydrophobic membrane modules available in the market, which have been extensively investigated during the past years.

A large variety of membrane configurations, including flat sheet (plate-and-frame and spiral wound modules) and tubular (tubular, capillary, and hollow fibre modules) have been tested in MD applications.

Most of the laboratory-scale modules are designed for use with flat-sheet membranes due to their versatility and simplicity of the preparation process; from an industrial standpoint, hollow fibre modules are more attractive due to their higher specific surface area. The choice of a module is usually determined by economic and operating conditions. An important criterion is based on an efficient control of concentration polarization and membrane fouling.

In plate-and-frame modules, the membranes, the porous support plates and the spacers are stacked between two endplates and placed in an appropriate housing. In this configuration, the packing density is about 100–400 m^2/m^3, depending on the number of membranes used.

In spiral-wound modules, the feed-flow channel spacer, the membrane and the porous support are enveloped and rolled around a perforated central collection tube. The feed solution moves in axial direction through the feed channel across the membrane surface. The permeate flows radially toward the central pipe.

A tubular membrane module consists of membrane tubes placed into porous stainless steel of fibre glass reinforced plastic pipes. The diameter of tubular membranes typically varies between 1.0 and 2.5 cm, with a packing density of about 300 m^2/m^3. In MD operations, such modules are used for highly viscous fluids; they also allow high feed-flow rates that reduce fouling tendency and polarization phenomena. In a capillary membrane module, a large number of capillary membranes (inner diameter of 0.2–3 mm) are arranged in parallel as a bundle in a shell tube; packing density is in the order of 600–1200 m^2/m^3.

Hollow-fibre modules are based on the same idea, but differ in the dimensions of the tubular membranes. In this case, the outer diameter typically ranges between 50 and 100 μm, and several thousands of fibres are installed in the vessel. This configuration has the highest packing density ($\sim$3000 m^2/m^3).

The main characteristic parameters of MD membranes are:

- Pore size. Sizes ranging from 100 Å to 1 μm.
- Porosity. This is one of the major factors affecting the mass transfer rate.
- Liquid entry pressure, because interfacial ΔP must be lower than the liquid entry pressure in order to avoid the wetting of membrane pores.

The materials used for MD membranes must be polymers with low surface tension and high chemical resistance. Materials such us PP (polypropylene), PTFE (polytetrafluoroethylene) and PVDF (polyvinylidenefluoride) meet these requirements.

Characteristic parameters of the different mass transport mechanisms (described in Section 11.3) may be estimated from gas permeation experiments as indicated previously. Gas permeation experiments using helium, air and argon and direct-contact membrane distillation (DCMD) experiments using distilled water were reported [62]. The characteristic parameters of the Knudsen and Poiseuille transport mechanisms were determined from gas permeation experiments. Such parameters were extrapolated in order to obtain the values corresponding to water vapour and these were used to estimate theoretical fluxes in DCMD processes employing two different models—one proposed previously by Schofield *et al.* (with some improvements) and another based on the 'dusty-gas' literature. In both models, the different transport mechanisms: ordinary diffusion, Knudsen flow and Poiseuille flow were taken into account. A very good agreement between the experimental fluxes and their theoretical predictions was found. A comparison between both models was also carried out. It was shown that in both models the viscous flow could be neglected under the operating conditions studied. The values of the parameter K_l, obtained from the fit of the experimental DCMD data using any of the models, were about 0.42. This parameter corrects the diffusion coefficient for ordinary diffusion transport. Therefore, it is possible to simulate MD experiments if such characteristic parameters are known from gas permeation experiments.

An experimental and theoretical investigation of the influence of concentration polarization and temperature polarization on the flux and selectivity of binary aqueous mixtures of ethanol was presented

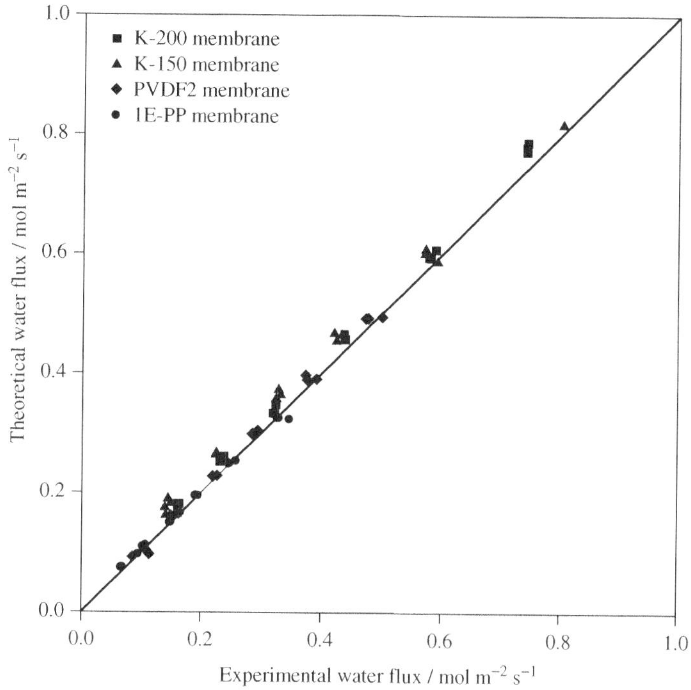

Figure 11.8 *Comparison of theoretical and experimental water molar fluxes in vacuum membrane distillation of aqueous binary mixtures of ethanol at different feed temperatures. Aqueous binary mixtures of ethanol (0.25 wt.%). Linear velocity through the cell: 2.65 m/s*

for vacuum membrane distillation processes [63]. Experimental results included changes in the following parameters: nature of solutions, membrane material and pore size, feed temperature, recirculation flow rate. One method was proposed in order to evaluate the concentration polarization effects from the fit of the experimental data. General models taking into account Knudsen and viscous flows were proposed, but viscous contribution was shown to be negligible under the operating conditions. Theoretical fluxes were therefore estimated using Knudsen model and a good agreement with the experimental data was found, as shown in Figures 11.8 and 11.9, where water and ethanol fluxes, respectively, are shown for the different membranes used and for the experiments changing feed temperature. The feed bulk temperature values used were: 20, 25, 30, 35, 40 and 45 °C.

11.5 Examples in biorefineries

Like the membrane pervaporation-bioreactor hybrid, the VMD-bioreactor hybrid process is also suitable for separation of ethanol and the other inhibitory compounds from fermentation broths [49, 50, 52, 63–65].

Continuous ethanol fermentation of concentrated glucose and molasses solutions was coupled with membrane distillation using a PTFE ethanol stripping module by Calibo *et al*. [49]. Experimental results indicated that the PTFE module can remove a high concentration of ethanol from the fermentation broth and thus maintain a low ethanol concentration in the broth, thereby alleviating the problem of product inhibition. Accordingly, the product yield and the specific ethanol production rate were increased. During the continuous fermentation runs, long-time operation using the PTFE module was found to be possible

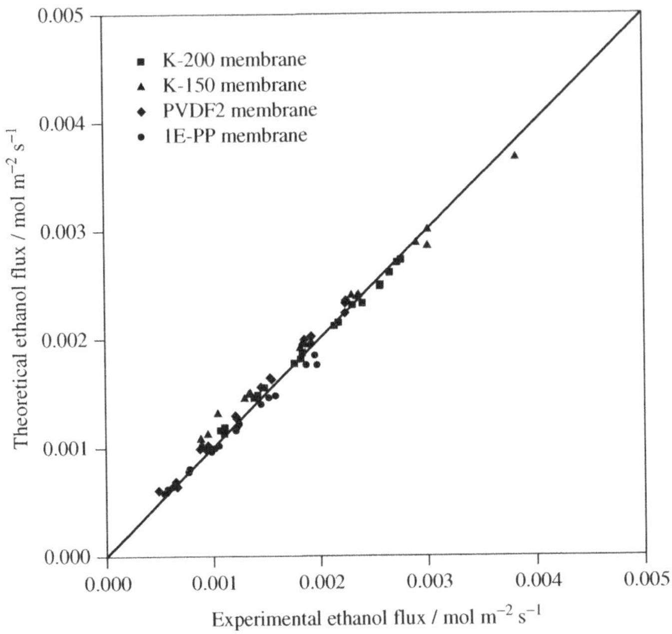

Figure 11.9 *Comparison of theoretical and experimental ethanol molar fluxes in vacuum membrane distillation of aqueous binary mixtures of ethanol at different feed temperatures. Aqueous binary mixtures of ethanol (0.25 wt.%). Linear velocity through the cell: 2.65 m/s*

(430 h using the glucose medium and 695 h using the molasses medium) and no significant change in the ethanol separation performance was observed. Although cell flocculation became undetectable when a concentrated molasses medium containing 316 g/l sugar solution was used, the ethanol separation performance of the ethanol stripper was not adversely affected by the presence of the free cells. This suggests that clogging of the membrane pores by cells or other particulates is not a major problem when using the PTFE module in continuous ethanol fermentation.

Banat and Simandl [52] examined air-gap membrane distillation as a possible technique for ethanol–water separation using PVDF membranes. The composition and flux of the permeate were monitored as a function of feed concentration, feed temperature, feed flow rate, cooling temperature and cooling flow rate. The effect of salt addition to the feed mixture was also examined. The upper feed concentration tested was 10 wt.% ethanol. Within the feed temperature range of 40–70 °C, ethanol selectivity of 2–3.5 was achieved. Two versions of a general mathematical model were solved numerically for the ethanol–water system; one did not include temperature and concentration polarization effects while the other did. Good agreement between experimental and predicted values was obtained with the latter version of the model. The operating variable that affected the permeate flux most significantly was the feed temperature. This is due to the exponential relationship between vapour pressure and temperature. Changes in other operating conditions had less effect upon flux. Modelling of the process indicated that concentration and temperature polarization occurred to a significant extent. The model version that neglected these effects did not adequately predict the experimental data at lower feed flow rates. There was a maximum selectivity achievable for each feed concentration and flow rate. The feed temperature requires optimization to achieve the maximum possible selectivity. The addition of salt increased ethanol selectivity significantly with only a slight decrease in total permeate flux.

Grypta *et al*. [50] produced ethanol in a membrane distillation bioreactor where porous capillary polypropylene membranes were applied for separating volatile compounds, including ethanol and other inhibitors, from the feed (broth), leading to an increase in the productivity and the sugar-to-ethanol conversion rate. The batch fermentation combined with the removal of ethanol from the broth using the membrane distillation process was investigated. The elimination of those compounds allows an increase in the productivity and the rate of conversion of sugar to ethanol, because they act as inhibitors. In the case of fermentation combined with MD, as was indicated above, the efficiency of 0.47–0.51 (g EtOH)/(g of sugar) and the production rate of 2.5–4 (g EtOH)/dm^3 h was achieved in relation to 0.35–0.45 (g EtOH) /(g of sugar) and 0.8–2 (g EtOH)/dm^3 h obtained in the classical batch fermentation. The ethanol flux obtained in membrane distillation varied in the range of 1–4 (kg EtOH)/m^2 per day and was dependent on the temperature and the feed composition. During several months of MD module work, a negative influence of separated broths on the hydrophobic polypropylene membrane was not observed. The performance of fermentation in the membrane bioreactor allows for a considerable acceleration of its course and increases its efficiency through the selective removal of fermentation products formed and hence a significant increase in the ethanol concentration compared to classical reactors. Vacuum membrane distillation is commercially competitive because of its high selectivity of ethanol over water, large flux, high thermal efficiency and low energy cost [51].

Later Grazyna *et al*. [65] applied membrane distillation for ethanol recovery during fuel ethanol production. The experimental system consisted of a bioreactor equipped with a capillary polypropylene microfiltration unit. Yeast cells' count and viability, assimilation of sugars, production of ethanol and fermentation of by-products (glycerol and lactic acid) were monitored during fermentations. The intercellular trehalose as well as Hsp70 and Hsp104 heat-shock protein contents were determined. It was concluded that membrane distillation can be regarded as a straightforward method, which leads to an increase in ethanol production, more complete fermentation of sugars, lowering the osmotic pressure in the fermentation broth, decreasing glycerol synthesis level and increasing yeast cells' number and viability. Continuous fermentation processes based on a bioreactor coupled with a membrane distillation unit have the potential to maximize the volumetric productivity and thus minimizing the production costs in the biofuel industry.

11.6 Economic importance and industrial challenges

Work has been reported on the techno-economic evaluation of fuel ethanol processes as well as on some proposed configurations [66]. This has analysed the most promising alternatives for compensating for ethanol production costs by the generation of valuable co-products and has outlined opportunities for the integration of fuel ethanol production processes and their implications. It has mainly studied ways of process intensification through reaction-reaction, reaction-separation and separation-separation processes for bioethanol production and presented some conclusions on current and future research trends in process design and integration in fuel ethanol production. It views obtaining a renewable, abundant, safe and effective source of energy as one of our main challenges. The biofuels, particularly bioethanol, are an environmentally clean source of energy. However, production costs of fuel ethanol are higher than production costs of gasoline in some cases, although this is strongly influenced by factors such as the price of oil and feed-stocks for ethanol production. Nevertheless, many groups and research centres in different countries are continuously carrying out studies aimed at reducing ethanol production costs for a profitable industrial operation. These research tendencies are related to the different steps of processing, the nature of the feed stocks used, and the process engineering tools, mainly process synthesis, integration and optimization.

Process engineering could provide the means to develop economically viable and environmentally friendly technologies for the production of fuel ethanol.

The various types of integration mentioned earlier allow process efficiency to increase through the improvement of reaction processes. However, separation is the step where major costs are generated in the process industry. Reaction–separation integration could therefore have the highest impact on the overall process in comparison with homogeneous integration of processes (reaction–reaction, separation–separation). Hence, the integration of reaction–separation processes plays a very important role in the production of fuel ethanol. Most integrated schemes of this type are oriented to the integration of the fermentation step and several separation unit operations.

Reaction-separation integration is particularly an attractive alternative for the intensification of alcohol fermentation processes. When ethanol is removed from the culture broth, its inhibition effect on growth rate is diminished or neutralized. The importance of this fact has been recognized in such early works as the one presented by Maiorella *et al*. [67], where membrane, extractive and flashing methods for ethanol removal were assessed for ethanol production from molasses and cellulose hydrolyzate.

Eleven alternative fermentation schemes for ethanol production were compared. Conventional batch, continuous, cell recycle, and immobilized cell processes, as well as membrane, extraction, and vacuum processes that remove ethanol from the broth selectively as it is produced, were considered. The processes were compared under identical conditions—that is, they were evaluated with the same molasses as the feed—using a consistent model for the yeast metabolism. Both molasses and cellulose hydrolyzate were considered as feeds. Optimized ethanol plants, including feed preparation, fermentation and product recovery sections were designed and total costs were projected. The concentration of cellulose hydrolyzate was proposed for reducing the costs of separation procedures.

Most of the proposed configurations using this type of reaction-separation integration are related to ethanol removal by different means including coupling different operation units to the fermentation, or simultaneous processes for the *in situ* removal of ethanol from culture broth. Ethanol removal from culture broth may be carried out by vacuum, by gas stripping, by membranes or by liquid extraction.

An important research trend in fuel ethanol production concerns the reduction of feedstock costs, especially through the use of less expensive lignocellulosic biomass. In general, most research efforts are oriented to the conversion of lignocellulosics into fermentable sugars and useful intermediates (due to the recalcitrance or resistance of the biomass to be converted). The key factor for enhancing the competitiveness of the biomass-to-ethanol process is the increase in the specific activity of cellulases and the decrease in their production cost. In addition, the technology of recombinant DNA (genetic engineering) will provide important advances for the development of fuel ethanol industry. The development of genetically modified microorganisms capable of converting starch or biomass directly into ethanol and with a proven stability under industrial conditions will allow the implementation of the consolidated bioprocessing of the feedstocks.

Synthesis will play a very important role in the evaluation of different technologies, especially those related to the integration of the reaction–separation processes, which could have the major effects on the economy of the global process. Similarly, the integration of different chemical and biological processes for the complete utilization of the feedstocks should lead to the development of big 'biorefineries' that allow the production of large amounts of fuel ethanol and many other valuable co-products at smaller volumes, improving the overall economical effectiveness of the conversion of a given raw material. Integration opportunities may provide ways to improve the process qualitatively and quantitatively so that both techno-economical and environmental criteria can be met.

The relatively higher production cost of ethanol is the main obstacle to be overcome. To undertake this, engineering plays a central role in the generation, design, analysis and implementation of technologies improving the indexes of the global process, or for the retrofitting of the bioprocesses used.

Undoubtedly, process intensification through integration of different phenomena and operation units as well as the implementation of consolidated bioprocessing of different feedstocks into ethanol (which requires the development of tailored recombinant microorganisms) will offer the most significant outcomes during the search for efficiency in fuel ethanol production. Efforts should focus on the development of the consolidated bioprocessing (CBP) of biomass as lignocellulosics is the most promising feedstock for ethanol production. The intensification of biological processes also implies better utilization of the feedstocks and the reduction of process effluents improving the environmental performance of the proposed configurations. Attaining this set of goals is a major challenge to be faced through collaboration between biotechnology and process engineering.

11.7 Comparisons with other membrane-separation technologies

Some laboratory membrane configurations for the removal of ethanol have shown interesting results, but their implementation on an industrial scale can be very difficult. The use of ceramic membranes has been proposed for the filtration of cell biomass and the removal of ethanol during the fermentation [68]. The removed ethanol is distilled and the obtained bottoms are recycled to the culture broth resulting in a drastic reduction in wastewater. This configuration uses a stirred ceramic membrane reactor (SCMR). In the same way, immobilized cells can be used to allow easier separation of ethanol and the recirculation of distillation bottoms to the reactor [69]. Kobayashi *et al*. [70] developed a mathematical model for the optimization of temperature profiling during the batch operation of a fermentor coupled with a hollow-fibre module; the temperature was kept initially at 30 °C, reducing later to 20 °C and attaining higher ethanol concentration and productivity. However, it is necessary to analyse the scalability of these configurations due to their complexities (immobilization, presence of membranes, recirculation, repeated batches). The utilization of liquid membranes (porous material with an organic liquid) in schemes involving the extraction of ethanol by the organic phase and the re-extraction with a liquid stripping phase used as an extractant (perstraction or membrane-aided solvent extraction) or gaseous stripping phase (pervaporation) have been also coupled to the fermentation process showing the increased effectiveness of the latter configuration [71].

Pervaporation has offered new possibilities for integration. The coupling of fermentation with pervaporation allows the removing of produced ethanol, reducing the natural inhibition of the cell growth caused by high concentrations of ethyl alcohol. Nomura *et al*. [72] observed that the separation factor of silicalite zeolite membranes used for continuous pervaporation of fermentation broth was higher than corresponding value for ethanol-water mixtures due to the presence of salts that enhance ethanol selectivity. Ikegami *et al*. [73, 74] used this same kind of membrane coated with two types of silicone rubber or covered with a silicone rubber sheet as a hydrophobic material for obtaining concentrated solutions of ethanol. The coupling of *C. thermohydrosulfuricum*, which directly converts uncooked starch into ethanol, with pervaporation has also been tested obtaining ethanol concentrations in the permeate of 27–32 %w/w [75].

O'Brien *et al*. [76] used process-simulation tools (Aspen Plus) to evaluate the costs of the global process involving fermentation-pervaporation in comparison to the conventional batch process from starch fermentation-pervaporation was simulated based on experimental data from tests carried out during more than 200 h employing commercial membranes of polydimethylsiloxane. Simulations showed costs to be slightly higher for the coupled fermentation-pervaporation process due to the capital and membrane costs. Nevertheless, fermentation costs were reduced by 75% and distillation costs decreased significantly. Sensitivity analysis indicated that few improvements in membrane flux or selectivity could make this integrated process competitive. Wu *et al*. [77] investigated the mass transfer coefficients for this type of membrane in the case of pervaporation of fermentation broths, showing that active yeast cells were favourable for ethanol recovery.

Sánchez *et al.* [78] carried out the modelling of SSF (simultaneous saccharification and fermentation) of lignocellulosic biomass coupled with a pervaporation unit. Ethanol removal allowed the reduction of the inhibition effect on cell growth by the accumulation of ethanol in the medium, whereas the SSF process permitted the reduction of the inhibition by glucose and cellobiose experimented by the enzymes during the cellulose hydrolysis. This reaction–reaction–separation integration configuration demonstrated the possibility of reaching higher productivities for the continuous process as well as the production of permeates with an elevated ethanol concentration in comparison to the broth; this implies the reduction in energetic costs during subsequent distillation. Kargupta *et al.* [79] carried out the simulation of continuous membrane fermentor–separator (CMFS) removing ethanol by pervaporation in a membrane reactor, which is coupled with a cell separator in order to increase the concentration of cells inside the reactor by recycling them. The proposed models predict an increase in productivity because this system could be operated at high dilution rates as a consequence of *in situ* product removal and higher cell concentrations.

Membrane distillation has been studied as well as pervaporation. In this type of distillation, aqueous solution is heated for the formation of vapours, which go through a hydrophobic porous membrane favouring the passage of vapours of ethanol (which is more volatile) compared to the vapours of water. The process driving force is the gradient of partial pressures mainly caused by the difference in temperatures across the membrane. Gryta *et al.* [50] implemented a batch fermentor coupled with a membrane distillation module leading to the ethanol removal from culture broth diminishing the inhibition effect and obtaining an increase in ethanol yield and productivities. Gryta [80] points out that when a tubular fermentor working in continuous regime is coupled with the membrane distillation module, higher increases in ethanol productivity can be achieved (up to 5.5 g EtOH/(l h)). This author determined that the number of yeast cells that are deposited on the membrane is practically zero during the operation of these modules [81]. Calibo *et al.* [49] also demonstrated the possibility of coupling the continuous fermentation with membrane distillation. They used a column fermentor, a cell settler and a membrane module. This system operated during almost 700 h with a feed of molasses. García-Payo *et al.* [82] studied the influence of different parameters for the case of air gap membrane distillation based on the model of temperature polarization. It was observed that permeate flux increases in a quadratic way when ethanol concentration increases in the membrane distillation module. A new multicomponent MD model based on the temperature and concentration polarization models was evaluated for AGMD processes of volatile alcohol-water mixtures. Two different equations were developed to estimate the equivalent film heat transfer coefficient and the membrane mass transfer coefficient. A theoretical approach was also developed to test the applicability of Sieder and Tate correlation for their AGMD module. Similarly, Banat and Simandl [52] indicate that the effects of concentration and temperature polarization should be taken into account during the modelling of this process and highlight the need to optimize it with respect to feed stream temperature.

Banat *et al.* [83] also analysed different models based on the Fick's law and on the solution of Maxwell–Stefan equations for this type of distillation. Likewise, the characteristics of vacuum membrane distillation [63] have been studied. An experimental and theoretical investigation of the influence of concentration polarization and temperature polarization on the flux and selectivity of binary aqueous mixtures of ethanol was presented for vacuum membrane distillation processes. Experimental results included changes of the following parameters: nature of solutions, membrane material and pore size, feed temperature, recirculation flow rate. A method was proposed to evaluate the concentration polarization effects from the fit of the experimental data. General models taking into account Knudsen and viscous flows were proposed, but the viscous contribution resulted to be negligible under the operating conditions. Theoretical fluxes were therefore estimated using the Knudsen model and a good agreement between them and the experimental ones was found.

Direct-contact membrane distillation was also studied for the concentration of aqueous solutions of ethanol [84, 85]. Selectivity for ethanol and other solvents and the separation characteristics of the membranes in DCMD type experiments were studied. For DCMD processes with PVDF hollow-fibre membranes, an ethanol solution of 5 wt.% was concentrated to 10 wt.% with an uncoated membrane and to 12 wt.% with a silicone-coated membrane. For all hollow-fibre membranes used, remarkably asymmetric permeation behaviour was observed: the flux was low but the selectivity was high in the normal mode and *vice versa* in the reverse mode.

In the case of the bioethanol production from sugar cane, the integration of fermentation with pervaporation or vacuum membrane distillation allows the recovery of a valuable product, the fructose. For this, mutant strains of yeasts without the capacity of assimilating this monosaccharide should be used. Continuous ethanol removal through the membranes coupled to the fermentor makes possible the accumulation of fructose in the culture medium, which can be recovered in an extraction column. According to Di Luccio *et al*. [86], the simulation of this process based on experimental data and semi-empiric models for the evaluation of the required membranes area allowed a preliminary economic analysis to be performed. This showed that variable costs involving membranes area influence the viability of the process to a higher degree. The process is only viable if the cost of membranes is not greater than US\$550/m^2 for a new plant or US\$800/m^2 for an adapted plant considering an internal return rate of 17%. (These figures were given in [86] and therefore relate to the year 2002.)

Compared to other membrane processes, such as ultrafiltration (UF), or reverse osmosis (RO), membrane distillation is more difficult to apply on an industrial scale because of some serious engineering problems. These include module design as well as heat loss during MD process, which may lead to uncertain economic costs.

11.8 Conclusions and future trends

Continuous fermentation processes based on a bioreactor coupled with a membrane distillation unit have the potential to maximize the volumetric productivity and thus minimizing the production costs in the biofuel industry. According to different research results [65], MD may be considered to be a straightforward method, which leads to an increase in ethanol production. However, the main problem of membrane distillation is that membrane selectivity is too low as well as the progressive wettability of the membranes observed during increasing concentration of volatile compounds [61]. Compared to other membrane processes, such as ultrafiltration (UF), or reverse osmosis, membrane distillation is more difficult to apply on an industrial scale because of some relevant engineering problems. These include module design as well as heat loss during MD process, which may lead to uncertain economic costs. Considerable effort should therefore be directed towards the design and construction of a novel membrane module to permit the successful industrial application of this separation technique. It is worth mentioning that a MD module design for a particular separation must be a trade-off of a number of factors, such as cost, and membrane performance including thermal stability, selectivity, and low thermal conductivity. However, there are many different hydrophobic membrane modules available on the market, which have been extensively investigated during the past years.

Despite their excellent potential, membrane distillation is still far from fulfilling expectations. To overcome the existing barriers, there is a need for more systematic analysis of possible advantages or drawbacks related to the introduction of an innovative membrane unit, clear protocols and comparison indexes for the choice of the best materials and operative conditions, accurate modelling for an easy scale-up or scale-down, and significant multidisciplinary research efforts.

According to this view, the problem of developing highly hydrophobic, stable, microporous membranes, with a narrow distribution of pore size and improved structural and morphological characteristics, is a crucial aspect to be addressed in research programmes devoted to the preparation of new specific membranes for MD and related operations. Moreover, all feasible approaches to increase the efficiency of these membrane contactor devices by reducing both concentration and temperature polarization phenomena, to enhance mass and heat transfer coefficients, to control the fouling problems and related drawbacks (such as clogging, loss of hydrophobicity) are expected to be investigated by. As the number of research papers published in MD is increasing continuously, it is reasonable to suppose that, in the near future, membranes with higher selectivity and improved ethanol resistance may become available.

Summing up, for the economic use of membrane distillation in ethanol fermentation process, future investigations should mainly concentrate on the systematic modelling of the proposed system as well as the development of high-flux and selective membranes. It is worth mentioning that the numerical results of mathematical models may be beneficial at the earlier stages of process scale-up.

References

1. H. Huang, S. Ramaswamya, U.W. Tschirner and B.V. Ramarao, A review of separation technologies in current and future biorefineries, *Sep. Purif. Technol.*, 62, 1–21 (2008).
2. K.W. Lawson, D.R. Lloyd, Membrane distillation, *J. Memb. Sci.*, 124, 1–25 (1997).
3. M.S. El-Bourawi, Z. Ding, R. Ma and M. Khayet, A framework for better understanding membrane distillation separation process, *J. Memb. Sci.*, 285, 4–29 (2006).
4. P.K. Weyl, Recovery of demineralized water from saline waters, United States Patent 3.340.186 (1967).
5. K. Schneider, W. Holz and R, Wollbeck, Membranes and modules for transmembrane distillation, *J. Membr. Sci.*, 39, 25–42 (1988).
6. R.W. Schofield, A.G. Fane, C.J.D. Fell and R. Macoun. Factors affecting flux in membrane distillation, *Desalination*, 77, 279–294 (1990).
7. G.C. Sarti and C. Gostoli, Use of hydrophobic membranes in thermal separation of liquid mixtures: Theory and experiments, in E. Drioli and M, Nakagaki (Eds.), *Membranes and Membrane Processes*, Plenum Press, New York, 1986, pp. 349–360.
8. V. Calabro, E. Drioli and F. Matera, Membrane distillation in the textile wastewater treatment, *Desalination*, 83, 209–224 (1991).
9. Y. Wu, Y. Kong, J. Liu, J. Zhang and J. Xu. An experimental study on membrane distillation: Crystallization for treating waste water in taurine production, *Desalination*, 80, 235–242 (1991).
10. P.P. Zolotarev, V.V. Ugrozov, I.B. Volkina and V.N. Nikulin, Treatment of waste water for removing heavy metals by membrane distillation. *J. Hazardous Mater.*, 37, 77–82 (1994).
11. M. Tomaszewska, Concentration of the extraction fluid from sulfuric acid treatment of phosphogypsum by membrane distillation, *J Membr. Sci.*, 78, 277–282 (1993).
12. V. Calabro, B.L. Jiao and E. Drioli, Theoretical and experimental study on membrane distillation in the concentration of orange juice, *Ind. Eng. Chem. Res.*, 33, 1803–1808 (1994).
13. S. Kimura and S. Nakao, Transport phenomena in membrane distillation, *J. Membr. Sci.*, 33, 285–298 (1987).
14. K. Sakai, T. Koyano and T. Muroi, Effects of temperature and concentration polarization on water vapour permeability for blood in membrane distillation, *The Chem. Eng. J.*, 38: B33–B38 (1986a).
15. F.A. Banat and J. Simandl, Theoretical and experimental study in membrane distillation, *Desalination*, 95, 39–52 (1994).
16. H. Kurokawa, K. Ebara, O. Kuroda and S. Takahashi, Vapor permeate characteristics of membrane distillation, *Sep. Sci. Technol.*, 25, 1349–1359 (1990).
17. S. Kubota, K. Ohta, I. Hayano, M. Hirai, K. Kikuchi, Y. Murayama, Experiments on seawater desalination by membrane distillation, *Desalination*, 69, 19–26 (1988).
18. C. Gostoli and G.C. Sarti, Separation of liquid mixtures by membrane distillation, *J. Membr. Sci.*, 33, 211–224 (1989).
19. C. Gostoli, G.C. Sarti and S. Bandini. Membrane distillation of ethanol-water mixtures. Proceedings of the International Conference on Separations for Biotechnology, Bologna. Italy, Horwood. UK, 1987, pp. 384–394.

20. H. Udriot, A. Araque and U. von Stockar, Azeotropic mixtures may be broken by membrane distillation, *Chem. Eng. J.*, 54, 87–93 (1994).
21. C.A. Rivier, M.C. García-Payo, I.W. Marison and U. von Stockar, Separation of binary mixtures by thermostatic sweeping gas membrane distillation. II. Theory and simulations, *J. Membr. Sci.*, 201, 1–16 (2002).
22. L. Basini, G. D'Angelo, M. Gobbi, G.C. Sarti and C. Gostoli. A desalination process through sweeping gas membrane distillation, *Desalination*, 64, 245–257 (1987).
23. G.C. Sarti, C. Gostoli and S. Bandini, Extraction of organic components from aqueous streams by vacuum membrane distillation, *J. Membr. Sci.*, 80, 21–33 (1993).
24. A. Saavedra, S. Bandini and G.C. Sarti, Separation of v.o.c. from aqueous streams through vacuum membrane distillation, Proc. of the Euromembrane Conf., Paris, France, 1992, pp. 119–204.
25. S. Bandini, G.C. Sarti and C. Gostoli, Vacuum membrane distillation: Pervaporation through porous hydrophobic membranes, Proceedings of the Third International Conference on Pervaporation Processes in the Chemical Industry, Nancy, France, Bakish Materials Corporation, 1988, pp. 117–126.
26. S. Bandini, C. Gostoli and G.C. Sarti, Separation efficiency in vacuum membrane distillation, *J. Membr. Sci.*, 73, 217–229 (1992).
27. H. Ohya *et al.*, Transport of mixed vapors in membrane distillation, Proceedings of the Third International Conference on Pervaporation Processes in the Chemical Industry, Nancy, France, Bakish Materials Corporation, 1988, pp. 501–507.
28. E.Curcio and E. Drioli, Membrane distillation and related operations—a review, *Separation and Purification Reviews*, 34, 35–86 (2005).
29. D.W. Gore, (1982) Gore-Tex Membrane Distillation, Proceedings of the Tenth Annual Conference of the Water Supply Improvement Association, July 25–29, 1982 (Honolulu, Hawaii).
30. S.I. Andersson, N. Kjellander and B. Rodesjo, Design and field tests of a new membrane distillation desalination process, *Desalination*, 56, 345–354 (1985).
31. W.T. Hambury and T. Hodgkiess, Membrane distillation—an assessment. *Desalination*, 56, 287–297 (1985).
32. N. Kjellander, Design and field tests of a membrane desalination system for seawater desalination, *Desalination*, 61, 237–243 (1987).
33. L. Basini, G. D'Angelo, M. Gobbi, G.C. Sarti and C. Gostoli, A desalination process through sweeping gas membrane distillation, *Desalination*, 64, 231–243 (1987).
34. M.P. Godino, L. Pena, C. Rincon and J.I. Mengual, Water production from brines by membrane distillation, *Desalination*, 108, 91–97 (1996).
35. F.A. Banat and J. Simandl, Desalination by membrane distillation: a parametric study, *Sep. Sci. and Tech.*, 33 (2), 201–226 (1998).
36. K.W. Lawson and D.R. Lloyd, Membrane distillation II. Direct contact MD, *J. Membrane Sci.*, 120, 123–133 (1996).
37. P.P. Zolotarev, V.V. Ugrozof, I.B. Yolkina and V.N. Nikulin, Treatment of waste water for removing heavy metals by membrane distillation. *J. of Hazardous Mat.*, 37, 7–82 (1994).
38. G. Zakrzewska-Trznadel, H. Harasimowicz and A.G. Chmielewski, Concentration of radioactive components in liquid low-level radioactive waste by membrane distillation, *J. Membrane Sci.*, 163, 257–264 (1999).
39. F.A. Banat and J. Simandl, Removal of Benzene Traces from contaminated water by vacuum membrane distillation, *Chem. Eng. Sci.*, 51, 1257–1265 (1996).
40. E. Drioli, B.L. Jiao and V. Calabro, The preliminary study on the concentration of orange juice by membrane distillation. Proceedings of the International Society of Citriculture, 3, 1140–1144 (1992).
41. F. Lagana, G. Barbieri and E. Drioli, Direct-contact membrane distillation: modelling and concentration experiments. *J. Membrane Sci.*, 166, 1–11 (2000).
42. S. Nene, S. Kaur, K. Sumod, B. Joshi and K.S.M.S. Raghavarao, Membrane distillation for the concentration of raw-cane sugar syrup and membrane clarified sugarcane Juice, *Desalination*, 147, 157–160 (2002).
43. M.B. Jensen, K.V. Christensen, R. Andrésen, L.F. Søtoft and B. Norddahl, A model of direct-contact membrane distillation for black currant juice, *Journal of Food Engineering*, 107, 3–4, 405–414 (2011).
44. S. Gunko, S. Verbych, M. Bryk and N. Hilal, Concentration of apple juice using direct-contact membrane distillation, *Desalination*, 190, 1–3, 117–124 (2006).

45. M.A. Izquierdo-Gil, Separation of Sugar Aqueous Solutions Using Hydrophobic Porous Membranes and Temperature Differences, Ph.D. thesis, University Complutense of Madrid, 1997.
46. K. Sakai, T. Muroi, K. Ozawa, S. Takesawa, M. Tamura and T. Makane, Extraction of solute-free water from blood by membrane distillation. *Trans. Am. Soc. Artif. Intern. Org.*, 32, 397–400 (1986).
47. E. Hoffmann, D.M. Pfenning, E. Philippsen, P. Schwahn, M. Sieber, R. When and D. Woermann, Evaporation of alcohol/water mixtures through hydrophobic porous membranes. *J. Membrane Sci.*, 34, 199–206 (1987).
48. H. Udriot, S. Ampuero, I.W. Marison and U. von Stockar, Extractive fermentation of ethanol using membrane distillation. *Biotechnology Letters*, 11 (7), 509–514 (1989).
49. R.L. Calibo, M. Matsumura and H. Kataota, Continuous ethanol fermentation for concentrated sugar solutions coupled with membrane distillation using a PTFE module. *J. Ferment. Bioeng.*, 67, 40–45 (1989).
50. M. Gryta, A.W. Morawski and M Tomaszewska, Ethanol production in membrane distillation bioreactor, *Catal. Today*, 56, 159–165 (2000).
51. Z. Lei, B. Chen and Z. Ding, *Special Distillation Processes*, first edn, Elsevier, Amsterdam, 2005.
52. F.A. Banat and J. Simandl, Membrane distillation for dilute ethanol. separation from aqueous streams. *J. Membrane Sci.*, 163, 333–348 (1999).
53. E.A. Mason and A.P. Malinauskas. *Gas Transport in Porous Media: The Dusty-Gas Model*, Elsevier, Amsterdam, 1983.
54. E.A. Mason, A.P. Malinauskas and R.B. Evans III, Flow and diffusion of gases in porous media, *J. Chem. Phys.* 46 (8) 3199 (1967).
55. M. Mulder, *Basic Principles of Membrane Technology*, Kluwer Academic Publishers, Dordrecht, 1991.
56. M.C. Porter, Concentration polarization with membrane ultrafiltration, *Ind. Eng. Chem. Prod. Res. Dev.* 11, 234 (1972).
57. A.J. Chapman, *Heat Transfer*, 3rd edn, Macmillan, New York, 1974.
58. L. Martínez-Díez, M.I. Vázquez-González and F.J. Florido-Díaz, Study of membrane distillation using channel spacers, *Journal of Membrane Science*, 144 (1–2), 45–56 (1998).
59. G.C. Sarti, C. Gostoli and S. Matulli, Low energy cost desalination processes using hydrophobic membranes, *Desalination*, 56, 277–286 (1985).
60. K. Schneider, W. Holz and R, Wollbeck, Membranes and modules for transmembrane distillation, *J. Membr. Sci.*, 39, 25–42 (1988).
61. M.C. García-Payo, M.A. Izquierdo-Gil and C. Fernández-Pineda, Wetting Study of Hydrophobic Membranes via Liquid Entry Pressure Measurements with Aqueous Alcohol Solutions, *J. Colloid Interface Sci.*, 230, 420–431 (2000).
62. C. Fernández-Pineda, M.A. Izquierdo-Gil and M.C. García-Payo, Gas permeation and direct-contact membrane distillation experiments and their analysis using different models, *J. Membr. Sci.*, 198, 33–49 (2002).
63. M.A. Izquierdo-Gil and G. Jonsson, Factors affecting flux and ethanol separation performance in vacuum membrane distillation (VMD), *J. Membr. Sci.*, 214, 113–130 (2003).
64. J. Bausa and W. Marquardt, Shortcut design methods for hybrid membrane distillation processes for the separation of nonideal multicomponent mixtures, *Ind. Eng. Chem. Res.*, 39, 1658–1672 (2000).
65. G. Lewandowicz, W. Biaas, B. Marczewski and D. Szymanowska, Application of membrane distillation for ethanol recovery during fuel ethanol production, *J. Membr. Sci.*, 375, 212–219 (2011).
66. C.A. Cardona and O.J. Sánchez, Fuel ethanol production: Process design trends and integration opportunities, *Bioresour. Technol.*, 98, 2415–2457 (2007).
67. B.L. Maiorella, H.W. Blanch and C.R. Wilke, Biotechnology report. Economic evaluation of alternative ethanol fermentation processes. *Biotechnol. Bioeng.*, 26, 1003–1025 (1984).
68. R. Ohashi, Y. Kamoshita, M. Kishimoto, T. Suzuki, Continuous production and separation of ethanol without effluence of wastewater using a distiller integrated SCM-reactor system. *J. Ferment. Bioeng.*, 86 (2), 220–225 (1998).
69. M. Kishimoto, Y. Nitta, Y. Kamoshita, T. Suzuki and K.I. Suga, Ethanol production in an immobilized cell reactor coupled with the recycling of effluent from the bottom of a distillation column. *J. Ferment. Bioeng.*, 84 (5), 449–454 (1997).

70. M. Kobayashi, K. Ishida and K. Shimizu, Efficient production of ethanol by a fermentation system employing temperature profiling and recycle. *J. Chem. Technol. Biotechnol.*, 63, 141–146 (1995).
71. P. Christen, M. Minier and H. Renon, Ethanol extraction by supported liquid membrane during fermentation, *Biotechnol. Bioeng.* 36, 116–123 (1990).
72. M. Nomura, T. Bin and S. Nakao, Selective ethanol extraction from fermentation broth using a silicalite membrane. *Sep. Purif. Technol.*, 27, 59–66 (2002).
73. T. Ikegami, D. Kitamoto, H. Negishi, K. Haraya, H. Matsuda, Y. Nitanai, N. Koura, T. Sano and H. Yanagishita, 2003. Production of high-concentration bioethanol by pervaporation. *J. Chem. Technol. Biotechnol.*, 78 (9), 1006–1010 (2003).
74. T. Ikegami, D. Kitamoto, H. Negishi, K. Iwakabe, T. Imura, T. Sano, K. Haraya and H. Yanagishita, Reliable production of highly concentrated bioethanol by a conjunction of pervaporation using a silicone rubber sheet-covered silicalite membrane with adsorption process. *J. Chem. Technol. Biotechnol.* 79 (8), 896–901 (2004).
75. Y. Mori and T. Inaba, Ethanol production from starch in a pervaporation membrane bioreactor using Clostridium thermohydrosulfuricum, *Biotechnol. Bioeng.*, 36, 849–853 (1990).
76. D. O'Brien, L. Roth and A. McAloon, Ethanol production by continuous fermentation-pervaporation: a preliminary economic analysis. *J. Memb. Sci.*, 166, 105–111 (2000).
77. Y. Wu, Z. Xiao, W. Huang and Y. Zhong, Mass transfer in pervaporation of active fermentation broth with a composite PDMS membrane, *Sep. Purif. Technol.*, 42, 47–53 (2005).
78. O.J. Sánchez, C.A. Cardona and D.C. Cubides, 2005. Modeling of simultaneous saccharification and fermentation process coupled with pervaporation for fuel ethanol production. Second Mercosur Congress on Chemical Engineering and Fourth Mercosur Congress on Process Systems Engineering, Rio de Janeiro, Brazil.
79. K. Kargupta, S. Datta and S.K. Sanyal, Analysis of the performance of a continuous membrane bioreactor with cell recycling during ethanol fermentation. *Biochem. Eng. J.*, 1, 31–37 (1998).
80. M. Gryta, The fermentation process integrated with membrane distillation. *Sep. Purif. Technol.*, 24, 283–296 (2001).
81. M. Gryta,. The assessment of microorganism growth in the membrane distillation system. *Desalination*, 142, 79–88 (2002).
82. M.C. García-Payo, M.A. Izquierdo-Gil and C. Fernández-Pineda, Air gap membrane distillation of aqueous alcohol solutions. *J. Memb. Sci.*, 169, 61–80 (2000).
83. F.A. Banat, F.A. Al-Rub and M. Shannag, Modeling of dilute ethanol–water mixture separation by membrane distillation. *Sep. Purif. Technol.*, 16, 119–131 (1999).
84. Y. Fujii, S. Kigoshi, H. Iwatani and M. Aoyama, Selectivity and characteristics of direct-contact membrane distillation type experiment. I. Permeability and selectivity through dried hydrophobic fine porous membranes. *J. Memb. Sci.*, 72, 53–72 (1992a).
85. Y. Fujii, S. Kigoshi, H. Iwatani, M. Aoyama and Y. Fusaoka,. Selectivity and characteristics of direct-contact membrane distillation type experiment. II. Membrane treatment and selectivity increase. *J. Memb. Sci.*, 72, 73–89 (1992b).
86. M. Di Luccio, C. Borges and T. Alves, Economic analysis of ethanol and fructose production by selective fermentation coupled to pervaporation effect of membrane costs on process economics. *Desalination*, 147, 161–166 (2002).

Part V

Solid-Liquid Separations

12

Filtration-Based Separations in the Biorefinery

Bhavin V. Bhayani and Bandaru V. Ramarao

Department of Paper and Bioprocess Engineering, Empire State Paper Research Institute, State University of New York College of Environmental Science and Forestry, USA

12.1 Introduction

Rapid global industrialization has resulted in an increased demand for transportation fuels and has led to the search for all kinds of fuel resources. The sharp increase in the price of oil is a result of the strength of global demand. Concerns about the environmental impact of increased fossil fuel use, and particularly worries about anticipated catastrophic and irreversible climate change caused by rapid increases in atmospheric CO_2 levels, drive the search for the replacement of oil by more sustainable fuels. Ethanol has served this purpose until now, primarily by replacing a fraction of oil consumption when it is blended into gasoline in smaller amounts. Ethanol blended to gasoline gives an oxygenated fuel, which burns cleaner with fewer emissions than gasoline alone, providing an additional environmental incentive. Currently, ethanol is produced from sugarcane and other starch-bearing agricultural or food crops such as corn and wheat. Its use as a biofuel has caused surging demand for these food crops and their prices have risen substantially in response. Increasing biofuels therefore presents a difficult choice for society between growing crops for food or fuel. Any opportunity to replace food grains as a source for ethanol is very attractive for the sustainability of future transportation and industrial development.

Biomass consisting of the waste from agricultural crops, such as the stems and other inedible portions of plants, is rich in carbohydrates and can serve as an excellent sustainable fuel resource. Other biomass found in forests and plantations based on wood, forest wastes, and waste materials from forest products industries such as pulp and paper mills presents an additional valuable resource for the production of transportation fuels. Carbohydrates in these biomass types occur in conjunction with lignin and are therefore collectively referred to as lignocellulosic biomass resources. Lignocellulosics comprise a vast variety, from stems and

Separation and Purification Technologies in Biorefineries, First Edition.
Edited by Shri Ramaswamy, Hua-Jiang Huang, and Bandaru V. Ramarao.

stalks of agricultural plants such as corn, sorghum, traditional pulping hardwoods, softwoods, and also plantation-based woody biomass such as eucalyptus.

12.2 Biorefinery

A biorefinery is a manufacturing facility that produces a spectrum of fuels, materials, and chemicals from lignocellulosic biomass resources such as wood, corn stover, wheat straw, and similar agricultural waste materials. Conceptually, it is similar to a modern petroleum refinery, producing gasoline and other liquid and gas fuels, plastics and petrochemicals. Biorefineries are expected to produce a mix of bioethanol, advanced biofuels such as butanol and jet fuels or different pyrolysis oils, by biochemical or thermochemical conversions. Bioplastics such as polyhydroxyalkanoates (PHA) and polylactic acid (PLA) can be produced by fermentation of biomass sugars. Biorefineries can also produce platform chemicals such as furfural, acetic acid and succinic acid, which can be converted to a variety of other chemicals.

Biorefineries can thus displace and extend the use of petroleum using renewable and widely distributed natural resources. They are therefore a key component for sustainable development in the future. Given the complex nature of any biomass, multiple products can be manufactured from its various constituents and thus the monetary yield can be maximized.

The process flow for a typical ethanol production from lignocellulosic biomass is shown in Figure 12.1. For purposes of separations analysis, the biorefinery processes can be divided into three sections. The first consists of pretreatment where biomass is rendered amenable to biochemical conversions using enzymes or acids. The second section consists of separation and purifications of different pretreatment output streams. Compounds that are toxic to fermentation microbes and inhibitors for enzyme activities are removed. In addition, some useful products such as lignin may be removed in this stage. The final stage is the downstream conversion of the liquors and biomass into fuel or polymeric products by fermentation or other processes. Each of these is considered in detail, from a solid–liquid separations point of view below.

12.2.1 Pretreatment

Biomass is a complex matrix composite of three components: cellulose, hemicellulose and lignin (Figure 12.2). Cellulose is a crystalline polymer of glucose units with a relatively high degree of polymerization, between 5000 to 100 000. Hemicellulose is a class of polysaccharides of six- and five-membered hydro-sugars with much lower degrees of polymerization in non-crystalline form. Lignin is a cross linked polymer of aromatic alcohols with a DP between 500 to 5000. Cellulose crystals in biomass typically occur in specific nanostructured forms called microfibrils, which are embedded in a matrix of hemicellulose and lignin, the lignin providing a protective and strongly binding sheath for the plant structure. The purpose of pretreatment is to loosen and break this internal structure of biomass so that the carbohydrate polymeric constituents, viz. cellulose and hemicellulose, are depolymerized into sugars for conversion. The crystalline structure of cellulose needs to be disrupted so that the enzymes can easily access and hydrolyze the cellulose and hemicellulose into its component sugars. Pretreatment is usually carried out by the application of a combination of heat, pressure, and chemical catalysts such as acids. Different pretreatment options are available, some of which are summarized in Table 12.1. Different types of liquors, including organic solvents such as ethanol and ionic liquids, have been proposed and investigated for pretreatment. However, water and aqueous solutions are the dominant ones used. Pretreatment can broadly be characterized by the pH conditions of the aqueous treatment liquor. Acidic solutions usually are dilute solutions of mineral acids (sulfuric) between 0.5 to 1.5% leading to a pretreatment solution pH below 2. Hot water is also used as pretreatment liquor. Although the initial

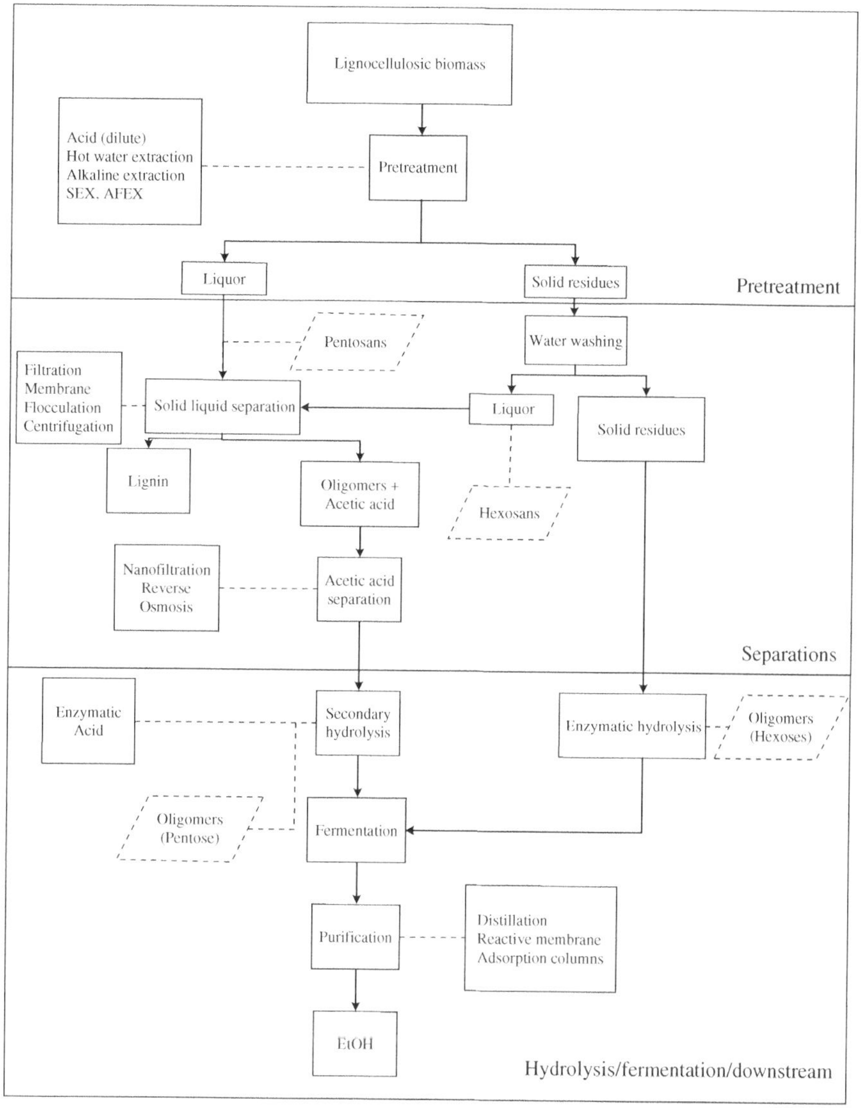

Figure 12.1 *Process flow diagram for ethanol production in a biorefinery*

water is at or near neutral pH conditions, rapid de-acetylation of the xylan and other hemicellulose constituents decreases the liquor pH to a final value between 3 and 4. Acetic acid predominates in the solution at this stage (small quantities of formic and other organic acids may be present). The organic acids are in protonated form in the liquors.

Alkaline conditions depolymerize lignin and liberate phenolic products (in dissolved form), whereas under acidic conditions, spherical lignin droplets are re-deposited onto the solid fractions [2]. Under any pretreatment technology thus employed it has been observed that certain fraction of lignin and other inhibitors are extracted along with cellulose and hemicellulose in the hydrolyzate. Thus to separate the inhibitors from the hydrolyzates, inclusion of a solid–liquid separation scheme is an essential process in

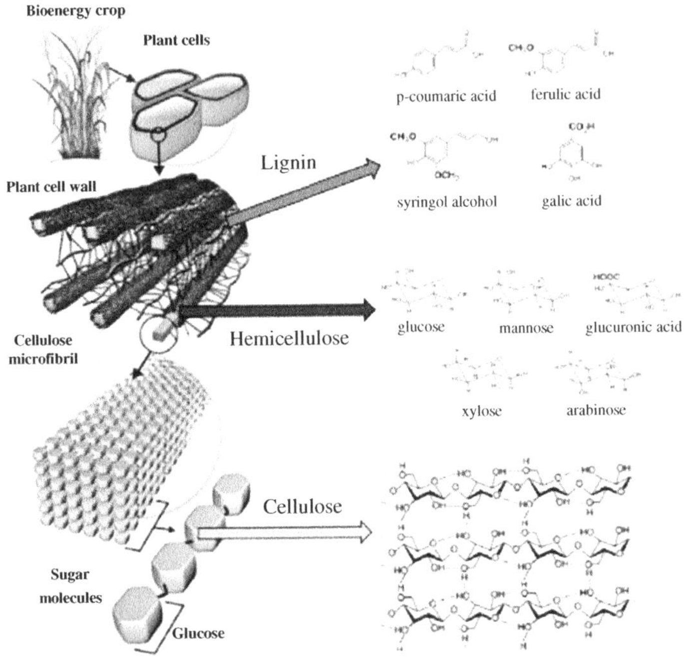

Figure 12.2 *Major components of lignocellulosic biomass [1]*

a biorefinery. Comparing Tables 12.2 and 12.3 we can conclude that the increase of pH leads to higher lignin removed in dissolved form as well as lower amount of other inhibitors while decrease of pH leads to insoluble lignin and some other inhibitors being released into the hydrolyzate.

The solid residues from pretreatment are separated by a simple screening or filtration operation and taken for further processing. In a typical biorefinery configuration, the solid residues rich in the cellulosic component are hydrolyzed enzymatically into glucose and other hexose sugars. Sometimes, acid hydrolysis is also used to depolymerize the cellulosic fraction. When collocated at a pulp and paper mill, the residual biomass (i.e. wood) is delignified by any of the traditional wood pulping and bleaching processes and the resulting wood pulp is made into paper and similar fibrous products.

Lignocellulosic hydrolyzate contains particulates, which are the main source of fouling in a membrane separation process. It has been seen that the extraction times determine the concentration (mass removal) of the particulate matter in hydrolyzate—i.e. as the extraction time increases the concentration increases and there are larger particle sizes. The particle charge has been determined to be strongly anionic by zeta potential measurements. It has also been observed that, as the mass removal increases, the fraction of suspended solids increases and the fraction of colloidal and dissolved solids decreases. The hydrolyzate obtained at lower extraction time is more electrostatically stable [3]. All these observations have been noted after various extraction severities were tested on a hardwood (Table 12.4).

12.2.2 Hydrolyzate separations

The liquor obtained from pretreatment will contain products from hemicellulose, lignin, and extractives to different extents. Under alkaline conditions, more lignin is removed from the biomass and it occurs

Table 12.1 *Pretreatment types*

Process	Conditions	Limitations and disadvantages
Steam explosion	Biomass pressurized with steam at about 1:1 ratio for short residence time (1– 10 min). Depressurization causes lignocellulose matrix to become porous.	Destruction of a portion of the xylan fraction; incomplete disruption of the lignin-carbohydrate matrix; generation of compounds inhibitory to microorganisms.
AFEX	Biomass pressurized with liquid anhydrous ammonia at moderate temperatures and high pressure (250–300 psi) for medium duration residence time. The combined chemical and physical effects of lignin solubilization, hemicellulose hydrolysis and cellulose decrystallization cause lignocellulose matrix to become porous.	Not efficient for biomass with high lignin content.
Hot water extraction	Biomass is treated with water at about 1:4 ratio for long duration residence time at temperatures exceeding 100 °C. High pressure and water penetration cause lignocellulose matrix to become porous.	Limited sugar extraction, some lignin and other inhibitor obtained in the extracts.
Acid hydrolysis	Biomass is either treated with dilute acid (longer residence times) or concentrated acid (shorter residence times). The residence times can be reduced by treating the biomass at higher temperature.	High cost; equipment corrosion; formation of toxic substances.
Organosolv (methanol, ethanol)	Biomass is treated with organic solvents at high temperatures (>100 °C) and duration times exceeding 50 minutes.	Solvents needed to be drained from the reactor, evaporated, condensed, and recycled; high cost.

Table 12.2 *Operating parameters and severity of extraction processes*

Pretreatment type	Temperature	pH	Time (min)	Severity log (Ro)
Lime	54–160	alkaline	60–4800	2.1–3.9
Ammonia fiber explosion	50–180	alkaline	5–30	0.4–3.5
Steam explosion	160–230	neutral	5–0	2.8–4.5
Acid hydrolysis	30	acidic	30	−0.6
Dilute acid hydrolysis	120–140	acidic	15–60	1.9–2.7
Hot water extraction	120–160	Neutral	90–120	3.28–3.85

Table 12.3 *Effect of process variables on extraction and production of inhibitors*

Pretreatment catalyst	pH initial	pH final	Lignin removed (%)	Acetic acid (g/kg)	5-HMF (g/kg)	Furfural (g/kg)
HCl	1	1.1	2	20	0.2	0.88
HCl	4	6.2	6	3	0	0
H_2SO_4	1	1.1	12	19	0.15	0.37
CH_3COOH	1	1.5	37	N/A	0.03	0.05
NaOH	13	10.6	68	16	0	0
NaOH	7	6.3	4	3	0	0
NaOH	10	6.4	4	3	0	0
NaOH	13	10.6	68	20	0	0

Table 12.4 *Effect of pretreatment (hot water extraction) time on particle size, zeta potential and mass removal*

Extraction time (hours)	Pretreatment severity (log Ro[a])	Particle size (nm)	Zeta potential (mV)	Mass removal (%)
0.5	3.28	220	−31	7.9
1	3.54	260	−23	13.7
1.5	3.73	340	−20.2	18.5
2	3.83	390	−17.4	20.6

[a]A measure of the intensity of overall reaction conditions in a chemical reaction.

in dissolved form. Hemicellulose is also in dissolved form and occurs as higher oligomers, close to its native form in the biomass. In acidic conditions, lignin precipitates into colloidal particles ranging in sizes from about a few hundred nanometers to the micrometer range. Some of the lignin may still be attached to hemicelluloses but the bulk of the hemicelluloses are dissolved as smaller molecular weight oligomers. Degradation products of hemicelluloses also occur with increasing abundance. Prominent among these are acetic acid, other short-chain carboxylic acids, acid degradation products of sugars such as furfural and 5-hydroxy methyl furfural (5-HMF). Hemicellulose also is dissolved and occurs mostly as oligomeric residues solid residues from pretreatment are separated by a simple screening or filtration operation and taken for further processing.

This complex liquor mixture containing hemicelluloses presents many opportunities for further use. It can be blended into the sugars stream obtained from the hydrolysis of the residual biomass after a secondary hydrolysis. The resulting sugar stream has hexose sugars from the cellulose decomposition and pentose sugars from hemicellulose decomposition, which can be fermented simultaneously into different liquid fuels by C6 and C5 fermenting microbes. For the general success of fermenting of all of these blends, it is necessary that the sugar streams are free of toxic and fermentation inhibitory compounds. A number of separation processes may be applied to these liquor streams to remove the inhibitory compounds. Primary separations include coagulation-flocculation of the particulate and colloidal lignin materials followed by simple cake filtration and microfiltration. The liquor hydrolyzates are usually highly turbid in these cases and the removal of the particulate matters results in significant drop in turbidity and clarification of the hydrolyzates [4].

12.2.3 Downstream fermentation and separations

The sugar-rich stream obtained from the solid–liquid separation process must be hydrolyzed to obtain sugar monomers prior to fermentation, which is accomplished by either acid (dilute or concentrated) or enzymatic hydolysis. Enzymatic hydrolysis can be combined in a process known as simultaneous saccharification and fermentation (SSF). Fermentation broth is a complex mixture containing unreacted sugars, fermentation products and byproducts, as well as microbial biomass. If the hydrolyzates were not purified prior to fermentation, they may also contain significant quantities of lignin products in particulate or colloidal form. The broth is usually separated by simple filtration and the liquor is purified/concentrated to produce bioethanol or other biofuels by distillation, adsorption or membrane pervaporation processes.

12.3 Solid–liquid separations in the biorefinery

The different mechanisms that develop solid–liquid separation (SLS) are [5]:

- increasing particle-size distribution and the concentration of particulate mass;
- settling and decanting of the liquor/supernatant;
- post-treatment by either washing or deliquoring.

The primary focus of this chapter is the development of filtration processes for the separation of solid phase from the liquids of the various streams encountered in the biorefinery. We will review some basic principles of filtration including dead-end and tangential-flow filtration. This will be followed by an exposition of typical methods to analyze both these processes. We will then provide some experimental setups that have been used to determine filtration characteristics, which allow the scale-up and optimization of these processes at an industrial scale. Typical experimental data are given to show the range of variation encountered in such streams. Experimental results of filtration studies carried out on hardwood hydrolyzates have been presented. The hydrolyzes were analyzed for particle size and zeta potential, which were measured using a BIC Particle Size and Zeta Potential Analyzer (90 Plus and ZetaPlus) (Brookhaven Instruments Corporation (BIC), Holtsville NY).

The processes in a biorefinery requiring SLS processes include:

- separation of non-fermentable constituents, for the dry biomass;
- separation of inhibitors from hydrolyzates (SLS);
- liquor obtained after washing of the solid residue after pretreatment (SLS);
- separation of solids from the fermentation broth (SLS).

Table 12.5 Different classes of separation process

Sr. no	Classes	Examples
1	Filtration	Dead end filtration, tangential flow filtration
2	Centrifugation	Centrifuges
3	Washing and leaching	
4	Precipitation and flocculation	Lime tanks, polymers
5	Adsorption	Char filters

Solid liquid separation can be carried out using various technologies as shown in Table 12.5.

12.4 Introduction to cake filtration

Cake filtration is one of the oldest separation technologies used in the chemical, processing, and mineral industries. It may be applied to recover either of the solid (cake) or liquid (filtrate) phases. The suspension is pushed, under a driving force (either pressure or vacuum), through a screen to obtain a stream of filtrate. The solid particles collect on the surface as a cake layer.

12.5 Basics of cake filtration

During cake filtration, the suspension flows normal to the screen, also called as the filter medium while the cake grows in the opposite direction. Another form of cake filtration is cross-flow filtration, in which the suspension to be treated flows tangentially to the filter medium surface. In this case the filter cake growth is limited and the resistance it poses to separation is substantially lower. Higher production rates are enabled. Thus the difference between cross-flow filtration and dead-end filtration is the direction of flow relative to the membrane and consequently the development of cake, as shown in Figure 12.3.

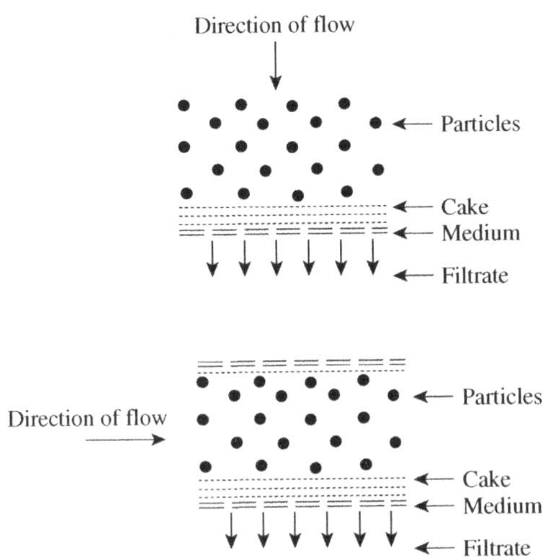

Figure 12.3 Comparison between dead end and cross flow filtration

Figure 12.4 shows a typical laboratory-scale dead-end filtration process. Dead-end filtration is inherently a batch process. The cake layer grows with time, increasing the hydraulic resistance. The program of operation of filtration usually takes one of the following three forms: (i) In constant-pressure filtration, the suspension pressure is held constant by means of a pump while the filtrate flux decreases with time, almost exponentially. Filtration is terminated when the incremental cost of obtaining filtrate exceeds its utility. (ii) The second method increases the suspension pressure in a programmed manner to keep the filtrate flow constant. This is also terminated when the incremental cost to increase the pressure exceeds the utility of the cake solids or the filtrate. (iii) A third method of operation is known as variable-rate filtration, where the pressure is varied according to the pump head flow characteristics to optimize energy expended in separation. Filtration performance is usually expressed in terms of the filtrate flux, J, defined as the filtrate flow rate per unit membrane area.

The filtrate flux is given by an application of Darcy's law, considering the medium resistance and the cake resistance to be in series.

$$J = \frac{\Delta P}{\mu \left(R_m + \alpha m\right)} \tag{12.1}$$

where J is the filtrate flux (m/s), ΔP the applied pressure (kPa), μ the fluid viscosity (Pa.s), Rm the membrane resistance (m/kg), α the specific cake resistance (m^2/kg) and m is the mass of cake per unit membrane area (kg/m^2). The specific cake resistance (α) is a measure of the "filterability" of the feed solution as it determines the flux dependence on pressure and thus the time of filtration. The pressure drop of a fluid flowing through a packed bed of solids (filter) can be quantified using the Kozeny–Carman equation (or Carman–Kozeny equation). The original equation is:

$$\frac{\Delta P}{L} = \frac{180\overline{V}_o \mu}{\Phi_s^2 D_p^2} \frac{(1-\varepsilon)^2}{\varepsilon^3} \tag{12.2}$$

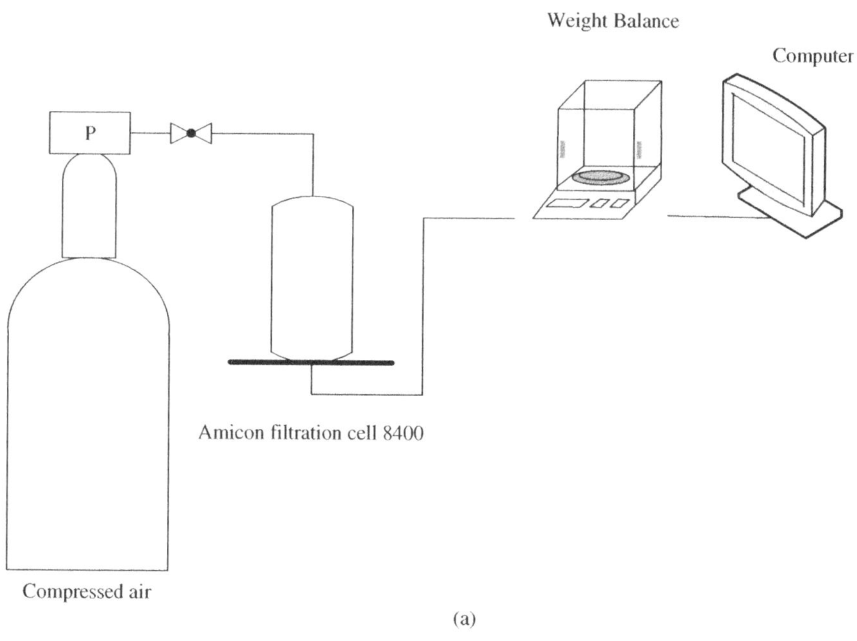

(a)

Figure 12.4 *(a) General schematic of the dead end filtration process (lab. scale). Reprinted from [7]; (b) Schematic of compressibility cell; (c) Photograph of a compressibility cell used for determining compression of fibrous suspensions*

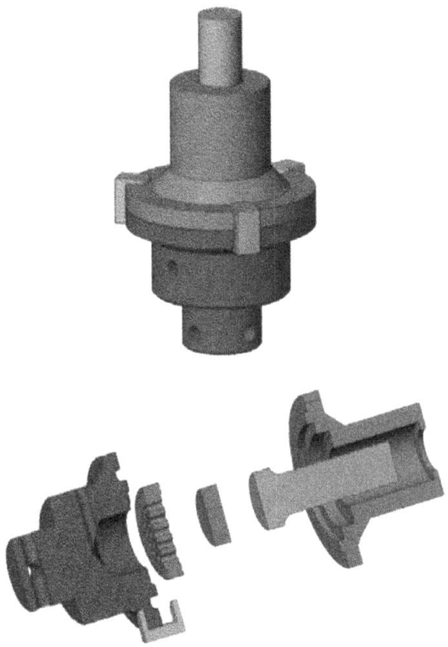

(b)

(c)

Figure 12.4 *(continued)*

The specific filtration resistance can be obtained as a function of the porosity and the specific surface area (volume basis) of the pores in the cake as follows:

$$\alpha = \frac{K(1-\varepsilon)S_v^2}{\varepsilon^3 \rho_c} \tag{12.3}$$

Dead-end filtration can be carried out in two modes depending upon the application of pressure. If the operating pressure is kept constant, the process is called constant-pressure filtration. In this type of filtration, the permeate flux decreases with time due to membrane fouling All the results presented in this chapter are for constant-pressure filtration. If the operating pressure is varied so as to maintain constant permeate flux the process is called constant-rate filtration.

$$\textit{Constant pressure filtration}: \quad \overline{\mu s \rho(1-\overline{m}s)^{-1}\left[\alpha_{av}\right]_{p_{sm}}} \frac{V^2}{2} + \mu R_{mV} = p_o t \tag{12.4a}$$

$$\textit{Constant rate filtration}: \quad q_{lm} = Q, V = QT$$

$$p_o = \mu Q\left\{s\rho(1-ms)^{-1}\left[\alpha_{av}\right]_{p_{sm}} Qt + R_m\right\} \tag{12.4b}$$

The test cell used by Grace can accommodate a suspension under higher pressure experiments (3446 kPa) and is made of Lucite acrylic resin and stainless steel. It provides the capability to record the filtrate continuously. This test cell is widely referred as the compression-permeability cell (C-P cell) and has been used for the measurement of permeability and solid concentration of filter cakes but the "Wall effects" and the dynamics of filtration have raised questions. Another interesting experimental set up [7] consisted of a Nutsche filter (a closed vessel filtration assembly used for applications designed to prevent evaporation of any liquid) with a computer controlled electro-pneumatic valves. This arrangement had the capability to measure cake thickness and cake solidosity while keeping the intrusion effects to the minimum.

The measurement of filtrate flow rate as a function of time has long been the standard method of evaluating filtration experiments. The filtrate flow rate gives an average value of the liquid velocity in the filter cake, thus providing limited insight into the filtration process.

12.5.1 Application in biorefineries

Dead-end filtration has been successfully used in pharmaceutical industries and it has been hypothesized that it can be used in a biorefinery to separate inhibitors from the hydrolyzate to obtain a sugar rich stream. Currently overliming is used to detoxify lignocellulosics in the biorefinery. Due to treatment time constraints (few hours to days to treat a feed-batch) the viability of the process has been questioned while dead-end filtration has not been investigated in detail. The instruments needed for a dead-end filtration set up are not complicated in geometric structure and can be easily and cheaply obtained. Therefore, the capital investment is small and may be used for solid–liquid separation problems encountered in biorefineries. Some of the applications of dead-end filtration to address the removal of inhibitors include:

- Separation of lignin and acetic acid from the hydrolyzate obtained after pretreatment of biomass. In this case a membrane with a higher molecular weight cut-off (MWCO) (i.e. >300 kDa) can be used.
- Separation of unclean feeding materials obtained in the washed liquor of solid residue after pretreatment.
- Separation of any inhibitors present in fermentation broth.

12.5.2 Specific points of interest

Lignin fragments or aromatics along with other inhibitors inhibit the fermentation process, but they have alternate uses. The use of a molecular-weight cut-off of 300 kDa is preferred because the sugar solutions will permeate but the lignin will be retained, thus achieving the separation of lignin and the rest of hydrolyzates. Along with lignin, furfural and hydroxylmethylfurfural (HMF), which are also detrimental to the fermentation process, are retained on the filter. Polyethersulfone filters can be used to remove solid fraction of the hydrolyzate and thus enhance the fermentability. It has been observed that the specific filtration resistance increases with pressure due to the higher solidosity—i.e. dryer cakes. There is a rule of thumb that the filterability based on the specific resistance can be classified as follows: easy to filter, moderately difficult to filter and hard to filter are 10^{10} and less, 10^{11} and higher than 10^{12} m/kg respectively. It has been observed [8] that the specific filtration resistance values range between 10^{12} and 10^{13}, although previous research [9] on similar material has determined average α-values in the order of magnitude of $10^{13}-10^{14}$ m/kg. Thus for application in a biorefinery, a detailed study should be undertaken to understand the specific filtration resistance of lignin and the hydrolyzate ($>10^{12}$ m/kg) and also the compressible nature of filter cake which influences the economics, viability and design of ethanol production from lignocellulosics.

12.6 Designing a dead-end filtration

12.6.1 Determination of specific resistance

The classic method for determining the specific cake resistance is to measure the volume, V, of filtrate as a function of time, t, during a batch filtration at constant pressure.

$$\frac{t}{V} = \frac{\alpha\mu c}{2A^2\Delta P}V + \frac{\mu R_m}{A\Delta P} \tag{12.5}$$

where A is the membrane area and c is the wet solid mass per unit volume of filtrate. Therefore, α can be determined from the slope of a plot of t/V versus V. The effect of pressure on α is seen in Figure 12.5, which shows that the filtration capacity changes with pressure and it can be deduced that the flux and specific resistance is affected by the pressure at which the process is operated.

12.6.2 Membrane fouling

When filtration is conducted over shorter periods of time such that the medium exerts a significant influence on the separation process, the dynamics of the filtrate flux can become quite complex. This is significant because it determines the quality of separation as well as the quantitative yield of solids and the liquid phase. The increase in flow resistance encountered by the suspension during filtration is referred to as fouling. Filter media fouling occurs by four mechanisms or different combinations of them (Table 12.6). Clogging of the pores within the medium as particles penetrate it and get retained at the walls, crevices and pore junctions. The second mechanism is the occlusion of the medium pores at the surface by particles. This can be either total or partial depending on the pore and the particle size distributions. Finally, the cake layer can be formed on top of the medium leading to increased resistance of flow and the development of filtration process as a conventional cake filtration. The cake layers can be compressible or incompressible depending upon a number of physic-chemical characteristics of the suspensions. Many models that are applicable for each of the above mechanisms for fouling can be found in the literature [6, 10–12]. The early filtration models [6, 11] assume that the formed filter cake is incompressible.

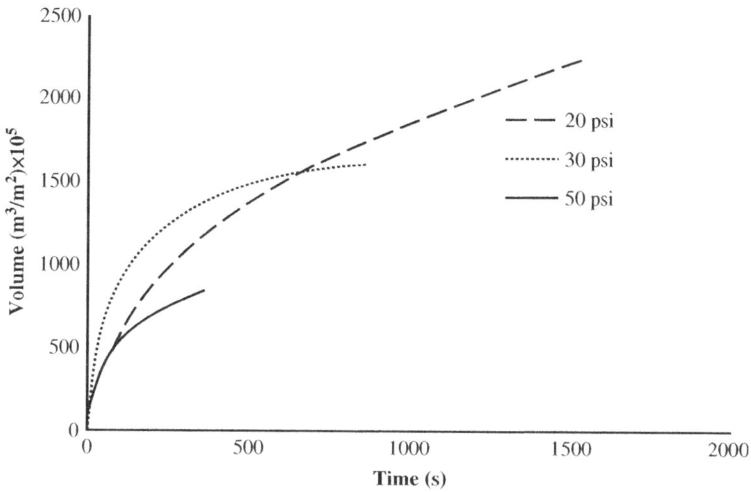

Figure 12.5 *Effect of pressure on cumulative flow*

Table 12.6 *Types of fouling and its characteristics*

Sr. no	Type of fouling	Fouling characteristic
1	Intermediate blocking	Particles deposit on other particles or block membrane pores (as in complete blocking)
2	Standard blocking	Porosity decreases proportionally to the volume of deposited particles due to the deposition of particles deposit within pores
3	Cake filtration	The membrane is covered by a layer of deposited solids from the feed suspension and the newer particles deposit on already deposited particles
4	Complete blocking	The membrane is completely clogged by the particles in the feed suspension

The laws of filtration first proposed by Hermans and Bredee consist of a set of power-law relationships corresponding to different particle-retention mechanisms in filtration. For understanding the fouling in membrane separations and the attendant flux reduction, there has been a resurgent interest in the laws of filtration [13]. Fouling is the process that results in a decrease in filtration capacity of a membrane due to the deposition of solids on the membrane surface, or within the membrane pores [14]. Thus the resulting resistance (Figure 12.6) to the flow offered can be classified as due to resistance of membrane (R_m), adsorption (R_a), pore blocking (R_{pb}), and the formation of cake layer (R_c).

Given the critical role of fouling, it is very important to understand the formation of cake and other mechanisms of membrane fouling for designing a membrane separation. The dynamics of filtration is given as [15]:

$$\frac{d^2t}{dV^2} = k_1\left(\frac{dt}{dV}\right)^{k_2} \tag{12.6}$$

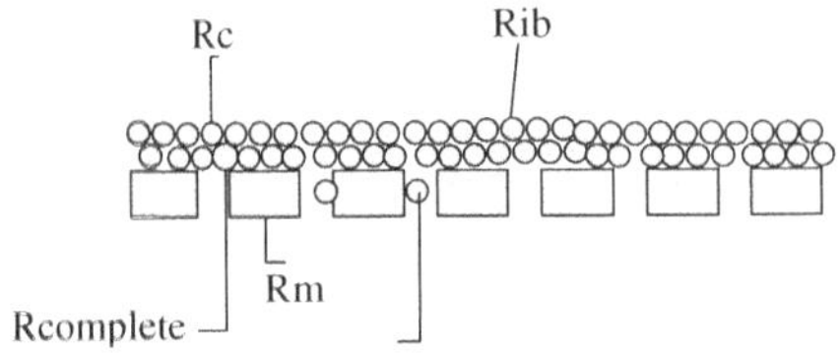

Figure 12.6 Membrane fouling mechanisms

Table 12.7 The "k" values for different types of fouling

Sr. no	"k"	Type of fouling
1	0	bridging (proper cake filtration)
2	1	intermediate blocking
3	1.5	standard blocking
4	2	complete blocking

where V represents the cumulative filtrate volume, t the time, and k_1 and k_2 are empirical constants. In the above equation, the value of k_2 characterizes the types of cake formation. The different k values that characterize the type of fouling are detailed in Table 12.7.

Equation 12.6 was developed by researchers based on assumptions including the uniform particle size, distribution of the solids in the feed, uniform deposition of cake, and different particle retention mechanisms. It was later subjected to modification by investigators [16]. The cake thus formed can be compressible or incompressible depends upon the solids in feed solution. Incompressible cakes tend to be more porous and thus have higher filtration capacity than compressible cakes, whose porosity decreases with time. The different variables that affect cake compressibility are pressure, particle charge, surface area, and the properties of the feed solution itself.

12.6.3 The effect of pressure on specific resistance—cake compressibility

Studies have shown that the specific cake resistance, α, changes with pressure (Table 12.8) and the correlation known as the power-law expression describes the relation:

$$\alpha = b_o P^n \text{ [17]} \tag{12.7}$$

where b_o is a constant and n is the compressibility index. A value of zero for n represents an incompressible filter cake and increasing values of n represents increasing filter-cake compressibility. In the filtration studies of non-microbial suspensions, it has generally been found to accurately represent the pressure-dependence of the specific resistance at high pressures [15].

12.6.4 Relating cake compressibility to cake particles morphology

It is a known fact that the permeability is higher in a medium with larger particle size. The permeability is also higher in incompressible cake wherein the porosity does not decrease. Filterability of feed solution is influenced by particle properties such as size, size distribution, shape, and interaction with the medium fluid [18–22]. As seen from Table 12.9 the particle shape influences the specific cake resistance. Flakes that have

Table 12.8 Effect of pressure on porosity and specific filtration resistance

Δp bar	$\varepsilon_{avg\ estimated}$ m^3/m^3	α m/kg
630	0.8	3.74E + 12
1200	0.78	5.12E + 12
2200	0.77	9.64E + 12
2200	0.77	8.06E + 12
3300	0.76	1.26E + 13
3300	0.78	1.55E + 13

Table 12.9 Effect of particle shape on specific resistance of the filter cake[a]. Reprinted with permission from [12] © 1935 WILEY-VCH Verlag GmbH & Co. KGaA, Weinheim

Particle shape	Specific surface (So) 1/m	Specific resistance m/kg
Fibrous	4/a	1.6×10^9
Cylinder	4.1/a	1.7×10^9
Sphere	6/a	3.6×10^9
Flake	104/a	1100×10^9

[a]Wakeman R. The influence of particle properties on filtration. *Separation and Purification Technology* 2007; 58 (2): 234–241.

the highest surface area have the highest "α" and thus have greater cake compressibility than ellipsoidal particles [23]. Similarly the suspension concentration also influences the filterability: lower concentration leads to streamline flow, in other words less particle-particle interaction and thus smaller particle sizes; in contrast, higher particle concentration leads to a greater possibility of bridging and thus a smaller possibility of particles entering or covering the pores, which results in a permeable cake. Pretreatment of feed solutions prior to membrane separations, with particle aggregating processes such as sedimentation, flocculation and centrifugation, has been widely practiced to decrease the specific surface area and thus increase the porosity.

Flocculation of hydrolyzates using organic compounds such as Poly-Dadmac, Polyethylene-imine and also alum can help reduce the fouling of membranes and thus increase the efficiency of filtration i.e. filtration capacity [4]. The mechanism by which flocculants separate particles can be one or combination of the following: selection of the organic flocculants can be based on the either charge neutralizations, interparticle bridging, double layer repression, floc entrapment. Studies [4] have shown that Polydadmac can successfully separate lignin (36%), leave most of the sugar in solution (96%), and give bigger flocs, which can be separated using membrane separation. To give a better understanding, a experiment of pretreating hydrolyzate of a hardwood with an organic polymer was conducted to study the impact of flocculation on filtration efficiency.

As illustrated in Figure 12.7, flocculated hydrolyzate has higher flux than the unflocculated hydrolyzate. But an optimum amount of flocculant must be added to induce flocculation or else the problem of membrane fouling due to excess amount of polymer, i.e. flocculant added, is seen. In the lab study carried out, it was

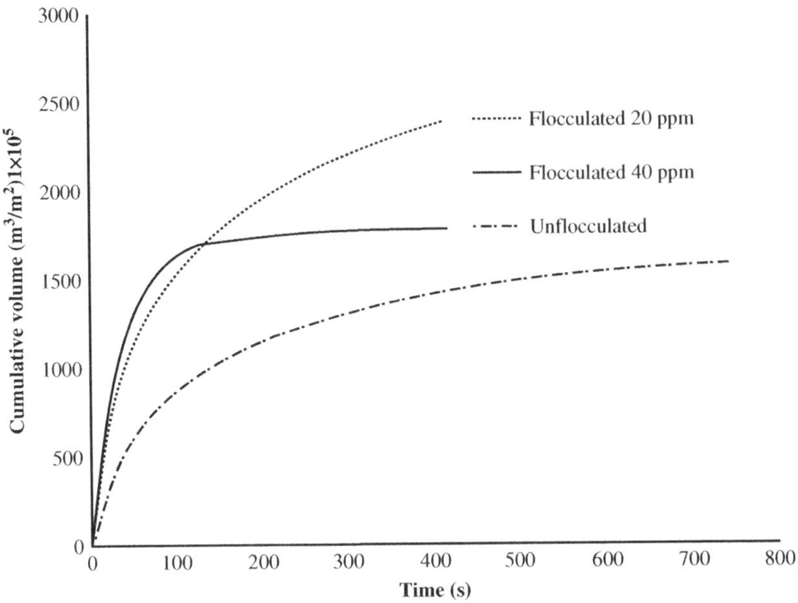

Figure 12.7 Effect of flocculation on permeate volume

observed that the 20 ppm and 40 ppm of a polymer flocculant produced the same quality of supernatant and the floc, but the 40 ppm fouled the membrane due to residual concentration of the flocculant in the suspension.

12.6.5 Effects of particles surface properties and the medium liquid

In addition to the physical properties, attributes of the feed such as its pH (Table 12.10), hydrophobicity, and surface charge [24–27] affect the specific cake resistance.

Similarly, the properties of the medium liquid, such as ionic properties and the composition, also influence the filterability. Investigations have shown that at constant pH and increasing ionic strength the specific resistance decreases. Particles tend to agglomerate at lower charge, thus having larger pores and thus better filterability [24–26, 28]. Using this property a filter aid/flocculant can be used to form permeable/incompressible cakes to achieve a zero zeta potential.

Table 12.10 Effect of [OH] on specific filtration resistances [2]

OH mol/kg	α m/kg
5.00E−04	3.00E + 10
1.00E−03	1.00E + 11
1.50E−03	2.50E + 11

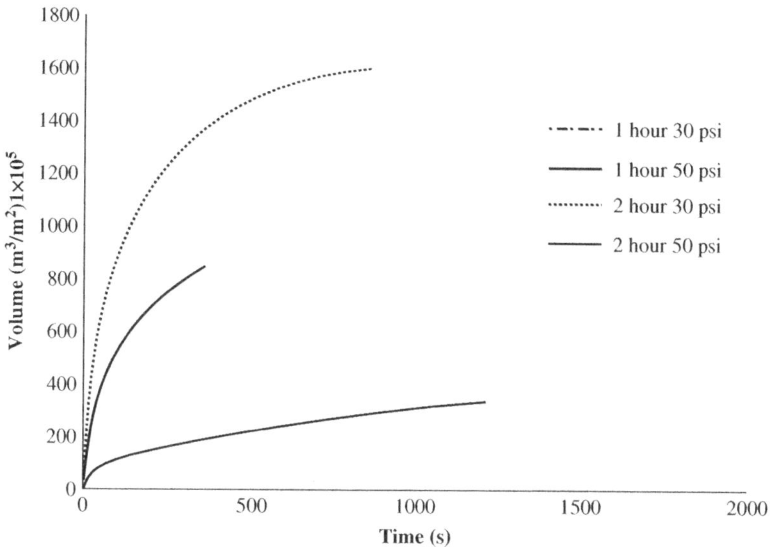

Figure 12.8 *Influence of extraction severity on permeate flow rate*

To summarize the experience of various research studies, the scale up of any filtration is a difficult process especially for compressible cake filtration which often require costly laboratory scale experiments. The information obtained from such experiments in combination with filtration models is very valuable.

As mentioned earlier, the mass removal, particle size, and thus the electrochemical stability of the hydrolyzate depends upon the extraction time. Higher the extraction time, larger is the particle size in the hydrolyzate [4]. The effect of different pretreatment severity/extraction time on mean particle size of a hardwood hydrolyzate can be seen in Table 12.4, while Figure 12.8 describes the effect of each of those variables on permeate volume. It can be seen that the smaller particle size fouls the membrane faster than the larger particle size. This is due to the lower floc formation capacity of particles at lower concentration, which have a larger surface area and thus lower porosity, which blocks membrane pores.

12.6.6 Fouling in filtration of lignocellulosic hydrolyzates

The hemicelluloses in wood extracts can be separated from inhibitory compounds. Ceramic microfilters of two different pore sizes, 0.2 μm and 0.01 μm, have been used to separate wood extracts [29] in a cross-flow configuration. This was the first separation step intended to prepare the permeate for subsequent nanofiltration where components will be separated. Large declines in the permeate fluxes were observed, indicating extensive fouling. The kinetics of the flux decay appeared to indicate pore blocking and the development of external fouling layers as the cause. Colloidal and particulate materials were separated from the extracts with turbidity reductions of 94 to close to 100% in most cases. No particulates were detected in the permeates. Significant separation of sugars assayed in xylose form was observed and could be correlated with the extent of fouling of the membranes. As the membrane pores are much larger than the xylo-oligomers, fouling layers built up during filtration seem to affect oligomeric separations.

The conditions that favor buildup of thicker fouling layers (larger pore size, higher TMP, and longer time into filtration) also seem to aid the retention of more hemicellulose sugars. In the most severe case of fouling, xylose concentrations in the permeate dropped to less than one-fifth of the feed values. Cleaning the membranes by soaking in alkali followed by alkali and de-ionized water rinse and back-flushing were effective in reversing most of the membrane fouling.

An experiment involving pretreating hydrolyzate of a hardwood with an organic polymer was conducted to study the impact of flocculation on filtration efficiency. Figure 12.5 shows the cumulative volume of the filtrate collected under constant pressure filtration at three different pressures. As pressure is increased, we observe the increase in filtrate fluxes initially (given by the slopes of these curves near the origin). However, the amount of filtrate produced decreases significantly with increased pressure, emphasizing that increased pressure leads to significant fouling of the membranes. For optimal filtration, lower pressure conditions are thus preferred. When the hydrolyzate was flocculated by a cationic polymer (in this case, PDADMAC at 47 ppm addition [30, 31] the flux is increased primarily due to decrease in the resistance of the aggregates.

12.7 Model development

One key advance in our ability to model fouling came about by combining the resistance of a fouling system due to the above factors linearly to describe the composite fouling evolution with time. The mathematical model developed by Ho and Zydney (2000) [32] (Eq. 12.8a and b) provides a comprehensive insight into the fouling behavior. According to the model, the initial drop in flow rate is a result of the pore blockage by the physical deposition of large aggregates on the membrane surface. Differing from previous models, this work accounted for the flow during pore blockage, with the specific cake resistance increasing with time as additional solid is deposited on the membrane surface. The Ho and Zydney model can be written for the volumetric flow rate of filtrate (Q, m^3/s) in terms of the time and specific filtration resistance characteristics of the cake layer and the occlusion of pores (α, R_p) as follows. The integral can be approximated resulting in the algebraic form of Eq (12.8.b). C_b represents the concentration of solids in the feed suspension, and the other symbols are as defined earlier.

$$Q = Q_0 \left[\exp\left(-\frac{\alpha \Delta P C_b}{\mu R_m} t \right) + \int_0^t \frac{\alpha \Delta P C_b}{\mu (R_m + R_p)} \times \exp\left(-\frac{\alpha \Delta P C_b}{\mu R_m} t_p \right) \right] dt_p \tag{12.8a}$$

Approximate solution

$$Q = Q_0 \left[\exp\left(-\frac{\alpha \Delta P C_b}{\mu R_m} t \right) + \frac{R_m}{R_m + R_p} \times \left(1 - \exp\left(-\frac{\alpha \Delta P C_b}{\mu R_m} t \right) \right) \right] \tag{12.8b}$$

The mathematical model developed by Ho and Zydney (2000) [32] (Eq. 12.8a) provides the most comprehensive insight into the fouling behavior. The model addresses the limitations of earlier models:

- The model accounts for initial flux decline due to pore blockage by physical deposition of large aggregates on the membrane surface.
- The model allowed flow during pore blockage and thus related the specific cake resistance to the amount of solid deposited and thus did not assume a uniform layer of cake.

The robust nature of this model made it possible for a smooth transition from pore blockage to cake filtration behavior during the course of the filtration thus eliminating the need to use completely separate mathematical descriptions in these fouling regimes, unlike the classical fouling models.

The predictions of the model derived by Ho and Zydney were compared with the laboratory observations for filtration of hardwood hydrolyzate. It has been observed (Figure 12.9) that the model predicts the filtrate

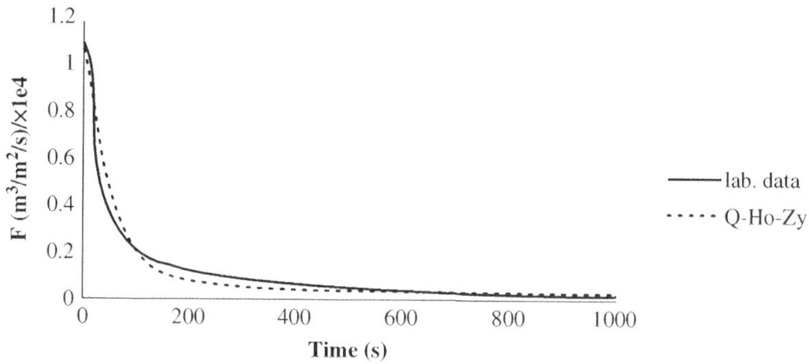

Figure 12.9 Comparison of prediction of Ho and Zydney model and actual data

flux and is suitable for application in a biorefinery project. The predictions of the model have been detailed in Table 12.11.

The parameters are described in Table 12.12.

The combination of amount of solids/particulate matter in the cake and the specific cake layer resistance of the hardwood hydrolyzate remains constant over different conditions but the resistance of a single solid aggregate changes. This is affected by extraction severity—lower resistance at higher severities—which can be attributed to the flocculation effect, as described in section 12.6.4. The effect of flocculation by an organic polymer has also been observed. Here, if an optimum amount of flocculant (20 ppm) was added, the Rpo was lower than the higher dose of flocculant, which blocked the membrane. The pore-blockage parameter was higher in a shorter extraction time due to standard blocking of the membrane resulting from shorter particle size in the feed stream.

Table 12.11 Variables quantified by Ho and Zydney model for different conditions

Extraction time (h)	Pressure	Rp0	f′R′	alfa
2	30psi	1E + 11	1E + 12	1
1	30 psi	1.00E + 12	1.00E + 12	1.4
2	Flocculated 40 ppm	5.00E + 12	1.00E + 12	1
2	Flocculated 20 ppm	1.00E + 11	2.00E + 12	1

Table 12.12 Ho and Zydney nomenclature

Rpo	Resistance of a single solid aggregate aggregate, m^{-1}
f′	Fractional amount of total solid content that contributes to deposit growth
R′	Specific cake layer resistance, m/kg
Rm	Resistance of the clean membrane, m^{-1}
alfa	Pore blockage parameter, m^2/kg

12.7.1 Requirements of a model

- The model calculations should be in excellent agreement with experimental data feed solutions over a range of bulk concentrations and transmembrane pressures.
- Models should be easy to use.
- Models should have good predictability.
- Models should have explanatory power.

12.8 Conclusions

Many studies [4, 8, 14–16, 19–30] have concluded that dead-end microfiltration is greatly affected by the specific resistance parameter. Research has shown that this parameter, α, is affected by the size and shape of solid loading, operating pressure, particles' surface properties, medium components, and by particle-particle and particle-medium interactions. The changes in loading solids orientation and deformation also have an affect on the mechanism of cake compressibility, which may lead to an interaction with membrane pore blocking.

The prediction of filtration rates, although difficult, is not impossible due to development of models such as Ho and Zydney (2000) [32], and Orsello (2006) [33], but laboratory experiments cannot be replaced. Thus the performance of a membrane filter solely based on modeling is very difficult to predict and model without actual laboratory experimental data, and currently dead-end microfiltration of biological suspensions largely remains an empirical science.

References

1. US DOE. 2005. Genomics:GTL Roadmap, DOE/SC-0090, U.S. Department of Energy Office of Science, p. 204.
2. Xiao, L., Sun, Z., Shi, Z. J., Xu, F., Sun, R. Impact of hot compressed water pretreatment on the structural changes of woody biomass for bioethanol production. *Bioresource Technol.* 2011; 6(2): 1576–1598.
3. Duarte, G. V., Ramarao, B. V., Amidon, T. E., Ferreira, P. T. Effect of hot water extraction on hardwood kraft pulp fibers (Acer saccharum, sugar maple). *Ind. Eng. Chem. Res.* 2011; 50(17): 9949–9959.
4. Duarte, G. V., Ramarao, B. V., Amidon, T. E. Polymer induced flocculation and separation of particulates from extracts of lignocellulosic materials. *Bioresour. Technol.* 2010; 101(22): 8526–8534.
5. Johansson, C. Pressure and Solidosity Profiles in Cake filtration. Ph.D. thesis, Chalmers University of Technology, Gothenburg, Sweden; 2005.
6. Grace, H. P. Resistance and compressibility of filter cakes. *Chem. Eng. Prog*. 1953; 49: 303.
7. Tarleton, E. S. *Determination of Cake Structure During Stepped and Variable Pressure Filtration*, Ninth World Filtration Congress, New Orleans, Louisiana, USA (2004).
8. Dingwell, K., Sedin, M., Theliander, H. *Filtration of lignin from a lignocellulose-based ethanol pilot plant. Environ. Eng. Sci.* 2011; 28: 775.
9. Sedin, P., Johansson, C., Theliander, H. *On the measurement and evaluation of pressure and solidosity in filtration. Chem. Eng. Res. Des.* 2003; 81: 1393.
10. Ruth, B. F. Correlating filtration theory with industrial practice. *Ind. Eng. Chem*. 1946; 38: 564.
11. Smiles, D. E. A theory of constant pressure filtration. *Chem. Eng. Sci*. 1970; 25: 985.
12. Chase, G. G., Willis, M. S., Kannel, J. *Averaging volume size determination of electroconductive porosity probes. Int. J. Multiphase Flow* 1990; 16: 103.
13. Hermans, P. H., Bredee, H. L. Zurkentnis der filtration sgesetze. *Recl. Trav. Chim. Pays-Bas* 1935; 54: 680–700.
14. Koros, W. J., Ma, Y. H., Shimidzu, T. Terminology for membranes and membrane processes (IUPAC Recommendations 1996). *Pure Appl. Chem.* 1996; 68: 1479.
15. Tien, C. *Introduction to Cake Filtration: Analyses, Experiments and Applications*. Amsterdam: Elsevier; 2006.

16. Tien, C., Ramarao, B. V. Revisiting the laws of filtration: An assessment of their use in identifying particle retention mechanisms in filtration. *J. Membr. Sci.* 2011; 383(1–2): 17–25.
17. McCarthy, A. A., Conroy, H., Walsh, P. K., Foley, G. The effect of pressure on the specific resistance of yeast filter cakes during dead-end filtration in the range 30–500 kPa. *Biotechnol. Tech.* 1998; 12(12): 909–912.
18. Wakeman, R. The influence of particle properties on filtration. *Sep. Purificat. Technol.* 2007; 58(2): 234–241.
19. McCarthy, A. A., O'Shea, D. G., Murray, N. T., Walsh, P. K., Foley, G. Effect of cell morphology on dead-end filtration of the dimorphic yeast Kluyveromyces marxianus var. marxianus NRRLy2415. *Biotechnol. Prog.* 1998; 14(2): 279–285.
20. Shimizu, Y., Shimodera, K., Watanabe, A. Cross-flow microfiltration of bacterial cells. *J. Ferment. Bioeng.* 1993; 76(6): 493–500.
21. Nakanishi, K., Tadokoro, T., Matsuno, R. On the specific resistance of microorganisms. *Chem. Eng. Commun.* 1987; 62(1–6): 187–201.
22. Tanaka, T., Abe, K., Asakawa, H., Yoshida, H., Nakanishi, K. Filtration characteristics and structure of cake in crossflow filtration of bacterial suspension. *J Ferment Bioeng* 1994; 78(6): 455–461.
23. Tanaka, T., Usui, K., Kouda, K., Nakanishi, K. Filtration behaviors of rod-shaped bacterial broths in unsteady-state phase of cross-flow filtration. *J. Chem. Eng. Japan* 1996; 29(6): 973–981.
24. Ohmori, K., Glatz, C. E. Effects of pH and ionic strength on microfiltration of C. glutamicum. *J. Membr. Sci.* 1999 2/3; 153(1): 23–32.
25. Ohmori, K., Glatz, C. E. Erratum: Effects of pH and ionic strength on microfiltration of C. glutamicum. *J. Membr. Sci.* 1999; 163(1): 151–152.
26. Ohmori, K., Glatz, C. E. Effect of carbon source on microfiltration of Corynebacterium glutamicum. *J Membr Sci* 2000; 171(2): 263–271.
27. Okamoto, Y., Ohmori, K., Glatz, C. E. Harvest time effects on membrane cake resistance of Escherichia coli broth. *J. Membr. Sci.* 2001; 190(1): 93–106.
28. Shirato, S., Esumi, S. Filtration of a culture broth of Streptomyces griseus. *J. Ferment. Technol.* 1963; 41: 87–94.
29. Hasan, A., Yasarla, L. R., Ramarao, B. V., Amidon, T. E. Separation of lignocellulosic hydrolyzates using ceramic microfilters. *J. Wood Chem. Tech.*, 2011; 31(4): 357–383.
30. Duarte, G.V., Ramarao, B. V., Amidon, T. E. Colloidal stability and flocculation of lignocellulosic hydrolyzates. *Bioresource Technol.* 2010; 101: 8526–8534.
31. Yasarla, L. R., Ramarao, B. V. Dynamics of flocculation of lignocellulosic hydrolyzates by polymers. *Ind. Eng. Chem. Res.* 2012; 51(19): 6847–6861.
32. Ho, C., Zydney, A. L. A combined pore blockage and cake filtration model for protein fouling during microfiltration. *J. Colloid Interface. Sci.* 2000; 232(2): 389–399.
33. Duclos-Orsello, C., Li, W., Ho, C. A three mechanism model to describe fouling of microfiltration membranes. *J. Membr. Sci.* 2006; 280(1–2): 856–866.

13

Solid–Liquid Extraction in Biorefinery

Zurina Zainal Abidin, Dayang Radiah Awang Biak, Hamdan Mohamed Yusoff and Mohd Yusof Harun

Department of Chemical and Environmental Engineering, Faculty of Engineering, Universiti Putra Malaysia, Malaysia

13.1 Introduction

Solid–liquid extraction (SLE) is a common separation method employed in industry. It is used for various applications including purification, enrichment, recovery, removal, and fractionation. Solid–liquid extraction, otherwise known as leaching, is a process that uses liquid solvent to extract the soluble substance (solute) residing within a solid matrix. The principle of SLE is similar to liquid–liquid extraction. Solid and liquid phases are in close contact, and the solutes from the solid diffuse out into the liquid. The partitioning of solutes between the two phases continues until equilibrium is attained. The traditional extraction method uses mechanical stress for expression of the solute by crushing the solid using a screw press. This method has a low yield, and the product requires post-processing treatment to achieve an acceptable quality.

The biorefinery industry produces bio-based products such as biofuels, bioenergy, and biochemicals [1]. Bioethanol, biodiesel and biogas are typical examples of biofuels. Bioenergy generation at a biorefinery plant is normally accomplished by harvesting the heat energy resulting from a particular process that generates steam to run turbines. Other bio-based products include carbohydrates, lignin, lipids, vegetable oil, fine chemicals, and bulk chemicals. Since the production of bioenergy is not directly related to SLE, this chapter will focus on bio-based products including biofuels.

In downstream processing, SLE usually appears at an early stage of product recovery. In a biorefinery, the raw materials are plant/lignocellulosic materials, which are mainly composed of cellulose, hemicelluloses, and lignin. Typical examples of biorefinery processes that employ SLE are leaching of sugar from sugar beet and sugarcane using hot water and extraction of oil from corn, rice bran, peanuts, rapeseed, *jatropha curcas* seed, soybeans and sunflower seeds using organic solvents. Solid–liquid extraction can also be

Separation and Purification Technologies in Biorefineries, First Edition.
Edited by Shri Ramaswamy, Hua-Jiang Huang, and Bandaru V. Ramarao.

used to recover active ingredients, metabolites, enzyme, and pharmaceutical products by leaching plants' roots, leaves and stems.

There have been major improvements in the design and modification of the extractors: there has also been progressive improvement in the SLE field through the use of other methods in combination with conventional ones. This intensifies the process, and includes the use of supercritical fluid, ultrasound, microwave, pulsed electric field-assisted processes, and others [2–4].

This chapter gives an overview of the current status of SLE in the biorefinery industry, and discusses SLE principles, process design, and modeling considerations, as well as elaborating on the industrial application of SLE by providing some examples. Finally, the economic importance and challenges of SLE are discussed.

13.2 Principles of solid–liquid extraction

Solid–liquid extraction allows soluble components (solute) to be removed from solids using a solvent. The SLE process can be considered in three stages:

1. Diffusion of the solvent through the pores of solid.
2. Dissolution of the solute in the solvent to form miscella, as the solvent enters the solid.
3. Transfer of the miscella back to the bulk solvent.

The basic principle behind SLE is the solute concentration difference between the solid and liquid, which causes solute molecules to diffuse from one to the other. However, the diffusion of the solute residing inside the solid is not the only mechanism involved in SLE process. Washing the solute from the solid surface, displacing the extract from inter-particle pores, and solubilization (or reaction-induced creation of soluble solutes from insoluble precursors) may also occur during SLE [5]. Figure 13.1 explains how the SLE process occurs. A good example is the sugar-refining process with hot water used as a solvent. The initial concentration of sugar in the hot water is zero, so there is a concentration driving force for the sugar to diffuse into the hot water. This mass transfer continues until the equilibrium concentrations in the solid and liquid phases are reached.

For plant materials, a pretreatment step is required to ensure efficient and rapid removal of the desired compounds. For example, nuts and seeds are ground or flaked to ensure rupture of the cell walls and

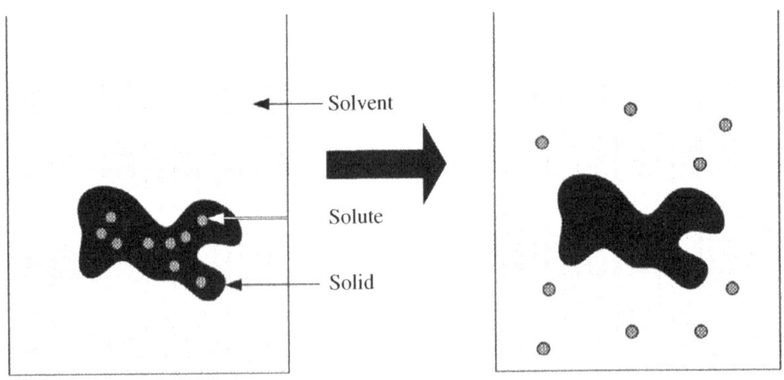

Figure 13.1 Schematic of extraction process

allow efficient mass transfer of desired compounds into the extraction liquid. The extraction of coffee from ground beans or tea from tea leaves is best accomplished after drying and grinding the raw materials. This pretreated solid has a larger mass transfer area and a shorter diffusion path compared to an untreated solid.

The distribution of solute between the raffinate—the liquid phase remaining in the solid after the extraction—and the extract can be expressed in terms of the partition coefficient, K:

$$K = \frac{C_E}{C_R}$$

where

C_E = equilibrium solute concentration in extracting solvent (kgm^{-3})
C_R = equilibrium solute concentration in raffinate (kgm^{-3})

The extraction factor, λ, is defined as amount of solute in extracting solvent divided by amount of solute in raffinate, where

$$\lambda = EC_E/RC_R = K\ (E/R)$$

E and R are the volume of extracting solvent and initial solvent, respectively. The value of K is often independent of the solute concentration, particularly at low solute concentrations. However, at higher solute concentrations, deviation from the linearity between C_R and C_E may be observed in some systems. The partition coefficient of a solute between two phases is usually determined by experimental methods.

13.2.1 Extraction mode

Most SLE equipment in biorefineries works on the following principles [6]:

1. Percolation
   In percolation, the solvent flows through solid fixture. In batch process, the solid is a fixed bed, whereas in continuous process, the solid is retained on a perforated conveyor or basket. This method is very popular for extraction of large amounts of solids [7].
2. Dispersed solids
   In this mode, the solid is retained in the solvent throughout the extraction process. The solid is usually pretreated, i.e. pulverized and/or dried, before being fed into the extractor. This method produces a higher yield compared to the percolation method but the operating cost is higher. There are two popular methods that fall under this category: immersion (immerse the solid particles in the stirred solvent), and maceration (soak the solid in the solvent without agitation) [8].

The liquid-to-solid contact area is where extraction occurs and maximizing this essential factor will reduce the corresponding equipment's size. Solvent selection plays an important role in solubility, as well as in the separation steps that follow SLE. Whether the SLE process is carried out by percolation or the dispersed-solids method, there are four important factors that can influence the process: temperature, contact area, extraction time, and solvent selection. Agitation is also employed to enhance mass transfer and to ensure proper mixing, as well as to prevent sedimentation of fine solid particles. Solid–liquid-extraction processes can be performed using either a batch or continuous system, depending on process requirements.

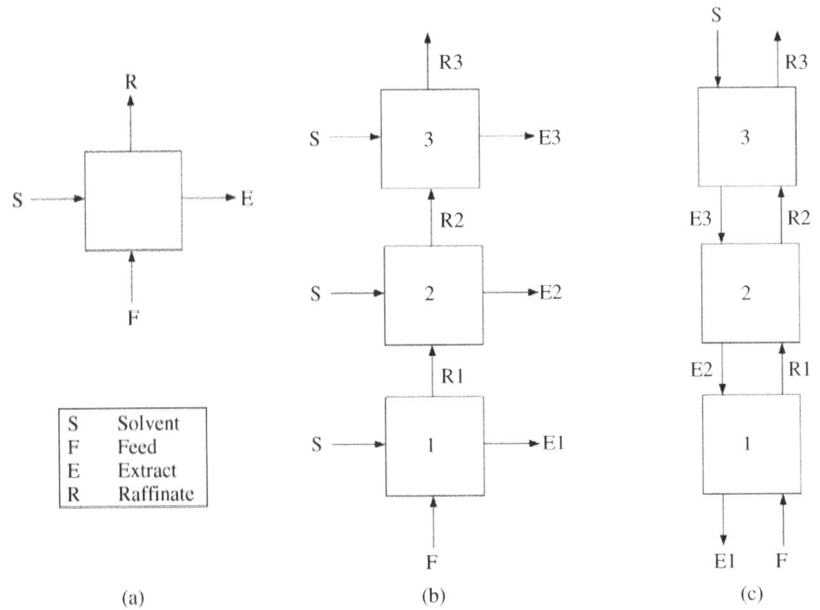

Figure 13.2 *Single batch, cross current and countercurrent SLE set-up*

13.2.1.1 Single-stage, batch

Solvent is allowed to come into contact with the prepared solids, and batch time is determined by rates of diffusion of the soluble components out of the solids. Once a batch of solids has been extracted, the vessel is emptied, cleaned, and refilled with a new charge of solids. This operation is illustrated in Figure 13.2(a).

13.2.1.2 Multistage crosscurrent flow

In this set-up, the solid comes in cross contact, for example in a perpendicular direction, with the solvent stream, as illustrated in Figure 13.2(b). The raffinate (solvent stream that remains after an extraction stage) from the first stage will be used to treat the solid in the subsequent stages. A two-stage crosscurrent operation is superior to a single stage that utilizes a similar amount of solvent but does not use the entire solvent capacity.

13.2.1.3 Multistage countercurrent flow

This setup is the most widely applied in industrial extractor. In the first stage, fresh solid enters while the final raffinate leaves. The fresh solvent is introduced at the vessel containing the most exhausted solid (the last stage).The extract retrieved from this vessel is passed through a battery of extractors successively until it arrives at the vessel that is most recently loaded with fresh solid (the first stage). This process is shown in Figure 13.2(c).

13.2.2 Solid–liquid extraction techniques

13.2.2.1 Solvent extraction

In this technique, a compatible solvent is added to a solid in order to extract the solute from the solids. The exhausted solid (solid discharge) will later be separated from the miscella (mixture of solvent and solute) by filtration or gravity. The extract in the miscella will be purified using evaporation, drying, or distillation.

13.2.2.2 High-pressure extraction

High-pressure extraction (HPE) is an alternative technique for extracting active compounds from plant materials, especially substances that are sensitive to high-temperature processing as in typical solvent extraction. This method utilizes high pressure, up to 1000 MPa. High pressure can cause some structural changes on the solids, thus improving the mass transfer rate, and subsequently providing faster and more effective extraction.

The most popular HPE is supercritical fluid extraction (SFE). The basis of SFE is the use of solvent that is near the supercritical phase, i.e. densified gas. At this state, the density of the solvent is very low, which would facilitate its penetration into the solid during the extraction process. In addition, in this high pressure condition, the solubility of the solute increases. Carbon dioxide is the common solvent for this kind of extraction since it is non-toxic, inflammable and available abundantly.

13.2.2.3 Ultrasonic-assisted extraction

Ultrasonic-assisted extraction (UAE) uses ultrasonic waves (18–20 kHz) to aid the extraction process by disrupting the solid structure [9]. The structural change is induced by the cavitation effects from the ultrasonic irradiation. Several investigations on the application of ultrasonication in pretreatment and extraction of plant materials have been reported [10, 11].

13.2.2.4 Microwave-assisted extraction

Microwave-assisted extraction (MAE) has many advantages over other extraction methods, such as low solvent usage and shorter extraction time. Numerous experiments have shown promising improvement in extraction of substances from plant material using MAE [2, 12, 13].

13.2.2.5 Heat reflux extraction

Reflux extraction is done by boiling the material in the solvent, and a chilled surface is used to condense the rising solvent vapors. The solvent vapors are returned to a liquid state in the container and continue to be boiled without the solvent boiling away. Then, the extract continues to concentrate in the solvent until it is reduced to essence later.

13.3 State of the art technology

The state-of-the-art technology for SLE in the biorefinery industry involves not only the equipment design and operation but also the process model. The most popular application for plant-based extraction are SFE, a process widely used for pharmaceutical and food products [14], and the solvent extraction process that includes accelerated solvent and high-pressure extraction [15]. Typically, the temperature of the two named processes can reach up to 100–200°C for a conventional solvent extraction process. Other new methods that are being investigated and tested on a laboratory scale include the microwave-assisted (MAE) and ultrasound-assisted extraction (UAE) processes [16]. However, the constraints in equipment and scale-up have hindered its application at industrial scale.

Feedstock for bio-based products synthesis including biofuels are mainly plant-based or of biomass origin. Conventional methods such as maceration, percolation and hydro-distillation are often the popular choice for processing such feedstock because these methods were fully understood and proven to produce satisfactory yield of product. The process was carried out in batch mode due to the required contact time for efficient extraction. The main constraint of these conventional methods is the long processing time and require large volumetric vessel to cater for a larger operation. Efficient leaching will require a multistage operation to provide a better product yield [14]. This multistage operation can be modified and configured accordingly in order to optimize the extraction time and solvent usage. Countercurrent has been preferred since it promises a highest rate of extract recovery. Here, the fresh solvent is brought into contact the extracted biomass, while the fresh biomass is fed to contact with the most concentrated solvent. The maceration process can also be enhanced through the application of microwave and ultrasound to reduce the processing time. However, as mentioned earlier, these supplementary methods are still under investigation [17] even though the current process can have a working volume up to 6000 L [14].

Prolonged processing period in maceration/immersion method makes the percolation process more favorable. Here, the solvent flows through a bed containing solid particles to be extracted. The liquid flows in down flow mode with no further requirement for solid-liquid separation like filtration or centrifugation [18]. Large quantity of solid can be treated with large volume of solvent due to the high concentration gradient across the extraction column. Perfect contact between the solid matrices and the solvent assists efficient solute transfer between the two phases to give an acceptable yield. Various process modifications like diffusion and osmosis principles [14], are implemented to improve the yield and reduce the extraction time. In addition, proper consideration on temperature and pressure is vital to determine the overall yield. Current industrial practices choose to either extract at a higher temperature or extend the contact time between the solid particles and the solvent.

The steam distillation and hydro distillation processes are mainly used to extract oil containing volatile components from plant materials. The use of water (steam) as solvent has long been used in essential oil industries, for example agarwood oleoresin, thyme, lavender, patchouli, and so forth [19–21]. This technique is also applied in the oilseed extraction process (vegetable oil production), mainly to remove the excess solvent contained in extracted oil products.

However, in industry, the vegetable oil extraction and fractionation processes are generally mechanical (boiling for fruits, pressing for seeds and nuts) or involve the use of solvents such as hexane [22, 23]. After boiling, the liquid oil is skimmed; after pressing, the oil is filtered; and after solvent extraction, the crude oil is separated, and the solvent is evaporated and then recovered. Sunflower, safflower, soybean, cottonseed, rapeseed, peanut, and palm oil are the potential renewable resources for biofuel applications [24]. For large-scale operations, continuous processes are favored. The design of equipment for continuous process is relatively compact and small [14]. Industrial units are available for a system with cross current and countercurrent flow of liquid (solvent) and solid. For SLE percolators, the operations (process designs) are considered pseudo-countercurrent due to the still position of the fixed bed column while the solvent flows through it. The commercially available SLE units include the tower extractor by Crown Iron Works, the screw extractor by GEA Niro, the Vatron Mau, Carousel™ extractor by Harburg-Freudenberger, the REFLEX™ and LM™ by De Smet, and the Sliding Cell™ extractor by Lurgi [14]. These units have a working capacity of between 100 to 17 000 tonnes per day [14].

The development of statistical models based on experimental design and the application of various process flows is one of the new modeling approaches in SLE. Distributed plug flow [25] and shrinking core models [26] were developed to imitate the actual extraction process. Here the phytology of the raw material, for example plant material, is also incorporated into the model. One of the most frequently used tools to predict such behavior is the response surface methodology (RSM). This method generates a statistical model to optimize the extraction processes. It is also capable of identifying significant process parameters that affect the extraction yield, as well as predicting the interactions between parameters (more than two). The targeted component yield, purity or cost, are set as a function of process variables, for example pressure, temperature, particle size, and extraction time. The effects of solvents, such as polarity, however, have to be addressed separately. Statistical models are then developed by fitting the attained values using polynomial regression. Later, pilot-scale tests can be performed based on the optimized result. Even though this model is able to demonstrate the process in a predefined condition, the mechanisms involved in the process are yet to be described. Thus, an in-depth understanding of the process related to the botanical aspects of the raw material is still needed. In the absence of this information, process intensification on the extraction cannot be undertaken.

There have also been attempts to integrate the process nature into the process model [21]—involving both the macroscale and microscale elements of the process. The macroscale elements are the equipment used in process design, whilst the microscale describes the particle size of the solid or the plant's cell wall [14]. The success of the integration between the process nature and the process model can be validated by the output of the pilot scale test.

It can be deduced that solid–liquid extraction for biorefinery operation is in need of further academic research in order to fully comprehend the process behavior. Whilst various process modifications and improvements have been made in the laboratory with the classical SLE approach, there is still a lack of transfer from academic research to feasible resolutions for large-scale industrial operations.

13.4 Design and modeling of SLE process

The success of any extraction process demands both theory and practical implementation especially during designing and modeling process parameters. Optimization in extraction process is an important step to achieve a reasonably high yield of the product within an acceptable time at a lower cost.

Solid–liquid extraction of biorefinery products generally consists of several processes, namely pretreatment of raw material, extraction, solute separation, and solvent recovery. These processes determine the technical and economic feasibility of plant material extraction. Each of these processes requires proper design, modeling and optimization to assist efficient product recovery without jeopardizing the quality in the most economical way. This discussion highlights important aspects to be considered for pretreatment, extraction, equipment and process operation.

13.4.1 Pretreatment of raw materials

As described in Section 13.2, any of the three steps in the extraction mechanism can limit the process rate. Several studies have indicated that, for plant material, the diffusion of solute into and out of the solid matrix is the rate-controlling factor [27, 28]. Hence, pretreatment of raw material prior to extraction process determines the way extraction proceeds and subsequently its efficiency. In designing and modeling of any pretreatment process, one may take into account the selection of suitable methods, nature and composition of solid, size of the solid feed, type of pretreatment techniques, moisture, and heat content.

Solid–liquid extraction in biorefinery deals with agricultural resources like corn, corn stover, straw, and others. A plant cell is protected by a cell wall that supports its rigid structure. The vacuole, nucleus and microbodies are present in the cytoplasmic area enclosed by a cell membrane. The solute is available inside the body of solid, on the surface of the solid, inside pores, or scattered within the plant structure, i.e. leaf, stem, flower and root. The extent of cell rupture and also the cell's complex structures, determine the solute's diffusivity.

There are two scenarios associated with pretreatment; (i) the cell wall is disrupted, and (ii) the cell wall remains intact. Typically, cell walls can be disrupted by mechanical, chemical, physical, thermal or biological methods [29]. For mechanical methods, a bead mill and homogenizer are used to break the cell wall. Small abrasive particles are used to disrupt the cell wall by high shear force resulting from grinding between beads and collision with beads. Some researchers have demonstrated the use of ultrasound to induce cell disruption for bio-product extraction at laboratory scale [2, 17, 30]. Osmotic pressure is sometimes adopted for cell lysis by increasing the volume of plant material and promoting cell-wall permeability. Figure 13.3 shows how a water hyacinth's cell wall becomes disrupted after being subjected to ultrasound and steaming treatment to assist the subsequent hydrolysis step for bioethanol

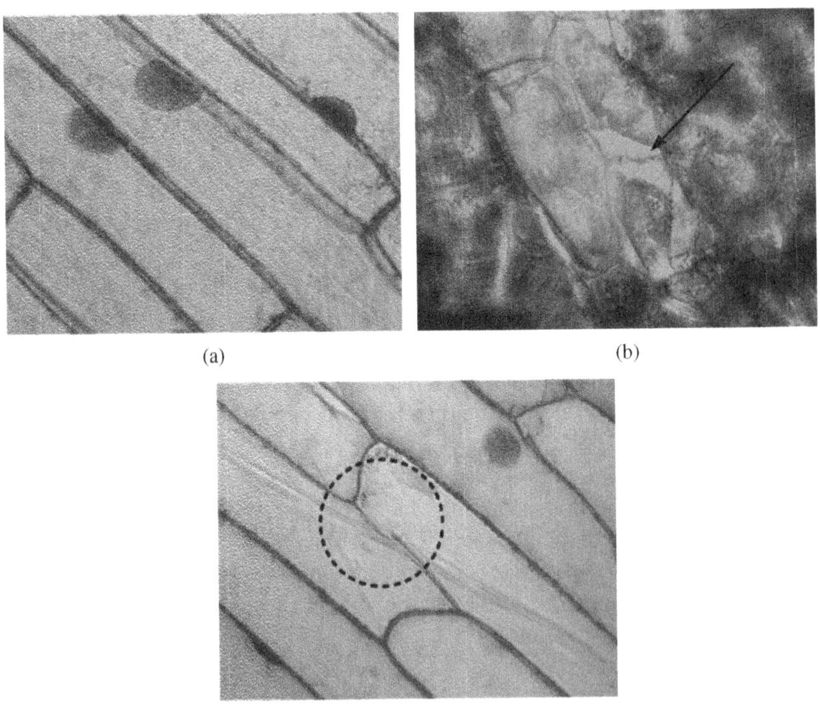

(a) (b) (c)

Figure 13.3 Cell morphology of water hyacinth plant at cellular level (1000x) for (a) untreated, (b) steamed sample and (c) ultrasonicated sample. Cell wall is seen to be disrupted after pretreated using osmotic shock and ultrasonication. Reprinted from [30] © 2011, with permission from Elsevier

production. Alternatively, for a thermally stable product, heat shock can be used to disrupt the cell. Enzymatic treatment using, for example, cellulose and pectinase, is good as it is very specific, harmless to the environment, poses a small risk to product damage and requires low energy [31]. However, this method suffers from high cost and may require subsequent treatment for removing the product from the enzyme. Additional pretreatment like hydrolysis can be conducted using acid or enzyme to breakdown the strong lignin and cellulosic structure to assist subsequent product recovery or extraction [32].

If the cell wall remains intact during contact with a solvent, the extraction process then depends solely on the diffusion of the solute through the cell walls, controlled by osmotic pressure. In such cases, size reduction can be one of the ways to reduce the distance that a solvent must travel to diffuse within the plant cell, in order to dissolve the solute and then diffuse back to the bulk solution for separation to occur. Size reduction can be achieved by crushing, grinding, flaking, milling, or cutting into small pieces. Shape modifications, like decortications, rolling, screw pressing, extruding, and pelletizing can improve the mass transfer in a process that employs the percolation method. It is vital that the solid is of a suitable size (surface area per unit volume) to increase the interfacial area for facilitating solute accessibility to the solvent. For instance, in the process of oilseed extraction, oil is retained in bodies that are enclosed in a phospholipid monolayer membrane, embedded with proteins such as oleosin, caleosins, and steroleosins [33, 34]. During the extraction, the oil travels from the enclosed bodies the to bulk solvent through many diffusion processes, which occur at various stages of extraction, including from the cell membrane to solvent to form a solution, from the inner solid to the surface of the solid, and finally from surface of solid to the bulk of the solvents. It is therefore very important to reduce the distance travelled by the oil solute (resistance) in reaching the solvents.

Typically, the main concern for the raw material lies in its size. Fine sizes can create undesirable dense solid packing and may contribute to the blockage of the interstices of larger particles. Good circulation of the solvent is hindered and this condition leads to a poor extraction process. Using a fine solid can not only cause difficulty in later solid separation and drainage of solid residue but also require high energy usage in solid preparation. The presence of impurities will further aggravate the situations. Besides size, other factors like thickness, shape, size homogeneity and internal structure of the solid particles are also important [35, 36].

13.4.2 Solid–liquid extraction

The pretreatment process is followed by the main leaching process. The SLE process is generally governed by a few main factors such as the chemical and physical properties of the separation process and product, solvent characteristics, operating conditions, prepared feed, and process setup. Proper selection and optimization of these factors can lead to better product yield and quality.

The selection of solvent for extraction is normally based on a number of characteristics, such as capacity, selectivity, chemical inertness, thermophysical properties, flammability, toxicity, cost, and availability. The most suitable solvent depends very much on the nature of the bioproducts to be extracted. A good solvent should be selective—it has higher affinity towards the solute of interest and is less soluble for solid matrix and other major components. Solvents are categorized according to the number of functional groups present in their molecules, solvent strength, and selectivity values [37]. Good solvents normally have similar functional groups as the solute. This characteristic determines the compatibility of the solute and solvent during the formations of physical and chemical interactions. Better product yield and quality can be attained through the use of the correct combination of solvent and solute. Non-polar solutes generally prefer non-polar solvents. Hexane is known to be the most widely selected solvent to be used for extracting non-polar solutes like oil.

The extraction process must be manipulated as efficiently as possible to get a high yield and high purity of desired components with minimal carry-over of undesired byproducts or impurities. For instance, in the process of vegetable oil extraction, the suitable candidate for solvent should have high triglyceride selectivity and low reactivity with the oilseeds. Low carry-over of byproducts such as free fatty acids is desirable if the oil is to be used for biodiesel production. The solvents should also possess low specific heat and low latent heat of vaporization to assist separation and recovery after the extraction process. Low viscosity of the solvent is desirable for easy flow with a good wetting ability on the solid matrix so that good penetration through seed capillaries and pores can be achieved. The initial use of pure solvent can establish a good driving momentum for the extraction process.

Other factors such as structure, boiling point, price, and safety are also among the important criteria in the selection of a compatible solvent. Hexane is used for the extraction of non-polar solute due to low cost and toxicity, good extraction efficiency, and ease of operation compared to other non-polar solvents such as petroleum ether, benzene, chloroform, isopropanol, and toluene. However, the concern with hexane is that it is flammable, explosive, and toxic, and this has prompted the industry to find a more environmentally friendly solvent. Other commonly used solvents for plant extraction are water, alcohols, glycol, acetates, and compressed/supercritical gas. The selection of solvents for extraction of the product with certain applications like pharmaceutical products, cosmetics, and foods, must comply with laws and regulations. Recently, research in using surfactant-water solution and ionic liquids as alternative solvents for the extraction process has been increasing due to issues like strict regulations, consumer acceptance, and environmental concerns [38]. Ionic liquids, for instance, have low vapor pressure to reduce the emission of volatile organic compounds. However, most ionic liquids are water soluble and can pose a threat to the environment if they leach out. Further research into the use of ionic liquids on the extraction of plant material should be undertaken. Issues concerning the recovery of ionic liquids from biomass and bioproducts have yet to be resolved.

The solubility of solvents also plays a major role in the extraction. Solubility is affected by the operating temperature. High temperature favors greater solubility of solute in solvent by increasing the diffusion coefficient and hence the extraction rate. A significant increase in bioproduct yield is obtained when the extraction temperature is near to the boiling point of the solvent. However, too high a temperature can cause solvent losses by evaporation, extraction of undesirable constituents, or damage to sensitive equipment. Normally, the upper limit of the temperature is determined by secondary considerations, such as enzyme attacks and product damage.

The solvent-to-solid ratio is another important parameter that affects the extraction process. The solvent-to-solid ratio is the ratio of solvent per feed used in the extraction process. Basically, a sufficient amount of solvent should be available for dissolving the solute and transferring it to the exterior of the solid matrix. There is no particular rule of thumb on the ratio of solvent to solid used. Many researchers have reported the use of different solvent to solid ratios. For instance, a 6:1 ratio was used to extract jatropha oil [2] and castor oil [39] respectively. A 4:1 ratio has been employed to extract olive foot cake [40] and 3:1 to 10:1 ratios were used to extract oil from tobacco [17]. The amount of solute in solid is fixed and thus requires a certain amount of solvent. Excessive use of solvents is commonly unfavorable. To improve SLE process, agitation of fluids can promote effective use of all the interfacial surface of solid, eddy diffusion, mass transfer of solute from surface of solid to bulk of solutions. This method can also prevent sedimentation of fine particles.

13.4.3 Equipment and operational setup

The efficiency of the extraction process partly depends on equipment selection, mode of operation (batch or continuous), solid handling (fixed bed, percolation, full immersion, intermittent drainage or dispersed/moving contact), arrangement of operation (single stage or multistage), and also dimension of extractor unit.

Percolation is often preferred for solid materials that have low internal resistance to mass transfer, while immersion is a better option for cases where the mass transfer in the solid is high. In percolation, a high velocity of solvent is necessary to wash the extract from the solid. Quite often the solid material is subjected to certain modifications, such as pelletization or granulation in order to sustain the high velocity of the solvent flow. For immersion, it is mostly sufficient to reduce the solid to smaller size in order to reduce diffusion length inside the solid. Immersion is also suitable for solids that tend to swell or disintegrate. However, in comparison to percolation, the immersion process requires an additional step for separating the solid from the solvent after the extraction. This is the reason why the industry preferred percolation over immersion methodology.

Batch extraction is not a common choice at industrial scale due to high hold-up time for the charging and discharging processes, and also high consumption of solvents. For low-volume products such as flavor and bioactive compounds, batch extraction is still applicable. Several improvements have been incorporated into the batch processing scheme, such as the combination of several batch extractors in a multistage operation, which give a better yield and also reduce the consumption of solvents that leads to lower energy usage for solvent recovery. In addition, this arrangement may back-up the production line in the event of scheduled maintenance or unplanned shutdown. Such operations require each extractor to be commissioned and maintained individually.

Cross-flow extraction is not preferred in industry, since large amounts of solvent are used without utilizing its solvent-extracting capacity efficiently, although higher product rates can be achieved. Continuous countercurrent systems are widely used as they are efficient in terms of yield, and solvent usage, and are also economical and convenient in operation. Several different type of industrial extractor-based countercurrent mode, either using immersion or percolation, have been developed for various applications. In countercurrent mode, the fresh solvent is placed in contact with the most depleted plant material while, conversely, the most enriched extract of plant material meets the fresh plant material. This produces a large concentration gradient of solute between solid and liquid for driving a good mass transfer process.

Different types of plant materials have different physical properties and may therefore require different types of extraction equipment. The quantities of the plant material to be handled at one time, such as single/multicomponents system, labor, cost, and also product changeover, are among other factors that need to be considered in the selection of the most suitable equipment to be used.

13.4.4 Process modeling

Process modeling is a systematic approach to describe a process and to predict the response of a system when subjected to various disturbances. Extraction performance is governed by mass transfer and also equilibrium conditions. Modeling of the extraction process requires knowledge of reaction kinetics, and of influencing chemical or physical factors by analyzing experimental/plant data. Extraction modeling is dealt with at the macroscale (equipment) level and also the microscale (plant cell) level [8, 14].

Extraction models proposed for the extraction of plant material can be divided into two main groups, based on empirical or differential mass balance models. The empirical model or mathematical equation is built based on the yield as a function of pressure, temperature, particle size, extraction time, solvent type, and also concentration. This is useful when the mass transfer mechanisms and equilibrium are unknown.

Differential mass-balance models start by assuming the mass transfer of solute from natural resources can be described by the desorption of the solute from the solid, diffusion of the solute into the solvent, and transfer of the solute in the solvent across the liquid boundary layer into the bulk solution. The two processes are often the rate-limiting step, which can be explained by Fick's Second Law of

Diffusion [28, 41]. Many assumptions and estimations have been made to simplify the model for a more realistic approach. Some investigators have modeled the process by only considering mass transfer resistance in the liquid phase, i.e. solvent and miscella [42, 43]. However, for the extraction of plant materials, the structure and also the location of the solute contribute significantly towards limiting the process and some models have attempted to include these factors [27, 44].

The adsorption/desorption equilibrium model considers the target product to be adsorbed onto the solid. The solvent diffuses through the pores of the solid to extract the product and then diffuses back into the bulk solution. The adsorption equilibrium can be described using linear [45], Langmuir [46], Freundlich, Henry or Brunauer–Emmett–Teller (BET) models depending on the assumptions made.

One of the widely used differential mass-balance models is the shrinking-core model, which depicts a sequence of irreversible desorption followed by diffusion through the pores within the solid. A particle is assumed to be divided into distinct extracted and non-extracted parts, having three zones—core, inner, and outer [46, 47]. The solute located within the core is continually extracted, resulting in the shrinkage of core size with time. When the solute concentration is greater than the solubility of the solute in the solvent phase, a constant concentration is achieved. The mass transfer in the outer zone is governed by effective diffusion.

With technological advancements, researchers have started to perform microanalysis of solid structures using scanning probes, electron microscopy, and optical methods. The extraction process was first visualized as the intact and broken-cells concept [47–50]. In this approach, for an easily accessible solute (when the cell walls are broken), the mass transfer resistance comes mainly from the fluid. For less accessible solutes—intact cells—the diffusion of solute in the solid phase is dominant. The diffusion rates of both regions differ significantly, where an initial fast extraction phase resulting from broken cells is followed by the slower extraction from intact cells. This reduces the model parameters to only the internal mass transfer coefficient, compared to that in the initial Fick's Law. Attempts have also been made to visualize the solute to be located in secretory structures, which can be cells, cavities, or ducts depending on the raw material [51, 52].

Other authors [46, 53] have attempted to modify and/or combine the previously described models to cater for different applications, such as a multi-component system and other systems. For instance, Fiori *et al.* (2009) described a combination of the shrinking core and the broken and intact cell model. Some researchers use a heat-transfer analogy of cooling a single sphere that is suitable for the idealized conditions of a fixed bed [14]. Modeling of industrial extractors such as De Smet [54], Rotocell™ [55] and the Crown Model [56] were performed by considering the mass transfer between phases, solute diffusion into the bulk phase, the presence of trays, the influence of drainage and loading zones, and other important system operational characteristics. A detailed review of these SLE (solid–liquid equilibrium) models is available in the literature [14].

So far, the discussion concerning process modeling has focused on mass transfer within a solid and its vicinity. The solute transfer in the fluid also contributes significantly to the performance of the extraction process. For this, the dynamics of the fluid are important, and depend on the type and operating conditions of the extractors. Each model produces an extraction curve that provides information about the variation of the concentration of solute with time, and hence the extraction equilibrium and kinetics that is important for extractor design. The model of the mass transfer in the fluid has its origin in the mass balance of the solute in the fluid. From this, the concentration profile of solute in the fluid is obtained. Useful models for an ideal stirred tank, ideal plug-flow extractor, combined tubular, and cascade of continuously stirred vessels are available. An ideal stirred tank is suitable for immersion while an ideal plug flow fits better with the percolation concept.

Extraction from various parts of different kinds of plants with different structures such as roots [57, 58], seeds [54, 59], leaves [41], flowers [60, 61] and tree bark [41, 62] have been investigated and described in the various extraction models mentioned earlier. However, it is clear that a good prediction of any extraction of biomass feedstock of various origins must not fail to consider the morphology of individual plant cells. Obviously, lignocellulosic feedstocks such as palm-oil fruit, wood and straw exhibit different cell morphologies compared to grain feedstock, for example corn, or even wet biomass such as green grass and alfalfa. The biomass feedstocks also respond differently to pretreatment. The extent of the cell disruption determines the changes that occur in the plant tissue, which will influence diffusion during the extraction process and hence the subsequent biorefinery process. Oil in seeds for instance, is known to reside in cavity cells with a low permeability wall and thus is often milled prior to extraction. Many oils, such as those of rape seed [63], sunflower seed [64], grape seed [53], and *jatropha curcas* [2], exhibit increasing extraction rates during the early hours of extraction and gradually decrease until the rate remains unchanged with time. The early fast extraction rate is driven by the fresh solvent dissolving the oil from the broken cells while the second slower rate is the diffusion of the remaining oil from the intact cells into the nearly saturated fluid phase. Similar extraction behavior was reported for extraction from agarwood [63], and wood chips [64]. Modeling work of the essential oil extraction from various plant families has also emphasized studying the plant cell morphology [51, 65]. For example, a detailed description of swelling and rupture of oil-bearing glands (glandular trichomes) during extraction, especially the similarity within *Lamiaceae* plant families, has been reported [66]. Pretreatment changes the plant tissue properties and disrupts a greater fraction of trichomes glands to yield more oil. This greatly influences the second part of the extraction, which is controlled by the rate of diffusion through the particles of plant material. Generally, the effect of pretreatment on the extraction is more pronounced during the earlier period and thus should not be neglected. These findings suggested that further investigation should be conducted at the cellular level to improve the existing models. This would be useful especially when dealing with challenging feedstocks such as lignocellulosic biomass to assist fractionation into lignin, hemicelluloses and cellulose, and further processing in a biorefinery. Moreover, the pretreatment methods and the corresponding extraction operational conditions should be optimized and modeled together to facilitate a better understanding for any specific plant material or plant family.

13.4.5 Scaling up

Designing and scaling up of extraction equipment is generally based on practical experience and experiments, whereby the experimental data are quantified to give information regarding the product and its reaction kinetics. Generally, a simple extraction process is performed using laboratory-scale equipment to quantify the effect of various parameters such as temperature, solid-to-solvent ratio, feed size, and the effect of solvent characteristics such as type, solubility, viscosity, concentration and other parameters on the extraction performance. The laboratory-scale data are then validated using pilot plant-scale equipment for further optimization.

13.5 Industrial extractors

The selection of an extraction unit for large volumetric production in industrial-scale production depends on several factors, which in general include (i) mode of operation—batch and continuous, (ii) extraction principles—immersion, maceration or percolation, (iii) production capacity—size of feed, target yield, and (iv) product specification. The following extractor examples are the common types of extractor used in biorefineries for the production of edible oil (vegetable, soybean, sunflower), essential oils, extracts (for perfumes, health supplements, Ayurvedic and traditional Chinese medication regiments), and special chemicals (phytosterols, polyphenols). The main classification of these examples is based on their mode of operation—batch or continuous.

13.5.1 Batch extractors

One of the classic examples of a batch extractor is the Soxhlet extraction apparatus that works on the percolation principle. As illustrated in Figure 13.4, this apparatus has a simple set-up, which includes a reservoir, condensing system, and a solid container. The reservoir holds the solvent as well as the extracts. The substrate is placed in a porous container located in the solvent recycle line. As the solvent passes through the solid during operation, active compounds and solutes are extracted and collected in the reservoir. From time to time, the solvent will be withdrawn from the system through a drain point in the condensing system, leaving concentrated extracts in the reservoir. These extracts are purified later by evaporation or distillation.

Another type of conventional batch extractor is the hydro/steam distiller, which is mainly used for the extraction of essential oil and volatile compounds from agarwood chips, patchouli, peppermint, eucalyptus, marjoram, and so forth. In the large-scale oilseed industry, this unit is used to desolventize (remove solvent from) the solid discharge. The design is similar to the Soxhlet apparatus, except that water is

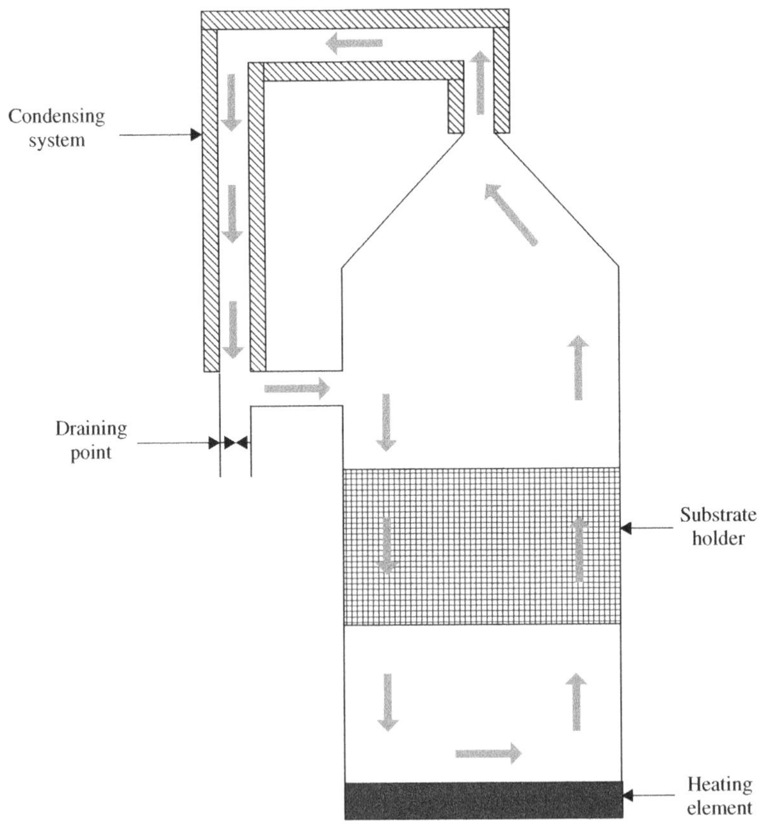

Figure 13.4 Soxhlet apparatus setup. Arrows showing solvent flow during operations

used as the main solvent, and the solid is immersed in the water throughout the process instead of being percolated through. As heat is continuously supplied to the unit, oil vapor, along with steam, travel into the condensing system. The vapor eventually condenses and collects in the separator. In this unit, the extract is collected from the separator in the condensing system, not from the reservoir as for the Soxhlet system. However, this method is associated with several drawbacks, such as slow process, product damage, or contamination [67].

Batch extraction is still favored in the production of fine chemicals and pharmaceutical products [68]. Even though it is associated with inefficient yield and processing time, this method provides flexibility in its operation and also versatility in its equipment [69]. To improve the batch-extraction process, various modifications and cutting-edge accessories have been introduced to meet the priorities of the extraction industry. These include faster processing capability, control modules, excellent solvent delivery system, easy solid loading, distribution and discharging, as well as flexibility of operation. For example, Figure 13.5 shows an improved batch extraction unit, DIG-MAZ, developed by Samtech Pte. Ltd. for extraction of active compounds, i.e. oleoresin, essential oil, colorant, and so forth, from plant materials. This compact equipment allows for cold and hot processing, combined with optional vacuum facilities and mechanical press. This system includes integrated unit operation, i.e. filtration and distillation, which eventually increases the efficiency of the process. DIG-MAZ is available in eight variants, which are based on its processing capacity, ranging from 10 L to 3000 L [70].

Figure 13.5 *DIG-MAZ batch extractor unit. Reprinted with permission from © B. Wolfsbauer, samtech Extraktlonstechnik GmbH*

13.5.2 Continuous extractors

From the success of developing practical extraction process models, which allow for approximate prediction of yield, improvement has been done on the process equipment to ensure good production in terms of yield per operation time and input. Continuous extractors have been adapted, especially in large-scale production, due to their capability to produce more products and higher yield in a shorter time compared to the batch extractors. Continuous extractor design is based on solvent flow, i.e. cross- or counter-current, positioning—horizontal or vertical—and type of raw material [14]. Counter-current operation provides a high concentration extract output with low residues [71].

Typically, a continuous extractor is in a horizontal position with a countercurrent flow of solvent and solid feed. The earliest type of continuous extractor is the belt- or basket-type extractor (Figure 13.6 and 13.7), which principally works by the percolation method. Solid is placed on a conveyor fixed with belt or basket, moving towards streams of spraying solvent. The contact area per shower head is proportional

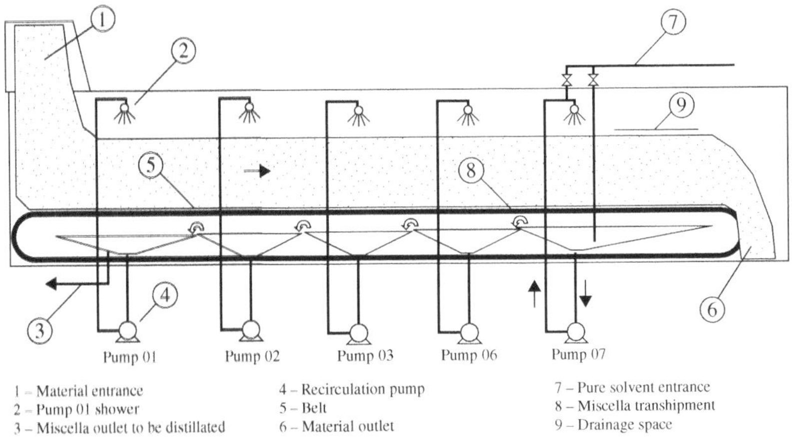

Figure 13.6 Belt-type extractor. Reprinted from [72] © 2010, with permission from Elsevier

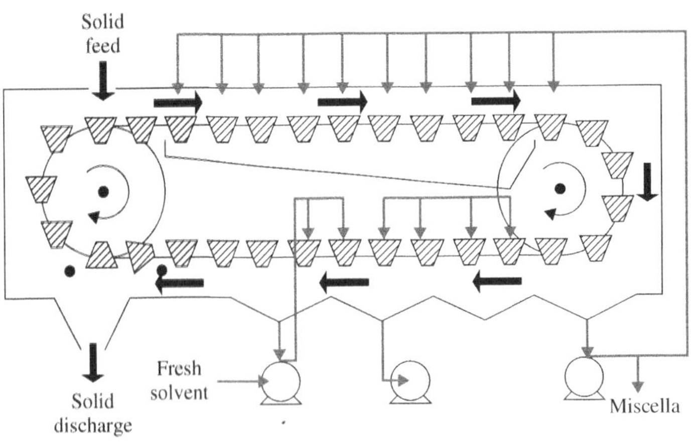

Figure 13.7 Basket-type extractor. Reference [7], with kind permission from Springer Science+Business Media © 1976

to the shower flow rate [72]. Concentrated solvent is distilled/evaporated to separate the extracts. These extractors are mainly used for the extraction of vegetable oil and sugar.

The designs of continuous extractors are improving to compensate for the drawbacks of earlier designs or to accommodate specific processing requirements. The Lurgi sliding-cell extractor was introduced by Lurgi GmbH in Germany. Its working principle is similar to those of classic continuous horizontal extractors, but it offers several processing advantages. These advantages include multiple solvent and solid inlets at different extraction phases, a shallow extraction bed to minimize solvent usage, improved hopper design to prevent clogging, and longer shelf life of assembly parts. Its processing capacity ranges from 100 to 5000 tonnes solid per day with a yield up to 99%. De Smet (Zaventem/Belgium) has produced a LM™ percolation belt-type extractor with processing capacity of between 500 and 5000 tonnes/day. This unit has a sloped conveying system equipped with raking attachment to distribute solids on the perforated belt for even extraction and to prevent solvent contamination. It allows high-pressure and high-temperature operation as it operates in a closed system.

The continuous extraction unit does not only come in horizontal position, which may take up a large amount of operation space, but is also available in vertical set-up. One example of a vertical extractor is the BMA tower extractor developed by BraunschweigischeMaschinebauanstalt AG for extraction of sugar from beet. The process starts with the beet slices being thermally treated in a counter-current cosette mixer; an integral part of the extraction system, before being fed into the extraction tower. Beet slices are not slurried to avoid formation of undesirable insoluble solids, which can clog the extraction line [73]. Two streams of water, i.e. fresh water and heated press water (80 °C), are supplied from the opposite direction of the solid feed in order to extract sugars from the beet slices. Solid remnants are screened out and the extracted juice is transported for further processing. The capacity of this unit can reach up to 17 000 tonnes/day.

The rotary extractor (Carousel™, Rotocell™, REFLEX™), in a horizontal setup, was introduced to solve mechanical problems associated with vertical-type reactors [8]. It occupies less space compared to an horizontal extractor with a similar capacity. This unit is currently developed by Harburg-Freudenberger (Hamburg/Germany), De Smet (Belgium) and PruessAnlagentechnik GmbH (Germany) with varying capacities (50–12 000 tonnes/day). This percolation extractor uses a rotating cell wheel (carousel) in single, double, or triple decks to hold and distribute solids during extraction. A typical number of cells per wheel is between 15 and 18 [73]. Each cell/chamber is filled with solid, and subjected to a solvent spray system designed to provide a countercurrent crossed flow across the extraction chamber (Figure 13.8). The solid

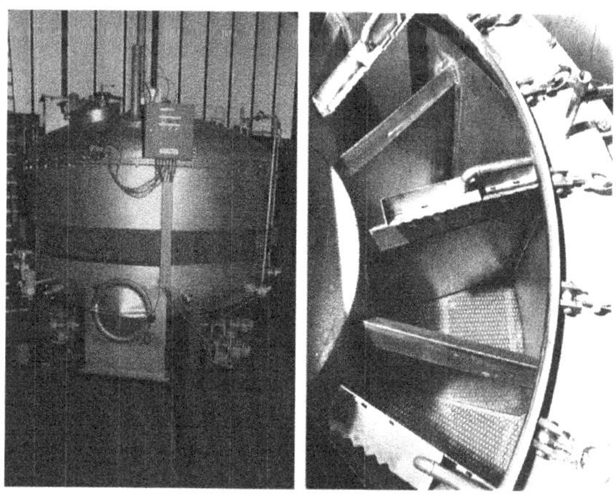

Figure 13.8 *Rotary extractor (left), solid chamber (right). Reproduced by permission of PruessAnlagentechnik GmbH*

cell's typical height is between 1.8 and 3.0 m, allowing for the formation of a deep solid bed [74]. The solvent is recycled in the system before being withdrawn for purification of the extracts.

The Bonotto extractor is a vertical column extractor installed with a stack of evenly spaced trays acting as dividers, forming several compartments for holding the immersed solid during extraction process (Figure 13.9). All the trays are connected to a rotating shaft. Each tray has an opening to allow aged solids from the upper plate to move to the lower plate as the center shaft rotates. Radial scrappers attached to the shaft helps to facilitate solid distribution and movement during the operation. The solvent is fed from the bottom of the column to give a countercurrent mixing between the solvent and the solid. The Bonotto extractor is mostly for the extraction of soybean oil [75].

The Hilderbrant extractor (Figure 13.10a) is an immersion-type extractor that has a combination of horizontal and vertical design. This extractor uses a screw press to transport the solid throughout the unit, which provides extra mechanical stress for better extraction yield. The solvent flows through the perforated feature of the press (Figure 13.10b). Since extra stress is applied by the press, some of the solid feeds may disintegrate into smaller particles and carried along in the solvent. These particles may contaminate the solvent and also choke the perforated press [73].

13.5.3 Extraction of specialty chemicals

The production of specialty chemicals by extraction methods differs from those examples explained previously like edible oil, due to (i) sensitivity of these substances to the fluctuation of extraction process variables, (ii) stringent requirements of solvent usage, and (iii) limited amount of solutes available in the solid feed. The preparation of these chemicals is therefore closely monitored in smaller scale operations and details of the manufacturer's operations such as types of equipment, process parameters, and so forth, are not published.

OmniChem Belgium, a subsidiary of AjinomotoOmniChem, had successfully developed a continuous extraction line for the production of oligomericproanthocyanidin (Tanal WG, Omnivin) and polyphenols (OmniCoa 35). In 1965, this company had its first continuous extraction line with a processing capacity of 1 to 3 tonnes/day. Five years later, the second production line materialized with an improved capacity of 20 tonnes/day. As of 2007, the two independent plants have a production capacity totaling 10 000 tonnes/year.

13.6 Economic importance and industrial challenges

The biobased economy is still in its infancy when compared to its competitor, the petroleum industry. Nevertheless, the development of the petroleum refining industry in recent years has provided some guidelines for the biobased economy. Technical, social and economic factors that can push the biobased industry into maturity [78] include yield, resources, and product diversifications, varying supplies of biomass resources, effects on agricultural activities, integration with the agricultural ecosystems, and the sustainability of the economy and its resources.

The four key factors that drive interest in bioenergy include rising prices for fossil fuels, energy security, climate change, and rural development [79]. Bioenergy markets are largely policy dependent in most parts of the world, as the production of biofuels is not currently competitive with that of fossil fuels. The cost of producing biobased products must compete favorably with the comparable petroleum-derived product to support the economic feasibility of the biobased production. To illustrate, the total production costs of ethanol and syndiesel from lignocellulosic feedstock in the United Kingdom for 2015 are estimated at USD 0.60/liter and USD 1.01/liter, respectively. These costs are expected to be reduced to USD 0.50/liter

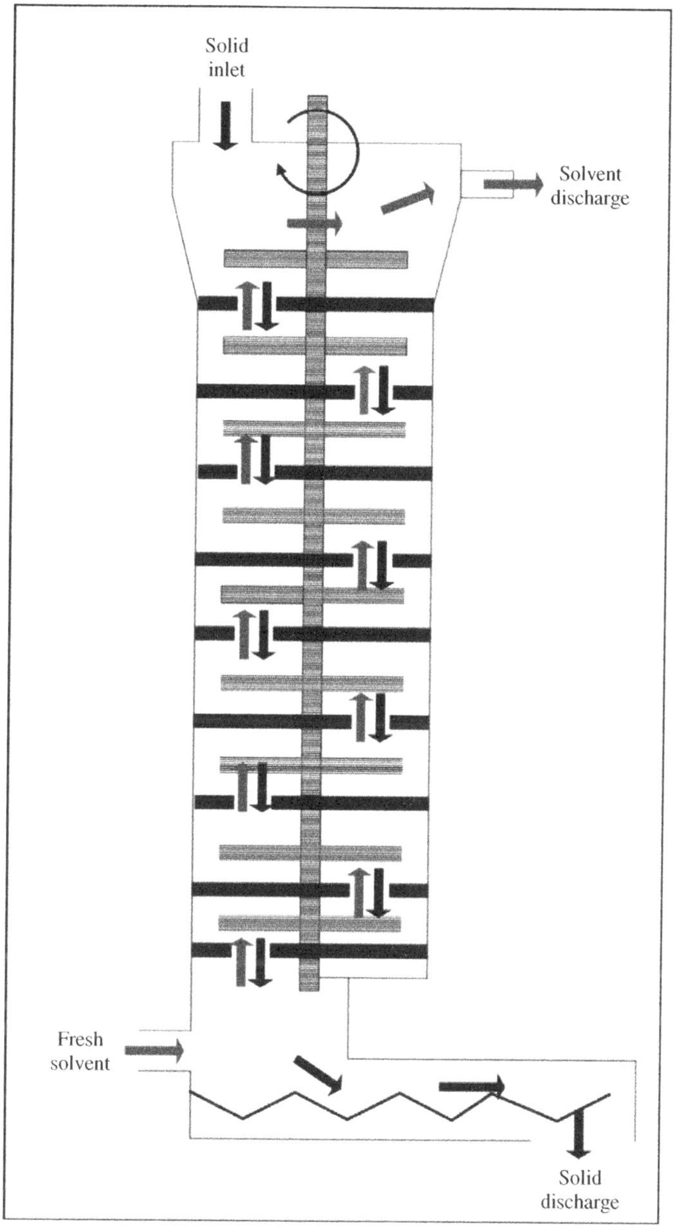

Figure 13.9 *Bonotto extractor. Reprinted with kind permission from Goss © 1946, Springer Berlin/Heidelberg*

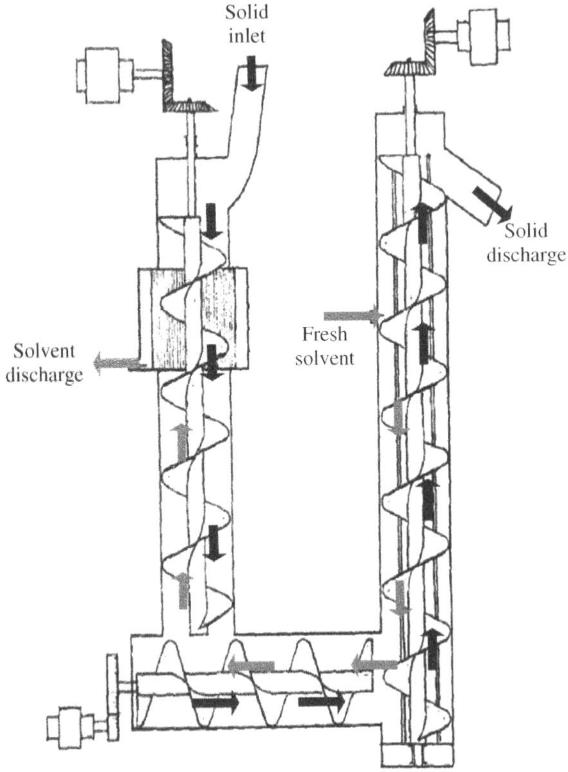

(a)

(b)

Figure 13.10 *(a) Hilderbrant extractor. Reference [76] adapted with kind permission from Springer Science+Business Media © 1953, (b) Perforated screw press. Reproduced with permission from Falcon Industries © 2012*

(ethanol) and USD 0.69/liter (syndiesel) as production is increased in 2022 [80]. The synthesis of novel bioproducts is the future alternative economic direction for the biorefinery industry.

The main application of the SLE process in supporting the biorefinery industry is in the production of first- and second-generation biofuels. In 2005, the world's biofuel production (both first- and second-generation biofuels) was approximately 38 300 liters/hectare, which is equivalent to 24 820 liters of gasoline/diesel equivalent per hectare and by 2050 this value is expected to increase to 42 210 liters of gasoline/diesel equivalent per hectare [80]. With this increase in the demand for biorefinery products, improved efficiency and performance in the extraction process together with development in biotechnology are needed. For the biorefinery industry to be sustained, the supporting production line should be in tandem to meet the current demands. With the current state–of-the-art technology it will be fairly difficult to achieve the targeted values. The industrial application of academic research is therefore important to ensure the sustainability of the biorefinery industry.

13.7 Conclusions

The role of SLE in the biorefinery industry is very important. It has been applied in various areas of biofuel production (such as vegetable oil from oil seed and sugar from starchy, sugary or lignocellulosic biomass), and bio-based and pharmaceutical compound production (such as extraction of fragrances and active compounds). The most popular SLE methods applied at an industrial scale are maceration, percolation, and steam distillation or hydro distillation. Larger units that operate continuously can be found mainly in vegetable oil production sector. Various modification methods, such as microwave assisted, ultrasonic assisted and high-pressure processes, have been introduced at the laboratory scale to enhance the production of the classical SLE system. Their success on an industrial scale still requires more integration of research. In addition, various models that predict the behavior of equipment, so that it can be applied in large-scale operations, have been developed. However, the effects of plants are not fully understood, making the models only partly successful. The demand for biorefinery products is increasing with more stringent requirements by governing bodies and depleting petrochemical resources. To ensure that the biorefinery industry is sustainable and able to meet the demands for biofuel and bio-based products, it is therefore imperative that the research findings are immediately converted into industrial applications. Close integration between academia and industry should take place. The main priority should be the development of systematic approaches to improve SLE processes by making them on a larger scale, with a shorter processing time and higher extraction yield.

References

1. H. J. Huang, S. Ramaswamy, U. Tschirner, B. Ramarao, A review of separation technologies in current and future biorefineries, *Separation and Purification Technology*, 62, 1–21 (2008).
2. S. Sayyar, Z. Z. Abidin, R. Yunus, A. Muhammad, Extraction of oil from *Jatropha* seeds–optimization and kinetics, *American Journal of Applied Sciences*, 6, 1390–5 (2009).
3. I. Sensoy, S. Sastry, Extraction using moderate electric fields, *Journal of Food Science*, 69, FEP7–FEP13 (2004).
4. J. T. Uhm, W. B. Yoon, Effects of high pressure process on kinetics of leaching oil from soybean powder using hexane in batch systems, *Journal of Food Science*, 76, E444–E9 (2011).
5. C. Geankoplis, *Transport Processes and Separation Process Principles (Includes Unit Operations)*, Prentice Hall, Upper Saddle River, NJ, USA, 2003.
6. W. Unterberg, *How to Prevent Spills of Hazardous Substances*, Noyes Data Corp., Park Ridge, NJ, 1988.
7. E. Milligan, Survey of current solvent extraction equipment, *Journal of the American Oil Chemists' Society*, 53, 286–90 (1976).

8. A. Pfennig, D. Delinski, W. Johannisbauer, H. Josten, in *Extraction Technology*, H.-J. Bart, S. Pilz (Eds.), Wiley-VCH Verlag GmbH, Weinheim, Germany, 2011.
9. N. Kardos, J. Luche, Sonochemistry of carbohydrate compounds, *Carbohydrate Research*, 332, 115–31 (2001).
10. H. Méndez, F. Alava, I. Lavilla, C. Bendicho, Ultrasonic extraction combined with fast furnace analysis as an improved methodology for total selenium determination in seafood by electrothermal-atomic absorption spectrometry, *Analytica Chimica Acta*, 452, 217–22 (2002).
11. Z. Hromadkova, A. Ebringerova, Ultrasonic extraction of plant materials—investigation of hemicellulose release from buckwheat hulls, *Ultrasonics Sonochemistry*, 10, 127–33 (2003).
12. X. Pan, G. Niu, H. Liu, Microwave-assisted extraction of tanshinones from Salvia *miltiorrhiza*bunge with analysis by high-performance liquid chromatography, *Journal of Chromatography A*, 922, 371–5 (2001).
13. D. P. Fulzele, R. K. Satdive, Comparison of techniques for the extraction of the anti-cancer drug camptothecin from *Nothapodytesfoetida*, *Journal of Chromatography A*, 1063, 9–13 (2005).
14. M. Kassing, U. Jenelten, J. Schenk, J. Strube, A new approach for process development of plant based extraction processes, *Chemical Engineering and Technology*, 33, 377–87 (2010).
15. M. Luque de Castro, L. Garcia-Ayuso, Soxhlet extraction of solid materials: an outdated technique with a promising innovative future, *Analytica Chimica Acta*, 369, 1–10 (1998).
16. F. Priego-Capote, M. Luque de Castro, Focused microwave-assisted Soxhlet extraction: a convincing alternative for total fat isolation from bakery products, *Talanta*, 65, 81–6 (2005).
17. I. T. Stanisavljevic, M. Lazic, V. Veljkovic, Ultrasonic extraction of oil from tobacco (*Nicotianatabacum L.)* seeds, *Ultrasonicssonochemistry*, 14, 646–52 (2007).
18. J. Singh, in *Percolation and Infusion Technique for the Extraction of Medicinal and Aromatic Plants*, S. S. Handa, S. P. S. Khanuja, G. Longo, D. D. Rakesh (Eds.), ICS-UNIDO, Trieste, 2008.
19. S. S. Hanc, S. Sahin, L. Y lmaz Isolation of volatile oil from thyme (*Thymbraspicata*) by steam distillation, *Food/Nahrung*, 47, 252–5 (2003).
20. F. Chemat, M. Lucchesi, J. Smadja, L. Favretto, G. Colnaghi, F. Visinoni, Microwave accelerated steam distillation of essential oil from lavender: A rapid, clean and environmentally friendly approach, *Analytica Chimica Acta*, 555, 157–60 (2006).
21. H. Sovová, S. A. Aleksovski, Mathematical model for hydrodistillation of essential oils, *Flavour and Fragrance Journal*, 21, 881–9 (2006).
22. V. Young, Processing of oils and fats, in *Fats and Oils: Chemistry and Technology*, R. J. Hamilton, A. Bhati (Eds.), Applied Science Publishers Ltd., London, 1980.
23. J. Rossell, Fractionation of lauric oils, *Journal of the American Oil Chemists' Society*, 62, 385–90 (1985).
24. A. Demirbas, Biodiesel production from vegetable oils via catalytic and non-catalytic supercritical methanol transesterification methods, *Progress in Energy and Combustion Science*, 31, 466–87 (2005).
25. J. Espinoza-Pérez, A. Vargas, V. Robles-Olvera, G. Rodriguez-Jimenes, M. Garcia-Alvarado, Mathematical modeling of caffeine kinetic during solid–liquid extraction of coffee beans, *Journal of Food Engineering*, 81, 72–8 (2007).
26. J. Ji, X. Lu, M. Cai, Z. Xu, Improvement of leaching process of *Geniposide* with ultrasound, *Ultrasonics Sonochemistry*, 13, 455–62 (2006).
27. E. Reverchon, Mathematical modeling of supercritical extraction of sage oil, *AIChE Journal*, 42, 1765–71 (1996).
28. J. M. Aguilera, Solid–liquid extraction, in *Extraction Optimization in Food Engineering*, C. Tzia, G. Liadakis (Eds.), Marcell Dekker, New York, 2003.
29. M. Negro, P. Manzanares, J. Oliva, I. Ballesteros, M. Ballesteros, Changes in various physical/chemical parameters of *Pinuspinaster* wood after steam explosion pretreatment, *Biomass and Bioenergy*, 25, 301–8 (2003).
30. M. Y. Harun, A. B. Dayang Radiah, Z. Zainal Abidin, R. Yunus, Effect of physical pretreatment on dilute acid hydrolysis of water hyacinth (*Eichhornia crassipes*), *Bioresource Technology*, 102, 5193–9 (2011).
31. Y. Sun, J. Cheng, Hydrolysis of lignocellulosic materials for ethanol production: a review, *Bioresource Technology*, 83, 1–11 (2002).
32. M. Balat, H. Balat, C. Öz, Progress in bioethanol processing, *Progress in Energy and Combustion Science*, 34, 551–73 (2008).

33. G. I. Frandsen, J. Mundy, J. T. C. Tzen, Oil bodies and their associated proteins, oleosin and caleosin, *Physiologia Plantarum*, 112, 301–7 (2001).
34. L. J. Lin, J. T. C. Tzen, Two distinct steroleosins are present in seed oil bodies, *Plant Physiology and Biochemistry*, 42, 601–8 (2004).
35. E. Reverchon, C. Marrone, Supercritical extraction of clove bud essential oil: isolation and mathematical modeling, *Chemical Engineering Science*, 52, 3421–8 (1997).
36. D. F. Othmer, J. Agarwal, Extraction of soybeans: Theory and mechanism, *Chemical Engineering Progress*, 51, 372–8 (1955).
37. S. Nyiredy, Solid–liquid extraction strategy on the basis of solvent characterization, *Chromatographia*, 51, 288–96 (2000).
38. X. Han, D. W. Armstrong, Ionic liquids in separations, *Accounts of Chemical Research*, 40, 1079–86 (2007).
39. O. Akaranta, A. Anusiem, A bioresource solvent for extraction of castor oil, *Industrial Crops and Products*, 5, 273–7 (1996).
40. S. Meziane, H. Kadi, Kinetics and thermodynamics of oil extraction from olive cake, *Journal of the American Oil Chemists' Society*, 85, 391–6 (2008).
41. E. Simeonov, I. Tsibranska, A. Minchev, Solid–liquid extraction from plants—experimental kinetics and modelling, *Chemical Engineering Journal*, 73, 255–9 (1999).
42. N. Bulley, M. Fattori, A. Meisen, L. Moyls, Supercritical fluid extraction of vegetable oil seeds, *Journal of the American Oil Chemists' Society*, 61, 1362–5 (1984).
43. A. K. K. Lee, N. Bulley, M. Fattori, A. Meisen, Modelling of supercritical carbon dioxide extraction of canola oilseed in fixed beds, *Journal of the American Oil Chemists' Society*, 63, 921–5 (1986).
44. O. J. Catchpole, J. B. Grey, B. M. Smallfield, Near-critical extraction of sage, celery, and coriander seed, *The Journal of Supercritical Fluids*, 9, 273–9 (1996).
45. M. Goto, M. Sato, T. Hirose, Extraction of peppermint oil by supercritical carbon dioxide, *Journal of Chemical Engineering of Japan*, 26, 401–7 (1993).
46. A. Tezel, A. Hortasu, Ö. Hortasu, Multi-component models for seed and essential oil extractions, *The Journal of Supercritical Fluids*, 19, 3–17 (2000).
47. H. Sovová, Rate of the vegetable oil extraction with supercritical CO_2—I. Modelling of extraction curves, *Chemical Engineering Science*, 49, 409–14 (1994).
48. C. Marrone, M. Poletto, E. Reverchon, A. Stassi, Almond oil extraction by supercritical CO2: experiments and modelling, *Chemical Engineering Science*, 53, 3711–8 (1998).
49. E. Reverchon, J. Daghero, C. Marrone, M. Mattea, M. Poletto, Supercritical fractional extraction of fennel seed oil and essential oil: experiments and mathematical modeling, *Industrial and Engineering Chemistry Research*, 38, 3069–75 (1999).
50. E. Reverchon, C. Marrone, Modeling and simulation of the supercritical CO_2 extraction of vegetable oils, *The Journal of Supercritical Fluids*, 19, 161–75 (2001).
51. I. Zizovic, M. Stamenic, A. Orlovic, D. Skala, Supercritical carbon dioxide essential oil extraction of *Lamiaceae* family species: Mathematical modelling on the micro-scale and process optimization, *Chemical Engineering Science*, 60, 6747–56 (2005).
52. M. Stamenic, I. Zizovic, A. Orlovic, D. Skala, Mathematical modelling of essential oil SFE on the micro-scale—Classification of plant material, *The Journal of Supercritical Fluids*, 46, 285–92 (2008).
53. L. Fiori, D. Basso, P. Costa, Supercritical extraction kinetics of seed oil: A new model bridging the broken and intact cells' and the shrinking-core models, *The Journal of Supercritical Fluids*, 48, 131–8 (2009).
54. G. Veloso, G. Thomas, V. Krioukov, A mathematical model of extraction in countercurrent crossed flows, *Chemical Engineering and Processing: Process Intensification*, 47, 1470–7 (2008).
55. G. Thomas, V. Krioukov, H. Vielmo, Simulation of vegetable oil extraction in counter-current crossed flows using the artificial neural network, *Chemical Engineering and Processing*, 44, 579–90 (2005).
56. M. L. M. N. Cerutti, A. A. U. de Souza, S. M. A. de Souza, Solvent extraction of vegetable oils: Numerical and experimental study, *Food and Bioproducts Processing*, 90, 199–204 (2011).
57. E. Simeonov, V. Koleva, Solid–liquid extraction from roots of *Geranium Sanquineum*, *Journal of the University of Chemical Technology and Metallurgy*, 43, 409–12 (2008).

58. I. Zizovic, M. Stamenic, J. Ivanovic, A. Orlovic, M. Ristic, S. Djordjevic, S. D. Petrovic, D. Skala, Supercritical carbon dioxide extraction of sesquiterpenes from valerian root, *The Journal of Supercritical Fluids*, 43, 249–58 (2007).
59. I. Papamichail, V. Louli, K. Magoulas, Supercritical fluid extraction of celery seed oil, *The Journal of Supercritical Fluids*, 18, 213–26 (2000).
60. M. Akgün, N. A. Akgün, S. Dinçer, Extraction and modeling of lavender flower essential oil using supercritical carbon dioxide, *Industrial and Engineering Chemistry Research*, 39, 473–7 (2000).
61. L. M. A. S. Campos, E. M. Z. Michielin, L. Danielski, S. R. S. Ferreira, Experimental data and modeling the supercritical fluid extraction of marigold (*Calendula officinalis*) oleoresin, *The Journal of Supercritical Fluids*, 34, 163–70 (2005).
62. A. Ibrahim, S. Al-Rawi, A. Majid, N. N. A. Rahman, K. Salah, M. O. A. Kadir, Separation and fractionation of Aquilaria Malaccensis oil using supercritical fluid extraction and the cytotoxic properties of the extracted oil, *Procedia Food Science*, 1, 1953–9 (2011).
63. M. Goto, B. C. Roy, T. Hirose, Shrinking-core leaching model for supercritical-fluid extraction, *The Journal of Supercritical Fluids*, 9, 128–33 (1996).
64. M. Perrut, J. Clavier, M. Poletto, E. Reverchon, Mathematical modeling of sunflower seed extraction by supercritical CO2, *Industrial and Engineering Chemistry Research*, 36, 430–5 (1997).
65. I. Zizovic, M. Stamenic, A. Orlovic, D. Skala, Supercritical carbon dioxide extraction of essential oils from plants with secretory ducts: Mathematical modelling on the micro-scale, *The Journal of Supercritical Fluids*, 39, 338–46 (2007).
66. M. Stamenic, I. Zizovic, R. Eggers, P. Jaeger, H. Heinrich, E. Roj, J. Ivanovic, D. Skala, Swelling of plant material in supercritical carbon dioxide, *The Journal of Supercritical Fluids*, 52, 125–33 (2010).
67. C. Grosso, M. A. T. Cardoso, A. Figueiredo, M. Moldão-Martins, J. Burillo, J. Urieta, J. Barroso, J. Coelho, A. Palavra (Eds.), Supercritical fluid extraction, Hydrodistillation and Soxhlet extraction of the aerial part of winter savory. Comparative evaluation of the extraction method on the chemical composition. Proceedings of European Congress of Chemical Engineering (ECCE-6); 2007; Copenhagen.
68. B. Grinbaum, An integrated method for development and scaling up of extraction processes, in *Exchange and Solvent Extraction*, *Vol*. 15, Y. Marcus, A. Sangupta (Eds.), Marcell Dekker, New York, 2002.
69. D. M. Roberge, L. Ducry, N. Bieler, P. Cretton, B. Zimmermann, Microreactor technology: A revolution for the fine chemical and pharmaceutical industries? *Chemical Engineering and Technology*, 28, 318–23 (2005).
70. S. E. GmbH. Extraction plants—samtech Extraction Technology. 2011; available from: www.samtech.at/en/extraktoren.php (accessed September 5, 2012).
71. F. Ullmann, W. Gerhartz, Y. S. Yamamoto, F. T. Campbell, R. Pfefferkorn, J. F. Rounsaville, *Ullmann's Encyclopaedia of Industrial Chemistry*, Wiley-VCH Verlag GmbH, Weinheim (2005).
72. R. Almeida, M. Ravagnani, A. Modenes, Soybean oil extraction in belt extractors with miscella recirculation, *Chemical Engineering and Processing: Process Intensification*, 49, 996–1005 (2010).
73. B. K. Dutta, *Principles of Mass Transfer and Separation Processes*, Prentice-Hall of India, New Delhi, 2007.
74. R. Eggers, P. Jaeger, C. Tzia, G. Liadakis, Extraction systems, *Extraction Optimization in Food Engineering*, 95–136 (2003).
75. N. Board, *Modern Technology of Oils*, Fats and Its Derivatives, NIIR Project Consultancy Services, Delhi, 2002.
76. W. Goss, Solvent extraction of oilseeds, *Journal of the American Oil Chemists' Society*, 23, 348–54 (1946).
77. N. H. Moore, Mechanics of solvent extraction with the immersion type extractor, *Journal of the American Oil Chemists' Society*, 30, 538–9 (1953).
78. B. E. Dale, S. Kim, in *Biomass Refining Global Impact–the Biobased Economy of the Twenty-First Century*, B. Kamm, P. R. Gruber, M. Kamm (Eds.), Wiley-VCH Verlag GmbH, Weinheim, 2006.
79. GBEP/FAO. *A Review of the Current State of Bioenergy Development in G8+5 Countries*, Natural Resources Management and Environment Department, FAO, Rome, 2007.
80. A. Bauen, G. Berndes, M. Junginger, M. Londo, F. Vuille, R. Ball, T. Bole, C. Chudziak, A. Faaij, H. Mozaffarian. *Bioenergy—A Sustainable and Reliable Energy Source: A Review of Status and Prospects*, International Energy Agency, Paris, 2009.

Part VI

Hybrid/Integrated Reaction-Separation Systems—Process Intensification

14

Membrane Bioreactors for Biofuel Production

Sara M. Badenes, Frederico Castelo Ferreira and Joaquim M. S. Cabral

Department of Bioengineering and Institute for Biotechnology and Bioengineering, Centre for Biological and Chemical Engineering, Instituto Superior Técnico, Technical University of Lisbon, Portugal

14.1 Introduction

Membrane bioreactors are efficient examples of process integration by coupling a biocatalytic process with an efficient membrane separation. Typically, the biocatalyst, a microorganism or an enzyme, is retained in a compartment by the membrane, allowing preferential permeation of smaller substrates or products. Membrane-based separations are usually highly selective and have relatively low energy requirements. Moreover their integration in biorefinery allows for process intensification. The concept of combining the unique catalytic features found in biological systems and the separating properties of membrane systems is widely explored for the treatment of industrial and domestic wastewaters. However, the application of membrane bioreactors is quite broad, including studies on systems so complex as artificial livers and pancreas (Giorno and Drioli, 2000). This chapter aims to describe the basic principles of membrane bioreactors in the context of biofuels, using the production of biodiesel, bioethanol, and biogas as illustrative examples.

Membrane technology offers a range of possible separations according to the selected membrane. Dense membranes are used for selective separation of gases or for reverse osmosis. Bioprocesses use large amounts of high-quality water and reverse osmosis units offer an effective route for *in situ* water recycling. Dense membranes may be also used in pervaporation to isolate ethanol from aqueous fermentation broths at lower temperatures and/or overcome water-ethanol azeotropes. Microfiltration, ultrafiltration and nanofiltration porous membranes are used to retain cells, proteins and small molecules, respectively. Notice, however, that more sophisticated membranes have been suggested in energy production: catalytic membranes can assist fuel gas reactions, supported liquid membranes may be used to promote difficult extractions, and ion exchange membranes can be used in biofuel cells for direct electricity production.

Separation and Purification Technologies in Biorefineries, First Edition.
Edited by Shri Ramaswamy, Hua-Jiang Huang, and Bandaru V. Ramarao.

Membrane bioreactors using porous and dense membranes have been studied for the production of bioethanol, biodiesel, fuel gases, such as methane or hydrogen, and harvesting algae biomass. Therefore, in this chapter, the case studies will focus on membrane bioreactors that use micro/ultrafiltration membranes to retain fermentation yeasts, improving bioethanol production yields; ultrafiltration membranes to retain transesterification enzymes in biodiesel production and hydrolytic enzymes used in lignocellulosic saccharification required for the second generation production of ethanol. Examples of the use of dense membranes in pervaporation to overcome ethanol/water azeotropes in ethanol downstream purification or for assisting fuel gases production are also described. This selection aims to provide a diverse array of combinations of biocatalyst and membrane types.

14.1.1 Opportunities for membrane bioreactor in biofuel production

Figure 14.1 shows, schematically, opportunities for production of biofuels with different calorific values according to the membrane pore size. Notice that enzymes and microorganisms can be immobilized in particles or incorporated in micelles, allowing the use of membranes with larger pore size for their retention. Membrane systems such as the use of catalytic membranes in gas fuel production or ion exchange membranes used in microbial fuel cells for direct electricity production are not described in this chapter. Functional membranes and applications are illustrated in Figure 14.2.

Membrane units not directly coupled to bioreactors can also be of particular importance in a biorefinery context. For example, the integration of reverse osmosis units in biofuel production processes offers an energy effective route for high-quality water purification and *in situ* recycling, which is of particular importance taking into account the large water volumes required in bioprocesses.

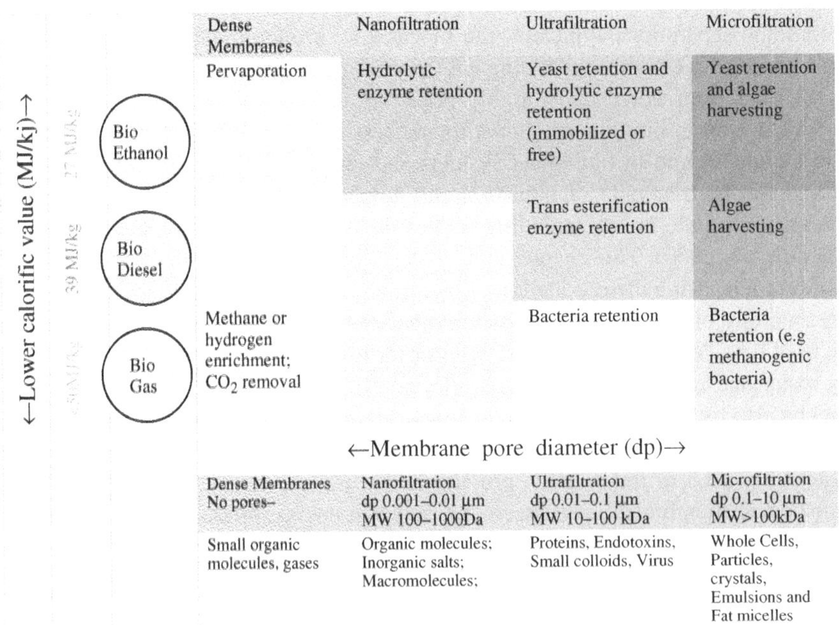

Figure 14.1 Opportunities for integrating membrane separations in the biofuel production in accordance with membrane cut off category and biofuel calorific value

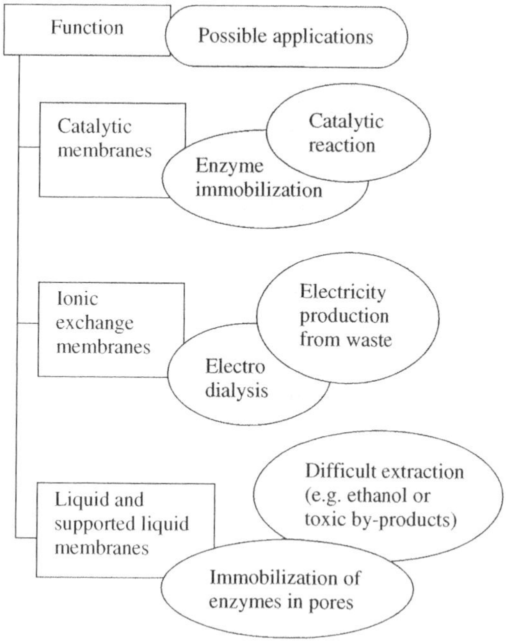

Figure 14.2 Functional membranes and their possible applications

14.1.2 The market and industry needs

Biofuel production has increased in recent years with attempts to provide sustainable and secure sources of energy. Petrol (gasoline), diesel, and natural gas are among the fossil fuels more commonly consumed worldwide, at a rate of about 8, 9 and 39 billion barrels a year, respectively (Figure 14.3). The combined consumption of the United States, European Union, and China represents 50% and 60% of the worldwide

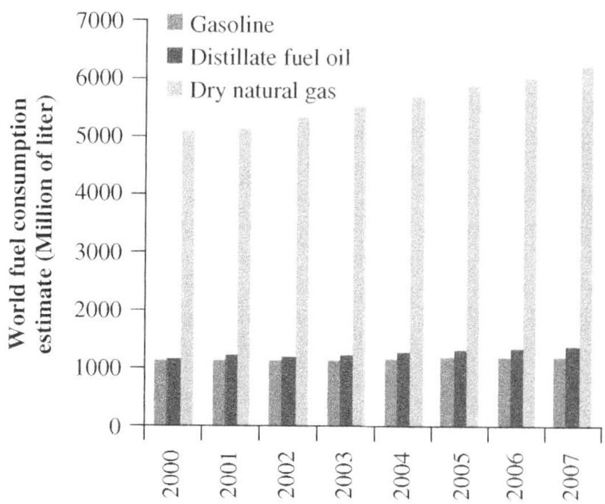

Figure 14.3 Worldwide fuels consumptions from 2000 to 2007 (values calculated using data from the site http://www.indexmundi.com/)

demand for diesel and petrol (gasoline), respectively. Different biofuels are used as drop-in fuels for the direct replacements of pure fossil fuels, according to their chemical-physical properties such as low calorific value, flashpoint, boiling point, density, and viscosity. Whereas bioethanol has been used as petrol replacement in spark-ignition engines, biodiesel, usually comprising fatty (m)ethyl esters (FA(M)EE), is used as a replacement for diesel in compression-ignition engines (Table 14.1).

In 2009, the consumption of bioethanol fuel and biodiesel had reached 4.64 and 1.03 M barrel a year, respectively (Figure 14.4). The US was leading bioethanol fuel consumption (57% worldwide share), whereas the EU was the champion in biodiesel use (71% share). This is not surprising since the same consumption trend is observed for fossil fuels, with four times more petrol being consumed in the US than in the EU and about more 30% of diesel consumed in the EU than in the US. Correspondingly, the current incorporation of bioethanol in gasoline is about 8.5% for the US and incorporation of FA(M)EE in diesel is of 3.6% to the EU. Brazil is the second main consumer of both bioethanol and biodiesel (shares worldwide of 31% and 10%, respectively), currently using more bioethanol than gasoline.

Biofuel consumption has been strongly driven by regulation. Biofuels are used in blends with fossil fuels. The use of bioethanol/gasoline blends, ranging in ethanol content from 5% to 25% (i.e. E5 to E25) has been adopted by many countries. The use of biofuel blends may be optional or mandatory, depending

Table 14.1 *Fuel properties*

	Gasoline	BioEthanol	Diesel	Biodiesel (FAME)	Natural Gas	Biogas
Density (g/cm^3)	0.737	0.79	0.856		0.625	0.863
Flash point (°C)	−72	−14	60–80	100–170	−188	
Melting point (°C)	−51	−114	~0	~0	−182	−182
Specific energy ($MJ \cdot kg^{-1}$)	47.3	27	44.8	39	63	50
Boiling point (°C)	121	78	253	344	161	−161
Flame temp (°C)					204	1911
Viscosity (cSt)	0.71	0.79	2.8	4.1	NA	NA

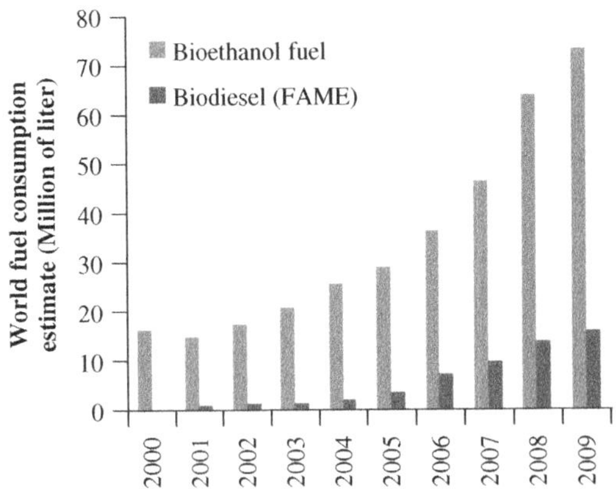

Figure 14.4 *Worldwide biofuels consumptions from 2000 to 2009, and consumption shares based in 2009 values (values calculated using data from www.indexmundi.com/)*

on local regulations. For example, the use of E25 and E10 is mandatory in Brazil and in about 10 states of US, respectively. Spark-ignited engines designed for petrol can use up to E10 without any adaptation, however, when using E25 these engines required specific alterations. Flex engines are specifically designed for using pure ethanol, as well as gasoline blends with higher ethanol contents. Notice that this is not a new concept, already in 1908, the well-known Ford Model T was able to be fuelled with petrol, ethanol or kerosene. Biodiesel/petrodiesel blends are available with biodiesel contents of 2% in fossil diesel (i.e. B2) up to pure biodiesel (i.e. B100). EU regulation established that conventional diesel can contain up to 7% of biodiesel, and the EU tendency is to push this limit to 10%. With exception of high performance pressure engines, it is believed that B20 can be used directly without or with minimal engine adaptation. B6 to B20 are certified in US by ASTM D7467 and B5 and B7 are certified in EU, respectively, by EN590 and EN 14214 norms.

The demand for natural gas should also not be neglected. US, EU and the Russia Federation are the three major consumers of natural gas, with a combined share of about 50%. This fuel consists mainly of methane but contains a fraction of up to 20% of other hydrocarbons, such as ethane. Natural gas is used directly for domestic and industrial heat/cooling or electricity production. Biogas consists mainly of methane (50–75%) and carbon dioxide (50–25%). Thus, it can be used as replacement for natural gas and burned directly for production of heat and electricity. Biogas can also be enriched to 95% methane and cleaned of other trace components, such as hydrogen sulfide and ammonia, to be either injected in the gas grid or used as transport fuel. Biogas use is still marginal when compared to natural gas, representing about 1% in countries such as US and United Kingdom; still, there are major efforts from many countries to improve biogas production, aiming at more sustainable agriculture and industry. In 2007, the market for industrial wastewater treatment systems for biogas generation was in Europe alone, 48.5 M Euros with an annual growth rate of 9.7%. Many of the industrial facilities that produce biogas re-use the obtained energy internally.

14.2 Basic principles

14.2.1 Biofuels: Production principles and biological systems

BIOETHANOL is produced from the fermentation of sugars and other hydrocarbonates by several yeasts, the most common being *Saccharomyces cerevisiae*. *Zymomonas mobilis*, a bacterium, is also among the more promising ethanol producer microorganisms, as it is able to metabolize sugars into ethanol and is tolerant to higher ethanol concentrations. Simple substrates used for fermentations are molasses, whose main component, sucrose requires to be broken down to glucose and fructose (Eq. (14.1)) before their conversion in ethanol (Eq. (14.2)); Invertase, an enzyme, has been used to achieve this purpose in mild conditions.

$$C_{12}H_{22}O_{11}\,(Sucrose) + H_2O \xrightarrow{Invertase} C_6H_{12}O_6\,(Glucose) + C_6H_{12}O_6\,(Fructose) \tag{14.1}$$

$$C_6H_{12}O_6\,(Glucose) \xrightarrow{S.cerevisiae\ \ or\ \ Z.mobilis} 2\times C_2H_5O_2\,(Ethanol) + \ 2\times CO_2 \tag{14.2}$$

Currently, bioethanol is obtained from fermentation of sugar and starch crops, such as sugar cane, corn, sugar beet and rice. In these crops, sugar monomers or dimers are easily accessible by microorganisms for fermentation into ethanol. However, these crops are raw materials in the food supply chain and thus their use for fuel production has raised criticism. The use of starch for ethanol production requires breaking down the polysaccharide into its sugar monomers using hydrolytic enzyme extracts, such as amylases, in a process of sequential hydrolysis and fermentation (SHF). To avoid enzymes inhibition by the accumulation of sugars, the use of simultaneous saccharification and fermentation (SSF) processes, where enzymes and yeasts are added in the fermentor, has also been suggested. As the enzymatic reaction yields sugars, these are continuously consumed in the fermentation, thus avoiding accumulation and inhibition of the microorganism by the product. Polysaccharide hydrolysis kinetics can be described by Eq. (14.3), based on the Michaelis–Menten kinetics, but also taking into account product (sugar monomers) inhibition in enzyme activity.

$$v = \frac{V_{\max} S_0}{(S_0 + K_m)\left(1 + \dfrac{I}{K_i}\right)} \tag{14.3}$$

Lignocellulosic materials are presented in many agro-forestry and industrial wastes, and it has been proposed that their use as raw materials yields a second generation of more sustainable bioethanol. Unfortunately, carbon sources in cellulosic materials are not so readily available to yeasts, and usually it requires a chemical pre-treatment, followed by SHF or SSF processes using adequate enzymes extracts, such as cellulase and glucosidase. These enzymes dramatically increase the cost of ethanol production and thus consolidated process had been idealized, where the yeast used is not only able to produce ethanol, but also produces *in situ* hydrolytic enzymes that are able to degrade lignocellulosic compounds.

Fermentation using adequate strains of *Saccharomyces cerevisiae* can only reach ethanol contents as high as 10–15% v/v. However, ethanol inhibition of fermentation is a main limitation when producing this fuel and as ethanol concentration increases, inhibition kinetics become significant. The Monod model is considered the basic equation and introduces the concept of growth-controlling (limiting) substrate, relating the growth rate to the concentration of a single growth controlling substrate via two parameters, the maximum specific growth rate ($\mu_{\max}$) and the substrate affinity constant (K_s).

$$\mu = \frac{\mu_{\max} S}{(K_s + S)} \cdot \left(\frac{K_p}{K_p - P}\right) \tag{14.4}$$

Equation (14.4) presents the Monod model corrected for product inhibition, where P is the product concentration and K_p is the product inhibition constant. Substrate consumption by microorganisms is used for cell growth, but also for cell maintenance. Therefore, when trying to establish a kinetic rate for ethanol (P) production and substrate (S) consumption, biomass ($Y_{X/S}$) and product ($Y_{P/S}$) yield coefficients are taken into account, as well as maintenance rates (m_q). The biomass yield coefficient rates attain its maximum value ($Y_{X/S}^{Max}$) when m_q is equal to zero. Several authors had established different equations; the ones shown below were initially defined by Pirt (1965) and modified by Roels (1980) (Eq. (14.5)):

$$-\frac{dS}{dt} = -\frac{1}{Y_{X/S}^{Max}} \frac{dX}{dt} \cdot + m_q \cdot X + \frac{1}{Y_{P/S}} \frac{dp}{dt}$$

where

$$\frac{1}{Y_{X/S}} = \frac{m_q}{\mu} \cdot + \frac{1}{Y_{X/S}^{Max}} \tag{14.5}$$

Using distillation, it is possible to recover the ethanol from the fermentation broth in the ethanol/water azeotrope at concentrations of about 96% v/v. Further dehydration of ethanol is required before preparing the final blend.

The different stages for bioethanol production and potential opportunities for application of membrane systems are briefly illustrated in Figure 14.5 and are further explored in Section 14.3.1.

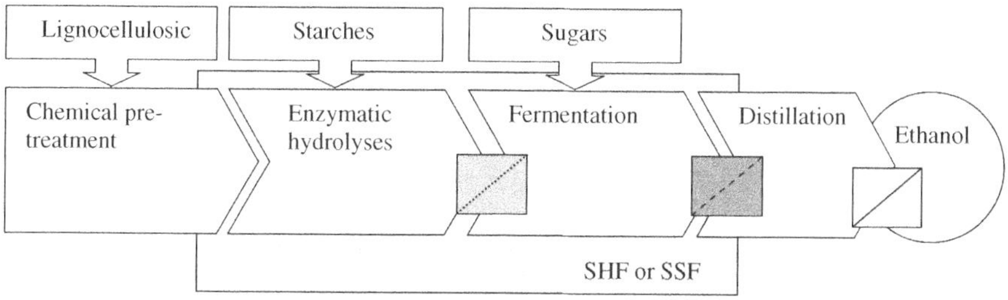

Figure 14.5 *Schematic diagram of stages required for production of bioethanol*

BIODIESEL is usually comprised by fatty acid (m)ethyl esters (FA(M)EE), which are obtained by chemical or enzymatic transesterification with methanol or ethanol of fatty oils, comprised mainly by triglycerides, but also mono-/di-glycerides and free carboxylic acids. The main by-product is glycerol, as can be seen in Eq. (14.6).

$$Triglyceride + 3 \times Methanol \xrightarrow{Transesterase} 3 \times FAME + Glycerol \tag{14.6}$$

Most commercial biodiesel is currently produced by transesterification of vegetable oils using a homogeneous base (sodium or potassium hydroxides, and sodium or potassium methoxide) as catalyst in a highly efficient process. Homogeneous alkali catalyzed biodiesel production processes show high yields and are about 4000 times faster than homogeneous acid catalysis. Acid catalysis commonly uses H_2SO_4, HCl, BF_3, H_3PO_4 and organic sulfonic acids, which are more corrosive than alkaline agents, causing damage to the equipment (Nagayama *et al*., 1999; Vyas *et al*., 2010). However, the use of alkaline-based continuous processes is limited as they are very sensitive to the presence of water and free fatty acids (FFA), requiring high-quality feedstock to avoid undesired side reactions. Acid catalyzed transesterification can be carried out, even with vegetable oils, crudes that contain high amounts of water and free fatty acids. However, in this type of catalysis high product yields are only obtained in reaction times fast enough for practical applications, when using high alcohol-to-oil molar ratios and temperatures (Lam *et al*., 2010). Moreover, both alkali and acid homogenous catalyzed transesterifications are still non-competitive when compared with petroleum diesel since the catalyst cannot be recovered and must be neutralized at the end of reaction, leading to disposal of large amount of wastewater during downstream process for isolation of products.

The enzymatic transesterification of oils for biodiesel production leads to high conversions percentages at mild operative conditions, and is environmentally more attractive alternative to the conventional chemical process. Also, contrary to alkaline catalysts, enzymes do not form soaps and can esterify both FFA and triglycerides in one step, becoming unnecessary a subsequent washing step, with reduction of production costs. Thus, enzymes are able to produce biodiesel in fewer process steps using less energy and with drastically reduced amount of wastewater (Ranganathan *et al*., 2008).

Nevertheless, if the enzymatic process is to compete with the chemical process in the market, the cost of enzymes has to be brought down. For long-term trials, reusing the enzyme for many cycles is needed to make correct evaluations of the industrial potential of the process. The longer the reuse of the same enzyme, the higher the productivity obtained with the enzyme batch and the lower the biodiesel production cost. Therefore, the drawback for the industrial application of the enzymatic processes related to the biocatalyst cost can be overcome by reusing the enzyme in a continuous operation or in a discontinuous batch mode membrane reactor capable of retaining the enzyme. The different stages for biodiesel production and

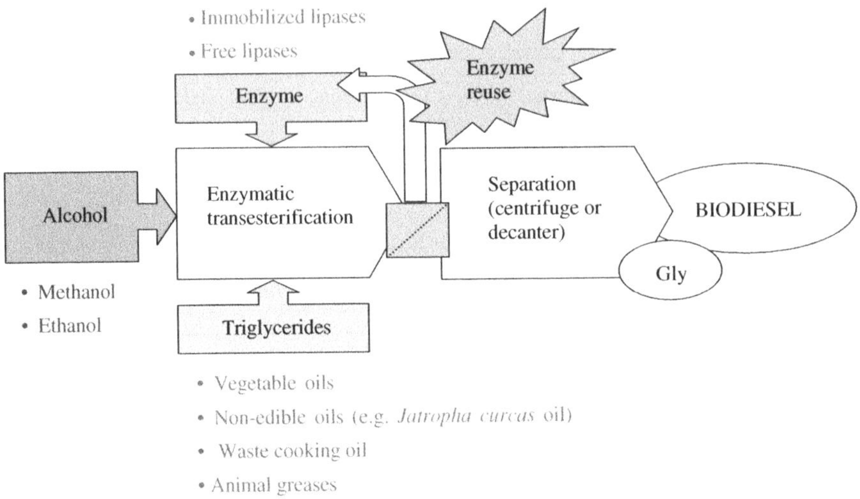

Figure 14.6 *Schematic diagram of the stages required for production of biodiesel*

potential opportunity for application of membrane systems are briefly illustrated in Figure 14.6 and are further explored in Section 4.3.2.

Lipases are among the enzymes found to be capable of catalyzing transesterification; these enzymes are produced by microorganisms such as *Rhizomucor miehei*, *Rhizopus oryzae*, *Candida antarctica*, *Chromobacterium viscosum*, *Thermomyces lagunisous*, *Pseudomonas fluorescencs* and *Pseudomonas cepacia*. Both extracellular and intracellular lipases are able to catalyze the transesterification of vegetable oils effectively. In both cases the enzyme can be immobilized and reused, which eliminates downstream operations and associated costs, and they are highly efficient compared with free enzymes processes. Lipase transesterification of triglycerides with an alcohol involves a two-step mechanism when looking at a single ester bond. In the first step, it occurs the formation of the active enzyme-substrate complex and the addition of an alcohol molecule to the complex. Then it occurs the separation of a molecule of the fatty acid alkyl ester and a glycerol moiety (di- or monoglyceride or glycerol), and the release of the active enzyme (Fjerbaek *et al*., 2009).

A kinetic model based on Ping Pong Bi Bi mechanism with competitive alcohol inhibition is the widely accepted model for transesterification of triglycerides. An example of an initial rate equation for this type of mechanism is shown above (Eq. (14.7)).

$$v_i = \frac{V_{\max}[TG][A]}{K_{m,TG}[A]\left(1+\frac{[A]}{K_{i,A}}\right)+K_{m,A}[TG]+[TG][A]} \tag{14.7}$$

The transesterification mechanism is actually more complex and the overall reaction can be divided in three consecutive reversible reactions to take into account the formation and consumption of intermediates (mono- and diglycerides). The kinetic model must be based on a material balance involving these three stepwise reactions characterized by the differential equations of each reaction component. Triglycerides are available in many vegetable oils such as soya bean and palma oil. However, criticism have been raised on the use of such oils as raw materials for energy since they are also needed for food products supply chains. Therefore, the use of oils of non-food crops, such as *Jatropha curcas* that is able to growth in non-agricultural lands or algae driven oils, have been proposed as second-generation biodiesels. Due to the high cost of crude and refined vegetable oil, cheaper animal fats and waste cooking oils, available from restaurants and households, are also attracting attention as possible substrates, often priced favorably, for conversion into biodiesel.

BIOGAS is usually produced by anaerobic digestion of different wastes. Membrane bioreactors have been used for wastewater treatment and biomass retention. When anaerobic systems are used, organic matter in the waste is converted by methanogenic bacteria to methane and CO_2, as described in Eq. (14.8):

$$C_{x1}H_{x2}O_{x3} \xrightarrow{\text{Methanogenic bacteria}} y1\ CH_4 + y2\ CO_2 \tag{14.8}$$

The biomass digestion process can be described as a sequence of reactions that leads to complete hydrolysis of cell polymers to monomers, fermentation of sugars, organic acids, amino acids, purines and pyrimidines to volatile organic acids, and conversion of C_5-C_3 organic acids to C_2 acid-acetate. Methanogens convert either acetate to CH_4 and CO, or H_2 and CO_2 to CH_4. The microorganisms involved in the anaerobic digestion can be mainly classified as hydrolytic, fermentative, acetogenic and methanogenic. Hydrolytic microorganisms excrete hydrolytic enzymes such as cellulose, cellobiase, xylanase, amylase, lipase and protease; they are responsible for reduce complex particulate compounds to soluble monomeric or dimeric substrates. Most of the bacteria are strict anaerobes such as *Bacteroides*, *Clostridium* and *Bifidobacteria*. Fermentative bacteria (for example, including species form the genera *Saccharomyces*, *Lactobacillus*, *Acetobacterium*) are responsible for conversion of the solutes obtained by hydrolysis into a mixture of short-chain volatile fatty acids, CO_2, H_2 and acetic acid. The volatile fatty acids are then converted to acetate and hydrogen by the acetogenic bacteria, typically belonging to the *Acetobacterium* and *Clostridium* genera. In the final step of the anaerobic digestion, two groups of methanogenic bacteria produce methane from acetate or H_2 and CO_2. Only few species are able to degrade acetate into CH_4 and CO_2, for example *Methanosarcina*, *Metanonococcus*, whereas all methanogenic bacteria are able to use H_2 to form CH_4 (Weiland, 2010).

The production of biogas by anaerobic digestion is one of the most energy-efficient and environmentally beneficial technologies for bioenergy production, mainly because of the high net energy yields per acreage and substrate flexibility. Biogas is produced from a variety of organic materials in plants, ranging from sewage treatment plants to organic waste utilization in landfill sites.

Mean biogas yields of various substrates are presented in Table 14.2:

Table 14.2 *Biogas yield for different sources (Weiland, 2010)*

Raw material category		Biogas yield (m^3/t FM)
Agriculture wastes	Cow manure	25
	Pig manure	30
Agriculture raw materials	Grasses	102
	Fodder beets	110
	Sudan grass	125
	Maize	200
	Wheat corn	630
Non-agricultural wastes	Biowaste	120
	Food residues	240
	Fat trap	400
	Used grease	800

14.2.2 Transport in membrane systems

Typically, in a membrane process, a membrane is used as a separating layer between two phases (liquid or gaseous), retaining specific compounds in one of the phases (feed or retentate phase), but allowing other compounds of the initial mixture to cross into the second phase (stripping or permeate phase). For transport to occur from the feed to the stripping phase, a driving force is required, such as differences in pressure, concentration, temperature or electric gradient. Different membrane processes are operated using different types of membranes, as schematically illustrated in Figure 14.7.

Porous membrane-based separations applied in biofuel production are usually pressure-driven processes with a membrane separating two liquid phases. As pressure is applied, the solvent (organic solvent or water) and specific smaller solutes are pushed across the membrane, and larger solutes are retained. Dense membranes are also widely used in reverse osmosis, usually to purify water, as virtually all the solutes are retained in a feed solution and the solvent (water) is pushed through the membrane. Pervaporation is also a pressure-driven process where a dense membrane is used, but vacuum is applied in the permeate promoting the preferential transport and phase transition of one of the species from a liquid feed to the gaseous permeate phase. Hydrophobic pervaporation is used for direct removal of ethanol from the fermentation broth and hydrophilic pervaporation is used to remove water from the ethanol/water azeotrope

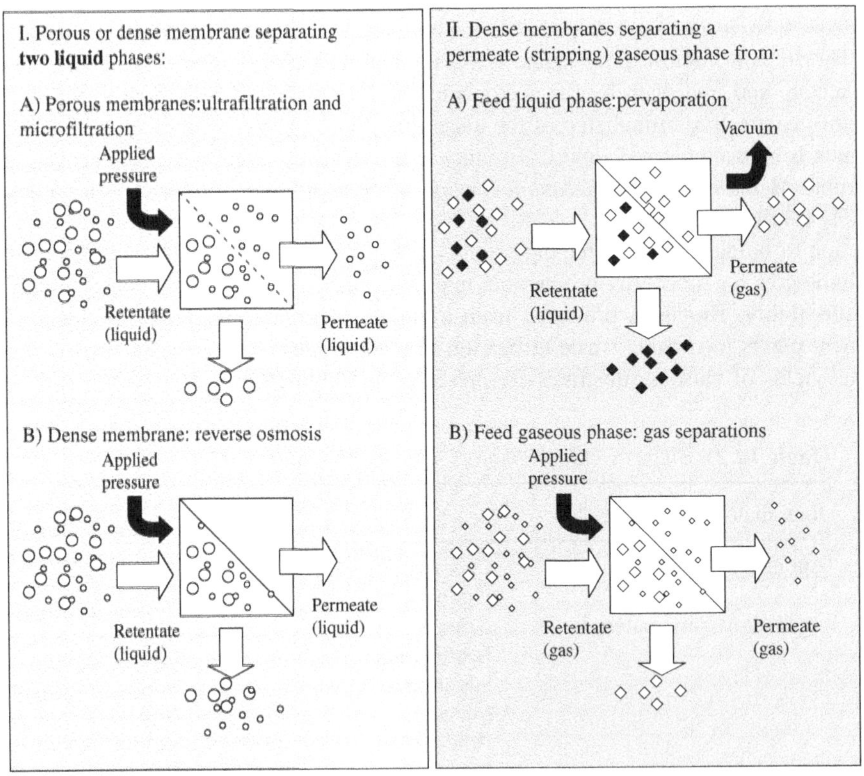

Figure 14.7 Schematic of membrane processes where (I) two liquid phases are separated by a porous membrane (IA: Micro and Ultra filtration) or dense membrane (IB: Reverse osmosis); A dense membranes (II) can also be used to separate a liquid from a gaseous phase (IIa: Pervaporation) or two gaseous phases (IIa: gas separations)

after distillation. Transport and separation through dense and porous membranes are usually described by solution-diffusion and pore-flow models, respectively. For nanofiltration, both of these two main theories were used to describe transport across the membranes, without consensus among membrane community.

Pressure-driven membrane processes are usually described by empiric parameters. Membrane rejection to solute ($R_{j,i}$), membrane permeability ($P_{m,i}$), and selectivity factors ($\alpha_{i,j}$) are commonly reported to assess membrane separation potential (Eqs. (14.9)–(14.11). These parameters determine process feasibility and economics. Membrane rejection to different solutes dictates separation efficiency and permeability to the solvent (or water) is used to select applied pressure and membrane area (A_m). Membrane permeability for pure water is >500, 50–500 and $<50 \, l \cdot m^{-2} \cdot hr^{-1} \cdot bar^{-1}$ for microfiltration, ultrafiltration and nanofiltration/reverse osmosis, respectively, implying application of 0.5–2 bar, 2–10 bar, 10–60 bar, for each of the respective operations.

$$Rj_i = 1 - \frac{C_P}{C_F} \tag{14.9}$$

$$J_{solvent} = A_m P_{m,solvent} (\Delta P - \Delta \pi) \tag{14.10}$$

$$\alpha = \frac{P_i}{P_j} \tag{14.11}$$

where J_i—flux, A_m—membrane area, $P_{erm,i}$—membrane permeability, P_{app}—applied pressure, $\Delta\pi$—osmotic pressure difference, Cp—concentration in permeate, C_F—concentration in feed solution, i—specie i.

Specifications for commercial available membrane often include a value of membrane molecular weight cut-off (MWCO) above which is expected solutes to be highly rejected (e.g. $Rj_i = 90\%$). Rejection curves are obtained by measuring rejection of solutes of different molecular weights (MW), in a given solvent (see examples in Figure 14.8). The underlying basis for the shape of rejection curves is the dispersion of pores size within each membrane, and the rejection curve can be seen as the cumulative curve of pore size distribution. Notice that transport phenomena across the membrane is a complex phenomenon (Table 14.3), and additionally to solute molecular size, other factors play important roles, including solute shape and polarity, as well as interactions between solute, solvent and membrane.

Dense membranes specifications are usually reported using specific gases permeability and selectivity factors (gas separations), and water permeability and salt rejections (reverse osmosis).

The description of transport through the membrane is usually attempted using parameters such as membrane permeability and selectivity, as mentioned earlier. However, there are additional phenomena, taking

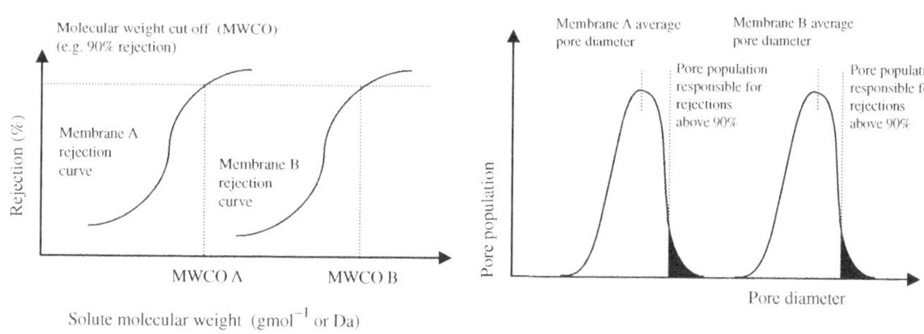

Figure 14.8 *Schematic of rejection curves for two membranes with different MWCO and corresponding pore size distribution*

Table 14.3 *Overview of pore flow and solution diffusion*

Pore-flow models	Solution diffusion model

Pore-flow models

Pore-flow models have been used to describe transport across porous membranes, linking solvent permeability (P_m, solvent) with membrane structure. Examples are the Hagen–Poiseuille (Eq. (14.12a)) and Kozeny–Carmen (Eq. 14.12b) correlations for parallel or close-packed spheres, respectively. Since membrane structure is not symmetric, more than one correlation type can be used in series.

$$P_{m,solvent} = \frac{\varepsilon . r^2}{\delta \eta \tau} \tag{14.12a}$$

$$P_{m,solvent} = \frac{\varepsilon^2}{K \eta s \left(1 - \varepsilon^2\right)} \tag{14.12b}$$

ε—surface membrane porosity, r—pore radius, τ—pore tortuosity (deviations to aligned pores), η —liquid viscosity, K—constant.

Solute transport across a membrane, neglecting ionic effects, can be described as function of a diffusion and convective term. The diffusion term can, in many cases, be neglected. The convective term is usually modeled taking into account the Poiseuille law and introducing a reflection coefficient (σ), according with a given model (e.g. Ferry, steric hindrance (SH) and Verniory) correlations based in differences between membrane pores and solute diameters. For example the SH model is based on effects of pore wall (H) and steric hindrance (S).

$$Js = P \Delta x \frac{dc}{dx} \pm (1 - \sigma) J_v c$$

$$\text{with } \sigma = 1 - HS$$

$$H = 1 + \frac{16}{9} \frac{d_s}{d_v};$$

$$S = 2\left(1 - \frac{d_s}{d_v}\right)^2 - \left(1 - \frac{d_s}{d_v}\right)^4 \tag{14.13}$$

In this condition, membrane rejection (R_j) of a given solute is a function of solvent and solute fluxes (J_V, J_{solute}), and corresponds to the reflective factor:

$$Rj = \frac{J_{solute}}{C_F \; J_V} \tag{14.14}$$

Solution diffusion model

In solution diffusion, transport of a given specie involves its partition from the feed solution into the membrane material; followed by its diffusion, described by Fick's law, across the membrane and its partition from the membrane material into the stripping phase. Partition coefficients (K_i) can be seen as a ratio of species activity coefficients for different phases γ_i. For gaseous phases, instead of K_i, gas solubility (S_i) into the membrane polymer is used.

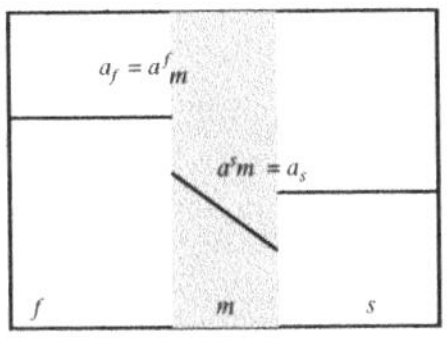

$$\mu^f_m = \mu_f \qquad \mu^s_m = \mu_s$$

$$a^f_m = a_f \qquad a^s_m = a_s$$

$$\gamma_m C^f_m = \gamma_f C_f \qquad \gamma_m C^s_m = \gamma_s C_s$$

$$K_{m/f} = \gamma_f / \gamma_m \qquad K_{m/f} = \gamma_s / \gamma_m$$

$$K_{m/f} = C^f_m / C_f \qquad K_{m/f} = C^s_m / C_s$$

μ—chemical potential, a—activity coefficient, γ—activity coefficients, C—concentration, m—membrane, f—feed solution, s—stripping solution, K—partition coefficient, D—Diffusion coefficient membrane, A_m—membrane area, J– flux

For feed and stripping solution comprised of same media, $\gamma_f = \gamma_f$ and $K_{m/f} = K_{m/s} \cdot P_{m,i}$ can be expressed as:

$$P_{m,i} = D \cdot K_m \text{or } P_{m,i} = D \cdot S_m \tag{14.15}$$

Concentration in the membrane phase is related to adjacent gas partial pressure as $C_{m,i} = k_{H,i}.P'_i$. In pervaporation, the feed phase is liquid and the stripping phase a gas at a pressure near zero, thus flux equations, driven by concentration gradients, can be used to describe transport through a dense membrane, separating two gaseous phases (gas separations) or a liquid from a gaseous phase (pervaporation), respectively.

$$J = A_m \frac{P_{m,1}}{\delta} \left(C_F - \frac{P'_V}{K_H}\right) \tag{14.16}$$

$$J = A_m \frac{P_{m,1}}{\delta} \left(P'_F - P'_p\right) \tag{14.17}$$

k_H corresponds to the Henri constant. Alternatively, the Langmuir isotherm or Freundlich adsorption equations could be used to relate concentration in the membrane with the respective partial pressure.

place into the adjacent phases and at membrane/phase interface, which can severely affect overall transport between phases, namely:

- Concentration polarization in the retentate solution takes place when mixing rates are not high enough and the solute concentration gradient increases abruptly from the bulk retentate solution to the membrane interface, as the solvent permeates through the membrane much faster than the solute. Therefore, as membrane rejection depends on solute concentration at the membrane interface, solute concentrations in the permeate are higher than those expected when predictions are made considering only retentate bulk solute concentration.
- Fouling of membrane surface builds up over longer operations, leading to solvent flux decrease, which increases operation costs as higher membrane areas or operation times are required. Membrane material, surface hydrophobicity and surface roughness influence transport and fouling.
- Clotting of the membrane surface is often disruptive of the separation operation and can be observed when handling solutions with high concentrations of suspended solids and particles or slurries.

Hydrodynamics play an important role in proper nutrient mixing and in the oxygenation of the bioreactor, but also in membrane operations, by avoiding stagnate zones in the fluid and additional liquid mass transfer resistances, concentration polarization, fouling and clotting. These phenomena can be mitigated as membrane module is operating in cross flow (i.e. pumping the solution tangential to the membrane surface), rather than in dead-end mode (representative schemes are presented in Figure 14.9), and carrying out periodic cleaning in place (CIP) operations. However, frequent cleaning also increases membrane material aging rate and the need to replace expensive membrane modules.

14.2.3 Membrane modules and reactor operations

There are several dense and porous membranes commercially available made of polymeric and ceramic materials, with different membrane structures. Many membranes have an asymmetric structure comprised of an active layer, which determines membrane selectivity, supported in a meso and then a macro porous layer. Ceramic membranes are usually made of metal oxides, such as silica, titania, alumina, zirconia, silicon carbide (SiC), and mixtures of oxides. Ceramic membranes are able to cope with a wider range of temperatures, pH and solvents; however, given their lack of flexibility, the module configurations are somewhat more limited. Usually, they are supplied as single membrane tubes or modules of parallel tubular membranes, as well as multi-channel monoliths. Polymeric membranes are available from a wide range of materials and, given their higher flexibility, they can be assembled in tubular hollow fibers, but also as flat sheets in plate and frame configurations or spiral wound modules. Several possible membrane are illustrated modules configurations in Figure 14.10.

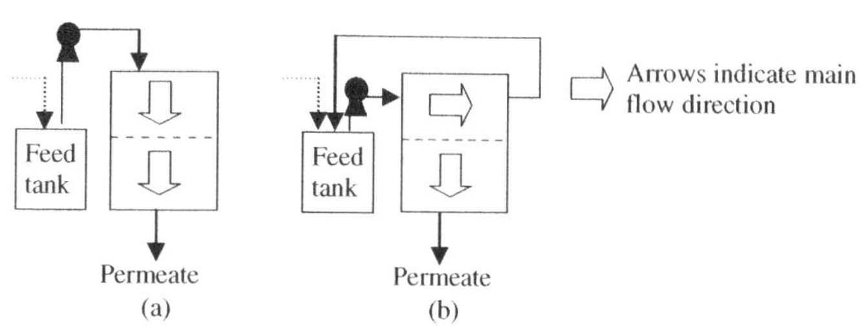

Figure 14.9 Operation modes: (a) dead-end cell operation, (b) cross-flow operation

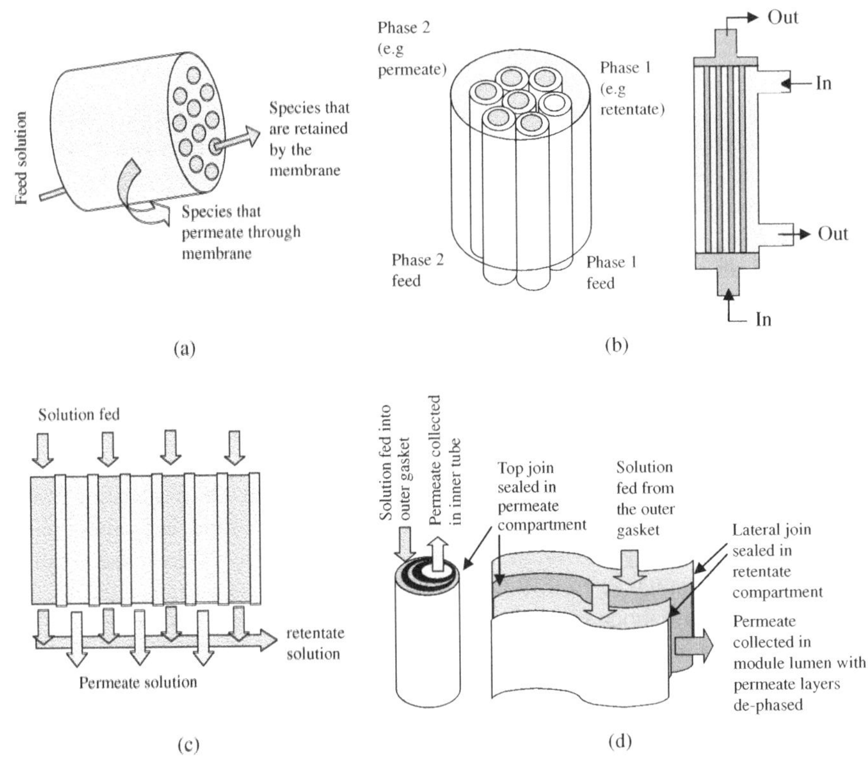

Figure 14.10 Membrane modules' different configurations: (a) multi-channel monolith membrane, (b) module of parallel hollow fibers, (c) plate and frame module of flat sheet membranes, and (d) spiral wound module of flat sheet membranes

14.2.4 Membrane bioreactor

Integration of membrane and bioreactor can be performed sequentially, with a side stream fed from the bioreactor to the membrane module, in one pass through or with optional recirculation of the retentate stream into the bioreactor, as permeate is collected (Figure 14.11a). Membrane modules can also be used between bioreactors in series. An alternative interesting configuration to minimize fouling is the submerged membrane bioreactor (Figure 14.11b), where the membrane is assembled inside the bioreactor and the solution pulled through the membrane.

14.3 Examples of membrane bioreactors for biofuel production

14.3.1 Bioethanol production

14.3.1.1 Overview

Since the 1970s' oil crisis and, more recently, with the rise of the carbon economy, several studies have focused on process intensification for improving bioethanol production cost efficiency (Cardona and Sanchez, 2007). Among several other technologies used to achieve such aim, membrane technology and

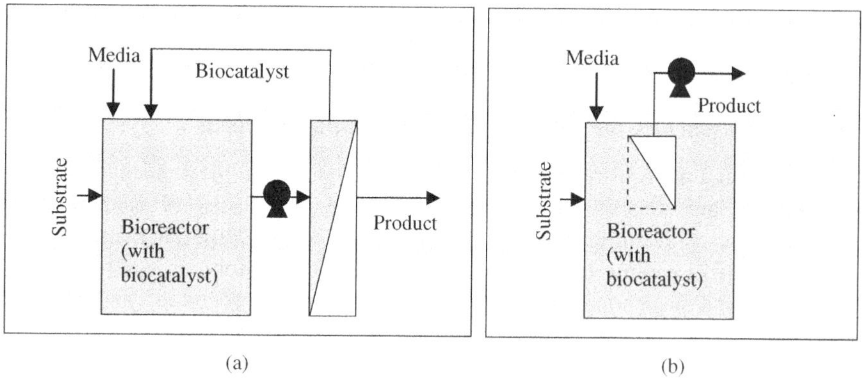

Figure 14.11 Schematic diagram of different membrane bioreactor configurations: (a) Bioreactor with recirculation in an external side stream cross-flow membrane module and (b) Submerged membrane bioreactor with internal membrane module

membrane bioreactors have the potential to play an important role (Lipnizki, 2010). This section comprises examples of studies in which the use of membrane bioreactors was suggested in order to:

- Achieve high cell concentration retention to improve productivities, with simultaneous removal of ethanol and by-products to avoid inhibition. Studies include microfiltration- and ultrafiltration-based processes.
- Enrich fermentation product to high grade ethanol >98%, using low energy intensive processes. Studies include downstream pervaporation for direct removal of ethanol or break down the azeotrope obtained after distillation.
- Reduce costs with hydrolytic enzymes in saccharification stage and remove solids and inhibitory compounds from the stream fed to the fermentation. Studies include upstream membrane bioreactors with ultrafiltration membranes retaining enzymes and polysaccharide compounds, but allowing sugar monomers to cross the membrane and to be fed into the fermentation.

The suggestion of using ultrafiltration for cell retention, and isolation of ethanol from the permeate stream with a pervaporation type process, was suggested as early as 1979 (Gregor and Jeffries, 1979). Figure 14.12

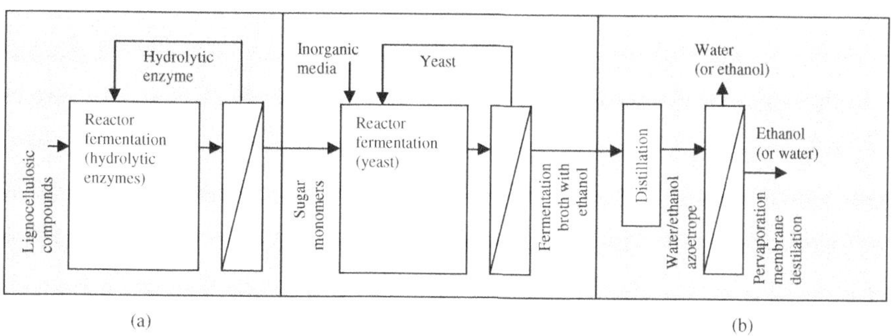

Figure 14.12 Membrane bioreactors for ethanol production: examples of (a) upstream saccharification stage—retention of hydrolytic enzymes and sugar permeation and (b) downstream ethanol purification stage—pervaporation

illustrates the opportunities for using membrane processes in ethanol production by fermentation as well as in upstream and downstream operations.

14.3.1.2 Membrane bioreactors for cell retention and ethanol removal

Membrane bioreactors using microfiltration and ultrafiltration membranes have been used for cell retention, with removal of ethanol and other by-products. In such strategies, the use of membrane bioreactor can offer advantages in terms of decoupling biomass and ethanol reactor dilution rates, (i) allowing higher biomass concentrations, (ii) avoiding ethanol and other by-product inhibitions, and (iii) providing a clear stream, free of biomass and suspended solids for further downstream processing and ethanol purification. In most of the trials for ethanol production with cell retention, an external membrane module with a recycle loop into the bioreactor was used, but bioreactors with internal submerged membrane modules had also been attempted.

In conventional *Saccharomyces cerevisiae*-based fermentations for ethanol production, the cell growth and ethanol production are severely inhibited at ethanol concentrations of $110\,g\cdot l^{-1}$. No ethanol inhibition is observed for values bellow $30\,g\cdot l^{-1}$ (Kargupta *et al.*, 1998). Ethanol concentration can be maintained at low levels by the continuous addition of lower sugar contents at higher dilution rates; however such a strategy leads to low cell density in the reactor, as cells are washed out and thus leading to low ethanol productivities. This problem was studied by Cysewski and Wilke (1977), who demonstrated that, by continuously removing ethanol by vacuum, ethanol productivities could be increased up to 82 and $40\,g\cdot l^{-1}\cdot h^{-1}$ with and without cell recycling, respectively. Indicative values for a conventional fermentation are the use of 10 wt% glucose solution to yield 50 dry $wt\cdot g\cdot l^{-1}$ of biomass and an ethanol productivity of $29\,g\cdot l^{-1}h^{-1}$.

Motivated by these findings, several studies have been carried out to attempt to improve ethanol productivity, following a strategy of continuous removal of ethanol and cell retention. Initial studies included the use of vacuum evaporation, solvent extraction for ethanol removal and settling tanks and centrifuges for cell retention. Aqueous/aqueous dialysis using a microfiltration membrane was suggested by Kyung and Gerhadt (1984) and evaluated experimentally, with cells being retained as ethanol was continuously removed; unfortunately, in this study, improvement in ethanol productivity was not observed. Additional studies also suggested the use of extractive systems, in which the role of porous membranes was to stabilize the interface between an aqueous phase and an organic phase; with inhibitory compounds being removed as partition favored their extraction from fermentation broth into the organic phase (Kang *et al.*, 1990; Chang *et al.*, 1992). The extracted compounds include not only ethanol but also by-products that tend to accumulate on the fermentation broth, such as acetic acid, glycerol, and succinic acid (Ben *et al.*, 2006), or unreacted compounds that are carried out from saccharification steps, such as furans or polyphenolic compounds (Maiorella *et al.*, 1983; Brandberg *et al.*, 2008). However, supported liquid membrane bioreactor industrial applications are usually hindered by stability issues and leakage of the organic phase solvent into the fermentation broth, which can affect negatively microorganism activity, even in small quantities.

Alternative approaches for cell retention include the immobilization of microbial cells (Ghose and Bandyopadhyay, 1980) and entrapment of *Saccharomyces cerevisiae* into the shell side of hollow fiber reactors (Mehaia and Cheryan, 1984), both strategies led to the improvement of ethanol production. Ethanol productivities of 2.1, 25 and $10\,g\cdot l^{-1}\cdot h^{-1}$ (corresponding to glucose consumptions of 100%, 74% and 85%) were observed for conventional batch conversion, carrier immobilization and membrane immobilization, respectively. Similar studies were performed for *Zymomonas mobilis* ethanol production.

Studies by Mehaia and Cheryan (1984), Hoffmann *et al.* (1985) and Lafforgue *et al.* (1987) coupled a fermentor with a cross-flow microfiltration unit, allowing ethanol productivities on the range of $33–44\,g\cdot l^{-1}\cdot h^{-1}$ at $90–140\,g\cdot l^{-1}$ cell concentration. Glucose consumption was not always completed,

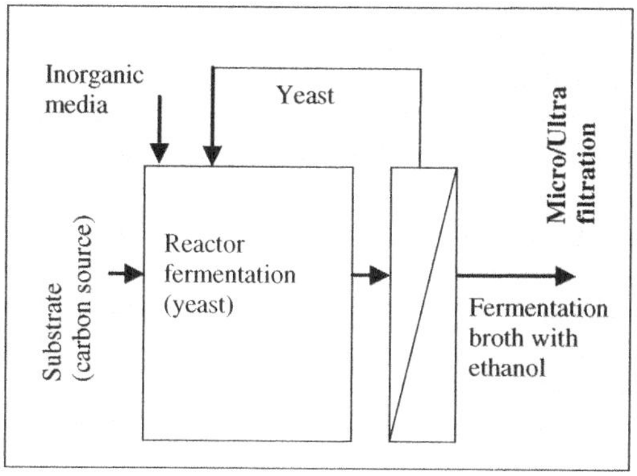

Figure 14.13 *Membrane bioreactors for ethanol production: bioreactor with recirculation with an external side stream cross-flow membrane module*

depending on dilution rates used, which implies non-optimized carbon economies. Still it is interesting to note that it was achieved cell concentrations and productivities up to $300\,g \cdot l^{-1}$ and $150\,g \cdot l^{-1} \cdot h^{-1}$, respectively. Notice that, in this membrane bioreactor configuration, the fermentation broth is recirculated between the fermentor and an external side stream cross-flow membrane unit tangentially to the membrane surface (Figure 14.13). Such a configuration has been used in several studies; usually a tubular ceramic membrane is used as a membrane module external to the fermentor, which allows for easy sterilization of the membrane, cleaning using alternate membrane units, and replacement when the membrane is damaged.

Further examples of trials using such configuration includes (i) the work by Melzoch and co-authors (Melzoch *et al.*, 1991), where tubular ultrafiltration membranes made of polyphenylenephthalamide with a MWCO of 25 kDa were used and it is reported the impact of such system on cell maintenance energy requirements; and (ii) the work by Escobar *et al.* (2001) in which the process was scaled up to a 7000 l membrane bioreactor, using a ceramic microfiltration membrane (Membralox, 19P19-40 modules from U.S. Filter, Warrendale, PA), with a transmembrane pressure kept constant at 140 kPa and a circulation cross flow rate of $1350\,l \cdot m^{-1}$. Such a membrane bioreactor, schematically illustrated in Figure 14.14, was fully controlled using pressure gages, flow meters and sensors with feedback loops to valves controlled by a programmable logic controller (PLC) for continuous adjust of flow rate, pressure, reactor levels, stream feeding and cells bleeding. Ethanol, glucose and cell concentrations over time were reported.

The membrane bioreactor was started with $23\,g \cdot l^{-1}$ cell concentration and after 18 hours of cell growth and pre-concentration, cell concentrations of $60{-}100\,g \cdot l^{-1}$ ($10^9{-}10^{10}$ cell $\cdot\, ml^{-1}$) were reached. Notice that in conventional batch systems without membranes, usual cell concentrations are limited to values of only $15{-}25\,g \cdot l^{-1}$. In this trial, at a cell concentration of $100\,g \cdot l^{-1}$, virtually all the glucose in the feed ($175\,g \cdot l^{-1}$) was consumed and an ethanol concentration in the broth of about 10.8% v/v was obtained. 50–70% of the cell population was viable, depending on nutrients feed and shear stress. Some key practical notes were highlighted in this pilot plant trial study:

- The cell concentration was maintained below $120\,g \cdot l^{-1}$ through cell bleed when necessary. This value was established to avoid excessively high viscosity in the fermentation broth, which would affect mass transfer of nutrients, oxygen and would also affect membrane permeation.
- As the cell concentration increased, possibly due to fouling, the flux through the membrane decreased. To reach and maintain a steady state at a constant dilution rate, small flux drops were compensated by increasing the transmembrane pressure, within reasonable values (>275 kPa in this study). Alternatively,

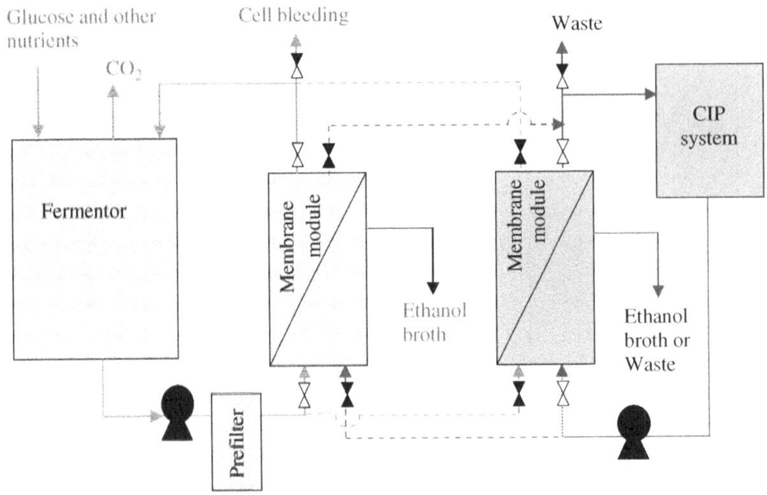

Figure 14.14 Membrane bioreactors for ethanol production: bioreactor with recirculation into an external side stream cross-flow membrane module equipped with a CIP system for membrane cleaning and an external spare membrane reactor. Adapted with permission from Escobar et al *© 2001, Humana Press Inc*

the residential time (i.e. the inverse of the dilution rate, D = F/V, where F is feed flow rate and V is the bioreactor volume) could be maintained by increasing the operational system volume. However, these two strategies can only be applied to a certain extent and thus are only suitable to correct small permeation flux drops.

- Cleaning in place (CIP) strategies were implemented using two membrane modules connected to the system. When the flux declined significantly and adjustments in pressure and volume were not able to maintain dilution rates, the flux recirculation was diverted to the second membrane module and pressure was decreased accordingly. The first membrane model was then submitted to the CIP protocol, in which the fermentation broth of this loop was firstly recovered, after the membrane was flushed first with water at 50 °C and then with cleaning chemical agents (dilute alkaline and chlorine solutions) until flux levels were restored. Permeate and retentate obtained in these operations were discharged. Membrane black flushing—inverting transmembrane pressure—is also possible. Interestingly, it was found that the antifoam used on the fermentation is a strong fouling agent (the antifoams normally used are polyoxyethylene/ polyoxypropylene oleyl ethers, polyethylene glycol and silicone-based compounds).

This study (Escobar *et al*., 2001) also includes an economic analysis for a plant with an ethanol annual capacity of 3.8 million m^3 (100 million $gal \cdot yr^{-1}$), which implies processing at $409\,m^3 \cdot h^{-1}$. To update this economic assessment to the current time it would be necessary to take into account changes in the costs of membranes and energy as well as fuel market prices. However, it is worth mentioning that membrane performance has an important effect in process economics. In this study, operating costs for ethanol production of $9\,\$/m^3$ (0.034 \$/gallon) or $4.5\,\$/m^3$ (0.017 \$/gallon) were estimated, when it was considered a membrane area of $5837\,m^2$ (flux $70\,l \cdot m^{-2} \cdot h^{-1}$) or $2724\,m^2$ (flux $150\,l \cdot m^{-2} \cdot h^{-1}$), respectively. Between 2000 and 2009, the drive for ethanol consumption increased about fivefold, with a higher growth rate in the period 2005–2009 and an average annual rate of about 20% for this five-year period. The expectations are that, with increasing demand and capacity limitations, the need for higher substrate conversion efficiencies became more stringent, leading to a call for process optimization, where membrane reactors can have an important role.

An additional interesting study (Ben *et al.*, 2006) tried to improve ethanol productivity using two-stage bioreactors with cell recycling, using an ultrafiltration module. The first reactor was used for cell growth and the second one for ethanol production. With complete consumption of glucose, it was possible to achieve an ethanol productivity of $42\,g \cdot l^{-1} \cdot h^{-1}$, with an ethanol concentration of $65\,g \cdot l^{-1}$. Studies targeting lignocellulosic hydrolysates are also described in the literature either using, again, the configuration where fermentation broth is recirculated through an external side stream cross-flow membrane module (Brandberg *et al.*, 2008) (Figure 14.11a) or an internal membrane bioreactor (Lee *et al.*, 2000) (Figure 14.11b).

14.3.1.3 Upstream saccharification stage: Retention of hydrolytic enzymes and sugar permeation

Saccharification of starch and lignocellulosic compounds into sugar monomers is required prior to their fermentation into ethanol. Chemical hydrolysis using sulfuric acid or alkaline solutions yields toxic compounds for the microorganisms used in the fermentation stage. The production of first generation bioethanol production from molasses and starches usually requires the mild enzymatic saccharification using, respectively, invertases or amylases. For second-generation bioethanol, lignocellulosic breakdown into sugars is better achieved with cellulases and β-glucosidases, however the costs with such enzyme extracts would account for 36%–45% of total costs in ethanol production.

Bringing down costs with enzymes is crucial to economic feasibility of second-generation ethanol production. The immobilization of enzymes is a possible strategy to facilitate enzyme recycling. However, immobilization can also hinder enzyme access to the insoluble lignocellulose and pose additional resistance to mass transfer. Then, retention of enzymes with membranes could be a possible alternative strategy for enzyme recycling. Cellobiose is an intermediate of lignocellulose break down in sugar monomers and its inhibition constant (K_I) for hydrolytic enzymes ranges from 0.01 to $6\,g \cdot l^{-1}$, whereas the glucose K_I ranges from 1 to $60\,g \cdot l^{-1}$. Therefore, cellobiose accumulation should be avoided, which was attempted by increasing its degradation rate into glucose adjusting amounts of beta-glucosidases, controlling addition of substrate and following SSF strategies. The use of nanofiltrations or narrower MWCO ultrafiltration could also be used to control cellobiose concentrations and increase the residence time of polysaccharide substrates, contributing to their complete hydrolysis (Cardona and Sanchez, 2007).

Conceptually, the retention of free enzymes using membrane reactors in SHF would bring tremendous advantages, since ultrafiltration membranes (see Figure 14.12a) will allow for (i) recycling enzymes, bringing down operation costs, (ii) maintaining a constant product flux, with continuously permeation of cellobiose and glucose formed, which are strong inhibitors of the hydrolytic enzymes, and (iii) retaining suspended solids and cellulosic compounds contributing to a fully substrate hydrolysis and a cleaner fermentation feeding stream, with reduced suspended solids and macromolecules toxic to microorganisms used in ethanol fermentation (Andric *et al.*, 2010).

Several studies had been performed, back from 1970 (Ghose and Kostick, 1970) up to day (Liu *et al.*, 2010; Zhang *et al.*, 2010) using cellulose acetate, polyethersulfone and polyamide ultrafiltration membranes with MWCO ranging from 5 to 50 kDa. The benefit of using a membrane bioreactor to remove inhibitory compounds was clearly demonstrated by Ohlson *et al.* (1984)) using a *T. reesei* enzyme and an ultrafiltration membrane. In this study, reactions of 20 h were carried out and the use of the membrane system allowed improvements in yields from 40% to 95%, with enzyme activity losses of only 5%. Ohlson and co-authors' results were confirmed by several other groups, including the studies by Yang and co-authors using cellulose from rice straw (Yang *et al.*, 2006) and corn stalk (Yang *et al.*, 2009), where product inhibition was overcome by increasing membrane bioreactor dilution rates. In the later system, after 40 h reaction, 85% and 40% conversions were obtained at dilution rates of 0.65 and zero respectively (zero corresponds to not using the membrane module). Some studies also suggested that, for given conditions, the reaction kinetics are the limiting step, rather than product inhibition, due to slow association/dissociation between substrate and enzymes (Gan *et al.*, 2002; Knutsen and Davis, 2004). In an attempt to facilitate kinetics and simultaneously allow for enzyme/inhibitory products separation, it was suggested (Bélafi-Bakó *et al.*, 2006) to use an immersed tubular composite membrane comprising a 30 kDa Nadir membrane and a non-woven textile layer to promote enzymes and substrate adsorption in the membrane walls. The introduction of the textile layer improved conversions in 10%.

The industrial application of membranes to retain enzymes in the saccharification step prior to fermentation is not trivial. Notice that polysaccharide solutions are usually very viscous slurries with high solid contents, which unfortunately results in severe membrane fouling and leads to quick operation disruption. Increasing mixing rates to improve mass transfer rates can also be somewhat limited, since it is reported that shear stress can deactivated the enzymes (Hong *et al.*, 1981). Several approaches and configurations of membrane reactor systems for saccharification (Andric *et al.*, 2010) had been evaluated:

- Feeding regimes, including fed batch and continuous substrate feeding, as well as bleeding of unconverted substrates and accumulated suspended solids, had been assayed to control viscosities and reaction rates in the bioreactor (Kinoshita *et al.*, 1986; Ishihara *et al.*, 1991; Lee and Kim, 1993).
- Several membrane configurations, including cascade membrane systems, with sequential microfiltrations and ultrafiltrations, had been assayed to minimize fouling and clotting (Ishihara *et al.*, 1991; Knutsen and Davis, 2004).
- Immobilization of beta-glucosidade in the membrane outer walls of a side stream membrane module with dissolved cellulases in the bioreactors was assayed (Klei *et al.*, 1981) to optimize relative cellobiose breakdown rate.

A large-scale trial was performed by Ishihara and co-authors (Ishihara *et al.*, 1991) using a 10 l bioreactor. The different studies mentioned, actually, didn't fully address the challenge of processing an extremely viscous solution through a membrane module, which is crucial to develop a robust industrial operation. In 2001 (Mores *et al.*, 2001), a trial was attempted in which sedimentation and membrane filtration were combined to remove larger particles from the settler. In this study, a microfiltration membrane was used for stream clarification and an ultrafiltration membrane was used for enzyme retention. A preliminary economic analysis presented in the same report pointed out that as long as 75% of the enzyme is recycled in the active form, the benefit of investing in a membrane system may reach as much as 0.18 $/gal of ethanol; costs and ethanol prices require update for actual values. Dynamic filtrations had been studied, where membrane separation is combined with a rotating disk or vibrating membranes in order to reach higher shear stresses and overcome mass transfer limitations. These systems had been applied, for example in food industry and have the potential to handle more viscous solutions (Jaffrin, 2008).

14.3.1.4 Downstream ethanol purification stage: Pervaporation

Pervaporation uses a dense membrane to separate a liquid phase from a gaseous phase submitted to negative pressure, promoting permeated species to pass to their gaseous state as they cross the membrane (see Figure 14.12b). There are hydrophobic and hydrophilic membranes available for pervaporation. In the context of ethanol production hydrophobic or hydrophilic pervaporation can be used, respectively, for (i) direct removal of ethanol from the fermentation broth allowing ethanol isolation and mitigating fermentation inhibition, or (ii) as an additional step after distillation for removal of the water to overcome water/ethanol azeotrope, allowing to reach 98% v/v ethanol grade without the use of chemical additives (O'Brien and Craig, 1996; Lipnizki, 2010; Lipnizki *et al.*, 2000). Pervaporation was applied to remove ethanol from fermentations catalyzed, not only by yeasts, but also by *Zymomonas mobilis* (Ikegami *et al.*,

2007) and *Clostridium thermohydrosulfuricum* (Mori and Inaba, 1990), a thermophile, able to produce ethanol at 66 °C.

14.3.2 Biodiesel production

14.3.2.1 Overview

Free or immobilized lipases have been used as biocatalysts in enzymatic transesterification of triglycerides and short-chain alcohols into fatty acids alkyl esters. Enzymes cost far more than chemical catalysts. In industrial applications of the enzymatic process, it is therefore desirable to include the retention of the lipase from the product stream for its recycling in successive operations. Usually, immobilized enzymes are preferred over free enzymes in enzymatic processes because immobilized enzymes are easily recovered, present higher stability, are insensitive to solvents, and reusable (Fjerbaek *et al.*, 2009). However, an enzyme-immobilization process increases costs and usually free enzyme catalyst is more effective, resulting in higher reaction efficiencies and yields. The use of membrane bioreactors is very promising as they are capable of retaining the free enzyme due to the nominal MWCO of the ultrafiltration membranes that are normally used (schematic diagram in Figure 14.15).

The optimization of enzymatic biodiesel production can be based on methods that focus on increasing lipase activity in organic media, like the use of known lipase-activating substances (such as cyclodextrins or crown ether) (Yasuda *et al.*, 2000; Van Unen *et al.*, 2002; Ghanem, 2003), the attainment of surfactant-coated lipase (using dialkyl glucosyl glutamates) (Okahata and Ijiro, 1998) or polyethylene glycol (PEG) (Persson *et al.*, 2002)) or the formation of homogeneous systems through lipase solubilization in the organic solvents (for example, in reversed micelles—Larsson *et al.*, 1990).

Biocatalysis in reversed micelles can be particularly advantageous in the conversion of water-insoluble substrates as a high interfacial area between water and organic solvent is possible. Moreover, the equilibrium can be shifted to the synthesis reaction by decreasing the water content (Carvalho and Cabral, 2000). The scale-up and development of reactor design that enables reactions in continuous mode or discontinuous batches with enzyme reutilization, taking full advantage of the reversed micellar medium, has been recognized as a critical demand in reversed micelles technology (Prazeres *et al.*, 1994; Carvalho *et al.*, 2001).

Ultrafiltration membrane reactors offer the most appropriate reactor configuration for the confinement of encapsulated enzymes in the retentate side of the membrane. The synthetic semipermeable membrane could retain the enzyme and selectively separate the products, while maintaining the activity of the biocatalyst.

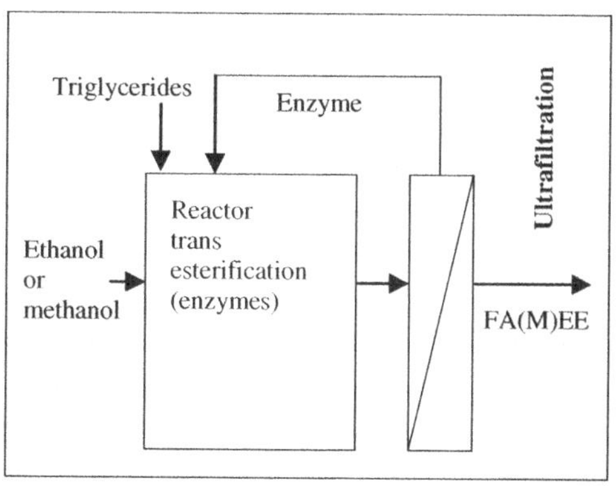

Figure 14.15 Membrane bioreactors for biodiesel production

Despite their cost, ceramic membranes are biologically compatible, and are highly resistant to chemicals and drastic operating conditions. Moreover, the filtration flux of these membranes is higher and more easily controlled than polymeric membranes.

14.3.2.2 Membrane bioreactor for biodiesel production

For the production of biodiesel, Badenes *et al*. (2011a) presented an enzymatic process using a membrane bioreactor in order to reuse the biocatalyst. The enzymatic transesterification of oils with an alcohol, using recombinant cutinase of *Fusarium solani pisi* microencapsulated in sodium bis(2-ethylhexyl) sulfosuccinate (AOT)/isooctane reversed micelles, was performed in a membrane reactor. Cutinase wild type and mutant T179C were tested for this process and the high long-term operational stability of the cutinase mutant demonstrates its potential as biocatalyst for the enzymatic continuous production of biodiesel. The production of fatty acid methyl esters (FAME) was tested departing from a commercial triolein mixture (oil), which contains a mixture of tri-, di- and monoglycerides with chains from oleic and linoleic acid, in order to approach real conditions, as far as possible. Cutinase was the biocatalyst used for the transesterification. This enzyme is an esterase that belongs to the α/β hydrolase family and present lipolytic activity, being its active site, constituted by the triad Ser, Asp and His, accessible to the solvent. Cutinase is a compact one-domain molecule (45 × 30 × 30 Å3 in size) constituted by 197 amino acids, with a molecular weight around 22 kDa and an isoelectric point of 7.6 (Carvalho *et al*., 1999).

The reversed micellar system provides a high interfacial area of contact, and hydrophobic substrates and products are solubilized in the continuous organic phase, whereas the enzyme remains active in the water core (Carvalho and Cabral, 2000). An ultrafiltration membrane reactor is an appropriate type of reactor to operate processes based on reversed micellar medium because the cutinase is confined in the reaction medium, while allowing the enzyme re-use and the removal of the products. The membrane reactor set-up is presented in Figure 14.16 and has a total reaction volume of 70 ml. A tubular ceramic membrane from Carbosep®, with a nominal MWCO of 15 KDa, 20 cm length, 6 mm internal diameter and 38 cm^2 interfacial area was arranged within a stainless-steel ultrafiltration module. The membrane is made of a zirconium

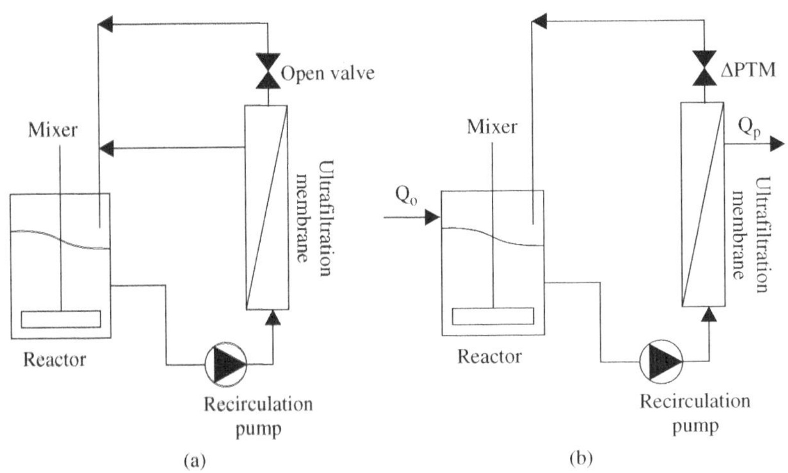

Figure 14.16 *Scheme of the membrane reactor (a) operating in a total recirculation batch mode and (b) for a continuous operation. ΔPTM is the transmembrane pressure, Q_p permeate flow rate and Q_0 feed flow rate*

layer over a porous carbon support and it is highly resistant to pressure, temperature, pH, and organic solvents. The recirculation of the reaction mixture was promoted by a gear pump with a high recirculation flow rate ($1000\,ml \cdot min^{-1}$) and tangential to the membrane surface reducing the deposition of solutes. A thermostated recirculation vessel was used in order to control the temperature at 30 °C.

Mutants of cutinase were evaluated in terms of their activity in the reversed micellar system and their stability in the presence of AOT (Badenes *et al.*, 2011b). It was found that the mutant T179C has a high resistance to the denaturing effect of AOT, and it has a higher stability in the presence of methanol when compared with cutinase wild type. The operational stability of the biocatalyst was evaluated and promising results were attained when cutinase mutant T179C was used.

The continuous operation of the membrane bioreactor was feasible due to the effective retention of cutinase within the reactor. An adsorption of 83% of protein was observed after 24 h of total recirculation and the real rejection of cutinase was higher than 96% when taking into account the adsorbed protein on the membrane retentate side. The membrane bioreactor could operate for more than 28 days; the conversion level is, as expected, dependent on the permeate flow rate—on the hydraulic residence time. High productivity levels (up to $500\,gproduct \cdot day^{-1} \cdot genzyme^{-1}$) were achieved indicating that the continuous production of biodiesel by cutinase mutant T179C in a membrane reactor has a high potential.

14.3.3 Biogas production

14.3.3.1 Overview

Anaerobic processes can be applied for treatment of a wide variety of industrial wastewaters with simultaneous generation of methane rich biogas. Biogas is a useful end product for energy production, and its production contributes to desirable diminution of biomass.

The membrane bioreactor process consists of a suspended growth biological reactor combined with a membrane unit either located external to the bioreactor (sidestream) or assembled directly within it (submerged or immersed) (Figure 14.11). The first generation of membrane bioreactor systems used in the 1980s for anaerobic wastewater treatment were mainly based on sidestream configuration with tubular membranes, where the biomass is circulated at high cross-flow velocity and the permeation is operated from inside to outside tubes. The second generation of membrane reactor systems was implemented in 1989, with the introduction of the submerged membrane configuration, which allowed a significant reduction of the capital and operational costs. In this type of bioreactor, the slight negative pressure imposed on the permeate side is responsible for the driving force that pulls clean water to the permeate through the membrane (Le-Clech, 2010). For the submerged membrane bioreactor processes, the membrane can be configured as vertical flat plates and vertical or horizontal hollow fibers, which are assemble in modules or cassettes including aeration ports. The circulation rate and operating transmembrane pressure used in external membrane systems are typically high ($1–5\,m \cdot s^{-1}$ and 2–7 bar, respectively), which is provided by a recirculation pump that reduces the deposition of suspended solids at the membrane surface. Although this configuration is simple and provides direct hydrodynamic control of fouling, the energy demand is relatively high. On the other hand, the cross-flow velocity and operating transmembrane pressure used in internal membrane systems are typically low (less than $0.6\,m \cdot s^{-1}$ and 0.2–1 bar, respectively) (Lei and Berube, 2004). Therefore, although submerged systems need less energy, they demand more membrane area. The submerged configuration relies on coarse bubble aeration to produce recirculation and suppress fouling. Microfiltration and ultrafiltration membranes are used in submerged configuration. The selection of the pore size remains a difficult task in membrane reactor applications because, with microporous membrane bioreactors, a high initial permeability is obtained, whereas with ultrafiltration membranes, a more stable performance from the early stage of filtration is achieved.

The main challenge for anaerobic membrane bioreactors has been the fouling of membranes. Membrane fouling is the result of adsorption of organic matter, precipitation of inorganic matter, and adhesion of microbial cells to the membrane surface. Therefore, fouling increases the mass transfer resistance to flux through a membrane. The use of submerged membrane bioreactor technology had been applied as a strategy to control the development of fouling by cake layer formation. In these reactors, a fraction of the biogas produced is recycled and sparged by a diffuser into the reactor below the membrane module (Lin *et al.*, 2009). The higher shear generated by the large bubbles scours the surface of the membrane.

The more common use of membrane bioreactors in fuel gases production is in cell retention by ultrafiltration membranes as wastewater is processed. However, upgrading of gases, just based on separation or combining separations and reactions, is also possible by membrane units. The selectivity of gases using porous membranes is usually poor and thus the membranes with more potential for gas separations are dense membranes. These membranes had been used in the biofuel production context for steam methane reforming in syngas (hydrogen and carbon monoxide), respective adjustment of carbon monoxide and hydrogen for Fischer–Tropsch processes, and biogas upgrading process, which deals primarily with the separations of CO_2 and CH_4.

Conventional methods of separating acid gases (CO_2 and H_2S) from mixtures with CH_4 include absorption of the acid gases in liquid solvents, adsorption on solids and chemical conversion to other compounds. These separation processes are energy intensive, require significant investment, and the process equipment typically occupies a large space. A potential technology is presented by a membrane separation process, which offers energy efficiency and a simple, compact, and modular equipment. The upgraded biogas, in the form of natural gas substitute, can be injected into existent natural gas grids or used as a vehicle fuel, using natural gas infrastructure (Makaruk *et al.*, 2010).

The case studies below illustrate the examples of detoxification of wastewater with simultaneous production of biogas, using either a submerged membrane bioreactor (Figure 14.17) or a bioreactor with an external sidestream membrane module (Figure 14.18), both for biomass and solid retention as permeation of detoxified wastewater is obtained. A study for biogas upgrading is also described (Figure 14.19).

14.3.3.2 Membrane bioreactor for biogas production

Submerged membrane bioreactor. Kanai *et al.* (2010) present a submerged anaerobic membrane bioreactor (SAnMBR) process for the production of biogas from distillation residues or food waste. Kanai and co-authors used the Kubota's submerged anaerobic membrane bioreactor (KSAnMBR) process that consists of

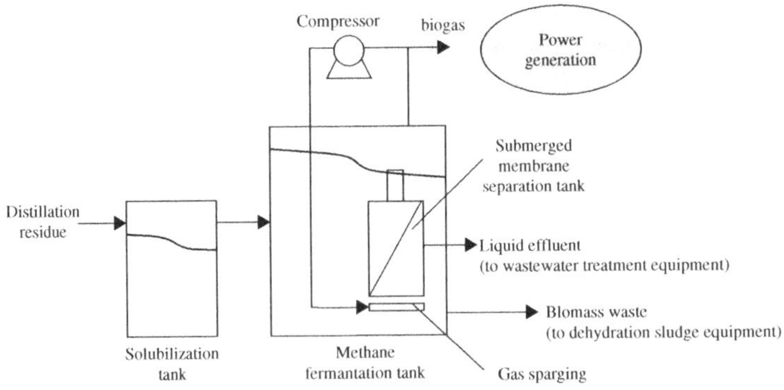

Figure 14.17 Submerged internal membrane bioreactors system for biogas production

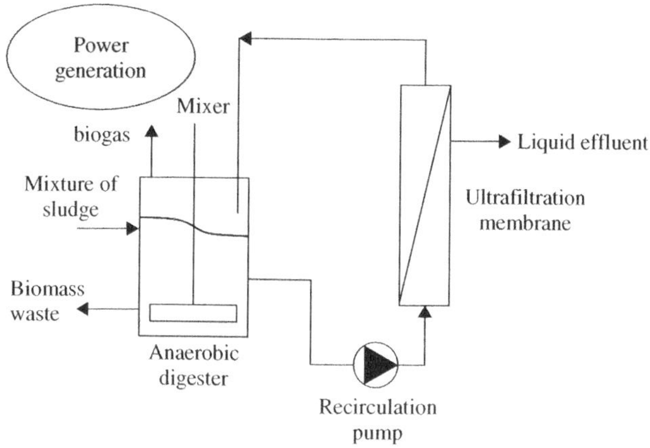

Figure 14.18 *Bioreactors with recirculation into external cross flow membrane module for biogas production*

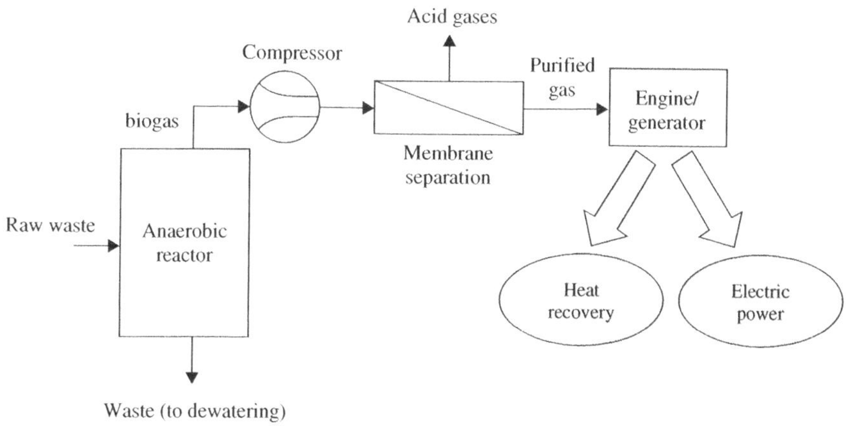

Figure 14.19 *Single pass membrane coupled to bioreactors for biogas upgrading*

a solubilization tank and a thermophilic digestion tank; the latter incorporating a submerged microfiltration membrane. Firstly, residues were induced to pre-treatment equipment, such as crushers or screens, and then stocked in the solubilization tank for equalization of nutrients and storage of the raw feed. After this, liquors are introduced into the thermophilic digestion tank. The anaerobic sludge was concentrated at the submerged membrane separation tank and then pumped to the sludge treatment line.

The biogas generated, consisting of ca. 60% methane, 40% CO_2 and a few minor components such as hydrogen sulfide, was collected for use at a power generation facility or boiler. In the KSAnMBR system, ammonia is filtered out through the membrane units and removed from the fermentation system, enabling the stable operation of the methane fermentation process. This system was successfully applied using the distillation residues of the Japanese distillery that produces Shochu from barley and sweet potato. An average of 12 GJ per day of energy was recovered from the KSAnMBR process and used for steam production. The generated energy is well above the electricity consumption and membrane bioreactor heating at site, enabling savings to the distillery. The KSAnMBR process was developed in the last decade and such patented technology has been successfully applied in 15 full-scale plants.

Another recent work by Huang *et al.* (2011) used a 6 l submerged anaerobic membrane bioreactor. This system consists of a 5 l well mixed anaerobic reactor coupled with 1 l gas lifter, in which a submerged plate and frame microfiltration membrane module were installed. The SAnMBR was seeded with biomass collected from a sludge digester of a local wastewater treatment plant. The maximum specific yield of biogas volume production per weight of mixed liquor volatile suspended solids (MLVSS) was $0.056\,l \cdot g^{-1} \cdot day^{-1}$, at an infinite solid retention time. Biogas production is increased using a shorter hydraulic retention time or longer infinite solid retention, due to increased organic loading rate or enhanced dominancy of methanogenics.

Bioreactor with external side stream membrane module. An example of an anaerobic digester coupled with a tubular ultrafiltration membrane for solid–liquid separation (Figure 14.18) was presented by Padmasiri *et al.* (2007). The AnMBR was inoculated with a mixture of sludge dredged from a swine lagoon, anaerobic granules from an upflow anaerobic sludge-blanket reactor treating brewery wastewater and anaerobic sludge from the primary anaerobic sludge digester of urban sanitary wastewater treatment plant. The specific biogas volume production per weight of volatile solids (VS) increased to $2\text{–}3\,l \cdot g^{-1} \cdot day^{-1}$ a few days after start-up and a two-fold increase was observed on day 53 and thereafter.

A membrane separation process for biogas upgrading. Membranes commonly used for separation of CO_2 are made from polymers, such as cellulose acetate and polyamides, because of their high selectivity for CO_2 relative to CH_4. The membranes can be used in the form of asymmetric or composite hollow fibers or flat sheets, which are assembled in hollow fiber or spiral-wound modules, respectively. A bench-scale membrane pilot plant trial was performed by Stern *et al.* (1998). This pilot plant was installed at a municipal wastewater treatment plant and was designed to process up to about $9.4 \times 10^{-4}\,m^3 \cdot s^{-1}$ of raw biogas at pressures up to 55 bar. The raw biogas contained 62 mol% CH_4, the balance being mainly CO_2, but also containing small amounts of a large number of organic compounds (aromatics, chlorinated and brominated hydrocarbons, and aromatics, alcohols, ketones, etc.).

Biogas, whether generated in wastewater treatment plants or on landfill sites, also contains small amounts of H_2S (less than 0.5 mol%). After being compressed, the raw biogas flowed through a cylindrical module containing a bundle of hollow-fiber membranes is then separated inside the membrane module into two fractions, a low-pressure (1.2–1.4 bar) permeate stream enriched in CO_2 and H_2S, and a high-pressure retentate stream enriched in CH_4, the desired product, i.e. the upgraded biogas. The hollow fiber modules used were obtained from Innovative Membrane Systems and consisted in cylindrical shells of 39 cm long and had an outside diameter of 7.3 cm. The hollow fibers in a module had an area of about $0.93\,m^2$. The permeate flowed axially through the bore of hollow fibers, whereas the feed and the retentate flowed outside the hollow fibers.

The performance of the membrane pilot plant was stable during periods over 1000 h, however it was necessary to pre-treat the raw biogas in order to prevent the condensation of organic impurities, which dissolved the hollow fibers. The pre-treatment consisted of the removal of a fraction of the impurities by condensation in a high-pressure heat exchanger; furthermore, the impurities could be removed by adsorption on solid adsorbents or by partial oxidation (Rautenbach and Welsch, 1993). Starting with raw biogas containing lower concentrations of H_2S (less than 500 ppm), this hollow fibers system could achieve the upgraded biogas to pipeline specifications ($\leq$2 mol% CO_2 and $\leq$4 ppm H_2S). An alternative is to use two different types of membranes simultaneously, one with high CO_2/CH_4 selectivity and another with high H_2S/CH_4 selectivity (e.g. polyurethane membranes—Chatterjee *et al.*, 1997).

This trial (Stern *et al.*, 1998) clearly demonstrates that the methane concentration in biogas produced at a municipal wastewater treatment plant can be enhanced from a value of 62 mol% to as much as 97 mol%, improving the biogas energy content. Therefore, the upgraded biogas could be used as a non-corrosive, more efficient fuel for generation of electricity. Even more, the permeate product stream, which is considered waste, is enriched in CO_2 but contains sufficient CH_4 to be used for heating applications.

14.4 Conclusions and future trends

In the next few decades, bioenergy will be a significant renewable energy source because it offers an economical, attractive alternative to fossil fuels. Future biofuel production systems should be integrated into existing technical biomass facilities and should have the potential to be implemented in the energy system of the existing infrastructure for fuel distribution and use. The examples brought into this chapter try to illustrate trials for the production of three different biofuels, namely bioethanol, biodiesel and biogas, using different membrane bioreactors. However, there is a large scope for integration of different catalysts and membrane bioreactors in the concept of biorefineries for fuel production. Just to mention three additional routes for fuel production that are in current research:

- *Biobutanol* has shown that it has the potential to play a significant role in a sustainable, non-petroleum-based, industrial system. Butanol fuel properties are considered to be superior to ethanol because of higher energy content and better air-to-fuel ratio. Several species of *Clostridium* bacteria are capable to metabolize different sugars, amino and organic aids, polyalcohols and other organic compounds to butanol (Harvey and Meylemans, 2011; Lee *et al*., 2008). Biobutanol production is possible in an anaerobic fermentation reactor, using different agricultural and industrial wastes and residues as substrates. Promising techniques for biobutanol recovery from fermentation broth are membrane-based methods, such as membrane evaporation, perstraction, pervaporation, and reverse osmosis. As an example, Izák *et al*. (2008) used an ionic liquid–polydimethylsiloxane ultrafiltration membrane for an efficient and stable removal of butanol out of a *Clostridium acetobutylicum* culture. The overall solvent productivity of fermentation, connected with continuous product removal by pervaporation, was $2.347 g \cdot l^{-1} \cdot h^{-1}$.
- *Algae harvesting for biodiesel*: algae offer great potential for biodiesel production because algae have higher productivities than land plants; some species can accumulate very large amounts of triacylglycerides, the major feedstock to produce biodiesel, and high quality agricultural land is not required to grow the biomass. Therefore, microalgae are a promising feedstock because of their lipid content and they help to decrease the use of land for non-food application. The first step of the process consists on the microalgae growth, followed by a harvesting process to concentrate it, which could be performed by a membrane system. Rios *et al*. (2011) presented dynamic microfiltration as a method for reducing fouling and concentration polarization at low transmembrane pressure in microalgae harvesting for biodiesel production. The scheme used was a rotational disk setup, which contains the membrane, to concentrate microalgae, thus removing water from the feed suspension.
- *Biohydrogen*: hydrogen is a clean and efficient energy carrier because it produces only water after combustion and can be directly converted to electricity via fuel cells. Biohydrogen is considered as the hydrogen produced through biological processes, such as fermentation, where wastewater streams and organic wastes are converted into H_2. In comparison with another biogas, hydrogen has more 58% of the energy content of methane (mass basis). Membrane bioreactors could be a promising alternative of bioprocesses for H_2 production from organic substrates. Oh *et al*. (2004) presented a cross-flow membrane coupled to an anaerobic reactor to produce hydrogen by fermentative bacteria grown at short retention times, being the produced biogas composition always at a range of 57% to 60% of hydrogen. Lee *et al*. (2007) used a hollow-fiber microfiltration membrane module connected to a continuous-flow stirred tank reactor to retain active biomass (sludge from a municipal sewage treatment plant) at a high organic loading rate to improve H_2 production ($1.27–1.39 \, molH_2 \cdot molhexose^{-1}$).

As the human population and demand for energy are dramatically increasing, it is important to develop new methods for the sustainable production of energy. A significant part of the answer can occur through the bioproduction of liquid and gaseous fuels needed for transportation, but also for industrial and households heating/cooling. The optimization of such production and the success of future biorefineries for energy production relies on the advances and combination of three main vectors:

- Efficient biocatalysts, where continuous improvement can be achieved unveiling natural features of microorganisms and enzymes, but also through advances in synthetic biology, where the microorganisms are engineered for specific catalytic functions.
- Efficient separations and process intensification, where membrane technology can have a key role, through its flexibility, wide range of possible separations, potential easy scale-up and low energy requirements.
- Efficient planning on access to raw materials and distribution of biorefinery products, avoiding misallocation of resources, which can disrupt different industries' supply chains and in the long term endanger the success of the biorefinery.

References

Andric P, Meyer AS, Jensen PA, Dam-Johansen K (2010) Reactor design for minimizing product inhibition during enzymatic lignocellulose hydrolysis: I. *Significance and mechanism of cellobiose and glucose inhibition on cellulolytic enzymes. Biotech Adv* 28 (3): 308–324.

Badenes SM, Lemos F, Cabral JMS (2011a) Performance of a cutinase membrane reactor for the production of biodiesel in organic media. *Biotechnol Bioeng* 108 (6): 1279–1289.

Badenes SM, Lemos F, Cabral JMS (2011b) Stability of cutinase, wild type and mutants, in AOT reversed micellar system—Effect of mixture components of alkyl esters production. *J Chem Technol Biotechnol* 86 (1): 34–41.

Bélafi-Bakó K, Koutinas AA, Nemestóthy N, Gubicza L, Webb C. (2006) Continuous enzymatic cellulose hydrolysis in a tubular membrane bioreactor. *Enzyme Microb Tech* 1–2: 155–161.

Ben Chaabane F, Aldiguier AS, Alfenore S, Cameleyre X, Blanc P, Bideaux C, Guillouet SE, Roux G, Molina-Jouve C (2006) Very high ethanol productivity in an innovative continuous two-stage bioreactor with cell recycle. *Bioproc Biosyst Eng 29*, no. 1 (4): 49–57.

Brandberg T, Sanandaji N, Gustafsson L, Franzén CJ (2008) Continuous fermentation of undetoxified dilute acid lignocellulose hydrolysate by *Saccharomyces cerevisiae* ATCC 96581 using cell recirculation. *Biotechnol Progr 21*, no. 4 (9): 1093–1101.

Cardona CA, Sanchez OJ (2007) Fuel ethanol production: Process design trends and integration opportunities. *Bioresource Technol* 98 (12): 2415–2457.

Carvalho CML, Aires-Barros MR, Cabral JMS (1999) Cutinase: From molecular level to bioprocess development. *Biotechnol Bioeng* 66 (1): 17–34.

Carvalho CML, Aires-Barros MR, Cabral JMS (2001) A continuous membrane bioreactor for ester synthesis in organic media: I. *Operational characterization and stability. Biotechnol Bioeng* 72 (2): 127–135.

Carvalho CML, Cabral JMS (2000) Reverse micelles as reaction media for lipases. *Biochimie* 82 (11): 1063–1085.

Chang HN, Yang JW, Park YS, Kim DJ, Han KC (1992) Extractive ethanol production in a membrane cell recycle bioreactor. *J Biotechnol* 24 (3): 329–343.

Chatterjee G, Houde AA, Stern SA (1997) Poly(ether urethane) and poly(ether urethane urea) membranes with high H_2S/CH_4 selectivity. *J Memb Sci* 135 (1): 99–106.

Cysewski GR, Wilke CR (1977) Rapid ethanol fermentations using vacuum and cell recycle. *Biotechnology and Bioengineering* 19 (8): 1125–1143.

Escobar JMK, Rane D, Cheryan M (2001) Ethanol production in a membrane bioreactor. *App Biochem Biotech* 91 (1): 283–296.

Fjerbaek L, Christensen KV, Norddahl B (2009) A review of the current state of biodiesel production using enzymatic transesterification. *Biotechnol Bioeng* 102 (5): 1298–1315.

Gan Q, Allen SJ, Taylor G (2002) Design and operation of an integrated membrane reactor for enzymatic cellulose hydrolysis. *Biochem Eng J* 12 (3): 223–229.

Ghanem A (2003) The utility of cyclodextrins in lipase-catalyzed transesterification in organic solvents: enhanced reaction rate and enantioselectivity. *Org Biomol Chem* 1: 1282–1291.

Ghose TK, Bandyopadhyay KK (1980) Rapid ethanol fermentation in immobilized yeast cell reactor. *Biotechnol Bioeng* 22 (7): 1489–1496.

Ghose TK, Kostick JA (1970) A model for continuous enzymatic saccharification of cellulose with simultaneous removal of glucose syrup. *Biotechnol Bioeng* 12 (6): 921–946.

Giorno L, Drioli E (2000) Biocatalytic membrane reactors: applications and perspectives. *Trends Biotechnol* 18 (8): 339–349.

Gregor HP, Jeffries TW (1979) Ethanolic fuels from renewable resources in the solar age. *Ann NY Acad Sci* 326 (1): 273–287.

Harvey BG, Meylemans HA (2011) The role of butanol in the development of sustainable fuel technologies. *J Chem Technol Biotechnol* 86: 2–9.

Hoffmann H, Kuhlmann W, Meyer HD, Schügerl K (1985) High previous productivity ethanol fermentations with crossflow membrane separation techniques for continuous cell recycling. J. Membr. Sci 22 (2): 235.

Hong J, Tsao GT, Wankat PC (1981) Membrane reactor for enzymatic hydrolysis of cellobiose. *Biotechnol Bioeng* 23 (7): 1501–1516.

Huang Z, Ong SL, Ng HY (2011) Submersed anaerobic membrane bioreactor for low-strength wastewater treatment: Effect of HRT and SRT on treatment performance and membrane fouling. *Water Res* 45: 705–713.

Ikegami T, Negishi H, Yanase H, Sakaki K, Okamoto M, Koura N, Sano T, Haraya K, Yanagishita H (2007) Stabilized production of highly concentrated bioethanol from fermentation broths by *Zymomonas mobilis* by pervaporation using silicone rubber-coated silicalite membranes. *J Chem Technol Biot* 82 no. 8 (8): 745–751.

Ishihara M, Uemura S, Hayashi N, Shimizu K (1991) Semicontinuous enzymatic hydrolysis of lignocelluloses. *Biotechnol Bioeng* 37 (10): 948–954.

Izák P, Schwarz K, Ruth W, Bahl H, Kragl U (2008) Increased productivity of *Clostridium acetobutylicum* fermentation of acetone, butanol, and ethanol by pervaporation through supported ionic liquid membrane. *Appl Microb Biotechnol* 78: 597–602.

Jaffrin MY (2008) Dynamic shear-enhanced membrane filtration: A review of rotating disks, rotating membranes and vibrating systems. *J Membr Sci* 324: 7–25.

Kanai M, Ferre V, Wakahara S, Yamamoto T, Moro M (2010) A novel combination of methane fermentation and MBR—Kubota submerged anaerobic membrane bioreactor process. *Desalination* 250: 964–967.

Kang W, Shukla R, Sirkar KK (1990) Ethanol production in a microporous hollow-fiber-based extractive fermentor with immobilized yeast. *Biotechnol Bioeng* 36 (8): 826–833.

Kargupta K, Datta S, Sanyal SK (1998) Analysis of the performance of a continuous membrane bioreactor with cell recycling during ethanol fermentation. *Biochem Eng J* 1 (1): 31–37.

Kinoshita S, Chua JW, Kato N, Yoshida T, Taguchi H (1986) Hydrolysis of cellulose by cellulases of Sporotrichum cellulophilum in an ultrafilter membrane reactor. *Enzyme Microb Tech* 8 (11): 691–695.

Klei HE, Sundstrom DW, Coughlin RW, Ziolkowski K (1981) Hollow-fiber enzyme reactors in cellulose hydrolysis. *Biotechnol Bioeng Symp* 11: 593–601.

Knutsen JS, Davis RH (2004) Cellulase retention and sugar removal by membrane ultrafiltration during lignocellulosic biomass hydrolysis. *Applied Biochem Biotech* 114 (1): 585–600.

Kyung KH, Gerhadt P (1984) Continuous production of ethanol by yeast "immobilize" in a membrane-contained fermentor. *Biotechnol Bioeng* 26: 252–256.

Lafforgue C, Malinowski J, Goma G (1987) High yeast concentration in continuous fermentation with cell recycle obtained by tangential microfiltration. *Biotechnol Letters* 9 (5): 347–352.

Lam M, Lee K, Mohamed A (2010) Homogeneous, heterogeneous and enzymatic catalysis for transesterification of high free fatty acid oil (waste cooking oil) to biodiesel: A review. *Biotechnol Adv* 28: 500–518.

Larsson K, Janssen A, Adlercreutz P, Matiasson B (1990) Three systems used for biocatalysis in organic systems. *A comparative study. Biocatal* 4: 163–175.

Le-Clech P (2010) Membrane bioreactors and their uses in wastewater treatments. *Appl Microbial Biotechnol* 88: 1253–1260.

Lee KS, Lin PJ, Fangchiang K, Chang JS (2007) Continuous hydrogen production by anaerobic mixed microflora using a hollow-fiber microfiltration membrane bioreactor. *Int J Hydrogen Ene* 32(8): 950–957.

Lee SG, Kim HS (1993) Optimal operating policy of the ultrafiltration membrane bioreactor for enzymatic hydrolysis of cellulose. *Biotechnol Bioeng* 42.(6): 737–746.

Lee SY, Park JH, Jang SH, Nielson LK, Kim J, Jung KS (2008) Fermentative butanol production by Clostridia. *Biotechnol Bioeng* 101(2): 209–228.

Lee WG, Park BG, Chang YK, Chang HN, Lee JS, Park C (2000) Continuous ethanol production from concentrated wood hydrolysates in an internal membrane-filtration bioreactor. *Biotechno Progr* 16 no. 2 (4): 302–304.

Lei E, Berube PR (2004) *Impact of membrane configurations and hydrodynamic conditions on the permeate flux in submerged membrane systems for drinking water treatment. Proceedings AWWA Water Quality technology Conference*, San Antonio, TX.

Lin HJ, Xie K, Mahendran B, Bagley DM, Leung KT, Liss, SN, Liao BQ (2009) Sludge properties and their effects on membrane fouling in submerged anaerobic membrane bioreactors (SAnMBRs) (2009) *Water Res* 43: 3827–3837.

Lipnizki F (2010) Membrane process opportunities and challenges in the bioethanol industry. *Desalination* 250 (3): 1067–1069.

Lipnizki F, Hausmanns S, Laufenberg G, Field R, Kunz B (2000) Use of pervaporation-bioreactor hybrid processes in biotechnology. *Chem Eng Technol* 23: 569–577.

Liu J, Lu J, Cui Z (2010) *Enzymatic hydrolysis of cellulose in a membrane bioreactor: assessment of operating conditions*. Bioproc Biosyst Eng 34, no. 5 (12): 525–532.

Maiorella B, Blanch HW, Wilke CR (1983) By-product inhibition effects on ethanolic fermentation by *Saccharomyces cerevisiae*. *Biotechnol Bioeng* 25 (1): 103–121.

Makaruk A, Miltner M, Harasek M (2010) Membrane biogas upgrading processes for the production of natural gas substitute. *Sep Purif Technol* 74: 83–92.

Mehaia MA, Cheryan M (1984) Ethanol production in a hollow fiber bioreactor using *Saccharomyces cerevisiae*. *Appl Microbiol Biot* 20 (2): 100–104.

Melzoch K, Rychtera M., Markvichov NS, Pospíchalová V, Basařová G, Manakov MN (1991) Application of a membrane recycle bioreactor for continuous ethanol production. *Appl Microbiol Biot* 34 (4): 469–472.

Mores WD, Knutsen JS, Davis RH (2001) Cellulase recovery via membrane filtration. *Appl Biochem Biotech* 91 (1): 297–309.

Mori Y, Inaba T (1990) Ethanol production from starch in a pervaporation membrane bioreactor using *Clostridium thermohydrosulfuricum*. *Biotechnol Bioeng* 36 (8): 849–853.

Nagayama K, Nishimura R, Doi T, Imai M (1999) Enhanced recovery and catalytic activity of *Rhizopus delemar* lipase in an AOT microemulsion system with guanidine hydrochloride. *J Chem Technol Biotechnol* 74: 227–230.

O'Brien DJ, Craig Jr, JC (1996) Ethanol production in a continuous fermentation/membrane pervaporation system. *Appl Microbiol Biot* 44 (6): 699–704.

Oh SE, Iyer P, Bruns MA, Logan BE (2004) Biological hydrogen production using a membrane bioreactor. *Biotechnol Bioeng* 87 (1): 119–127.

Ohlson I, Trägaardh G, Hahn-Hägerdal B (1984) Enzymatic hydrolysis of sodium-hydroxide-pretreated sallow in an ultrafiltration membrane reactor. *Biotechnol Bioeng* 26 (7): 647–653.

Okahata Y, Ijiro K (1998) A lipid-coated lipase as new catalyst for triglyceride synthesis in organic media. *J Chem Soc Chem Commun* 20: 1392–1394.

Padmasiri SI, Zhang J, Fitch M, Norddahl B, Morgenroth E, Raskin L (2007) Methanogenic population dynamics and performance of an anaerobic membrane bioreactor (AnMBR) treating swine manure under high shear conditions. *Water Res* 41: 134–144.

Persson M, Mladenoska I, Wehtje E, Adlercreutz P (2002) Preparation of lipases for use in organic solvents. *Enzyme Microb Technol* 31: 833–841.

Pirt SJ, (1965) The maintenance energy of bacteria in growing cultures. *P R Soc B* 163, 224–231.

Prazeres DMF, Garcia FAP, Cabral JMS (1994) Continuous lipolysis in a reversed micellar membrane bioreactor. *Bioprocess Eng* 10: 21–27.

Ranganathan SV, Narasimhan SL, Muthukumar K (2008) An overview of enzymatic production of biodiesel. *Bioresour Technol* 99 (10): 3975–3981.

Rautenbach R, Welsch K (1993) Treatment of landfill gas by gas permeation-pilot plant results and comparison to alternatives. *Desalination* 90: 193–207.

Rios SD, Clavero E, Salvado J, Farriol X, Torras C (2011) Dynamic microfiltration in microalgae harvesting for biodiesel production. *Ind Eng Chem Res* 50: 2455–2460.

Roels JA (1980) The application of macroscopic principles to microbial metabolism. *Biotechnol Bioeng* 22: 2457–2514.

Stern SA, Krishnakumar B, Charati SG, Amato WS, Friedman AA, Fuess DJ (1998) Performance of a bench-scale membrane pilot plant for the upgrading of biogas in a wastewater treatment plant. *J Memb Sci* 151: 63–74.

Van Unen D, Engbersen J, Reinhoudt D (2002) Why do crown ethers activate enzyme in organic solvents? *Biotechnol Bioeng* 77: 248–255.

Vyas AP, Verma JL, Subrahmanyam N (2010) A review on FAME production processes. *Fuel* 89: 1–9.

Weiland P (2010) Biogas production: current state and perspectives. *Appl Microbiol Biotechnol* 85: 849–860.

Yang S, Ding W, Chen H (2006) Enzymatic hydrolysis of rice straw in a tubular reactor coupled with UF membrane. *Process Biochem* 41 (3): 721–725.

Yang S, Ding W, Chen H (2009) Enzymatic hydrolysis of corn stalk in a hollow fiber ultrafiltration membrane reactor. *Biomass Bioenerg* 33 (2): 332–336.

Yasuda M, Kiguchi T, Kasahara H, Ogino H, Ishikawa H (2000) Effect of additives on transesterification activity of Rhizopus chinensis lipase. *J Biosci Bioeng* 90: 681–683.

Zhang M, Su R, Li Q, Qi W, He Z (2010) Enzymatic saccharification of pretreated corn stover in a fed-batch membrane bioreactor. *Bioenerg Res* 4, no. 2 (12): 134–140.

15

Extraction-Fermentation Hybrid (Extractive Fermentation)

Shang-Tian Yang[1] and Congcong Lu[2]

[1]*William G. Lowrie Department of Chemical and Biomolecular Engineering, Ohio State University, USA*
[2]*Dow Chemical Company, Midland, MI, USA*

15.1 Introduction

Many biobased chemicals and fuels are produced, or can be produced, from renewable biomass via microbial fermentation. After fermentation there are usually two approaches for product recovery from the fermentation broth: "end-of-pipe" and "slip-stream" (Vane, 2008). The end-of-pipe approach refers to product recovery after the fermentation is completed, and the product-depleted broth is sent to the next step for processing. This approach is usually employed in ethanol recovery from fermentation due to the high end-product concentration present in the feed stream. The slip-stream approach refers to product recovery while the fermentation is still on-going in the bioreactor, and the product-depleted stream is returned to, or never leaves the bioreactor. This process is also known as the ***extractive fermentation*** process, or integrated process, meaning that the separation technology is integrated with fermentation and the desired product can be extracted *in situ*. The end-of-pipe approach is most widely used in industry for its process simplicity and is easy to use with batch processes. In contrast, the slip-stream approach is mostly studied at a laboratory scale for its potential to alleviate product inhibition and to improve productivity and substrate conversion.

This chapter reviews recent advances in separation technologies that can be used in the slip-stream approach for *in situ* recovery of fermentation products, focusing on carboxylic acids and alcohols. An introduction to industrial needs for integrated fermentation-separation processes and their benefits to the biorefinery industry will be provided first, followed with a brief description of the basic principles of extractive fermentation and more detailed discussion of the design and operation of various *in situ* product recovery techniques. Finally, two extractive fermentation examples will be illustrated, one for biobutanol

Separation and Purification Technologies in Biorefineries, First Edition.
Edited by Shri Ramaswamy, Hua-Jiang Huang, and Bandaru V. Ramarao.

production and the other for butyric acid production by *Clostridia*. Both of them are still at the laboratory-scale research and development stage but hold great promise for industrial application. Lastly, a conclusion and some perspectives about the future trend in this technology area are provided.

15.2 The market and industrial needs

Bioethanol is the largest industrial fermentation product with a projected annual production of more than 30 billion gallons by 2015. Butanol is an important industrial solvent that can be produced in the acetone-butanol-ethanol (ABE) fermentation that was once the second largest industrial fermentation next to ethanol fermentation before the rise of the modern petroleum industry. Several carboxylic acids, including citric, itaconic, and lactic acids, are currently produced almost exclusively from carbohydrate-based feedstock by fermentation. Many more carboxylic acids such as succinic, fumaric, butyric and propionic acids can also be produced from renewable biomass via microbial fermentation (Yang *et al*., 2006). However, the potential of these biobased alcohols and carboxylic acids as chemical intermediates or feedstock chemicals replacing the petroleum-based chemicals depends on their production costs, of which a large portion, 20% to 50%, can be attributed to their separation and purification costs.

Conventional organic acid and alcohol fermentations are subject to end-product inhibition, which significantly reduces cell growth and limits its ability to produce metabolites such as lactic acid and ethanol. Product inhibition limits product titer, which increases the recovery costs. It also limits substrate utilization and requires the use of dilute substrates in the fermentation process, which increases process water usage and wastewater generated. The accumulation of an acidic product would also require the addition of base to maintain the fermentation pH, which not only increases the material cost but may also make the downstream processing more complicated. In addition, overproduction of a metabolite could prompt cells to shift to produce other byproducts, thus reducing the product yield. Overall, the reduced reactor productivity, product yield, and product titer would significantly increase the production cost and could make the biobased chemicals uneconomical in competition with their petroleum-based counterparts. It is therefore desirable to remove the fermentation end product while it is being produced by cells using online product recovery techniques in an integrated process, also known as extractive fermentation.

Carboxylic acids produced in a fermentation process are usually recovered either by distillation (for volatile products), precipitation (for non-volatile products), or solvent extraction. Adsorption and electrodialysis have also been studied for some carboxylic acids. A comparison of these separation methods for carboxylic acid recovery from fermentation broth is given in Table 15.1. Distillation, which can be used for compounds such as volatile organic acids and esters, is energy intensive. Precipitation has thus been the most widely used separation method in the production of several carboxylic acids, including citric, fumaric, and lactic acids, which have a low solubility when present in their calcium salt form. However, the acidification of the calcium salt with sulfuric acid to produce the free acid product also generates large amounts of $CaSO_4$, a solid waste expensive to be disposed of by landfill. Current US industrial production of citric and lactic acids uses solvent extraction. Extraction, adsorption, and electrodialysis can be used for online recovery of carboxylic acids in an integrated fermentation-separation process.

For the production of ethanol and butanol, multi-column distillation followed with dehydration by molecular sieve adsorption has been the standard separation process used in the biorefinery industry. Distillation offers a wide range of advantages, such as high alcohol recovery, and multi-stage operation; it is also easy to scale up, and relatively energy-efficient when alcohol concentration in the feed stream is high. There are also many less attractive facts about recovering alcohol using distillation. It is energy-intensive for low alcohol concentration feeds. It requires a high-temperature operation, which is lethal to microorganisms. An additional dehydration step is necessary in order to reach the fuel-grade specification

Table 15.1 *Separation methods for recovery of carboxylic acids from fermentation broth. Reprinted from* Yang et al. © *2006, with permission from Elsevier*

Method	Principle	Advantages	Disadvantages
Precipitation	$CaCO_3$ or CaO is added in the medium to neutralize acid. The calcium carboxylate solution is concentrated by evaporation, then crystallized and separated from the mother liquor	Low impurities in product; low capital costs; high yield	Requires the use of H_2SO_4 to release carboxylic acid, which generates $CaSO_4$, a solid waste requiring landfill disposal
Distillation	NH_3 is used to neutralize acid. Ammonia carboxylate then reacts with alcohol to form ester, which is separated by distillation	High product purity; the byproduct $(NH_4)_2SO_4$ can be used as a fertilizer	Requires hydrolysis of ester and distillation to separate the alcohol from carboxylic acid. High capital and energy costs associated with distillation. Requires economy of scale
Extraction	Uses organic solvents to extract carboxylic acid from the broth	Low costs, high yield, best for carboxylate salt production	The solution needs to be acidified to allow efficient extraction of the free carboxylic acid. Extractant needs to be regenerated by distillation or back-extraction (stripping).
Adsorption	Normally using ion-exchange resins to adsorb carboxylate ions from the broth	Easy to operate	Low adsorption capacities, high resin costs, requires energy-intensive resin regeneration, separation is not highly selective
Electrodialysis	An electric current is applied to move negatively charged carboxylate ions through an anion-exchange membrane towards the anode in the electrodialyzer	Carboxylate is concentrated in aqueous solution; does not require acid addition to adjust the solution pH	Product purity is low and may require further purification; high energy input; membrane fouling; difficult to scale up

(Vane, 2008). Because the ethanol concentration is usually high at the end of the fermentation process (~10%), distillation is favorable for ethanol recovery.

Butanol recovery is the most energy-intensive and costly step in the whole biobutanol production process (Ezeji *et al*., 2004a; 2007a). In ABE fermentation, the final concentration of butanol is usually below 1.5% (w/v) in the fermentation broth. Recovering butanol using distillation is thus extremely energy-intensive and costly. Unlike ethanol, butanol has a low vapor pressure and high boiling point (118 °C), which pose further challenges in distillation and require more energy. Alternative separation technologies that are energy-efficient and suitable for recovering low-concentration alcohol in the fermentation broth are in demand.

Table 15.2 *Alternative separation techniques for butanol recovery from ABE fermentation*

Method	Principle	Advantages	Disadvantages
Adsorption	Adherence of solvents to silicalite resin, clay, activated carbon, or other adsorptive materials	Easy to operate, low energy requirement	High adsorbent cost, low selectivity, low adsorption capacity, require heating for adsorbent regeneration
Gas stripping	Volatile solvents being stripped out by gases and then condensed	Easy to operate, no harm to the culture, strips only the volatiles, no fouling	Low selectivity among volatile compounds
Liquid–liquid extraction	Using the solubility difference of solvent in extractant and aqueous phase for separation	High selectivity, easy to operate and scale up	High extractant cost, forming emulsion, toxic to the culture, low distribution coefficients for conventional extractants
Perstraction	Membrane-based extraction, separating the fermentation broth from the extractive solvents	High selectivity, low toxicity to the culture compared to liquid–liquid extraction	High membrane cost, emulsion and membrane fouling
Pervaporation	Using membrane to selectively let the vaporous solvents pass through, permeate side is under vacuum	High selectivity, relatively high mass flux for butanol, low energy input	High membrane cost, membrane fouling, difficult to operate and scale up

Over the years, many novel, energy-efficient and economically feasible separation techniques, including gas stripping, liquid–liquid extraction, adsorption, pervaporation, and perstraction (see Table 15.2 for their operational principles), have been developed to recover volatile compounds such as solvents from the fermentation broth. These technologies are more energy-efficient than the traditional distillation approach and can significantly lower the process cost. Table 15.2 summarizes and compares the advantages and disadvantages of these alternative butanol recovery methods.

15.3 Basic principles of extractive fermentation

Various extractive fermentation processes have been widely studied for the production of organic acids and alcohols (Fernandes *et al*., 2003; Stark and von Stockar, 2003; Yang *et al*., 2006). In an extractive fermentation process, the product is separated either *in situ* (within the bioreactor) or *ex situ* by circulating the fermentation broth through an external separation unit (see Figure 15.1). The integrated separation process removes the inhibitory fermentation product as it is being produced and thus prevents its accumulation to a level that is toxic to the microorganism. For example, ABE fermentation is severely limited by its low product concentration and productivity due to strong inhibition by butanol on the microorganism. By employing extractive fermentation with continuous gas stripping to recover butanol from the fermentation broth, the butanol concentration can be controlled at a low level in the fermentor to relieve butanol inhibition, thus increasing the reactor productivity and allowing more substrates to be converted to the final product.

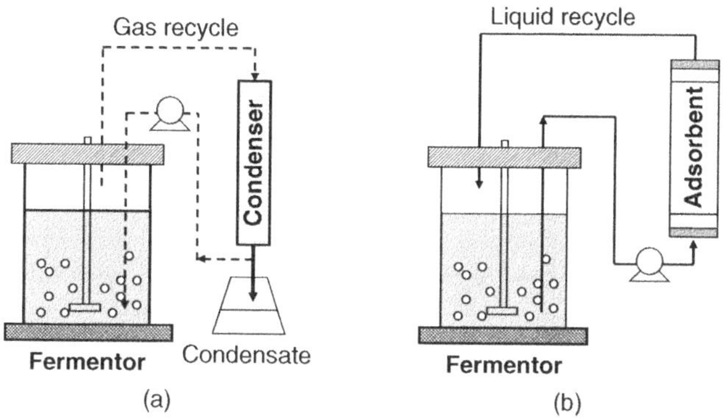

Figure 15.1 *Integrated fermentation process with in situ gas stripping (a) or ex situ adsorption (b) for online product recovery. In situ recovery occurs within the fermentor, while ex situ recovery takes place outside the fermentor*

Depending on the product and process, the separation method can be gas stripping, pervaporation, adsorption, liquid–liquid extraction, or electrodialysis. However, not all of them are suitable for *in situ* product recovery. The design and application of these separation methods in extractive fermentation are discussed in detail in the following section. Distillation employing a high temperature cannot be used in extractive fermentation, and thus will not be discussed.

15.4 Separation technologies for integrated fermentation product recovery

15.4.1 Gas stripping

Gas stripping is an easy-to-operate technique for butanol recovery from fermentation broth. Figure 15.2a shows a schematic diagram of a typical gas-stripping process. Gas stripping can either be integrated with fermentation in the bioreactor, or performed in an individual stripping column. Therefore, the gas stripper shown in Figure 15.1a can either be a bioreactor or a separate stripping column. In ABE fermentation, either nitrogen or fermentation gases (H_2 and CO_2) can be used as stripping gases (Ezeji *et al*., 2004a) to ensure the anaerobic condition. In the integrated scenario, stripping gas is introduced into the fermentation broth in the bioreactor to capture the volatile solvents in the broth, and the gas containing solvents is subsequently passed through a condenser where the solvents are condensed and enriched in the condensate stream. In the separate gas-stripper scenario, feed stream (broth) is sent to the stripper where the solvents are captured by the stripping gas, and the feed low in solvents is recycled back to the bioreactor. Gas flow can also be operated in either single-pass mode or recycle mode. In the single-pass mode, once gas passes the condenser it is released into open air, which may result in solvent loss depending on the efficiency of the condenser. In the recycle mode, gas low in solvents after condensing is recycled back into the stripper/bioreactor to capture more solvents, and the process is a closed loop, which prevents any solvent loss.

Gas stripping offers many advantages as an integrated product recovery technology with fermentation, including utilization of fermentation gases as stripping gases, the ability to operate at fermentation temperature, and flexibility with or without the removal of solids from the fermentation broth (Vane, 2008). This technology makes use of the solvent-to-water ratio in the inert gas at equilibrium, which is strongly governed by temperature. The partial pressure of a volatile component i in the gas phase (P_i) is proportional to its saturated vapor pressure (P_i^{sat}), liquid-phase mole fraction (x_i), and activity coefficient (γ_i),

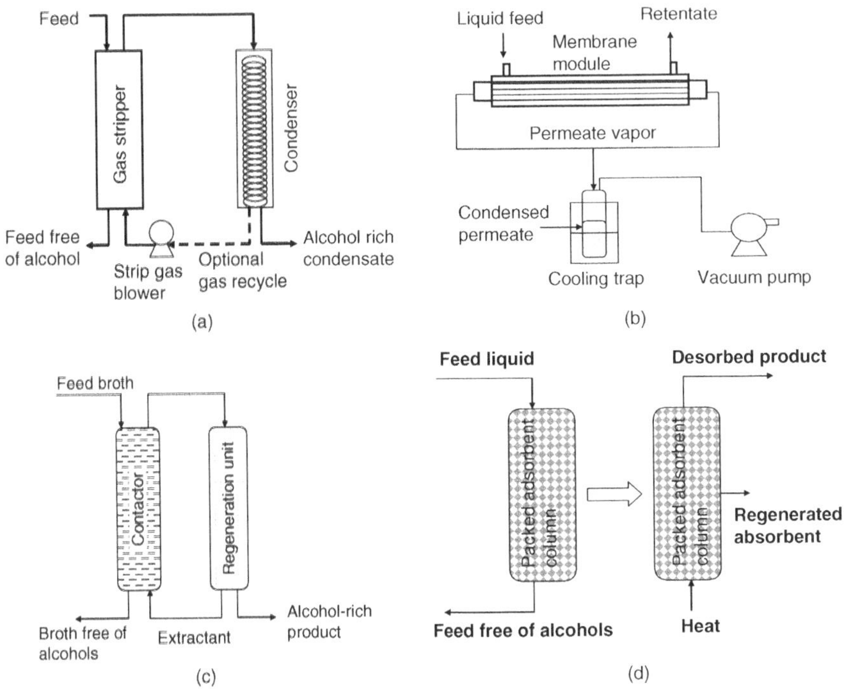

Figure 15.2 Alternative butanol recovery processes: (a) Gas stripping, (b) Pervaporation, (c) Liquid–liquid extraction, (d) Adsorption

as follows:

$$P_i = y_i P_{total} = x_i \gamma_i P_i^{sat}$$

where P_{total} is the total pressure of gas phase, P_i is the partial pressure, and y_i is the mole fraction of component i in the liquid phase. P_i^{sat} is strongly dependent on temperature. In general, increasing temperature also increases P_i^{sat} and thus P_i in the gas phase. Although increasing temperature favors higher vapor pressure and gas-stripping rate, this principle applies to both volatile solvents and water. Therefore, the optimal temperatures for gas stripping and condensation depend on the selectivity of solvents over water.

Many other factors also affect the performance of gas stripping, such as bubble size, mass transfer coefficient, interfacial contact area and contact time, cooling temperature, and gas flow rate. Ezeji *et al.* (2005a) studied the effects of bubble size and gas flow rate on butanol removal, and reported that a bubble size between 0.5 to 5 mm had no effect on butanol removal rate under the condition tested, whereas increasing the flow rate from 43 cm^3/s to 80 cm^3/s resulted in a 2.51-fold increase in gas-stripping rate. They reported that in a 2 L reactor the gas bubbles had sufficient contact time and reached 95% saturation with butanol within 0.14 s, and thus smaller bubbles (<0.5 mm) were not necessary. They also mentioned that further reducing the bubble size had no impact on increasing the solvent stripping rate, but actually reduced the reactor productivity.

Ezeji *et al.* (2003) studied butanol removal using model solution and fermentation broth, and they reported that gas stripping was highly selective towards butanol over acetone, and the presence of cells in the fermentation broth adversely affected butanol removal. No acids were taken out by gas stripping during the process, indicating that gas stripping was only selective towards volatile solvents. It has also

Table 15.3 *Solvent selectivities and operating conditions for butanol recovery in gas stripping processes*

Process	Stripping temp. (°C)	Condensation temp. (°C)	Stripping gas and flow rate[a]	Selectivity	References
Integrated with batch reactor	34	−60	$N_2$2.7 L/min	ABE 23.4	Ennis *et al.*, 1986b
Separate stripper, continuous fermentation	30	−40 to −5	$N_2$10 L/L·min	ABE 4.0	Groot *et al.*, 1989
Separate stripper, continuous fermentation	65–67	3–4	$N_2$2.5 L/min	ABE 30.5	Qureshi and Maddox, 1991
Integrated with fed-batch reactor	35	0–3	H_2 and CO_2,3–3.2 L/L·min	ABE 6–23	Qureshi *et al.*, 1992
Integrated with batch reactor	34	−0.8	H_2 and CO_2,1.5–3.3 L/L·min	ABE 9.5–13	Maddox *et al.*, 1995
Model solution	35	−2	$N_2$4.6 L/min	A 4.1–6.4 B 10.3–13.8 E 4.9–7.9	Ezeji *et al.*, 2003
Integrated with batch reactor	33–35	−2	H_2 and $CO_2$3 L/min	A 4.7–10.5 B 6.7–13.2 E 4.7–9.3	
Integrated with fed-batch reactor	33–35	−2	H_2 and $CO_2$6 L/min	B 10.3–22.1	Ezeji *et al.*, 2004b

[a] L/min: liter gas per minute; L/L·min: liter gas per liter broth per minute.
A: acetone; B: butanol; E: ethanol

been reported that gas stripping did not harm cells or remove any nutrients from the broth when integrated with fermentation (Qureshi and Blaschek, 2001).

Gas stripping has been successfully demonstrated and applied in many fermentation processes and improved overall butanol production and productivity (Ezeji *et al.*, 2003, 2004b, 2005b, 2007b). Table 15.3 summarizes performance and solvent selectivities of gas stripping processes for butanol recovery under various operating conditions. With simultaneous product removal, concentrated substrate can be utilized by microorganism in an integrated fermentation process, which would otherwise cause product inhibition. It was reported (Ezeji *et al.*, 2003) that 161.7 g/L glucose was utilized and 75.9 g/L ABE were obtained in a batch process integrated with gas stripping, whereas only 17.7 g/L ABE were produced from 45.4 g/L glucose consumed in the control or non-integrated batch process. If operated in fed-batch mode, highly concentrated substrate can be used because product accumulation and inhibition can be avoided. Ezeji *et al.* (2004b) studied fed-batch fermentation with gas stripping, where 500 g/L glucose was periodically added into the reactor to replenish depleted sugar. They reported that, in total, 500 g glucose were utilized by the bacteria in a 1 L reactor and 232.8 g/L ABE with an enhanced productivity of 1.16 g/L · h were obtained from this integrated fed-batch process. A continuous ABE fermentation with solvents recovered by gas stripping was also reported to utilize 1163 g/L glucose, resulting in a total of 460 g/L ABE production with 0.91 g/L·h productivity (Ezeji *et al.*, 2004a).

Currently, gas stripping has not been commercially used for ethanol or butanol recovery. The alcohol-rich condensate from gas stripping requires at least one additional step for alcohol dehydration in order to meet fuel-grade specifications. Vane (2008) suggested that, in case of butanol recovery by gas stripping, phase separation is a feasible choice due to the high alcohol (butanol) concentration in the condensate. Besides process design and unit fabrication, he also suggested that improvement on mass and energy integration schemes for gas stripping is needed in order to make this process economically feasible and attractive.

15.4.2 Pervaporation

Pervaporation is a membrane-based separation technique. Liquid feed containing volatile species flows on one side of the membrane, while the other side of the membrane is under vacuum. Volatile components of the liquid stream penetrate and diffuse through the membrane and evaporate into permeate vapor under vacuum. The permeate vapor is then condensed in a cooling trap as condensed permeate (Vane, 2005, 2008; Thongsukmak and Sirkar, 2007). Figure 15.2b provides a schematic diagram of pervaporation.

Pervaporation is a selective separation process based on the membranes employed. Components in the liquid feed have different chemical and physical properties; some components have similar properties to the selective membrane material, and can diffuse through the membrane and enrich in the permeate side, whereas others stay on the other side of the membrane. The concentration of solvents on the permeate side is a function of feed concentration, and depends on the composition and selectivity of the membrane used (Ezeji *et al*., 2004a; Vane, 2005, 2008). When the selected components diffuse to and enrich in the permeate side, the concentration of these components is reduced in the liquid feed, and the retentate leaving the module is low in concentration of the selected components, completing the separation process. Due to the selective nature of the membrane and diffusion rates of different components, the concentration ratio of one component in permeate to feed can range from single digit to over 1000 (Vane, 2005). If the membrane is hydrophobic, the permeate side will become rich in organic compounds relative to water. If the membrane is hydrophilic, the feed liquid will be dehydrated as water permeates through the membrane, which is the primary commercial use of pervaporation (Jonquieres *et al*., 2002).

The chemical activity difference (concentration gradient) on the feed side and the permeate side is the driving force for a component to transport across the membrane, and the flux is inversely proportional to the overall resistance and proportional to the concentration gradient. The resistance to transporting across the membrane includes diffusion in the stagnant feed liquid to the membrane, diffusion through the membrane, and diffusion in the permeate vapor. The primary factors affecting separation by pervaporation are membrane materials and feed species, whereas feed temperature, composition and permeate pressure are only secondary factors (Vane, 2005).

Pervaporation has been widely studied for butanol recovery from water or fermentation broth (Geng and Park, 1994; Jonquieres and Fane, 1997; Qureshi and Blaschek, 1999a; 1999b; Qureshi *et al*., 1999; 2001; Fadeev *et al*., 2000, 2001). In general, a hydrophobic membrane is needed in order to get butanol-rich condensate on the permeate side. Table 15.4 presents various membranes that have been applied in the pervaporation process for butanol recovery and their performances.

Currently, the polydimethyl siloxane membrane, which is also known as the ***PDMS*** or ***silicone rubber*** membrane, is the benchmark of hydrophobic membrane commonly used in alcohol/water separation by pervaporation (Vane, 2005, 2008). The PDMS membrane offers a separation factor of 4.4–10.8 for the ethanol/water system, and 40–60 for butanol/water separation (Vane, 2005). Many factors, such as operating temperature, feed concentration, thickness of the membrane, and membrane source and fabrication procedure, affect the performance of PDMS membranes. There have been research efforts trying to improve the performance of pervaporation using PDMS. Recently, Li *et al*. (2010) reported using a tri-layer PDMS composite membrane (PDMS/PE/brass support) to recover butanol by pervaporation, and a separation factor of 34 was obtained.

Table 15.4 *Comparison of membrane performance for butanol recovery in pervaporation processes*

Membrane	Membrane thickness (μm)	Total flux (g $m^{-2}h^{-1}$)	Selectivity	Temp. (°C) Feed, condensate	Feed C_{BuOH} (g/L)	References
Poly (dimethyl siloxane) (PDMS)	25	282–1000	15–35	50–198	5–7	Hickey *et al.*, 1992
	50	70	37	50, −198	10	Boddeker *et al.*, 1990
	190	300	26.8	40, cold trap	10–50	Jonquieres and Fane, 1997
PDMS filled with zeolite	210	100–230	36–45	40, cold trap	10–50	Jonquieres and Fane, 1997
PDMS filled with silicalite	306	90–237	55–105	78, −198	7–78	Qureshi and Blascheck, 1999a
Polytetrafluoroethylene (PTFE)	25–40	35–2100	2.7–4.8	30, −55, dry ice	3–30	Vrana *et al.*, 1993
Poly (methoxy siloxane) (PMS)	N/A	150–400	10–15	50, −198	10–70	Hickey *et al.*, 1992
Polyurethane (PU)	50	7–88	9	50, −198	10	Boddeker *et al.*, 1990
Polyether block amide (PEBA)	50	60–800	20	50, −198	10–52.5	Boddeker *et al.*, 1990
Polypropylene (PP)	N/A	1400–1600	6.3	36, 5	3.5–14	Gapes *et al.*, 1996
Silicone	1000	4.42–11.5	46–58	37, −30	14–17.5	Larrayoz and Puigjaner, 1987
	400	12.9–19.5	45–47	37, −60	4.3–17	Groot *et al.*, 1984
Silicone (thin film)	50	52.8	42	30, cold trap	10	Huang and Meagher, 2001
Silicone filled with silicalite	19	62.8–607	85.9–111.3	30 – 70, cold trap	10	Huang and Meagher, 2001
Zeolite (Ge-ZSM-5)	30	9.6	19	30, N/A	50	Li *et al.*, 2003
Liquid membrane (oleyl alcohol)	25	25–450	180	30, −20 to −100	2.5–37.5	Matsumura and Kataoka, 1987
Liquid membrane (trioctylamine)	N/A	8.3–10.7	71–104	54, −198	16.4–19.7	Thongsukmak and Sirkar, 2007

N/A: not available

Poly[1-(trimethylsilyl)-1-propyne], also known as PTMSP, is another polymeric/organic membrane that offers a good alcohol/water separation factor. It has a high free volume in the membrane, offering more void spaces for higher permeability than PDMS (Volkov *et al.*, 1997, 2004). It has been reported that the butanol/water separation factor in PTMSP can reach as high as 70 (Fadeev *et al.*, 2001). However, due to the high free volume, which attracts foulants inside the membrane, the performance of PTMSP is not as stable as PDMS, and the flux and selectivity of PTMSP gradually decrease over time (Schmidt *et al.*, 1997; Fadeev *et al.*, 2003). Other polymeric materials, including polypropylene (PP) and polytetrafluoroethylene (PTFE), have also been studied as potential membranes for pervaporation, but they had relatively low separation factors ranging from 3 to 9.5 for butanol/water separation (Qureshi *et al.*, 1992; Vrana *et al.*, 1993).

In addition to polymeric membranes, inorganic zeolite materials, such as silicalite and Ge-ZSM-5, have also been studied as hydrophobic membranes in pervaporation applications (Sano *et al.*, 1994; Li *et al.*, 2003). These inorganic materials are usually supported by a solid frame, such as stainless steel, in order to act as a membrane. Li *et al.* (2003) studied the stainless steel supported Ge-ZSM-5 membrane on ethanol, methanol, butanol, and 2-propanol separation through pervaporation. The ethanol/water separation factor was reported at 47, which was at least fourfold to fivefold higher than PDMS, but the butanol/water separation factor was lower than PDMS, only at 19. Silicalite also delivered excellent separation factor for ethanol/water separation, with 60 reported by Sano *et al.* (1994) and an average of 40 widely reported in the literature (Vane, 2005). However, the biggest downside associated with the inorganic membrane is the fabrication cost. Therefore, it has also been proposed that silicalite can be dispersed in PDMS to fabricate a mixed matrix membrane to incorporate the advantages of both zeolite and PDMS. A wide range of butanol/water separation factors of 50–111 (Huang and Meagher, 2001), 55–209 (Qureshi and Blaschek, 1999a), 70–97 (Qureshi *et al.*, 2001), and 100–108 (Qureshi *et al.*, 1999) have been reported in the literature using ABE model solution and fermentation broth. Compared with typical 40–60 separation factor for butanol/water separation in PDMS membrane, the addition of these inorganic silicalite improved the performance of PDMS. The fabrication process of the mixed matrix silicalite/PDMS membrane is similar to that for PDMS, and the cost is expected to be close to PDMS and significantly lower than the inorganic membrane (Vane, 2005).

In addition to the membranes mentioned above, which are solid membranes, ***supported liquid membranes*** have also been studied in the pervaporation process to recover alcohol from dilute aqueous solutions (Matsumura and Kataoka, 1987; Thongsukmak and Sirkar, 2007; Izak *et al.*, 2008). Oleyl alcohol is the common material employed in supported liquid membrane, and a high 180 butanol/water separation factor was reported using a porous PP supported oleyl alcohol liquid membrane in the pervaporation process (Matsumura and Kataoka, 1987). The general requirement for liquid membranes is that the organic solvent must be biocompatible with the microorganisms in the fermentation and must be stable under the operating conditions otherwise a solvent toxic to the culture would hinder the fermentation, and loss of the solvent into the fermentation broth would decrease the life of the liquid membrane. In general, the solvent concentration in permeate from the liquid-membrane based pervaporation is higher than that from the polymeric and ceramic membrane-based pervaporation process. A major problem associated with the liquid membrane is that the liquid leaks into the fermentation broth over time and the liquid membrane has to be regenerated. Thongsukmak and Sirkar (2007) employed a novel nanoporous coating (fluorosilicone) on the polypropylene hollow fiber as the support material to minimize or prevent the migration of liquid membrane into the fermentation broth, and used trioctylamine (TOA) as a liquid membrane for butanol recovery through pervaporation. A butanol selectivity of 108–141 was reported in this porous PP hollow fiber supported TOA liquid membrane using model butanol solution, and a selectivity of 71–104 was reported using the ABE mixture model solution. Pervaporation via a supported ionic liquid membrane integrated with ABE fermentation enhanced the solvent productivity to 2.34 g/L·h (Izak *et al.*, 2008). Tetrapropylammonium tetracyano-borate was the ionic liquid used in the study, which was supported by PDMS as a supported liquid membrane. The butanol enrichment factor was reported to be 11.23 in this study.

The employment of a membrane in pervaporation makes it very efficient and highly selective; even a low concentration species in the feed mixture can be enriched significantly through pervaporation if using a suitable membrane. However, due to the employment of the membrane, pervaporation performance is very sensitive and can be affected by many factors when integrated with an ongoing fermentation process. Fouling is the most common problem with any membrane-based separation technology (Qureshi and Blaschek, 1999c ; Fadeev *et al*., 2000). Vane (2005) summarized a list of factors that impede the performance of pervaporation by fermentation broth, including dead cells, suspended solids, cell metabolites, sugars, organic acids, and fatty acids. Dead cells and suspended solids are most likely to accumulate in the pervaporation membrane module, clog the pores, and block the flow path. Organic acids are the second group that impacts on the pervaporation performance, and competitive sorption with alcohol has been proposed as a potential mechanism. Since acids are in the undissociated form at pH lower than their pK_a values, increasing the pH to 4–6 can significantly reduce the impact of acids on the membranes.

In summary, pervaporation is an emerging membrane-based technology with high selectivity for efficiently recovering alcohol from dilute aqueous solutions. Many obstacles are still to be overcome in order to develop a process suitable for commercial application for butanol recovery, such as membrane fouling and high fabrication cost. Membranes that are highly permeable to alcohol with good alcohol/water separation factor are desired for pervaporation, and the stability of membrane over extended operation period is required. It has thus been suggested that a silicone rubber (PDMS) membrane coupled with efficient vapor condensation and dehydration system is a good choice for butanol recovery (Vane, 2008).

15.4.3 Liquid–liquid extraction

Liquid–liquid extraction is another alternative separation technique proposed for recovering butanol from dilute aqueous solutions. Extraction is also widely used in the recovery of several carboxylic acids, including citric and lactic acids. Extractant liquid is placed in contact with the fermentation broth, and solvents are extracted from the fermentation broth into the extractant phase due to the solubility difference, thus being separated from the aqueous solution. This broth/extractant contact can be either done in a direct way, i.e. mixing, or an indirect way, i.e. using a membrane to separate the two phases. The latter procedure is often referred as ***perstraction*** (Ezeji *et al*., 2007a; Vane, 2008). The employment of membrane in perstraction to separate the two phases is to avoid problems usually associated with traditional liquid–liquid extraction, including toxicity to microorganisms, emulsion, loss of extractant, and transfer of cells from broth to extractant phase (Ezeji *et al*., 2007a). After the extractant is enriched with alcohols, these alcohols must be removed and recovered in a regeneration unit in order to get the desired product and recycle the extractant back into the process. Common extractant regeneration methods include: distillation, vacuum evaporation, and pervaporation for volatile product such as butanol (Ezeji *et al*., 2004a; Vane, 2008) and back extraction or stripping with hot water, strong acid or base for non-volatile organic acids (Yang *et al*., 2006). The schematic design of a liquid–liquid extraction process is depicted in Figure 15.2c.

There are many requirements a solvent must meet in order to be considered as a suitable extractant for liquid–liquid extraction, including:

- high selectivity for the product to water (separation factor);
- high distribution coefficient, which reduces the volume of extractant needed to recover the same amount of the product;
- immiscible, non-emulsifying, clear phase separation from aqueous solution;
- it should be non-toxic to microorganisms, non-reactive with fermentation components, and non-flammable to ensure safety when operating;
- inexpensive to use and easily available.

Table 15.5 shows the performance and toxicity of some solvents evaluated for butanol recovery by liquid–liquid extraction. In general, hydrocarbon is non-toxic, and has a high selectivity for butanol

Table 15.5 *Comparison of solvents for butanol extraction and their toxicity towards* Clostridium beijerinckii

Solvent	Toxicity	Distribution coefficient	Selectivity	Solvent	Toxicity	Distribution coefficient	Selectivity
Hexane	N	0.5	2700	Hexanol	T	12	160
Heptane	N	0.5	3300	Heptanol	T	11	180
Octane	N	0.3	4100	Octanol	T	10	130
Decane	N	0.3	4300	Oleyl alcohol	N	3.6	ND
Dodecane	N	0.3	2900	Decanol	T	8	200
Gasoline	N	0.3	ND	Dodecanol	T	6	140
Butyl acetate	T	~3	ND	Methyl laurate	N	1.8	7
Hexyl acetate	N	3.6	5	Ethyl laurate	N	1.7	7
Dibutyl phthalate	N	1.4	3	Methyl oleate	N	1.3	6
Dibutyl adipate	T	2.5	3	Ethyl oleate	N	1.3	6
Dibutyl maleate	T	2.0	3	Ethyl stearate	N	0.8	7
Tributyl citrate	N	2.4	2	Butyl stearate	N	1.2	ND
Ethyl oenanthate	N	2.0	4	Isopropyl myristate	N	1.4	7
Oleic acid	N	3.9	6	Corn oil	N	0.7	440
Isophytol	N	3.2	ND	Olive oil	N	0.7	470
				Sesame oil	N	0.3	220

N: non-toxic; T: toxic; ND: not determined;

Distribution coefficient: ratio of butanol concentrations in the solvent phase to the aqueous phase at equilibrium

Selectivity: distribution coefficient of butanol/distribution coefficient of water

References: Barton and Daugulis, 1992; Groot *et al.*, 1990

(against water) but low distribution coefficient for butanol. Long-chain aliphatic alcohols have a good selectivity and high butanol distribution coefficient, but most of them, except for oleyl alcohol, are toxic. Most of the esters of short-chain and long-chain fatty acids are non-toxic, and have a moderate distribution coefficient but a poor selectivity.

Thirty-one commonly used solvents were evaluated as extractants in an extractive ABE fermentation process by *C. acetobutylicum* (Barton and Daugulis, 1992), and some of the good candidates reported were poly(propylene glycol) (PPG), oleyl alcohol, isophytol, eutanol G and triethyl citrate, based on butanol partition coefficient and biocompatibility. They also reported that an extractive ABE fermentation using PPG 1200 resulted in 58.6 g/L acetone and butanol, which was threefold higher than the production in the control.

Other solvents including n-decanol, dibutyl phthalate, and oleyl alcohol have also been reported as suitable extractants to recover butanol with high partition coefficients and low toxicity (Eckert and Schügerl, 1987; Wayman and Parekh, 1987; Roffler *et al.*, 1988). Oleyl alcohol is the most often used and investigated extractant in butanol recovery (Ezeji *et al.*, 2004a; Roffler *et al.*, 1987a; 1987b). It was reported that oleyl alcohol was the most effective candidate in extracting butanol and the least in reducing the productivity (Qureshi and Maddox, 1995). Roffler *et al.* (1987a) studied six solvents/solvent mixtures as extractants in extractive ABE fermentation, including kerosene, tetradecanol, oleyl alcohol, dodecanol, benzyl benzoate, and reported that oleyl alcohol or oleyl alcohol and benzyl benzoate mixture resulted in the best result in batch fermentation. Glucose consumption was improved from 80 g/L to over 100 g/L, with a 60% increase in volumetric butanol productivity; 19.7 g/L and 19.3 g/L butanol was produced in oleyl alcohol and oleyl alcohol with benzyl benzoate extractive fermentations, respectively, compared to 14.6 g/L butanol obtained in control batch fermentation. In an extractive fed-batch ABE fermentation, oleyl alcohol was mixed with broth at a ratio of 1, 1.5, and 2.3, and the final butanol production achieved in each process was 32 g/L,

45 g/L and 63 g/L, respectively (Roffler *et al.*, 1987b). This indicated that with a high extractant/broth ratio, more butanol was recovered in the extractant phase and end-product inhibition was relieved on microorganisms, resulting in higher total butanol production. In each scenario, fermentation stopped when butanol concentration reached 30–35 g/L in the extractant phase, indicating that the saturation point of butanol in oleyl alcohol is about this concentration. Oleyl alcohol can also be mixed with other extractants that have higher partition coefficients but are toxic to cells to obtain an extractant mixture with overall high partition coefficient and relatively low toxicity (Evans and Wang, 1988).

In addition to the more traditional extractants like long-chain alcohols, alkanes, esters, fatty acids, and oils, some novel materials including ionic liquid and biodiesel have also been suggested as potential extractant for butanol extraction. ***Ionic liquid*** (IL) is a group of salts that exist in the liquid form at low temperature (<100 °C) or room temperature, and is considered as a green and safe solvent due to its thermally and chemically stable properties (Seddon, 1997; Huddleston *et al.*, 1998; Earle and Seddon, 2000; Hagiwara and Ito, 2000; Fadeev and Meagher, 2001; Zhao *et al.*, 2005; Toh *et al.*, 2006). The miscibility and hydrophobicity of ILs can be adjusted by manipulating the structure of anions and cations. It was reported that anions determine the water miscibility of ILs, whereas cations have more influence on the hydrophobicity of ILs (Zhao *et al.*, 2005). ILs have been used as extractants in many application areas, including metal ions (Wei *et al.*, 2003), carbohydrates (Liu *et al.*, 2005), organic acids (Matsumoto *et al.*, 2004), and biofuels (Fadeev and Meagher, 2001). $[PF_6]^-$ based ILs are usually water-immiscible, and 1-butyl-3- methyl-1*H*-imidazol-3-ium ($[BMIM][PF_6]$) has been identified as a suitable extractant for butanol recovery (Fadeev and Meagher, 2001).

In addition to ILs, biodiesel is another exotic extractant proposed for butanol recovery (Li *et al.*, 2010). Biodiesel can be utilized as diesel fuel; with butanol added into biodiesel via extraction the fuel properties of biodiesel can be enhanced. Li *et al.* (2010) reported that biodiesel preferentially extracted butanol with a partition coefficient of 1.23, and the fuel properties of ABE-enriched biodiesel were significantly improved, with the cetane number increasing from 48 to 54, and cold filter plugging point decreasing from 5.8 to 0.2 °C.

The alcohol-rich product recovered by liquid–liquid extraction usually requires additional steps for dehydration and purification, and the concentration of alcohol in the extractant strongly depends on the selectivity of the extractant. The regeneration step is the most energy-intensive procedure in liquid–liquid extraction. It was suggested that a butanol/water separation factor of 30–50 is needed in order to reduce the energy demand significantly (Vane, 2008).

Unlike butanol, carboxylic acids are hydrophilic and conventional solvents, including most alcohols, ketones, ethers, and aliphatic hydrocarbons, are not efficient extractants for their separation. Reactive extraction with long-chain aliphatic amines such as tricaprylyl amine (Alamine 336) and ditridecyl amine (Adogen 283), which have low solubility in water and high distribution coefficients for carboxylic acids, has therefore been more commonly used to recover carboxylic acids from dilute solutions (Yang *et al.*, 1991; King 1992; Eyal and Canari, 1995; Tik *et al.*, 2001). Depending on the carboxylic acid and amine extractant, the distribution coefficient usually ranges from ~3 to over 20. The reactive extraction of carboxylic acids by amines is a complexation reaction between undissociated acids (HA) and amines (R_3N), as follows (Kertes and King, 1986; Prochazka *et al.*, 1994):

$$m\left[R_3N\right]_{org} + n\,[HA]_{aqu} \Leftrightarrow \left[(R_3N)_m\,(HA)_n\right]_{org}$$

Long-chain aliphatic amines are hydrophobic, so a polar diluent, such as 1-octanol, is required to improve the solvation of hydrophilic organic acid in the solvent. The diluent can better solvate the amine-acid complex and avoid its precipitation and the formation of a separate phase (Hong and Hong, 2000; Wasewar *et al.*, 2003; Senol, 2004, 2006). Diluent also lowers the viscosity of the solvent phase.

The extraction of carboxylic acids with aliphatic amines is greatly affected by pH and temperature (Yang *et al*., 1991; Eyal and Canari, 1995). The complexation reaction takes place between the undissociated acid and amine molecules, so the extraction works better at an acidic pH while back extraction can be done with a strong base to regenerate the solvent (Yang *et al*., 1991). Since the complexation reaction is exothermic, the distribution coefficient decreases with increasing temperature. Back extraction or stripping with hot water or temperature swing can therefore also be used to recover the acid product and regenerate the solvent (Tamada and King, 1990; Prochazka *et al*., 1994). The amine extractant can also be regenerated by distillation or heating if the acid is volatile and the diluent is non-volatile.

15.4.4 Adsorption

Adsorption is another separation process that has been widely studied for fermentation product recovery. The fermentation product, butanol or an organic acid, is first adsorbed by selected adsorbent materials in a packed column from dilute solution during the loading cycle, and then desorbed to obtain a concentrated product solution during the regeneration cycle. Like extractant used in the liquid–liquid extraction process, the adsorbent also needs to be regenerated to recover the adsorbed product and for continued reuse. Desorption of a volatile compound such as butanol is usually done by heating the adsorbent, whereas eluting with a stripping solution or solvent is normally used for desorbing non-volatile organic acids. High separation factor and distribution coefficient are two key parameters in selecting proper adsorbent materials. A typical adsorption process involving adsorption (loading) and desorption (regeneration) is illustrated in Figure 15.2d.

Anion exchange resins have been widely studied for adsorption of carboxylic acids, including lactic, citric, fumaric, and acetic acids (Cao *et al*., 1996; 2002; Wang *et al*., 2000; Anasthas and Gaikar, 2001). In general, the undissociated acid is adsorbed onto weak or strong base polymer resins containing tertiary or quaternary amine groups. The adsorbed acid molecules are then eluted or desorbed with methanol, ammonia, or H_2SO_4. The main disadvantages of the adsorption process for carboxylic acid separation are its low adsorption capacity (usually less than 100 mg/g resin) and the requirement of additional chemicals for acidifying the feed broth and for eluting/desorbing the acid molecules from the adsorbents. The presence of other anions, such as SO_4^{2-} and Cl^-, in the fermentation broth can significantly reduce the adsorption efficiency due to competition for active sites on the ion-exchange resin. Other materials including activated carbon, polyvinyl pyridine, and silicalite (zeolite) molecular sieves have also been studied for the adsorption of lactic acid (Ju and Chen, 1998; Aljundi *et al*., 2005).

The most commonly used adsorbent materials for alcohol recovery are hydrophobic zeolites, especially silicalite-1 (Milestone and Bibby, 1981; Groot *et al*., 1992; Holtzapple and Brown, 1994; Qureshi *et al*., 2005; Oudshoorn *et al*., 2009). Other materials such as resin, activated carbon, and polyvinylpyridine have also been suggested and studied as adsorbent materials for alcohol recovery (Groot and Luyben, 1986; Qureshi *et al*., 2005; Nielsen and Prather, 2009). Table 15.6 lists the butanol adsorption capacities and performances of various adsorbent materials for butanol recovery by adsorption. Milestone and Bibby (1981) reported that using silicalite as adsorbent in butanol recovery from a dilute 0.5% solution, a highly concentrated butanol (~98% pure) was obtained by heating the adsorbent to 150 °C after preliminary drying at 40 °C to remove bulk water. However, this result has not been duplicated by other researchers. Oudshoorn *et al*. (2009) evaluated three zeolites of different structures and SiO_2/Al_2O_3 ratios, and reported that the zeolite with a higher SiO_2/Al_2O_3 ratio also had a higher butanol adsorption capacity (higher distribution coefficient). Zeolite with a ZSM-5 structure and a high SiO_2/Al_2O_3 ratio showed excellent affinity for butanol even when butanol concentration was low, indicating that the affinity for butanol was associated with the hydrophobicity of the zeolite. It was mentioned that the presence of cells did not affect the butanol adsorption behavior of all zeolites investigated.

Table 15.6 Comparison of adsorption capacities of some adsorbents used for butanol recovery

Adsorbent	Feed C_{BuOH} (g/L)	Adsorbent loading (g/L)	Adsorption capacity (mg/g)	References
Activated carbon	15.0	10	252	Groot and Luyben, 1986
Silicalite	21.5	40	97	Milestone and Bibby, 1981
	10.0	200	48	Meagher *et al.*, 1998
	11.7–16.8	168	64–85	Maddox, 1982
	8.3	85	63.5	Ennis *et al.*, 1987
XAD-16	9.2	85	75	
XAD-2	16.5	10	78	Groot and Luyben, 1986
XAD-4	14.4	10	100	
XAD-8	15.5	10	66	
Amberlite XAD-4	4.0–20.0	100–200	27–83	Nielsen *et al.*, 1988
Amberlite XAD-7	4.0–20.0	100–200	22–69	
Bonopore	4.0–20.0	100–200	23–74	
Bonopore, nitrated	4.0–20.0	100–200	13–55	
Polyvinylpyridine	14.9	100	68	Yang *et al.*, 1994
Zeolite (CBV811)	4.8–9.0	7–25	98–117	Oudshoorn *et al.*, 2009
Poly(styrene-co-DVB)	5.0	100	22.3–56.3	Nielsen and Prather, 2009
Poly(methacrylate)	5.0	100	34.7	
Poly(butrylene phthalate)	5.0	100	7.4	

Nielsen and Prather (2009) investigated and compared the performance of several commercially available resins for butanol adsorption. They identified two resins of poly(styrene-co-divinylbenzene) that showed the best n-butanol affinity, and concluded that the butanol partition coefficient of resins was determined by the specific surface areas. Due to the high specific loadings of resin (266–403 g butanol/kg resin), butanol was recovered by vacuum evaporation at 100 °C with 78–85% recovery efficiency, and this process was predicted to be economically favorable.

Qureshi *et al*. (2005) reviewed butanol adsorption studies of various adsorbent materials, including silicalite, resins, bone charcoal, activated charcoal, bonopore, and polyvinylpyridine, and provided a systematic comparison of butanol adsorption efficiency using these materials. Bone charcoal and activated carbon were reported to have the highest butanol adsorption capacity using ABE model solution, but the adsorbed butanol on these materials could not be completely recovered during the desorption process. Silicalite was suggested to be the most appealing adsorbant, concentrating butanol to 810 g/L from a 5 g/L dilute feed solution with complete butanol recovery in the desorption process. Qureshi *et al*. also made a comparison of energy inputs for ABE separation by various technologies, and concluded that adsorption was the most energy-efficient one, followed by liquid–liquid extraction, pervaporation, gas stripping, and steam-stripping distillation. However, the actual energy consumption in butanol recovery would be highly dependent on the butanol concentration present in the fermentation broth, and the best separation technology will have to be determined based on the fermentation butanol titer and other process factors.

Adsorbent fouling by cells and adsorption of other fermentation components, such as nutrients, substrates and acids, have been the major concerns with applying adsorption technology with fermentation to recover alcohols. In order to avoid fouling by cells, it was suggested that a membrane-assisted cell recycle or cell removal by centrifuge could be considered in the integrated fermentation process with adsorption (Nielsen *et al*., 1988; Yang and Tsao, 1995).

15.4.5 Electrodialysis

Electrodialysis (ED) is a membrane separation process in which ions driven through an ion-selective membrane under an electric field are separated and concentrated (Xu, 2005; Nagarale *et al.*, 2006; Huang *et al.*, 2007). There are mainly two types of ED available for separating carboxylic acids from fermentation broth. Conventional ED with cation- and anion-exchange membranes stacked between anode and cathode can concentrate and partially purify carboxylates (Figure 15.3a), whereas a three-compartment bipolar membrane electrodialysis (BMED) with additional bipolar membranes (BM), which split water into H^+ and OH^-, stacked between cation exchange membranes (CAM) and anion exchange membranes (AEM), can produce concentrated free acid and base from salt (Figure 15.3b) (Bailly, 2002; Wiśniewski *et al.*, 2004). A simple two-compartment BMED with either CAM or AEM can also be used to convert carboxylates to carboxylic acids if the feed is relatively pure. More details can be found in a review article by Huang *et al.* (2007),

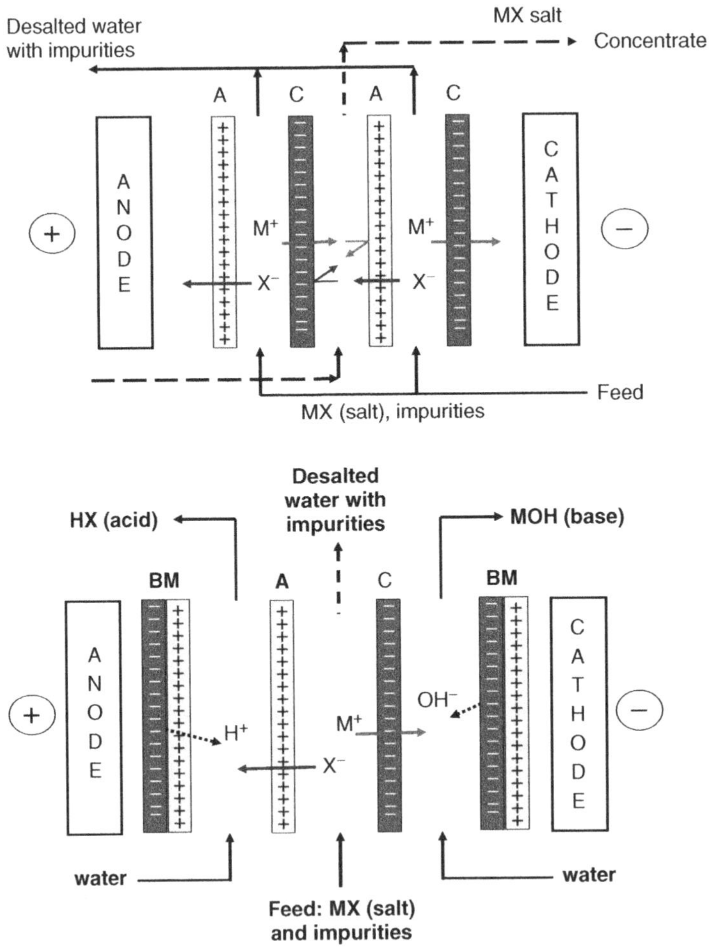

Figure 15.3 Electrodialysis. A. Principle of desalting electrodialysis. B. Three-compartment water-splitting electrodialysis with bipolar membranes (BM) for producing free organic acid and base from organic salt. A: anion exchange membrane; C: Cation exchange membrane; M^+: cation; X^-: anion. Reprinted from Yang et al. *© 2006, with permission from Elsevier*

which also discussed additional types of EDs, including electrometathesis (EMT), electro-ion substitution (EIS), electro-electrodialysis (EED), electrodeionization (EDI), and two-phase electrodialysis (TPED).

Various types of electrodialysis processes have been widely studied for the separation of carboxylic acids, including acetic, butyric, citric, formic, gluconic, itaconic, lactic, malic, propionic, pyruvic, succinic, and tartaric acids (Belafi-Bako *et al*., 2004; Godjevargova *et al*., 2004; Fidaleo *et al*., 2005; Ferrer *et al*., 2006; Fidaleo and Moresi, 2006; Groot 2011; Luo *et al*., 2004; Strathmann 2004; Thang *et al*., 2005; Wee *et al*., 2005; Zelić and Vasić-Rački, 2005; Wang *et al*., 2006; Wang *et al*., 2011b; Zhang *et al*., 2011). They have the advantage of producing free acid from salt without using acidifying chemicals. Base recovered in ED can also be recycled for use in fermentation to control the pH. However, membrane fouling by proteins, amino acids, and divalent ions Ca^{2+} and Mg^{2+}, which form insoluble hydroxides at the interface of the bipolar membrane, can cause operation problems (Ren *et al*., 2008). The fermentation broth may therefore have to be clarified and partially purified before sending to ED, which could be difficult to do in an integrated fermentation-separation process. Another problem is the low current efficiency (down to ~25%) when ED is used for separating carboxylic acids, which have low electrical conductivity, resulting in high energy (electricity) costs (Nagarale *et al*., 2004). The energy consumption was ~0.22 kWh/kg for concentrating sodium lactate and sodium propionate, whereas much more energy (>1 kWh/kg) would be required for making organic acids, for example 2–5 kWh/kg for citric acid. Adding granular cation exchange resins (e.g., Amberlite IR 120 Plus) in the weak acid compartment can increase the conductivity and thus reduce the energy consumption. Zhang *et al*. (2012) used conductive spacers to facilitate the transport of hydrogen ions in the product compartment and thus decrease the energy consumption of the process. The current efficiency and final acid concentration are negatively affected by back diffusion of acid molecules and osmosis, which can be minimized by maintaining an overpressure in the concentrated compartment (Luo *et al*., 2004).

Wang *et al*. (2011a) compared the economics of BMED and ion exchange for the production of gluconic acid. In general, ion exchange is much cheaper than BMED ($0.057/kg versus $0.085–0.407/kg) at 2011 bipolar membrane prices. However, acidifying gluconate with ion-exchange also generates wastes, which could have a high environmental cost. They therefore suggested that the best strategy is to integrate BMED and ion-exchange. They also suggested using conventional ED to concentrate the organic salts in the feed stream first, and then feed the concentrated organic salts to BMED to produce the organic acids (Wang *et al*., 2010). The integrated process could achieve a high current efficiency (higher than 100%) and lower the energy consumption and overall process cost.

Electrodialysis fermentation (EDF) uses an electrodialyzer with ion-exchange membranes to remove ionized product carboxylates from the fermentation broth, thus providing good pH control without requiring a base and reducing chemical use and waste generation. Electrodialysis fermentation has been extensively studied for lactic acid production from glucose (Yao and Toda, 1990; Siebold *et al*., 1995; Gao *et al*., 2004; 2005; Li *et al*., 2004; Arora *et al*., 2007). Although EDF can improve reactor productivity and produce a concentrated and relatively pure product stream, it usually gives lower product yields due to product loss and low conversion rate of the substrate. The final product concentrations from EDF are also lower than those of other fermentation processes. Membrane fouling by cells and proteins is a main concern in EDF. Cell immobilization or removal by filtration can be applied to reduce fouling and to extend the operating period. However, the removal of nutrients (e.g., amino acids) and inorganic ions (phosphates, calcium, etc.) by electrodialysis, and changes in the redox potential of the fermentation medium caused by the hydrogen gas produced during electrodialysis (Vonktaveesuk *et al*., 1994) present additional challenges to EDF. These problems and the process complexity greatly limit EDF for the production of carboxylic acids.

Although ED, and especially BMED, is a promising technology for organic acids production, there are several challenges, including high membrane costs, membrane fouling, and low ion selectivity, that need to be addressed with further research and development efforts before its industrial applications can be economically competitive.

15.5 Examples in biorefineries

15.5.1 Extractive ABE fermentation for enhanced butanol production

Butanol is the most desired product from ABE fermentation and yet the most toxic product to the culture. Severe butanol inhibition exists in ABE fermentation, which results in low final butanol concentration, low yield, and low productivity (Qureshi and Ezeji, 2008). All these limitations hamper the economic application of deriving biobutanol from ABE fermentation.

In a typical batch ABE fermentation, only 12–15 g/L butanol and 20–25 g/L total ABE can be obtained through a period of 40–60 h until the fermentation stops due to inhibition (Woods, 1995). The *in situ* recovery of butanol is therefore crucial in improving the reactor performance. Simultaneous butanol recovery can relieve the product inhibition and leads to a more complete conversion of the carbon source. It allows the use of a concentrated feed and extends the fermentation period (Groot *et al.*, 1990; Dürre, 1998; Ezeji *et al.*, 2004a). Moreover, online butanol recovery also simplifies the downstream separation process, which lowers the energy consumption and brings down the whole process cost. In the past, distillation was widely employed to recover butanol, and it was proved to be costly due to the low butanol concentration in the broth (Ezeji *et al.*, 2004a). In recent years, new advances in butanol recovery techniques, including liquid–liquid extraction, adsorption, pervaporation and gas stripping, have allowed them to be integrated with fermentation in an effort to develop a commercial process for biobutanol production (Groot *et al.*, 1990; Qureshi and Maddox, 1995; Ezeji *et al.*, 2003; 2005a; Vane, 2005; Izak *et al.*, 2008). These integrated fermentation processes have been shown to be superior in aspects of sugar consumption, ABE final concentrations, and productivity. Some of the reported research on integrated ABE process are summarized and compared in Table 15.7.

From Table 15.7, it is clear that online butanol recovery can increase the final ABE concentration and the reactor productivity significantly. As butanol is continuously removed from the fermentation broth, butanol in the reactor never reaches the inhibitory level, thus allowing a higher sugar utilization rate. Compared to conventional batch fermentation, online butanol recovery allows the use of highly concentrated feed solution in fed-batch and continuous processes in an extended period, which further leads to high ABE production. Some of the integrated processes also exhibit the potential of commercializing the ABE fermentation at the industry scale. Ezeji *et al.* (2005b) reported that gas stripping integrated with continuous fermentation used 1163 g/L glucose and produced 460 g/L ABE in total.

More recently, Lu *et al.* (2011) studied fed-batch fermentation with continuous gas stripping for butanol production from cassava bagasse hydrolysate using a hyper-butanol-producing *Clostridium acetobutylicum* strain JB200 in a fibrous bed bioreactor. The fed-batch fermentation was operated for 263 h, producing a total of 108.5 g/L ABE (butanol: 76.4 g/L, acetone: 27 g/L, ethanol: 5.1 g/L) with an average ABE productivity of 0.47 ± 0.06 g/L·h. The gas stripping kept the butanol concentration in the fermentation broth between 8 g/L and 12 g/L, and generated a condensate containing 10% to 16% (w/v) of butanol, ~4% (w/v) of acetone, a small amount of ethanol (<0.8%) and almost no acids. After phase separation, a highly concentrated butanol solution of ~64% (w/v) was obtained in the upper organic phase, which can be easily further purified and dehydrated by distillation. This work illustrated that integrating gas stripping with ABE fermentation can greatly enhance the fermentation efficiency and final product titer and purity that would be more amenable for further purification at a much lower energy input.

Table 15.7 *Integrated fermentation-separation processes for ABE production by Clostridia from various substrates*

Recovery technique	Substrate	Strain	Fermentation mode	ABE (g/L)	ABE yield (g/g)	Productivity (g/L·h)	Reference
Gas stripping	Whey permeate	*C. acetobutylicum* P262	Batch	70.0	0.35	0.32	Maddox *et al.*,1995
			Continuous	69.1	0.38	0.26	Qureshi *et al.*, 1992
	Glucose	*C. beijerinckii* BA101	Batch	79.5	0.47	0.60	Ezeji *et al.*, 2003
			Fed-batch	232	0.47	1.16	Ezeji *et al.*, 2004b
	Wheat straw	*C. beijerinckii* P260	Batch	47.6	0.37	0.36	Qureshi *et al.*, 2007
	Soluble corn starch	*C. beijerinckii* BA101	Batch	23.9	0.43	0.31	Ezeji *et al.*, 2007b
	Saccharified corn starch		Batch	26.5	0.41	0.40	
			Fed-batch	81.3	0.36	0.59	
	Cassava bagasse	*C. acetobutylicum* JB200	Fed-batch	108.5	0.37	0.47	Lu *et al.*, 2011
Pervaporation	Whey permeate	*C. acetobutylicum* P262	Continuous	42.0	0.34	0.14	Qureshi *et al.*, 1992
	Glucose	*C. beijerinckii* BA101	Fed-batch	165	0.43	0.98	Qureshi and Blaschek, 2000
		C. acetobutylicum ATCC 824	Fed-batch	155	0.35	0.18	Qureshi *et al.*, 2001
		C. beijerinckii NRRL B592	Continuous	13.1	0.28	1.72	Gapes *et al.*, 1996
Liquid–liquid extraction	Glucose	*C. acetobutylicum* ATCC 824	Batch	22.5–34.3	0.26–0.33	N/A	Roffler *et al.*, 1987a
			Fed-batch	50.5–96.5	0.33–0.36	1.4–2.3	Roffler *et al.*, 1987b
	Whey permeate	*C. acetobutylicum* P262	Continuous	23.8	0.35	0.14	Qureshi *et al.*, 1992
Perstraction	Whey permeate + lactose	*C. acetobutylicum* P262	Batch	136.6	0.44	0.21	Qureshi and Maddox, 2005
	Whey permeate		Continuous	57.8	0.37	0.24	Qureshi *et al.*, 1992
Adsorption	Glucose	*C. acetobutylicum*	Batch	23.2	0.32	0.92	Yang and Tsao, 1995
			Fed-batch	59.8	0.32	1.33	
			Repeated fed-batch	387.3	0.32	1.69	

15.5.2 Extractive fermentation for organic acids production

Integrated fermentation-separation processes have been extensively studied for the production of various carboxylic acids from sugars (Table 15.8). Compared to the same fermentation without simultaneous product removal, online product separation significantly increased the final product concentration and reactor volumetric productivity. Extractive fermentation selectively removing the desirable product also significantly increased the product yield due to reduced byproducts formation. However, solvent toxicity and nutrients removal from the fermentation media by solvent extraction or adsorption could significantly lower cell productivity and thus reduce reactor productivity. Extraction and adsorption of carboxylic acids also usually work better at an acidic pH value, whereas most of the carboxylic acid fermentations have an optimal pH value around neutral or greater than the pK_a value of the acid. Consequently, extractive fermentation may have to be operated at a suboptimal pH, lowering its reactor productivity.

Wu and Yang (2003) developed an extractive fermentation process using 10% (v/v) Alamine 336 in oleyl alcohol as the extractant contained in a hollow-fiber membrane extractor for butyric acid production from glucose by *Clostridium tyrobutyricum*. The membrane-based extraction or pertraction process selectively removed butyric acid from the fermentation broth. In the pertraction process, the extractant was simultaneously regenerated by stripping with NaOH in a second membrane extractor. The solvent was toxic to cells if they were in direct contact. So cells were immobilized in a fibrous bed bioreactor (FBB) to protect them from solvent toxicity. Figure 15.4 illustrates the extractive fermentation process with a FBB and two hollow-fiber membrane extractors for extraction and back extraction. During the extractive fermentation, pH was self-regulated and maintained at ~pH 5.5, which resulted from a balance between acid production by cells and removal by extraction. The extractive fermentation gave a high product concentration of 300 g/L with a high butyric acid yield of 0.45 g/g and product purity of 91%. It also gave a high reactor productivity of 7.4 g/L h. In contrast, fermentation without on-line extraction gave the final butyric acid concentration of ~43.4 g/L, with a butyric acid yield of 0.42 g/g and productivity of 6.8 g/L h at the optimal pH of 6.0. These values were much lower at pH 5.5 (see Table 15.9). They concluded that the extractive fermentation reduced product inhibition by selectively removing butyric acid from the fermentation broth and thus improved butyric acid production in both its final titer and yield, and reactor productivity.

However, in order to produce butyric acid, instead of butyrate salts, further purification and acidification of the product from the extractive fermentation process are necessary. This can be done by using the water-splitting electrodialysis with bipolar membrane (BMED) discussed earlier. The integration of the solvent extractive fermentation with BMED may be the best strategy to produce carboxylic acids in terms of energy and environmental costs. This remains to be investigated.

15.6 Economic importance and industrial challenges

The energy cost for the recovery and purification of a fermentation product is high, usually ranging from around 20% to 50% of the final product cost. The recovery cost usually increases inversely proportional to the product concentration in the fermentation broth. It is therefore important to have the fermentation product produced at a sufficiently high titer before final recovery and purification. For example, distillation is the most energy-intensive unit operation in producing butanol from the ABE fermentation. The conventional distillation process requires more than ~79 MJ per kg of butanol produced from a binary butanol-water solution containing 0.5% (w/v) butanol (Matsumura *et al.*, 1988), which is more than the energy content of butanol (36 MJ/kg). However, the energy requirement in distillation decreases dramatically to ~36 MJ/kg and 24 MJ/kg as the butanol concentration in the feed solution increased to 1% and 1.5%, respectively. Extractive fermentation can increase fermentation productivity and can give a higher concentration product

Table 15.8 *Integrated fermentation-separation processes for production of carboxylic acids. Reprinted from Yang et al. © 2006, with permission from Elsevier*

Product	Microorganism / fermentation conditions	Yield (g/g sugar)	Productivity (g/L·h)	Final conc. (g/L)	Reference
Product removal by adsorption with ion exchange resins					
Lactic acid	*L. delbrueckii* Batch fermentation	1.17 (0.95)	0.9 (1.3)	1.84 (105)	Chen and Ju, 2002
	L. casei / Continuous fermentation with cell recycle	(0.98)	(138)	(80)	Gonzalez-Vara *et al.*, 2000
	L. delbrueckii Batch fermentation	1.12 (0.93)	5.3 (1.7)	1.25 (100)	Srivastava *et al.*, 1992
Fumaric acid	*Rhizopus oryzae* Rotary biofilm reactor	1.13 (0.85)	1.12 (4.25)	(85)	Cao *et al.*, 1996; 1997
Citric acid	*Aspergillus niger* Batch fermentation	1.15 (0.95)	1.6 (0.54)	1.2 (78)	Wang *et al.*, 2000
Electrodialysis fermentation					
Lactic acid	*L. lactis* / batch fermentation with periodic electrodialysis	~1.0 (0.78)	4.7 (3.4)	~1.0 (52)	Vonktaveesuk *et al.*, 1994
	R. oryzae / Ca-alginate beads fluidized bed bioreactor	0.96 (0.71)	1.7 (14.8)	5.2 (52)	Xuemei *et al.*, 1999
Propionic acid	*P. freudenreichii* Batch fermentation	1.2 (0.14)	1.8 (0.22)	2.0 (38)	da Costa *et al.*, 1999
Extractive fermentation					
Citric acid	*A. niger* / batch fermentation	–	1.7 (0.15)	5.2 (232)	Wieczorek and Brauer, 1998
Butyric acid	*C. tyrobutyricum* / batch fermentation with immobilized cells	1.1 (0.45)	1.1 (7.37)	6.9 (300)	Wu and Yang, 2003
Lactic acid	*R. oryzae* / rotating fibrous bed bioreactor, fed-batch fermentation	1.02 (0.92)	0.3 (0.73)	2.4 (293)	Tay, 2002
Propionic acid	*P. acidipropionici* / batch fermentation	1.3 (0.66)	8.3 (1.0)	3.9 (75)	Jin and Yang, 1998

Note: The numbers in the table indicate the relative performance as compared with the same fermentation without simultaneous product removal (= 1). The numbers in parentheses show the actual performance data for the integrated fermentation-separation processes.

before final purification. As discussed in the previous two examples, ABE fermentation with online gas stripping can produce a condensate with more than 15% butanol and a final product with more than 60% butanol after phase separation. Conventional ABE fermentation can only produce ca. 12–15 g/L butanol and 20–30 g/L ABE. Butyric acid fermentation with integrated pertraction can produce a product with more than 30% butyric acid, whereas conventional fermentation without pertraction can only produce 5% to 8% butyric acid. The final product purification cost thus can be reduced by more than 75% in these two cases.

Table 15.9 Comparison of butyric acid fermentations with and without online extraction of butyric acid. Adapted with permission from Wu et al. © 2003 Wiley Periodicals, Inc

	Batch fermentation		Extractive fermentation
	pH 6.0	pH 5.5	~pH 5.5
Final butyrate concentration (g/L)	43.4 ± 0.8	20.4 ± 0.8	301 ± 8
Butyrate yield (g/g)	0.42 ± 0.02	0.38 ± 0.02	0.45 ± 0.01
Acetate yield (g/g)	0.095 ± 0.010	0.115 ± 0.004	0.111 ± 0.004
Butyrate productivity (g/L·h)	6.77 ± 0.23	5.11 ± 0.34	7.37 ± 0.42
Product selectivity	0.81 ± 0.02	0.77 ± 0.03	0.80 ± 0.01
Product purity			0.91 ± 0.02

Product selectivity: butyric acid yield/total acid yield from the fermentation
Product purity: fraction of butyric acid in the final acid product

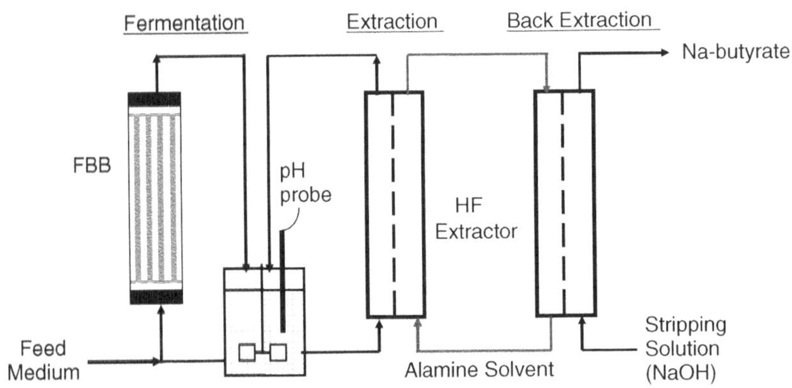

Figure 15.4 An extractive fermentation process with continuous product separation by extraction using hollow-fiber (HF) membrane extractors. The fermentation product, butyric acid, is extracted in the first HF unit and then stripped (back extracted) in the second HF unit. The final product is a concentrated sodium butyrate solution with NaOH as the stripping solution

In addition to energy consumption, water usage in a fermentation process is also a major concern because of limited water resources. Extractive fermentation allows the use of highly concentrated substrates and thus can reduce water usage in fermentation by a factor of 5–10. The high-titer product from extractive fermentation also reduces the amount of wastewater to be disposed of or treated. For example, fed-batch ABE fermentation with online gas stripping allows the utilization of highly concentrated substrate and can reduce the net water use by over 90%. About 2.88 gallons of water per gallon of ABE production will be required in the integrated process without recycling the fermentation water. For comparison, a typical corn ethanol plant uses 3–4 gallons of water per gallon of ethanol produced with almost all fermentation water being recycled (IATP, 2006). Similar benefits can be realized in the production of butyric acid and other carboxylic acids by employing the extractive fermentation technology.

Although extractive fermentation with online product recovery offers an energy-efficient and environmentally friendly process for the production of many biochemicals that potentially can be economically competitive to the petroleum-based counterparts, the biorefinery industry has yet to adopt this technology. One main reason for this is the lack of industrial experience in operating this kind of process at a large scale or even a smaller pilot scale. The fermentation industry is very conservative and skeptical in adopting a new or "untried" technology. Additional challenges include: increased process complexity that may not only complicate the process design but also may increase operation difficulty and risks of culture contamination and process failure, and increased equipment and material costs that may not give a justifiable return of investment (ROI). Further optimization to improve efficiency and reduce costs, and scale up of novel separation technologies that are more friendly for integration with existing fermentation processes are also imperative to the development and commercialization of extractive fermentation technology.

15.7 Conclusions and future trends

Extractive fermentation with integrated online product recovery can alleviate product inhibition, increase reactor productivity, and allow the use of concentrated feedstock. It can also produce the product at a higher titer, purity, and yield. It offers a great opportunity for the biorefinery industry to produce biofuels and biobased chemicals such as butanol and butyric acid at an economically competitive cost. However, the industry is slow to adopt this technology because of the lack of experience and the increased complexity. Nevertheless, as oil prices continue to go up and petroleum-based fuels and chemicals become increasingly expensive, and biomass or biorenewable feedstock, including agricultural and forestry residues and energy crops, is becoming increasingly important in supplying the future fuels and chemicals, the biorefinery industry must search for and develop new fermentation technologies, including extractive fermentation and novel online separation technologies, which are not only green and technically advantageous, but also economically competitive.

Among the available separation technologies, gas stripping appears to be the most promising one for separating butanol from ABE fermentation broth, whereas solvent extraction or pertraction is the most energy-efficient method for recovering fermentation-produced carboxylic acids. Membrane technologies, such as pervaporation and electrodialysis, also have great potential for recovering fermentation products, but their costs are high and must be reduced significantly before they can be applied widely in industry.

References

I.H. Aljundi, J.M. Belovich, and O. Talu. Adsorption of lactic acid from fermentation broth and aqueous solutions on Zeolite molecular sieves, *Chem. Eng. Sci.*, 60, 5004–5009 (2005).

H.M. Anasthas, and V.G. Gaikar, Adsorption of acetic acid on ion-exchange resins in non-aqueous conditions, *Reactive Functional Polymers*, 47, 23–35 (2001).

M.B. Arora, J.A. Hestekin, S.W. Snyder, E.J. Martin, Y.J. St. Lin, M.I. Donnelly, and C.S. Millard, The separative bioreactor: a continuous separation process for the simultaneous production and direct capture of organic acids, *Separation Sci. Technol.*, 42, 2519–2538 (2007).

M. Bailly, Production of organic acids by bipolar electrodialysis: realizations and perspectives, *Desalination*, 144, 157–162 (2002).

W.E. Barton, and A.J. Daugulis, Evaluation of solvents for extractive butanol fermentation with *Clostridium acetobutylicum* and the use of poly(propylene glycol) 1200, *Appl. Microbiol. Biotechnol.*, 36, 632–639 (1992).

K. Belafi-Bako, N. Nemestothy, and L. Gubicza, A study on applications of membrane techniques in bioconversion of fumaric acid to l-malic acid, *Desalination*, 162, 301–306 (2004).

K.W. Boddeker, G. Bengtson, and H. Pingel, Pervaporation of isomeric butanols, *J. Memb. Sci.*, 54, 1–12 (1990).

N. Cao, J. Du, C. Chen, C.S. Gong, and G.T. Tsao, Simultaneous production and recovery of fumaric acid from immobilized *Rhizopus* with a rotary biofilm contactor and an adsorption column, *Appl. Environ. Microbiol.*, 62, 2926–2931 (1996).

N. Cao, J. Du, C. Chen, C.S. Gong, and G.T. Tsao, Production of fumaric acid by immobilized *Rhizopus* using rotary biofilm contactor, *Appl. Biochem. Biotechnol.*, 63/65, 387–394 (1997).

X. Cao, H.S. Yun, and Y.-M. Koo, Recovery of L-(+)-lactic acid by anion exchange resin Amberlite IRA-400, *Biochem. Eng. J.*, 11, 189–196 (2002).

C.C. Chen, and L.-K. Ju, Coupled lactic acid fermentation and adsorption, *Appl. Microbiol. Biotechnol.*, 59, 170–174 (2002).

J.P.L.C. da Costa, C. Schorm, A. Quesada-Chanto, K.W. Boddeker, and R. Jonas, On-line dialysis of organic acids from a *Propionibacterium freudenreichii* fermentation: evaluation of a new pH control strategy, *Appl. Biochem. Biotechnol.*, 76, 99–106 (1999).

P. Dürre, New insights and novel developments in clostridial acetone/ butanol/ isopropanol fermentation, *Appl. Microbiol. Biotechnol.*, 49, 639–648 (1998).

M.J. Earle, and K.R. Seddon, Ionic liquids. Green solvents for the future, *Pure Appl. Chem.*, 72, 1391–1398 (2000).

G. Eckert, and K. Schügerl, Continuous acetone-butanol production with direct product removal, *Appl. Microbiol. Biotechnol.*, 27, 221–228 (1987).

B.M. Ennis, N.A. Gutierrez, and I.S. Maddox, The acetone-butanol-ethanol fermentation: a current assessment, *Process Biochem.*, 21, 131–147 (1986).

B.M. Ennis, N. Qureshi, and I.S. Maddox, In-line toxic product removal during solvent production by continuous fermentation using immobilized Clostridium acetobutylicum, *Enzyme Microbial. Technol.*, 9, 672–675 (1987).

P.J. Evans, and H.Y. Wang, Response of Clostridium acetobutylicum to the presence of mixed extractants, *Appl. Environ. Microbiol.*, 54, 175–192 (1988).

T.C. Ezeji, P.M. Karcher, N. Qureshi, and H.P. Blaschek, Improving performance of a gas stripping-based recovery system to remove butanol from *Clostridium beijerinckii* fermentation, *Bioprocess Biosyst. Eng.*, 27, 207–214 (2005a).

T.C. Ezeji, N. Qureshi, and H.P. Blaschek, Production of butanol by *Clostridium beijerinckii* BA101 and in-situ recovery by gas stripping, *J. Microbiol. Biotechnol.*, 19, 595–603 (2003).

T.C. Ezeji, N. Qureshi, and H.P. Blaschek, Butanol fermentation research: upstream and downstream manipulations, *The Chemical Record*, 4, 305–314 (2004a).

T.C. Ezeji, N. Qureshi, and H.P. Blaschek, Acetone-butanol-ethanol production from concentrated substrate: reduction in substrate inhibition by fed-batch technique and product inhibition by gas stripping, *Appl. Microbiol. Biotechnol.*, 63, 653–658 (2004b).

T.C. Ezeji, N. Qureshi, and H.P. Blaschek, Process for continuous solvent production, *United States Patent Application Publication*, US patent 20050089979A1 (2005b).

T.C. Ezeji, N. Qureshi, and H.P. Blaschek, Bioproduction of butanol from biomass: from genes to bioreactors, *Current Opinion in Biotechnol.*, 18, 220–227 (2007a).

T.C Ezeji, N. Qureshi, and H.P. Blaschek, Production of acetone butanol (AB) from liquefied corn starch, a commercial substrate, using *Clostridium beijerinckii* coupled with product recovery by gas stripping, *J. Ind. Microbiol. Biotechnol.*, 34, 771–777 (2007b).

A. M. Eyal, and R. Canari, PH dependence of carboxylic and mineral acid extraction by amine-based extractants: effects of pK_a, amine basicity, and diluent properties, *Ind. Eng. Chem. Res.*, 34, 1789–1798 (1995).

A.G. Fadeev, S.S. Kelley, J.D. McMillan, YaA Selinskaya, V.S. Khotimsky, and V.V. Volkov, Effect of yeast fermentation by-products on poly(1-trimethylsilyl-1-propyne) pervaporative performance, *J. Membr. Sci.*, 214, 229–238 (2003).

A.G. Fadeev, and M.M. Meagher, Opportunities for ionic liquids in recovery of biofuels, *Chem. Commun.*, 295–296 (2001).

A.G. Fadeev, M.M. Meagher, S.S. Kelley, and V.V. Volkov, Fouling of poly-[1-(trimethylsilyl)-1-propyne] membranes in pervaporative recovery of butanol from aqueous solutions and ABE fermentation broth, *J. Membr. Sci.*, 173, 133–144 (2000).

A.G. Fadeev, Y.A. Selinskaya, S.S. Kelley, M.M. Meagher, E.G. Litvinova, V.S. Khotimsky, and V.V. Volkov, Extraction of butanol from aqueous solutions by pervaporation through poly(1-trimethylsilyl-1-propyne), *J. Membr. Sci.*, 186, 205–217 (2001).

P. Fernandes, D.M.F. Prazeres, and J.M.S. Cabral, Membrane-assisted extractive bioconversions, *Adv. Biochem. Eng./Biotechnol.*, **80**, 115–148 (2003).

J.S.J. Ferrer, S. Laborie, G. Durand, and M. Rakib, Formic acid regeneration by electromembrane processes, *J. Membrane Sci.*, 280, 509–516 (2006).

M. Fidaleo, and M. Moresi, Modeling of sodium acetate recovery from aqueous solutions by electrodialysis, *Biotechnol. Bioeng.*, 91, 556–568 (2005).

M. Fidaleo, and M. Moresi, Assessment of the main engineering parameters controlling the electrodialytic recovery of sodium propionate from aqueous solutions, *J. Food Eng.*, 76, 218–231 (2006).

M. Gao, M. Hirata, M. Koide, H. Takanashi, and T. Hano, Production of L-lactic acid by electrodialysis fermentation (EDF), *Process Biochem.*, 39, 1903–1907 (2004).

M. Gao, M. Koide, R. Gotou, H. Takanashi, M. Hirata, and T. Hano, Development of a continuous electrodialysis fermentation system for production of lactic acid by *Lactobacillus rhamnosus*, *Process Biochem.*, 40, 1033–1036 (2005).

J.R. Gapes, D. Nimcevic, and A. Friedl, Long-term continuous cultivation of *Clostridium beijerinckii* in a two-stage chemostat with on-line solvent removal, *Appl. Environ. Microbiol.*, 62, 3210–3219 (1996).

Q. Geng and C.-H. Park, Pervaporative butanol fermentation by *Clostridium acetobutylicum* B18, *Biotech. Bioeng.*, 43, 978–986 (1994).

T. Godjevargova, R. Dayal, and S. Turmanova, Gluconic acid production in bioreactor with immobilized glucose oxidase plus catalase on polymer membrane adjacent to anion-exchange membrane, *Macromol. Biosci.*, 4, 950–956 (2004).

Y.R.A. Gonzalez-Vara, G. Vaccari, E. Dosi, A. Trilli, M. Rossi, and D. Matteuzzi, Enhanced production of L-(+)-lactic acid in chemostat by *Lactobacillus casei* SM 20011 using ion-exchange resins and cross-flow filtration in a fully automated pilot plant controlled via NIR, *Biotechnol. Bioeng.*, 67, 147–156 (2000).

W.J. Groot, Process for manufacturing succinic acid, PCT International Patent Application, *WO* 2011098598 A1 (2011).

W.J. Groot and K.C.A.M. Luyben, In situ product recovery by adsorption in the butanol/isopropanol batch fermentation, *Appl. Microbiol. Biotechnol.*, 25, 29–31 (1986).

W.J. Groot, H.S. Soedjak, P.B. Donck, R.G.J.M. van der Lans, K. Ch. A.M. Luyben, and J.M.K. Timmer, Butanol recovery from fermentations by liquid–liquid extraction and membrane solvent extraction, *Bioprocess Eng.*, 5, 203–216 (1990).

W.J. Groot, R.G.J.M. van der Lans, and K.Ch.A.M. Luyben, Batch and continuous butanol fermentations with free cells: integration with product recovery by gas- stripping, *Appl. Microbiol. Biotechnol.*, 32, 305–308 (1989).

W.J. Groot, R.G.J.M. van der Lans, and K. Ch. A.M. Luyben, Technologies for butanol recovery integrated with fermentations, *Proc. Biochem.*, 27, 61–65 (1992).

W.J. Groot, C.E. Van der Oever, and N.W.F. Kossen, Pervaporation for simultaneous product recovery in the butanol/isopropanol batch fermentation, *Biotechnol. Lett.*, 6, 709–714 (1984).

R. Hagiwara and Y. Ito, Room temperature ionic liquids of alkylimidazolium cations and fluoroanions, *J. Fluor. Chem.*, 105, 221–227 (2000).

C. Huang, T. Xu, Y. Zhang, Y. Xue, and G. Chen, Application of electrodialysis to the production of organic acids: State-of-the-art and recent developments, *J. Membrane Sci.*, 288, 1–12 (2007).

P.J. Hickey, F.P. Juricic, and C.S. Slater, The effect of process parameters on the pervaporation of alcohols through organophilic membranes, *Separation Sci. Technol.*, 27, 843–861 (1992).

M.T. Holtzapple, and R.F. Brown, Conceptual design for a process to recover volatile solutes from aqueous solutions using silicalite, *Sep. Technol.*, 4, 213–229 (1994).

Y.K. Hong, and W.H. Hong, Equilibrium studies on reactive extraction of succinic acid from aqueous solutions with tertiary amines, *Bioprocess Eng.*, 22, 477–481 (2000).

J. Huang, and M.M. Meagher, Pervaporative recovery of n-butanol from aqueous solutions and ABE fermentation broth using thin-film silicalite-filled silicone composite membranes, *J. Membr. Sci.*, 192, 231–242 (2001).

J.G. Huddleston, H.D. Willauer, R.P. Swatloski, A.E. Visser, and R.D. Rogers, Room temperature ionic liquids as novel media for "clean" liquid-liquid extraction, *Chem. Commun.*, 1765–1766 (1998).

Institute for Agriculture and Trade Policy (IATP), Water use by ethanol plants potential challenges, Minneapolis, MN (2006), www.iatp.org/files/258_2_89449.pdf (accessed September 28, 2012).

P. Izak, K. Schwarz, W. Ruth, H. Bahl, and U. Kragl, Increased productivity of *Clostridium acetobutylicum* fermentation of acetone, butanol, and ethanol by pervaporation through supported ionic liquid membrane, *Appl. Microbiol. Biotechnol.*, 78, 597–602 (2008).

Z. Jin, and S.T. Yang, Extractive fermentation for enhanced propionic acid production from lactose by *Propionibacterium acidipropionici*, *Biotechnol. Prog.*, 14, 457–465 (1998).

A. Jonquieres, R. Clement, P. Lochon, J. Neel, M. Dresch, and B. Chertien, Industrial state-of-art of pervaporation and vapour permeation in the western countries, *J. Membr. Sci.*, 206, 87–117 (2002).

A. Jonquieres, and A. Fane, Filled and unfilled composite GFT PDMS membranes for the recovery of butanol from dilute aqueous solutions: influence of alcohol polarity, *J. Membr. Sci.*, 125, 245–255 (1997).

L. Ju, and C. Chen, Adsorption characteristics of polyvinylpyridine and activated carbon for lactic acid recovery from fermentation of *Lactobacillus delbrueckii*, *Sep. Sci. Technol.*, 33, 1423–1437 (1998).

A.S. Kertes, and C.J. King, Extraction chemistry of fermentation product carboxylic acids, *Biotechnol. Bioeng.*, 28, 269–282 (1986).

C.J. King, Amine-based systems for carboxylic acid recovery. *CHEMTECH, May* (1992) 285–291 (1992).

M.A. Larrayzo, and L. Puigjaner, Study of butanol extraction through pervaporation in acetobutylic fermentation, *Biotechnol. Bioeng.*, 30, 692–696 (1987).

H. Li, R. Mustacchi, C.J. Knowles, W. Skibar, G. Sunderland, I. Dalrymple, and S.A. Jackman, An electrokinetic bioreactor: using direct electric current for enhanced lactic acid fermentation and product recovery, *Tetrahedron*, 60, 655–661 (2004).

S. Li, R. Srivastava, and R.S. Parnas, Separation of 1-butanol by pervaporation using novel tri-layer PDMS composite membrane, *J. Membr. Sci.*, 363, 287–294 (2010).

S. Li, V.A. Tuan, J.L. Falconer, and R.D. Noble, Properties and separation performance of Ge-ZSM-5 membranes, *Micropor. Mesopor. Mater.*, 58, 137–154 (2003).

Q. Liu, M.H.A. Janssen, F. van Rantwijk, R.A. Sheldon, Room temperature ionic liquids that dissolve carbohydrates in high concentrations, *Green Chemistry*, 7, 39–42 (2005).

C. Lu, J. Zhao, S.T. Yang, and D. Wei, Fed-batch fermentation for butanol production from cassava bagasse hydrolysate in a fibrous bed bioreactor with continuous gas stripping, *Bioresources Technol.*, 104, 380–387 (2012).

G.S. Luo., X.Y. Shan, X. Qi, and Y.C. Lu, Two-phase electro-electrodialysis for recovery and concentration of citric acid, *Separation Purification Technol.*, **38**, 265–271 (2004).

I.S. Maddox, Use of silicalite for the adsorption of n-butanol from fermentation *liquors*, *Biotechnol. Lett.*, 4, 759–760 (1982).

I.S. Maddox, The acetone-butanol-ethanol fermentation: recent progress in technology, *Genet. Eng. Rev.*, 7, 189–220 (1989).

I.S. Maddox, N. Qureshi and K. Roberts-Thomson, Production of acetone-butanol-ethanol from concentrated substrates using *Clostridium acetobutylicum* in an integrated fermentation-product removal process, *Process Biochemistry*, 30, 209–215 (1995).

M. Matsumoto, K. Mochiduki, K. Fukunishi, K. Kondo, Extraction of organic acids using imidazolium-based ionic liquids and their toxicity to *Lactobacillus rhamnosus*, *Separ. Purif. Technol.*, 40, 97–101 (2004).

M. Matsumura, and H. Kataoka, Separation of dilute aqueous butanol and acetone solutions by pervaporation through liquid membranes, *Biotechnol. Bioeng.*, 30, 1991–1992 (1987).

M. Matsumura, H. Kataoka, M. Sueki, and K. Araki, Energy saving effect of pervaporation using oleyl alcohol liquid membrane in butanol purification, *Bioprocess Eng.*, 3, 93–100 (1988).

M.M. Meagher, N. Qureshi, and R.W. Hutkins, Silicalite membrane and method for the selective recovery and concentration of acetone and butanol from model ABE solutions and fermentation broth, *U.S. Patent 5,755,967* (1998).

N.B. Milestone, and D.M. Bibby, Concentration of alcohols by adsorption on silicalite, *J. Chem. Technol. Biotechnol.*, 31, 732–736 (1981).

R.K. Nagarale, G.S. Gohil, and V.K. Shahi, Recent developments on ion-exchange membranes and electro-membrane processes, *Adv. Colloid Interface Sci.*, 119, 97–130 (2006).

R.K. Nagarale, G.S. Gohil, V.K. Shahi, G.S. Trivedi, S.K. Thampy, and R. Rangarajan, Studies on transport properties of short chain aliphatic carboxylic acids in electrodialytic separation, *Desalination*, 171, 195–204 (2004).

L. Nielsen, M. Larsson, O. Holst, and B. Mattiasson, Adsorbents for extractive bioconversion applied to the acetone-butanol fermentation, *Appl. Microbiol. Biotechnol.*, 28, 335–339 (1988).

D.R. Nielsen, and K.J. Prather, In situ product recovery of n-butanol using polymeric resins, *Biotechnol. Bioeng.*, 102, 811–821 (2009).

A. Oudshoorn, L.A.M. van der Wielen, and A.J.J. Straathof, Adsorption equilibria of bio-based butanol solutions using zeolite, *Biochem. Eng. J.*, 48, 99–103 (2009).

J. Prochazka, A. Heyberger, V. Bizek, M. Kousova, and E. Volantova, Amine extraction of hydroxycarboxylic acids. 2. Comparison of equilibria for lactic, malic, and citric acids, *Ind. Eng. Chem. Res.*, 33, 1565–1573 (1994).

N. Qureshi, and H.P. Blaschek, Butanol recovery from model solution/ fermentation broth by pervaporation: evaluation of membranes performance, *Biomass and Bioenergy*, 17, 175–184 (1999a).

N. Qureshi and H.P. Blaschek, Production of acetone butanol ethanol (ABE) by a hyper-producing mutant strain of *Clostridium beijerinckii* BA101 and recovery by pervaporation, *Biotechnol. Prog.*, 15, 594–602 (1999b).

N. Qureshi, and H.P. Blaschek, Fouling studies of a pervaporation membrane with commercial fermentation media and fermentation broth of hyper-butanol-producing *Clostridium beijerinckii* BA101, *Sep. Sci. Technol.*, 34, 2803–2815 (1999c).

N. Qureshi, and H.P. Blaschek, Butanol production using *Clostridium beijerinckii* BA101 hyper-butanol producing mutant strain and recovery by pervaporation, *Appl. Biochem. Biotechnol.*, 84, 225–235 (2000).

N. Qureshi, and H.P. Blaschek, Recovery of butanol from fermentation broth by gas stripping, *Renewable Energy*, 22, 557–564 (2001).

N. Qureshi, and T.C. Ezeji, Butanol, "a superior biofuel" production from agricultural residues (renewable biomass): recent progress in technology, *Biofuels, Bioprod. Bioref.*, 2, 319–330 (2008).

N. Qureshi, S. Hughes, I.S. Maddox, and M.A. Cotta, Energy-efficient recovery of butanol from model solutions and fermentation broth by adsorption, *Bioprocess Biosyst. Eng.*, 27, 215–222 (2005).

N. Qureshi and I.S. Maddox, Integration of continuous production and recovery of solvents from whey permeate: use of immobilized cells of *Clostridium acetobutylicum* in a fluidized bed reactor coupled with gas stripping, *Bioproc. Eng.*, 6, 63–69 (1991).

N. Qureshi, and I.S. Maddox, Continuous production of acetone-butanol- ethanol using immobilized cells of *Clostridium acetobutylicum* and integration with product removal by liquid-liquid extraction, *J. Ferment. Bioeng.*, 80, 185–189 (1995).

N. Qureshi, and I.S. Maddox, Reduction in butanol inhibition by perstraction: Utilization of concentrated lactose/whey permeate by *Clostridium acetobutylicum* to enhance butanol fermentation economics, *Trans. IChemE, Part C, Food and Bioproducts Processing*, 83, 43–52 (2005).

N. Qureshi, I.S. Maddox, and A. Friedl, Application of continuous substrate feeding to the ABE fermentation: relief of product inhibition using extraction, perstraction, stripping and pervaporation, *Biotechnol. Prog.*, 8, 382–390 (1992).

N. Qureshi, M.M. Meagher, J. Huang, R.W. Hutkins, Acetone butanol ethanol (ABE) recovery by pervaporation using silicate-silicone composite membrane from fed-batch reactor of *Clostridium acetobutylicum*, *J. Membrane Sci.*, 187, 93–102 (2001).

N. Qureshi, M.M. Meagher, and R.W. Hutkins, Recovery of butanol from model solutions and fermentation broth using silicalite/silicone membrane, *J. Membr. Sci.*, 158, 115–125 (1999).

N. Qureshi, B.C. Saha, and M.A. Cotta, Butanol production from wheat straw hydrolysate using *Clostridium beijerinckii*, *Bioprocess Biosyst. Eng.*, 30, 419–427 (2007).

H. Ren, Q. Wang, X. Zhang, R. Kang, S. Shi, and W. Cong, Membrane fouling caused by amino acid and calcium during bipolar membrane electrodialysis, *J. Chem. Technol. Biotechnol.*, 83, 1551–1557 (2008).

S.R. Roffler, H.W. Blanch, and C.R. Wilke, In-situ recovery of butanol during fermentation, Part 1: batch extractive fermentation, *Bioprocess Eng.*, 2, 1–12 (1987a).

S.R. Roffler, H.W. Blanch, and C.R. Wilke, In-situ recovery of butanol during fermentation, Part 2: fed-batch extractive fermentation, *Bioprocess Eng.*, 2, 181–190 (1987b).

S.R. Roffler, H.W. Blanch, and C.R. Wilke, In situ extractive fermentation of acetone and butanol, *Biotechnol. Bioeng.*, 31, 135–143 (1988).

T. Sano, H. Yanagishita, Y. Kiyouzumi, F. Mizukami, and H. Haraya, Separation of ethanol/water mixture by silicalite membrane on pervaporation, *J. Membr. Sci.*, 95, 221–228 (1994).

S.L. Schmidt, M.D. Myers, S.S. Kelley, J.D. McMillan, and N. Padukone, Evaluation of PTMSP membranes in achieving enhanced ethanol removal from fermentations by pervaporation, *Appl. Biochem. Biotechnol.*, 63–65, 469–482 (1997).

K.R. Seddon, Review: Ionic liquids for clean technology, *J. Chem. Tech. Biotechnol.*, 68, 351–356 (1997).

A. Senol, Influence of diluent on amine extraction of pyruvic acid using Alamine system, *Chem. Eng. Processing: Process Intensification*, 45, 755–763 (2006).

A. Senol, Effect of diluent on Amine extraction of acetic acid: modeling considerations, *Ind. Eng. Chem. Res.*, 43, 6496–6506 (2004).

M. Siebold, P. Vonfrieling, R. Joppien, D. Rindfleisch, K. Schugerl, and H. Roper, Comparison of the production of lactic acid by three different Lactobacilli and its recovery by extraction and electrodialysis, *Process Biochem.*, 30, 81–95 (1995).

A. Srivastava, P.K. Roychoudhury, and V. Sahai, Extractive lactic acid fermentation using ion-exchange resin, *Biotechnol. Bioeng.*, 39, 607–613 (1992).

D. Stark, and U. von Stockar, In situ product removal (ISPR) in whole cell biotechnology during the last twenty years, *Adv. Biochem. Eng./Biotechnol*. **80**, 149–175 (2003).

H. Strathmann, *Ion-exchange Membrane Separation Processes*, Elsevier, Amsterdam, 2004.

J.A. Tamada, and C.J. King, Extraction of carboxylic acids with amine extractants. 3. Effect of temperature, water coextraction, and process considerations, *Ind. Eng. Chem. Res.*, 29, 1333–1338 (1990).

A. Tay, Production of L(+)-lactic acid from glucose and starch by fermentations with immobilized cells of *Rhizopus oryzae*. Ph.D. thesis, Ohio State University, Columbus, Ohio, 2002.

V.H. Thang, W. Koschuh, K.D. Kulbe, and S. Novalin, Detailed investigation of an electrodialytic process during the separation of lactic acid from a complex mixture, *J. Membrane Sci.*, 249, 173–182 (2005).

A. Thongsukmak, and K.K. Sirkar, Pervaporation membranes highly selective for solvents present in fermentation broth, *J. Memb.Sci*., 302, 45–48 (2007).

N. Tik, E. Bayraktar, and Ü. Mehmetoglu, In situ reactive extraction of lactic acid from fermentation media, *J. Chem. Technol. Biotechnol.*, 76, 764–768 (2001).

S.L.I. Toh, J. McFarlane, C. Tsouris, D.W. DePaoli, H. Luo, and S. Dai, Room-temperature ionic liquids in liquid-liquid extraction: effects of solubility in aqueous solutions on surface properties, *Solvent Extraction and Ion Exchange*, 24, 33–56 (2006).

L.M. Vane, Review: A review of Pervaporation for product recovery from biomass fermentation processes, *J. Chem. Technol. Biotechnol.*, 80, 603–629 (2005).

L.M. Vane, Separation technologies for the recovery and dehydration of alcohols from fermentation broths, *Biofuels, Bioprod. Bioref.*, 2, 553–588 (2008).

V.V. Volkov, A.G. Fadeev, V.S. Khotimsky, E.G. Litvinova, YaA Selinskaya, J.D. McMillan, and S.S. Kelley, Effects of synthesis conditions on the pervaporation properties of poly[−1-(trimethylsilyl)-1-propyne] useful for membrane bioreactors, *J. Appl. Polym. Sci.*, 91, 2271–2277 (2004).

V.V. Volkov, V.S. Khotimsky, E.G. Litvinova, A.G. Fadeev, YaA Selinskaya, N.A. Plate, J.D. McMillan, and S.S. Kelley, Development of novel poly[−1-(trimethylsilyl)-1-propyne] based materials for pervaporation separation in biofuel production, *Polym. Mater. Sci. Eng.*, 77, 339–340 (1997).

P. Vonktaveesuk, M. Tonokawa, and A. Ishizaki, Simulation of the rate of L-lactate fermentation using *Lactococcus lactis* IO-1 by periodic electrodialysis, *J. Ferment. Bioeng.*, 77, 508–512 (1994).

D.L. Vrana, M.M. Meagher, R.W. Hutkins, and B. Duffield, Pervaporation of model acetone-butanol-ethanol fermentation product solutions using Polytetrafluoroethylene membranes, *Sep. Sci. Technol.*, 28, 2167–2178 (1993).

J.L. Wang, X.H. Wen, and D. Zhou, Production of citric acid from molasses integrated with in-situ product separation by ion-exchange resin adsorption, *Bioresource Technol.*, 75, 231–234 (2000).

Y. Wang, C. Huang, and T. Xu, Which is more competitive for production of organic acids, ion-exchange or electrodialysis with bipolar membranes? *J Membrane Sci.*, 374, 150–156 (2011a).

Y. Wang, N. Zhang, C. Huang, and T. Xu, Production of monoprotic, diprotic, and triprotic organic acids by using electrodialysis with bipolar membranes: effect of cell configurations, *J Membrane Sci.*, 385–386, 226–233 (2011b).

Y. Wang, X. Zhang, and T. Xu, Integration of conventional electrodialysis and electrodialysis with bipolar membranes for production of organic acids, *J Membrane Sci.*, 365, 294–301 (2010).

Z. Wang, Y. Luo, and P. Yu, Recovery of organic acids from waste salt solutions derived from the manufacture of cyclohexanone by electrodialysis, *J. Membrane Sci.*, 280, 134–137 (2006).

K.L. Wasewar, V.G. Pangarkar, A.B.M. Heesink, and G.F. Versteeg, Intensification of enzymatic conversion of glucose to lactic acid by reactive extraction, *Chem. Eng. Sci.*, 58, 3385–3393 (2003).

M. Wayman, and R. Parekh, Production of acetone-butanol by extractive fermentation using dibutylphthalate as extractant, *J. Ferment. Technol.*, 65, 295–300 (1987).

Y.-J. Wee, J.-S. Yun, Y.Y. Lee, A.-P. Zeng, and H.-W. Ryu, Recovery of lactic acid by repeated batch electrodialysis and lactic acid production using electrodialysis wastewater, *J. Biosci. Bioeng.*, 99, 104–108 (2005).

G.T. Wei, Z. Yang, and C.J. Chen, Room temperature ionic liquid as a novel medium for liquid/liquid extraction of metal ions, *Analytica Chimica Acta*, 488, 183–192 (2003).

S. Wieczorek, and H. Brauer, Continuous production of citric acid with recirculation of the fermentation broth after product recovery, *Bioprocess Eng.*, 18, 75–77 (1998).

J. Wiśniewski, G. Wiśgniewska, and T. Winnicki, Application of bipolar electrodialysis to the recovery of acids and bases from water solutions, *Desalination* 169, 11–20 (2004).

D.R. Woods, The genetic engineering of microbial solvent production, *Trends Biotechnol.*, 13, 259–264 (1995).

Z. Wu, and S.T. Yang, Extractive fermentation for butyric acid production from glucose by *Clostridium tyrobutyricum*, *Biotechnol. Bioeng.*, 82, 93–102 (2003).

T. Xu, Ion exchange membranes: State of their development and perspective, *J Membrane Sci.*, 263, 1–29 (2005).

L. Xuemei, L. Jianping, L. Mo'e, and C. Peilin, L-lactic acid production using immobilized *Rhizopus oryzae* in a three-phase fluidized-bed with simultaneous product separation by electrodialysis, *Bioprocess Eng.*, 20, 231–237 (1999).

S.T. Yang, H. Huang, A. Tay, W. Qin, L. De Guzman, and E.C. San Nicolas, Extractive fermentation for the production of carboxylic acids, in S.T. Yang (ed.), *Bioprocessing for Value-Added Products from Renewable Resources*, Elsevier, pp. 421–446 (2006).

X. Yang, G.J. Tsai, and G.T. Tsao, Enhancement of in situ adsorption on the acetone-butanol fermentation by *Clostridium acetobutylicum*, *Separation Technol.*, 4, 81–92 (1994).

X. Yang, and G.T. Tsao, Enhanced acetone-butanol fermentation using repeated fed-batch operation coupled with cell recycle by membrane and simultaneous removal of inhibitory products by adsorption, *Biotechnol. Bioeng.*, 47, 444–450 (1995).

S.T. Yang, S.A. White, and S.T. Hsu, Extraction of carboxylic acids with tertiary and quaternary amines: effect of pH, *Ind. Eng. Chem. Res.* 30, 1335–1342 (1991).

P.X. Yao, and K. Toda, Lactic acid production in electrodialysis culture, *J. Gen. Appl. Microbial.*, 36, 111–125 (1990).

B. Zelić, and D. Vasić-Rački, Process development and modeling of pyruvate recovery from a model solution and fermentation broth, *Desalination* 174, 267–276 (2005).

K. Zhang, M. Wang, and C. Gao, Tartaric acid production by ion exchange resin-filling electrometathesis and its process economics, *J Membrane Sci.*, 366, 266–271 (2011).

K. Zhang, M. Wang, and C. Gao, Ion conductive spacers for the energy-saving production of the tartaric acid in bipolar membrane electrodialysis, *J Membrane Sci.*, 387–388, 48–53 (2012).

H. Zhao, S. Xia, and P. Ma, Review: use of ionic liquids as "green" solvents for extractions, *J. Chem. Technol. Biotechnol.*, 80, 1089–1096 (2005).

16

Reactive Distillation for the Biorefinery

Aspi K. Kolah, Carl T. Lira and Dennis J. Miller

Department of Chemical Engineering and Materials Science, Michigan State University, USA

16.1 Introduction

Process intensification is a paradigm shift from conventional process design that leads to lower manufacturing costs, smaller equipment size and capital investment, higher energy efficiency, safer plant design, and reduced waste production that results in a lower carbon emission footprint. At its core, process intensification combines two or more unit operations that require novel process equipment and strategies with the above advantages over conventional chemical manufacturing techniques. Combining reactions with separation leads to reactive separation processes that include reactive distillation, reactive extraction, membrane reactors, oscillating flow reactors, spinning tube reactors, and fuel cells. Generally, new chemical conversion pathways or catalyst development are not considered as process intensification, but more efficient ways of carrying out chemical conversions are.

Reactive distillation is a unit operation in which chemical reaction and distillation occur simultaneously in a single piece of equipment. The reaction can be either homogeneously or heterogeneously catalyzed. The subset of heterogeneously catalyzed reactive distillation processes are commonly referred to as catalytic distillation processes. This technology of integrating reactors and separators has made enormous progress since it was first reported by Backhaus [1] in 1921. The first known industrial commercialization of reactive distillation was reported by Shell Chemicals in 1953, followed by Eastman Chemicals methyl acetate process in 1980 [2]. There are currently more than 150 reactive distillation processes operating on a commercial scale worldwide [2].

16.1.1 Reactive distillation process principles

Excellent reviews on reactive distillation have been presented by Terrill *et al.* [3], Sharma [4], Doherty and Buzad [5], Stichlmair and Frey [6], Sakuth *et al.* [7], Taylor and Krishna [8], Malone and Doherty [9],

Separation and Purification Technologies in Biorefineries, First Edition.
Edited by Shri Ramaswamy, Hua-Jiang Huang, and Bandaru V. Ramarao.

Mahajani and Chopade [10], Sundmacher and Kienle [11], Hiwale *et al*. [12], Harmsen [2] and Luyben and Yu [13]. Sundmacher *et al*. [14] have attempted to classify reactive distillation systems.

16.1.2 Motives for application of reactive distillation

The motives for applying reactive distillation to a particular chemical system may stem from either the properties of the chemical reaction or those of the phase equilibriums of the species involved. The most common properties of reaction or separation that drive use of reactive distillation are given in Sections 16.1.2.1 and 16.1.2.2.

16.1.2.1 Reaction properties

Equilibrium limitations—Reactions that are thermodynamically limited can be carried out to give a high conversion by continuously removing one product from the system as the reaction progresses. This is the main reason for the use of reactive distillation in a large number of chemical systems such as esterification, where water or ester can be selectively removed in the distillate stream as products of reaction.

Reactive products—Selectivity to the desired product of a reaction may be increased by removing the desired product from the reaction mixture as it forms in order to prevent further reaction.

Exothermicity—The use of reaction heat to drive distillation can lead to significant energy savings. The dissipation of such heat can control reaction temperature and thus avoid an unfavorable shift in the chemical reaction equilibrium, a reduction in selectivity, accelerated catalyst degradation, and reactor hot spots. Endothermic reactions are not precluded from reactive distillation, but the heat for reaction and vaporization must be introduced via the reboiler or preheated feeds.

16.1.2.2 Separation properties

Close boiling mixtures: Two species that have very close boiling points can be separated by reaction of one of them to a product with different volatility. An example is the separation of cyclohexane and cyclohexene, which have a relative volatility close to unity and are thus difficult to separate by conventional distillation. Cyclohexene reacts with a carboxylic acid to form an ester, which is easily separated from cyclohexane.

Presence of azeotropes: Many conventional processes lead to formation of azeotropes between the various constituents involved. Reactive distillation can facilitate one component of an azeotrope being reacted away, thus leading to considerably simpler phase equilibrium for the system and achievement of high-quality purification of a product.

High-purity product requirement: In some systems, small quantities of impurities can be difficult to separate by conventional means. Reactive distillation can remove these impurities efficiently via reaction, giving a high-purity product stream. An example of this approach is the hexamethylene diamine/water separation (in the Nylon 6,6 process) using adipic acid as the reactive entrainer.

16.1.3 Limitations and disadvantages of reactive distillation

There are several limitations of reactive distillation as a process strategy. For reactive distillation to be effective, reactions of interest must occur at temperatures and pressures that are compatible with the desired distillation. In the ideal case, the volatilities of the reactants and products should be such that the concentration of the reactants in the liquid phase inside the reaction zone should be as high as possible and the concentration of the products should be minimized. Second, reactive distillation cannot be applied for gas–liquid reactions that require very high temperature and pressure. Third, reactive distillation is difficult to implement for very slow reactions; the extended catalyst contact times required result in large column sizes that are expensive to build and challenging to operate. Reduction in overall reaction selectivity can also occur as a result of undesired removal of one of the reactive components because of volatility, leading to side reactions not observed in single fluid phase reactors. Finally, maintaining proper chemical gradients to drive reactions forward can be challenging because of mass transfer or volatility issues.

From a catalyst standpoint, catalyst life should be long in solid-catalyzed reactive distillation processes in order to avoid frequent plant shutdowns for catalyst replacement. Finally, extended development time for design, laboratory- and pilot-scale testing is typically required because of the complexity of simultaneous reaction and separation. The implementation of reactive distillation requires corporate commitment to tolerate extended development time and additional development costs, and to overcome inherent reluctance to initiate new approaches in processing.

16.1.4 Homogeneous and heterogeneous reactive distillation

Reactive distillation can be carried out using either homogeneous catalysts such as sulfuric acid dissolved into the reaction media, or using solid heterogeneous catalysts such as ion exchange resins (e.g., Amberlyst-15). The use of heterogeneous catalysts for reactive distillation has the inherent advantages of avoiding the need for catalyst recovery and recycling, and of precisely defining the position and height of the reaction zone(s) in a reactive distillation column. Homogeneous catalysts may also give rise to other challenges such as corrosion or environmental issues. The primary challenge in using heterogeneous catalysis is the requirement to maintain catalyst activity for extended time periods (6 months to 2 years)—this is because the replacement of the heterogeneous catalyst is expensive and requires shutdown of the column. If catalyst life expectancy is in question, homogeneous catalysts are used because they are continuously introduced and removed from the column.

16.2 Column internals for reactive distillation

The means by which solid, heterogeneous catalysts are held in a reactive distillation column remain among the most important considerations in the application of a viable reactive distillation system. In most cases, the catalyst is incorporated into the structure responsible for ensuring adequate gas–liquid and liquid–solid contact within the column. This structure should exhibit low pressure drop per unit length, high separation efficiency (small height of a theoretical plate (HETP)) via extensive gas–liquid interfacial contact, and good liquid-phase radial distribution, holdup, and solid wetting to optimize reaction rate on the catalyst. In addition to these requirements, the structure containing the catalyst must have high mechanical, thermal, and chemical stability, and the cost must be reasonable.

Within the catalyst structure, individual catalyst particles must have properties that are generally desirable for all heterogeneous chemical reactions. They must be small enough (1–3 mm) to avoid intraparticle diffusion effects, must possess appropriate pore structure and surface area, must have sufficient activity and selectivity for the desired reaction, and must be easily handled to facilitate installation and change-out. For resin-type catalysts, particles should also exhibit low osmotic swelling forces in order to limit breakage of the polymer structure and subsequent reduction of void fraction in the catalyst bed.

The open literature is replete with methods describing techniques for immobilizing the heterogeneous catalyst in the distillation column. Towler and Frey [15] have presented an excellent review of catalyst packing used in RD columns. A similar review by Taylor and Krishna [8] also discusses in detail various hardware alternatives for catalytic packings. In addition to the catalytic packings described in this article, numerous proprietary packings are available from corporate entities.

Keeping in view the requirements posed on catalytic packings for reactive distillation, a general structure of three scales of porosity for packings has been described by Krishna and Sie [16]: (i) micron scale—pores inside the catalyst particles providing access to active sites and the high surface area desired for reaction; (ii) millimeter scale—pores between the catalyst particles held inside the catalytic packing structure, providing channels for liquid access to the catalyst by gravity flow and (iii) centimeter scale—channels between or built into the catalyst packing structure, providing channels for gas flow and gas–liquid contact in the reactive distillation column.

The catalyst and catalyst support structure must be properly designed and constructed in order for the reactive distillation column to operate properly. A summary of the different types of catalyst structures used in reactive distillation are given below.

16.2.1 Random or dumped catalyst packings

Random packings for reactive distillation take one of two forms. In the first, the actual catalyst is formed into macroscopic structures that are placed in the column as the catalytic section. This form is essentially the same as a fixed catalyst bed in a heterogeneous reactor. The second form is the use of small containers or "bags" made of plastic or metal mesh into which catalyst is placed and sealed. These bags are then dumped or placed randomly into the column to form the reactive bed. Random packings are often used for reactive distillation columns because they are cheaper than structured packings and easier to handle. Some types of random packings achieve higher separation efficiencies than structured packings.

Different manufacturing technologies exist for the synthesis of random packings, such as sintering of polymers [17], melting of polymers [18], block polymerization [19, 20] and precipitation polymerization [21, 22]. Figure 16.1 shows Raschig rings constructed of ion exchange resin for use in reactive distillation [23].

16.2.2 Catalytic distillation trays

Another way of introducing catalyst into a reactive distillation column is through use of catalytic distillation trays. An essential challenge of catalytic distillation trays is providing a quantity of catalyst sufficient for

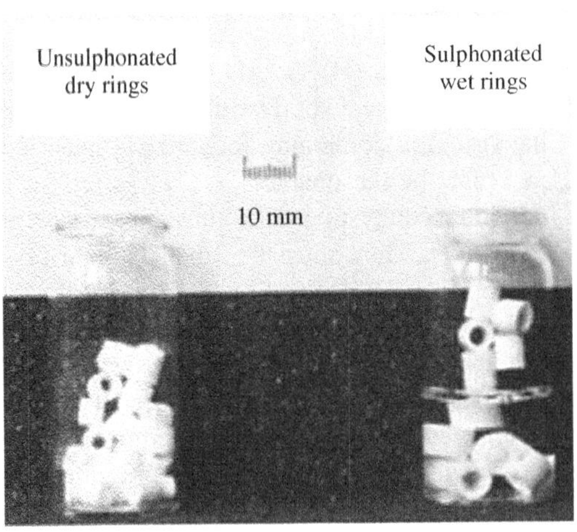

Figure 16.1 Raschig ring dumped packings constructed of ion-exchange resin polymer. Reproduced with permission from [23] © 1992 VCH Verlagsgesellschaft mbH

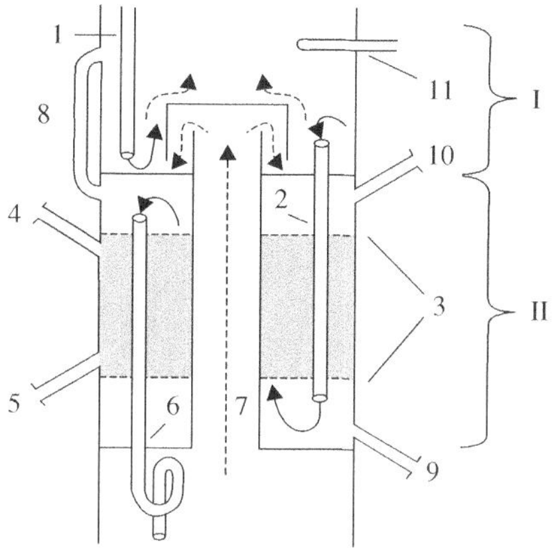

Figure 16.2 Novel D+R reactive distillation column tray. Reproduced with permission from [33] © 2011 Pergamon, Institution of Chemical Engineers. (I) Distillation section (D), (II) reactive section (R), (1), (2) downcomers, (3) sieves, (4) upper catalyst port, (5) lower catalyst port, (6) downcomer, (7) gas chimney, (8) venting pipe, (9) sampling port before the catalyst bed, (10) sampling port after the catalyst bed, (11) temperature measurement. The grey area indicates the catalyst bed

reaction without compromising pressure reduction or liquid flow capacity within the column. The design of catalytic trays for reactive distillation processes is very different from that used for normal distillation, and several configurations have been reported by various authors including Carland [24], Jones [25, 26], Marion *et al.* [27], Sanfilippo *et al.* [28], and Yeoman *et al.* [29–32]. A recent publication by Harbou *et al.* [33] describes a novel hybrid tray, referred to as a "D + R" tray, in which a conventional bubble cap tray is combined with a reactive distillation section (Figure 16.2).

16.2.3 Catalyst bales

Catalyst bales, also referred to as "teabag" packings, have been described by Smith [34, 35] and are widely used in industry for reactive distillation processes (C.D.Tech, now ABB Lummus) because they are inexpensive and readily replaceable. To produce bales, catalyst particles are sewn into fiberglass cloth pockets. The cloth pockets are rolled up and wrapped with alternating wire gauze layers to form cylindrical catalyst bales as shown in Figure 16.3 [36, 37]. These catalyst bales can be positioned onto trays of distillation columns and placed on top of each other to achieve a desired catalyst packing height. Fluid dynamic and mass transfer characteristics of catalytic bales have been presented by Manduca *et al.* [37] and Caetano *et. al.* [36] respectively.

16.2.4 Structured packings

Structured packings represent the most advanced means of placing catalysts within the reactive distillation column. Here, catalyst particles are sealed within corrugated wire gauze sheets to form envelopes that are bundled together in blocks (for large columns) or cylinders (for small columns) that fill the entire cross-section of the column. The bundles typically contain features such as wall "wipers" or additional corrugated metal sheets to ensure uniform liquid distribution and provide for the highest possible gas–liquid interfacial area and liquid–solid catalyst contact. Structured packings therefore provide the most desirable features of catalyst support combined with effective distillation characteristics.

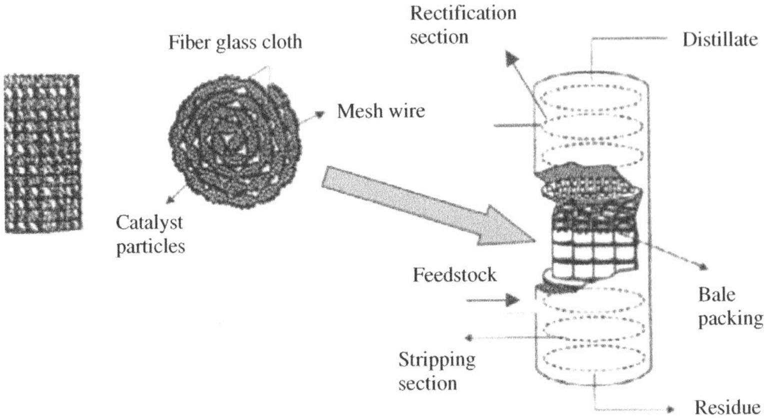

Figure 16.3 *Catalyst bales for supporting solid catalysts in reactive distillation [36]*

Katapak-S and Katapak-SP are commercially available from Sulzer Chemtech, Switzerland. The Katapak-SP11 packings for a 100 mm diameter column have the following geometrical properties—specific surface area of 210 m^2/m^3, a packing porosity of 0.74, porosity of spheres inside catalytic baskets of 0.385 and a volume fraction of catalyst baskets in the packing of 0.42 [38]. Figure 16.4 shows the structure of Katapak-SP11 and Katapak-SP12 [39].

Technical details of Katapak-S packings are given by Stringaro [39]. Hydrodynamic and mass transfer behavior for Katapak-S has been studied by Moritz and Hasse [40], Moritz *et al.* [41], Kolodziej *et al.* [42], and van Baten and Krishna [43]. For Katapak-SP packing, hydrodynamic and mass transfer characteristics have been reported by Behrens *et al.* [44–46], Ojulic and Beherns [47], Kolodziej *et al.* [48], Gotze *et al.* [49], and Viva *et al.* [50]. Computational fluid dynamics has been used to

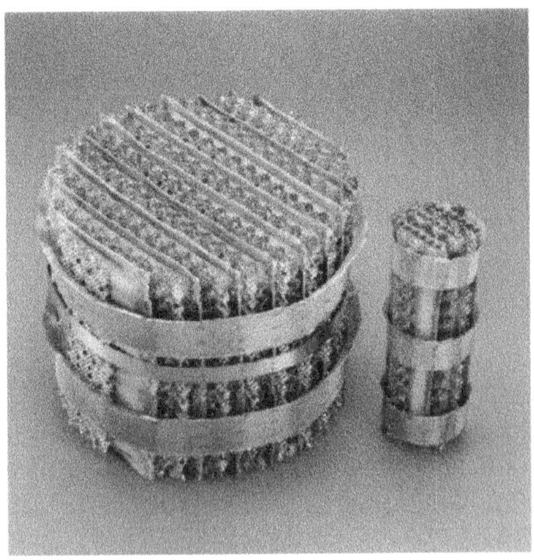

Figure 16.4 *Katapak-SP11 and Katapak-SP12 structured packing. Reprinted from [49] © 2001, with permission from Elsevier*

study radial and axial liquid dispersion in Katapak-S [51, 52] and in Katapak-SP [53]. Recently, x-ray tomography and CT scans have been utilized for investigation of hydrodynamics of Katapak-SP packings in reactive distillation columns [54–58]. Figure 16.5 shows the static and dynamic liquid holdups for the Katapak SP-11 and Katapak-SP12 packings [50]. Static liquid holdup indicates liquid loading inside the catalyst bed without gas flow.

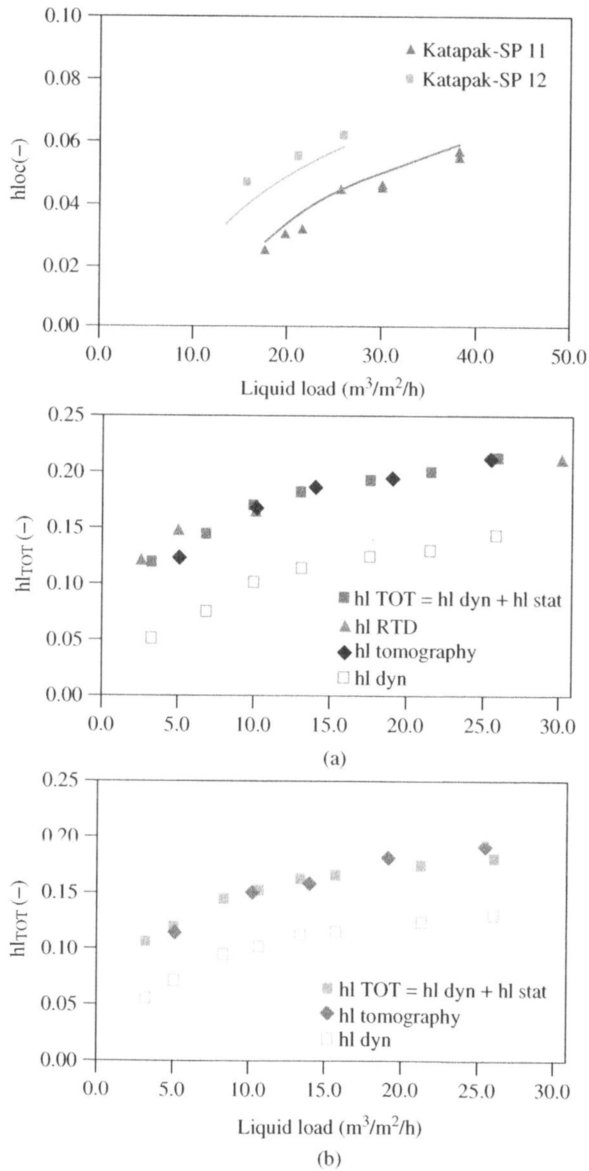

Figure 16.5 Open channel holdup (top) for Katapak-SP11 and Katapak-SP12. Total liquid holdup (hlT_{OT}) and dynamic holdup (hl dyn) for Katapak-SP11 (a, middle) and Katapak-SP12 (b, bottom). Reproduced with permission from [50] © 2011 Institution of Chemical Engineers

Figure 16.6 *An internally finned monolith catalyst for reactive distillation. Reprinted from [60] © 2005, with permission from Elsevier*

16.2.5 Internally finned monoliths

The use of internally finned monolith in reactive distillation is a relatively new concept. Details about internally finned monoliths can be obtained from Lebens *et al*. [59] and Schildhauer *et al* [60]. Figure 16.6 shows an internally finned monolith packing [60]. Schildhauer *et al*. [60], and Brinkmann *et al*. [61] describe the hydrodynamic modeling for reactive stripping columns using monolith packings.

16.3 Simulation of reactive distillation systems

The feasibility of reactive distillation as a process option is best explored with initial simulation of the process using modern software packages such as CHEMCAD and AspenPlus. Such simulations require significant effort to characterize the physical properties and reaction rates of the species present, and to choose the appropriate model for the process design. The steps involved in the mathematical simulation and design analysis of a reactive distillation process are broadly classified and described below.

16.3.1 Phase equilibria

Feasibility of separation in reactive distillation depends ultimately on relative volatility and/or azeotrope behavior. Phase-equilibrium characterization is essential early in the design process, first as a screening tool, and then in parallel with kinetics measurements described below. The most efficient reactive distillations involve reactants of intermediate volatility with one product that is more volatile and another that is less volatile than the reactants. This permits reaction in the center of a column, with one product leaving the top and the other leaving the bottom. This configuration is most efficient because the reactants tend to stay in the center of the column rather than exiting, and the desired product is separated by distillation from co-products, thus maximizing the effect of Le Châtelier's principle. However, other arrangements are feasible and are readily practiced. For example, in esterification with ethanol product water forms a low boiling ethanol-rich azeotrope that must be taken out the top of the column. Azeotropes can sometimes be broken inside the column by judicious choices. For example, in the Eastman Chemicals methyl acetate process [62], acetic acid is fed near the top of the column to "react away" the low-boiling azeotrope between methyl acetate and methanol, enabling a methyl acetate-rich distillate.

Screening of volatility and azeotropic behavior is facilitated by process design software. Data packages often contain parameters fitted to vapor–liquid equilibrium (VLE) or liquid-liquid equilibrium (LLE) data. Frequently, multifunctional bio-derived molecules have polar and non-polar functionality that make them only partially miscible with other species involved in the reactive distillation. For many bio-based species, the phase equilibrium data used to fit the parameters are not provided in the software package, so the researcher typically needs to obtain literature data for verification. A good open resource for data is the IUPAC-NIST solubility data series [63]. In cases where literature data are not available, UNIFAC can be used to screen compounds. UNIFAC should only be used for screening because the errors can be significant for unusual multifunctional compounds. If data are not available and the screening calculations are promising, then experimental phase equilibrium measurements are necessary to provide confidence in the property data. Activity coefficients fitted to phase equilibrium data are important for modeling distillation behavior in the column. The systems are typically non-ideal enough that activity coefficients must be incorporated into the kinetic rate laws (activity-based kinetics) and reaction equilibrium constants. Vapor–liquid equilibrium data can be collected with either static cells or circulating stills. Static cells operate at a fixed temperature, and the pressure is measured as the composition is varied [64]. Circulating stills operate at fixed pressure, and the bubble temperature is measured as composition is varied. Static cells are often user-built because of simplicity but are slower to produce data because the apparatus and liquids must be carefully degassed prior to measurement. Circulating stills are run at controlled pressure [65] with an inert gas blanket obviating the need for degassing, but stills have two limitations: (i) they must be carefully designed to eliminate superheating such as by use of an internal Cottrell pump and jacketed disengaging region; (ii) they often are unreliable when measuring systems with large relative volatilities, near 10 or above. Static cells have an advantage for thermally sensitive compounds or reactive systems because they can be operated at low temperature, often avoiding measureable reaction. For reactive systems, preventing reaction in a circulating still is more difficult because the apparatus must usually be heated to operate above about 150 mmHg for stable boiling. Liquid–liquid equilibrium data can typically be collected rapidly by submersing small vials in a bath. The phase equilibrium data collected by various methods can be fitted with activity coefficient models. The NRTL model is often the most flexible model, but even that model is challenged by simultaneous VLE and LLE, unless a large number of temperature-dependent parameters are introduced, or different parameters are used in different composition ranges.

16.3.2 Characterization of reaction kinetics

Following phase-equilibrium behavior, the estimation of the reaction rate at reactive distillation conditions for the chemical system of interest must be carried out. Most of the time this means liquid-phase experimental measurements of kinetics in stirred batch reactors with either dissolved homogeneous catalyst or suspended heterogeneous catalyst at conditions reflecting the range of bubble point temperatures and pressures expected in the column. For heterogeneous catalysts, the key outcome of these kinetic studies is specific rate—moles of key species converted per unit catalyst mass per unit time. The magnitude of this quantity defines the opportunity for reactive distillation. As liquid flow in distillation is the discontinuous phase and is gravity driven, there is a limited practical range of liquid velocity ($m^3/m^2/s$) that defines the range of contact time of liquid with solid catalyst at a given distillation "stage." If the rate of reaction is so slow that relatively little reaction occurs at a stage, then an RD column would have to have a number of stages, to accomplish the desired reaction, which far exceeds that required for separation. In this case, higher pressure and thus higher temperature is a first option; if this is not possible then conventional reactive distillation is impractical and the reaction should either take place in dedicated reactors with intermediate separation, or possibly using a novel approach such as sidedraw reactors on a conventional distillation column that could provide the required reaction time with a much larger quantity of catalyst than could be fit within a conventional distillation column.

If, on the other hand, kinetic measurements show that rate is very rapid, then it becomes reasonable to assume that chemical reaction approaches equilibrium on a given stage within the column. With this assumption, process simulation becomes relatively straightforward as long as the chemical equilibrium can be appropriately modeled, and reactive distillation is very attractive as a process option.

In many cases, a reaction may be fast enough to make reactive distillation feasible but not so rapid as to be able to assume chemical equilibrium at a stage. In this case, careful characterization of reaction kinetics under the range of conditions where reactive distillation will be run is required in order to develop a proper simulation of the column. When incorporating chemical kinetics into process simulators such as AspenPlus, either concentration, mole fraction, or activity-based kinetic models can be used to describe rates. For systems where thermodynamic nonidealities exist, activity-based models are essential to properly relate reaction behavior.

Reaction kinetic studies conducted in stirred batch systems must avoid external (liquid–solid) and intraparticle mass transport influences whenever possible. Liquid–solid mass transport may be minimized by ensuring complete catalyst suspension and demonstrating independence of rate from agitation speed, and intraparticle diffusion limitations are minimized at low temperature and small particle sizes. It is necessary to evaluate mass transport via calculation of effectiveness factor or the observable modulus as defined by Weiss [66]. One criterion for ensuring intrinsic kinetic measurements is that rate per unit mass catalyst is independent of the quantity of catalyst in the reacting phase.

When applying a kinetic model derived from a batch reactor to a reactive distillation column, there are several cautions that must be addressed as the column design emerges. In addition to making sure kinetics are evaluated in the range of operation of the column, incomplete catalyst wetting can lead to lower than expected reaction rates, and side reactions that are unimportant in batch systems can lead to production of species of intermediate volatility that become trapped in the column and build up over time.

16.3.3 Calculation of residue curve maps

The characterization of simultaneous chemical reaction and phase equilibrium behavior, recently reviewed by Frey and Stichlmair [67, 68], Seider and Widagdo [69], Marcelino *et al.* [70], is essential for any theoretical study of reactive distillation systems. Four possible modeling approaches can be taken in reactive distillation: both vapor–liquid equilibrium (VLE) and chemical equilibrium (CE) are assumed (equilibrium controlled reactions); VLE is not assumed but CE is assumed; VLE is assumed but CE is not assumed; and neither VLE nor CE is assumed. The concept of distillation line diagrams, commonly called residue curve maps (RCM), is often used. In an RCM, the change of liquid composition with time is plotted. Residue curve maps for reactive systems are often referred to as reactive residue curve maps, but are also denoted as RCM in this paper.

Developing an RCM, which maps out distillation boundaries and identifies feasible operating zones, is invariably the first step in determining whether a given system is suitable for reactive distillation. Once the zone of feasible operation is mapped out, it is possible to predict the distillate and reboiler product compositions.

Barbosa and Doherty [71–74] and Ung and Doherty [75] have studied RCM for the first case (above) where both VLE and CE are reached. The authors make use of transformed variables defined by

$$X_i = \frac{\left(x_i/\nu_i - x_k/\nu_k\right)}{\left(\nu_k - \nu_T x_k\right)} \quad Y_i = \frac{\left(y_i/\nu_i - y_k/\nu_k\right)}{\left(\nu_k - \nu_T y_k\right)} \tag{16.1}$$

where X_i and Y_i are the transformed liquid and vapor compositions, respectively, ν_i is the stoichiometric coefficient of each component, ν_T is the sum of stoichiometric coefficients in the reaction, and k denotes the reference component. Knowing the actual x_i and y_i compositions, the transformed compositions, X_i and Y_i, can be easily calculated. The reverse calculations require search algorithms. Singular or azeotropic points are defined by

$$X_i = Y_i \tag{16.2}$$

These points are commonly referred to as "reactive azeotropes" as they are points in a reactive distillation system where the concentrations of vapor and liquid phase cannot be changed via distillation.

The advantage of using transformed variables is that X_i has the same value before and after chemical reaction. For a simple batch distillation, the governing equations reduce to

$$\frac{dX_i}{d\tau} = X_i - Y_i \tag{16.3}$$

Integration of the earlier equations gives the RCM for the system.

Residual curve maps for slow, kinetically controlled reactions where CE is not reached but VLE is reached have been studied by Venimadhavan *et al.* [76] and Thiel *et al.* [77,78] for heterogeneously catalyzed systems. For these systems, the Damkohler number [79] is defined as the ratio of the characteristic liquid residence time to the characteristic reaction time and given as

$$Da = \frac{W/FC_{cat}}{1/kC_{cat}} \tag{16.4}$$

where W is the total amount of catalyst in the column (kg), F is the total feed rate to the column (kmol/hr), k is the equivalent first-order forward rate constant evaluated at a reference temperature, usually the boiling point of the lowest boiling pure component in the system (kmol/hr kg cat), and C_{cat} is the concentration of the catalyst (kg catalyst per kmol). A low Damkohler number indicates that reaction is kinetically limited, while a high Damkohler number indicates that the system approaches chemical equilibrium.

Tang *et al.* [80] have reported RCMs for five esterification systems, including ethyl acetate and n-butyl acetate formation. Daza *et al.* [81] have reported RCMs for the synthesis of ethyl lactate. Jiminez *et al.* [82] have reported RCMs for the transesterification of methyl acetate with n-butanol, and Orjuela *et al.* [83] have reported RCMs for butyl acetate synthesis. A residue curve map for the ternary system in biodiesel synthesis has been generated by Simasititkul *et al.* [84] and is shown in Figure 16.7.

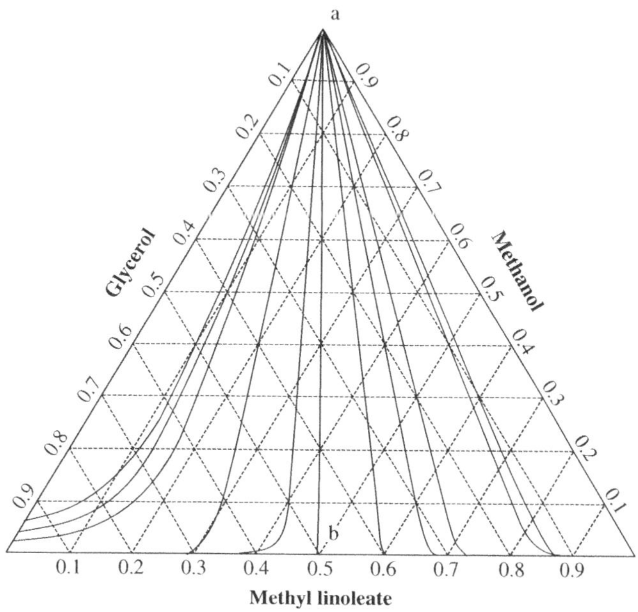

Figure 16.7 *Residue curve map for ternary system in biodiesel synthesis. Reference [84], with kind permission from Springer Science+Business Media © 2011*

16.3.4 Simulation and design of reactive distillation systems

Steady-state process design of reactive distillation column can be performed by either a simulation approach, where the column configuration is specified and the goal is to calculate the outlet product streams, or a design approach, where the desired outlet product streams are specified and the goal is to calculate a feasible column configuration. Simulation calculations for reactive distillation are based on either equilibrium stage modeling or rate-based modeling.

16.3.4.1 Equilibrium stage model

In this approach, the distillation column is treated as a series of equilibrium stages where the vapor from the stage below and the liquid from the stage above are brought into contact on the stage together with fresh or recycle feeds. The vapor and liquid streams leaving each stage are in thermodynamic equilibrium and finite reaction rates are accounted for in each reactive stage. Height equivalent to a theoretical plate (HETP) is used to translate the number of equilibrium stages to the actual column height. For steady-state equilibrium-based models, various methodologies have been adapted for solving the model equations, ranging from manual hand calculations performed tray-to-tray, to short cut methods, theta methods, tearing algorithms, relaxation methods, Newton's method, and homotopy continuation methods. Equilibrium-based models have been widely used to model reactive distillation processes; see Baur *et al*. [85], Peng *et al*. [86], Taylor and Krishna [8], Sundmacher and Kienle [11], and Taylor [87] for reviews on modeling of reactive distillation columns.

Commercially available software simulation packages such as ASPEN Plus, particularly the RADFRAC inside-out algorithm, Pro/II, and SpeedUp, *etc*. are used for obtaining steady-state behavior of reactive distillation columns.

READYS, a dynamic simulator for equilibrium based models, is described by Scenna *et al*. [88] Taylor and Krishna [8] provide a general overview of other approaches to dynamic simulation of reactive distillation systems using equilibrium models.

16.3.4.2 Rate-based model

For the modeling of slow reactions in reactive distillation, it is necessary to use kinetic expressions that are dependent on the temperature and concentration of the components involved. Normal distillation and reactive distillation processes almost never operate in complete equilibrium. To overcome the problems associated with use of distillation efficiencies, non-equilibrium rate-based models are used.

Rate-based models are set up differently for heterogeneously or homogeneously catalyzed systems. Using rate-based models, heterogeneous reactions can either be considered as pseudo-homogeneous, where a lumped term is considered for catalyst diffusion and reaction, or rigorous, with explicit treatment of diffusion and reaction, as with the dusty gas model [8].

Rate-based models make use of finite mass transfer coefficients that are either calculated from fundamental mass transfer models or are determined experimentally for reactive distillation packings [8]. In rate-based models, mass transfer coefficients, interfacial area, liquid holdup, and so forth, are hardware dependent. Moreover, calculations of mass transfer driving forces in rate-based models require thermodynamic properties of the reaction system.

Commercial software such as Aspen Plus has the RATEFRAC module for solving rate-based reactive distillation problems such as those described by Pinjala *et al*. [89]. Zheng and Xu [90] have used a MERQ rate-based model from the work of Taylor and Krishna [91], where the vapor and liquid phases have individual material and energy balances. The model equations have been solved using the Newton–Raphson method. ChemSep, a non-equilibrium based computational model that can be used to simulate reactive distillation processes, is available for academic and commercial use from Clarkson University [92].

Sundmacher and Hoffmann [93] present a detailed non-equilibrium model for packed columns, taking into account transport processes inside the porous catalytic packings. The porous catalytic packings are totally wetted internally and externally [94]. Mass and energy transport occur across vapor–liquid and solid–liquid interfaces.

Most non-equilibrium models cannot account for flow patterns of vapor and liquid on tray columns or for liquid maldistribution in packed columns. Higler *et al.* [95–97] developed a non-equilibrium cell model to describe the mass transfer, reaction, and flow patterns on a catalytic distillation tray. On each tray, the two-phase mixture is split up into a number of contacting cells with flexibility of interconnection. The connection pattern and backmixing characteristics can be chosen to study flow patterns and maldistribution adequately. Mass transfer resistances are located in the vapor–liquid and liquid–solid films. The Stephan-Maxwell equations are used for calculation of mass transfer coefficients. Mass transfer inside the porous catalyst is characterized using the dusty gas model.

16.3.4.3 Design of reactive distillation systems

Design algorithms for reactive distillation systems are mainly based on equilibrium stage models. Barbosa and Doherty [98, 99] developed the fixed-point method for the design of single- and double-feed reactive distillation columns. Mahajani and Kolah [100] extended the methods of Barbosa and Doherty for design of packed reactive distillation columns. An excellent review on design methods has been given by Taylor and Krishna [8].

16.4 Reactive distillation for the biorefinery

The emergence of biomass as a raw material for the production of fuels and chemicals provides many opportunities for application of reactive distillation. This is, in part, because the feedstocks, intermediates, and products of biomass refining have high oxygen content and thus significant functionality. Many of these functional groups undergo chemical reactions that are reversible or thermodynamically limited, making them candidates for reactive distillation. In this section, we review recent activities in the open literature pertaining to reactive distillation processes for the biorefinery. The review is broken down according to the types of functional groups and reactions taking place, and describes both the accomplishments and potential opportunities for further work in the area.

There are several classes of chemical reactions for which reactive distillation has found significant application. These are (i) esterification of carboxylic acids to alcohols and transesterification to exchange the alcohol functionality on the ester—without question the most widely investigated reaction systems to which reactive distillation has been applied; (ii) etherification, upon which much of the reactive distillation technology existing today has been built through the large-scale production of the now-banned gasoline additive methyl tert-butyl ether (MTBE); and (iii) acetal formation, which has proven useful for recovering and purifying polyhydroxyl compounds and for forming fuel additives as substitutes for ethers. These reactions are reviewed in the sections that follow, along with a brief mention of recent reactive distillation applications in other areas including thermochemical conversion routes.

16.4.1 Esterification of carboxylic acids and transesterification of esters

Carboxylic acids are the largest single class of chemicals made from biomass resources and provide opportunities for use as platform chemicals for synthesis of numerous intermediates and products.

Esterification is the reaction of a carboxylic acid with an alcohol to form an ester and liberate water (Eq. 16.1). This reaction is very commonly carried out industrially and is characterized by a reaction equilibrium constant that is typically on the order of 0.1 to 10.

$$R - COOH + R' - OH = R - COO - R' + H_2O \tag{16.5}$$

Transesterification is the exchange of the alcohol functionality of an ester with another alcohol; the reaction is again used industrially and typically has an equilibrium constant on the order of unity.

$$R - COOR' + R'' - OH = R - COOR'' + R' - OH \tag{16.6}$$

There are numerous fermentation processes in existence and under development that produce organic acids from carbohydrate sugars in either free acid form, or more often, in salt form in concentrations ranging from 10 to 150 g/L. Often, a part of the recovery of acids from the fermentation broth involves esterification, through which the acids are volatilized, separated from the fermentation residue, and purified by distillation. A second major sources of carboxylic acids and esters (referred to in this case as fatty acids and esters) is from plant oils or triglycerides, the fatty acid tri-ester of glycerol (1,2,3-propanetriol), which are the primary starting points for biodiesel production via transesterification. From either source, organic acid esters have commodity-scale applications as fuels, monomers, solvents, and food/flavor/cosmetic ingredients, and thus play a large and varied role in the biorefinery.

Many of the alcohols that are commonly used for esterification can also be produced from biomass resources. These routes include fermentation to ethanol, butanol, and isobutanol, and thermochemical conversion via synthesis gas to methanol and higher alcohols.

The following paragraphs highlight recent work in esterification using reactive distillation. The review is not intended to be exhaustive but is up to date, so older work may be identified from references listed in the papers included.

16.4.1.1 Biodiesel production

Kiss *et al*. [101–104] have presented a broad perspective on the potential for biodiesel production via transesterification with methanol in continuous reactive distillation. Their approach involves the use of acid catalysts, including metal oxides [104] to facilitate the transesterification reaction. A number of other authors [105–108] have also investigated and proposed designs for biodiesel production via continuous reactive distillation processes using acid catalysts. Figure 16.8 shows one such design proposed by He *et al*. [105]

With acid catalysts, the challenge is achieving high enough reaction rates to attain nearly complete conversion in a reasonably sized distillation column. Qui *et al*. [107] present a summary review of the various continuous processing strategies proposed for biodiesel production. Two experimental studies, by Mueanmas *et al*. [109] and Salis *et al*. [110], have examined the transesterification reaction using triolein (triglyceride of oleic acid (C18:1)), and achieved reasonable conversions in both cases. In another study by Simasititkul *et al*. [84], simulation of the RD column showed good conversion to the methyl ester product.

In an attempt to circumvent the slow reaction rate of triglyceride transesterification with solid acid catalysts, an alternative two-step approach to biodiesel production has been proposed [111–114] that involves (i) hydrolysis of triglycerides to liberate free fatty acids and glycerol from triglycerides, and then (ii) esterification of free fatty acids with methanol to form biodiesel. Essentially quantitative conversion can be obtained via reactive distillation through proper column design with this approach. Bart *et al*. [115], in a recent book on biodiesel production, review the various reactive distillation technologies reported in

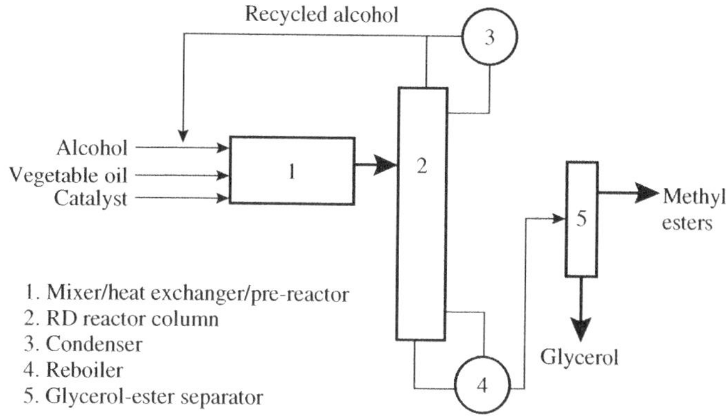

Figure 16.8 Flow diagram for biodiesel synthesis via reactive distillation [105]

the literature. Cossio-Vargas *et al.* [116] have mathematically studied the thermal coupling of reactive distillation sequences using side rectifier, side stripper, and fully thermally-coupled Petlyuk columns.

16.4.1.2 Esterification of long-chain fatty acids

The direct esterification of free fatty acids is an important pretreatment step when using waste oils as feedstocks for conventional (base-catalyzed) biodiesel production [117], and for the production of fatty-acid esters in general that have applications in consumer and industrial products. While earlier work on fatty-acid esters defined approaches, both in experiments and simulations [118–120], more recent studies of fatty-acid esterification have focused on energy efficiency and yield. Nghi [121] reported a 27% reduction in process energy requirements via heat integration of the esterification column and methanol recovery column in fatty acid methyl laurate production. Hernandez *et al.* [122] and Castro *et al.* [123] report a 54% reduction in energy requirements relative to a conventional process by integrating two reactive distillation columns in a supercritical methanol esterification process. Kiss *et al.* [124] used simple internal heat integration to achieve up to a 45% reduction in energy use in fatty acid esterification.

Fatty acid esterification in reactive distillation has been examined for several other alcohols as well. Dimian *et al.* [111] studied the simultaneous fatty acid esterification with mixed ethanol and 2-ethylhexanol. They found that the arrangement of feed alcohols is important to achieving high conversion and efficient use of column trays—in particular, the heavy alcohol, 2-ethylhexanol, is fed to the condenser of the column to produce an organic phase that partially dissolves some of the ethanol coming overhead with water. This phase is refluxed into the column, while the aqueous phase is withdrawn. Isopropyl palmitate synthesis via reactive distillation [125–127] has also been examined as an attractive biofuel component, because branch-chained esters have better cold-weather properties than methyl esters. Both experimental and simulation studies were conducted; simulations consistently over predicted experimental conversions but predicted trends with process parameters well. Over 99% yield of isopropyl palmitate can be achieved in a continuous reactive distillation process.

16.4.1.3 Lactate esterification

Lactic acid is the prototypical bio-renewable monomer. It is produced in fermentation at concentrations of 100–150 g/liter; recovery and purification costs can be as much as 50% of total lactic acid cost. One route to recovery of lactic acid is esterification via reactive distillation, followed by hydrolysis in a second

RD column [128–131]. Methanol is the alcohol of choice for the purification process, because it does not form an azeotrope with water and therefore can be purified and recycled. Significant energy conservation is possible via appropriate heat integration [131] of the esterification, hydrolysis, and methanol purification columns, and appropriate control schemes have been proposed by Liu *et al*. [130]. Both batch [128] and continuous [129–131] processes have been shown to be viable for lactic acid recovery and purification.

In addition to the esterification of lactic acid with methanol for purification, ethanol [132, 133] and butanol [134] have been investigated. Because lactic acid oligomerizes as its concentration in aqueous solution is increased, the formation of oligomers must be accounted for in esterification via reactive distillation. An example of a reactive distillation process for ethyl lactate production is given in Figure 16.9.

Lactic acid is fed to the top of the reactive zone in aqueous solution ranging from 20 wt% to 88 wt%. Ethanol is fed near the bottom of the reactive zone of the column, and moves up the column as the most volatile species. As lactic acid is esterified, the water of reaction as well as water fed with lactic acid moves to the vapor phase and exits the top of the column along with excess ethanol. The more highly concentrated lactic acid tends to oligomerize as it moves down the column, leading to a distribution of lactic acid oligomers and their ethyl esters, which exit the bottom of the RD column. These oligomers can be broken down via alcoholysis with ethanol in a subsequent RD column, leading to an essentially 100% yield of ethyl lactate [132]. Esterification with butanol is carried out in a similar fashion, with one important advantage being that a decanter is used with the condenser at the top of the column [134]. Butanol and water form two phases upon condensation; the heavier water phase is removed from the column and the butanol-rich phase is refluxed. In this way, stoichiometric quantities of butanol can be fed to the column and a high concentration of butanol can be maintained in the column.

16.4.1.4 Short-chain organic acid esterification

There are many examples of reactive distillation applied to esterification of short-chain carboxylic acids produced from renewable feed stocks. Many of the esters formed have commercial importance in the near term, and others are of particular interest because of their phase equilibrium properties. A brief overview of key systems is provided.

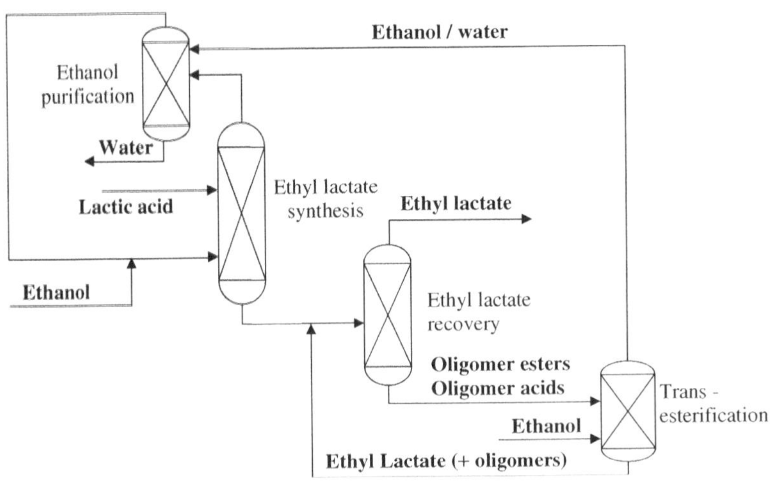

Figure 16.9 Process concept for ethyl lactate production via reactive distillation. Reprinted with permission from [132] © 2005, American Chemical Society

In alkanoic (monocarboxylic) acids, methyl acetate production via reactive distillation has been practiced commercially by Tennessee Eastman, where a traditional process with 13 unit operations was economized into a single reactive distillation column. As with methyl acetate formation, in which the methyl acetate–water azeotrope was cleverly broken by introduction of acetic acid at the top of the reactive distillation column, the formation of short chain acetate esters is challenged by several azeotropes in each system. Thus, Lai *et al.* [135] combined a decanter-condenser with a novel side-stream stripper in formation of ethyl acetate and isopropyl acetate. They introduced both acid and alcohol at the bottom of the column, and the organic phase in the decanter was taken to the stripping column where pure ester was recovered, thus circumventing the low-boiling azeotrope. With longer-chain alcohols, Brehelin *et al.* [136] obtained partial acetic acid conversion to propyl acetate in experimental studies, and several authors [137–140] have reported reactive distillation studies for butyl acetate, an important industrial solvent. These studies include reactive distillation in a packed-bed column [140], an economic analysis of various reactive separation approaches using dilute acetic acid as feed stock [138], the design of a column to achieve 99.5% yield of butyl acetate with acetic acid containing 20% water [137], and simultaneous esterification with butanol and amyl (C5) alcohols [139]. In all these cases, the RD column is characterized by a decanter following the condenser at the top of the column to take advantage of the heterogeneous butanol-water azeotrope, where the butanol-rich organic phase is refluxed to the column and the water-rich aqueous phase is withdrawn as distillate.

Several groups have applied a decanter at the condenser to take advantage of heterogeneous azeotropes in similar systems. For example, in the production of propyl propionate [141, 142] water and propyl propionate have very limited mutual solubility below about 30 wt% n-propanol content. Thus, the water-propyl propionate azeotrope gives two phases when the column is operated to limit the amount of propanol in the distillate. The water phase, which is nearly pure, is discarded, and part of the organic phase is refluxed with the remainder being taken as distillate product. In the production of butyl propionate, Lee *et al.* [143] examine the splitting of feed streams to achieve optimal conversion and low reboiler duties. Finally, the production of butyl levulinate, an attractive biofuel oxygenate for diesel or gasoline engines, is examined in a conceptual study [144].

The production of dicarboxylic acid esters is of significant importance because of their role in forming linear polyesters such as nylon-6,6 and renewable analogs such as polybutlene succinate, a polymer of 1,4-butanediol and succinic acid (1,4-butanedioic acid). The formation of diethyl succinate in the presence of acetic acid (a fermentation co-product) has been demonstrated in reactive distillation [145], with 98% yield of nearly pure diethyl succinate produced in the bottoms of a pilot-scale reactive distillation column. The acetate ester produced was taken from the distillate stream along with excess ethanol and product water.

The esterification of adipic acid via reactive distillation has been examined [146, 147]. The tricarboxylic citric acid has been esterified in reactive distillation with ethanol to give triethyl citrate by Kolah *et al.* [148] Because of the slow reaction kinetics of the third esterification step of producing triethyl citrate from diethyl citrate, a large column is required to achieve the target (98%) product yield.

16.4.1.5 Reactive distillation for glycerol esterification

Glycerol has become a renewable feedstock of great interest in the past ten years, because it is the primary byproduct of biodiesel production. Because of its reactive nature, it has been converted via reactive distillation processes to several products. One product of interest is triacetin, the triacetate ester of glycerol, of interest as a biofuel additive. For this system, the use of an entraining solvent has proven valuable, just as the use of entrainers for other esterification reactions has proven useful [149]. In triacetin production [150], ethylene dichloride (EDC) has been identified as a desirable entrainer. A process concept for triacetin production using ethylene dichloride is given in Figure 16.10.

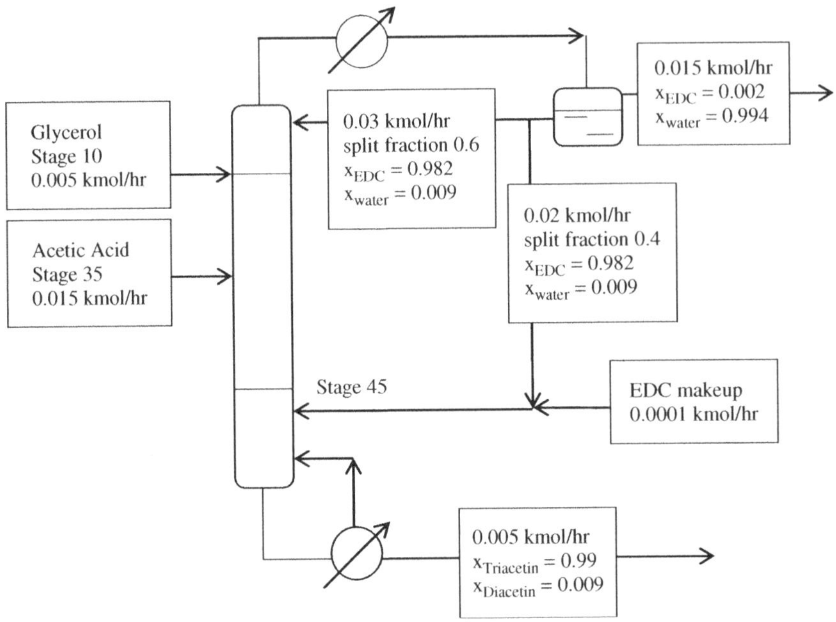

Figure 16.10 *Reactive distillation configuration for triacetin production using ethylene dichloride as entrainer. Reprinted with permission from [150] © 2010 American Chemical Society*

Ethylene dichloride forms a heterogeneous azeotrope with water that carries water to the distillate of the reactive distillation column and, more importantly, facilitates efficient separation of water from acetic acid. Thus, addition of EDC along with acetic acid near the bottom of an RD column, with glycerol being fed near the top of the column, drives product water of esterification upward, allowing the reaction to proceed to completion, while acetic acid becomes an intermediate boiler and remains within the column. This is another example of how a reactant in an RD system (acetic acid in this case) can be maintained at a high concentration in the column while being fed at a stoichiometric ratio, thus implementing Le Chatelier's principle. A similar approach has been used to form ethylene glycol diacetate, the C2 analog of triacetin [151].

Two other reactions of glycerol—not esterification per se but routes to important intermediates for chemicals from glycerol—have been examined in reactive distillation. The first is formation of dichloropropanol, a key intermediate in producing epichlorohydrin for epoxy resins, from glycerol [152], in which yields in excess of 92% are reported. The second is the dehydration of glycerol to hydroxyacetone (acetol), the key intermediate in a route to forming 1,2–propanediol from glycerol [153]. Glycerol can also be converted to other species by etherification and acetalization chemistry via reactive distillation; these routes are covered in the following sections.

16.4.2 Etherification

The single largest commercial application for reactive distillation was the production of methyl tert-butyl ether (MTBE) as an oxygenated gasoline additive. With MTBE production now ceased, there is interest in alternative ethers in this family of chemicals as renewable fuel additives. These include ethyl tert-butyl ether [154] and tert-amyl methyl ether [155], both of which are acceptable replacements for MTBE and can be efficiently produced via reactive distillation. There are also considerable energy savings to be realized in formation of ethers in this family, as illustrated by Huang *et al.* [156] in their study of MTBE formation. They found that the energy requirements in the MTBE formation column can be reduced substantially by proper location of feed streams into the column.

Glycerol ethers have also attracted attention as biofuel oxygenate additives; for example, reactive distillation to produce tert-butyl glycerol ethers has been proposed by Kiakittipong *et al.* [157]. They achieved about 60% conversion of glycerol in experimental studies, with the mono t-butyl ether as the predominant product (85%). Steric hindrance plays a significant role in achieving high yields of the di- and tri- ether products of glycerol.

16.4.3 Acetal formation

Acetals are formed by reaction of an aldehyde or ketone with two alcohols to produce a carbon atom with two ether linkages. An example is the production of 1,1-diethoxybutane from reaction of butyraldehyde with ethanol [158, 159], where approximately 50% conversion of butyraldehyde was achieved. The process is challenged by removing water liberated in reaction.

The formation of acetals from vicinal diol compounds constitutes an effective approach to recovering these compounds from aqueous solution. Glycerol can be recovered from the aqueous phase of biodiesel production by acetalization in a reactive distillation column [160]. Similarly, 1,3-propanediol can be reacted with an aldehyde to form a six-membered cyclic acetal, and can be simultaneously distilled out of aqueous solution [161]. In both of the above cases, the acetals can be hydrolyzed to recover the polyol, and the aldehyde can be recycled to the acetal formation step. This approach has also been successfully used to recover ethylene glycol and propylene glycol from aqueous solutions produced in sugar alcohol hydrogenolysis [162], as shown in Figure 16.11. In this case, the acetals of PG and EG were formed by reaction with acetaldehyde and then the separated by simple distillation. The individual acetals were then separately hydrolyzed to recover the individual polyols.

16.4.4 Reactive distillation for thermochemical conversion pathways

An alternative to fermentation for processing biomass is thermochemical or pyrolytic conversion, in which high temperatures are used to break down the constituents of biomass into small molecules. The advantage of this approach is that the entire biomass is converted, the disadvantage is that several hundred compounds are typically formed in detectable quantities. These compounds are typically classified into gaseous, liquid, or solid products—the primary product of interest for fuel use is the liquid phase, also known as bio-oil, which can constitute up to 70% of the original biomass weight. This bio-oil, when first formed, contains a large fraction of unsaturated, reactive compounds, significant oxygen, and up to 30 wt% water. It must therefore be stabilized, upgraded, and dried to become suitable for fuel use.

Reaction of raw bio-oil with alcohols in reactive distillation has been attempted as a first step in stabilizing and upgrading [163–165]. Both esterification [163, 165] and acetalization reactions [164] take place, reducing the degree of unsaturation and the quantity of oxygen present, and increasing the overall volume of the liquid product. It remains uncertain, however, whether the degree of stabilization achieved via reactive distillation warrants investment in and operation of a reactive distillation system.

Finally, reactive distillation has been suggested as a vehicle for carrying out Fischer–Tropsch synthesis of hydrocarbons from synthesis gas [166]. Gasification of biomass gives a mixture of CO and H_2, which can be advantageously converted in reactive distillation to separate product water as it is formed.

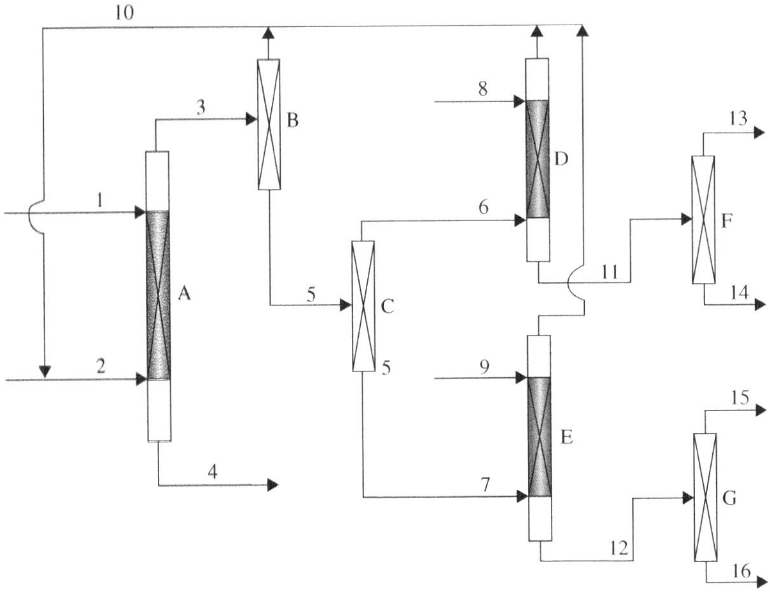

Figure 16.11 Process concept for EG/PG recovery from mixed glycol stream. Reprinted with permission from [162] © 2004 Pergamon. Columns: (A) acetal formation; (B) acetaldehyde recovery; (C) 2MD/2,4-DMD separation; (D) 2MD hydrolysis; (E) 2,4-DMD hydrolysis; (F,G) water removal. (Columns F and G required for dry EG and PG only.) Streams: (1) mixed glycol aqueous feed solution; (2) acetaldehyde feed; (3) acetaldehyde/acetal distillate; (4) residual aqueous glycol solution; (5) 2MD/2,4-DMD; (6) 2MD; (7) 2,4-DMD; (8,9) water; (10) acetaldehyde recycle; (11) EG/water; (12) PG/water; (13,15) water; (14) EG; (16) PG

16.5 Recently commercialized reactive distillation processes for the biorefinery

ZeaChem Inc. and Sulzer Chemtech have jointly worked on a commercial process to convert acetic acid into ethyl acetate. The annual market for ethyl acetate is approximately \$2.2 billion globally and \$115 million in the U.S. ZeaChem's biological pathway provides a lower cost route for the production of ethyl acetate as compared to the currently used natural gas feed stock based processes. The company intends to build a commercial biorefinery upon successful operations at its 250 000 gallon-per-year facility, which is proposed to be built in Boardman, Oregon [167].

Sulzer Chemtech has developed, together with Chemopetrol, a process for the production of ethyl acetate and butyl acetate [168]. Sulzer Chemtech announced in 2008 that it has commercialized the world's first fatty-acid esterification plant based on a hybrid reactive distillation / membrane separation technology in Asia, using a proprietary heterogeneous catalyst which has a predicted catalyst life of five years [169].

16.6 Conclusions

Reactive distillation is a commercially established process intensification technology that has significant applications in the emerging biorefinery. It offers the opportunity for reduced capital investment, energy savings, and minimization of waste production. These advantages are offset by greater complexity in design and operation arising from simultaneous reaction and separation, challenges in the reliability of process simulation, and a lack of familiarity with its operation among engineering staff. For renewable chemicals and fuels, such challenges are met by careful characterization of chemical and physical properties and by use of appropriate hardware and catalyst structures to optimize column efficiency. Reactive distillation will continue to expand into biorefining capability as the industry matures and becomes more competitive.

References

1. A.A. Backhaus, Continuous process for the manufacture of esters, US Patent 1400849 (1921).
2. G.J. Harmsen, Reactive distillation: The front-runner of industrial process intensification—A full review of commercial applications, research, scale-up, design and operation. *Chem. Eng. Proc.*, 46, 774–780 (2007).
3. D.L. Terrill, L.F. Sylvestre and M.F. Doherty, Separation of closely boiling mixtures by reactive distillation. *Ind. Eng. Chem. Pro. Des. Dev.*, 24, 1062–1071 (1985).
4. M.M. Sharma, Separations through reactions. *J. Sep. Pro. Technol.*, 6, 9–16 (1985).
5. M.F. Doherty and G. Buzad, Reactive Distillation by Design, *Chem. Eng. Res Des., Trans. of the Inst. Chem. Eng. Part A*, 70, 448–458 (1992).
6. J.G. Stichlmair and T. Frey, Reactive distillation processes. *Chem. Eng. Technol.*, 22, 95–103 (1999).
7. M. Sakuth, D. Reusch and R. Jawnosky, Reactive distillation in Ullmann's Encyclopedia of Industrial Chemistry. Wiley-VCH Verlag GmbH, Wienheim, http://www.onlinelibrary.wiley.com/book/10.1002/14356007 (accessed 18 September, 2012), ISBN: 9783527306732.
8. R. Taylor and R. Krishna, Modelling reactive distillation. *Chem. Eng. Sci.*, 55, 5183–5220 (2000).
9. M.F. Malone and M.F. Doherty, Reactive distillation. *Ind. Eng. Chem. Res.*, 39, 3953–3957 (2000).
10. S. M. Mahajani and S. P. Chopade, Reactive distillation: Processes of commercial importance, in *Encyclopedia of Separation Science* by D.I. Wilson, T.R. Edlard., C.A. Poole and M. Cooke (Eds.), Academic Press, London, 4075–4082 (2001).
11. K. Sundmacher and A. Kienle (Eds.), *Reactive Distillation: Status and Future Directions*, Wiley-VCH Verlag GmbH, Wienheim (2002).
12. R.S. Hiwale, N.V. Bhate, Y.S. Mahajan and S.M. Mahajani, Industrial applications of reactive distillation: Recent trends. *Int. J. Chem. Reac. Eng.*, 2, 1–52 (2004).
13. W.L. Luyben and C.C. Yu (Eds.), Reactive Distillation and Control, John Wiley & Sons, Inc., New York (2008).
14. K. Sundmacher, L.K. Rihko and U. Hoffmann, Classification of reactive distillation processes by dimensionless numbers. *Chem. Eng. Comm.*, 127, 151–167 (1994).
15. G.P. Towler and S.J. Frey, Reactive distillation. in S. Kulprathipanja, *Reactive Separation Processes*, Taylor & Francis, Philadelphia, PA (Chapter 2) (2000).
16. R. Krishna and S.T. Sie, Strategies for multiphase reactor selection. *Chem. Eng. Sci.*, 49, 4029–4065 (1994).
17. Y. Fuchigami, Hydrolysis of methyl acetate in distillation column packed with reactive packings of ion exchange resins. *J. Chem. Eng. Japan* 3, 354–359, (1990).
18. D.N. Chaplits, V.P. Kazakov, E.G. Lazariants, V.F. Chebotaev, M.I. Balashov and L.A. Serafimov, Ion exchange moulded catalyst and method of its preparation. US Patent 3965039 (1976).
19. U. Hoffmann, H. Bruderreck, K. Gottlieb, K. Schädlick, A. Rehfinger and J. Flato, Formkoerper aus makroporoesen ionenaustauscherharzen sowie verwendung der formkoerper. German Patent 3930515 (1989).
20. L.A. Smith. Catalyst structure and a process for its preparation. US Patent 4250052 (1981).
21. U. Hoffmann, U. Kunz, H, Bruderreck, K. Gottlieb, K. Schädlick and S. Becker, Tragerkatalysator und verwendung desselben, German Patent 4234779 (1992).
22. U. Kunz and U. Hoffmann, Method of chemically reacting substances in reaction column. European Patent EP0859653 B1 (1996).
23. J. Flato and U. Hoffmann, Development and start-up of a fixed bed reaction column for manufacturing antiknock enhancer MTBE. *Chem. Eng. Technol.*, 15, 193–201 (1992).
24. R.J. Carland., Fractionation tray for catalytic distillation. US Patent 5308451 (1994).
25. E.M. Jones, Contact structure for use in catalytic distillation. US Patent 4536373 (1985).
26. E.M. Jones, Distillation column reactor with catalyst replacement apparatus. US Patent 5133942 (1992).

27. M.C. Marion, J.C. Vitard, P. Travers, I. Harter and A. Forestiere, Process and apparatus for reactive distillation with particular distribution of liquid and vapour phases. US Patent 5776320 (1996).
28. D. Sanfilippo, M. Lupieri and F. Ancillotti, Process for preparation of tert-alkyl ethers and process for reactive distillation. US Patent 5493059 (1996).
29. N. Yeoman, R. Pinaire, M.A, Ulowetz, J.O. Berven, T.P. Nace and D.A. Fruse, Catalytic reaction and mass transfer process. US Patent 5447609 (1995).
30. N. Yeoman, R. Pinaire, M.A, Ulowetz, T.P. Nace and D.A. Furse, Internals for distillation columns including those for use in catalytic reactions. US Patent 5454913 (1995).
31. N. Yeoman, R. Pinaire, M.A, Ulowetz, T.P. Nace and D.A. Furse. Internals for distillation columns including those for use in catalytic reactions. US Patent 5496446 (1996).
32. N. Yeoman, R. Pinaire, M.A, Ulowetz, T.P. Nace and D.A. Furse, Method for concurrent reaction with distillation. US Patent 5593548 (1997).
33. E. von Harbou, M. Schmitt, S. Parada, C. Grossmann and H. Hasse, Study of heterogeneously catalyzed reactive distillation using the D+R tray—A novel type of laboratory equipment. *Chem. Eng. Res. Des.*, 89, 1271–1280 (2011).
34. L.A. Smith, Process for separation of butenes from C4 streams. US Patent 4242530 (1980).
35. L.A, Smith, Catalytic distillation structure. US Patent 4443559 (1984).
36. M.G. Caetano, J.C. Gonzalez and R.B. Solari, Flowdynamic modeling of bale-type catalytic distillation packing. *Sep. Sci. Technol.*, 39, 855–877 (2004).
37. E. Manduca, J.C. Gonzalez and H. Elman, Mass transfer characteristics of bale-type catalytic distillation packing. *Sep. Sci. Technol.*, 38, 3535–3552 (2003).
38. A. Viva, S. Aferka, D. Toye. P. Marchot, M. Crine and E. Brunazzi, Measurement of liquid hold-up in catalytic structured parkings: Comparison of different experimental techniques. Proceedings of the Distillation and Absorption Conference 2010 (Eds. A.B. de Haan, H. Kooijman and A. Gorak), September 12–15, Eindhoven, Netherlands (2010).
39. J.P. Stringaro, Catalyzing fixed bed reactor. US Patent 5470542, (1995).
40. P. Moritz, H. Hasse, Fluid dynamics in reactive distillation packing Katapak-S. *Chem. Eng. Sci*. 54, 1367–1374 (1999).
41. P. Moritz, B. Bessling, G. Schembecker, Fluid-dynamic consideration of catalyst supports in reactive distillation. *Chem. Ing. Technol.* 71, 131–135 (1999).
42. A. Kołodziej, M. Jaroszyński and I. Bylica, Mass transfer and hydraulics for KATAPAK-S, *Chem. Eng. Pro.: Process Intensification*, 43, 457–464 (2004).
43. J.M. van Baten and R. Krishna, Liquid phase mass transfer within Katapak-S structures studied using computational fluid dynamics simulations. *Cat. Today*, 69, 371–377 (2001).
44. M. Behrens, Ž. Olujić and P.J. Jansens, Combining reaction with distillation: Hydrodynamic and mass transfer performance of modular catalytic structured packings, *Chem. Eng. Res. Des.*, 84, 381–389 (2006).
45. M. Beherns, Z. Olujic and P.J. Jansens, Liquid flow behavior in catalyst-containing pockets of modular catalytic structured packing Katapak-SP. *Ind. Eng. Chem. Res.*, 46, 3884–3890 (2007).
46. M. Behrens, Z. Olujic and P.J. Jansens, Liquid holdup in catalyst containing pockets of a modular catalytic structured packing. *Chem. Eng. Technol.*, 31, 1630–1637 (2008).
47. Z. Ojulic and M. Berhens, Holdup and pressure drop of packed beds containing a modular catalytic structured packing. *Chem. Eng. Technol.*, 29, 979–985 (2006).
48. A. Kolodziej, M. Jaroszynski, H. Schoenmakers, K. Althaus, E. Geissler, C. Ubler, and M. Kloeker, Dynamic tracer study of column packings for catalytic distillation. *Chem. Eng. Proc.* 44, 661–670 (2005).
49. L. Gotze, O. Bailer, P. Moritz and C. von Scala, Reactive distillation with Katapak. *Cat. Today*, 69, 201–208 (2001).
50. A. Viva, S. Aferka, D. Toye, P. Marchot, M. Crine and E. Brunazzi, Determination of liquid hold-up and flow distribution inside modular catalytic structured packings, *Chem. Eng. Res. Des.* 89, 1414–1426 (2011).
51. J.M. van Baten, J. Ellenberger and R. Krishna, Hydrodynamics of reactive distillation tray column containing envelopes: Experiments vs CFD simulations. *Chem. Eng. Sci.* 56, 813–821 (2001).

52. J.M. van Baten, J. Ellenberger and R. Krishna, Hydrodynamics of reactive distillation tray column with catalyst containing envelopes: experiments vs. CFD simulations. *Cat. Today*, 66, 233–240 (2001).
53. J. Chen, C. Liu, X. Yuan and G. Yu, CFD simulation of flow and mass transfer in structured packing distillation columns. *Chin. J. Chem. Eng*., 17, 381–388 (2009).
54. S. Aferka, M. Crine, A.K. Saroha, D. Tove and P. Marchot, In situ measurement of the static liquid holpup in Katapak-packed column using X-ray tomography. *Chem. Eng. Sci.* 62, 6076–6080 (2007).
55. S. Aferka, A. Viva, E. Brunazzi, P. Marchot, M.. Crine and D. Toye, Liquid load point determination in a reactive distillation packing by x-ray tomography. *Can. J. Chem. Eng.*, 88, 611–617 (2010).
56. S. Aferka, P. Marchot, M. Crine and D. Toye, Interfacial area measurement in a catalytic distillation packing using high energy X-ray CT. *Chem. Eng. Sci.* 65, 511–516 (2010).
57. S. Aferka, A. Viva, E. Brunazzi, P. Marchot, M. Crine and D. Toye, Tomographic measurement of liquid hold up and effective interfacial area distributions in a column packed with high performance structured packings. *Chem. Eng. Sci.*, 66, 3413–3422 (2011).
58. A. Viva, S. Aferka, E. Brunazzi, P. Marchot, M. Crine and D. Toye, Processing of X-ray tomographic images: A procedure adapted for the analysis of phase distribution in MellapakPlus 752.Y and Katapak-SP packings. *Flow Meas. Instr.*, 22, 279–290 (2011).
59. P.J.M. Lebens, F. Kapteijn, S.T. Sie and J.A. Moulijn, Potential of internally finned monoliths as a packing for multifunctional reactors. *Chem. Eng. Sci.*, 54, 1359–1363 (1999).
60. T.J. Schildhauer, S. Tromp, I. Muller, A. Schilkin, E.Y. Kennig, F. Kapteijn and J.A. Moulijn, Modeling of reative stripping in monolith reactors. *Cat. Today*, 105, 414–420 (2005).
61. U. Brinkmann, T.J. Schildhauer and E.Y. Kennig, Hydrodynamic analogy approach for modeling of reactive stripping with structured catalyst supports. *Chem. Eng. Sci.*, 65, 298–303 (2010).
62. V.H. Agreda, L.R. Partin and W.H. Heise, High purity methyl acetate via reactive distillation. *Chem. Eng. Prog.*, 86, 40–46 (1990).
63. See http://srdata.nist.gov/solubility/ (accessed September 15, 2012).
64. S.I. Sandler, *Chemical, Biological, and Engineering Thermodynamics*, John Wiley & Sons, Inc., Hoboken, NJ, p. 539 (2006).
65. S.I. Sandler, *Chemical, Biological, and Engineering Thermodynamics*, John Wiley & Sons, Inc., Hoboken, NJ, p. 531 (2006).
66. P.B. Weisz and C.D. Prater, Interpretation of measurements in experimental catalysis. *Adv. in Catalysis*, 6, 143–196 (1954).
67. T. Frey and J. Stichlmair, Thermodynamic fundamentals of reactive distillation. *Chem. Eng. Technol.*, 22, 11–18 (1999).
68. T. Frey and J. Stichlmair, Reactive azeotropes in kinetically controlled reactive distillation. *Chem. Eng. Res. Des., Trans. Inst. Chem. Eng., Part A*, 77, 613–618 (1999).
69. W.D. Seider and S. Widagdo, Multiphase equilibria of reactive systems. *Fluid Phase Equil.*, 123, 283–303 (1996).
70. C.R. Marcelino, G.S.H. Juan and B.P. Adrian, Short-cut method for the design of reactive distillation columns. *Ind. Eng. Chem. Res.*, 50, 10730–10743 (2011).
71. D. Barbosa and M.F. Doherty, Theory of phase diagrams and azeotropic conditions for two-phase reactive systems. *Proc. Royal Society London A*, 413, 443–458 (1987).
72. D. Barbosa and M.F. Doherty, A new set of composition variables for the representation of reactive phase diagrams. *Proc. Royal Society London A*, 413, 459–464 (1987).
73. D. Barbosa and M.F. Doherty, The influence of equilibrium chemical reactions on vapor liquid-phase diagrams. *Chem. Eng. Sci.*, 43, 529–540 (1988).
74. D. Barbosa and M.F. Doherty, The simple distillation of homogeneous reactive mixtures. *Chem. Eng. Sci.*, 43, 541–550 (1988).
75. S. Ung and M.F. Doherty, Calculation of residue curve maps for mixtures with multiple equilibrium chemical reactions. *Ind. Eng. Chem. Res.*, 34, 3195–3202 (1995).
76. G. Venimadhavan, G. Buzad, M.F. Doherty and M.F. Malone, Effect of kinetics on residue curve maps for reactive distillation. *Amer. Inst. Chem. Eng. J.*, 40, 1814–1824 (1994).

77. C. Thiel, K. Sundmacher and U. Hoffmann, Residue curve maps for heterogeneously catalyzed reactive distillation of fuel ethers MTBE and TAME. *Chem. Eng. Sci.*, 52, 993–1005 (1998).
78. C. Thiel, K. Sundmacher and U. Hoffmann, Synthesis of ETBE: Residue curve maps for the heterogeneously catalyzed reactive distillation process. *Chem. Eng. J.*, 66, 181–191 (1998).
79. G. Damköhler, Strömungs und Wärme übergangsprobleme in Chemischer Technik und Forschung, *Chem. Eng. Technol.*, 12, 469 (1939).
80. Y.T. Tang, Y.W. Chen, H.P. Huang, C.C. Yu, S.B Hung and M.J. Lee, Design of reactive distillations for acetic acid esterification, *AIChE J.* 51, 1683–1699 (2005).
81. O.S. Daza, G.V. Escobar, E.M. Zarate and E.O. Munoz, Reactive residue curve maps—A new study case, *Chem. Eng. J.*, 117, 123–129 (2006).
82. L. Jimenez and J. Costa-Lopez, The production of butyl acetate and methanol via reactive and extractive distillation. II. Process modeling, dynamic simulation, and control strategy, *Ind. Eng. Chem. Res.*, 41, 6735–6744 (2002).
83. A.L. Orjuela, F.L. Leiva, L.M. Alejandro, G.N. Rodriguez and L.C. Maria, Constructing residue curve maps for butyl acetate synthesis. *Revista Ingeneiria E Investigacion*, 26, 26–34 (2006).
84. L. Simasatitkul, P. Siricharnsakunchai and Y. Patcharavorachot, Reactive distillation for biodiesel production from soybean oil. *Korean J. Chem. Eng.* 28, 649–655 (2011).
85. R. Baur, A.P. Higler, R. Taylor and R. Krishna, Comparison of equilibrium stage and nonequilibrium stage models for reactive distillation. *Chem. Eng. J.*, 76, 33–47 (1999).
86. J.J. Peng, S. Lextrait, T.F. Edgar and R.B. Eldridge, A comparison of steady-state equilibrium and rate-based models for packed reactive distillation columns. *Ind. Eng. Chem. Res.*, 41, 2735–2744 (2002).
87. R. Taylor, (Di)Still modeling after all these years: A review of the state of the art, *Ind. Eng. Chem. Res.* 46, 4349–4357 (2007).
88. N.J. Scenna, C.A. Ruiz and S.J. Benz, Dynamic simulation of startup procedures of reactive distillation columns. *Comp. Chem. Eng.* 22, S719–S722 (1998).
89. V. Pinjala, J.L. DeGarmo, M.A. Ulowetz, T.L. Marker and C.P. Luebke, Rate based modeling of reactive distillation operations. Proceedings of the Annual Meeting of the American Institute of Chemical Engineering (Session 3, Distillation with reaction), Miami Beach, Florida, November 1–6, 1992.
90. Y. Zheng and X. Xu, Study on catalytic distillation processes, Part II: Simulation of catalytic distillation processes, Quasi homogeneous and rate based model. *Chem. Eng. Res. Des. Trans. Inst. Chem. Eng., Part A*, 70, 465–470 (1992).
91. R. Taylor and R. Krishna, *Multicomponent Mass Transfer*, John Wiley & Sons, Inc., New York (1993).
92. See http://www.chemsep.org/program/index.html (accessed September 15, 2012).
93. K. Sundmacher and U. Hoffmann, Development of a new catalytic distillation process for fuel ethers via a detailed nonequilibrium model. *Chem. Eng. Sci.*, 51, 2359–2368 (1996).
94. K. Sundmacher, *Reaktivdestillation mit katalytischen fuellkoerperpackungen—ein neuer Process zur Herstellung der Kraftsto*wkomponente *MTBE*. Ph.D thesis, Universitat Clausthal (1995).
95. A.P. Higler, R. Taylor and R. Krishna, Modeling of a reactive separation process using a nonequilibrium stage model. *Comp. Chem. Eng.*, 22, S111–S118 (1998).
96. A. Higler, R. Krishna and R. Taylor, A nonequilibrium cell model for multicomponent (reactive) separation processes. *Amer. Inst. Chem. Eng. J.*, 45, 2357–2370 (1999).
97. A.P. Higler, R. Taylor and R. Krishna, Nonequilibrium Modelling of reactive distillation: Multiple steady states in MTBE synthesis. *Chem. Eng. Sci.* 54, 1389–1395 (1999).
98. D. Barbosa and M.F. Doherty, Design and minimum reflux calculations for single-feed multicomponent reactive distillation columns. *Chem. Eng. Sci.*, 43, 1523–1537 (1998).
99. D. Barbosa and M.F. Doherty, Design and minimum reflux calculations for double-feed multicomponent reactive distillation columns. *Chem. Eng. Sci.*, 43, 2377–2389 (1998).
100. S.M. Mahajani and A.K. Kolah, Some design aspects of reactive distillation columns (RDC). *Ind. Eng. Chem. Res.*, 35, 4587–4596 (1996).
101. A.A. Kiss, A.C. Dimian and G. Rothenberg, Solid acid catalysts for biodiesel production—towards sustainable energy, *Adv. Synth. Catal.*, 348, 75–81 (2006).

102. A.A. Kiss, A.C. Dimian and G. Rothenberg, "Green catalysts" for enhanced biodiesel technology, catalysis of organic reactions, *Chemical Industries Series*, 115, 405–414 (2006).
103. A.A. Kiss, F. Omota, A.C. Dimian and G. Rothenberg, The heterogeneous advantage: biodiesel by catalytic reactive distillation, *Topics in Catalysis*, 40 (1–4), 141–150 (2006).
104. A.A. Kiss, A.C. Dimian and G. Rothenberg, Biodiesel by catalytic reactive distillation powered by metal oxides, *Energy & Fuels*, 22, 589–604 (2008).
105. B.B. He, A.P. Singh and J.C. Thompson, A novel continuous-flow reactor using reactive distillation for biodiesel production. *Trans ASABE* 49, 107–112 (2006).
106. E. Lotero, Y.J. Liu, D.E. Lopez, K. Suwannakarn, D.A. Bruce and J.G. Goodwin, Synthesis of biodiesel via acid catalysis, *Ind. Eng. Chem. Res.* 44, 5353–5363 (2005).
107. Z. Qiu, L. Zhao and L. Weatherley, Process intensification technologies in continuous biodiesel production. *Chem. Engg. Proc.*, 49, 323–330 (2010).
108. N.L. Silva, C.M.G. Santander, C.B. Batistella, R.M. Filho and M.R.W. Maciel, Biodiesel production from integration between reaction and separation systems: Reactive distillation process, *Appl Biochem. Biotechnol.*, 161, 245–254 (2010).
109. C. Mueanmas, K. Prasertsit and C. Tongurai, Transesterification of triolein with methanol in reactive distillation column, simulation studies, *Int. J. Chem. Reac. Engg.* 8, A141 (2010).
110. A. Salis, M. Pinna, M. Monduzzi and V. Solinas, Biodiesel production from triolein and short chain alcohols through biocatalysis, *J. Biotechnol.*, 119, 291–299 (2005).
111. A.C. Dimian, C.S. Bildea, F. Omota and A.A. Kiss, Innovative process for fatty acid esters by dual reactive distillation, *Comp. Chem. Engg.*, 33, 743–750 (2009).
112. F.I. Gomez-Castro, V. Rico-Ramirez, J.G. Segovia-Hernandez and S. Hernandez, Feasibility study of a thermally coupled reactive distillation process for biodiesel production, *Chem. Eng Processing.*, 49, 262–269 (2010).
113. A.A. Kiss, Separative reactors for integrated production of bioethanol and biodiesel, *Comp. Chem. Eng.*, 34, 812–820 (2010).
114. G.D. Machado, D.A.G. Aranda and M. Castier, Computer Simulation of Fatty Acid Esterification in Reactive Distillation Columns, *Ind. Eng. Chem. Res.*, 50, 10176–10184 (2011).
115. C.J. Bart, N. Palmeri and S. Cavallaro, Industrial process technology for biodiesel production in C.J. Bart, N. Palmeri and S. Cavallaro (Eds.). Biodiesel science and technology: From Soy to Oil. **7**, 761–766, CRC Press LLC, Boca Raton, FL, 2010.
116. E. Cossio-Vargas, S. Hernandez, J.G. Segovia-Hernandez and M.I. Cano-Rodriguez, Simulation study of the production of biodiesel using feedstock mixtures of fatty acids in complex reactive distillation columns. *Energy*, 36, 6289–6297 (2011).
117. B.M.E. Russbeuldt and W.F. Hoelderich, New sulfonic acid ion-exchange resins for the preesterification of different oils and fats with high content of free fatty acids. *App. Catal. A-Gen.*, 362, 47–57 (2009).
118. F. Omota, A.C. Dimian and A. Bliek, Fatty acid esterification by reactive distillation. Part 1: equilibrium-based design, *Chem. Eng. Sci.*, 58, 3159–3174 (2003).
119. C. von Scala, P. Moritz, and P. Fassler, Process for the continuous production of fatty acid esters via reactive distillation. *Chimia*, 57, 799–801 (2003).
120. S. Steinigeweg, and J. Gmehling, Esterification of a fatty acid by reactive distillation. *Ind. Eng. Chem. Res.*, 42, 3612–3619 (2003).
121. N. Nghi and Y. Demirel, Using thermally coupled reactive distillation columns in biodiesel production. *Energy* 36(8), 4838–4847 (2011).
122. S. Hernandez, J.G. Segovia-Hernandez and L. Juarez-Trujillo, Design study of the control of a reactive thermally coupled distillation sequence for the esterification of fatty organic acids. *Chem. Eng. Commun.*, 198, 1–18 (2010).
123. F. Israel Gomez-Castro, V. Rico-Ramirez and J. G. Segovia-Hernandez, Esterification of fatty acids in a thermally coupled reactive distillation column by the two-step supercritical methanol method, *Chem. Eng. Res. Des.*, 89, 480–490 (2011).
124. A. Kiss, Heat-integrated reactive distillation process for synthesis of fatty esters. *Fuel Proc. Technol.*, 92, 1288–1296 (2011).

125. S. Bhatia, A.L. Ahmad and A.R. Mohamed, Production of isopropyl palmitate in a catalytic distillation column: Experimental studies, *Chem. Eng. Sci.*, 61, 7436–7447 (2006).
126. S. Bhatia, A.R. Mohamed, A.L. Ahmed and S.Y. Chin, Production of isopropyl palmitate in a catalytic distillation column: Comparison between experimental and simulation studies, *Comp. Chem. Engg.*, 31, 1187–1198 (2007).
127. S.Y. Chin, A.R. Mohamed and A.L. Ahmad, Esterification of palmitic acid with iso-propanol in a catalytic distillation column: Modeling and simulation studies. *Int. J. Chem. React. Eng.*, 4, #A32 (2006).
128. E. A. Edreder, I.M. Mujtaba and M. Emtir, Optimal operation of different types of batch reactive distillation columns used for hydrolysis of methyl lactate to lactic acid, *Chem. Eng. J.*, 172, 467–475 (2011).
129. R. Kumar, H. Nanavati, S. Noronha and S. Mahajani, A continuous process for the recovery of lactic acid by reactive distillation. *J. Chem. Technol. Biotechnol.*, 81, 1767–1777 (2006).
130. M. Liu, S.T. Jiang and L.J. Pan, Design and control of reactive distillation for hydrolysis of methyl lactate. *Chem. Eng. Res. Des.* 89, 2199–2206 (2011).
131. I. M. Mujtaba, E. A. Edreder and M. Emtir, Significant thermal energy reduction in lactic acid production process. *App. Energy.*, 89, 74–80 (2012).
132. N. Asthana, A. Kolah, D.T. Vu, C.T. Lira and D.J. Miller, A continuous reactive separation process for ethyl lactate formation. *Org. Proc. Res. Dev.*, 9, 599–607 (2005).
133. J. Gao, X.M. Zhao and L.Y. Zhou, Investigation of ethyl lactate reactive distillation process. *Chem. Eng. Res. Des.*, 85, 525–529 (2007).
134. R. Kumar and S. Mahajani, Esterification of lactic acid with n-butanol by reactive distillation. *Ind. Eng. Chem. Res.*, 46, 6873–6882 (2007).
135. I.K. Lai, S.B. Hung and W.J. Hung, Design and control of reactive distillation for ethyl and isopropyl acetates production with azeotropic feeds. *Chem. Eng. Sci.*, 62, 878–898 (2007).
136. M. Brehelin, F. Forner and D. Rouzineau, Production of n-propyl acetate by reactive distillation—Experimental and theoretical study. *Chem. Eng. Res. Design*, 85, 109–117 (2007).
137. A. Arpornwichanop, C. Wiwittanaporn and S. Authayanun, The use of dilute acetic acid for butyl acetate production in a reactive distillation: Simulation and control studies. *Korean J. Chem. Eng.*, 25, 1252–1266 (2008).
138. A. Arpornwichanop, K. Koomsup and W. Kiatkittipong, Production of n-butyl acetate from dilute acetic acid and n-butanol using different reactive distillation systems: Economic analysis. *J. Taiwan Inst. Chem. Eng.*, 40, 21–28 (2009).
139. H.Y. Lee, L.T. Yen, and I.L. Chien, Reactive distillation for esterification of an alcohol mixture containing n-butanol and n-amyl alcohol, *Ind. Eng. Chem. Res.*, 48, 7186–7204 (2009).
140. E. Sert, and F.S. Atalay, Esterification of acetic acid with butanol: operation in a packed bed reactive distillation column. *Chem. Biochem. Eng. Quart.*, 25, 221–227 (2011).
141. E. Altman, P. Kreis and T. van Gerven, Pilot plant synthesis of n-propyl propionate via reactive distillation with decanter separator for reactant recovery. Experimental model validation and simulation studies. *Chem. Eng. Proc.*, 49, 965–972 (2010).
142. M. Kotora, C. Buchaly and P. Kreis, Reactive distillation—experimental data for propyl propionate synthesis. *Chem. Papers*, 62, 65–69 (2008).
143. H.Y. Lee, C.H. Jan, and I.L. Chien, Feed-splitting operating strategy of a reactive distillation column for energy-saving production of butyl propionate. *J. Taiwan. Inst. Chem. Eng.*, 41, 403–413 (2010).
144. A. Harwardt, K. Kraemer and B. Ruengeler, Conceptual design of a butyl-levulinate reactive distillation process by incremental refinement. *Chinese J. Chem. Eng.*, 19, 371–379 (2011).
145. A. Orjuela, A. Kolah, C.T. Lira and D.J. Miller, Mixed Succinic Acid/Acetic Acid Esterification with Ethanol by Reactive Distillation. *Ind. Eng. Chem. Res.*, 50, 9209–9220 (2011).
146. K.W. Chan, Y.T. Tsai and H. Lin, Esterification of adipic acid with methanol over Amberlyst 35. *J. Taiwan Inst. Chem. Eng.*, 41, 414–420 (2010).
147. S.B. Hung, I.K. Lai and H.P. Huang, Reactive distillation for two-stage reaction systems: Adipic acid and glutaric acid esterifications. *Ind. Eng. Chem. Res.*, 47, 3076–3087 (2008).
148. A.K. Kolah, N.S. Asthana, D.T. Vu, C.T. Lira and D.J. Miller, Triethyl citrate synthesis by reactive distillation. *Ind. Eng. Chem. Res.*, 47 (4), 1017–1025 (2008).

149. M.C. de Jong, E. Zondervan and A.C. Dimian, Entrainer selection for the synthesis of fatty acid esters by entrainer-based reactive distillation. *Chem. Eng. Res. Design*, 88, 34–44 (2010).
150. A. Hasabnis and S. Mahajani, Entrainer-based reactive distillation for esterification of glycerol with acetic acid. *Ind. Eng. Chem. Res.*, 49, 9058–9067 (2010).
151. T. Suman, S. Srinivas and S. Mahajani, Entrainer based reactive distillation for esterification of ethylene glycol with acetic acid. *Ind. Eng. Chem. Res.*, 48, 9461–9470 (2009).
152. Z.H. Luo, X.Z. You and J. Zhong, Design of a reactive distillation column for direct preparation of dichloropropanol from glycerol. *Ind. Eng. Chem. Res.*, 48, 10779–10787 (2009).
153. C.W. Chiu, M.A. Dasari and G.J. Suppes, Dehydration of glycerol to acetol via catalytic reactive distillation. *AIChE. J.*, 3543–3548 (2006).
154. R. Khaledi and B.R. Young, Modeling and model predictive control of composition and conversion in an ETBE reactive distillation column. *Ind. Eng. Chem. Res.*, 44, 3134–3145 (2005).
155. A.E. Plesu, J. Bonet and V. Plesu, Residue curves map analysis for tert-amyl methyl ether synthesis by reactive distillation in kinetically controlled conditions with energy-saving evaluation. *Energy*, 33, 1572–1589 (2008).
156. K. Huang, S.J. Wang and W. Ding, Towards further internal heat integration in design of reactive distillation columns—Part III: Application to a MTBE reactive distillation column. *Chem. Eng. Sci.*, 63, 2119–2134 (2008).
157. W. Kiatkittipong, P. Intarachareon, N. Laosiripojana, C. Chaisuk, P. Praserthdam and S. Assabumrungrat, Glycerol ether synthesis from glycerol etherification with tert-butyl alcohol in reactive distillation. *Comp. Chem. Engg.* 35, 2034–2043 (2011).
158. I. Agirre, V.L. Barrio, B. Guemez, J.F. Cambra and P.L. Arias, Bioenergy II: The development of a reactive distillation process for the production of 1,1 diethoxy butane from bioalcohol: Kinetic study and simulation model. *Int. J. Chem. React. Eng.*, 8, A86 (2010).
159. I. Agirre, V.L. Barrio, B. Guemez, J.F. Cambra and P.L. Arias, Catalytic reactive distillation process development for 1,1 diethoxy butane production from renewable resources. *Bioresource Technol.*, 102, 1289–1297 (2011).
160. L. Vijaya, U. Ravuru and V. Kotra, Novel Route for Recovery of Glycerol from Aqueous Solutions by Reversible Reactions. *Int. J. Chem. React. Eng.*, 7, A38 (2009).
161. J. Hao, F. Xu and H.J. Liu, Downstream processing of 1,3-propanediol fermentation broth. *J. Chem. Technol. Biotechnol.* 81, 102–108 (2006).
162. A.D. Dhale, L.K. Myrant, S.P. Chopade and D.J. Miller, Propylene glycol and ethylene glycol recovery from aqueous solution via reactive distillation, *Chem. Eng. Sci.*, 59, 2881–2890 (2004).
163. X. Li, R. Gunawan, and C. Lievens, Simultaneous catalytic esterification of carboxylic acids and acetalization of aldehydes in a fast pyrolysis bio-oil from mallee biomass. *Fuel*, 90, 2530–2537 (2011).
164. F.H. Mahfud, I. Melian-Cabrera and R. Manurung, Biomass to fuels—Upgrading of flash pyrolysis oil by reactive distillation using a high boiling alcohol and acid catalysts. *Proc. Safety Env. Protection*, 85, 466–472 (2007).
165. J. Xu, J. Jiang and Y. Sun, Bio-oil upgrading by means of ethyl ester production in reactive distillation to remove water and to improve storage and fuel characteristics. *Biomass Bioenergy*, 32, 1056–1061 (2008).
166. S. Srinivas, S.M. Mahajani and R.K. Malik, Reactive distillation for Fischer–Tropsch synthesis: feasible solution space. *Ind. Eng. Chem. Res.*, 49, 6350–6361 (2010).
167. See http://www.zeachem.com/press/pressrelease042110.php (accessed September 15, 2012).
168. See http://www.sulzerpumps.com/portaldata/7/Resources/03_NewsMedia/STR/2001/2001_03_moritz_e.pdf (accessed September 15, 2012).
169. P. Moritz, P. Faessler, C. von Scala and O. Bailer, Method for the esterification of a fatty acid. U.S. Patent 7 091 367 B2, August 15, 2006.

17

Reactive Absorption

Anton A. Kiss[1] and Costin Sorin Bildea[2]

[1]*Arnhem, The Netherlands*
[2]*University 'Politehnica' of Bucharest, Romania*

17.1 Introduction

In recent decades, the chemical process industry showed more interest than ever in the development of reactive separation processes that combine reaction and separation into a single, integrated unit. Compared to conventional processes, reactive separations bring several important benefits such as increased selectivity and reaction yield, overcoming thermodynamic limitations, lower energy requirements and significant reduction in water and solvent consumption (Noeres *et al*., 2003). The most important and well known examples of reactive separations are reactive distillation (RD) and reactive absorption (RA). By integrating absorption and chemical reactions into a single operating unit with enhanced performance, RA qualifies as a process-intensification technique. Just as reactive distillation, RA is typically carried out in tray or packed columns—but unlike reactive distillation, RA requires neither a reboiler nor a condenser (Kenig and Seferlis, 2009).

The separation and/or purification of a gas mixture by the absorption of part of the mixture (e.g., CO_2, H_2S, NOx and SOx) in a solvent, which is regenerated afterwards, is the most commonly encountered use of RA as an industrial process. However, apart from gas cleaning, RA is also applied in the production of bulk chemicals, such as nitric and sulfuric acid (Kiss *et al*., 2010; Yildirim *et al*., 2012).

More recently, reactive separations (RD and RA) using solid catalysts offer excellent opportunities for manufacturing fatty esters, involved in specialty chemicals and biodiesel production. Integrating reaction and separation into one production unit brings key benefits such as simplified operation, no waste, reduced capital investment and low operating costs (Kiss, 2009, 2010, 2011).

This chapter introduces the basics of reactive absorption and then presents a case study that is relevant to biofuel production. The novel heat-integrated reactive absorption process described here eliminates all the conventional catalyst-related operations, efficiently uses the raw materials and equipment and considerably reduces the energy requirements for biodiesel production—over 85% lower as compared to the RA and

Separation and Purification Technologies in Biorefineries, First Edition.
Edited by Shri Ramaswamy, Hua-Jiang Huang, and Bandaru V. Ramarao.

RD base cases. Rigorous simulations based on experimental results were carried out using Aspen Plus and Aspen Dynamics. Despite the high degree of integration, the process is easily controllable using an efficient control structure proposed in this work. The main results are provided for a plant producing 10 KTPY (10000 tons per year) fatty acid methyl esters (FAME) from methanol and waste vegetable oil with high free fatty acid content, using sulfated zirconia as solid acid catalyst.

17.2 Market and industrial needs

Reactive absorption is essentially a mature process that is known since the foundation of modern chemical industry. More recently, the role of RA as a core environmental protection process has grown up significantly, and nowadays RA is the most widely applied reactive separation process.

The current market for reactive absorption processes is growing, mainly due to extensive efforts in the area of CO_2 capture and storage (Rahimpour and Kashkooli, 2004), as well as due to novel proposed applications in biofuel production (Kiss, 2009; Kiss and Bildea, 2011). Most of the current research focuses on development of new solvents with high capacity and easy regeneration properties (Yildirim *et al.*, 2012). The main room for improvement is in the regeneration of the solvent, responsible of up to 70–80% of the operating costs. Hence, either the existing processes must be improved or different solvents have to be considered to reduce the energy requirements. In addition, the absorption can be integrated into the total plant, which could significantly reduce the total energy requirements.

Due to tighter legislation on greenhouse gas emissions, industrial needs are also shifting towards more efficient, less expensive, sustainable and eco-friendly reactive separation processes. However, the main industrial applications of reactive absorption remains focused on purification of gas streams and the production of bulk chemicals (Kenig and Gorak, 2005; Yildirim *et al.*, 2012):

- removal of harmful substances (e.g. coke oven gas purification, CO_2 and NOx removal);
- retrieval of valuable substances or non-reacted reactants (e.g. solvent regeneration);
- production of chemical products (e.g. sulfuric and nitric acid, formaldehyde synthesis);
- water removal (e.g. water removal from natural gas, air drying);
- conditioning of gas streams (e.g. synthesis gas conditioning);
- separation of substances (e.g. olefin/paraffin separations).

17.3 Basic principles of reactive absorption

Absorption may be either a physical or a chemical process. Absorption is defined as a process by which a substance included in one state is transferred into the bulk volume of another substance in a different state—typically a gas being absorbed in a liquid. This should not be confused with adsorption, which is the physical adherence/bonding of chemicals on the surface of another substance.

Physical absorption of a gas or part of a gas mixture in a liquid solvent involves the mass transfer that occurs at the interface between the gas and the liquid and the diffusion of the gas into liquid. Physical absorption of gases in a liquid solvent depends on the solubility of the gases and the pressure and temperature conditions. A classic example of physical absorption of a gas into a liquid is the absorption of carbon dioxide (CO_2) into water (H_2O)—well known from the beverage industry.

Chemical absorption, known also as reactive absorption, involves a chemical reaction between the substance being absorbed and the bulk liquid. It depends upon the stoichiometry of the reaction and the concentration of the reactants. An illustrative example of chemical absorption is the purification of natural gas by passing it through an aqueous solution of mono-ethanolamine (MEA) in which the acid gases (e.g. H_2S, CO_2) are removed by reacting with MEA. Note that reactive absorption may be reversible or irreversible, depending on the reaction type (e.g. equilibrium or irreversible).

17.4 Modelling, design and simulation

The modelling of reactive absorption received considerable attention, reflected in several publications (Kenig *et al.*, 2003; Noeres *et al.*, 2003; Kenig and Gorak, 2005; Kenig and Seferlis, 2009). The optimal design of reactive absorption processes requires adequate models covering column hydrodynamics, mass and heat transfer, as well as reaction kinetics. Large-scale applications are modelled by dividing the columns into smaller segments, called stages. Each stage corresponds to a single tray or to a segment of packed column. A general overview of the reactive absorption modelling approaches was given by Kenig and Gorak (2005). Several models were developed along the history of reactive absorption—the main ones being described hereafter (Yildirim *et al.*, 2012).

Equilibrium stage model. This assumes that the gas and liquid streams leaving a stage are in thermodynamic equilibrium. This simple model was used in the past decades for a variety of applications, especially for non-reactive systems. However, reactive systems require further extensions for a proper description. The chemical reactions are taken into consideration using a source term in the mass and energy balances. Yet, in real absorption processes, the thermodynamic equilibrium is usually not attained within a stage. For this reason, tray efficiencies or HETP (height equivalent of a theoretical plate) values were introduced to build a link to real columns. This method often fails for RA processes, due to the specific features of multi-component mixtures (Taylor and Krishna, 1993).

HTU/NTU-concepts and enhancement factors. A simple approach to take mass transfer kinetics into account for the determination of transfer units was developed by Chilton and Colburn (1934). The column height is determined as a product of the number and height of the transfer units (NTU and HTU, respectively). The height of the transfer unit is estimated via empirical Sherwood correlations, whereas the number of transfer units is obtained by numerical or analytical integration of the inverse of the driving force over the column height. Chemical reactions are considered via enhancement factors, which are defined as the quotient of mass transfer rate with reaction and the mass transfer rate without reaction (Danckwerts, 1970). Since the complexity of the occurring reactions is described using one single parameter, this method often leads to inaccurate results.

Rate-based stage model. The rate-based approach is a method that directly takes into account the multicomponent mass and heat transfer and the chemical reaction. The mass transfer between the phases can be described by different theories, such as the two-film model or the penetration/surface renewal theory. The corresponding model parameters are determined using empirical correlations. For many applications, the two-film model parameters can be found in the literature and, therefore, this method is often preferred (Kenig *et al.*, 2003).

For the design and simulation of reactive absorption processes, adequate software tools are required. Nowadays, the rate-based models are mostly used, being typically implemented in AspenTech AspenPlus (RATEFRAC or RateSep unit), Aspen Custom Modeller (Kenig *et al.*, 2003; Kucka *et al.*, 2003), gPROMS (Kiss *et al.*, 2010), Mathworks Matlab (Gabrielsen *et al.*, 2006), FORTRAN (Thiele *et al.*, 2007), ProTreat or even CFD software (e.g. ANSYS CFX).

Reactive absorption can be performed in a variety of equipment types, which provide a continuous flow of both contacting phases. Just as reactive distillation, RA is typically carried out in tray or packed columns. However, RA requires no reboiler or condenser (Kenig and Seferlis, 2009; Kiss, 2009). The RA process is characterized by independent flow of both gas and liquid phases, and allows co-current (down-flow and up-flow) and more typical counter-current operating regimes. Reactive absorption operating units can be conveniently classified according to which phase is in a continuous or a disperse form. Using this criterion, the classification of the reactive absorption equipment can be summarized as:

- *both phases in a continuous form*: packed columns, thin-film contactors, wetted-wall columns, contactors with flat surface, laminar jet absorber, disc (sphere) columns;
- *a disperse gas phase and a continuous liquid phase*: plate columns, plate columns with packing, bubble columns, packed bubble columns, mechanically agitated columns, jet absorbers;
- *a disperse liquid phase and a continuous gas phase*: spray columns, Venturi scrubbers.

17.5 Case study: Biodiesel production by catalytic reactive absorption

Fatty esters are important chemicals used mainly in cosmetics, pharmaceuticals, cleaning products and in the food industry. However, since the 1990s the main interest has shifted to the large-scale production of biodiesel—hence the stronger market drive for more innovative and efficient processes. Biodiesel is a biodegradable and renewable alternative fuel with properties similar to petroleum diesel (Bowman *et al*., 2006; Balat and Balat, 2008; Knothe, 2010). Unlike petroleum diesel, which is a mixture of hydrocarbons, biodiesel consists of fatty acid methyl esters (FAME). Typically, it is produced from green sources such as vegetable oils, animal fat or even waste cooking oils from the food industry (Encinar *et al*., 2005; Kulkarni and Dalai, 2006). The increasing worldwide interest in biodiesel is illustrated by the exponential increase of the production, mostly in Western Europe, USA, and Asia (Figure 17.1), with Germany and France as top EU consumers (Kiss, 2009).

Nowadays, employing waste and non-edible raw materials is mandatory to comply with the economical, ecological and also ethical requirements for biofuels. Basically, the 'food-versus-fuel' competition can be conveniently avoided when the raw materials are waste vegetable oils, non-food crops such as Jatropha (Kumar and Sharma, 2005; de Oliveira *et al*., 2009) and Mahua (Puhan *et al*., 2005; Jena *et al*., 2010) or castor oil (da Silva *et al*., 2009; Canoira *et al*., 2010). However, waste raw materials can contain a substantial amount of free fatty acids (FFA), up to 100%.

Fatty esters are currently produced by acid/base catalysed esterification or trans-esterification with methanol or ethanol (Kiss, 2010). There are several processes currently in use at pilot or industrial scale: batch, continuous, supercritical, enzymatic, and two-step processes involving hydrolysis and esterification (Kiss and Bildea, 2011). At present, the most common manufacturing technologies employ homogeneous

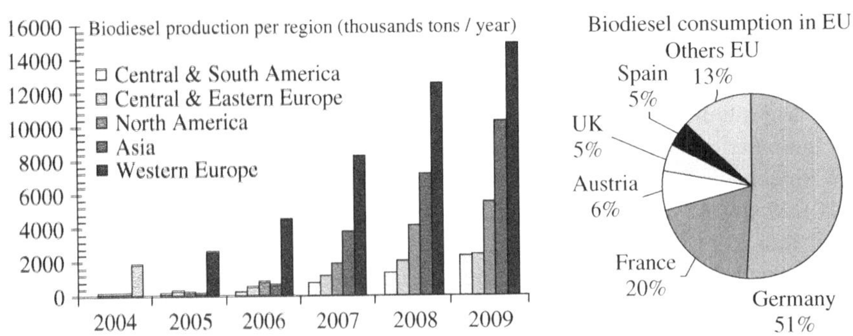

Figure 17.1 Biodiesel production in the world (left) and biodiesel consumption in the EU (right)

catalysts, in batch or continuous processes. The reaction is followed by several neutralization and product purification steps (Hanna *et al.*, 2005; Meher *et al.*, 2006), the use of liquid catalysts leading to severe economical and environmental penalties. Obviously, this has to change considering the huge impact of the large-scale production of biofuels. A continuous biodiesel process that uses a solid catalyst (Kiss *et al.*, 2006a; Yan *et al.*, 2010) in order to suppress the costly downstream processing steps and waste treatment appears as a very interesting alternative.

The recent literature is quite abundant in studies on integrated processes, such as reactive distillation (Kiss *et al.*, 2006b, 2008) and reactive absorption (Noeres *et al.*, 2003). However, the vast majority of the reported studies on reactive separation processes for biodiesel production are based solely on conventional reactive distillation (He *et al.*, 2006; Kiss *et al.*, 2006b, 2008; Suwannakarn *et al.*, 2009), or alternatives such as entrainer-based reactive distillation (RD) and dual RD (Dimian *et al.*, 2009).

This section presents a novel energy-efficient integrated reactive absorption (RA) process for biodiesel production, which is very well controllable in spite of the high degree of integration. The integration of reaction and separation into one unit combined with the use of a heterogeneous catalyst offers major advantages such as: reduced capital investment, low operating costs, simplified downstream processing steps as well as no catalyst-related waste streams and no soap formation (Kiss and Bildea, 2012). The main results are given for a plant producing 10 KTPY biodiesel by esterification of methanol with free fatty acids, using sulfated zirconia as solid acid catalyst.

17.5.1 Problem statement

Biodiesel is a very appealing but still expensive alternative fuel. The common problem of all conventional processes is the use of liquid catalysts that require neutralization and costly separation steps that generate salt waste streams (van Gerpen, 2005; Meher *et al.*, 2006; Narasimharao *et al.*, 2007). To solve these problems, the process presented here makes use of solid acids applied in an esterification process based on catalytic reactive separation. Such an integrated process is able to shift the chemical equilibrium to completion and preserve the catalytic activity of the solid acid by continuously removing the products (Kiss *et al.*, 2006b). Moreover, compared to conventional processes, the investment and operating costs are much reduced. However, the market pressure demands further decrease of the operating costs by replacing the raw materials with inexpensive waste oils (high FFA content), and reducing the energy requirements per ton of product (Janulis *et al.*, 2004; Kiss, 2009). Several reactive separation processes based on fatty acids esterification were reported recently, aiming at high productivity performance along with low energy requirements:

- reactive distillation: 191.2 kW·h/ton biodiesel (Kiss *et al.*, 2008);
- dual reactive distillation: 166.8 kW·h/ton biodiesel (Dimian *et al.*, 2009);
- reactive absorption: 138.4 kW·h/ton biodiesel (Kiss, 2009);
- heat-integrated reactive distillation: 108.8 kW·h/ton biodiesel (Kiss, 2011);
- heat-integrated reactive absorption: 21.6 kW·h/ton biodiesel (Kiss and Bildea, 2011).

Figure 17.2 shows a comparison of the energy requirements for a conventional two-step process involving acid and basic catalysis (Vlad *et al.*, 2010) versus recently reported reactive separation processes (Kiss, 2009, 2010; Kiss *et al.*, 2008; Dimian *et al.*, 2009; Kiss and Bildea, 2011, 2012).

17.5.2 Heat-integrated process design

In this work, rigorous computer simulations were used to evaluate the integrated process for the synthesis of fatty esters by reactive absorption. Starting from a base case design described in previous work (Kiss, 2009),

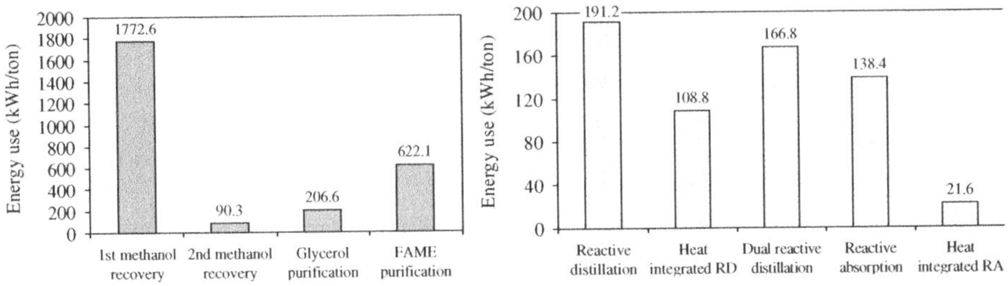

Figure 17.2 *Energy requirements for conventional process (left) and reactive separations (right)*

pinch analysis is applied to identify the possibilities of heat recovery by internal process/process exchange, as well as the optimal selection and placement of utilities. Then, steady-state process simulation (Dimian and Bildea, 2008) is employed to obtain a description of material and energy streams in the process with the scope of evaluating the plant and understanding its steady state behaviour. Finally, dynamic simulations are used to evaluate controllability of the plant and to demonstrate the effectiveness of a proposed plant-wide control strategy.

The conceptual design of the process is based on a reactive absorption column that integrates the reaction and separation steps into one operating unit. The chemical equilibrium is shifted towards product formation by continuous removal of the reaction products, instead of using an excess of a reactant—typically the alcohol. An additional flash vessel and a decanter are used to guarantee the high purity of both products. Since methanol and water are much more volatile than the fatty ester or fatty acid, these will separate easily in the top of the column. Figure 17.3 (left) presents the flowsheet of this process based on conventional reactive absorption, as reported by Kiss (2009).

The column has 15 theoretical stages with the reactive zone located between stages 3–12, the solid catalyst loading being 6.5 kg per stage. The fatty acid is pre-heated then fed as hot liquid in the top of

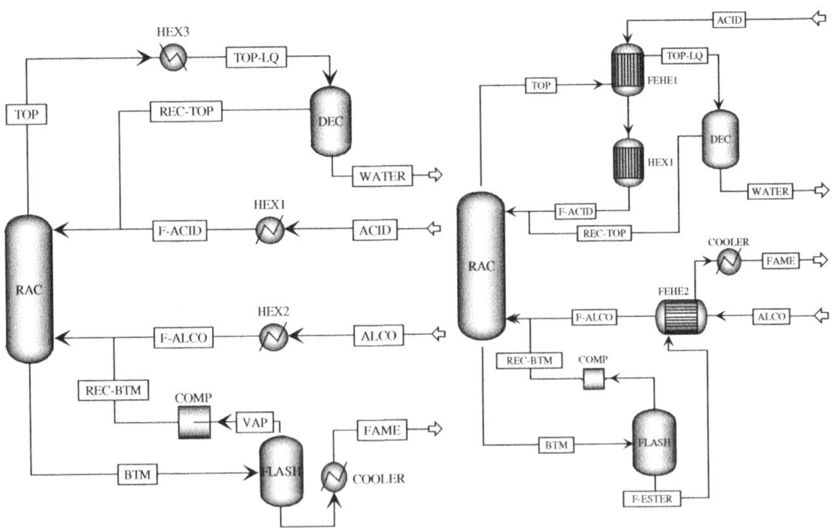

Figure 17.3 *Synthesis of fatty esters by reactive absorption base case flowsheet (left) and heat-integrated process (right)*

the reactive column while a stoichiometric amount of alcohol is injected as vapour into the bottom of the column, thus creating a counter-current flow regime over the reactive zone. Water by-product is removed as top vapour, then condensed and separated in a decanter from which the fatty acids are recycled back to the column while water by-product is recovered at high purity. The fatty esters are delivered as a high-purity bottom product of the reactive column. The hot product is flashed first to remove the remaining methanol, and then it is cooled down and stored.

The reference flowsheet presented in Figure 17.3 (left) is relatively simple, with just a few operating units, two cold streams that need to be pre-heated (fatty acid and alcohol) and two hot streams that have to be cooled down (top water and bottom fatty esters). The heat-integration was therefore performed by applying previously reported heuristic rules (Hamed *et al*., 1996; Chen and Yu, 2003; Dimian and Bildea, 2008). Consequently, two feed-effluent heat exchangers (FEHE) should replace partially or totally each of the two heat exchangers HEX1 and HEX2. Figure 17.3 (right) illustrates the improved process including heat integration around the RA column (Kiss and Bildea, 2011). The top vapour stream is used to pre-heat the fatty acid feed stream (FEHE1). An additional heat exchanger (HEX1) is required to finally heat the fatty acids to the desired reaction temperature. When changes of the production rate are desired, the duty of this heat-exchanger can be changed to ensure the required temperature of the column-inlet acid stream (F-ACID). The hot liquid product of the FLASH, a mixture of fatty esters, is used to pre-heat and vaporize the alcohol feed stream. If changes of the production are expected, the nominal design should include a bypass of the hot stream that can be used for control objectives.

Note that other heat-integration alternatives are possible. For example, the top outlet stream could be used to vaporize the methanol feed in FEHE1, while the liquid product from the FLASH could be used to preheat the acid feed FEHE2. This alternative has slightly higher energy requirements compared to the flowsheet presented in Figure 17.3, right (29 kW instead of 27 kW) while the duty of the FAME cooler decreases to −3.3 kW. Still, an additional heat exchanger (13 kW) is necessary to cool the top stream to the decanting temperature (50 °C), which is a drawback of this alternative.

17.5.3 Property model and kinetics

Process simulations embedding kinetics experimental results were performed using AspenTech Aspen Plus and Aspen Dynamics as powerful CAPE tools (Aspen Technology, 2009a, b, c). The RA column was simulated using the rigorous RADFRAC unit with RateSep (rate-based) model, and considering three phase balances. The physical properties required for the simulation and the binary interaction parameters for the methanol–water and acid-ester pairs were available in the Aspen Plus database of pure components, while the other interaction parameters were estimated using the UNIFAC–Dortmund modified group contribution method (Aspen Technology, 2009a, b, c). Using other state-of-the-art estimation methods, such as UNIFAC and UNIFAC–Lyngby modified, leads to similar results. Note that, at ambient pressure, the boiling points of fatty acids and their corresponding fatty acid methyl esters are quite high (Yuan *et al*., 2005; Kiss, 2009).

The residue curve maps (RCM) and the ternary diagram of the mixture of fatty acid, methanol and water is presented in Figure 17.4 (Kiss, 2011). As phase splitting may occur (Kiss, 2009), the RA column is modelled considering VLL equilibrium data. Phase splitting must be accounted for as the free water phase can deactivate the solid acid catalyst. Nevertheless, as revealed by the column composition profiles, the liquid molar fraction of water does not exceed 0.1 at any given stage. Catalyst deactivation therefore does not occur here, under the designed process conditions.

The fatty components were conveniently lumped into one fatty acid and its fatty ester—according to the reaction: R-COOH + CH_3OH ↔ R-COO-CH_3 + H_2O. Dodecanoic (lauric) acid/ester was selected as a lumped component due to the availability of experimental results, kinetics and VLLE parameters

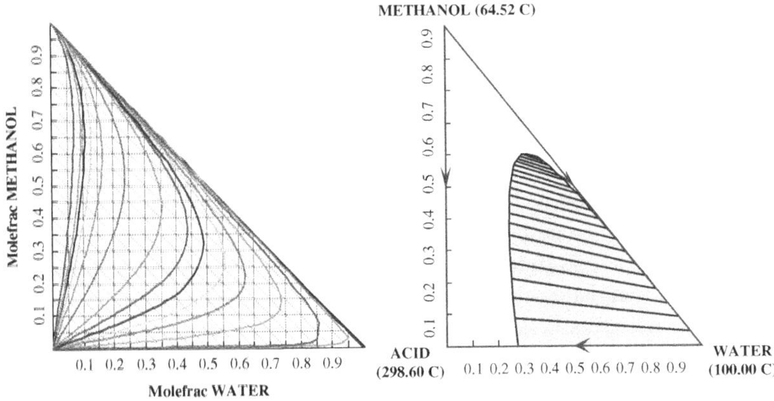

Figure 17.4 *Residue curve map (left) and ternary diagram (right) for the mixture fatty acid-methanol-water*

for this system (Kiss *et al*., 2006a, 2006b; Dimian *et al*., 2009). Moreover, as lauric acid is among the lightest fatty acids, this represents the worst-case scenario in terms of separation from the alcohol. The assumption of lumping components is very reasonable since fatty acids and their corresponding fatty esters have similar properties. This approach was already reported to be successfully used to simulate other fatty ester production processes (Kiss *et al*., 2006b, 2008; Dimian *et al*., 2009, Kiss, 2009).

In this work, sulfated zirconia is considered as green catalyst. Kinetic data for the esterification with methanol is available from previous work (Kiss *et al*., 2008; Dimian *et al*., 2009; Kiss, 2009). The esterification reaction is reversible, hence the reaction rate accounts for both reactions:

$$r = (k_1 W_{cat})\, C_{Acid} C_{Alcohol} - (k_2 W_{cat})\, C_{Ester} C_{Water} \tag{17.1}$$

where k_1 and k_2 are the kinetic constants for the direct (esterification) and reverse (hydrolysis) reactions, W_{cat} is the weight amount of catalyst, and $C_{Component}$ is the molar concentration of the components present in the system. Note that by changing the weight amount of catalyst used (W_{cat}) the reaction rate can be similar for different catalysts. As water is continuously removed from the system, the reverse hydrolysis reaction is extremely slow, hence the second term of the reaction rate can be neglected. Therefore, a simple kinetic expression can be used:

$$r = (k_1 W_{cat})\, C_{Acid} C_{Alcohol} = k C_{Acid} C_{Alcohol} \tag{17.2}$$

$$k = A \exp\left[-E_a/(RT)\right] \tag{17.3}$$

where C_{Acid} and $C_{Alcohol}$ are mass concentrations, A is the pre-exponential (Arrhenius) factor ($A = 250\,\text{kmol·m}^3\text{·kg}^{-2}\text{·s}^{-1}$) and E_a is the activation energy ($E_a = 55\,\text{kJ/mol}$).

17.5.4 Steady-state simulation results

The integrated reactive absorption process was designed according to previously reported process synthesis methods for reactive separations (Noeres *et al*., 2003). The next simulation results are given for a plant producing 10 KTPY ($1250\,\text{kg h}^{-1}$) fatty acid methyl esters (FAME) from methanol and waste vegetable oil with the highest free fatty acids (FFA) content, using solid acids as green catalysts. Table 17.1 presents the complete mass balance of the process, while Table 17.2 lists the main design parameters, such as column size, catalyst loading, and feed condition (Kiss and Bildea, 2011).

Table 17.1 Mass balance of a 10 KTPY FAME process based on integrated reactive-absorption.

	F-ACID	F-ALCO	BTM	REC-BTM	REC-TOP	TOP	WATER	FAME
Temperature / (°C)	160	65.4	136.2	146.2	51.8	162.1	51.8	30
Pressure / (bar)	1.05	1.05	1.03	1.216	1	1	1	0.203
Vapour frac	0	1	0	1	0	1	0	0
Mole flow / (kmol h^{-1})	5.824	5.876	6.125	0.252	0.059	5.886	5.828	5.873
Mass flow / (kg h^{-1})	1166.7	188.3	1261.3	11.3	9.369	114.4	105.06	1250
Mass flow / (kg h^{-1})								
METHANOL	0	188.3	9.125	7.544	0.002	0.103	0.101	1.581
ACID	116.74	0	Trace	Trace	9.218	9.233	0.016	Trace
WATER	0	0	Trace	Trace	0.24	105.2	104.93	Trace
FAME	0	0	1252.2	3.764	0.846	0.846	Trace	1248.4
Mass fraction								
METHANOL	0	1	0.007	0.667	172 ppm	894 ppm	965 ppm	0.001
ACID	1	0	Trace	Trace	0.894	0.08	148 ppm	Trace
WATER	0	0	Trace	10 ppb	0.023	0.912	0.999	Trace
FAME	0	0	0.993	0.333	0.082	0.007	513 ppb	0.999
Mole fraction								
METHANOL	0	1	0.046	0.931	873 ppm	546 ppm	0.001	0.008
ACID	1	0	Trace	Trace	0.726	0.008	13 ppm	Trace
WATER	0	0	Trace	26 ppb	0.211	0.992	0.999	Trace
FAME	0	0	0.954	0.069	0.062	670 ppm	43 ppb	0.992

Table 17.2 Design parameters for simulating the reactive absorption column.

Parameter	Value	Units
Total number of theoretical stages	15	—
Number of reactive stages	10 (from 3 to 12)	—
Column diameter	0.4	m
HETP	0.6	m
Valid phases	VLL	—
Volume liquid holdup per stage	18	L
Mass catalyst per stage	6.5	kg
Catalyst bulk density	1050	kg/m^3
Fatty acid conversion	>99.99	%
Fatty acid feed (liquid, at 160 °C)	1167	kg h^{-1}
Methanol feed (vapour, at 65 °C)	188	kg h^{-1}
Production of biodiesel (FAME)	1250	kg h^{-1}
RA column productivity	19.2	kg FAME kg cat^{-1} h^{-1}

A high level of conversion of the reactants is achieved, with the productivity of the RA unit exceeding 19 kg fatty esters per kg catalyst hour. The purity specification is higher than 99.9 %wt for the final biodiesel product (FAME stream). Note that the total amount of the optional recycle streams (REC-BTM) is not significant, representing less than 0.9% of the biodiesel production rate.

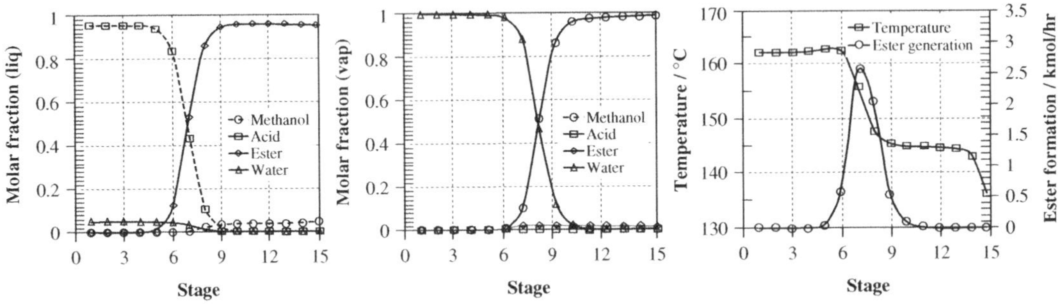

Figure 17.5 *Liquid-vapour composition, temperature and reaction rate profiles along the column*

The liquid-vapour composition profiles along the reactive absorption column are plotted in Figure 17.5, along with the temperature and reaction rate profiles in the reactive column (Kiss, 2009; Kiss and Bildea, 2011). The concentration of fatty ester and methanol increases from the top to bottom, while the concentration of the fatty acid and water increases from the bottom to the top of the column. Therefore, the top product of the reactive separation column is water vapours with small amounts of fatty acids, while the bottom liquid stream is the fatty esters product (biodiesel) with a limited amount of methanol. In particular cases, the concentration of methanol in the liquid phase could be further increased by working at higher pressure. The RA column is operated in the temperature range of 135–165 °C, at ambient pressure. The reaction rate exhibits a maximum in the middle of the column, where the reactive zone is placed.

17.5.5 Sensitivity analysis

Despite the recent progress in understanding the feasibility, design and control of reactive separations, the conceptual design of RA may lead to several different process configurations and operating parameters. In this work, sensitivity analysis was used to evaluate the range of the operating parameters: reactant ratios, temperature of feed streams, decanting temperature, flashing pressure and recycle rates. Note that the optimal molar ratio of the reactants (alcohol:acid) is very close to the stoichiometric value of 1 (Kiss, 2009). If this ratio is higher than 1 (i.e. if there is an excess of alcohol) then the fatty acid is completely converted to fatty ester but the excess of alcohol becomes a significant impurity in the top stream and thereafter in the water by-product. On the other hand, when the ratio is less than 1 (i.e. there is an excess of fatty acids) then the purity of the water by-product remains high but the conversion of fatty acid is incomplete—hence the bottom product contains unreacted fatty acid that cannot be removed easily from the final product by simple flashing. The separation of fatty acids from fatty esters is more difficult than the separation of fatty acids from water, so this situation should be avoided. In practice, using a very small excess of methanol (up to 1%), or an efficient control structure that can ensure the stoichiometric ratio of reactants, is sufficient for the complete conversion of the fatty acids and prevention of the difficult separation previously mentioned. Note that integrated reactive separations can be easily controlled by PID controllers in a multi-loop framework (Dimian *et al.*, 2009; van Diggelen *et al.*, 2010) or by using more advanced techniques such as model predictive control (Nagy *et al.*, 2007).

The most significant results of the sensitivity analysis are presented in Figure 17.6 and Figure 17.7, for a constant acid flow rate of 1166.7 kg h^{-1} (5.824 kmol h^{-1}) and variable alcohol flow rate (Kiss and Bildea, 2011). Figure 17.6 shows the duty of heat exchangers versus the molar reactants ratio (alcohol:acid), for different temperatures of the acid feed. Vaporization of an increasing flow rate of alcohol requires more energy to be exchanged in FEHE2, which is realized by increasing the amount of hot liquid that passes through the heat exchanger. This can be accomplished if—in the base case design—only a certain fraction

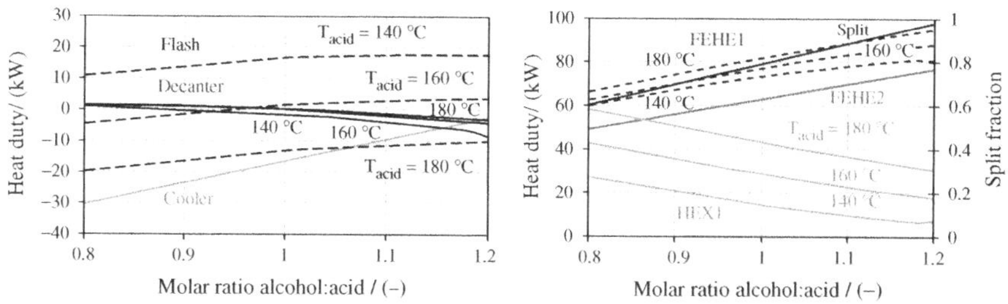

Figure 17.6 *Duty of heat exchangers (coolers and heaters) versus molar reactants ratio (alcohol:acid)*

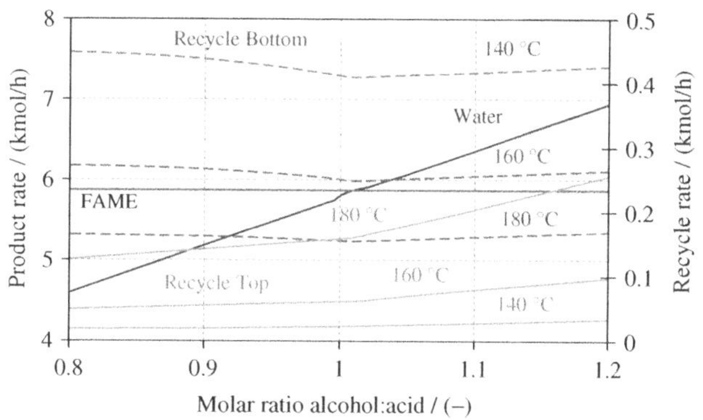

Figure 17.7 *Product and recycle flow rates versus molar reactants ratio (alcohol:acid)*

of the hot stream (80% in our base case design, variable "Split") is used for alcohol vaporization. The fraction of hot liquid that is necessary for alcohol vaporization increases to almost 100% when the alcohol flow rate is increased to 120% of the base case value. The dependence of FEHE2 heat duty versus the alcohol:acid ratio is insensitive with respect to the temperature of the acid feed. Increasing the alcohol flow rate leads to a larger flow rate at the top outlet of the column. More energy is exchanged in FEHE1 with the result of lower duty for the heater HEX1. In general, higher acid temperatures require larger heat duties in both FEHE1 and HEX1 heat-exchangers.

Figure 17.7 illustrates the flow rates of products and recycle streams versus the molar reactants ratio (alcohol:acid), at different temperatures of the fatty acid feed stream (Kiss and Bildea, 2011). The flow rate of the FAME stream is unchanged when the reactants ratio changes. However, below the stoichiometric ratio of 1, this stream contains large quantities of unreacted acid. The flow rate of the WATER product stream increases with the alcohol flow rate. Above the stoichiometric ratio of 1, this stream contains water and the excess of methanol. A higher temperature acid stream leads to increasing the top recycle—because more acid is vaporized—and reduces the bottom recycle.

Table 17.3 shows a head-to-head comparison of the novel heat-integrated RA process described here (Kiss and Bildea, 2011) against the previously reported reference RD and RA processes (Kiss, 2008; Kiss *et al.*, 2008; Kiss, 2009). The heating and cooling requirements are figures that ultimately translate into equipment size and cost. Remarkably, the energy demand is less than 22 kW·h / ton biodiesel—equivalent to only 34 kg steam per ton of biodiesel produced. Also, compared to the reference reactive absorption base case (Kiss, 2009), the heating and cooling requirements are reduced by 85% and 90%, respectively.

Table 17.3 *Comparison between integrated reactive-absorption versus reactive-distillation processes (at a production rate of 1250 kg h^{-1} fatty esters).*

Equipment / parameter / units	RD	HI-RD	RA	HI-RA
Reactive column—reboiler duty (heater), kW	136	136	n/a	n/a
HEX-1 heat duty (fatty acid heater), kW	95	0	108	27
HEX-2 heat duty (methanol heater), kW	8	0	65	0
Reactive column—condenser duty (cooler), kW	−72	−72	n/a	n/a
HEX-3 water cooler/decanter, kW	−6	−6	−77	0
COOLER heat duty (biodiesel cooler), kW	−141	−38	−78	−14
FLASH heat duty (methanol recovery), kW	0	0	0	0
Compressor power (electricity), kW	0.6	0.6	0.6	0.6
Reactive column, number of reactive stages	10	10	10	10
Feed stage number, for acid / alcohol streams	3/10	3/10	1/15	1/15
Reactive column diameter, m	0.4	0.4	0.4	0.4
Reflux ratio (mass ratio R/D), kg/kg	0.10	0.10	n/a	n/a
Boil-up ratio (mass ratio V/B), kg/kg	0.12	0.12	n/a	n/a
Productivity, kg ester / kg catalyst / h	20.4	20.4	19.2	19.2
Energy requirements per ton biodiesel, kW·h/ton FAME	191.2	108.8	138.4	21.6
Steam consumption, kg steam / ton FAME	295	168	214	34

17.5.6 Dynamics and plantwide control

Heat-integrated reactive absorption offers significant advantages such as minimal capital investment and operating costs, as well as no catalyst-related waste streams and no soap formation. However, the controllability of the process is just as important as the capital and operating costs savings (Bildea and Kiss, 2011). In processes based on reactive distillation or absorption, feeding the reactants according to their stoichiometric ratio is essential to achieve high products purity (Dimian *et al.*, 2009; Bildea and Kiss, 2011). Thus, the fatty acid is completely converted to fatty esters when there is an excess of methanol, but the excess of methanol becomes an impurity in the top stream and thereafter in the water by-product. On the contrary, when there is an excess of fatty acids, the purity of water by-product remains high, but the conversion of fatty acids is incomplete. In the latter case the bottom product contains unreacted fatty acids that cannot be easily removed from the final product by simple flashing. As the separation of fatty acids from fatty esters is more difficult than the separation of fatty acids from water, this situation should be avoided. This constraint must be fulfilled not only during the normal operation, but also during the transitory regimes arising due to planned production-rate changes or unexpected disturbances.

However, integrated biodiesel processes based on reactive absorption have fewer degrees of freedom compared to reactive distillation. This makes it challenging to set the reactant-feed ratio correctly, consequently avoiding impurities in the products. In the following, we will prove that this reactive absorption process is very controllable, despite the high degree of integration. A key result of this study is an efficient control structure that can ensure the stoichiometric ratio of reactants and fulfils the operating constraint of using an excess of methanol. The excess of methanol is sufficient for the total conversion of the fatty acids, and consequently for the prevention of difficult separations (e.g. fatty acid—fatty ester).

Figure 17.8 presents the proposed plantwide control structure (Kiss and Bildea, 2011). Compared to Figure 17.3 (right), more details concerning the practical implementation are given. The production rate is set by the flow rate of the acid stream (controller FC1). After mixing with the recycle and passing the

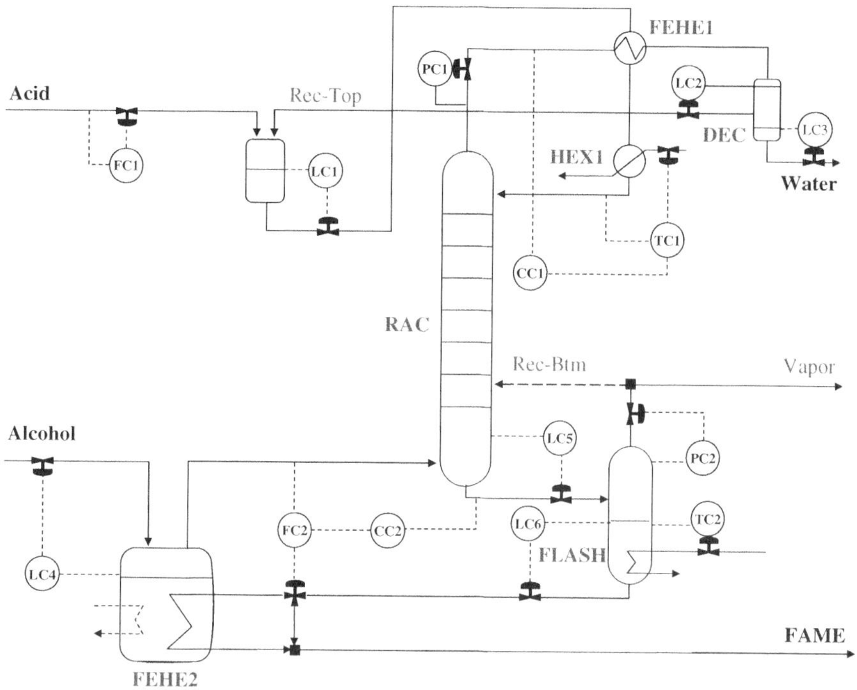

Figure 17.8 Plantwide control structure: concentration of acid in bottom stream is controlled by manipulating the alcohol flow rate. The temperature of the column-inlet acid stream is set by a methanol concentration controller

feed-effluent heat exchanger FEHE1, the temperature of the acid fed to the column is controlled (by TC1) by manipulating the duty of the heat exchanger HEX1. The flow rate of alcohol evaporated and fed to the column is set by changing the split of the hot stream from the flash-outlet stream (controller FC2). The temperature of the alcohol stream entering the column is therefore the boiling temperature. Because the hot FAME stream is not available during the start-up of the plant, the evaporator should include the option of using another heat source. This is shown by the dashed drawing in Figure 17.8.

It should be noted that setting the flow rates such that the stoichiometric ratio is fulfilled is not possible using only the flow controllers FC1 and FC2 because of unavoidable measurement or control implementation errors. For this reason, an additional concentration controller is necessary. An excess of alcohol will have as a result the complete consumption of acid and a drop in the acid concentration at the bottom of the column. On the other hand, large quantities of acid will be present when this reactant is in excess. We conclude that the imbalance in the alcohol:acid ratio can be detected by measuring the concentration of acid in the bottom outlet of the column.

The concentration controller CC2 is therefore used to give, in a cascade manner, the correct setpoint to the alcohol flow-rate controller FC2. In this way, when production rate changes are implemented by changing the acid flow rate, the correct ratio between reactants is achieved. However, for large production rate changes the overall reaction rate is not sufficiently large to guarantee the complete conversion of both reactants and high purity cannot be achieved. In particular, large amounts of methanol will be present in the top product of the absorption column. The concentration controller CC1 avoids this by increasing the setpoint of the temperature controller TC1. The result is higher temperatures inside the column, and therefore increased reaction rates. Control of the material inventory is also achieved by conventional level

Table 17.4 Controller tuning parameters.

Controller	Control action	Range of manipulated variable	Range of controlled variable	Setpoint	Bias	Gain (%/%)	Integral time (min)
Level LC1 (Buffer vessel)	Direct	0–2350 $kg\,h^{-1}$	0–1 m	0.5 m	1180 $kg\,h^{-1}$	1	—
Level LC2 (Dec., org. lq.)	Direct	0–40 $kg\,h^{-1}$	0–1 m	0.5 m	14.2 $kg\,h^{-1}$	1	—
Level LC3 (Dec., aq. lq.)	Direct	0–210 $kg\,h^{-1}$	0–1 m	0.45 m	105 $kg\,h^{-1}$	1	—
Level LC4 (Alc. evap.)	Reverse	0–400 $kg\,h^{-1}$	0–2	1	197 $kg\,h^{-1}$	1	—
Level LC5 (Col. sump)	Direct	0–2500 $kg\,h^{-1}$	0–1.75 m	0.875 m	1259 $kg\,h^{-1}$	10	60
Level LC6 (Flash)	Direct	0–2490 $kg\,h^{-1}$	0–1 m	0.5 m	1245 $kg\,h^{-1}$	1	—
Pressure PC1 (Column)	Direct	0–12 $kmol\,h^{-1}$	0.95–1.05 bar	1 bar	5.91 $kmol\,h^{-1}$	1	12
Pressure PC2 (Flash)	Reverse	0–1.22 $kmol\,h^{-1}$	0.18–0.22 bar	0.2 bar	0.29 $kmol\,h^{-1}$	1	12
Temp. TC1 (Acid inlet)	Reverse	0–0.2 $GJ\,h^{-1}$	140–180 °C	Output of CC1	0.12 $GJ\,h^{-1}$	1	20
Mole fraction CC1 (Methanol, column top)	Direct	140–180 °C	$1–5 \times 10^{-3}$	3×10^{-3}	167.6 °C	0.1	40
Flow rate FC2 (Methanol inlet)	Reverse	0.5–1	4–8 $kmol\,h^{-1}$	Output of CC2	0.9	1	20
Mole fraction CC2 (Acid, col. btm.)	Direct	2–8 $kmol\,h^{-1}$	$1–2 \times 10^{-3}$	1×10^{-3}	6.15 $kmol\,h^{-1}$	0.2	40
Temperature TC2 (Flash)	Reverse	0–0.1 $GJ\,h^{-1}$	130–140 °C	137 °C	0.01 $GJ\,h^{-1}$	1	20

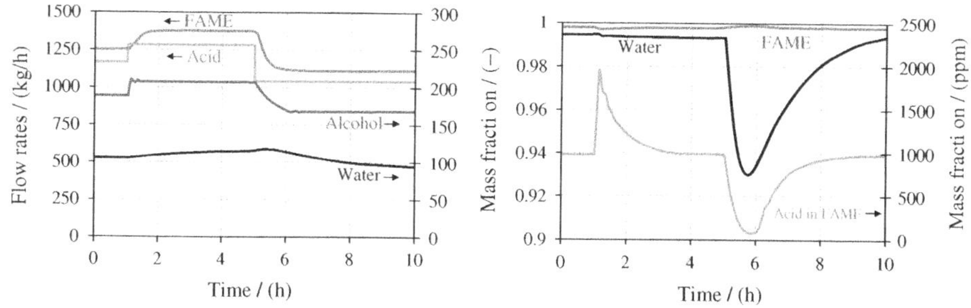

__Figure 17.9__ Dynamic simulation results for the flowsheet with methanol recycle: acid flow rate disturbance of +10% at 1 h, and –10% at 5 h. Production rate changes are easily achieved and the products purity is maintained at high values

and pressure control loops. Details of the controller action and controller tuning parameters are presented in Table 17.4 (Kiss and Bildea, 2011).

Figure 17.9 depicts the dynamic simulation results for the flowsheet presented in Figure 17.8, when the recycle of methanol vapours is considered (Kiss and Bildea, 2011). The simulation starts from the steady state. At time $t = 1$ h, the acid flow rate is increased by ~10%, from 1166.7 kg h^{-1} to 1282 kg h^{-1} (5.824 kmol h^{-1} to 6.4 kmol h^{-1}). Then, at time $t = 5$ h, the acid flow rate is decreased to 1041.7 kg h^{-1} (5.2 kmol h^{-1}), representing a ~10% decrease with respect to the nominal value. The new production rate is achieved in about 1 hour. The dynamic response of the water product stream is slower. The purity of FAME remains high throughout the dynamic regime, with the acid concentration below the 2000 ppm requirement of the ASTM D6751-08 standard (i.e. acid number lower than 0.50 mg KOH/g biodiesel).

Similarly, Figure 17.10 shows the dynamic simulation results for the same flowsheet presented in Figure 17.8, but without the bottom recycle of methanol vapours (Kiss and Bildea, 2011). The same scenario is assumed for the dynamic simulation. Again, the production rate responds quickly to changes of acid flow rate (Figure 17.10, left). However, a degradation of dynamic performance is observed when the purity of the product streams is taken into account (Figure 17.10, right). The alternative with methanol recycle is therefore preferable and recommended.

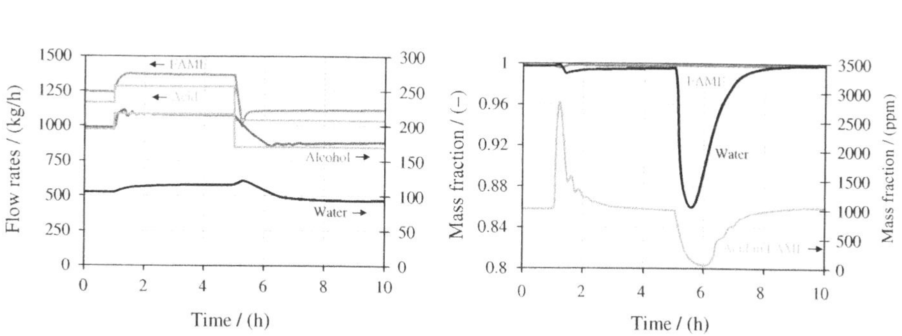

__Figure 17.10__ Dynamic simulation results for the flowsheet without methanol recycle: acid flow rate disturbance of +10% at 1 h, and –10% at 5 h. Production rate changes are easily achieved and a high degree of product purity is maintained

17.6 Economic importance and industrial challenges

Reactive absorption is essentially an established technology, known and used since the foundation of the chemical industry. Recently, the role of RA as a core environmental protection process has grown up significantly, and nowadays RA is the most widely applied reactive separation process. As illustrated by the case study presented in this chapter, reactive absorption can lead to significantly lower investment and operating costs as compared to reactive distillation or the rest of conventional processes for biodiesel production. Moreover, it can also bring significant green advantages such as less waste, lower energy requirements, cleaner gas emissions and drastically reduced greenhouse gas (GHG) pollution.

Industrial challenges depend mainly on the nature of the process. In the case of CO_2 absorption and sour gas treatment, the main room for improvement in the industrially used RA processes is in the regeneration of the solvent (e.g. amines). Although the RA step is not energy demanding, solvent recovery by desorption typically requires 70–80% of the total operating costs. Current research focuses on development of new solvents (e.g. amine mixtures or hindered amines) with high capacity and easy regeneration properties. In general, either existing processes or equipment must be improved or different solvents have to be considered in order to reduce the energy usage. Moreover, absorption can also be integrated into the total plant, which could significantly reduce the overall energy requirements. For NOx removal, the latest research activities were dedicated to aqueous alkaline solutions, since valuable by-products can be obtained (e.g. nitric acid, nitrates and nitrites). Recent research on SOx removal focuses on investigating wet processes and significant progress was made (e.g. with limestone technology, Wellman-Lord process, ammonia or seawater scrubbing). For biodiesel production the main challenges could be the stability of the solid catalyst (Bildea and Kiss, 2011), the use of various mixtures of fatty acids, and the integration of esterification and trans-esterification steps.

17.7 Conclusions and future trends

The novel heat-integrated reactive absorption process presented here eliminates all conventional catalyst-related operations, improves efficiency and considerably reduces the energy requirements for biodiesel production—85% lower as compared to the base case. Another important result is an efficient control structure that ensures the required reactant ratio and fulfils operating constraint of having an excess of methanol. This is, in fact, sufficient for the total conversion of the fatty acids and consequently for prevention of difficult separations. Remarkable, despite the high degree of integration, this reactive absorption process is very controllable as illustrated by the results of rigorous dynamic simulations.

Given tighter legislation, future work will aim to apply integrated RA technologies to the eco-efficient removal of contaminants from gas streams, as well as the production of biofuels—for example, biodiesel production by green processing of waste materials with high FFA content.

References

Aspen Technology, *Aspen Dynamics—Reference and User Guide*, Aspen Technology, Burlington MA, 2009a.

Aspen Technology, *Aspen Physical Property System—Physical Property Models*, Aspen Technology, Burlington MA, 2009b.

Aspen Technology, *Aspen Plus: User Guide*. Volumes 1 and 2, Aspen Technology, Burlington MA, 2009c.

Balat M., Balat H., A critical review of bio-diesel as a vehicular fuel, *Journal of Energy Conversion and Management*, 49 (2008), 2727–2741.

Bildea C.S., Kiss A.A., Dynamics and control of a biodiesel process by reactive absorption, *Chemical Engineering Research and Design*, 89 (2011), 187–196.

Bowman M., Hilligoss D., Rasmussen S., Thomas R., Biodiesel: a renewable and biodegradable fuel, *Hydrocarbon Processing*, 85 (2006), 103–106.

Canoira L., Galean J.G., Alcantara R., Lapuerta M., Garcia-Contreras R., Fatty acid methyl esters (FAMEs) from castor oil: Production process assessment and synergistic effects in its properties, *Renewable Energy*, 35 (2010), 208–217.

Chen Y.H., Yu C.C., Design and control of heat-integrated reactors, *Industrial and Engineering Chemistry Research*, 42 (2003), 2791–2808.

Chilton T.H., Colburn A.P., Mass transfer (absorption) coefficients prediction from data on heat transfer and fluid friction, *Industrial and Engineering Chemistry*, 26 (1934), 1183–1187.

Danckwerts P.V., *Gas Liquid Reactions*, McGraw-Hill, New York, 1970.

Da Silva N.D., Batistella C.B., Filho R.M., Maciel M.R.W., Biodiesel production from castor oil: Optimization of alkaline ethanolysis, *Energy and Fuels*, 23 (2009), 5636–5642.

De Oliveira J.S., Leite P.M., De Souza L.B., Mello V.M., Silva E.C., Rubim J.C., Meneghetti S.M.P., Suarez P.A.Z., Characteristics and composition of Jatropha gossypiifolia and Jatropha curcas L. oils and application for biodiesel production, *Biomass & Bioenergy*, 33 (2009), 449–453.

Dimian A.C., Bildea C.S., *Chemical process design—Computer-Aided Case Studies*, Wiley-VCH Verlag GmbH, Weinheim, 2008.

Dimian A.C., Bildea C.S., Omota F., Kiss A.A., Innovative process for fatty acid esters by dual reactive distillation, *Computers and Chemical Engineering*, 33 (2009), 743–750.

Encinar J.M., Gonzalez J.F., Rodriguez-Reinares A., Biodiesel from used frying oil. Variables affecting the yields and characteristics of the biodiesel, *Industrial and Engineering Chemistry Research*, 44 (2005), 5491–5499.

Gabrielsen J., Michelsen M.L., Stenby E.H., Kontogeorgis G.M., Modelling of CO2 absorber using an AMP solution, *AIChE Journal*, 52 (2006), 3443–3451.

Hamed O.A., Aly S., AbuKhousa E., Heuristic approach for heat exchanger networks, *International Journal of Energy Research*, 20 (1996), 797–810.

Hanna M.A., Isom L., Campbell J., Biodiesel: Current perspectives and future, *Journal of Scientific and Industrial Research*, 64 (2005), 854–857.

He B.B., Singh A.P., Thompson J.C., A novel continuous-flow reactor using reactive distillation for biodiesel production, *Transactions of the ASAE*, 49 (2006), 107–112.

Janulis P., Reduction of energy consumption in biodiesel fuel life cycle, *Renewable Energy*, 29 (2004), 861–871.

Jena P.C., Raheman H., Kumar G.V.P., Machavaram R., Biodiesel production from mixture of mahua and simarouba oils with high free fatty acids, *Biomass and Bioenergy*, 34 (2010), 1108–1116.

Kenig E.Y., Gorak A., Reactive absorption, in Sundmacher K., Kienle A., Seidel-Morgenstern A. (Eds.), *Integrated Chemical Processes—Synthesis, Operation, Analysis, and Control*, Wiley-VCH Verlag GmbH, Weinheim, 2005.

Kenig E.Y., Kucka L., Gorak A., Rigorous modelling of reactive absorption processes, *Chemical Engineering and Technology*, 26 (2003), 631–646.

Kenig E.Y., Seferlis P., *Modelling reactive absorption*, Chemical Engineering Progress, January (2009), 65–73.

Kiss A.A., Novel process for biodiesel by reactive absorption, *Separation and Purification Technology*, 69 (2009), 280–287.

Kiss A.A., Separative reactors for integrated production of bioethanol and biodiesel, *Computers and Chemical Engineering*, 34 (2010), 812–820.

Kiss A.A., Heat-integrated reactive distillation process for synthesis of fatty esters, *Fuel Processing Technology*, 92 (2011), 1288–1296.

Kiss A.A., Bildea C.S., Integrated reactive absorption process for synthesis of fatty esters, *Bioresource Technology*, 102 (2011), 490–498.

Kiss A.A., Bildea C.S., A review on biodiesel production by integrated reactive separation technologies, *Journal of Chemical Technology and Biotechnology*, 87 (2012), 861–879.

Kiss A.A., Bildea C.S., Grievink J., Dynamic modelling and process optimization of an industrial sulfuric acid plant, *Chemical Engineering Journal*, 158 (2010), 241–249.

Kiss A.A., Dimian A.C., Rothenberg G., Solid acid catalysts for biodiesel production—towards sustainable energy, *Advanced Synthesis and Catalysis*, 348 (2006a), 75–81.

Kiss A.A., Dimian A.C., Rothenberg G., Biodiesel by reactive distillation powered by metal oxides, *Energy and Fuels*, 22 (2008), 598–604.

Kiss A.A., Rothenberg G., Dimian A.C., Omota F., The heterogeneous advantage: biodiesel by catalytic reactive distillation, *Topics in Catalysis*, 40 (2006b), 141–150.

Knothe G., Biodiesel: Current trends and properties, *Topics in Catalysis*, 53 (2010), 714–720.

Kucka L., Muller I., Kenig E.Y., Gorak A., On the modelling and simulation of sour gas absorption by aqueous amine solutions, *Chemical Engineering Science*, 58 (2003), 3571–3578.

Kulkarni, M.G., Dalai A.K., Waste cooking oil-an economical source for biodiesel: A review, *Industrial and Engineering Chemistry Research*, 45 (2006), 2901–2913.

Kumar N., Sharma P.B., Jatropha Curcus—A sustainable source for production of biodiesel, *Journal of Scientific and Industrial Research*, 64 (2005), 883–889.

Meher L.C., Vidya Sagar D., Naik S., Technical aspects of biodiesel production by transesterification: A review, *Renewable & Sustainable Energy Reviews*, 10 (2006), 248–268.

Nagy Z.K., Klein R., Kiss A.A., Findeisen R., Advanced control of a reactive distillation column, *Computer Aided Chemical Engineering*, 24 (2007), 805–810.

Narasimharao K., Lee A., Wilson K., Catalysts in production of biodiesel: A review, *Journal of Biobased Materials and Bioenergy*, 1 (2007), 19–30.

Noeres C., Kenig E.Y., Gorak A., Modelling of reactive separation processes: reactive absorption and reactive distillation, *Chemical Engineering and Processing*, 42 (2003), 157–178.

Puhan S., Vedaraman N., Rambrahamam B.V., Nagarajan G., Mahua (Madhuca Indica) seed oil: A source of renewable energy in India, *Journal of Scientific and Industrial Research*, 64 (2005), 890–896.

Rahimpour M.R., Kashkooli A.Z., Enhanced CO2 removal by promoted hot potassium carbonate in a split-flow absorber, *Chemical Engineering and Processing*, 43 (2004), 857–865.

Suwannakarn K., Lotero E., Ngaosuwan K., Goodwin J.G., Simultaneous free fatty acid esterification and triglyceride transesterification using a solid acid catalyst with in situ removal of water and unreacted methanol, *Industrial and Engineering Chemistry Research*, 48 (2009), 2810–2818.

Taylor R., Krishna R., *Multicomponent Mass Transfer*, Wiley, New York, 1993.

Thiele R., Faber R., Repke J. U., Thielert H., Wozny G., Design of industrial reactive absorption processes in sour gas treatment using rigorous modelling and accurate experimentation, *Chemical Engineering Research and Design*, 85 (2007), 74–87.

Van Diggelen R.C., Kiss A.A., Heemink A.W., Comparison of control strategies for dividing-wall columns, *Industrial and Engineering Chemistry Research*, 49 (2010), 288–307.

Van Gerpen J., *Biodiesel processing and production, Fuel Processing Technology*, 86 (2005), 1097–1107.

Vlad E., Bildea C.S., Plesu V., Marton G., Bozga G., Design of biodiesel production process from rapeseed oil, *Revista de Chimie*, 61 (2010), 595–603.

Yan S.L., DiMaggio C., Mohan S., Kim M., Salley S.O., Ng K.Y.S., Advancements in heterogeneous catalysis for biodiesel synthesis, *Topics in Catalysis*, 53 (2010), 721–736.

Yildirim O., Kiss A.A., Huser N., Lessmann K., Kenig E.Y., Reactive absorption in chemical process industry: A review on current activities, *Chemical Engineering Journal*, 213 (2012), 371–391.

Yuan W., Hansen A.C., Zhang Q., Vapour pressure and normal boiling point predictions for pure methyl esters and biodiesel fuels, *Fuel*, 84 (2005), 943–950.

Part VII

Case Studies of Separation and Purification Technologies in Biorefineries

18

Cellulosic Bioethanol Production

Mats Galbe, Ola Wallberg and Guido Zacchi

Department of Chemical Engineering, Lund University, Sweden

18.1 Introduction: The market and industrial needs

Due to ever increasing demand for energy, both for heating and transportation, society is facing an enormous challenge in the near future: how will it be possible to supply an increasing population with its needs? Oil shortage is an imminent threat, which no doubt will increase energy prices and is likely to affect living expenses to some extent. Another reason for introducing alternatives to fossil fuels is to secure the supply (at least to some extent) of transportation fuels. The search for alternative energy sources has been ongoing for several decades, and has become even more intense since the early to mid-2000s. In an effort to reduce our dependence on fossil fuels for transportation, a number of alternative fuels can be produced from renewable sources such as agricultural residues, forest residues, and food residues. Some of the potential fuels are alcohols such as methanol, ethanol, or butanol. These have the advantage that they are fairly easy to include in the existing infrastructure for gasoline, without totally replacing existing cars if mixtures of alcohols and gasoline are used. In a similar way, biodiesel, produced from various vegetable oils, can be mixed with diesel to reduce the amount of fossil fuels used in diesel engines. A somewhat different concept is the use of biogas, which requires more modification of cars, and completely new systems for storage and filling at filling stations. Biogas can be produced from various types of organic waste such as agricultural residues or household waste, but also from dedicated energy crops, as is the case with bioethanol.

Currently, production of bioethanol is carried out using starch- or sugar-containing raw materials, such as wheat or sugar cane. The industrial processes are mature and work well from a technical point of view. Utilization of wheat involves enzymatic liquefaction of the starch content at temperatures around 90–100 °C and then enzymatic hydrolysis to glucose followed by fermentation to ethanol. In many cases hydrolysis and fermentation are, to a large extent, performed simultaneously. In addition to ethanol, a valuable co-product is obtained, rich in proteins, which is used as animal feed. The sugar-cane process involves extraction of

Separation and Purification Technologies in Biorefineries, First Edition.
Edited by Shri Ramaswamy, Hua-Jiang Huang, and Bandaru V. Ramarao.

sucrose from the crushed sugar cane. The cane leftovers are then incinerated to produce mostly electricity and / or some process heat. Most of the bioethanol is produced in the United States and Brazil—almost 90% of the annual production in 2009. Globally, about 77 million m^3 of bioethanol was produced in 2009 [1].

The starch- and sugar-containing raw materials are also used for human needs and animal feed and it is therefore not acceptable that the increasing demand for fuel ethanol will be based on these feedstocks. Furthermore, the reduction of greenhouse gases resulting from use of sugar- or starch-based ethanol is not as high as desirable [2]. The future expansion of ethanol production must thus be based on lignocellulosic materials. These raw materials are sufficiently abundant and also available worldwide [3].

The production of alternative fuels from lignocellulose sources is often not straightforward. The intricate structure of plants and trees make them difficult to utilize "as is." Plant and tree materials in general comprise three polymers: cellulose, hemicellulose, and lignin, which make up most of the structure. The polymers are heavily entangled, thus creating a structure that is difficult to degrade without advanced chemical processing. Also, the degradation of the materials results in a multitude of chemical components. To be able to convert lignocellulosic materials to bioethanol biologically, advanced process technology has to be employed. The conversion process involves an enzymatic saccharification step where the sugars in the raw material are released as monomers. However, because of the recalcitrance of lignocellulosic materials it is necessary to perform some kind of pretreatment to make the structure accessible for efficient enzymatic attack. The released sugars are fermented to bioethanol, while the solid residue, containing mostly lignin, can be used for internal heat generation, or for production of solid fuel pellets. Excess heat is used internally in the plant or, if possible, utilized for district heating. It is of utmost importance to maximize the utilization of the raw material, since the cost for the starting materials has an impact on the minimum selling price of the fuel. Depending on the production process, some of the by-products may be of high value, which adds to the revenue. In some cases chemicals that may be detrimental to microorganisms are released from the material or may be formed in the pretreatment, such as aldehydes and organic acids, which may inhibit or even kill fermenting organisms.

The production of bioethanol from lignocellulosic materials involves several separation operations, which include material streams in various states. Not only are the starting materials heterogeneous, but the intermediate process streams consist of a vast number of compounds—and proteins and living cells—that need to be taken care of in an efficient manner. The minimum-selling price (MSP) of bioethanol is affected by separation steps that yield co-products, such as lignin for production of steam or pellets, to add to the value of the bioethanol. In this chapter, some aspects of various separation technologies applied in bioethanol production from lignocellulosic materials are discussed.

18.2 Separation procedures and their integration within a bioethanol plant

18.2.1 Process configurations

Bioethanol can be produced from lignocellulosic biomass using several different process configurations. In this section one possible process, which is frequently described in literature [4–7], is briefly described—see Figure 18.1. The process is based on dilute-acid pretreatment and simultaneous saccharification (enzymatic hydrolysis) and fermentation, so-called SSF. First the material is chipped or milled to sizes suitable for feeding into the pretreatment reactor. The feedstock is then impregnated with water or acid in either liquid (e.g., H_2SO_4, H_2SO_3, H_3PO_4, Acetic acid (HAc)) or in gaseous form (e.g., SO_2, CO_2). The feedstock will have a total dry-solids content, prior to pretreatment, between 35% and 65%. Depending on the raw material, different conditions (i.e., type of acid and concentration, residence time, temperature) during pretreatment will result in various feeding strategies for enzymatic hydrolysis and fermentation [8]. Typical pretreatment conditions are 180 °C to 220 °C for 5–10 minutes. The resulting solution (slurry) after pretreatment has a pH typically between 1.5 and 3.5.

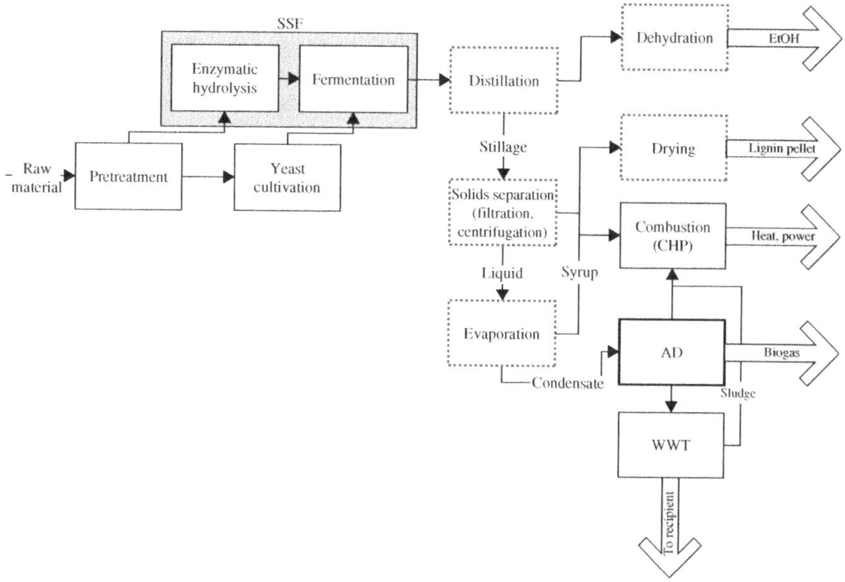

Figure 18.1 General process description of bio-ethanol production from lignocellulosic biomass. Important separation challenges are in boxes with dashed lines. AD: Anaerobic digestion; WWT: Waste-water treatment

After pretreatment, a large part of the hemicellulose sugars will have been hydrolyzed into water-soluble monomer or oligomer form. The structure of the remaining solid part has become more accessible to enzymatic attack by, for example, the effect of increased pore size, making the material suitable for enzymatic hydrolysis of the cellulose and the remaining fraction of hemicellulose. The enzymatic hydrolysis and fermentation steps can be performed either separately (SHF), or simultaneously (SSF). When SSF is used, as shown in Figure 18.1, the pretreated material may be prehydrolyzed prior to SSF to reduce the viscosity. Yeast for fermentation (or SSF) can be cultivated on-site using a part of the liquid hydrolyzate as substrate, which also to some extent adapts the microorganism to the toxic nature of the hydrolyzate and helps the yeast cells to cope with the harsh environment.

When SSF is completed, the liquid fraction of the fermentation broth contains mainly ethanol, but also a large number of other organic compounds, such as residual sugars, acetic acid and fusel oil (i.e. mainly amyl alcohols). The concentration of ethanol depends on the concentration of solids used in the SSF and on the conversion yield. In most cases the solid concentration is chosen to result in at least 4 to 5 wt% ethanol. The broth also contains a solid residue, consisting mainly of lignin, some unreacted cellulose, and yeast.

The main challenges in terms of separation processes in the bioethanol production are downstream of the fermentation, although in some cases concentration of the hydrolysates and removal of fermentation inhibitors may also be required, which involves separation steps. The interesting products in the broth are ethanol, solids, and residual low molecular-weight organic substances. Some selected processes for various separation operations are described in more detail in the following sections.

In brief, separation is employed at various stages of the process. The fermentation broth is first distilled in a stripper column, in order to separate the ethanol content from the broth. The general idea behind distilling the entire broth still containing the solids is that the ethanol losses will be negligible compared with the situation if the broth is filtered and then the liquid fraction is distilled. After the stripper, the ethanol stream is further concentrated in a rectifier column. As ethanol and water form an azeotrope, the distillate from the rectifier has to be dehydrated by means of azeotropic distillation, adsorption or pervaporation in order to yield water-free ethanol, which could be blended with gasoline.

The stillage from the stripper column is fractionated by filtration or centrifugation. The solids represent a large part of the energy content of the raw material and this energy will in part be utilized internally in the ethanol production by burning in a heat and power plant to supply the ethanol plant with steam and electricity. When only part of the solid lignin is required to supply the plant with energy the remaining solids can be dried, pelletized, and sold as a solid fuel co-product for combined heat and power plants (CHP).

The liquid part of the stillage represents one of the major challenges in lignocellulosic ethanol production. The liquid contains a large quantity of water-soluble material, which must be treated in some way before the wastewater is discharged to the recipient. Some of this liquid can be recycled back to be used for dilution purposes in the process. The recycled fraction, however, cannot be too large as non-process elements—for example, inorganic components would build up in the plant, which would lead to scaling and fouling problems. Some of the non-process elements also act as inhibitors in the biological processes upstream in the process. This puts a limit on the amounts that can be recycled. One previously suggested method [5] of handling this liquid is to evaporate to a high-enough dry-solids content to allow for combustion of the material in the heat and power plant (as indicated in [1]). This method requires a large energy input, as well as a large capital investment. An alternative method, showing promising results at laboratory scale, is to produce biogas from the organic compounds in the liquid part of the stillage by anaerobic digestion (AD). In terms of process equipment, nothing would be added in this case as biogas production from the evaporator condensate is part of the base-case process configuration and the added amount of liquid when treating the entire liquid part of the stillage is minute compared to when only condensates are used. The energy efficiency for this process alternative is also much better than when evaporation is used [6]. However, evaporation and combustion are mature technologies that would certainly work, whereas biogas production and subsequent aerobic waste water treatment are untested in this application even though research is ongoing. The main difference would also be that the inorganic materials are concentrated and in the evaporation configuration while they have to be coped with in the waste water treatment in the AD case.

18.3 Importance and challenges of separation processes

Separation challenges in the ethanol production process can be divided into two categories: energy-demanding separations and technically difficult separations. The main energy-demanding process steps in lignocellulosic ethanol production are distillation to concentrate the ethanol, adsorption to remove the final content of water, and, optionally, evaporation either of the sugar solution before fermentation or of the stillage stream, as an option to anaerobic digestion, and drying of the solid residue (mainly lignin) if this is required. In these cases the main challenge is to reduce the energy demand. One technical challenge is the filtration and washing of the solid material, either after enzymatic hydrolysis (in the SHF configuration) or after distillation (in the SSF configuration) because the solid material has a very high filtration resistance. The challenge is probably larger in the SHF case as it is important to recover all sugars present in the liquid, to minimize losses, without diluting the sugars too much.

Figure 18.2 shows the energy demand levels in a process for production of ethanol from spruce based on a yearly capacity of 200 000 ton spruce as dry matter (DM) [9]. The overall energy demand is about 25.7 MJ l^{-1} of ethanol, of which 74% is for the separation processes. The net energy demand, after recovery of the secondary steam, is about 15.2 MJ l^{-1} ethanol.

18.3.1 Distillation

The distillation unit comprises several columns, for example one or more strippers to concentrate the fermentation broth from a typical 4–6 wt% to above 20 wt% and a rectifier (Figure 18.3) to concentrate the ethanol to near azeotropic concentrations (93–94 wt%). There could also be other columns to purify the

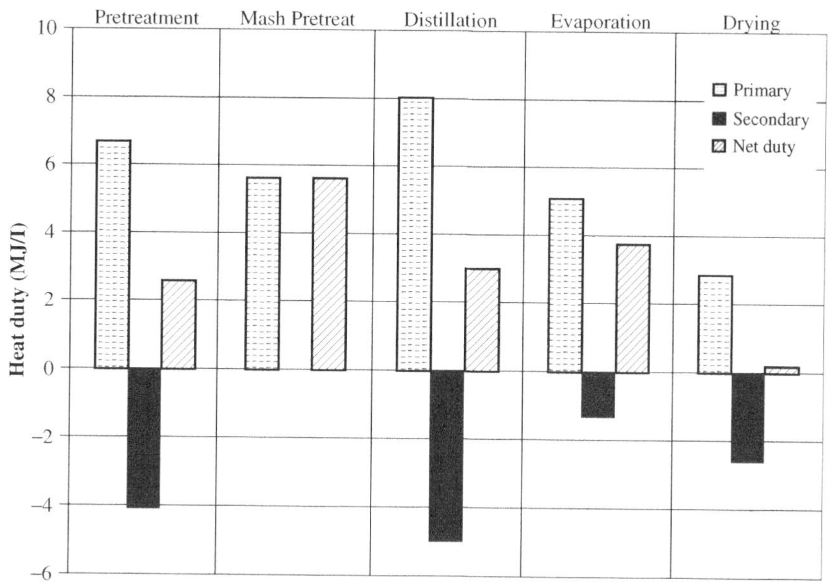

Figure 18.2 *Heat duty of the energy-demanding process steps in the proposed ethanol production process. The dotted bars represent the primary steam demand while the black bars represent the amount of secondary steam that is generated. The striped bars are the difference between the primary steam demand and the generated secondary steam.*

ethanol further, for example for removal of small amounts of aldehydes and methanol. The main columns are usually run in an energy-integrated way, with the various columns working at different pressures so that the overhead vapor from one column is used to provide heat in the reboiler of the next column when the vapor is condensed. Figure 18.4 shows the energy demand for a distillation unit comprising two stripper columns and a rectifier working in series as function of the ethanol concentration in the feed [9]. This shows the importance of reaching ethanol concentration in the fermentation (or SSF) at least above 4 wt%.

In process configurations where SSF is applied, the whole slurry after SSF is fed to the stripper units to avoid ethanol losses or dilution, which would occur in case the solid material is removed by filtration. This means that stripper columns must be capable of handling solid material, which requires special type of plates, for example, disc-and-donut trays, in order not to clog. This is an additional cost as the open structure of these trays usually has lower efficiency than normal trays, like valve trays. Further research on the development of more efficient equipment for stripping high solid containing streams may improve this.

Figure 18.4 shows that the energy demand for the distillation is significant, particularly for dilute solutions, even if multiple column distillation is performed. This leaves room for newer, more energy-efficient systems to be developed, especially based on membrane-separation techniques. In a study on a hybrid distillation-vapor permeation [10] using a hydrophilic membrane to separate the ethanol-water vapor from the stripper into a water-rich permeate and ethanol-enriched retentate vapor, the energy demand was reduced significantly. For a feed with a concentration of 5 wt% ethanol the energy demand for a stripper was 6 MJ kg^{-1} ethanol to reach a vapor of 40 wt% while it was reduced to 2.2 MJ kg^{-1} ethanol to reach an ethanol concentration of about 80 wt%. Other alternatives to distillation, especially various membrane-based techniques, are also promising but there is a need for development of membranes with higher ethanol-water selectivity and also to be resistant to fouling materials present especially from lignocellulosic based ethanol production. The alternatives to distillation are described in a review by Vane [11].

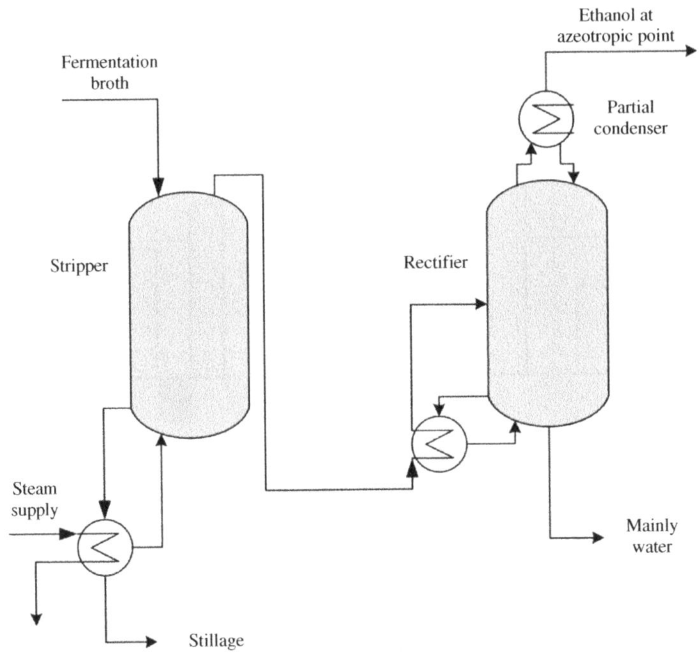

Figure 18.3 *Distillation using one stripper and a rectifier*

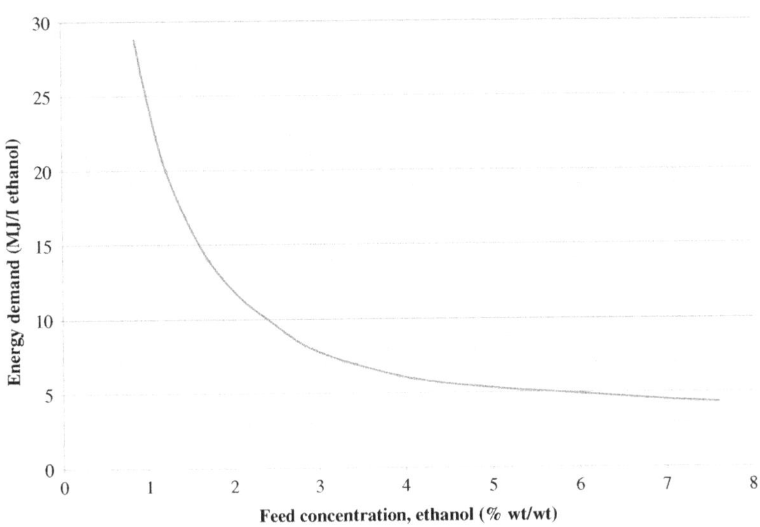

Figure 18.4 *Energy demand in the distillation step, where ethanol is concentrated to 94 wt- %, as a function of the ethanol feed concentration. The step was assumed to consist of two stripper columns (25 trays each) and a rectification column (35 trays) heat integrated by operating at different pressures. The inlet feed temperature was increased from 80°C to the boiling temperature before entering each stripper column*

18.3.2 Dehydration of ethanol

When ethanol is to be blended with gasoline it must be water free. With normal distillation the ethanol concentration that can be achieved is up to near azeotropic concentration, about 93–94 wt% so other methods have to be used to remove the final 6–7 wt% of water. In the past this was usually performed by azeotropic distillation using addition of a third component, such as benzene or cyclohexane, to form a heterogeneous azeotrope. However, this was rather energy demanding and has now in most cases been replaced by adsorption using zeolites. Another low-energy demanding method, based on membrane separation, is also available now.

18.3.2.1 Adsorption on zeolites

The dehydration of ethanol using adsorption on zeolites is performed in two packed columns, where one is adsorbing water while the other is regenerated (Figure 18.5). The adsorption column is fed with the near-azeotropic ethanol distillate from the rectifier after being superheated in a superheater. The water in this vapor stream is adsorbed on the zeolite material while anhydrous ethanol comes out as a product stream. The main part is condensed and cooled down to ambient temperature for storing while a smaller part is used to regenerate the zeolite in the second column. The regeneration is performed at lower pressure where the water desorbs and comes out together with the ethanol used for regeneration. This stream is then taken to the rectifier column to recover the ethanol and concentrate it up to near azeotropic concentration.

The two columns operate sequentially and are cycled so that one is under regeneration while the other is under operation and well before the adsorption column is saturated with water they are switched so the adsorption column is regenerated while the regenerated column works as adsorption column.

The adsorption of water in the zeolites, which have pores of about 3 Å, is exothermic, so the packed bed is heated. This heat is then used when the zeolites are regenerated. This makes dehydration by zeolites a much less energy-demanding process than azeotropic distillation. The energy that has to be provided is the heat to superheat the vapor to the adsorption column operating temperature and electricity to operate the pumps as well as the energy required to distill the ethanol/water stream from the regeneration. According to

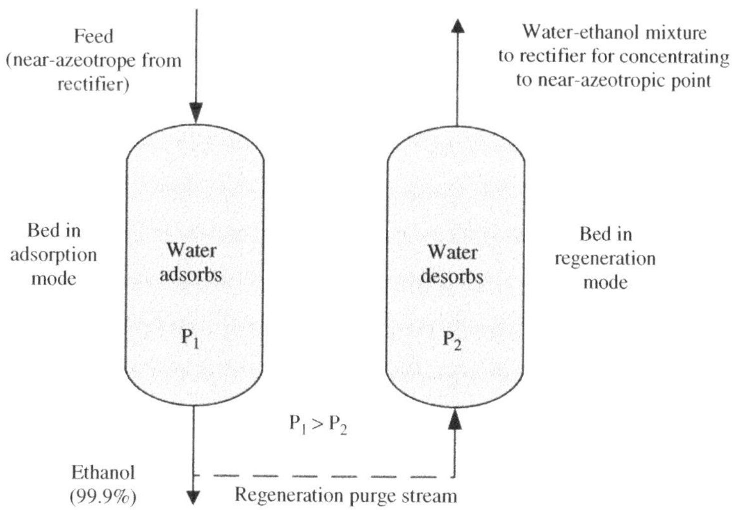

Figure 18.5 *Dehydration of azeotropic ethanol*

Swain [12], the total energy demand for the dehydration is about 1.1 MJ l^{-1} of anhydrous ethanol, where the main part is for the distillation/rectification of the stream from the regeneration column. Depending on how the rectifier is heat integrated with the other distillation columns, the actual energy demand can be lower.

18.3.2.2 Pervaporation and vapor permeation

Pervaporation and vapor permeation are two membrane processes that use differences in solubility of water and ethanol in the membrane (Figure 18.6). The main difference between the two processes from an outside view is that the feed solution is liquid in the case of pervaporation and gaseous in the case of vapor permeation. In both processes the driving force for the separation is a difference in partial pressure (or chemical potential) across the membrane. This difference can be achieved either by high temperature on the feed (or retentate) side of the membrane or by decreasing the partial pressure on the permeate side of the membrane by keeping the total pressure low or by using a sweep gas to remove the vapors. In either case a low condensation temperature is required on the permeate side of the membrane.

At some point in the process, energy needs to be added into the system to supply the heat of vaporization. In the case of vapor permeation this energy is supplied from the process while for pervaporation the energy is taken from the sensible heat of the solution, thereby lowering the temperature of the feed solution. This energy then needs to be replenished in order to maintain the driving force for the separation. This then adds complexity to the design as more heat exchangers are needed.

Separation between ethanol and water is achieved because water is preferentially dissolved into the hydrophilic membrane over the slightly more hydrophobic ethanol molecule. It is therefore only a small

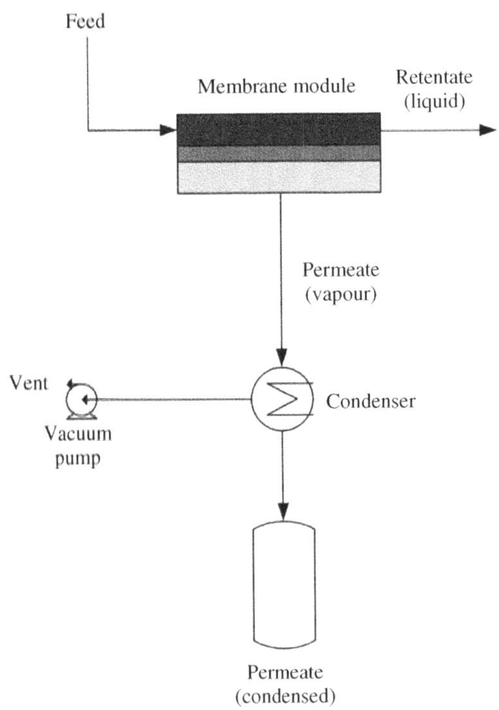

Figure 18.6 Pervaporation

amount of the feed solution (6–7 wt%) that needs to pass the membrane, leaving the ethanol on the concentrate side. The choice of membrane material will be a tradeoff between flux and selectivity and generally satisfactory performance—that is both high flux and high selectivity—will be difficult to achieve. This will lead to a product loss to the permeate, and the permeate then needs to be returned to the rectifying column, similar to the case with zeolite adsorption, in order to not lose too much ethanol.

As described earlier, pervaporation and vapor permeation can be utilized in hybrid systems together with distillation. One alternative is to use the membrane only to transcend the azeotrope and distill the ethanol solution concentrated above the azeotropic point, whereby the ethanol would become the heavy (less volatile) component collected from the bottoms of the distillation column, while the azeotropic concentration is obtained as distillate.

18.3.3 Evaporation

Evaporation is one method proposed for concentration of the soluble non-volatile compounds in the stillage stream for burning in the CHP. This could be considered as a technically rather safe method to take care of the waste water treatment in the process, at least the non-volatile compounds. However, part of the volatile compounds ends up in the evaporation condensate and needs to be taken care of by traditional waste-water treatment. The volatile compounds are, however, much less toxic and the condensate streams can be re-used in the process without problems for the yeast in the fermentation or SSF step [13]. Evaporation is an energy-demanding process, although it is usually performed as multiple-effect evaporation with typically five to six effects in series. As shown in Figure 18.4 the energy demand for a softwood-based ethanol process was estimated to be in the range of 4 to 5 MJ l^{-1} of ethanol produced, based on a five-effect evaporation unit. The energy required for the evaporation is in the same range as that obtained by burning the concentrated residue, which means that the evaporation is primarily a waste water treatment and not a method to generate more energy output from the process.

There are several ways of reducing the energy demand in the evaporation step, an example is shown in Figure 18.7. Wingren *et al.* [6] investigated the effect of increased number of effects, from 5 to 8,

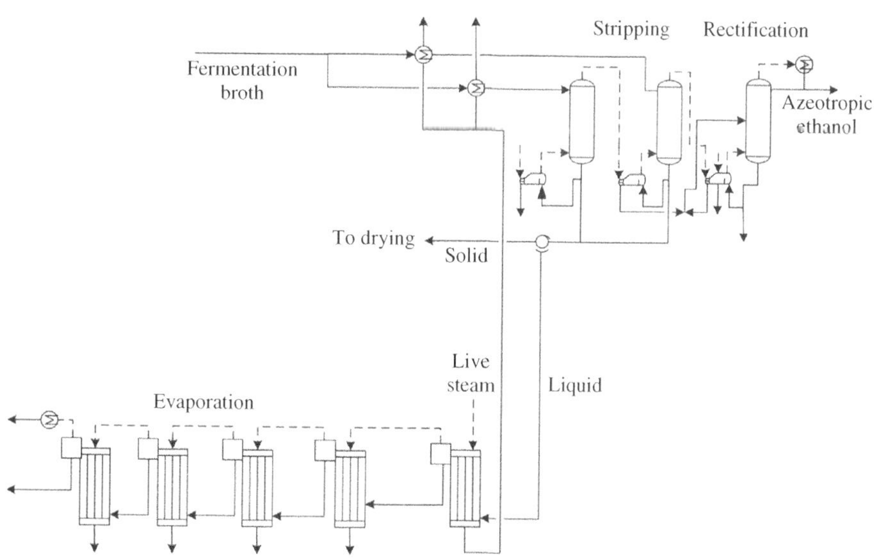

Figure 18.7 Integration of evaporation with distillation

the use of mechanical vapor recompression and also of replacing the evaporation step by an anaerobic digestion of the stillage stream in a process for production of ethanol from spruce based on 200 000 ton of raw material annually on a DM basis. The increased number of effects resulted in a decrease in the overall energy demand from 19 to 16.1 MJ l^{-1} of ethanol but had marginal effect on the production cost. Both anaerobic digestion and mechanical vapor recompression applied to the evaporation step seem more promising and resulted in both lower energy demands (9.8 and 10.2 MJ l^{-1}, respectively) and in a decreased ethanol production cost by about 7%. However, experimental work is still needed to determine if anaerobic digestion is sufficient as treatment of the stillage and if the treated water is suitable for recycling.

18.3.4 Liquid–solid separation

18.3.4.1 Filtration of solid residue (lignin)

The solid fraction remaining after enzymatic hydrolysis or after SSF has to be recovered as a solid fuel used in the CHP plant or sold as a solid fuel co-product. The separation of the solid fraction from the liquid is usually performed by filtration, for example in a filter press. These solids are rather difficult to filter. The average specific filtration resistance, α_{av}, for the solid residue after SSF of steam pretreated SO_2 impregnated spruce is in the range of 10^{13}–10^{14} m kg^{-1} [14]. The challenge is probably larger in the SHF case as it is important to recover all sugars present in the liquid, to minimize losses, without diluting the sugars too much. In this case it is important that the solid cakes have not been compressed during filtration in order to avoid low wash-flow rates [15].

In the SSF configuration the filtration is performed after the stripper. This means that the material is heated before filtration, which has a positive effect. Filter media could also be added, such as saw dust or other untreated biomass, but the economic criteria need to be considered. There is certainly room for improvement of this separation step and to obtain a better understanding on how the material properties of these materials influence the cake formation and the filtration resistance.

18.3.4.2 Recovery of yeast

When fermentation is performed on the liquid fraction—in the SHF concept—the yeast can be recovered and used again. It is common practice to separate the yeast from the fermentation broth by centrifugation. Centrifuges are rather expensive and require regular maintenance. An alternative to centrifugation is to use filtration—either conventional plate-and-frame filters or membrane-based micro filtration. Matta and Medronho [16] investigated separation of yeast from fermentation broth by filtration in a stainless steel plate-and-frame filter assisted by the use of filtration aids. The filter aid, perlite, was then recovered from the yeast using hydro cyclones. This resulted in a yeast cell recovery of about 85%. The recovery of yeast can probably be improved in the future but it is, at least currently, a real challenge to do so from the SSF step.

New methods have been suggested to combine the benefits of SHF and SSF by performing the SSF in a hybrid mode. Figure 18.8 shows a schematic flow sheet for such a process. In this case the enzymatic hydrolysis (EH) is performed in a separate vessel from the fermentation but the liquid is circulated from the EH to the fermentor and back again. In this way sugars released in the EH are removed, thus avoiding end-product inhibition, and the sugars are consumed in the fermentor, without being in contact with the solid material in the EH, which makes it possible to recycle the yeast. In order to achieve these advantages, the liquid from the EH has to be continuously separated from the solid, and the liquid from the fermentor has to be separated from yeast. This needs development of filters and membranes to work well with this type of mixtures.

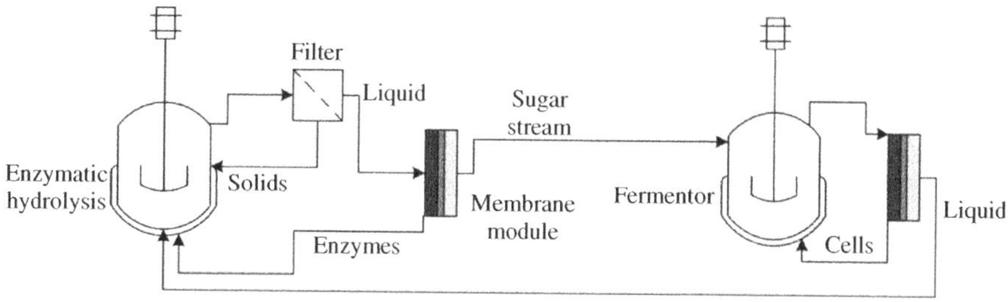

Figure 18.8 *Hybrid enzymatic hydrolysis and SSF*

18.3.5 Drying of solids

As mentioned in Section 18.3.4.1, the solids can be used to produce process heat and electricity. However, in most process simulations [6, 7] the solid residue, with a large margin, covers the energy needed for the ethanol processing. One option for the remaining solid fraction would be to market it as a solid fuel for additional byproduct income. The solids from the filtration step usually have a dry-matter content of 40–50 wt%, which means that in order to obtain a solid fuel product they should be dried to at least 85 wt% dry matter and then be pelletized. When used internally for heat and power production there could also be a benefit to dry the solids if this can be made in an energy-efficient way.

Drying is a process that uses a large amount of energy. The energy input is needed to evaporate the water in the substrate (2.0–2.5 MJ kg^{-1} water depending on pressure). It is also technically difficult (but not entirely impossible) to reduce the energy needed by connecting several dryers in series, in a similar way to multiple-effect evaporation discussed earlier. There are means that could be employed to reduce the specific energy demand of the drying process. Preferably, the substrate should be preheated as much as possible employing low-grade heat; the substrate should also be fed at a constant rate and the drying temperature should be high (in air drying, if permitted by the dried material) to allow for a large moisture uptake by the air.

In a second-generation bioethanol plant, several different drying technologies could be considered. The most important consideration is the potential for energy integration of the drying operation with other energy-intensive processes in an ethanol plant, and the way this integration influences the product mix of the ethanol plant. There are three main options to consider:

- air dryer heated to low temperature by waste heat;
- air dryer heated by back-pressure steam;
- superheated steam dryer heated by high-pressure steam.

These systems individually have different benefits and drawbacks [17].

18.3.5.1 Air dryer heated to low temperature by waste heat

The major benefit of low-temperature drying using waste heat streams is that no extra energy input is required. However, it relies on a sufficient amount of non-utilized waste heat being available within the production system. As a lower temperature is used, the drying time will be longer; thus, the equipment size becomes larger, which will affect the capital cost for the equipment. Other alternatives may be available for the waste heat, such as district heating, which will be negatively influenced by this type of drying operation.

18.3.5.2 Air dryer heated by back-pressure steam

This type of dryer relies on a primary energy supply in the form of live steam and consequently the amount of solid fuel that needs to be dried is reduced (as a somewhat larger part of the solids is incinerated to provide the additional energy). Electricity production will increase somewhat as steam is withdrawn after the steam turbine, which shifts the energy co-product from being a solid fuel towards electricity. A higher temperature is used, so the drying equipment can be made smaller. There might be a possibility to utilize the waste heat from the dryer as low-grade heat for preheating purposes or district heating. However, in most cases this type of dryer will increase the net heat energy demand of the process considerably.

18.3.5.3 Superheated steam dryer heated by high pressure steam

In superheated steam drying, the drying medium is superheated steam, instead of air. The moisture in the substrate evaporates into steam, which at a point reaches the saturation temperature. Saturated steam can therefore be extracted from the dryer and the latent heat can be used in different processes in the ethanol plant as an energy source, provided that the steam pressure in the dryer is sufficiently high—for example, 4 bar. According to Mujumdar, by utilizing a superheated steam dryer, the net energy consumption can be lowered to 1000–1500 kJ kg^{-1} water evaporated compared with 4000–6000 kJ kg^{-1} for more conventional dryers [18].

The equipment for superheated steam drying is more complicated than for a conventional dryer as it is constructed to work at higher pressures. Thus, the risk for leaks in the system is enhanced, which may result in a quite significant energy loss and may also contribute to local environmental pollution and smell problems. It is also more difficult to feed in the moist material, against a higher pressure, without causing leaks from the steam atmosphere. Superheated steam drying usually takes place at higher temperatures compared to the other two alternatives, which can affect the dried product if it is heat sensitive. The superheated steam dryer will also affect electricity production as steam is withdrawn from the turbines at higher pressures compared with the other two alternatives. On the other hand, the net energy demand of the process will be reduced (see Figure 18.2) as the secondary steam from the drier will be used to replace back-pressure steam from the turbine. This also relies on there being sufficient demand for low-grade energy, which could use this steam. There is certainly room for improvements in this type of dryer, including the possibility to use vapor recompression of steam from an atmospheric dryer and by this way utilize the secondary heat with diminished leakage problems.

18.3.6 Upgrading of biogas

As mentioned earlier, AD of the stillage stream to biogas is a promising alternative to evaporation. If the biogas from the AD is used as transportation fuel or distributed in the natural gas grid, it has to be upgraded to increase the heating value of the product gas. This means removal of undesired compounds in the gas mix, mainly carbon dioxide, which can constitute 30–60% of the gas from the AD. However, several other impurities, such as hydrogen sulfide, and siloxanes, may be present in the gas, which, depending on their concentrations, need to be removed [19].

The upgrading can be performed in several ways, such as by physical and chemical absorption (e.g., amine scrubbing), by adsorption (e.g., pressure swing adsorption), and by permeation (e.g., membrane separation).

18.4 Pilot and demonstration scale

One of the most important issues is to verify all process steps in an integrated way at a pilot and/or demonstration scale (Table 18.1). These plants should provide data for full-scale ethanol production and

Table 18.1 Pilot and demonstration plants for ethanol production from cellulosic materials in operation in Europe and USA [22]

Company	Location	Raw material	Production ethanol (m^3/year)
Abengoa Bioenergy	Spain, Salamanca	Cereal straw	4000
AE Biofuels	USA, Butte	Switchgrass + other	600
DuPont Danisco Cellulosic Ethanol	USA, Vonore	Corn residues, Switchgrass	900
Etanolpiloten i Sverige AB	Sweden, Örnsköldsvik	Wood + sugar cane bagasse	100
Iogen Corporation	Canada, Ottawa	Wheat, barley and oat straw	2000
Inbicon (Dong Energy)	Denmark, Kalundborg	Wheat straw	5000
KL Energy Corporation	USA, Upton	Wood waste	6000
Lignol Energy Corporation	Canada, Burnaby	Wood residues	100
Mascoma Corporation	USA, Rome	Wood, Switchgrass + other	600
MossiGhisolfi—Chemtex Italia	Italy, Tortona	Corn stover, straw and wood material	60
POET	USA, Scotland	Corn residues	75
Verenium	USA, Jennings	Sugar cane bagasse, energy cane	200
Verenium	USA, Jennings	Sugar cane bagasse, energy crops, wood and switchgrass	5000

at the same time define new challenges for further research. The step from pilot- and demonstration-scale production of lignocellulosic ethanol to competitive full-scale production requires further reduction of the production cost. At pilot and demonstration scale it is important to prove the whole process for longer periods of time with variations in raw material, which can be normal variations in crops due to where they are grown and the climate as well as changes during storage. High accessibility and verification of yields and productivities obtained at lab and bench scale has also to be achieved, especially for critical process steps such as pretreatment and SHF or SSF. There are also more technical issues like separation of lignin and the influence of process integration and recycling of process streams on fouling. The qualities of products and co-products, and waste water treatment, are other important factors.

In recent years several companies have moved from research to pilot and demonstration-scale plants where all steps are integrated into a continuous process, which gives more reliable data for economic analysis and scale up to industrial plants. One of the most rigorous pilot plants that operates in a continuous mode and where all steps are integrated is the pilot plant in Örnsköldsvik, Sweden [20], which was put into operation in 2005. It is optimized to process wood but has lately also been used to process agricultural residues. The demonstration plant that has been in operation the longest time is the Iogen plant in Ottawa, which went into operation in 2004. It has a capacity to convert about 1600 ton per year of wheat, barley or oat straw to ethanol [21]. Since then several other pilot and demonstration plants have started operation, mainly designed for the use of agricultural residues, like corn stover and different types of straw. In Europe demonstration plants are currently in operation in Salamanca, Spain (owned by Abengoa) and in Kalundborg, Denmark (owned by Inbicon, a subsidiary of Dong Energy); several demonstration plants are also operated in the United States. These and also planned new facilities, including commercial-scale plants, are presented in a report from the IEA/Task39 group, which is available on the web [22]. Most available pilot and demonstration plants are designed for steam explosion or acids for pretreatment, followed by enzymatic hydrolysis and fermentation

18.5 Conclusions and future trends

So far, the main efforts to make lignocellulosic ethanol production feasible from an economic point of view have been focused on conversion steps—pretreatment, enzymatic hydrolysis and fermentation. Considerable resources have been spent to reduce the cost of enzymes and to develop microorganisms that can ferment all sugars in hydrolysates, both hexoses and pentoses, in an efficient way. However, although these steps have been improved, attention must also be directed to the remaining steps in the process. Although the separation steps are more mature and in most cases proven at large scale, they are in general very energy demanding; thus, improvements are required in the future to increase the energy efficiency of the whole process to improve the economics. This includes both improvements to current separation methods, including energy integration, and the development of alternative, less energy-demanding processes, such as the use of membrane technologies where applicable. The development of the separation steps has to be considered as an integrated part of the improvement of the whole process and must take into consideration specific aspects of each process configuration.

A future trend is also the combined production of several products from biomass in so-called biorefineries. This could comprise conversion to a variety of energy products in biomass based energy combines, like ethanol, biogas, electricity, solid fuels and district heating [23], or a combination of fuels and chemicals for the production of a variety of products and materials based on fossil raw materials. Examples are the use of hemicelluloses as precursors for polymers [24] or the use of lignin for bulk chemicals [25], or as materials replacing fossil materials [26] or extractives. These types of compounds will all require an increased fractionation of the raw material and more separation steps to purify the various products.

Future biorefineries have significant potential for improvement and development in terms of products, individual process steps, process configurations, and process integration options. This makes development very complex. A huge amount of knowledge will be required to develop optimal production systems, which are not yet available. This would include improved separation methods, more adapted to this kind of material, as most conventional separation methods have been developed for fossil-based raw materials.

References

1. Swedish Energy Agency (2011) Analys av marknaderna för etanol och biodiesel, Swedish Energy Agency, Sweden, ER 2011:13, ISSN 1403–1892.
2. Farrell AE, Plevin RJ, Turner BT, Jones AD, O'Hare M and Kammen DM (2006) *Science* 311, 506–508.
3. Kamm B, Gruber PR and Kamm M (2006) In: *Biorefineries—Industrial Processes and Products, Volume 1*, Eds.: Kamm B, Gruber PR, Kamm M, Wiley-VCH Verlag GmbH, Weinheim, Germany, pp. 1–33.
4. Lynd LR, Elander TE and Wyman CE (1996) Appl Biochem Biotechnol 57/58, 741.
5. Aden A, Ruth MF, Ibsen K, Jechura J, Neeves K, Sheehan J and Wallace B NREL/TP-510-32438 2002.
6. Wingren A, Galbe M and Zacchi G (2008) *Bioresource Technol* 99, 2121–2131.
7. Sassner P and Zacchi G (2008) *Biotechnol Biofuels*, 1, 4 (doi:10.1186/1754-6834-1-4).
8. Galbe M and Zacchi G (2007) Pretreatments of lignocellulosic materials for efficient bioethanol production. In: *Adv Biochem Eng/Biotechnol* (108) Ed.: Olsson L., Springer Verlag, Berlin-Heidelberg, Germany, pp.41–65.
9. Galbe M, Sassner P, Wingren A and Zacchi G (2007) In: *Adv Biochem Eng/Biotechnol* (108) Ed.: Olsson L. Springer Verlag, Berlin-Heidelberg, 303–327.
10. Vane LM, Alvarez FR, Huang Y and Baker RW (2010), *J Chem Technol Biotechnol* 88, 502–511.
11. Vane, ML (2008) Biofuels, *Bioprod Bioref* 2, 553–588.

12. Swain RLB (2003) *In: The Alcohol Textbook*, 4th edition. Ed.: Jacques KA, Lyons TP, Kelsall DR, Nottingham Univ. Press, Nottingham, UK, pp. 337–342.
13. Alkasrawi M, Galbe M and Zacchi G (2002) *Appl Biochem Biotechnol*, 98–100, 849–861.
14. Dingwell K, Sedin M and Theliander H (2010) *Proc 13th Nordic Filtration Symp* Ed.: Antti Häkkinen, June 10–11, 2010, Lappeenranta, Finland, pp. 58–65.
15. Johansson C (2005), Pressure and Solidosity Profiles in Cake Filtration. *An Experimental Study on Filtration Properties of Lignin from Bioethanol Production and Model Materials*. PhD Thesis, ISBN/ISSN:91-7291-709-1, Chalmers University of Technology, Gothenburg, Sweden.
16. Matta MV and Medronho A (2000) *Biosep*, 9, 43–53.
17. Wimmerstedt R (2007) In: *Handbook of Industrial Drying*, 3rd edition, Ed.: Mujumdar AS, CRC Press, Boca Raton, FL, 2007, pp. 743–753.
18. Mujumdar AS (2007) In: *Handbook of Industrial Drying*, 3rd edition, Ed: Mujumdar AS, CRC Press, Boca Raton, FL, 2007, pp. 439–452.
19. Greer D (2010) *BioCycle*, 51(2), 27–30.
20. SEKAB E-Technology AB, http://www.sekab.com/cellulose-ethanol/demo-plant (accessed September 19, 2012).
21. Iogen Corp., http://www.iogen.ca/company/demo_plant/index.html (accessed September 17, 2012).
22. Bacovsky D, Dallos M and Wörgetter M (2010) Status of 2nd Generation Biofuels Demonstration Facilities T39-P1b 27 http://www.ascension-publishing.com/BIZ/IEATask39-0610.pdf (accessed September 17, 2012).
23. Barta Zs, Reczey K and Zacchi G, (2010) Biotechnol Biofuels 3, 21.
24. Persson T, Ren JL, Joelsson E and Jönsson A-S (2009), *Bioresour Technol* 100, 3906–3913.
25. Van Haveren J, Scott EL and Sanders J (2007) *Biofuels Bioprod Biorefin* 2, 41–57.
26. Lora JH and Glasser WG (2002) *J Polym Environ* 10, (1/2), 39–48.

19

Dehydration of Ethanol using Pressure Swing Adsorption

Marian Simo

Praxair Technology Center, New York, USA

19.1 Introduction

Fuel-grade ethanol is the fastest growing market for the ethanol today. Fermentation-derived ethanol comprises about 10% of all motor vehicle fuels in the United States. During the production process, the ethanol contained in the fermentation broth is enriched up to 92–95 wt% by distillation. Further enrichment of ethanol must obviate the azeotropic point. Conventionally, the final purification was done by azeotropic distillation [1]. With the development of adsorption processes and the invention of molecular sieves, the pressure swing adsorption (PSA) process completely replaced the azeotropic and extractive distillation routes for ethanol dehydration due to performance, cost, and environmental reasons [2].

In the original PSA process invented by Skarstrom [3], the two steps of adsorption and desorption/purge are carried out in two adsorbent beds operated in tandem, enabling processing of a continuous feed. A similar Guerin-Domine cycle, basis for the modern vacuum swing cycle, was patented later [4]. The adsorbent is regenerated by rapidly reducing the partial pressure of the adsorbed component by lowering the total pressure and/or by using a purge gas.

The first adsorption-based patents on the ethanol dehydration process were filed very early. In 1938, Derr described a method of producing anhydrous and absolute alcohol through a bed of freshly reactivated alumina moistened with liquid alcohol [5]. Oulman *et al*. described a process of concentrating relatively dilute aqueous solutions of ethanol by passing them through a bed containing granules of crystalline silica polymorph, such as silicate, which preferentially adsorbs ethanol [6].

Ethanol can also be dehydrated by adsorption with biomaterials such as cornmeal, cellulose or cornstarch [7]. The liquid phase adsorption process with cornmeal, employing a countercurrent moving bed or a simulated moving bed countercurrent flow system, was described in a patent by UOP Inc. [8].

Separation and Purification Technologies in Biorefineries, First Edition.
Edited by Shri Ramaswamy, Hua-Jiang Huang, and Bandaru V. Ramarao.

The gaseous phase version of UOP process was proposed later by Ladisch and coworkers [9]. Their cornmeal adsorption/distillation process did not succeed on a large scale probably due to problems associated with handling and transport of solid materials. Also, an energy-intensive temperature swing adsorption (TSA) process with air or nitrogen as the heating medium was used to regenerate the adsorption beds.

The use of zeolites for ethanol drying was first suggested in patents by Ginder and Greenbank [10, 11]. Fornoff describes a process for dehydration of ethanol comprising distilling a crude aqueous ethanol feedstock to produce a gaseous ethanol-water mixture containing about 90 wt% of ethanol, drying the mixture in the presence of carbon dioxide with a crystalline zeolite 3Å and allowing the product ethanol to condense at ambient temperatures [12].

The liquid phase adsorption with a 3Å zeolite was investigated experimentally by Teo and Ruthven [13]. It was concluded that the adsorbent exhibited high-water adsorption capacity; however, the adsorption kinetics was very slow (possibly due to slower diffusion in the liquid phase). The study by Carton *et al.* compared the separation performance for both vapor and liquid adsorption system utilizing a 3Å zeolite [14]. The breakthrough curves were more favorable in the vapor phase due to faster adsorption kinetics and the treatment capacity in the vapor phase was superior to the one obtained in the liquid phase.

The PSA process with a 3Å zeolite for the ethanol production was introduced to the industrial practice in 1980s and practically did not change since [15]. The pressure swing adsorption process utilizes a 3Å molecular sieve that preferentially adsorbs water [14, 16]. The column pressure can be changed much faster than the column temperature. The fast changes in the pressure enable short cycle times and thus much higher throughput is possible in the case of the PSA compared to the TSA systems. As a result, the productivity of a PSA system defined as the amount of dry ethanol produced per hour per ton of adsorbent is much higher. The PSA process is attractive due to a low energy consumption, its capability of producing very dry product, and proven commercial experience (Brown C. J. personal communication, 2005). No solvent, entrainer or salt is needed; and thus the expenses for the regeneration, transportation or storage do not apply here and the final product is not contaminated. The PSA is now the technology of choice for the ethanol dehydration for both small and large plants [17].

19.2 Ethanol dehydration process using pressure swing adsorption

19.2.1 Adsorption equilibrium and kinetics

In the ethanol PSA process the water is removed with appropriately sized molecular sieve adsorbent. Zeolite 3A is a synthetic crystalline potassium aluminosilicate obtained by ion exchange from the sodium form of zeolite A (NaA also known as 4A zeolite). Synthetic zeolites are highly ordered structures. The main interconnecting 3D pore system in zeolite 3A consists of α cages separated by 3Å restrictions [18]. The collision diameters for water and ethanol are 2.7 Å and 4.3 Å respectively. As a result, water molecules are allowed to enter the structure and eventually adsorb on the surface while the ethanol molecules are excluded due to steric hindrance.

The thermodynamic relationship between the adsorbent equilibrium loading capacity and the component partial pressure at constant temperature is the adsorption isotherm. The water-3A zeolite adsorption equilibrium can be obtained through various established experimental techniques [19] and the data is available in the open literature [18, 20–22].

The water adsorption isotherm on 3A zeolite is shown in Figure 19.1 [21]. The adsorption isotherm approaches the rectangular shape at low temperatures, which translates to high adsorption capacity and very strong adsorption forces between the water molecules and the adsorbent surface. At high temperatures, the isotherm approaches linear shape, the adsorption capacity decreases and so does the interaction energy.

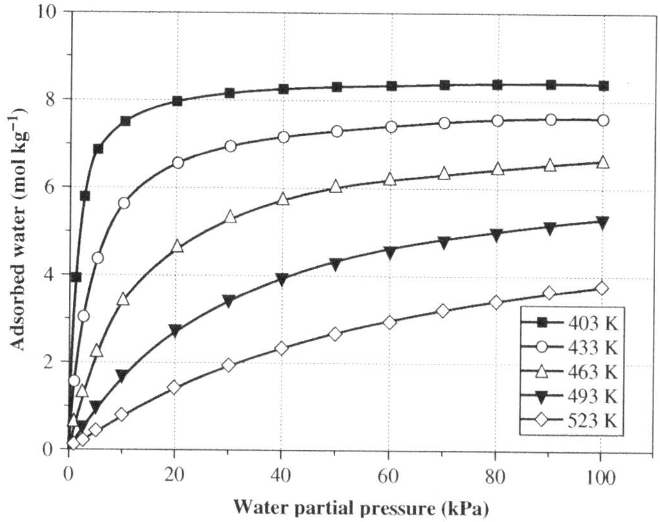

Figure 19.1 Water equilibrium isotherms on 3A zeolite. The amount of water adsorbed in mol kg^{-1} is plotted as a function of water partial pressure in kPa at different temperatures

Water adsorption isotherm on 3A zeolite is classified as "favorable" [23] meaning that the system exhibits higher water adsorption capacity at low fluid phase concentrations. However, a very favorable adsorption isotherm is "unfavorable" for the desorption process. The linear isotherm is equally favorable for both adsorption and desorption. Thus, the selection of proper operating temperature for the cyclic process is critical when optimizing process performance and energy efficacy.

The adsorption of ethanol on 3A zeolite in the ethanol dehydration studies is usually neglected [20, 22]. The sorption capacity of ethanol on a 3A zeolite was measured using microcalorimetry [24] and breakthrough apparatus [21]; the reported ethanol sorption capacity was 0.044 mol kg^{-1} at 100 °C and 0.03 mol kg^{-1} at 167 °C, respectively. In both cases, it was concluded that the extent of ethanol adsorption was minimal. Recent computational study has confirmed that the ethanol co-adsorption can be safely neglected with regard to the PSA process design and performance [25].

The adsorption equilibrium isotherm, used to calculate the sorption capacity and also the heats of adsorption [19], is the most important characteristic of the adsorption system. The adsorption kinetics represents the second necessity required to predict the process performance through the mathematical modeling. In general, the zeolite particle (bead) uptake rate can be controlled by the mass transfer resistances in series due to: (i) laminar fluid film separating the particle from the bulk fluid, (ii) macropores, acting as conduit to transport the gas molecules from the surface to particle interior, (iii) surface barrier, where molecules are adsorbed on the surface of zeolite crystals (pore mouth) and further (iv) diffusion within the zeolite crystals (activated micropore diffusion). The last step is the fast physical adsorption on the surface and concomitant liberation of the heat of adsorption. The micropore resistance was previously identified as the sole diffusion mechanism for the adsorption of water on 3A zeolite [13, 20, 22]. A recent study revealed that above 100 °C more than ~40% of the overall diffusion resistance is due to fluid film and macropore diffusion and ~60% is due to micropore resistance. The contribution from the micropore diffusion decreases as the temperature increases and above 200 °C becomes completely negligible [21].

Precise design of the PSA unit is a difficult task because of many interacting operational parameters characterizing this separation process. Pilot-scale experiments are time consuming and economically demanding. These reasons have lead to the development of mathematical models which are used to guide the design of the PSA process and to optimize the process performance [17]. For any model, it is very important to use reliable and accurate equilibrium and kinetic data.

19.2.2 Principle of pressure swing adsorption

The pressure swing adsorption is a cyclical process consisting typically of two or more vessels packed with the adsorbent [26]. The high partial pressure of component to be removed translates to high adsorption capacity according to adsorption isotherms shown in Figure 19.1. After the adsorber is saturated, the pressure in the vessel is reduced (sometimes using a vacuum) resulting in low partial pressure of the impurity and as a consequence the molecules desorb from the surface. The purge gas is often used to further dilute the gas phase thus enhancing the regeneration process.

The temperature-swing adsorption process is based on the fact that, for a fixed partial pressure, the sorption capacity is high at low temperatures (adsorption) and low at high temperatures (regeneration). The TSA processes use much longer cycles since it takes a considerable amount of time (hours or days) to heat up tons of adsorbent material packed in the vessels. As a result, the TSA systems are typically larger and more expensive then the PSA systems. The TSA systems are almost exclusively used for the trace separations and not for the bulk separations such as ethanol PSA process.

Each process has its own characteristic advantages and disadvantages. The TSA is energy intensive because of the need to supply heat for the regeneration. It is preferred for very strongly adsorbed components, because a modest change in temperature produces a large change in the gas-solid adsorption equilibrium. On the other hand, the PSA technology is best suited for components that are not too strongly adsorbed. However, the short cycle time (seconds or minutes) characteristic for the PSA systems creates unique challenges in other aspects of process design and development, for example, the adsorbent mechanical properties [26].

19.2.3 Ethanol PSA process cycle

Currently, two- and three-bed PSA ethanol dehydration processes are common in the industry. There are some differences in the efficiency, cost and performance; however, the overall trade-off between the two and three bed processes is only marginal due to the fact that both processes use essentially the same PSA cycle, see Tables 19.1 and 19.2 respectively.

The pressure history for a single adsorber in two bed process is depicted in Figure 19.2. The sequence of elementary steps and interactions between the beds are depicted in Figure 19.3 [17].

19.2.3.1 Two-bed ethanol PSA cycle steps

1. *Adsorption step*. The water-ethanol vapor stream is fed from the top of the bed at 379.2 kPa and 440 K. The high pressure product stream (preferably dry ethanol) is collected at the bottom of the bed. The adsorption step takes about 345 s. During this step significant amount of heat is released upon water adsorption.
2. *Countercurrent blowdown*. During the blowdown the pressure is decreased from 379.2 kPa to approximately 137.9 kPa in about 60 s and the pressure decrease with time is approximately linear. The bed pressure is further reduced by applying vacuum from the top of the vessel (~13.8 kPa). The whole step takes approximately 210 s.
3. *Purge*. The bed is purged counter currently by a product stream at the lowest possible pressure. This step is very short; in the ethanol plant, it takes only 15 s.
4. *Pressurization*. The bed needs to be pressurized to high pressure in order to enter the adsorption step. The product stream is used for a countercurrent pressurization of the bed from the bottom of the vessel. The pressure is increased from 13.8 to 379.2 kPa in approximately 120 s.

Table 19.1 *Two-bed ethanol PSA cycle*

Bed 1	ADSORPTION			BLOWDOWN	PURGE	PRESSURIZATION
Bed 2	BLOWDOWN	PURGE	PRESSURIZATION	ADSORPTION		

Table 19.2 *Three-bed ethanol PSA cycle. Reprinted from [18] © 2008, with permission from Elsevier*

Bed 1	ADSORPTION	BLOWDOWN	PURGE PRESSURIZATION
Bed 2	PURGE PRESSURIZATION	ADSORPTION	BLOWDOWN
Bed 3	BLOWDOWN	PURGE PRESSURIZATION	ADSORPTION

It is important that the rate of pressure change during the blowdown and pressurization steps is performed in a controlled fashion; high upward velocities could "lift the bed" owing to the onset of fluidization. The rate of pressure changing steps is restricted due to limited mechanical stability of the adsorbent material; e.g. faster pressurization can result in the occurrence of zeolite dust particles in the product stream. For depressurization steps, the size of the exhaust condensers and vacuum level used can be additional rate-limiting factors.

19.2.4 Process performance and energy needs

The ethanol PSA is continuously processing wet ethanol feed stream containing 7–9 wt% water delivered at elevated pressure (275–450 kPa, absolute pressure). The mixture is super-heated using a heat exchanger upstream of the PSA unit. The inlet temperature depends on the adsorption pressure; typical temperatures are in the range from 420 to 450 K. The feed stream is passed through one of the vessels available and the dry ethanol product is obtained at the adsorber outlet (>99.5 wt% ethanol). The other bed(s) is undergoing the regeneration stage comprised of several steps. The water adsorbed in the adsorption step

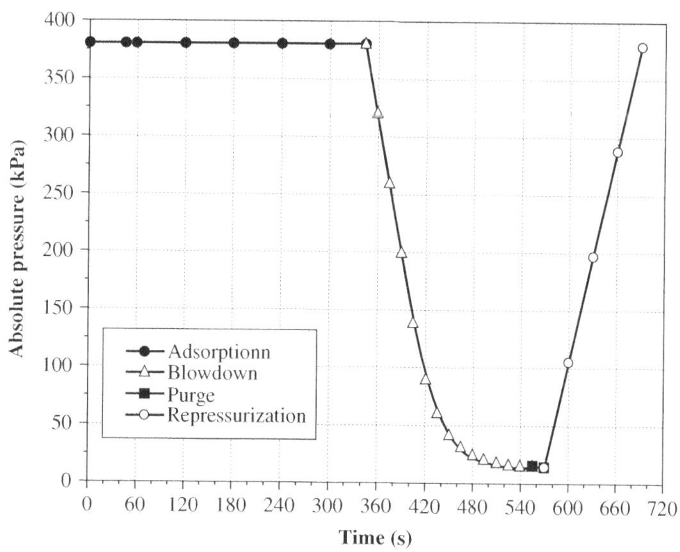

Figure 19.2 *Pressure history for two-bed ethanol PSA cycle. Reprinted from [18] © 2008, with permission from Elsevier*

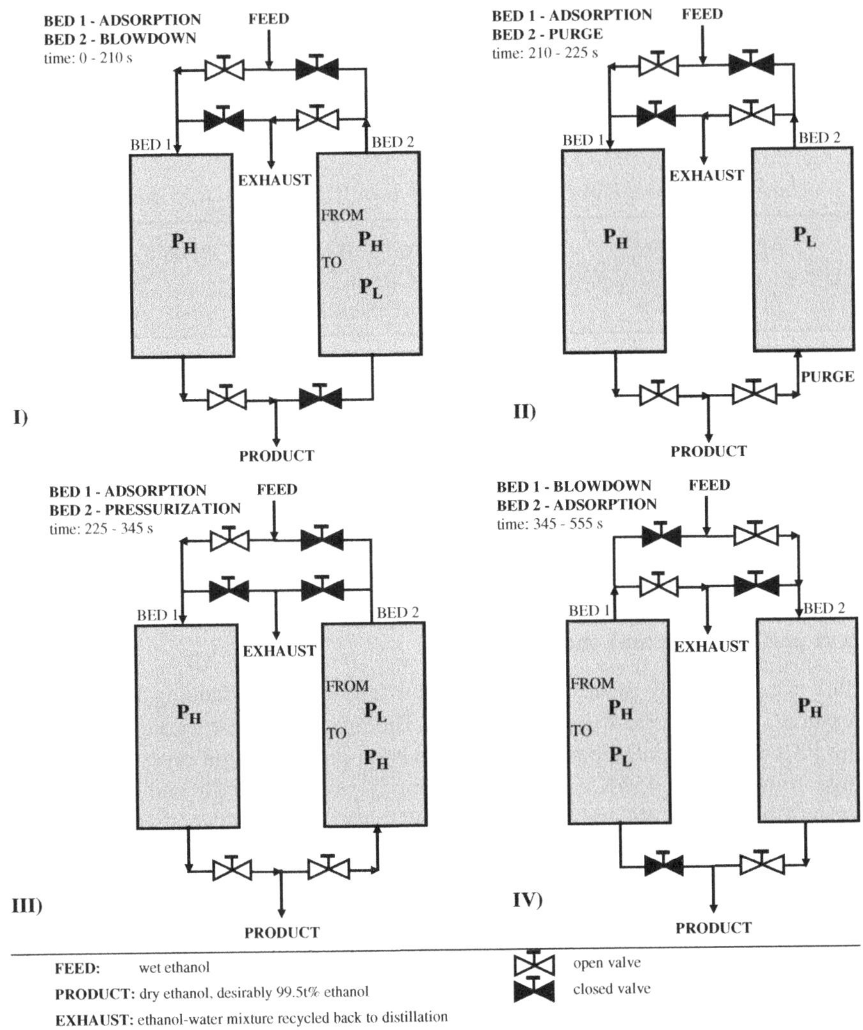

Figure 19.3 *Half cycle for ethanol two-bed PSA cycle (I) to (III). Reprinted from [18] © 2008, with permission from Elsevier*

leaves the PSA process during the regeneration stage carrying significant amount of ethanol as well. This stream is usually recycled back to the rectifier. Simplified flowsheet for the two bed ethanol PSA process is shown in Figure 19.4.

The ethanol recovery of the PSA process as reported by industry is typically between 85% to 95% (Brown C. J. personal communication, 2005). The recovery is defined as the ratio of the amount of ethanol in the final product to the amount of ethanol in the feed. The actual value depends on many factors such as the operating pressure, temperature, number of beds, size of the beds, cycle time, and the amount of purge gas used, to name a few.

Ethanol is used to purge and pressurize the bed. The only step where the ethanol is lost is during the blowdown and purge steps. The ethanol-rich condensate from the initial blowdown step is recycled and introduced to the top of the rectifying column while the water rich condensate collected towards the end

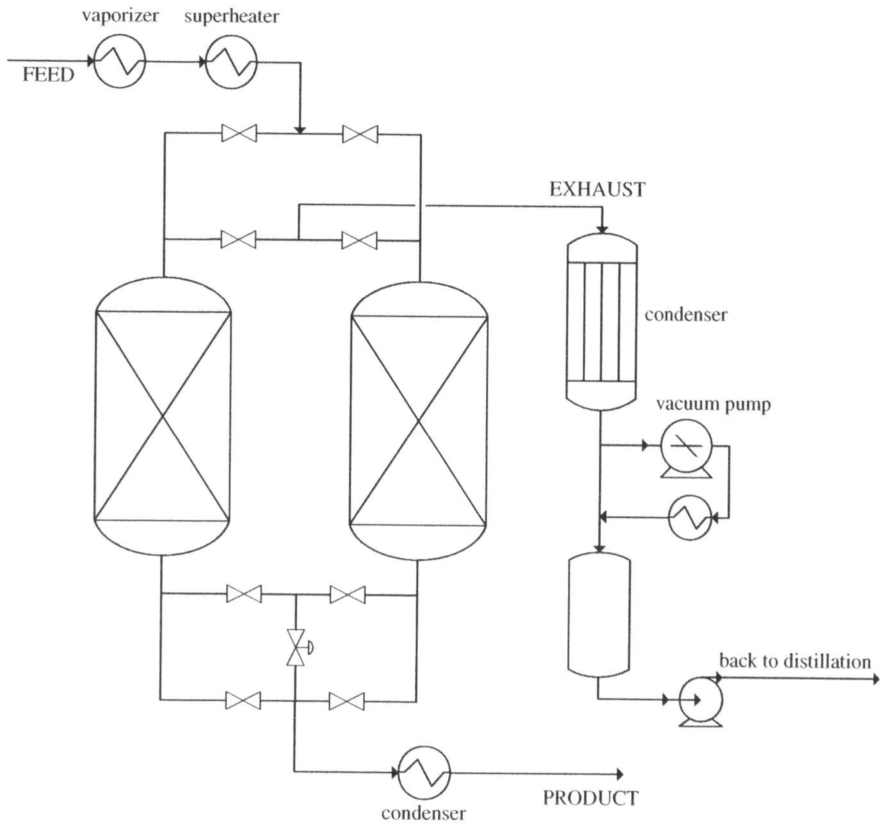

Figure 19.4 *Two-bed ethanol PSA process flowsheet*

of blowdown and purge is mixed with the rectifier (or beer column) feed stream. It is desired that the amounts of recycled PSA condensates are minimal, since they perturb the operation of the distillation columns. Ethanol recovery also increases as the amount of condensates decreases. The reasons discussed above led to the elimination of the purge step in some plants (Brown C. J. personal communication, 2005).

The production capacity of a typical ethanol plant is ~50 million gallons per year (~148 778 tons) assuming 8400 hours per year plant availability. This value corresponds to the PSA unit feed of ~20.4 tons hr^{-1} assuming 8 wt% water content. Typical productivity of the present ethanol PSA process is 0.3–0.35 h^{-1} (kg ethanol per hour per kg of zeolite). The energy required to run the PSA process includes the high-pressure steam for the super-heater, cooling water for the exhaust stream (regeneration) condenser and power for the vacuum pump. The latter two are much smaller compared to steam use for the super-heater.

In modern ethanol plants, the feed stream is already available as a vapor at high pressure since the rectifier operates at elevated pressure. The PSA super-heater requires approximately 305 BTU Gal^{-1} (~108 kJ kg^{-1}). The total energy consumption for the ethanol PSA process for a 50 MM gallon plant translates to 370 lbs hr^{-1} of high-pressure steam. This fact makes the PSA process very efficient in terms of the energy consumption within the thermally integrated ethanol plant; in addition, the thermal energy content of the anhydrous ethanol product is recovered. A stand-alone PSA unit would use additional 3900 BTU Gal^{-1} (~1384 kJ kg^{-1}) to vaporize the feed.

The amount of energy needed to break the ethanol-water azeotrope represents only a fraction of the energy consumed by the ethanol plant. The dry mill ethanol plant consists of: the fermentation; distillation (typically comprised of two or three distillation or stripping columns); the dehydration section utilizing a molecular sieve PSA system, and the stillage evaporation section processing the beer column bottoms into animal feed (DDGS). The energy consumption for the distillation, dehydration and evaporation (DD&E) for a typical commercial plant can be between 15 000 to 20 000 Btu Gal^{-1} of ethanol. The recent improvements in the process efficiency, pressure cascaded distillation systems and high level of the thermal integration resulted in the thermal energy use of 10 000 BTU Gal^{-1} (Brown C. J. personal communication, 2005). The overall plant energy including the DD&E section—10 000 BTU Gal^{-1}, the cooking process—7000 BTU Gal^{-1} and the dryer—14 000 BTU Gal^{-1} adds up to the 31 000 BTU Gal^{-1} of ethanol (11 000 kJ kg^{-1}). Where in fact more than 45% of the energy used is associated with the production of DDGS—a high value revenue stream for the ethanol plant.

The energy required for the production of fuel-grade ethanol 17 000 BTU Gal^{-1} on a large scale compares very favorably with the higher heating value of ethanol 84 000 BTU Gal^{-1}.

19.3 Future trends and industrial challenges

The ethanol PSA process introduced in the 1980s utilized a 3A zeolite material and essentially the same adsorbent is in use today. Some work was done on the optimum level of the cation exchange with potassium in the 3A zeolite manufacturing process to reduce the adsorbent cost. There is a trade-off between the adsorbent cost and the ethanol and water adsorption capacities for various contents of potassium in the zeolite structure.

Ladish and co-workers investigated the use of bio-based adsorbents for the dehydration of ethanol [7]. They proposed dehydration agents such as cellulose, cornmeal, cracked corn, corn cobs, wheat straw, bagasse, starch, hemicellulose and wood chips to remove the moisture from the aqueous alcohol. In the latter study, a cyclical TSA process was successfully operated for several months using corn grits [9]. The energy required to run the process was 530 kJ kg^{-1} of ethanol when processing a vapor feed compared to 108 kJ kg^{-1} needed by the commercial PSA process. In order to reduce the energy demand for the corn grit process the PSA cycle should be adopted by eliminating the external heat source required for the regeneration gas. In addition, the TSA system will always be larger resulting in higher installation capital cost and adsorbent volume owing to the longer cycle time characteristic for the TSA process.

While the PSA system avoids many of the disadvantages associated with the TSA process, the short cycle time that characterizes the PSA has its own consequences. In each cycle of the operation, the adsorbent is subjected to rapid changes in pressure causing attrition, fluidization, and particle movement, which can eventually lead to the formation of dust. This presents a great challenge for the bio-based adsorbents. The mechanical properties of synthetic zeolites are far more superior to those exhibited by biomass adsorbents.

The major advantages of bio-based adsorbents is the low cost and the fact that the materials are renewable. The adsorption equilibrium and kinetic properties are comparable with current synthetic adsorbents; however, there is still a lot to do with regard to the process design and development.

The growth in the biofuel market since the early 2000s can be almost entirely credited to corn or sugarcane based ethanol. This trend is expected to change with the advent of bio-refineries and lignocellulosic ethanol technologies. The larger production scale will result in the reduction of the production costs through higher process efficiency, lower capital cost and higher level of process integration. Pressure swing adsorption is mature technology ready to take on all of these challenges.

The higher production capacities will justify the use of multiple adsorption vessels. The advanced PSA cycles and steps are very well established in the industry. The basic principles were analyzed, understood, and explained by academia [26, 27]. The product recovery of the PSA process typically increases with an increasing number of adsorption vessels and so does the process throughput and productivity. The novel high performance ethanol PSA processes can emerge very quickly given the right economic and technological circumstances.

The ethanol PSA process is quite flexible in terms of operating conditions, such as pressures, temperatures, and feed-stream composition. This allows for numerous process arrangements with regard to the PSA process integration within the biorefinery complex resulting in more efficient plant flowsheet. For example, by changing the process cycle and operating pressures in the ethanol plant the PSA was able to deliver fuel grade ethanol product utilizing the feed stream containing 20 wt% water instead of a typical value of 8 wt% [28]. As a consequence, the distillation train upstream of the PSA unit can be redesigned to achieve lower capital cost and energy consumption.

19.4 Conclusions

Pressure swing adsorption technology is the preferred and industrially established method for the separation of water-ethanol mixtures for ethanol production. Single-train PSA units have been successfully installed and operated with annual capacities exceeding 100 million gallons ($\sim 3.8 \times 10^5$ m^3). A further growth in the ethanol production scale is pending the commercialization of technologies for converting lignocellulosic residues as well as energy crops to ethanol. With the increase in the plant size and capacity, the advanced ethanol PSA process will offer lower specific capital investment, greater energy efficiency and superior process performance. In this aspect the adsorption technologies present an attractive solution to the separation of the final biorefinery products due to flexibility in the choice of adsorbents, operating conditions, process design, and cycles.

Despite the technological advancements described earlier, there remains an urgent need for the development of more effective and economic methods for the separation of water from ethanol on a large scale. The promising candidates besides the PSA are membrane technologies and highly integrated hybrid processes.

References

1. Katzen R, Moon GD, Jr., Kumana JD, inventors; Katzen Raphael Associates, USA, assignee. Distillation method and apparatus for making motor fuel grade anhydrous ethanol. European patent 11147. 1980.
2. Wheals AE, Basso LC, Alves DMG, Amorim HV. Fuel ethanol after 25 years. *Trends in Biotechnology*. 1999; 17(12): 482–7.
3. Skarstrom CW, inventor; Esso Research and Engineering Co., assignee. Fractionating gas mixtures by adsorption. United States patent 2944627. 1960.
4. de Montgareuil PG, Domine D, inventors; Air Liquide SA pour l'Etude et l'Exploitation des Procedes Georges Claude, assignee. Process for separating a binary gaseous mixture by adsorption. United States patent 3155468. 1964.
5. Derr RB, inventor Aluminum Co. of America, assignee. Drying alcohol. United States patent US 2137605. 1938.
6. Oulman CS, Chriswell CD, inventors; Iowa State University Research Foundation, Inc., USA, assignee. Concentrating ethanol from dilute aqueous solutions. United States patent 4277635. 1981.
7. Ladisch MR, Tsao GT, inventors; Purdue Research Foundation, USA, assignee. Vapor phase dehydration of aqueous alcohol mixtures. United States patent 4345973. 1982.

8. Priegnitz JW, inventor UOP Inc., USA, assignee. Separation of water from ethanol. United States patent 4333740. 1982.
9. Ladisch MR, Voloch M, Hong J, Bienkowski P, Tsao GT. Cornmeal adsorber for dehydrating ethanol vapors. *Industrial and Engineering Chemistry Process Design and Development.* 1984; 23(3): 437–43.
10. Ginder WF, inventor; AD Pro Industries, Inc., USA, assignee. Removing water vapor from gaseous ethanol. United States patent 4407662. 1982.
11. Greenbank M, Rosene MR, inventors; Calgon Carbon Corp., USA, assignee. Dehydrating ethanol. United States patent 4465875. 1984.
12. Fornoff LL, inventor; CE Lummus, USA, assignee. Dehydrating ethanol and production of gasohol therefrom. United States patent 4273621. 1981.
13. Teo WK, Ruthven DM. Adsorption of water from aqueous ethanol using 3-.ANG. molecular sieves. Industrial and Engineering Chemistry Process Design and *Development.* 1986; 25(1): 17–21.
14. Carton A, Gonzalez G, Iniguez de la Torre A, Cabezas JL. Separation of ethanol-water mixtures using 3A molecular sieve. *Journal of Chemical Technology and Biotechnology.* 1987; 39(2): 125–32.
15. Tindall BM, Natarajan RS. Production of anhydrous ethanol by pressure swing adsorption. Paper presented at: AICHE annual meeting in Minneapolis, MN, USA: AIChE, New York, NY, USA; 1987.
16. Carmo MJ, Gubulin JC. Ethanol-water separation in the PSA process. *Adsorption.* 2002; 8(3): 235–48.
17. Simo M, Brown CJ, Hlavacek V. Simulation of pressure swing adsorption in fuel ethanol production process. *Computers and Chemical Engineering.* 2008; 32(7): 1635–49.
18. Llano-Restrepo M, Mosquera MA. Accurate correlation, thermochemistry, and structural interpretation of equilibrium adsorption isotherms of water vapor in zeolite 3A by means of a generalized statistical thermodynamic adsorption model. *Fluid Phase Equilibria.* 2009; 283: 73–88.
19. Do D.D. *Adsorption Analysis: Equilibria and Kinetics*. London: Imperial College Press, 1998.
20. Kupiec K, Rakoczy J, Mirek R, Georgiou A, Zielinski L. Determination and analysis of experimental breakthrough curves for ethanol dehydration by vapor adsorption on zeolites. *Inzynieria Chemiczna i Procesowa.* 2003; 24(2): 293–310.
21. Simo M, Sivashanmugam S, Brown CJ, Hlavacek V. Adsorotion/Desorption of Water and Ethanol on 3A Zeolite in Near-Adiabatic Fixed Bed. *Industrial Engineering and Chemistry Research.* 2009; 48: 9247–60.
22. Sowerby B, Crittenden BD. A vapor-phase adsorption and desorption model for drying the ethanol-water azeotrope in small columns. *Chemical Engineering Research and Design.* 1991; 69(A1): 3–13.
23. Ruthven DM. *Principles of Adsorption and Adsorption Processes*. New York: Wiley Interscience, 1984.
24. Lalik E, Mirek R, Rakoczy J, Groszek A. Microcalorimetric study of sorption of water and ethanol in zeolites 3A and 5A. *Catalysis Today.* 2006; 114(2–3): 242–7.
25. Simo M, Leo DM, Safavieh M, Brown CJ, Hlavacek V. Modeling of PSA Process in Fuel Ethanol Production: Effect of Ethanol Co-Adsorption on the Performance. Paper presented at: AIChE Annual Meeting, Conference Proceedings Paper 329w. AIChE, New York, NY, 2007.
26. Ruthven DM, Farooq S, Knaebel KS. *Pressure Swing Adsorption*. New York: VCH Publishers, Inc., 1994.
27. Yang RT. *Gas Separation by Adsorption Processes*. London: Imperial College Press, 1997.
28. Simo M. *Pressure Swing Adsorption Process for Ethanol Dehydration*. Dissertation: University at Buffalo, 2009.

20

Separation and Purification of Lignocellulose Hydrolyzates

G. Peter van Walsum

Forest Bioproducts Research Institute, Department of Chemical and Biological Engineering, University of Maine, USA

20.1 Introduction

20.1.1 Sugar platform

There are many process configurations possible for converting biomass to liquid fuels. Some of the most recently developed conversion technologies, as well as others that have been studied for years, use sugars as their "platform" for producing fuels and chemicals. Once derived from biomass, sugars can be converted into liquid fuels using fermentation or a series of chemical conversions catalyzed by inorganic catalysts. In both cases, it is advantageous to have the sugars in a solution that is free from inhibitory levels of other compounds derived from the biomass feedstock. The sugar platform can be implemented using different sources of sugars. Sugar crops, such as sugar cane or sugar beets are the most direct source of sugars. Starch crops can also produce sugars relatively easily using a cooking step and enzymatic hydrolysis. Producing sugars from lignocellulose is the most challenging but holds the greatest promise for the potential scale of the resource availability. Different simplified processing pathways for the sugar platform are illustrated in Figure 20.1. The separation, purification, and detoxification of biomass hydrolyzates are of particular interest in this chapter.

20.1.2 Biomass hydrolysis

Plant-derived sugars, such as sucrose or fructose, do not require hydrolysis to be rendered fermentable by microorganisms. Starch feedstocks require hydrolysis of the long-chain starch molecules in order to

Separation and Purification Technologies in Biorefineries, First Edition.
Edited by Shri Ramaswamy, Hua-Jiang Huang, and Bandaru V. Ramarao.

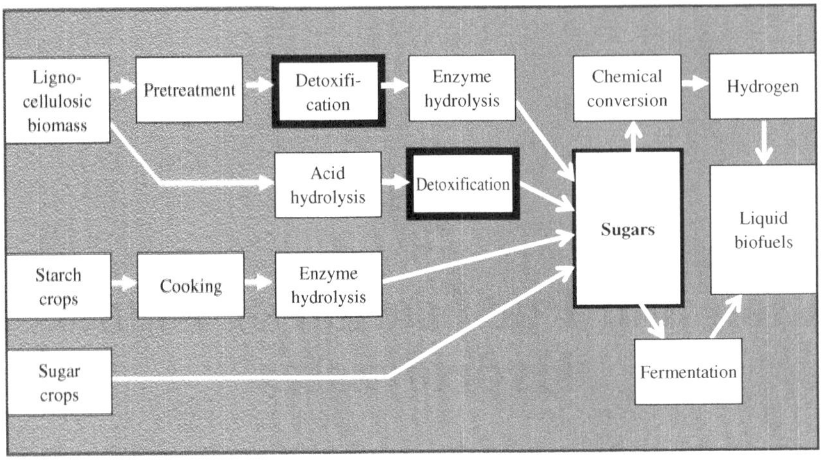

Figure 20.1 *Sugar platform conversion pathways*

release the glucose monomers. This is achieved relatively easily as amylases are very effective on starch and are relatively inexpensive to produce. In contrast to starch, depolymerization of lignocellulose into sugars is considerably more challenging. This is because the β 1–4 glycosidic bond in cellulose is difficult to break, much of the cellulose chains are arranged in a crystalline structure that impedes ready access to the bond linkages, and because the greater structure of lignocellulose, which incorporates hemicellulose and lignin, also impedes hydrolysis. Hydrolysis of lignocellulose can be catalyzed by strong acids or cellulytic enzymes. In general, acid hydrolysis is faster but results in some degradation of the sugars, while enzymatic hydrolysis progresses slowly but can give good yield of sugars. On its own, enzymatic hydrolysis progresses too slowly to be commercially viable. To accelerate enzymatic hydrolysis to a viable rate requires pretreatment of the lignocellulose prior to hydrolysis. Pretreatment is usually a thermochemical treatment of some sort that disrupts the physical and chemical structure of biomass in order to render them more accessible and amenable to enzymatic hydrolysis.

In some situations it may be advantageous to produce sugars from only a fraction of the carbohydrate content of biomass. For example, extracting the hemicellulose sugars from pulp wood will generate sugars that can be used for conversion to fuels and chemicals, while leaving the cellulose largely intact so that it can be used to make pulp and paper products. In this case, only partial hydrolysis of the wood carbohydrates is desired. Pretreatment and wood extraction through partial hydrolysis are similar processes and both can be carried out using a variety of chemical environments.

20.1.3 Biomass pretreatment

In order to facilitate enzymatic hydrolysis of biomass, a pretreatment is necessary to render the biomass amenable to rapid hydrolysis. Raw biomass, without pretreatment, is highly recalcitrant to enzymatic hydrolysis. Many approaches have been investigated for pretreating biomass. Most systems are aqueous and span the range of pH from acidic to neutral to alkaline. Some systems involve solvents other than water, such as ionic liquid or organosolv pretreatments. In general, more severe pretreatment conditions (smaller particle size, higher temperature, longer reaction times, more aggressive catalysts) result in greater reactivity of the fiber to enzymatic hydrolysis, but also generate more degradation products that can be inhibitory to enzymes or fermenting organisms. Table 20.1 lists some of the different pretreatment technologies that have been investigated.

Table 20.1 *Different biomass pretreatment technologies*

Pretreatment environment	Technology approach	Sample references
Low pH	Dilute sulfuric acid	Lloyd and Wyman (2003), Galbe and Zacchi (2002)
	Dilute nitric acid	Su *et al.* (1980)
	SO_2 catalyzed steam explosion	Bura *et al.* (2003)
	Phosphoric acid	Zhang *et al.* (2007)
	Formic acid	Grous *et al.* (1986)
	Carbonic acid	McWilliams and van Walsum (2002), van Walsum (2001)
Neutral pH	Steam explosion	Clark and Mackie (1987)
	Liquid Hot Water	van Walsum *et al.* (1996)
	Buffered pH	Weil *et al.* (1998)
	Green liquor	Walton *et al.* (2010)
High pH	Aqueous ammonia	Kim *et al.* (2000a), Du *et al.* (2010)
	Ammonia fiber expansion (AFEX)	Holtzapple *et al.* (1991)
	Lime	Chang *et al.* (1998)
Solvent enhanced	Phosphoric acid + solvent	Zhang *et al.* (2007)
	SO_2 + solvent	Iakovlev *et al.* (2011)
	Supercritical CO_2	Kim and Hong (2001)
	Organosolv	Pan *et al.* (2006)
Oxidative	Wet oxidation	Skammelsen and Thomsen (1998)
	Oxidative lime	Chang *et al.* (2001)
	Peroxide	Azzam (1989)
	Ozone	Bjerre *et al.* (1996)

Direct comparisons between pretreatment technologies have shown that most well developed pretreatment technologies, when applied at optimal conditions, work well on herbaceous feedstocks. Feedstocks with high lignin content are more challenging, and soft woods in general respond poorly compared to hardwoods and herbaceous feedstocks (Mosier *et al.* 2005).

20.1.4 Wood degradation products and potential biological inhibitors

Through the pretreatment or extraction processes, wood components change their physical structures and are broken down into smaller molecules. The goal of pretreatment is to enable the release of sugars into solution while minimizing the degradation of these sugars and other wood components that could result in compounds inhibitory to cellulolytic enzymes or fermenting organisms. Degrading the sugars also reduces the desired yield. Investigation of pretreatment-induced degradation products dates back at least as far as 1938 (Sjolander *et al.* 1938). Several review articles have summarized findings on microbial inhibitors in biomass hydrolysates (McMillan 1994, Olsson and Hahn-Hagerdahl 1996, Palmqvist and Hahn-Hagerdahl 2000a, 2000b, Klinke *et al.* 2004, Pienkos and Zhang, 2009). Determining the chemical makeup of pretreatment hydrolysates has also received attention. The principal difficulties in analysis of hydrolysates are the very wide variety of compounds present and the complex nature of the samples, which often results in fouled analytical equipment. HPLC, GC-MS and LC-MS have all been applied to determining the composition of biomass hydrolysates (Chen *et al.* 2006, van Walsum *et al.* 2007, Du *et al.*

2010). The colloidal properties of extracts such as particle size distributions and their zeta potentials were determined by Duarte *et al*. (2010).

20.1.5 Detoxification of wood hydrolysates

Several methods have been used to reduce the inhibitory nature of biomass hydrolysates. Inhibitors can affect the enzymatic hydrolysis of cellulose or the microbial growth in the fermentation. Methods that have been suggested to reduce the inhibition include: evaporation of volatiles, overliming (raising the pH to above 10, and lowering it again), ion exchange, adsorption, liquid–liquid extraction, dialysis, treatment with microbes or enzymes, flocculation, and development of organisms more tolerant to the inhibitory environment.

20.2 The market and industrial needs

20.2.1 Microbial inhibition by biomass degradation products

Several studies have investigated microbial response to pretreatment degradation products (Wilson *et al*. 1989, Frazer and McCaskey 1991, Larsson *et al*. 1998, Ranatunga *et al*. 2000). However, there has not been a consistently applied method for determining inhibition. Different toxicity tests have been carried out using different protocols and measuring different output variables, inhibitor concentrations, and organisms. The effects of inhibition have been measured by quantifying different indicators of microbial activity. For example, some studies have quantified ethanol (or other product) production (Larsson *et al*. 1998, Jennings and Schell 2011), others have measured sugar consumption (Alriksson *et al*. 2005, Yourchisin and van Walsum, 2004), and yet others have quantified cell-growth rates (Um *et al*. 2011). Others simply quantify the components in the pretreated material without testing fermentation (Chen *et al*. 2006). Batch growth is the preferred process configuration (Walton *et al*. 2010), yet some use continuous culture (Lynd *et al*., 2001). Thus, comparing results between studies is difficult, though some general trends can be discerned.

With the exception of acetic acid, most of the commonly quantified products, such as furfural and 5-hydroxymethylfurfural (HMF), have been reported to show relatively low to moderate toxicity to fermentative organisms on their own, at the concentrations at which they occur (Clark and Mackie 1984, Tran and Chambers 1986, Palmqvist *et al*. 1999a, 1999b, 2000a, 2000b). It has been reported that some of the less commonly measured products, typically present in lower concentrations, can contribute substantially to the inhibitory nature of the prehydrolysate (Frazer and McCaskey 1991, Pfeifer *et al*. 1984, Ando *et al*. 1986, Tran and Chambers 1986, Nishikawa *et al*. 1988). For example, Tran and Chambers (1986) found that products from degradation of the extractive fraction of wood were among the most inhibitory on a per-mole basis. Clark and Mackie (1987) concluded that low molecular-weight phenolics, derived from lignin, were ten times more inhibitory than carbohydrate-derived products. Frazer and McCaskey (1991) also concluded that the phenolic compounds were the most growth inhibitory. In a review of several studies, McMillan (1994) concurred with this conclusion, adding that, on a molar basis, the most inhibitory compounds were "aromatic aldehydes and acids and C6-to-C9 straight-chain organic acids." Results from different studies are not always in agreement. In some cases, compounds that are reported as inhibitory in one study are found to be less so in another. For example, Tran and Chambers (1986) reported that the relative toxicity of furfural was low, whereas Pfeifer *et al*. (1984) reported that it was decisive.

Inhibitors can be classified by their defining functional group, for example classification as aldehydes (including the furans), ketones, organic acids, alcohols or phenols. In addition, inorganic materials can also be inhibitory, such as metals or salts. Among the organic inhibitors, a relatively consistent observation is that the inhibitory quality of a compound increases with the hydrophobicity of the compound (McMillan 1994).

Some mechanisms of microbial inhibition have been identified. Banerjee *et al.* (1981) determined that cell growth rate and yields were reduced because of glycolysis inhibition by furfural and HMF. Aldehyde inhibition is also correlated with hydrophobicity, although this effect was not found to be related to disruption of cell membranes (Zaldivar and Ingram 1999a). Organic acids at low pH will collapse pH gradients and disrupt cellular energy generation. This effect is exacerbated for longer chain acids that are more hydrophobic, and thus able to more easily penetrate the cell wall (Zaldivar and Ingram, 1999b). Alcohols demonstrate relatively low toxicity to microbes compared with their acid or aldehyde counterparts. Alcohols soften the cell membranes and thus tend to be more inhibitory when more hydrophobic. (Zaldivar and Ingram 2000). In alcohol fermentations, butanol, for example, is more inhibitory than ethanol. Phenolics also reduce membrane integrity and interfere with sugar transport (Heipieper *et al.* 1994).

Interactions between inhibitors are not yet well understood. In one study, Palmqvist *et al.* (1999b) demonstrated a synergistic inhibition between acetic acid and furfural, with the two compounds producing an enhanced inhibition on *S. cerevisiae*.

20.2.2 Enzyme inhibition by biomass degradation products

While the study of inhibitors in wood hydrolysates is often directed at the inhibition of fermentative microorganisms, it has also been reported that pretreatment products can inhibit the production or action of cellulase enzymes (Mes-Hartree and Saddler 1983, Szengyel and Zacchi, 2000, Tengborg *et al.* 2001, Cantarella *et al.* 2004). Tengborg *et al.* (2001) found up to 37% inhibition of cellulolytic enzymes in the presence of softwood hydrolysate. Cantarella *et al.* (2004) found that acetic acid, furfural, 5-HMF, syringaldehyde, 4-hydroxybenzaldehyde and vanillin did not significantly affect enzymatic activity, while formic acid and levulinic acid did. Berlin *et al.* (2006) demonstrated that lignin inhibits cellulase enzymes. They found that cellulases are the most susceptible, xylanases less so and β-glucosidase the least inhibited of the enzymes that hydrolyze carbohydrate bonds.

20.3 Operation variables and conditions

20.3.1 Effects of pretreatment conditions on enzymes and microbial cultures

There have been several reports seeking to correlate pretreatment severity (R_0) (Overend and Chornet 1987) or combined severity (CS) (Chum *et al.* 1990) to fermentability of pretreated hydrolysates. These two functions are defined as:

$$R_0 = te\left(\frac{T - 100}{14.75}\right) \qquad (20.1)$$

for which t is time in minutes and T is temperature in degrees Celcius. Thus, severity acts as a consolidation variable for effects of time and temperature. Severity is often reported as $\log(R_0)$.

$$cs = \log(R_0) - pH \qquad (20.2)$$

By including the pH with the severity function, the combined severity takes into account differing chemical environments resulting from different levels of acidification of the pretreatment environment.

Tengborg *et al.* (1998) found that sulfuric acid pretreatment of sprucewood gave optimal sugars near CS 3.0 but that fermentability declined at this CS. Larsson *et al.* (1998) conducted an extensive study of dilute acid hydrolysis of sprucewood at 76 different conditions, over a combined severity range of 1.4 to 5.4. Their study looked at concentrations of glucose, mannose, xylose, furfural, HMF; acetic, formic and levulinic acids and the fermentability of the hydrolysates by *S. cerevisiae*. Fermentability, as measured by ethanol yield and productivity, decreased with increasing CS, with the greatest decreases occurring at approximately CS 3. A similar study by Bura *et al.* (2003) confirmed these results, in which they found optimal sugar accumulation from SO_2-catalysed pretreatment of corn stover to be at a log severity of 3 to 3.3 while accumulations of furfural and HMF showed increasing concentrations up to log(Ro) = 4.3. Acid-catalyzed steam explosion is now routinely carried out in two steps, using different temperatures or acid concentrations, which have been shown to improve fermentability while maintaining good enzymatic hydrolysis rates (Kim *et al.* 2000b, Nguyen 2000, Shevchenko *et al.* 2000, Bura *et al.* 2003, Kim *et al.* 2005). This two-step approach to reducing inhibition underscores the complexity of the kinetic reactions at work in the pretreatment process.

Several studies have contributed to the development of kinetic models for the hydrolysis of the major polysaccharides. Decomposition kinetics of xylose, galactose, mannose, glucose, 2-furfural, and HMF have been investigated over varying severities with the aim of enhancing methane fermentation (Baugh *et al.* 1988a, 1988b). The degree of deacetylation of hemicellulose has been shown to correlate well with the severity factor (Garrote *et al.* 2002). Chen *et al.* (2007) reported kinetic parameters for accumulation of different decomposition products resulting from dilute acid pretreatment. Most significantly, this work demonstrated that accumulation trends for monitored degradation products do not follow the classical severity relationship as defined by Overend and Chornet (1987). The Ladisch research team has investigated neutral pH pretreatment with the intent of reducing microbial inhibition (Weil *et al.* 1998). They demonstrated that the neutral-pH pretreatment method produced a prehydrolysate that was fermentable by a genetically modified cofermenting yeast (Mosier *et al.* 2004).

20.3.2 Quantification of microbial inhibitors in pretreatment hydrolysates

Bouchard *et al.* (1991) presented an analysis that characterized the general chemical properties of the pretreatment products, without identifying the individual compounds. Results were presented as characterizing qualities such as molecular-weight distribution, abundance of O-acetyl groups or the relative distribution of chemical-bond types as determined by Fourier-transformed infrared (FTIR) analysis. An increasingly detailed picture is developing of the accumulated degradation products in pretreatment hydrolysates derived from different substrates employing HPLC and LC-MS methods (Chen *et al.* 2006, 2007, Du *et al.* 2010).

20.3.3 Separations challenges posed by biomass degradation products

The complexity of biomass hydrolysates results in several challenges for performing separations work, both for chemical analysis and processing applications.

For analytical work, hydrolysates are notorious for their fouling of chromatography columns and analysis for minor degradation products typically requires a sample extraction prior to analysis. The variation in hydrophobicity of target analytes results in varying degrees of extraction efficiency during sample preparations. The wide variety of analytes often necessitates more than one analytical technique for analysis of all target analytes, though the Chambliss group has demonstrated the ability to quantify on the order of 40 compounds in a single liquid chomatograpy run (Chen *et al.* 2007, 2008, Du *et al.* 2010).

At a commercial scale, the functioning of separation techniques to remove inhibitory compounds is often at odds with sugar retention. Several methods for detoxification are effective at reducing inhibition, but they can also reduce sugar concentrations, reducing process yield. Some methods, such as liquid-liquid extraction, are simply difficult to perform with complex aqueous mixtures. In this case, the phase separation and recycling of the extractant necessary for continuous operation can perform poorly in the presence of complex hydrolysate mixtures. The use of adsorption beds is also hampered by the complex nature of the hydrolysate solutions. Simple procedures such as evaporation, which may be desired to increase the concentration of sugars in solution, will concentrate some of the inhibitors. This is particularly true for salt-laden media such as pre-pulping wood extracts, which require concentration several-fold to attain suitable sugar concentration for most fermentations.

20.4 The hydrolyzates detoxification and separation processes

Several methods have been suggested for detoxification of pretreated or acid-hydrolyzed biomass, including: overliming, ammonium hydroxide, use of adsorbents such as activated carbon or polymeric adsorbents, solvent extraction, ion exchange, flocculation, treatment with microbes, treatment with enzymes. The following sections describe some of the findings reported on these methods.

20.4.1 Evaporation, flashing

Flashing of hot hydrolysates or heated evaporation, sometimes under vacuum, can reduce the concentration of volatiles such as furfural, acetic acid or vanillin. (Larsson *et al*. 1999, Converti *et al*. 2000, Rodrigues *et al*. 2001). Wilson *et al*. (1989) reported improved the fermentability by *P. stipitis CBS 5776* of steam-pretreated aspen wood hydrolysates after rotoevaporation, compared to the untreated hydrolysate. However other no-volatile inhibitors, such as salts or some lignin derivatives, will increase through evaporation, which can increase inhibition. Thus the net benefit of evaporation depends on the inclusive composition of the hydrolysate. Walton *et al*. (2010) showed that evaporation and concentration of green liquor wood extracts resulted in increased inhibition of *E. coli* KO11. This was attributed to the increased levels of acetate and sodium in the evaporated green liquor. On the other hand, acetic acid can also be removed from solution by evaporation at low pH, followed by re-dilution of the medium with fresh water.

20.4.2 High pH treatment

Raising the pH of biomass hydrolysate, often called "overliming," has been discussed for use in reducing toxicity of hydrolysate at least since the WWII era (Sjolander *et al*. 1938, Leonard and Hajny 1945). High pH treatment consists of raising the pH of the hydrolysate to a level of about pH 10, and then reducing it again to a level appropriate for fermentation. $Ca(OH)_2$ is the base usually associated with pH adjustment, hence the term "overliming," but the treatment can also be effective with other bases as well. Although it has been recognized as effective for many years, the mechanisms of inhibition reduction through pH adjustment remain unclear and the process continues to attract attention from researchers. The results presented below can be summarized as showing that pH close to 10 is generally optimal, since lower pH gives too little improvement in detoxification and higher pH results in too much degradation of sugars. Different bases have different levels of effectiveness, with $Ca(OH)_2$ being one of the better choices and NH_4OH often showing the best subsequent fermentation performance.

20.4.2.1 Cation effects in overliming

High pH treatment is simple, effective, and relatively inexpensive. When lime, a commonly used base, is used after a sulfuric acid in pretreatment, it results in the formation of insoluble gypsum. This precipitate must be removed from the process stream, requiring a costly separations step. It also requires solid disposal and eliminates the possibility of recycling the pretreatment catalyst—which add to the chemical and operating costs of the plant.

A study by Persson *et al*. (2002a). looked at four different bases (NaOH, $Ca(OH)_2$, KOH, and NH_4OH) to treat acid hydrolysate of spruce. The treatment raised the pH 10 followed by a pH reduction back to 5.5 using either HCl or H_2SO_4. Of the four bases tested, the $Ca(OH)_2$ and NH_4OH treatments gave the greatest improvements of fermentability. These samples performed five times better than the untreated hydrolysate and were comparable with, even superior to, the reference fermentation. The treatments showed little effect on concentrations of formic, acetic, or levulinic acids or on the concentration of glucose and mannose. Ammonia treatment resulted in the largest changes in likely inhibitory compounds, such as hydroxymethylfurfural (HMF), furfural, phenol, vanillin, coniferyl aldehyde and cinnamic acid.

Subsequent work by Alriksson *et al.* (2005) tested another collection of bases, including: $Ba(OH)_2$, $Ca(OH)_2$, $Mg(OH)_2$, NaOH, and NH_4OH. This study also tested spruce hydrolysate with pH changes to pH 10 and back down to 5.5. Contrary to the study by Persson, in this case acid levels were found to increase while furans and phenolics declined. In this study, NH_4OH was not the most effective base for reducing concentration of degradation products. $Mg(OH)_2$ caused the largest decrease in furan concentrations while $Ca(OH)_2$ caused the largest effect on phenolics. However, in agreement with the Persson study, it was observed that baker's yeast fermentations showed much higher glucose consumption rates after NH_4OH conditioning than lime, which was second in this group of bases. The high glucose consumption resulted in improved volumetric ethanol productivity and ethanol yields. Hypothesizing that the NH_4^+ ions could have been assisting the fermentation, the researchers tested to see if comparable benefits were derived from addition of NH_4Cl to other hydrolysates. This enrichment did not result in improved fermentations in these other cases.

Zymomonas mobilis 8b was also observed to grow better in ammonia-treated hydrolysate than in lime-treated hydrolysate. When *Z. mobilis* was used to ferment overlimed corn stover hydrolysate, ethanol yields fell from 79% to 23% when the hydrolysate concentration was increased from 40% to 85% (Jennings and Schell 2011). When corn stover hydrolysate was conditioned with ammonium hydroxide, ethanol yields were much higher overall, and fell from 90% to 50% when the hydrolysate concentration was raised from 40% to 85% (Dutta *et al.* 2010). Jennings and Schell (2011) studied the benefits of NH_4OH in more detail on corn stover hydrolysate. They confirmed that ethanol production in fermentors was greatly enhanced when ammonia conditioning was employed in place of overliming. Some of that improvement came from enhanced sugar recovery with ammonia conditioning, but the metabolic ethanol yields were also higher with ammonia conditioning, which indicates that there is a biological effect to ammonia conditioning as well. In this study they found that that *Z. mobilis* reached peak ethanol production in half the time with ammonia-conditioned hydrolysate as compared with lime-treated hydrolysate.

At high levels of pH, Nilvebrant *et al.* (2003) found that degradation of furans and lignin was more pronounced with $Ca(OH)_2$ than with NaOH.

20.4.2.2 *pH and temperature effects*

The study by Persson *et al.* (2002a) using four different bases tested two methods of adjusting the pH. One was the conventional method of adjusting the pH to 10 and then lowering it to 5.5. They also tested simply adjusting the pH directly to 5.5. Neutralization to pH 5.5 with any of the bases tested had very little effect on the concentration of the toxic compounds measured. Thus, they confirmed the necessity of "over" liming. Martinez *et al.* (2001) reported that the optimal addition of $Ca(OH)_2$ for overliming can be determined by a simple titration of the hydrolysate.

Nilvebrant *et al.* (2003) explored going to higher pH with dilute acid hydrolysate of spruce. It was shown that at higher pH and temperature, degradation of furan and lignin increased, leading to higher levels of formic, acetic and levulinic acid as well as higher levels phenolics such as vanillic acid. Purwadi *et al.* (2004) found similar results with forest residue hydrolysate overlimed to different pH levels with $Ca(OH)_2$ at various temperatures. More severe conditions increased sugar degradation. For fermentation results, baker's yeast fermentation produced the highest ethanol level with hydrolysate that had been overlimed to pH 10 at 45 °C, giving a yield of 0.47 of the fermentable sugars. While this condition gave the most ethanol, faster growth rates were achieved with hydrolysates that had been treated at higher temperatures and pH. Thus, the higher severity conditions resulted in less inhibition of growth, but reduced the amount of sugar available to support ethanol production. Sugar yields went down as pH increased from 11 to 12, and the amount of ethanol decreased as well.

In work using NH_4OH, optimum fermentation conditions were found after treatment to only pH 9 at 55 °C (Alriksson *et al*. 2006). Under these conditions, furans were reduced by 30%, phenols by 13%, glucose and mannose by 6%, and baker's yeast fermentation led to an ethanol120% yield higher than that of the control sugar solution. Thus, this also supports the benefit of NH_4OH pH treatment for the fermentation process downstream. NaOH treatment was also optimized with best results at 80 °C, pH 9, yielding about the same reductions in furans and phenols but slightly higher sugar losses. When the optimal conditions were used for NH_4OH and NaOH conditioning and compared with overliming (at 30 °C, pH 11), all conditioned hydrolysates supported similar levels of ethanol productivity with a range of 90–110% of sugar solution control compared to about 15% for unconditioned hydrolysate.

Zymonomas mobilis 8b was also used to determine the optimum conditions for overliming of dilute acid hydrolysate of corn stover (Mohagheghi *et al*. 2006). As in other studies, it was found that glucose, arabinose, and xylose sugar losses increase with increased pH. More specifically, they reported that xylose was the most sensitive sugar (35% loss) and glucose the least (15% loss). Arabinose was in between at 20% loss, all at pH 11. As was found in other studies, detoxification appears to be most effective at high pH, evidenced by the increasing ethanol yield on sugar from pH 9 to 11, but overall yields peak at pH 10 because of the increase in sugar losses at higher pH.

20.4.2.3 Different fermentative organisms

The studies reported above made use of baker's yeast and *Z. mobilis* as their fermentation test organisms. Results appear to be similar for these two very different organisms, suggesting that the optimization of pH adjustment is mostly related to the chemistry of the solution rather than the peculiar natures of the organisms.

20.4.3 Adsorption

Many of the compounds found to be inhibitory can be adsorbed onto appropriate media. Since hydrophobic molecules are typically the most toxic, adsorbents that have a high affinity for such compounds are effective at reducing inhibition.

Several solid phase extraction methods have been tested for conditioning of hydrolysates. Activated carbon treatment of corn stover dilute acid hydrolysate was shown to reduce acids without a significant reduction in sugar concentration (Berson *et al*. 2005). Alves *et al*. (1998) reported that pH variations combined with activated charcoal provided optimal production of xylitol from bagasse hydrolysate. Schirmer-Michel and co-workers (2008) found that treatment of hydrolysate with activated charcoal reduced furfural by 95%, acetic acid by 37%, and phenolics by 75% in sulfuric acid hydrolysates of soy-bean hulls. They found minimal reduction of sugars but unfortunately the treatment had no impact on fermentability by *C. guilliermondii* NRRL Y-2075. They reported identical growth on treated and untreated hydrolysate, both of which were much lower than the synthetic medium control.

Weil *et al*. (2002) found that the styrene-based polymeric adsorbent XAD-4 enhanced fermentability of *E. coli* KO11 on dilute acid pretreated corn fiber hydrolysates. They noted a dramatic decrease of furfural in the medium and suggested that the adsorbent would be useful for reducing concentrations of aldehyde compounds in general.

20.4.4 Liquid–liquid extraction

Solvent extraction can also be effective at removing inhibitory compounds from hydrolysates. Several studies have demonstrated benefits of such an operation. Wilson *et al.* (1989) reported improved the fermentability by *P. stipitis CBS 5776* of steam-pretreated aspen wood hydrolysates after liquid extraction with ethyl acetate. Prepulping green liquor extract derived from hardwoods was found to have reduced inhibition of *Pichia stipitis* when first extracted with trioctylphosphine oxide (TOPO) (Um *et al.* 2010). The TOPO was demonstrated to reduce acetic acid and it is a known extractant of phenolic compounds. An MTBE extract of a yellow poplar acid hydrolysate was shown by Ranatunga *et al.* (2000) to inhibit *Z. mobilis* CP4(pZB5) fermentation when added to a synthetic hydrolysate made up of glucose, xylose, acetic acid and H_2SO_4. Unfortunately the authors did not report whether the extracted hydrolysate showed improved fermentation performance.

In a different approach to liquid extraction, supercritical CO_2 extraction of an acid hydrolysate of spruce removed a number of potentially toxic compounds, resulting in improved fermentation yields and productivity with baker's yeast (Persson *et al.* 2002b). Furfural was reduced by 93%, coniferyl aldehyde by 91%, but other compounds were less affected, such as HMF which was only reduced by 10%, acetic acid by 19% and levulinic acid by 6%.

20.4.5 Ion exchange

In their comparative study of different inhibition reduction methods, Larsson *et al.* (1999) found that ion exchange was the most effective of the methods tested, but it also reduced sugar yields. This finding has been replicated by several other studies by different research groups. Many studies looked at comparing different types of resins. In general, it was found that most resins improved fermentability, and anion exchange resins were usually the most effective.

DeMancilha and Karim (2003) studied production of xylitol from xylose using dilute acid cornstover hydrolysate and *C. mogi* ATCC 18364. In this study, several types of resin were tested. Cation exchangers removed 40–60% of HMF, 50–80% of furfural, 0% of acetic acid, and 70–80% of color, while losing less than 8% of the xylose. A weak base anion exchanger performed better, removing all or nearly all the HMF, furfural, color and acetic acid with only 6% of the xylose removed. The strong anion exchangers did not remove any acetic acid but did remove nearly all the furfural and 50– 65% of the HMF and 40–70% of the color. In this case, less than 4% of the xylose was lost. The weak base anion exchanger and strong cation exchanger were used to condition hydrolysate for fermentation, but performance was not as good as in the reference fermentation.

Nilvebrandt *et al.* (2001) also tested several resins: an anion exchanger (AG1-X8), a cation exchanger (AG50 W-X8), and a plain resin (XAD-8). They conditioned dilute acid hydrolysate of spruce for fermentation by baker's yeast. For biomass yield and ethanol productivity, the performance of the different resins followed the pattern of anion exchange > plain resin > cation exchange. AG1-X8 removed nearly all aliphatic acids as well as more than 60% of the furfural and HMF, 80% of the phenols and 90% of the Hibbert's ketones. XAD-8 and AG50-X8 removed no acids and very little Hibbert's ketones, but did remove a significant amount of furans and phenols. One drawback to the anion resin was that it was the only resin that sequestered glucose, though it could be recovered in the presence of added sulfate. Adding onto this work, the group tested additional strong base (Dowex 1X4, Dowex 2X8, IRA458) and weak base (IRA67, IRA92, and Duolite A7) anion exchangers (Horvath *et al.* 2004). All improved fermentation relative to untreated hydrolysate, with the strong base resin being the most effective at improving fermentation rates. All the treated hydrolysates did eventually achieve comparable yield of ethanol, though at different rates. All the resins but IRA67 returned close to 100% glucose and mannose. The IRA67 resin returned slightly lower yields. Dowex 1X4 was best at removing levulinic and acetic acids and phenolics, and was among the best in removing formic acid and furfural.

In another example of ion exchange chromatography, Luo *et al*. (2002) started with a dilute nitric acid hydrolysis of green hybrid poplar. It was conditioned by anion exchange chromatography using Dowex MWA-1, a weak base resin. The composition of the extractant and treated hydrolysate were analyzed. An ion chromatography run with an ethyl acetate extract of the untreated hydrolysate generated more than 70 separate peaks. Analysis of the treated hydrolysate demonstrated that all the aromatic acids and much of the aliphatic acids were gone. Eliminated compounds included: 2-furancarboxylic acid, 2-furanacetic acid, 5-hydroxymethyl-2-furancarboxylic acid, and ferulic acid, which are considered to be likely inhibitors. Most of the aromatic acids and aliphatic acids were recovered by regenerating the anion exchange resin. This treatment of the hydrolysate resulted in enhanced fermentation by *S. cerevisiae* D_5A.

20.4.6 Polymer-induced flocculation

Investigation of colloidal particles in maple wood hot water extracts found them to be in the size range from ~220 nm to 270 nm and negatively charged. By treating the extracts with a cationic flocculating agent poly-DADMAC and kaolin, it was possible to preferentially precipitate out the colloidal fraction containing lignin and lignin derived compounds. (Duarte *et al*. 2010, Hasan *et al*. 2011). This reduced the lignin concentration while the sugar content remained unchanged. This method is likely to reduce inhibition, although no fermentation results on the treated material have yet been reported.

20.4.7 Dialysis

Electrodialysis can be used to remove ions from hydrolysates. A common ion of concern is acetate, which is often one of the more inhibitory compounds due to its high concentration in most hydrolysates. In addition, salt inhibition resulting from pH neutralization can cause inhibition in acid-generating fermentations (Lynd *et al*., 2001). In a study by Sreenath and Jeffries (2000) electrodialysis was used to remove acids from mixed wood hydrolysate.

Several batches were analyzed in which the fermentable sugars (glucose, galactose, mannose, and xylose) ranged in concentration from 10–121 g/L; acetic acid from 0.43 to 6.2 g/L and HMF from below detection limit to 2.2 g/L. Different batches supported varied fermentation results with *C. shehatae* strain PFL-Y-049, although results were not compared with an unconditioned hydrolysate control.

20.4.8 Microbial detoxification

In another approach, studies (Lopez *et al*. 2004, Nichols *et al*. 2005, 2008) have demonstrated that biological detoxification of hydrolysates can be achieved through the addition of selected microorganisms in the pretreatment phase. In particular, Lopez *et al*. isolated novel soil microorganisms capable of using ferulic acid, furfural and HMF. They showed that the fungus, *Coniochaeta lignaria*, successfully eliminated inhibitors including furfural and HMF. Five bacteria were also shown to reduce inhibitors, as well. These bacteria included *Methylobacterium extorquens*, *Pseudomonas sp.*, *Flavobacterium indologenes*, *Acinetobacter sp.*, and *Arthrobacter aurescens*. Nichols *et al*. (2005, 2008) continued this work and isolated a strain of *C. ligniaria* NRRL30616 that was capable of removing more than 99% of the furfural, 85% of the HMF, and 20% of the acetate from hydrolysate. Compounds representing all of the classes of inhibitory side-products were removed during the course of fungal growth. Unfortunately, the fungus also consumed 50% of the glucose and 20% of the xylose. Abatement of corn stover hydrolysate was followed by fermentation with *Saccharomyces sp*. LNH-ST in the presence of cellulases and showed a vast improvement over fermentation of unabated hydrolysate.

Acetate is a common inhibitor, and a novel isolate of *S. cerevisiae* YGSDC was found that metabolizes acetate but not sugars (Schneider 1996). This strain could reduce acetic acid from 0.68 to 0.04% in stream stripped hardwood-spent sulfite liquor. After treatment, growth of *Pachysolen tannophilus* and *P. stipitis* increased by approximately an order of magnitude, as did ethanol yields.

The thermophilic bacterium *Ureibacillus thermosphaericus* was investigated for the biological detoxification of hydrolysate of waste house wood (Okuda *et al.* 2008). In synthetic hydrolysates, *U. thermosphaericus* oxidizes furfural and HMF to 2-furancarboxylic acid and 5-hydroxymethyl furancarboxylic acid, respectively, which are less toxic to yeast in synthetic hydrolysate than their parent molecules. Hydrolysate that had been treated by this bacterium over 24 h resulted in markedly increased ethanol production rate by *S. cerevisiae*, which was comparable to overlimed samples. In biomass-derived hydrolysates, however, the concentrations of these compounds were not decreased markedly by the bacterium. The combination of bacterial and overliming treatments of hydrolysates helped significantly to maintain ethanol production rate by *E. coli* KO11. *U. thermosphaericus* has a significant advantage of not consuming sugars, thus preserving product yields.

In addition to reducing microbial inhibition, inhibition of cellulase activity has also been shown to diminish after fermentation of hydrolysates (Tengborg *et al.* 2001),

20.4.9 Enzyme detoxification

Some reduction of inhibition can be achieved by application of enzymes that will break down the various products derived from lignin. These enzymes remove monoaromatic phenolics by catalyzing oxidative polymerization. In one example, willow extract prepared by steam and SO_2 treatment was conditioned by treatment with laccase and lignin peroxidase from *T. versicolor* (Jonsson *et al.* 1998). This treatment improved the rate of glucose consumption and ethanol production.

20.4.10 Microbial accommodation of inhibitors

Another approach to improving fermentation of hydrolysates has been to develop or isolate microbial strains more adept at growing amidst inhibitors (Fein *et al.* 1984, Jeon *et al.* 2002). Positive results have been reported from genetic modification of *S. cerevisiae* (Larsson *et al.* 2001). Immobilization of yeast in a fermentation reactor, which diminished the need for cell propagation and also enhances acclimation to the environment, has been shown by Taherzadeh *et al.* (2001) to be effective at enhancing conversion. Directing the evolution of fermentative organisms has also been demonstrated. Gilbert *et al.* (2009) demonstrated rapid (five days) evolution of *S. cerevisiae* to develop greater tolerance to acetic acid through use of a cytostat.

20.5 Separation performances and results

Results from the preceding discussion are summarized in the following Table 20.2. Several methods for detoxifying hydrolysates have been developed and applied. Issues that would direct decision of an appropriate technology include the particular sensitivities of the fermentative organisms or hydrolytic enzymes to be used, the feedstock used, the pretreatment conditions used and nature of the resulting hydrolysate. In addition, integration with other aspects of the biorefinery could direct choice of technology. For example, conducting bioconversion at a kraft pulp mill equipped with a lime kiln may make options utilizing lime more attractive than at another site with no such infrastructure in place.

Table 20.2 *Summary of chemical methods for detoxification of biomass hydrolysates*

Method	Compounds removed	Inhibition reduction	Sugar retention
Evaporation, flashing	Volatiles	Some reduction, can also exacerbate	Very good
pH adjustment	Aldehydes, phenols. Aliphatic acids not much affected	Effective	Good up to pH 10
Carbon or polymer adsorption	Furans, phenolics, acids	Fair	Good
LLE	Organic acids, phenolics	Moderate	Fair
Ion exchange	Organic acid anions, metal ions	Effective	Fair
Dialysis	Salts, organic ions	Effective	Good
Polymer flocculation	Lignin, lignin derivatives	Promising, not demonstrated	Good
Biodetoxification	Organics	Effective	Poor

20.6 Economic importance and industrial challenges

20.6.1 Cost of slow enzymes

In 2010, two major international enzyme manufacturers, Novozymes and Genencor, each announced that they had developed enzyme cocktail systems that would reduce the cost of enzymea for hydrolysis of pretreated biomass to about 50 cents/gallon of ethanol. Assuming a stoichiometric yield of ethanol on glucose, this suggests a cost of about 8 cents/kg of glucose derived from cellulose. Thus, if these prices hold in the eventual commercialization of lignocellulose-derived sugars, there will remain strong economic incentives to improve hydrolysis rates and reduce enzyme inhibition. Reaction durations for enzymatic hydrolysis of cellulose are commonly modeled to last for tens of hours. Thus, the volume of hydrolysis equipment and their associated capital and operating costs can be substantially lowered by increasing the rate of hydrolysis. As a result, for any process making use of enzymatic hydrolysis of lignocellulose, inhibition reduction is a key lever for cost reduction.

20.6.2 Cost of slow fermentations

Similar to the hydrolysis costs associated with inhibited enzymes, reducing inhibition of cells has strong economic incentives. Higher conversion rates will allow shorter residence times and smaller process volumes. In addition, some cell conversion systems are capable of converting the variety of wood sugars found in lignocellulose to the desired products, say ethanol, but they do so with inhibited growth, meaning that the cells do not propagate within the ethanol fermentation. This situation requires that the cells be grown in separate sugar-fed bioreactors, which dramatically increases capital and operating costs. Reducing or eliminating cell inhibition thus increases both the productivity of the conversion process and also improves the propagation of the cells, enabling further reduced costs.

The most effective way to reduce the effects of inhibitory environments on microorganisms is likely to be to develop microbes that are more tolerant of inhibitors. Developing such organisms allows for process simplification, with benefits of lower capital and operating costs. Developing improved strains of organisms also has the benefit of being something that can be practiced continually while a plant is operational. As improved strains are developed, they can be implemented in the existing process with little or no required physical plant changes. This is one area in which biorefinery technology has an edge over conventional petroleum processing, since the powerful and still rapidly developing tools of genetic engineering and directed evolution can be applied to improve commodity-scale processes. A second approach to improving the robustness of microbial conversion is to make use of mixed cultures. While it is possible, it may not be optimal to use one organism to perform all the biological conversions required in a biorefining process. Thus, it may be possible to grow in one vessel, microbes that ferment sugars to the desired products, and microbes that consume inhibitory compounds that impede the production microbe. This could achieve the benefits of detoxification without requiring a separate unit operation.

20.6.3 Benefits of co-products

Many biorefinery scenarios envision producing a high-volume, low-value product, such as a biofuel, in conjunction with production of smaller volume, higher value co-products. This business strategy makes good sense, and emulates how the well-established oil-refining and petrochemical industries operate. The corn wet mill is another example of a commodity scale processing facility that makes optimal use of the diverse components available in its feedstock. However, in most cases, selling co-products requires that they be relatively pure, and thus separation issues become involved in the adding of value to the process co-products. To the extent that inhibitory compounds could be removed from the sugar platform process and resold as co-products, biorefineries could achieve a twofold improvement in their profitability. For example, neutral pH prepulping extracts from hardwoods contain high levels of acetic acid. Removing the acetic acid in a form sufficiently pure to have market value achieves the benefits of aiding fermentation of the extract while also generating an additional revenue stream. Other potential co-products, such as furfural or formic acid, might also be separated from the sugar stream in such a scenario.

20.6.4 Material consumption

Much of the research into pretreatment methods has sought to reduce consumption of chemical inputs, notably acids and bases. Accordingly, several methods have been proposed that make use of catalysts that can be recycled. Much work has focussed on lime- or ammonia-catalyzed pretreatment, for example. Both of these catalysts can, in theory at least, be recycled back to the process, much in the way that Kraft pulp mills can recycle its pulping chemicals. Alkaline pretreatment also offers the benefit of generating hydrolysates that are less inhibitory than acid-pretreated materials, though the drawback is generally lower yields of final fermentable sugars than the acid routes. The challenge for such recycle operations is the added complexity of the product recycle stream and the associated need for larger economy of scale to justify the added complexity.

For separations using alternate phases, such as adsorbents or liquid extraction, an important issue is the recycling and reconditioning of the extractant material. For example, liquid extraction of acetic acid from solution has been done commercially using ethyl acetate or trialkylphoshineoxides (TOPO). In the former case, recycling the solvent by distillation is capital and energy intensive, as the solvent is volatile and thus requires a large distillation column and high energy input. Using a high boiling-point solvent such as TOPO reduces the distillation cost but in this case the solvent becomes contaminated with lignin-derived phenolics and other organics and needs to be back-extracted for reuse. Thus, need for separations operations propagates as additional materials are added to the processing technology.

20.6.5 Complexity: Capital and operating cost

When the pretreatment process and its recycle stream start to look like a refinery of their own, then it is likely that the cost of process complexity is getting too high. The innate attributes of biomass processing systems associated with the heterogeneous nature of biomass, the solids handling and mass transfer limitations make separations a challenging issue. Most biorefineries are likely to handle aqueous streams with relatively dilute concentrations of target compounds, be they valuable co-products or troublesome molecules that need to be removed. The result is that separations will likely involve large process equipment and relatively complex collections of unit operations. The magnitude of the challenge can be appreciated when comparing a kraft pulp mill to an oil refinery. In an oil refinery, the raw material, crude oil, is a liquid that requires no dilution for processing. In contrast, a pulp mill receives a feedstock that is solid and 50% water. To process the fiber into desired products requires size reduction and further dilution, relatively slow processing (hours instead of minutes) and then separation of the products from water downstream. Thus, while the size of the process equipment can be comparable to those found in oil refineries, the volume of the final product is an order of magnitude, or two, smaller. Developing separations technologies that can process the complex mixtures of compounds found in biorefinery process streams at low cost is therefore a critical issue.

20.6.6 Waste reduction

Because biorefineries process organic materials of recent biological origin, most of the byproducts emanating from a biorefinery are also biodegradable or at least combustible. Thus, the volume of waste derived from the initial raw material need not be very high, provided that adequate waste treatment or combustion processes are in place. However, the chemical processing of the biomass, which can often include mineral acids and bases, does generate waste streams that have little value and can accumulate considerable volume. The most pressing example is the accumulation of gypsum resulting from sulfuric acid pretreatment (or hydrolysis) of biomass and subsequent neutralization (or over liming) with calcium hydroxide. As mentioned above, this is a major incentive for developing pretreatment and detoxification methods that make use of materials that can be recycled.

20.7 Conclusions

The sugar platform is a promising approach to producing commodity-scale products from renewable biomass. Sugars can be converted through biological or chemical means to a wide assortment of valuable products. The challenge of realizing this approach is to overcome the recalcitrant nature of cellulose. Pretreatment and enzymatic hydrolysis together can produce high yields of sugars from biomass, but the pretreatment process needs to be optimally designed so as to enable the biologically catalyzed process downstream. Many methods of reducing the inhibitory nature of biomass hydrolysates have been developed, though all add cost and complexity to the operations of a biorefinery. The optimal solution will likely be specific to the biomass feedstock, downstream processing needs, and local opportunities for synergy with other biomass processing facilities.

References

Alriksson, B., Horváth, I.S., Sjöde, A., *et al*. (2005) Ammonium hydroxide detoxification of spruce acid hydrolysates. *Appl. Biochem. Biotechnol.* 121–124: 911–922.

Alriksson, B., Sjöde, A., Nilvebrant, N.O., *et al*. (2006) Optimal conditions for alkaline detoxification of dilute-acid lignocellulose hydrolysates. *Appl. Biochem. Biotechnol.* 129–132: 599–611.

Alves, L.A., Felipe, M.G.A., Silva, J.B. *et al*. (1998) Pretreatment of sugar cane bagasse hemicellulose hydrolysate for xylitol production of Candida guilliermondu. *Appl. Biochem. Biotechnol.* 70–72: 89–98.

Ando, S., Arai, I., Kiyoto, K., and Hanai, S. (1986) Identification of aromatic monomers in steam-exploded poplar and their influence on ethanol fermentation. *J. Ferment. Technol.* 64: 567–570.

Azzam, A.M. (1989) Pretreatment of cane bagasse with alkaline hydrogen peroxide for enzymatic hydrolysis of cellulose and ethanol fermentation. *J. Environ. Sci. Health B* 24: 421–433.

Banerjee, N., Bhatnagar, R., and Viswanathan, L. (1981) Inhibition of glycolysis by furfural in *Saccharomyces cerevisiae*. *Eur. J. Appl. Microbiol. Biotechnol.* 11: 226–228.

Baugh, K.D., Levy, J.A., and McCarty, P.L. (1988) Thermochemical pretreatment of lignocellulose to enhance methane fermentations: II. *Evaluation and application of pretreatment model. Biotechnol. Bioengin.* 31: 62–70.

Baugh, K.D., and McCarty, P.L. (1988) Thermochemical pretreatment of lignocellulose to enhance methane fermentations: 1. *Monosaccharide and furfurals hydrothermal decomposition and product formation rates. Biotechnol. Bioengin.* 31: 50–61.

Berlin, A., Balakshin, M., Gilkes, N. *et al*. (2006) Inhibition of cellulase, xylanase and beta-glucosidase activities by softwood lignin preparations. *J. Biotechnol.* 125(2): 198–209.

Berson, R.E., Young, J.S., Kamer, S.N., and Hanley, T.R. (2005) Detoxification of actual pretreated corn stover hydrolysate using activated carbon powder. *Appl. Biochem. Biotechnol.* 121–124: 923–934.

Bjerre, A.B., Olesen, A.B., Fernqvist, T., *et al*. (1996) Pretreatment of wheat straw using combined wet oxidation and alkaline hydrolysis resulting in convertible cellulose and hemicellulose. *Biores. Technol.* 49: 568–577.

Bouchard, J., Nguyen, T.S., Chornet, E., and Overend, R.P. (1991) Analytical methodology for biomass pretreatment. Part 2: Characterisation of the filtrates and cumulative product distribution as a function of treatment severity. *Biores. Technol.* 36: 121–131.

Bura, R., Bothast, R., Mansfield, S., and Saddler, J.N. (2003) Optimization of SO_2-catalyzed steam pretreatment of corn fiber for ethanol production. *Appl. Biochem. Biotechnol.* 105–108: 319–335.

Cantarella, M., Cantarella, L., Gallifuoco, A., *et al*. (2004) Effect of inhibitors released during steam explosion treatment of poplar wood on subsequent enzymatic hydrolysis and SSF. *Biotechnol. Prog.* 20(1): 200–206.

Chang, V.S, Nagwani, M, and Holtzapple, M.T (1998). Lime pretreatment of crop residues bagasse and wheat straw. *Applied Biochemistry and Biotechnology* 74(3): 135–159.

Chang, V.S., Nagwani, M., Kim, C., and Holtzapple, M.T. (2001) Oxidative lime pretreatment of high-lignin biomass. *Appl. Biochem. Biotechnol.* 94: 1–26.

Chen, S.-F., Mowery, R.A., Castleberry, V. A. *et al*. (2006) High performance liquid chromatography method for simultaneous determination of aliphatic acid, aromatic acid and neutral degradation products in biomass pretreatment hydrolysates. *J. Chrom. A.* 1104: 54–61.

Chen, S.-F., Mowery, R.A., Chambliss, C.K., and van Walsum, G.P. (2007) Pseudo reaction kinetics of organic degradation products in dilute-acid-catalyzed corn stover pretreatment hydrolysates. *Biotech. Bioeng.* 98(6): 1135–1145.

Chum, H.L., Johnson, D.K., Black, S.K., and Overend, R.P. (1990) Pretreatment-catalyst effects and the combined severity parameter. Appl. *Biochem. Biotechnol*. 24/25: 1–14.

Clark, T. and Mackie, K.L. (1984) Fermentation inhibitors in wood hydrolysates derived from the softwood *Pinus radiata*. *J. Chem. Biotechnol.*, B34: 101–110.

Clark, T. and Mackie K.L. (1987) Steam explosion of the softwood Pinus radiata with sulphur dioxide addition. *I. Process optimization. J. Wood Chem. Technol.* 7: 373–403.

Converti, A., Dominguez, J.M., Perego, P., *et al*. (2000). Effects of lignocellulose degradation products on ethanol fermentations of glucose and xylose my *Saccharomyces cerevisiae*, *Zymomonas mobilis*, *Pichia stipitis*, and *Candida shehatae*. *Enzyme Microb. Technol.* 19: 220–225.

deMancilha, I.M., Karim, M.N. (2003) Evaluation of ion exchange resins for removal of inhibitory compounds from corn stover hydrolyzate for xylitol fermentation. *Biotechnol Prog* 19:1837–1841.

Du, B., Sharma, L.N., Becker, C., *et al*. (2010) Effect of varying feedstock-pretreatment chemistry combinations on the production of potentially inhibitory degradation products in biomass hydrolysates. *Biotechnol. Bioengineer.* 107(3): 430–440.

Dutta, A., Dowe, N., Ibsen, K., *et al*. (2010) An economic comparison of different fermentation configurations to convert corn stover to ethanol using *Z. mobilis and Saccharomyces*. *Biotechnology Progress*, 26(1): 64–72.

Duarte, G.V., Ramarao, B.V., and Amidon, T.E. (2010) Colloidal stability and flocculation of lignocellulosic hydrolyzates. *Biores. Technol.* 101: 8526–8534.

Fein, J.E., Tallim, S.R., and Lawford, G.R. (1984) Evaluation of D-xylose fermenting yeasts for utilization of a wood-derived hemicellulose hydrolysate. *Can. J. Microbiol.* 30: 682–690.

Frazer, F.R. and McCaskey, T.A. (1991) Effect of components of acid-hydrolysed hardwood on conversion of D-xylose to 2,3-butanediol by *Klebsiella pneumoniae*. *Enz. Microb. Technol.* 13(Feb): 110–115.

Galbe, M. and Zacchi, G. (2002) A review of the production of ethanol from softwood. *Applied Microbiology and Biotechnology*. 59(6): 618–628.

Garrote, G., Dominguez, H., and Parajo, J.C. (2002) Interpretation of deacetylation and hemicellulose hydrolysis during hydrothermal treatments on the basis of the severity factor. *Process Biochem.* 37: 1067–1073.

Gilbert, A., Sangurdekar, D.P. , and Srienc, F. (2009) Rapid strain improvement through optimized evolution in the cytostat. *Biotechnol. Bioeng*. 103(3): 500–512.

Grous, W.R., Converse A.O., and Grethlein, H.E. (1986) Effect of steam explosion pretreatment on pore size and enzymatic hydrolysis of poplar. *Enz Microb Technol* 8: 274–280.

Hasan, A., Yasarla, L.R., Ramarao, B.V., and Amidon, T.E. (2011) Separation of Lignocellulosic Hydrolyzate Components Using Ceramic Microfilters. *J. Wood Chem. Tech.* 31(4): 357–383.

Heipieper, H.J., Weber, F.J., Sikkema, J. *et al*. (1994) Mechanism of resistance of whole cells to toxic organic solvents. *TIBTECH* 12: 409–415.

Holtzapple, M.T., Jun, J.H., Ashok, G. *et al*. (1991) The ammonia freeze explosion (AFEX) process: a practical lignocellulose pretreatment. *Appl. Biochem. Biotechnol*. 28/29: 59–74.

Horvath, I.S., Sjode, A., Nilvebrant, N.-O., *et al*. (2004) Selection of anion exchangers for detoxification of dilute-acid hydrolysates from spruce. *Appl. Biochem. Biotechnol.* 113–116: 525–538.

Iakovlev, M., Sixta, H., and Van Heiningen, A. (2011) SO2-Ethanol-Water (SEW) Pulping: II. Kinetics for Spruce, Beech, and Wheat Straw. *J. Wood Chem. Technol.* 31(3): 250–266.

Jennings, E. and Schell, D. (2011) Conditioning of dilute-acid pretreated corn stover hydrolysate liquors by treatment with lime or ammonium hydroxide to improve conversion of sugars to ethanol. *Bioresource Technology*. 102(2): 1240–1245.

Jeon, Y.J., Svenson, C.J., Joachimsthal, E.L.H., and Rogers, P.L. (2002) Kinetic analysis of ethanol production by an acetate-resistant strain of recombinant *Zymomonas mobilis*. *Biotech. Letters* 24: 819–824.

Jonsson L.J., Palmqvist E., Nilvebrant, N.O., and Hahn-Hägerdal, B. (1998) Detoxification of wood hydrolysates with laccase and peroxidase from the white rot fungus Trametes versicolor. *Appl. Microbiol. Biotechnol.* 49: 691–697.

Kim, J.S., Lee, Y.Y., and Park, S.C. (2000a). Pretreatment of Wastepaper and pulp mill sludge by aqueous ammonia and hydrogen peroxide. *Appl. Biochem. Biotech.* 84–86: 129–179.

Kim, K.H., and Hong, J. (2001). Supercritical CO_2 pretreatment of lignocellulose enhances enzymatic cellulose hydrolysis. *Biores. Tech.* 77(2): 139–144.

Kim, K.H., Tucker, M.P., and Nguyen, Q.A. (2005) Conversion of bark-rich biomass mixture into fermentable sugar by two-stage dilute acid-catalyzed hydrolysis. *Biores. Tech* 96(11): 1249–1255.

Kim, S.B., Yum, D.M., and Park, S.C. (2000b) Step-change variation of acid concentration in a percolation reactor for hydrolysis of hardwood hemicellulose. *Bioresource Technol.* 72: 289–294.

Klinke, H.B., Ahring, B.K., Schmidt, A.S., and Thomsen, A.B. (2002) Characterization of degradation products from alkaline wet oxidation of wheat straw. *Biores. Technol.* 82: 15–26.

Klinke, H., Thomsen, A., and Ahring, B.K. (2004) Inhibition of ethanol-producing yeast and bacteria by degradation products produced during pre-treatment of biomass. *Appl. Microbiol. Biotechnol.* 66: 10–26.

Larsson, S., Cassland, P., and Jonsson, L.J. (2001) Development of a *Saccharomyces cerevisiae* strain with enhanced resistance to phenolic fermentation inhibitors in lignocellulose hydrolysates by heterologous expression of laccase. *Appl. Environ. Microbiol.* 67(3): 1163–1170.

Larsson, S., Palmqvist, E., Hahn-Hägerdal, B. *et al*. (1998) The generation of fermentation inhibitors during dilute acid hydrolysis of softwood. *Enz. Microb. Technol.* 24: 151–159.

Larsson, S., Reimann, A., Nilvebrant, N.-O., and Jonsson, L.J. (1999) Comparison of different methods for the detoxification of lignocellulose hydrolyzates of spruce. *Appl. Biochem. Biotechnol.* 77–79: 91–103.

Leonard, R.H. and Hajny, G.J. (1945) Fermentation of wood sugars to ethyl alcohol. *Ind. Eng. Chem.* 37: 390–395.

Lloyd, T. and Wyman, C.E. (2003) Application of a depolymerization model for predicting thermochemical hydrolysis of hemicellulose. *Appl. Biochem. Biotechnol.* 105–108: 53–67.

Lopez, M.J., Nichols, N.N., Dien, B.S. *et al.* (2004) Isolation of microorganisms for biological detoxification of lignocellulosic hydrolysates. *Appl. Microbiol. Biotechnol.* 64: 125–131.

Luo, C., Brink, D.L., and Blanch, H.W. (2002) Identification of potential fermentation inhibitors in conversion of hybrid poplar hydrolyzate to ethanol. *Biomass Bioenergy* 22: 125–138.

Lynd, L.R., Baskaran, S., and Casten, S. (2001) Salt accumulation resulting from base added for ph control, and not ethanol, limits growth of thermoanaerobacterium thermosaccharolyticum HG-8 at elevated feed xylose concentrations in continuous culture. *Biotechnol. Prog.* 17: 118–125

Martinez, A., Rodriguez, M.E., Wells, M.L. *et al.* (2001) Detoxification of dilute acid hydrolysates of lignocellulose with lime. *Biotechnol. Prog.* 17: 287–293.

McMillan, J.D. (1994) Conversion of hemicellulose hydrolyzates to ethanol, in Enzymatic Conversion of Biomass for Fuels Production, Himmel, M.E., Baker, J.O., and Overend, R.P., Editors. *ACS Symposium Series* 566: 411–437.

McWilliams, R.C. and van Walsum, G.P. (2002) comparison of aspen wood hydrolysates produced by pretreatment with liquid hot water and carbonic acid. Appl. Biochem. and *Biotechnol.* 98: 109–121.

Mes-Hartree, M., and Saddler, J.N. (1983) The nature of inhibitory materials present in pretreated lignocellulosic substrates which inhibit the enzymatic hydrolysis of cellulose. *Biotechnol. Lett.* 5(8): 531–536.

Mohagheghi, A., Ruth, M., and Schell, D. (2006) Conditioning hemicellulose hydrolysates for fermentation: effects of overliming pH on sugar and ethanol yields. *Process Biochem.* 41:1806–1811.

Mosier, N., Hendrickson, R., Ho, N.W.Y., *et al.* (2004) Optimization of pH controlled liquid hot water pretreatment of corn stover, in Presentation 31d at the AIChE 2004 Annual Meeting, Austin TX, Nov. 7–12.

Mosier, N., Wyman, C., Dale, B., *et al.* (2005) Features of promising technologies for pretreatment of lignocellulosic biomass. *Biores Technol* 96: 673–686.

Nguyen, Q.A., Tucker, M.P., Keller, F.A., and Eddy, F.P. (2000) Two-stage dilute-acid pretreatment of softwoods. *Appl. Biochem. Biotechnol.* 84–86: 561–576.

Nichols, N.N., Dien, B.S., Guisado, G.M., and Lopez, M.J. (2005) Bioabatement to remove inhibitors from biomass-derived sugar hydrolysates. *Applied Biochem. Biotechnol.* 121–124: 379–390.

Nichols, N.N., Sharma, L.N., Mowery, R.A., *et al.* (2008) Fungal metabolism of pretreatment side-products present in corn stover dilute acid hydrolysate. *Enzyme and Microbial Technol.* 42: 624–630.

Nilvebrant N.O., Persson P., Reimann, A., *et al.* (2003) Limits for alkaline detoxification of dilute-acid lignocellulose hydrolysates. *Appl. Biochem. Biotechnol.* 105–108: 615–628.

Nilvebrant, N.O., Reimann, A., and Larsson, S., *et al.* (2001) Detoxification of lignocellulose hydrolysates with ion exchange resins. *Appl. Biochem. Biotechnol.* 91–93: 35–49.

Nishikawa, N.K., Sutcliffe, R., and Saddler, J.N. (1988) The influence of lignin degradation products on xylose fermentation by *Klebsiella pneumoniae*. Appl. Microbiol. Biotechnol. 27: 549–552.

Okuda, N., Soneura, M., Ninomiya, K., *et al.* (2008) Biological detoxification of waste house wood hydrolysate using Ureibacillus thermosphaericus for bioethanol production. *J. Biosci. Bioeng.* 106:128–133.

Olsson, L. and Hahn-Hagerdahl, B. (1996) Fermentation of lignocellulosic hydrolysates for ethanol production. *Enyme Microb. Technol.* 18: 312–331.

Overend, R.P. and Chornet, E. (1987) Fractionation of lignocellulosics by steam-aqueous pretreatments. Phil. Trans. R. Soc. Lond. A321: 523–536.Palmqvist, E., Almeida, J., and Hahn-Hagerdahl, B. (1999) Influence of furfural on anaerobic glycolytic kinetics of *Saccharomyces cerevisiae* in batch culture. *Biotechnol. Bioeng.* 62: 447–454.

Palmqvist, E. and Hahn-Hagerdahl, B. (2000a) Fermentation of lignocellulosic hydrolysates. *I: inhibition and detoxification. Biores. Technol.* 74: 17–24.

Palmqvist, E. and Hahn-Hagerdahl, B. (2000b) Fermentation of lignocellulosic hydrolysates. *II: inhibitors and mechanisms of inhibition. Biores. Technol.* 74: 25–33.

Palmqvist, E., Meinander, N.Q., Grage, H., and Hahn-Hagerdahl, B, (1999) Main interaction effects of acetic acid, furfural and p-hydroxybenzoic acid on growth and ethanol productivity of yeasts. *Biotechnol. Bioeng.* 63: 46–55.

Pan, X., Gilkes, N., Kadla, J., *et al.* (2006). Bioconversion of hybrid poplar to ethanol and co-products using an organosolv fractionation process: Optimization of process yields. *Biotechnol. Bioeng.* 94(5): 851–861.

Persson, P., Andersson, J., Gordon, L., *et al*. (2002a) Effect of different form of alkali treatment on specific fermentation inhibitors and on the fermentability of lignocellulose hydrolysates for production of fuel ethanol. *J. Agric. Food Chem.* 50: 5318–5325.

Persson, P., Larsson, S., Jönsson, L.J. *et al*. (2002b) Supercritical fluid extraction of a lignocellulosic hydrolysate of spruce for detoxification and to facilitate analysis of inhibitors. *Biotechnol. Bioeng.* 79: 694–700.

Pfeifer, P.A., Bonn, G., and Bobleter, O. (1984) Influence of biomass degradation products on the fermentation of glucose to ethanol by *Saccharomyces carlbergensis* W 34. *Biotechnol. Letters* 6(8): 541–546.

Pienkos, P.T. and Zhang, M. (2009) Role of pretreatment and conditioning processes on toxicity of lignocellulosic biomass hydrolysates. *Cellulose* 16: 743–762.

Purwadi, R., Niklasson, C., and Taherzadeh, M.J. (2004) Kinetic study of detoxification of dilute-acid hydrolyzates by $Ca(OH)_2$. *J Biotechnol* 114:187–198.

Ranatunga, T.D., Jervis, J., Helm, R.F., *et al*. (2000) The effect of overliming on the toxicity of dilute acid-pretreated lignocellulosics: the role of inorganics, uronic acids and ether-soluble organics. *Enz. Microb. Technol.* 27: 240–247.

Rodrigues, R.C.L.B., Felipe, M.G.A., Almeida e Silva, J.B., *et al*. (2001) The influence of pH, temperature and hydrolysate concentration on the removal of volatile and nonvolatile compounds from sugarcane bagasse hemicellulosic hydrolysate treated with activated charcoal before or after vacuum evaporation. *Braz. J. Chem. Eng.* 18: 299–311.

Schirmer-Michel, A.C., Flores, S.H., Hertz, P.F. *et al*. (2008) Production of ethanol from soybean hull hydrolysate by osmotolerant Candida guilliermondii NRRL Y-2075. *Biores. Technol.* 99: 2898–2904.

Schneider, H. (1996) Selective removal of acetic acid from hardwood-spent sulfite liquor using a mutant yeast. *Enz. Microb. Technol.* 19: 94–98.

Shevchenko, S.M., Chang, K., Robinson, J., and Saddler, J.N. (2000) Optimization of monosaccharide recovery by post-hydrolysis of the water-soluble hemicellulose component after steam explosion of softwood chips. *Bioresource Technol.* 72: 207–211.

Sjolander, N.O., Langlykke, A.F., and Peterson, W.H. (1938) Butyl alcohol fermentation of wood sugar. *Indust. Eng. Chem.* 30: 1251–1255.

Skammelsen, A.S. and Thomsen, B.A. (1998) Optimization of wet oxidation pretreatment of wheat straw. *Biores. Technol.* 64: 139–151.

Sreenath, H. and Jeffries, T.W. (2000) Production of ethanol from wood hydrolyzate by yeasts. *Biores. Technol.* 72: 253–260.

Su, T.M, *et al*. (1980) Final Report to the United States Department of Energy, Subcontract No. XR-9-8271-1 by General Electric Co.

Szengyel, Z. and Zacchi, G. (2000) Effect of acetic acid and furfural on cellulase production of *Trichoderma reesei RUT C30*. *Appl. Biochem. Biotechnol.* 89: 31–42.

Taherzadeh, M.J., Millati, R., and Niklasson, C. (2001) Continuous cultivation of dilute-acid hydrolysates to ethanol by immobilized *Saccharomyces cerevisiae*. *Appl. Biochem. Biotechnol.* 95: 45–57.

Tengborg, C., Galbe, M., and Zacchi, G. (2001) Reduced inhibition of enzymatic hydrolysis of steam-pretreated softwood. *Enz. Microb. Technol*. 28(ER9-10): 835–844.

Tengborg, C., Stenberg, K., Galbe, M. (1998) Comparison of SO_2 and H_2SO_4 impregnation of softwood prior to steam pretreatment on ethanol production. *Appl. Biochem. Biotechnol.* 70–72: 3–15.

Tran, A.V. and Chambers, R.P. (1986) Ethanol fermentation of red oak acid prehydrolysate by the yeast *Pichia stipitis* CBS 5776. *Enz. Microb. Technol.* 8(July): 439–444.

Um, B.-H., Freedman, B., and van Walsum, G.P. (2011) Conditioning hardwood-derived pre-pulping extracts for use in fermentation through removal and recovery of acetic acid using trioctylphosphine oxide (TOPO). *Holzforschung* 65: 51–58.

van Walsum, G.P. (2001) Severity function describing the hydrolysis of xylan using carbonic acid. *Applied Biochem. Biotechnol.* 91–93: 317–329.

van Walsum, G.P., Allen, S.G., Laser, M.S. *et al*. (1996) Conversion of lignocellulosics pretreated with hot compressed liquid water to ethanol. *Applied Biochem. Biotechnol*. 57/58: 157–170.

van Walsum, G.P., Garcia-Gil, M., Chen, S.F., and Chambliss, K. (2007) Effect of dissolved carbon dioxide on accumulation of organic acids in liquid hot water pretreated biomass hydrolysates. *Applied Biochem. Biotechnol*. 137–140: 301–311.

Walton, S., Van Heiningen, A., and van Walsum, P. (2010) Inhibition effects on fermentation of hardwood extracted hemicelluloses by acetic acid and sodium. *Bioresource Technol.* 101(1): 1935–1940.

Weil, J., Brewer, M., Hendrickson, R. *et al*. (1998) Continuous pH monitoring during pretreatment of yellow poplar wood sawdust by pressure cooking in water. *Appl. Biochem. Biotechnol.* 70–72: 99–111.

Weil, J.R., Dien, B., Bothast, R., *et al*. (2002) Removal of fermentation inhibitors formed during pretreatment of biomass by polymeric adsorbents. *Ind. Eng. Chem. Res.* 41(24): 6132–6138.

Wilson, J.J., Deschatelets, L., and Nishikawa, N.K. (1989) Comparative fermentability of enzymatic and acid hydrolysates of steam-pretreated aspenwood hemicellulose by *Pichia stipitis* CBS 5776. *Appl. Microbiol. Biotechnol.* 31: 592–596.

Yourchisin, D. and van Walsum, G.P. (2004) Comparison of the microbial inhibition and enzymatic hydrolysis rates of liquid and solid hydrolysates produced from pretreatment of biomass with carbonic acid and liquid hot water. *Appl. Biochem. Biotechnol.* 115(1–3): 1073–1086.

Zacchi, G. (2002) A review of the production of ethanol from softwood. *Appl. Microbiol. Biotechnol.* 59: 618–628.

Zaldivar, J. and Ingram, L.O. (1999a) Effect of selected aldehydes on the growth and fermentation of ethanologenic Escherichia coli. *Biotechnol. Bioeng.* 65: 24–33.

Zaldivar, J. and Ingram, L.O. (1999b) Effect of organic acids on the growth and fermentation of ethanologenic *Escherichia coli* LY01. *Biotechnol. Bioeng.* 66: 203–210.

Zaldivar, J. and Ingram, L.O. (2000) Effect of alcohol compounds found in hemicellulose hydrolysate on the growth and fermentation of ethanologenic Escherichia coli. *Biotechnol. Bioeng.* 68: 524–530.

Zhang, Y.H., Ding, S.Y., Mielenz, J.R., *et al*. (2007) Fractionating recalcitrant lignocellulose at modest reaction conditions. *Biotech. Bioeng.* 97: 214–223.

21

Case Studies of Separation in Biorefineries—Extraction of Algae Oil from Microalgae

Michael Cooney

Hawaii Natural Energy Institute, USA

21.1 Introduction

According to the International Energy Agency's (IEA) report on biodiesel production which looked at the 21 leading biofuel producing countries, global biodiesel production has increased tenfold from 2000 to 2008 and could double again to 21.8 billion liters by 2012 (Sustainable Production of Second-Generation Biofuels; IAE Information Paper; http://www.iea.org/publications/freepublications/publication/second_generation_biofuels-1.pdf). Although the motivations for governments to pursue biodiesel development aggressively are complex and multidimensional [1], key driving forces behind the government policies can be divided into three principle points. Firstly, with the shadow of a plateau or decline in world peak oil production [2–4], the global energy crisis approaching, biodiesel fuel will play a more important role in strengthening US energy security. Secondly, as a renewable energy, biodiesel is derived from plant materials which can contribute to the reduction of greenhouse gas (GHG) emissions when replacing fossil oil—assuming they are sustainably managed. Thirdly, the increased demand for oil crops for biodiesel production clearly can have a positive effect on net farm income whilst also reducing government outlays to farmers by raising the market price of oil crops.

The production of biodiesel, however, needs a quality feedstock and must use far less energy than that used to produce the fuel. The principal source of biodiesel feedstock has been and remains oil seed crops although microalgae have gained attention as a collaborative source due to limitations observed with oil seed crop production. The production of microalgae oil as a feedstock, however, remains a most challenged industry [5, 6] with nearly no commercial operation deployed anywhere in the world despite an

Separation and Purification Technologies in Biorefineries, First Edition.
Edited by Shri Ramaswamy, Hua-Jiang Huang, and Bandaru V. Ramarao.

impressive array of activity in terms of patents, start-up companies, and large-scale government investment [7]. This is due, in part, to the fact that the production of microalgae oil is an energy-complex combination of multiple energy-consuming unit operations, which include growth of biomass in open ponds (or perhaps closed photobioreactors), harvesting, dewatering, drying, extraction, and purification of the oil prior to its transesterification into biodiesel. In response, many have proposed unit operations that integrate two or more of the harvesting, dewatering, drying, and extraction steps in an effort to simplify and reduce energy loads but, in general, the more that are combined the less robust the process becomes. This chapter focuses mostly on the extraction step and discusses the most recent activity and proposed technologies, with some commentary on the market for biodiesel. As a recent review [8] extensively covered the academic literature, this chapter focuses heavily on the patent literature.

21.2 The market and industrial needs

21.2.1 Feedstock markets

United States biofuel policy includes a biodiesel use mandate that rises to almost 4 hm^3 (1 $hm^3 = 1\,000\,000\,m^3$) by 2012, calling for an analysis of biomass feedstock that recognizes the complex interdependence among potential feedstocks, as well as competition for food and industrial uses [9]. In their study summarizing supply and demand of 13 sources of oils and five sources of fats and butter, Thompson *et al.* (2010) found that in a period between 2006 and 2009 the largest US bio-oil production came from corn and soybean at 1113 and 8622 Gg per year (1.113×10^6 and 8.622×10^6 metric tonne, respectively) and for fats and butter the largest contributor was from inedible tallow at 2890 Gg (2.890×10^6 Mt) per year. Of all these sources, only soybean oil was reported as a feedstock for biodiesel production at 280 Gg (0.28×10^6 metric tonne) per year, with the rest going for food and industrial use. Thompson *et al.* also reported imports of these products to the US, with the largest contributors being coconut and rapeseed oil at 442 and 571 Gg (0.442 and 0.571×10^6 metric tonne) per year. Globally conventionally grown edible oils included rapeseed, soybean, sunflower and palm with about 7% of global vegetable oil supplies used for biodiesel production in 2007 [10]. Extensive use of edible oils for fuels, however, may cause other significant problems such as starvation in developing countries as growing crops for fuel squanders land, water, and energy resources vital for the production of food for people [11, 12]. In recent years additional sources of waste oil have entered the market, including waste oil from restaurants, whether it is the oil recovered from used cooking oil or waste trap grease [13]. In fact, used cooking oil has become so valuable that one can find reports of its theft from restaurants.

Fats and oils can be used directly by consumers or indirectly through industrial uses (here industrial use being defined broadly to include not only processed goods such as soap but also as animal feed). Soybean oil is among the cheaper oils commonly used for food consumption, certainly much cheaper than olive oil, but fats tend to be cheaper than oils with the cheapest, according to these data, being poultry fat. Different fats and oils have varying degrees of suitability for different uses, leading to imperfect substitutability. In the extreme case, some fats and oils are put exclusively to one of these two broad types of uses (e.g. such as castor oil, inedible tallow, and poultry fat) and are not directly consumed whereas corn, cottonseed, olive, and peanut oils tend to be used only for direct consumption. This distinction is judged important: biodiesel demand may tend to be made more cheaply from low priced oils and fats that are used for industrial processes rather than from higher priced goods that are typically consumed directly. Moreover, elasticities of substitution are presumably important within either category of use—more so than between the two categories.

It is sometimes overlooked that growing biofuel production is perceived to have important effects on agricultural commodity markets through processors' demand for feedstocks. The potential stress that sourcing biofuel feedstocks would place on agricultural commodities otherwise used for food production has also led to a broad interest in biofuel policies and provisions intended to limit the scope for certain feedstocks to be diverted from food use. Another stress on biodiesel production in the United States is the mandate to reduce greenhouse-gas emissions imposed by the Energy Independence and Security Act of 2007 (Pub. L. No. 110–40, Title II; Dec 19 2007), which sets greenhouse gas reduction thresholds that must be met for different feedstocks or processes in order to count towards the mandates. This effectively means that sourcing biomass feedstocks must include more screening criteria than just biomass productivities and oil content. For example, preliminary rules established by the EPA suggest that biodiesel from soybean oil does not meet greenhouse gas reduction requirements necessary to qualify for the mandate [14].

Analyzing the impact of biodiesel policy on feedstock markets introduces complications relating to substitutability among these goods in consumption, their production response, and trade flows. A recent study that took into account the potential that the rapid increase in biodiesel production could lead to climbing prices for the feedstock and for other competing fats and oils cautioned against overly simplifying feedstock markets by holding prices constant when considering the economics of a particular feedstock [9]. Specifically, they used a simulation model to explore the tradeoffs between uses caused by the expanding US mandate for biodiesel use both with and without soybean oil counted as an eligible feedstock. Their findings suggested a hierarchy of price effects that tends to be largest for cheaper fats and oils typically used for industrial and feed purposes and smallest for fats and oils traditionally used exclusively for direct consumption, with the cross-commodity effects and other key economic parameters playing a critical part in determining the scale in each case. Any analysis that condenses the market into a broad aggregate or that focus on a single good might give useful broad brush approaches that meet certain needs, but does not capture how price impacts are transmitted throughout the animal fats and vegetable oils complex.

These factors have led to an increase in the production of biodiesel from different non-edible oilseed crops over the last few years [10]. These reports suggest that non-edible plant oils have been found to be promising crude oils for the production of biodiesel. The use of non-edible oils when compared with edible oils is very significant in developing countries because of the tremendous demand for edible oils as food, which can be far too expensive for use as fuel [15]. Throughout the world, large amounts of non-edible oil plants are available in nature [16], including the jatropha tree (*Jatropha curcas*) [17, 18], karanja (*Pongamia pinnata*) [19, 20] tobacco seed (*Nicotiana tabacum L.*) [21, 22], rice bran [23, 24], mahua (*Madhuca indica*) [25, 26], neem (*Azadirachta indica*) [27], rubber plant (*Hevea brasiliensis*) [28], castor [29], linseed [30], and microalgae [31]. Microalgae have also been proposed [32], and are often defended from a well distributed list of the oil productivity of microalgae versus various oil seed plants as presented in Table 21.1 [33].

Unfortunately, the numbers for microalgae have been misused. In a recent paper the thermodynamic basis for the calculation of the upper photosynthetic limit for the amount of biodiesel that can be produced per acre per year has been outlined [5]. To reach the upper limit of 15 000 gallons per acre per year one must make a large number of highly unrealistic assumptions including (i) sunlight at an intensity available at the equator, (ii) no more than 10% cloud cover year round, (iii) photosynthetic efficiency of 21.4%, (iv) 100% efficiency in harvesting, dewatering, and extraction, and (v) roughly 50% composition of lipids. While all of these assumptions are unrealistic, perhaps the most problematic is the assumption of photosynthetic efficiency of 21.4%. This describes the efficiency at which the cells can convert the energy contained in the photons arriving at the earth's surface to energy in the form of biomass. This value is the thermodynamic upper limit and assumes 100% efficiency of photon transmission, 100% utilization of all photons by the cells (i.e. all photons are taken up by cells), and 100% efficiency of converting the energy in

Table 21.1 Oil productivity of different crops

Oil crop	Productivity (gallons $acre^{-1}$ $year^{-1}$)
Corn	18
Soybeans	48
Safflower	83
Sunflower	102
Rapeseed	127
Oil palm	635
Microalgae	5000–15 000

the photons to biomass. In reality, an overall photosynthetic efficiency of around 6% is far more realistic, particularly for the conditions of growth in open ponds. Using this value, and the measure photon intensity at Honolulu, one gets just at 4279 gallons per acre per year of biodiesel (and this still assumes 100% efficiency in harvesting, dewatering, and extraction). This is in line with recent reports that microalgae are the fastest growing photosynthesizing organisms, completing an entire growth cycle every few days. Approximately 46 tons of oil/hectare/year (~4100 gallons per acre per year) can be produced from diatom algae [1]. When one compares, however, the water load when growing microalgae [5], along with the land and nutrient loads [6], as well as extensive obstacles to harvesting, dewatering, and extraction [8], bio-oil from microalgae is not as superior to oil seeds as once thought.

The discussion on markets above suggests that biodiesel production facilities will need to source a variety of feedstocks, either processed independently or as blends. In other words, biodiesel will be manufactured from a wide range of feedstocks and any attempt to estimate its demand on commodity prices of feedstocks that would otherwise go to food is too complicated to predict, and will ultimately be affected by complex pricing schemes that take into account local markets, process difficulties with respect to handling blends and changing feedstocks during the year, trade flows, and international supplies. Moreover, the blending of feedstocks, or even varying quality in a single feedstock, can cause significant processing problems that add to the final cost of biodiesel. As such, extraction technologies put in place commercially will likely need to be sufficiently robust to handle a wide range of feedstocks.

21.2.2 Biodiesel markets

Biodiesel production expanded from about 5 million gallons in 2001 to 1100 million gallons in 2011 (*Source*: National Biodiesel Board under "Production Statistics" accessed at http://www.biodiesel.org/production/production-statistics). Even though high oil prices will always tend to reduce biodiesel production, several forces will contribute to long-term expansion in the biodiesel industry: (i) High petroleum prices will raise petroleum diesel prices and as we have passed peak world oil production the future costs of oil production will continue to increase over time as per Hotelling's model of depleted resources [34]; (ii) Government mandates, such as the provisions of the Energy Independence and Security Act of 2007 (H.R. 6 of the 110 U.S. Congress: Energy Independence and Security Act of 2007. Accessed at Library of Congress http://thomas.loc.gov/cgi-bin/bdquery/z?d110:HR00006:@@@X) that includes mandates of up to 36 million gallons of biofuels; (iii) The public and government's concern over global warming will provide additional value for biodiesels' CO_2 recycling characteristics. For example, the US government has discussed the use of GHG emission price in a cap-and-trade system, as in the Lieberman-Warner Climate

Security Act of 2008 (The Library of Congress. Lieberman-Warner Climate Security Act of 2008 (S3036). Washington, DC; 2008. Available at: http://thomas.loc.gov/cgi-bin/query/D?c110:4:./temp/_c1106UdsR6, accessed 21 October, 2008). However, there are negative forces that will hinder the expansion of the biodiesel industry, such as (i) cost of feedstocks have risen rapidly threatening industry viability; (ii) The expiration of biofuel subsidies at the end of 2009; and (iii) large energy requirements for ethanol may push soybean oil prices above breakeven points. Despite these factors, the US agricultural model, FASOMGHG, has been used to predict that the maximum biodiesel market penetration will only reach 9% in 2030 with a wholesale diesel price of $4 per gallon [35].

21.2.3 Algae products

Given the relative difficulty of commercializing bio-oil from microalgae [5, 6], with the possible exception of those linked to wastewater [7], the bio-fuel from microalgae thrust has focused heavily on the production of co-products [36]. Algae products include, but are not limited to, food, fertilizers, pharmaceuticals, dyes, bio-plastics, lipids, feedstock for chemical production, and feedstock for energy production [37]. Algae products are collected through application of precipitation, chromatography, adsorption, electrophoresis, crystallization, binding, foam, fractionation, osmotic shock, and combinations thereof [37]. Historically, the post-lipid fractionation (i.e. the waste algae biomass) are typically discarded or made into biogas (methane) using digesters without further fractionation of the remaining valuable products [38]. Despite this wealth of byproducts, there still remains no commercially viable biofuels from microalgae process, due in part to the high costs of even the simplest algal production processes and, in even larger part, the presently undeveloped nature of algal mass culture technology [7].

21.2.4 Industrial needs

The production of biodiesel starts with the acquisition of feedstock. Despite extensive research information on transesterification [39–42] the production technology and process optimization for various biomaterials, the major challenge to the biodiesel production industry remains a lack of a consistent quality and supply of feedstock [43]. As discussed earlier, the feedstocks for biodiesel production are primarily vegetable oils and animal fats although the residual fats and oils of domestic, commercial, and industrial processing can also be used as feedstocks [44]. Different vegetable oils with various compositions of fatty acids will be found; among them soybean, sunflower, rapeseed, and palm being the most studied, and less common or unconventional oilseeds being tobacco, pongamia, jatropha, and rubber seeds [45]. Though animal fats have chemical structures similar to vegetable oils, with the only differences being in the distribution of fatty acids, they remain potentially useful sources of biodiesel that are not studied as extensively as vegetable oils [43].

Biodiesel is produced from these sources using transesterification, a reaction by which the fats or oils are reacted with alcohols (usually methanol or ethanol) to form fatty acid alcohol (i.e. methyl, ethyl) esters (also termed biodiesel esters such as fatty acid methyl or ethyl esters), and glycerol (Figure 21.1). That being said, the transesterification is not as straightforward as one would hope, owing to the fact that the reaction depends upon the chemical structure of the feedstock. Specifically, if applied to triglycerides (i.e., triacylglycerols or TAGs), transesterification is a stepwise process where monoacylglycerols (MAGs) and diacylglycerols (DAGs) are produced as intermediates. The TAGs are first converted to DAGs, and since they are in methanol they have a greater chance of reacting to product MAGs and glycerol rather than moving back into the oil phase. Although catalysts, alkaline or acidic, are employed to increase the reaction rate and yield, unwanted side reactions producing undesired byproducts are caused by the presence of undesired physical and chemical characteristics (free fatty acid (FFA) content, moisture content and other

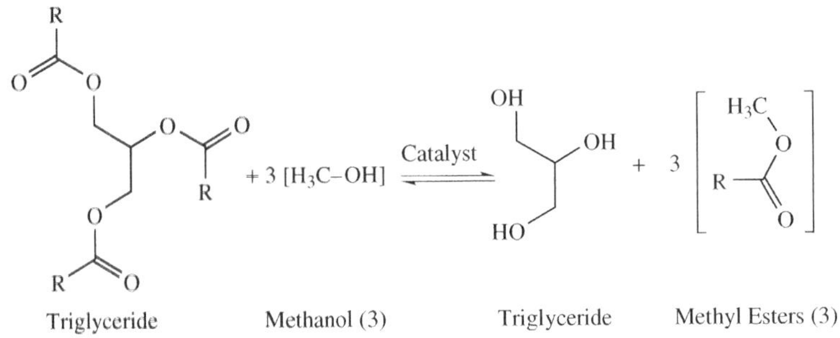

Figure 21.1 *Transesterification reaction*

impurities) present in the oils used as feedstock. For example, besides transesterification, TAG and free fatty acids (FFA) can undergo the side-reactions of hydrolysis or saponification. In hydrolysis, water reacts with the TAG to produce glycerol and a free fatty acid as opposed to the desired fatty acid methyl ester. In saphonication the free fatty acid reacts with sodium hydroxide (a common base catalyst used in the transesterification of TAGs) to form soap (a sodium salt of the free fatty acid) and glycerol.

Of the several variables that can affect the yield/conversion of the transesterification reaction, one of the most difficult to control at commercial scale is the quality of the feedstocks (i.e. presence of impurities) [46]. As discussed earlier, impurities in biodiesel, either due to side-reactions, unreacted feedstock, or non-fatty acid constituents, will increase pollutants and/or limit reaction rates and yields [46]. For example, the base-catalyzed biodiesel process is sensitive to water and free fatty acids. Water (as low as 1%) can hydrolyze the triglycerides to diglycerides and form a FFA and this FFA then reacts with the alkali catalyst to forms soaps, resulting in separation inefficiencies downstream of the reactor [43]. Waste cooking oil is generally high in free fatty acids (2 wt%) and therefore more prone to the saponification side reaction. To avoid this, a pretreatment step such as those involving reaction with methanol in presence of acid catalyzed esterification is applied because, although the acid-catalyzed transesterification is significantly slower than alkaline catalysis, there is no side reaction converting fatty acids to soaps [47]. Consequently, acid catalysis has become an accepted pretreatment step for esterification of feedstocks with high FFA presence in the feed, although alternative treatments such as refining the waste oil using steam treatment has also been suggested to reduce the free fatty acid and moisture content of oil [48]. While moisture and free fatty acids are major examples, other impurities include sand, dirt, seed fragments, glucosides, unsaponifiables (organics which do not react with base to form soaps such as sterols, higher molecular weight alcohols, pigments, waxes, and hydrocarbons), metals, and gums [43]. Their presence requires other pretreatment steps such as washing with a polar solvent (e.g. water).

The oils with high FFA content decreases biodiesel yield and increases production cost. Contaminated feedstocks require extra processing steps like filtration, centrifuging and heating [43]. In the United States, soybean oil is the primary feedstock for biodiesel. The fuel industry competes with the food industry for soy and, as discussed earlier, in order to keep biodiesel production costs low and to meet additional energy demand, biodiesel producers will look for alternative feedstocks. Because of the requirement that biodiesel production processes will require multiple feedstocks, manufactures will need to carefully consider processing conditions, continuous versus batch operations, methanol and glycerol separation techniques, and catalysis as a function of the varying quality of the feedstocks they source and potentially blend. Ever stricter quality demands on biodiesel before its sale to the end user mean that there remains a need to develop inexpensive and rapid fuel quality tests to characterize biodiesel as well as the feedstocks. As such, perhaps the greatest industrial need will be for the methods to pretreat the feedstocks.

21.3 The algae oil extraction process

Overall, the extraction processes requires the coordinated integration of multiple unit operations such as: (i) harvesting/isolation of the cells from a growth unit such as a bioreactor or pond, (ii) cell wall lyses/disruption, (iii) fully drying the cells, (iv) extraction of the oil from the cells, and finally (v) recovery and refining of the oil. In general the most straightforward and efficient extraction technology has been accomplished through the application of solvent systems (methanol/chloroform, hexane, hexane/alcohols, alcohol-ionic liquids) to dried cells that have been exposed to some sort of cell-wall rupturing technique (e.g. grinding) [8]. The classic solvent system would be the methanol-chloroform system of Bligh and Dyer while the more practical is hexane [8]. These approaches, however, suffer severe limitations in terms of energy loads associated with drying and solvent distillation and are thus not commercial viable [8]. In response to this there has been a general trend towards the development of extraction systems that work in the presence of water and thus avoid the need for drying and extraction of the oil into a miscible solvent (i.e. hexane) that must then be distilled away [8]. In general these techniques require the application of energy intensive techniques to make the cell walls more permeable (i.e. some application of sonication or equivalent wave energy) while suspended in water such that the oil is both extracted from the cells and easily separate (downstream) from the immiscible water phase. Although appropriately addressing the need to limit the dewatering/drying pretreatment step, such processes have generally suffered from scale up problems. In summary, when reviewing extraction processes one must also consider the impact of and interaction with the upstream processes as the two are dependent upon each other, with the choice of one impacting the other [5].

21.3.1 Harvesting/isolation

The isolation or harvesting of microorganisms involves the separation of the cells from the cultivation medium and their subsequent concentration through application of dewatering techniques. Many solid–liquid separation technologies have been employed to harvest algae cells from the fermented medium including precipitation, filtration, centrifugation and flocculation [49–51]. A simple example would include initial harvesting through flocculation, concentration through membrane filtration coupled with centrifugation to a paste. Usually a minimum of two steps are applied although this is dependent upon the extraction technology applied.

21.3.2 Drying

When lipids are not extracted immediately, or if the extraction technology is inhibited by the presence of water, the isolated microorganisms are typically dried and stored under nitrogen (or treated with an antioxidant) to avoid exposure of microorganisms to air which can further degrade the lipids. They can also be sealed under vacuum, to prevent degradation of lipids, although this would generally be applied for the production of nutraceuticals or other high-value products. In the case of biofuels the more viable option would be the rapid extraction, purification, and chemical transformation of the bio-oil to fuel. Drying technologies include drum drying, freeze-drying, spray drying, and solar drying [50]. The drying process can expose the microorganisms to heat, which can damage (degrade) the quality of, the lipid if done incorrectly.

21.3.3 Cell wall lyses/disruption

Cells can be lysed/disrupted using a variety of methods including the application of enzymes, chemical treatment, thermal treatment, and mechanical pressing [37]. Enzymes are applied to degrade the cell wall, with usually more than one applied (e.g. lytic, proteases, cellulases, hemicellulases, chitinases and/or pectinases) [52–54]. Cells can also be genetically engineered to express a protein that will lyse the microorganism's cell wall [54]. Termed "autolysis," the process requires inoculums with the microorganisms generally grown in a bioreactor to a desired cell density followed by induction to produce the lytic protein to lyse the cells. The cells can also be infected with a lytic virus that invade and rupture algal cells [55], including Chlorella virus, and cyanophages [52, 53]. Chemical lysing agents can include base or acid salts [54], and/or one or more surfactants or detergents. The application of high salt concentration for osmotic shock has been proposed [55]. Thermal lysing comprises heating (for example, by addition of steam [56]) the cells until their walls degrade or break down and this is particularly useful for microorganisms whose cell walls are composed of proteins [57]. Heating the broth also denatures proteins and helps to solubilize organic material, including proteins. The temperature can range from between 50 °C to 130 °C, although higher temperatures are generally more effective. Mechanical lysis disrupts cell walls by physical means, which are designed to impart shear forces [54]. Methods for mechanical lysis can include use of the French press, ball or bead mills, ultrasonication, spray drying, cold pressure, and pressure homogenizers (e.g. Gaulin.TM. homogenizer performed at pressures that can vary from 300 to 900 bar) [54, 55]. Homogenization will require multiple passes, usually from 1 to 3, depending upon the pressure applied, which can range from low (100 bar) to high (900 bar) although optimal performance is achieved at 500 to 700 bar [58].

21.4 Extraction

21.4.1 Organic-solvent based

Organic-based extraction processes apply to the use of nonpolar hydrophobic solvents (such as hexane or alcohol) with which the extracted bio-oil are partially or wholly immiscible. In general, the cells are first dried (dry extraction) after which a cell-disruption technique (such as freeze drying or grinding) is applied prior to their dissolution in the organic solvent such as hexane [50, 59, 60]. Hexane primarily extracts triglycerides, diglycerides, monoglycerides, and esterified sterols, although other components of the total lipid fraction such as phospholipids, free sterols, and carotenoids are also extracted to a lesser degree [61]. One problem with the use of hexane is that in addition to extracting the desired triglycerides, all other fat-soluble components are extracted. These can include fat soluble pigments, which, if present, will contaminate or alter the quality of the extracted oil.

There are many approaches used to contact the hexane with the cells, such as Soxhlet extraction [50] or high-pressure techniques (e.g., the Dionnex accelerated solvent extraction system). Other approaches have proposed dimethyl ether [60] although in the presence of this solvent, gums between complex lipids (if they are present) will form, which need to be separated from the neutral lipids later (usually via phase separation aided by centrifugation) [60]. Once the lipids have been extracted into the solvent, the lipid-containing solvent will need to be separated from the solid phase, which is best accomplished with centrifugation. Once separated, the extracted lipids must then be further separated from the solvent. This can be accomplished by application of super or subcritical CO_2 [60] to the solvent (with the lipids carried with the CO_2 phase that is later evaporated away to leave behind the lipids) or by vacuum distillation of the solvent [50].

In some instances the algae remain wet to some extent and need not be fully dewatered before the extraction process [59]. For example, the microalgae can be wet milled in hexane [61]. In this process the milling serves to break through the water interface and initiate contact between the lipids and the hexane phase, which is immiscible with the water phase and able to substantially extract triglycerides, diglycerides, monoglycerides, and esterified sterols [61]. Although conceptually attractive, large-scale application to wet samples requires large volumes of hexane, which promotes application to fully dry biomass in order to minimize the volumes of solvent used. After the extraction step is completed, one is left with a miscella (i.e. the solvent containing the extracted oil or grease) and extracted biomass [61] that can be separated by centrifugation [61]. The oil can then be separated from the hexane (or equivalent) via techniques such as distillation under vacuum. In other applications, the lipid content of the clarified miscella can be adjusted to about 45% using n-hexane, and then chilled to approximately $-1\,^{\circ}$C for 8 to 12 hours. The last chilling step crystallizes any saturated fats or high melting point components, which helps to improve performance of oil in winter climates. Finally, the miscella can then be filtered to remove the crystallized steering phase while the hexane is removed from the miscella, leaving behind the winterized lipid [61].

Direct transesterification (DT) has emerged as a method that avoids the initial use of hexane (or similar solvent). The biomass is first dried and then contacted with an alcohol and catalyst [55]. In some cases the biomass is first disrupted prior to the addition of alcohol and catalyst. The alcohol is typically added at a 3:1 mass ratio (relative to the oil in the cell) [50]. Typical catalysts include, but are not limited to, concentrated sulfuric acid or enzymes such as lipases [50, 62]. If the direct transesterification is done solely in the presence of methanol as solvent and reactant, with added catalyst, the FAMES must then be separated from the alcohol phase, either using by extraction into an organic solvent, such as hexane, or by addition of water to modify the two phase properties such that the methanol and water partition to their own phase and leave the FAMEs to their own immiscible phase.

21.4.2 Aqueous based

In aqueous systems the extraction of microalgae oil is applied to the cells while still suspended or concentrated in the presence of water. In addition to potentially avoiding the need to distill large quantities of organic solvent, it also avoids the high energy cost (and potential damage to the lipids) associated with drying. With some modifications at key points, directed for the needs of a specific process, the general process starts with cells that have first been are cultured in a bioreactor to a given concentration. The final concentration can range from as low as 0.3 $\text{gdw}\,\text{l}^{-1}$ for microalgae grown photosynthetically to as high as 80 $\text{gdw}\,\text{l}^{-1}$ for microalgae grown heterotrophically [57, 58]. Even at 80 $\text{gdw}\,\text{l}^{-1}$, however, there is a significant amount of water present (e.g., the water content of cells is at least 60%). If necessary, the cells can be sterilized at this point (usually through pasteurization) to kill the cells and/or to inactivate any undesirable enzymes that can degrade the isolated lipids.

Next the cells are lysed, mechanically disrupted, or some combination thereof. If desired, the cells can be concentrated, prior to this step, through application of dewatering techniques such as filtration or centrifugation. For example, the concentration of cells facilitates additional washing steps with water that may be applied in order to remove any undesired extracellular water-soluble or water-dispersible compounds that may be present prior to the lysing step. It is also possible that the selected cell-disruption technique may work best on concentrated cells. Prior to cell disruption, a separation inducer (to encourage separation of oil from cell debris) may be added. This inducer can comprise addition of one or more salts (e.g. NaCl at concentrations of 10–150 $\text{g}\,\text{l}^{-1}$), alkali, and/or one or more cell-degrading enzymes, surfactants or emulsifiers [58]. Nonetheless, the cell lysing or disruption will generally leave a poorly defined two-phase system (a lighter oil containing phase and a heavier aqueous phase containing the cells) with an emulsion at the interface. In some cases the emulsion can incorporate the entire light lipid carrying phase although it is believed that the oil–water interface of the emulsion is stabilized by residual cellular debris [57].

To further separate the oil phase from the aqueous cell debris containing phase, an additional separation step is needed. This is usually carried out with two or three phase centrifugation, while one or more surfactants or detergents may be added to assist with this process [58]. In the case of two-phase centrifugation, the lighter oil phase will contain emulsified lipids while the water and cell debris will separate to the heavier phase. In three-phase centrifugation, the aqueous phase comprises mainly water plus any water-soluble constituents (proteins, inducer agents, for example) while the cell debris will concentrate to a solid paste. A third, lipid rich phase, is also taken off and subsequently washed until a substantially non-emulsified oil phase is obtained. It is believed that the oil–water interface of the emulsion is stabilized by residual cellular debris, which is removed by this washing process [57]. The lighter lipid phase can be further refined as discussed above to meet the required specifications of its target use. The solid phase cells may be further washed, for example, using an aqueous solution (such as water) to remove any extracellular water-soluble or water-dispersible compounds.

Other methods include the formation of a hydrolysate, assuming the cells can first be ruptured [63]. In this approach an aqueous slurry of microalgae is formed, the cell walls are ruptured, sufficient acid is added to the algal slurry to form an acid concentration of 2 to 3 M and to partially hydrolyze the proteins in the algae, the acid-insoluble fraction is discarded from the acid-soluble fraction of the resultant hydrolysate, removing the acid from the soluble fraction until the soluble fraction has a pH of at least 1, titrating the hydrolysate with a base to convert any remaining acid in the hydrolysate to a salt and adjust the pH to within the range of about 6.5 to about 7.0.

Similarly, acid hydrolysis can be applied to an aqueous suspension of algal biomass and the lipids extracted by addition of at least one non-polar organic solvent and at least one inorganic solvent under atmospheric pressure to the aqueous suspension of algae biomass [64]. After these steps, three phases are obtained: (i) a semisolid phase comprising a slurry of algae biomass, (ii) an aqueous phase comprising inorganic compounds and hydrophilic organic compounds, and (iii) an organic phase comprising fatty acids and hydrophobic organic compounds other than fatty acids. The solvent and acid have to be added at the same time and the process executed at temperatures below 100 °C.

More indirect methods in aqueous media include the use of inorganic materials to absorb fat soluble compounds selectively in the presence of ruptured (or rupturing) cells. If the particles are heavy enough to fall out of solution in the absence of agitation, a conceptually easy downstream separation can be achieved, either with centrifugation of lower energy settling techniques. The fat-soluble compounds can be recovered easily by contacting the absorbed particles with a wash solution to desorb the fat-soluble compounds. A purification step to isolate the fat-soluble compound from the solution would follow. Challenges to this technology include the ability to lyse the cells with the fluidized particles and to bind lipids selectively from the other fat-soluble compounds that are produced in photosynthetically grown microalgae. Microalgae cells such as *D. salina*, which lack a protective cell wall, may serve this process well. In terms of selective isolation, this method may be most suitable for cells grown heterotrophically on synthetic media that limits the production of pigments and fat-soluble compounds other than triglycerides [65]. A slightly different version to this would be the grinding of dilute aqueous dispersions of microalgae in the presence of grinding media and then applying adsorptive bubble separation [66]. Theoretically the lipids absorb to gas bubbles and rise to the surface where they separate into an immiscible phase as the gas bubbles pop [67]. The potential limitation to this type of process is the presence of emulsions that can develop when the lipid-laden cells are ground in the presence of water and then followed with aeration.

Some researchers have proposed the separation of lipids from microalgae suspended in water through the application of supercritical CO_2, with a downstream separator used to partition the treated algae composition into an organic phase, which includes the lipids, an aqueous phase, and a solid phase with the biomass residue [68]. Compounds other than carbon dioxide such as methane, ethane, or propane have also been used [49].

An emerging method gaining attention is the secretion of lipids into the medium from selected species of microalgae. One scheme proposed is the use of lipophilic particulates or ion-exchange resins [69]. They can be circulating in a separate medium and then collected, or the medium may be passed over a fixed-bed column. The fatty acids are then eluted from the absorbents by use of an appropriate solvent, from which they must then be separated, presumably using low-pressure distillation techniques. Initially appealing, this method is challenged by recovery of the excreted lipids before bacteria consume them, or their forming a surface layer that deflects incoming light for growth.

Others have proposed to lyse algae concentration (from centrifugation) with a live steam at a mass flow rate of steam being roughly 2–20%, or more preferably 2–5% of the mass flow rate of the algae cell concentrate [56]. The steam is applied at low pressure (between 3–5 bar) and the unbound oil is then recovered after the lysed cell concentrate is passed to a oil separator. Some energy saving is gained by passing the heated lysed concentrate (now diluted with water) through a heat exchanger that heats up the incoming algae to be lysed.

The proposed Westfalia Separator® Friolex process for obtaining oil uses an aqueous base, not hexane, to extract by displacement [70]. This process enables high-quality oil or fat to be obtained directly and efficiently from vegetable and animal raw materials containing oils, where the non-oil phase (e.g. proteins) remains "native." In the form of a slurry, the fruit flesh is separated into an oil fraction and pulp in a specially developed decanter centrifuge. This is only possible if there is no emulsion between the oil and the aqueous phase. With most natural raw materials such as seeds, fish or eggs, however, this emulsion is the case. In order to break this emulsion, therefore, aqueous alcohol is added, not as a solvent for extraction, but as a purely physical aid to break the emulsion. This simple method enables a clear distinction to be made between the oil phase and the solids slurry. The entire Friolex process has a nitrogen blanket. This ensures protection for operating in an alcoholic environment and has a positive impact on product quality. Any oxidation of oil is avoided. Oils with valuable polyunsaturated fatty acids (PUFAs) are preserved, as are vitamins. The raw material is first mixed with water before going into a milling apparatus. This slurry is then mixed with more water, additives, and alcohol in a separate mixing unit. This then goes to a decanter to separate into oil. The oil is dried (with the water coming off returned upstream) and then fed to a separator for further improvement of the oil quality. The sludge from this step is dried with the solids disposed off, and the water/alcohol, after evaporated, is collected and sent back upstream to the mixing tank.

21.4.3 Combined aqueous and organic phases

Recently, many processes have been proposed that combine extraction across both aqueous and organic phases. In general, these processes will first apply techniques (usually through the use of sound wave energy) to lyse the cells in the presence of water and then use an organic solvent to extract the desired triglycerides. One example involves the microalgae containing broth being passed through a reservoir where low-energy sound waves are provided (<2 MHz on 120 V) to produce an algal slurry (a completely dispersed emulsion of broken algae cells, lipids, and growth media) [71]. The solution is composed of three layers: a lipid/oil emulsion on top, a thin green layer of cell debris at the interface, and a lower heavy water layer. The top lipid/oil layer is then separated using traditional techniques with the addition of acid and/or heat in the presence of an organic solvent [72, 73]. The oil is then subjected to further processing to produce a fuel.

There have been recent advances with organic solvents that promote the selective extraction of only triglycerides and that work in the presence of water. In general this method applies solvents, such as dodecane, which have biocompatible octanol numbers (the logarithm of the octanol-water partition coefficient). A biocompatible solvent will generally have an octanol number greater than 5 [74]. The technique is generally applied in the presence of sonication, high-energy sound or acoustical radiation, for the disruption of aggregates of molecules in order to either separate them or permeabilize them [75]. This process, termed "hydrocarbon milking," mixes at least a portion of the culture with a solvent to obtain a heterogenous mixture of solvent, cells, and water. The mixture is then passed into a partitioning chamber to obtain an extracted aqueous fraction containing an extracted viable organism and a solvent-oil fraction; and a recycle step, in which at least a portion of the extracted viable organism is recycled into a culturing system [75]. Although the process eliminates the need for centrifugation/flocculation and the destructive solvent (methanol) or mechanical disruption steps typically used to extract oil from microalgae, the process ultimately requires the separation of the extracted lipids from the organic solvent, a process that will still require vacuum or pressure reduced distillation. The application of the sonication step, which is used to permeabilize the cells in the presence of the water-organic co-solvent, is also difficult to scale up and can induce emulsions that challenge the downstream separation step.

Energy-efficient membrane systems have been proposed for the separation of alcohol and oil from water solvent as an efficient algae oil extraction process [76]. This approach requires that the bio-oil is efficiently extracted from the cells using an alcohol, such as ethanol, with downstream processing only required to remove the water from the alcohol/oil mixture, as the alcohol (e.g. methanol or ethanol) that remains is then simply used as a substrate in transesterification reactions required to make biodiesel. The membranes are composed of perfluoropolymer membranes modified with functional groups that transform the polymer into hydrophilic NF membranes that, under pressure, will facilitate transport of the hydrophilic water whilst holding back the oil and a significant percentage of the alcohol. Although attractive in concept, key issues arise—for example, the fouling of the membrane pores by the oil or alcohol or cell debris. There is also the issue of designing energy-efficient ethanol extraction units.

21.4.4 Supercritical fluids

A recent review has shown at least 20 articles devoted to the application of supercritical CO_2 to extract molecules of interest from microalgae [77]. The most extracted compounds are neutral lipids and antioxidants. Of all the parameters that affect the extraction performance (e.g., temperature, pressure, extraction time) it is noteworthy that the greatest effect is from algae pretreatment. The first step removes all water through two stages of centrifugation followed by low temperature or freeze drying. The algae are then crushed. Supercritical CO_2 is the most often recommended solvent as it allows the avoidance of toxic solvents. However, if polar lipids are desired, the presence of a polar solvent (e.g. ethanol) is necessary. However, in addition to the extreme energy requirements of the pretreatment, the extraction process also carries along liposoluble pigments such as antioxidants, chlorophyll, hydrocarbons, and vitamin E, all of which can be considered contaminants of the final oil. Supercritical fluid extraction is also a highly energy intensive operation owing to the high pressures needed to maintain supercritical conditions (e.g. 50 °C and 240 to 380 bar). Real problems occur with scale up in terms of high pressure rated equipment as well.

Low-pressure CO_2 can also be used to avoid the cost of high-maintenance pump systems, such as the proposed BoSonX system, which uses CO_2 in a liquid phase that filters through the algae in order to extract the necessary oils [78]. The process involves roughly seven unit operations that begin with the premixing of an algae slurry, where the "as-received" algae oil slurry is combined with the recycled algae oil extract and CO_2 gas. This step is generally required to bring the incoming slurry to a predetermined composition that can be pumped. The combined slurry is then pressurized with CO_2 as it passes through an ultrasonic reactor heated to elevated temperatures—a step that was previously established for CO_2-based degreasing (i.e. extraction of oils and greases using pressurized CO_2). The extracted slurry is then passed sequentially through three unit operations designed to remove, first, the algal cell solids, then the carbonated water, and finally the algae oil. The solids separation is a three-stage filtration process, operated at elevated pressure (~500 psig) that uses pressurized CO_2 to push the oil through a filter bed that retains the defatted cells. The carbonated oil/water slurry is then transported to an electrocoagulation separator designed to separate water, carbonated oil and trailing residues from each other. Following the water separation system, the carbonated oil is then passed to an oil/CO_2 separator where the slurry is slowly depressurized until the

CO_2 is removed from the oil. The recovered CO_2 is then returned to the final unit operation, a CO_2-processing unit that repressurizes the CO_2 for its return to the upstream process units. Although obviously a competent system, the energy load for pressurization to move the slurry through the system is unlikely to be successful on a commercial scale.

21.4.5 Solventless extraction

Solventless extraction technologies traditionally used for oil seeds, such as expeller presses and extruders, are inefficient for microalgae, in part because the microalgae cells are much smaller than the dimensions of mechanical surfaces used to press them. Nonetheless high-velocity adiabatic impact compaction has been proposed to recover biofuel from algae efficiently [79]. This technique is implemented by placing dried algae in a casting die and compacting it with a battering ram at a controlled velocity to deliver a sufficient impact to disrupt the outer wall structure. The power-ram velocity (and thus power of impact) is usually varied with the microalgae feedstock. The disruption is delivered by shock waves that are distributed throughout the biomass upon impact of battering rams traveling at velocities around 2–10 meters per second prior to impact.

21.4.6 Emerging technologies

Some researchers have taken the view that the issues presented earlier are so challenging as to suggest the bio-fuels from microalgae will remain severally limited at commercial scale for the next decade, if not much longer [7]. As such, some have suggested the growth of microalgae to support the growth of higher organisms (shrimp or fish) from which bio-oil can then be extracted using much simpler and less energy intensive processes, such as mechanical pressing or grinding in the presence of an extracting solvent such as hexane, benzene, isopropanol, and the like [80].

Advanced materials and processes (AMP) have developed non-dispersive, static fiber reactors that can be used, among other uses, to replace the need for centrifuges when processing vegetable oils and making biodiesel [81]. These processes comprise a fiber reaction process whereby reactive components contained in immiscible streams are brought into contact to effect chemical reactions and separations. The conduit reactor utilized contains wettable fibers onto which one stream is substantially constrained and a second stream is flowed over to create continuously a new interface to bring about efficiently contact between the reactive species and thus promote reactions or extractions. Co-solvents and phase-transfer catalysts may be employed to facilitate the process. Such reactors are efficient at contacting two immiscible liquid phases without dispersion of one phase into the other. The separation of the two phases is quick, clean, and simple with fiber having 560X times the surface area of contact between the two liquids as compared to conventional stirred tank reactors. These types of systems open up the possibility of processing aqueous-phase (paste) microalgae extracts, and catalyzing the transesterification reaction in the presence of water, with the FAMES solubilizing in the phase immiscible with water. Although the process is proposed for vegetable oils, the same technology could be applied to microalgae paste.

One of the more intriguing approaches in recent years proposes to culture algae that are consumed by fish, which are then captured and extracted for oil [82]. The extraction of lipids from fish is a step in the commercial process for producing fish meal which is the main product. Standard hydrothermal processes can be applied to extract residual lipids from the fish meal. Because harvesting and processing the fishes does not require removing and heating large volumes of water, as is necessary in conventional methods that directly process the algae, a reported net gain of energy can be obtained from the system (estimated at a net gain of $+22\,250\,\text{kwhr acre}^{-1}$ in contrast to a net gain of $-174\,000\,\text{kwhr acre}^{-1}$ for conventional methods applied to algae). As such, this approach attempts to solve the problems involved with conventional processes that must separate the algae from the pond water and then dewater the algae. Dewatering is problematic because water has a high heat capacity and thus, requires a large amount of energy to evaporate. The use of fish is also advantageous because it allows for the growth of marine strains of microalgae without adding salt to the processing. The presence of salt is a major impediment to extraction of oil from marine strains of microalgae—fish are natural salt filters.

21.4.7 Refining lipids

The isolated lipid can be refined further using a process similar to that used to refine standard vegetable oils [57]. Such a lipid-refining process generally involves hydrating phospholipids by adding phosphoric acid followed by adding sodium hydroxide to neutralize free fatty acids. These compounds are then removed via centrifugation. This is then followed by washing with water to further remove any remaining amounts of hydrated phospholipids ("gums") and neutralized fatty acids ("soapstock") from the lipid. The washed lipid is then bleached with an appropriate agent (e.g. Trysil.TM.), and a standard bleaching clay. Citric acid can also be added to remove divalent metal ions by chelation. The bleaching agent and bleaching clay are then removed via filtration to produce refined lipid. The bleached lipid can be further cold filtered to remove high melting-point compounds that may be present in the lipid; however, this step is less often required. The resulting lipid can also be further refined by removing any low molecular weight components that may be present through sparging with steam at high temperatures, under high vacuum. This process also destroys any peroxide bonds that may be present and reduces or removes off odors and helps improve the stability of the oil. An antioxidant may then be added to the resulting deodorized lipid to improve product stability.

Prior to the refining process, the isolated lipid can be winterized to remove high melting compounds, such as saturated fatty acids. In general, winterization is the process of removing sediment that appears in the vegetable oils at low temperature [61]. The winterization process generally involves dissolving the isolated lipid in an organic solvent, such as hexane, cooling the resulting organic solution, and filtering the solution to remove the high melting-point components of the lipid phase. It can also involve the contacting of the lipid composition with a more polar solvent (such as acetone or isopropyl alcohol) to remove more polar fatty acids that would otherwise form crystals of triglycerides at lower temperatures (these crystals can have melting points as high as 50–50 °C) [61]. Upon precipitation of the contaminants from the lipid composition, a separation can be conducted to remove the precipitated material from the lipid composition [61]. The winterization process generally produces a clear lipid, especially when the isolated lipid is cloudy or opaque.

21.5 Separation performance and results

It is particularly difficult, if not misleading, to define the performance of results for the majority of separation techniques discussed above because, with the exception of those applied for the production of nuetraceutical oils, none of them has been applied on a commercial scale. It is also difficult to extract accurate extraction information from patents or patent applications as they are generally vague in specifics but long on potential. For example, one of the more conceptually pleasing technologies is that proposed by Benamann and Oswald in their 1996 report to the DOE [83]. In that work the authors proposed a relatively simple and low-cost process that requires no drying of the biomass: cell breakage followed by emulsification with recycle oil and then separation with a three-phase centrifuge. Like most proposed processes, however, this has never been tried at large scale and is also limited to strains of microalgae (e.g. *Duniaella*) that lack a cell wall and thus can break open in the presence of hot oil with the application of most modest shear forces. Clearly, this process is limited to very specific strains of microalgae and thus very challenged if applied to strains with cell walls.

As discussed in our previous review, most liquid-phase solvent extraction technologies are ultimately limited by solvent access to intracellular contained lipids, contact of the solvent with the lipid, and the solvent's carrying capacity [5]. Perhaps the best example of this is the classic technique of Bligh and Dyer [84], which used a co-solvent mixture containing methanol and chloroform to extract lipids from tissues. While quoted and used extensively, this method's accuracy was recently shown to be limited by the carrying capacity of the chloroform [85, 86]. The basic message to be taken from these works is that if the cell mass is dried, appropriately ground open to allow solvent penetration and contact, and if enough of the solvent is used, nearly complete extraction of bio-oil can be expected. In this sense, liquid-phase solvent extraction using organic solvents can be expected to produce yields approaching 100 percent but not necessarily in a commercially viable way.

Direct transesterification techniques can also be quite successful, particularly when under conditions where enough acid catalyst is added to help digest the biomass [87, 88]. That said, the FAMES produced are still miscible in methanol and must be separated by addition of water to create a two-phase system where the lipids are isolated into a hydrophobic solvent, such as hexane, from which they must later be recovered using evaporative distillation of the hexane. Again, these processes can be scaled up to produce very high yields but not on a commercially successful basis.

The technique of milking lipids from microalgae has gained some attention in the commercial sector but it has not yet been proven successful on a commercial scale. The concept is to selectively extract triglycerides from algal cells using a solvent of appropriate hydrophobicity and in the presence of sonication or similar energy that partially permeabilizes the cell wall. In practice the method requires the mixing of the aqueous solution of cells with the appropriate organic solvent (e.g. dodecane) and then passing the mixture through a pipe fitted with some kind of sonication energy distribution device, during which time the solvent selectively extracts only TAGs. Theoretically the mixture is then passed to a holding tank wherein the still live cells separate with the water and the extracted TAGs separate into the organic solvent phase. While conceptually attractive, and proposed to have nearly 100% extraction efficiency (private communications), the process had not passed pilot trials and would still require, if successful, extraction of the lipids from the organic phase. Related strategies to utilize cells genetically modified to secrete lipids will suffer degradation and consumption of the lipids before they can be harvested.

Techniques using supercritical fluids such as CO_2, and even accelerated solvent extraction using solvents such as methane or hexane, can be considered very efficient. In the case of CO_2 as the solvent, one advantage is that the CO_2 is easily separated from the extracted materials upon lowering its pressure. The extraction efficiency, however, can actually be so good that all fat-soluble compounds (such as chlorophyll) will likewise be extracted. This can be problematic if such compounds are considered contaminants that must be removed prior to the transesterification of the lipids to biodiesel or more complicated catalysis to jet fuel.

Of all technologies, the most likely processes to have commercial success are those that extract the lipids from larger biomass sources such as those that propose to grow microalgae to grow fish or shrimp. The advantage is that dewatering and mechanical extraction (via pressure compacting or extrusion) are far simpler to apply to fish or shrimp. Both also act as natural salt filters, which is great advantage supporting the use of brackish or salt water—salt being a tremendous problem undermining most extraction processes. While both methods will still deliver a relatively "dirty" bio-oil that will need further refining, the method of extracting oils from fish is a proven industry with commercially viable unit operations [50, 68, 82]. There are serious issues, however, that need to be worked out. Among them are the challenge of cultivation microalgae in the wild at high abundance and cheaply, and whether or not they are suitable for growth of the shrimp or fish. These kinds of open ponds are challenged by disturbances such as sudden heavy rainfall (particularly when grown in brackish or marine water), which can adjust pH or salinity, as well as crashes due to the presence of predators. That said, these challenges are relatively equivalent to those already faced by the bio-oil from microalgae industry. As such, the superior extraction options would seem to make this approach the most likely to be successful in the long term. The sole exception would be the dark fermentation industry in which microalgae are grown on simple sugars in sterile bioreactors to high concentration [89].

21.6 Economic importance and industrial challenges

The biofuel from microalgae industry, to date, is one of projection as opposed to reality. There is a global paucity of production sites dedicated to the growth of microalgae for fuels. To date, the industry is largely one of unrealized projections, often with projects being swayed by a desire to meet current consumption loads of petroleum-based liquid fuels. In doing so the economical importance of bio-oil from microalgae has dominated the culture of the entire industry, often at the exclusion of a realistic observation of industrial challenges.

The most noteworthy of industrial challenges has to do with the availability of natural resources to support algal biofuels production at scale up levels. Significant displacement of petroleum fuels will require that algae feedstock production will reach large volumes that will put demand on key resources (e.g., CO_2, nitrogen and phosphorous nutrients, and water [5, 6, 90]). A recent study has shown that significant resource supply challenges occur as biofuel production capacity approaches 10 billion gallons, let alone 20, 50, or even 100 billion gallons per year [6]. Another issue is that the specific details of resource challenges depend upon the geographic region, the target feedstock production volume. The implications are that the supply of CO_2, nutrients, and water, in particular, can be expected to severely limit the extent to which the US production of algae biofuel can be sustainably expanded. It appears that the most restrictive limitation is the source of CO_2 [6].

The latter statement is particularly important because it indicates that of all the industrial challenges facing production of microalgae biofuels (nitrogen and phosphorous and water limitations, the biology of large-scale commercial growth of strains in open ponds, dewatering and extraction), the single biggest problem is a source of CO_2. This statement underlines the correlation between the requirement of feeding large-scale microalgae ponds a concentrated source of CO_2 in order to achieve the kind of biomass productivities that supports biofuel productivities that can subsidize the US consumption of petroleum fuels. Sources of concentrated CO_2 are typically from power-plant flue gas and, beyond the irony of predicating a renewable source of energy on the consumption of a byproduct of petroleum combustion, the majority of power plants are not in locations suitable for the installation of large-scale microalgae farms. It also underscores the reality that large-scale commercial production of microalgae biofuels is uncompetitive when grown in the absence of a concentrated source of CO_2. The response to this has been a significant focus on the production of byproducts and the application of carbon substrates at night (e.g. Heteroboost™).

The production and recovery of byproducts, as well as the application of carbon feedstocks at night, only add to the industrial challenges. The most common byproduct is the use of defatted microalgae biomass as an animal feed [91]. Although reasonable at first glance, these production avenues introduce additional processing challenges. Lipids, particularly those with nutraceutical value, are an important value component of animal or fish feeds; however, the extraction process lowers its value. Moreover, there is concern that commercial ponds can suffer invasion of microalgae species that are toxic to animals [92]. The addition of sugar-based carbon, for example, increases the opportunity for invasion by predatory bacteria and higher organisms. Perhaps more relevant is the fact that such sugars themselves are a consequence of an agricultural growth process, ironically suggesting that one must grow, harvest, and process crops to produce enough sugar to sustain biomass productivities of another crop (microalgae) to produce a liquid-phase biofuel. This becomes all the more problematic considering that, if sugars are used to spurt heterotrophic growth at night, one could simply produce the microalgae from purely heterotrophic high-density growth in pure culture.

In industrial-scale production, a large amount of volatile organic solvent is typically used, creating hazardous operating conditions. The use of organic solvent in the ***extraction*** process may also necessitate using an explosion-proof oil recovery system, thereby adding to the cost of lipid recovery. In addition, if the lipid is to be used as a foodstock, it is important that certain solvents, such as hexane, are removed completely, or only remain in very small quantities. If the hexane is removed by evaporation then this may involve heating, and that not only adds to costs but can cause lipid degradation. Furthermore, with increasing environmental considerations, the use of solvents for the *extraction* of lipids is becoming increasingly expensive and unpopular.

21.7 Conclusions and future trends

The number one impediment to the extraction of bio-oils from microalgae is the presence of water [8]. This becomes more challenging if saltwater is used to grow the microalgae—an approach that a growing number of people are suggesting as the only path forward for the bio-oil from microalgae industry [93]. In a typical photosynthetic pond, the maximum biomass concentration that one can likely assume is around 0.3 gdw l^{-1}. This means that the water column is 99.7% water. Beyond the realization that this represents a production process wherein 99.7% of the reactor space is composed of an inert material (water) that must be processed [5], it is often overlooked that water is a material that infers with extraction of the bio-oils. The natural oil–water separation that early researchers proposed would happen if the cells were simply shocked open (e.g. by osmotic shock, pressure release, and the like) does not occur because the lipids either remain attached to the biomass material or form a cloudy emulsion in water (data not shown). To some extent water will either form solvent shells around the lipids or free fatty acids, thus inhibiting the formation of larger micelles that will separate via gravity differences, or inhibit access of a miscible hydrophobic solvent from contacting the lipids and thus solubilize them. Thus the presence of water has proven to be one of the most significant hurdles in the extraction of biomass.

Early research recognized this fact by simply drying (heat or freeze drying) the cells prior to the application of solvent extraction techniques (liquid phase extraction, accelerated solvent extraction, or sub/supercritical fluid extraction). These techniques [77, 85, 87, 94–99] have not proven appropriate for lipid extraction from algae on a commercial scale due to relatively high energy requirements in the drying the biomass, as well as the distillation costs of separating the extracted solvent from the organic solvent it was extracted into (i.e. hexane). These roadblocks have led to a large number of technologies developed to work in the presence of water, including the milking strategies [75] and those using hydrophilic ionic liquids [100].

To date, despite the evolution of extraction approaches, no single commercially successful extraction process has been developed or put in place. The obvious need to move cultivation of microalgae to seawater (excepting niche uses like in remediation of wastewater) for access to nutrients such as phosphorous or nitrogen, will further complicate future developments as virtually none of the extraction technologies discussed above have dealt with the presence of salt. Salt, even more so than the presence of water, looms as one of the most significant challenges to extraction of bio-oils from water. As such, it would appear that future trends in the development of extraction technologies around bio-oils from microalgae will need to address the presence of salt. The reason for this is fairly clear: the removal of salt will require washing steps that either need the cells first to be fried and then washed and then dried, or washed through dilution with fresh waters—both approaches that are not feasible on a large scale. Given these constraints, the most logical and likely future approach is the extraction of oil from larger organisms (such as fish) that feed off of the marine algae. The extraction of this oil can be accomplished using well established techniques although the extracted oil will still be in the form of an emulsion that will need to be cleaned up prior to its use as a feedstock for transesterification of the oil to fatty acid methyl esters (biodiesel). Future work should focus on the treatment of pressed oil from fish for use as a feedstock for transesterification into biodiesel or chemical transformation into jet fuel.

References

1. L. Lin, Z. Cunshan, S. Vittayapadung, S. Xiangqian, and D. Mingdong, Opportunities and challenges for biodiesel fuel, *Applied Energy*, 88, 1020–1031 (2011).
2. L. Hughes, and J. Rudolph, Future world oil production: growth, plateau, or peak? *Current Opnion in Environmental Sustainability*, 3, 225–234 (2011).
3. P. Almeida, and P. D. Silva, The peak of oil production—timings and market recognition, *Energy Policy*, 37, 1267–1276 (2009).
4. A. Kjell, M. Höök, K. Jakobsson, M. Lardelli, S. Snowden, and B. Söderbergh, The peak of the oil age—analyzing the world oil production Reference Scenario in World Energy Outlook 2008 *Energy Policy*, 38, 1398–1414 (2010).
5. M. J. Cooney, G. Young, and R. Pate, Bio-oil from photosynthetic microalgae: Case study, *Bioresource Technology*, 102, 166–177 (2010).
6. R. Pate, G. Kilse, and B. Wu, Resource demand implications for US algae biofuels production scale-up, *Applied Energy*, 88, 3377–3388 (2011).
7. J. R. Benemann, Microalgal Biofuels: A Brief Introduction, Available from: http://advancedbiofuelsusa.info/wp-content/uploads/2009/03/microalgae-biofuels-an-introduction-july23-2009-benemann.pdf (accessed September 26, 2012) (2009).
8. M. J. Cooney, G. Young, and N. Nagle, Extraction of Bio-oils from Microalgae: A Review, *Journal of Separation and Purification Reviews*, 38, 291–325 (2009).
9. W. Thompson, S. Meyer, and T. Green, The US biodiesel use mandate and biodiesel feedstock markets, *Biomass and Bioenergy*, 34, 883–889 (2010).
10. M. Balat, Potential alternatives to edible oils for biodiesel production—a review of current work, *Energy Conversion and Management*, 52, 1479–1492 (2011).
11. D. Pimentel, A. Marklein, M. A. Toth, M. Karpoff, G. S. Paul, and R. McCormack, Food versus biofuels: environmental and economic costs, *Human Ecology*, 37, 1–12 (2009).
12. D. Pimentel and M. Pimentel, Corn and cellulosic ethanol cause major problems, *Energies*, 1, 35–37 (2008).
13. M. Canakci, The potential of restaurant waste lipids as biodiesel feedstocks, *Bioresource Technology*, 98, 183–190 (2007).
14. Regulation of fuels and fuel additives: changes to renewable fuel standard program, Environmental Protection Agency, *Federal Register*, 74(99), 24904–25142, (2009).
15. K. Pramanik, Properties and use of Jatropha curcas oil and diesel fuel blends in compression ignition engine, *Renewable Energy*, 28, 239–248 (2003).
16. A. Demirbas, Potential resources of non-edible oils for biodiesel, *Energy Source Part B*, 4, 310–314 (2009).
17. T. Ganapathy, K. Murugesan, and R. P. Gakkhar, Performance optimization of Jatropha biodiesel engine model using Taguchi approach, *Applied Energy*, 86, 2476–2486 (2009).
18. A. K. Tiwari, A. Kumar, and H. Raheman, Biodiesel production from Jatropha oil (Jatropha curcas) with high free fatty acids: an optimized process, *Biomass and Bioenergy*, 31, 569–575 (2007).
19. M. Naik, L. C. Meher, S. N. Naik, and L. M. Das, Production of biodiesel from high free fatty acid Karanja (Pongamia pinnata) oil., *Biomass and Bioenergy*, 32, 354–357 (2008).
20. A. K. Agarwal, and K. Rajamanoharan, Experimental investigations of performance and emissions of Karanja oil and its blends in a single cylinder agricultural diesel engine, *Applied Energy*, 86, 106–112 (2009).

21. N. Usta, An experimental study on performance and exhaust emissions of a diesel engine fuelled with tobacco seed oil methyl ester, *Energy Conversion and Management*, 46, 2373–2386 (2005).
22. V. B. Veljkovic, S. H. Lakicevic, O. S. Stamenkovic, Z. B. Todorovic, and M. L. Lazic, Biodiesel production from tobacco (Nicotiana tabacum L.) seed oil with a high content of free fatty acids, *Fuel Processing Technology*, 85, 2671–2675 (2006).
23. S. Sinha, A. K. Agarwal, and S. Garg, Biodiesel development from rice bran oil: transesterification process optimization and fuel characterization, *Energy Conversion and Management*, 49, 1248–1257 (2008).
24. L. Lin, D. Ying, S. Chaitep, and S. Vittayapadung, Biodiesel production from crude rice bran oil and properties as fuel, *Applied Energy*, 86, 681–688 (2009).
25. S. V. Ghadge, and H. Raheman, Biodiesel production from Mahua (Madhuca indica) oil having high free fatty acids, *Biomass and Bioenergy*, 28, 601–605 (2005).
26. H. Raheman, and S. V. Ghadge, Performance of compression ignition engine with Mahua (Madhuca indica) biodiesel., *Fuel Processing Technology*, 86, 2568–2573 (2007).
27. T. V. Rao, G. P. Rao, and K. H. C. Reddy, Experimental investigation of Pongamia, Jatropha and Neem methyl esters as biodiesel on CI engine, *Jordan Journal of Mechanical and Industrial Engineering*, 2, 117–122 (2008).
28. A. S. Ramadhas, S. Jayaraj, and C. Muraleedharan, Biodiesel production from high FFA rubber seed oil, *Fuel Processing Technology*, 84, 335–340 (2005).
29. L. L. Sousa, I. L. Lucena, and F. Fernandes, Transesterification of castor oil: effect of the acid value and neutralization of the oil with glycerol, *Fuel Processing Technology*, 91, 194–196 (2010).
30. A. Demirbas, Production of biodiesel fuels from linseed oil using methanol and ethanol in non-catalytic SCF conditions, *Biomass and Bioenergy*, 33, 113–118 (2009).
31. Y. Chisti, Biodiesel from Microalgae, *Biotechnology Advances*, 25, 294–306 (2007).
32. T. M. Mata, A. A. Martins, and N. S. Caetano, Microalgae for biodiesel production and other applications: A review, *Renewable and Sustainable Energy Reviews*, 14, 217–232 (2010).
33. E. W. Becker. *Microalgae: Biotechnology and Microbiology*, Cambridge University Press, Cambridge, 1994.
34. H. Hotelling, The economics of exhaustible resources, *The Journal of Political Economy*, 39, 137–175 (1931).
35. K. R. Szulczyk, and B. A. McCarl, Market penetration of biodiesel, *International Journal of Energy and Environment*, 1, 53–68 (2010).
36. R. Harun, M. Singh, M. F. Gareth, and M. K. Danquah, Bioprocess engineering of microalgae to produce a variety of consumer products, *Renewable and Sustainable Energy Reviews*, 14, 1037–1047 (2010).
37. L.-K. Ju, Multi-Step Method for Producing Algae Products, *USPTO*, 20100297714, (2010).
38. T. J. Czartoski, R. Perkins, J. L. Villanueva, and G. Richards, Algae Biomass Fractionation, *USPTO*, 20100233761, (2010).
39. P. Felizardo, J. Machado, D. Vergueiro, M. J. N. Correia, J. Pereira Gomes, and J. M. Bordado, Study on the glycerolysis reaction of high free fatty acid oils for use as biodiesel feedstock, *Fuel Processing Technology*, 92, 1225–1229 (2011).
40. B.-X. Peng, Q. Shu, J.-F. Wang, G.-R. Wang, D.-Z. Wang, and M.-H. Han, Biodiesel production from waste oil feedstocks by solid acid catalysis, *Process Safety and Environmental Protection*, 86, 441–447 (2008).
41. L. F. Bautista, G. Vicente, R. Rodriguez, and M. Pacheco, Optimisation of FAME production from waste cooking oil for biodiesel use, *Biomass and Bioenergy*, 33, 862–872 (2009).
42. C. Kaya, C. Hamamci, A. Baysal, O. Akba, S. Erdogan, and A. Abdurrahman-Saydut, Methyl ester of peanut (Arachis hypogea L.) seed oil as a potential feedstock for biodiesel production, *Renewable Energy*, 34, 1257–1260 (2009).
43. A. Karmakar, S. Karmakar, and S. Mukherjee, Properties of various plants and animals feedstocks for biodiesel production, *Bioresource Technology*, 101, 7201–7210 (2010).
44. S. D. Sanford, J. M. White, P. S. Shah, C. Wee, M. A. Valverde, and G. R. Meier, *Feedstock and Biodiesel Characteristics Report*, Renewable Energy Group, Ames, IA (2009).
45. H. Fukuda, A. Kondo, and H. Noda, Review: biodiesel fuel production by transesterification of oils, *Journal of Bioresource and Bioengineering*, 92, 405–416 (2001).
46. S. Saraf, and B. Thomas, Influence of feedstock and process chemistry on biodiesel quality, *Process Safety and Environmental Protection*, 85, 360–364 (2007).

47. H. Lepper, and L. Friesenhagen, Process for the production of fatty acid esters of short chain aliphatic alcohols from fats and/or oils containing free fatty acids, *USPTO*, 4608202 (1986).
48. B. Supple, R. Howard-Hilige, E. Gonzalez-Gomez, and J. J. Leahy, The effect of steam treating waste cooling oil on yield of methyl ester, *Journal of the American Oil Chemists Society*, 79, 175–178 (2002).
49. G. A. Deluga, V. Zamansky, K. Liu, and T. E. P. Westendorf, Integrated System and Method for Producing Fuel Composition from Biomass, *USPTO*, 20090259082 (2009).
50. Q. Wu, and W. Xiong, Method for producing biodiesel from an alga, *USPTO*, 20090298159 (2009).
51. S. Salim, R. Bosma, M. H. Vermue, and R. H. Wijffels, Harvesting of microalgae by bio-flocculation, *Journal of Applied Phycology*, 23, 849–855 (2011).
52. J. R. Oyler, Process of Producing Oil from Algae Using Biological Rupturing, *USPTO*, 20090269839 (2009).
53. J. R. Oyler, Integrated Processes and Systems for Production of Biofuels Using Algae, *USPTO*, 20090081748 (2009).
54. D. E. Trimbur, C.-S. Im, H. F. Dillon, D. A. G., F. S., and A. Coragliotti, Renewable Diesel and Jet Fuel from Microbial Sources, *USPTO*, 20090047721 (2009).
55. A. G. Day, B. G., and S. Franklin, Soaps Produced from Oil-Bearing Microbial Biomass and Oils, *USPTO*, 20090305942 (2009).
56. E. H. Dunlop, and D. A. Hazlebeck, High Efficiency Separations to Recover Oil from Microalgae, *USPTO*, 20090081742 (2009).
57. C. M. Ruecker, S. P. Adu-peasah, B. S. Engelhardt, and G. T. Veeder, Solventless extraction process, *USPTO*, 7,781,193 (2010).
58. H. L. Bijl, and A. Schaap, Isolation of microbial oils, *USPTO*, 7,431,952 (2008).
59. A. M. Aravanis, B. L. Goodall, M. Mendez, J. L. Pyle, and J. E. Moreno, Methods and Systems for Biofuel Production, *USPTO*, 20100297749 (2010).
60. O. J. Catchpole, J. B. Grey, A. D. MacKenzie, and S. J. Tallon, Extraction of Highly Unsaturated Lipids with Liquid Dimethyl Ether, *USPTO*, 20100160659 (2010).
61. D. G. Deuppen, S. G. Zeller, S. I. Diltz, and R. H. Driver, Extraction and Winterization of Lipids from Oilseeds and Microbial Biomass, *USPTO*, 7695626 (2010).
62. Q. Wu, W. Zhou, and W. Xiong, Method for producing biodiesel using high-cell-density cultivation of microalgae Chlorella protothecoides in bioreactor, *USPTO*, 20090211150 (2009).
63. J. C. Cox, H. Chen, and C. Kabacoff, Media for Cell Growth and Method for Making Them, *USPTO*, 5324658 (1994).
64. S. P. A. Eni, E. M. Piazzale, E. N. D'addario, F. Capuano, E. D'angeli, and R. Medici, Process For The Extraction of Fatty Acids From Algal Biomass, *WIPO*, WO/2010/000416 (2010).
65. P. J. Keating, Method for recovering pigments from algal cultures, *USPTO*, 6805800 (2004).
66. R. Lemlich, and J. Arod. *Adsorptive Bubble Separation Techniques*, Academic Press, New York, 1972.
67. R. L. Clayton, S. N. Falling, and J. S. Kanel, Process For Microalgae Conditioning and Concentration, *USPTO*, 20100167339 (2010).
68. B. C.-P. Wu, C. A. DeLuca, and E. K. Payne, Systems and Methods for Hydrothermal Conversion of Algae into Biofuel, *USPTO*, 20100050502 (2010).
69. P. G. Roessler, Y. Chen, B. Liu, and C. N. Dodge, Secretion of Fatty Acids by Photosynthetic Microorganisms, *USPTO*, 20090298143 (2009).
70. J. C. Lippmeier, J. W. Pfeifer III,, J. M. Hansen, A. K. E., W. R. Barclay, P. W. Behrens, and D. C. Martin, Biological Oils and Production and Uses Thereof, *USPTO*, 20090064567 (2009).
71. C. Russel, H. Calvin, and J. Rodriguez, Novel Process for Separating Lipids from a Biomass, *USPTO*, 20100261918 (2010).
72. T. P. Hensarling, and T. J. Jacks, Solvent extraction of lipids from soybeans with acidic hexane, *Journal of the American Oil Chemists' Society*, 60, 783–784 (1983).
73. Z. Dubinsky, and S. Aaronson, Increase of lipid yields from some algae by acid extraction, *Phytochemistry*, 18, 51–52 (1979).

74. J. Frenz, C. Largeau, E. Casadevall, F. Kollerup, and A. J. Daugulis, Hydrocarbon recovery and biocompatibility of solvents for extraction of culture of Botryococcus braunni, *Biotechnology and Bioengineering Symposium*, 34, 755–762 (1988).
75. R. T. Sayre, Optimization Of Biofuel Production, *USPTO*, 20090181438 (2009).
76. S. M. Nemser, S. Majumdar, and K. J. Pennisi, Removal of Water and Methanol from Fluids, *USPTO*, 20080099400 (2008).
77. C. Crampon, O. Boutin, and E. Badens, Supercritical carbon dioxide extraction of molecules of interest from microalgae and seaweeds, *Industrial and Engineering Chemistry Research*, 50, 8941–8953 (2011).
78. S. Christy, T. Dallas, M. Alva, and W. Wikstrom, editors. Algae-Based Biofuel System for Military Use. *Proceedings of the 2011 National Security Innovation Competition*; 2011; Colorado Springs, Colorado.
79. G. Thomas, and L. Lindell, Adiabatic Compaction of Algae Biomass for Extraction of Biofuel, *USPTO*, 20090298158 (2009).
80. L. Beecher, and D. E. Brune, Concentration and Separation of Lipids from Renewable Resources, *USPTO*, 20090181436 (2009).
81. J. L. Massingill Jr.,, Fiber Film Reactors to Effect Separation and Reaction Between Two Immiscible Reaction Components, *USPTO*, 7,618,544 (2009).
82. B. C.-P. Wu, D. Stephen, G. E. Morgenthaler, and D. V. Jones, Systems and Methods for Producing Biofuels from Algae, *USPTO*, 20100081835 (2010).
83. J. R. Benemann, and W. J. Oswald, Systems and Economic Analysis of microalgae ponds for conversion of CO2 to biomass, Report No. DOE/PC/93204–T5 (1996).
84. E. G. Bligh, and W. J. Dyer, A rapid method for total lipid extraction and purification, *Canadian Journal of Biochemistry and Physiology*, 37, 911–917 (1959).
85. S. J. Iverson, S. L. C. Lang, and M. H. Cooper, Comparison of the Bligh and Dyer and Folch Methods for Total Lipid Determination in a Broad Range of Marine Tissue, *Lipids*, 36, 1283–1287 (2001).
86. F. Smedes, and T. K. Askland, Revisiting the development of the Bligh and Dyer total lipid determination method, *Marine Pollution Bulletin*, 38, 193–201 (1999).
87. R. Lepage, and C. C. Roy, Improved recovery of fatty acid through direct transesterification without prior extraction or purification, *Journal of Lipid Research*, 25, 1391–1396 (1984).
88. G. Young, F. Nippen, S. Titterbrandt, and M. J. Cooney, Direct transesterification of biomass using an ionic liquid co-solvent system, *Biofuels*, 2, 261–266 (2011).
89. F. Chen, High Density culture of microalge in heterotrophic growth, *Trends in Biotechnology*, 14, 421–426 (1996).
90. D. P. van Vuuren, A. F. Bouwman, and A. H. W. Beusen, Phosphorous demand for the 1970 –2100 period: A scenario analysis of resource depletion, *Global Environmental Change*, 20, 428–439 (2010).
91. P. Sporalore, C. Joannis-Cassan, E. Duran, and A. Isambert, Commercial applications of microalgae, *Journal of Bioscience and Bioengineering*, 101, 87–96 (2006).
92. M. Cerejo, and J. M. Dias, Tidal transport and dispersal of marine toxic microalgae in a shallow, temperate coastal lagoon, *Marine Environmental Research*, 63, 313–340 (2007).
93. L. Thomsen, How "green" are algae farms for biofuel produciton? *Biofuels*, 1, 515–517 (2010).
94. P. C. K. Chueng, Temperature and pressure effects on supercritical carbon dioxide extraction of n-3 fatty acids from red seaweed, *Food Chemistry*, 65, 39–403 (1999).
95. M. Herrero, A. Cifuentes, and E. Ibanez, Sub- and supercritical fluid extraction of functional ingrediants from different natural sources: Plants, food-by-products, algae and microalgae. A review, *Food Chemistry*, 98, 136–148 (2006).
96. T. Lewis, P. D. Nichols, and T. A. McMeekin, Evaluation of extraction methods for recovery of fatty acids from lipid-producing microheterotrophys, *Journal of Microbial Methods*, 43, 107–116 (2000).
97. M. D. Luque de Castro, M. Valcarcel, and M. T. Tena. *Supercritical Fluid Extraction*, Springer Verlag, Heidelberg, 1994.

98. E. Molina Grima, A. Robels Medina, A. Gimenez Gimenez, J. A. Sanchez, F. Garcia Camacho, and J. L. Garcia Sanchez, Comparison between extraction of lipids and fatty acids from microalgal biomass, *Journal of the American Chemical Society*, 71, 955–959 (1994).
99. J. Rodriguez-Ruiz, E. H. Belarbi, J. L. G. Sanchez, and D. L. Alonso, Rapid simultaneous lipid extraction and transesterification for fatty acid analysis, *Biotechnology Techniques*, 12, 689–691 (1998).
100. G. Young, F. Nippen, S. Titterbrandt, and M. J. Cooney, Extraction of biomass using an ionic liquid co-solvent system, *Separation and Purification Technology*, 72, 118–121 (2009).

22

Separation Processes in Biopolymer Production

Sanjay P. Kamble, Prashant P. Barve, Imran Rahman and Bhaskar D. Kulkarni

Chemical Engineering and Process Development Division, CSIR—National Chemical Laboratory, Pune, India

22.1 Introduction

The chemical and allied process industries all over the world are facing the challenge of developing sustainable products and processes in an environment of highly globalized market competition and fast-growing environmental constraints. Process intensification through the revolutionary development of eco-friendly new or modified products and processes, which can be produced with reduced material and energy consumption, is urgently required (Pal *et al.*, 2009). Recently lactic acid (LA) and its polymer, polylactic acid (PLA), have attracted the attention of world researchers due to their versatile commercial applications. Traditionally approximately 85% of lactic acid produced has been used in the food industry. The multifunctional nature of lactic acid means that it can be used as a starting material for the synthesis of various organic compounds through reactions such as esterification, condensation, polymerization, reduction, and substitution. This has contributed to its potential as a platform chemical for a whole range of products that are used in very large volumes for industrial and consumer products. The production of many specialty chemicals, solvents (acetone, acrylic acid, propionic acid), biodegradable thermoplastics (PLA), green solvents (ethyl, propyl, butyl lactates) and oxygenated chemicals (propylene glycol) are a few examples of lactic acid-derived products. Market demand for lactic acid-derived products are growing exponentially over the years (Vijayakumar *et al.*, 2008). Polymers and copolymers of LA are known to be environmentally benign because of their biodegradability into harmless products. Polylactic acid is the most sought-after plastic among biodegradable polymers because of its low carbon footprint (it is entirely derived from renewable resources and is also fully biodegradable) (Farrington *et al.*, 2005; Sauer *et al.*, 2010). Several applications of PLA, such as in textiles (shirts and upholstery), non-wovens (diapers),

Separation and Purification Technologies in Biorefineries, First Edition.
Edited by Shri Ramaswamy, Hua-Jiang Huang, and Bandaru V. Ramarao.

packaging (cups, bottles, films, and trays), biomedical products (various types of screws, barrier films, drug-delivery devices), electronics (mobile, laptop housings), cutlery (in fast food joints), foams (as a replacement for expanded polystyrene (EPS)), and auto-components (spare tire cover, door handles), have been reported in the literature (Lunt, 1998; Vaidya *et al*., 2005).

However, whether LA can be used to its full potential largely depends on whether it can be produced in a cost-effective manner with high purity and optical activity. The major technology barrier in the cost-effective production of high-purity lactic acid is its downstream separation and purification. In this case study, different techniques such as electrodialysis, adsorption, solvent extraction, membrane separation and reactive distillation were analyzed for separation of lactic acid from fermentation froth. Each technique has some advantages and disadvantages, which are briefly reviewed. Among these methods, reactive distillation was found to be quite promising at an industrial scale, with some cost incentives. This work suggests a promising novel autocatalytic, cost-effective, eco-friendly, process concept involving a countercurrent operation for the recovery of high-quality lactic acid. This mode of operation gives several advantages, which are also discussed in detail later. Autocatalytic reactive distillation of lactic acid with methanol using a packed bed column followed by hydrolysis of methyl lactate with lactic acid catalyst, using three CSTRs in series, has been reported for purification of lactic acid at pilot plant scale. The optical purity and purity of the synthesized lactic acid were also estimated. The cost of purification of lactic acid (>99%) was also estimated and compared for both autocatalytic and ion-exchange resin processes based on their operating cost.

22.2 The market and industrial needs

The world market for lactic acid is growing every year. Key market leaders include Natureworks LLC (US), Purac (Netherlands), Galactic (Belgium), ADM (US), and Pyramid Bioplastics (Germany). The level of production is around 350 million kg /year in 2011 and the worldwide growth is believed by some observers to be 12–15%, the growth rate can go well above the predicted values due to change in governmental policies, subsidies, etc. The world market for lactic acid is forecasted to reach 367.3 thousand metric tons by the year 2017. The world market for lactic acid is forecasted to reach 367.3 thousand metric tons by the year 2017. The total global production capacity for bio-plastics was 327 000 tonnes in 2010 and the production capacity for bio-plastics will increase to a predicted 2.1 million tonnes by 2013 as reported by the UK's National Centre for Biorenewable Energy, Fuels and Materials (Williams, 2010). Tables 22.1 and 22.2 provide more details of the market size for bio-plastics and biodegradable polymers. Although bio-plastics are between two and ten times more expensive than petroleum-derived plastics at the present time, bio-plastics will reduce the petroleum consumption for plastic by 15–20% in 2025. There were over 500 bio-plastics processing companies in 2012; more than 5000 are expected by 2020. (Bio-plastics

Table 22.1 *Global market for biodegradable polymers in million kg*

Application	2006	2007	2012	CAGR %
Compost bags	78.47	109.77	266.26	8.80
Loose-fill packaging	68.95	73.03	97.07	2.59
Other packaging[a]	23.13	26.76	21.77	10.61
Miscellaneous[b]	14.97	24.49	77.56	11.34
Total	185.52	234.05	544.31	7.85

[a]Includes medical/hygiene products, agricultural, paper coatings, etc.
[b]Unidentified biodegradable polymers CAGR-compound annual growth rate.
Source: Adapted from Table A in Sarnacke and Wildes (2008) and million lbs have been converted into million kgs.

Table 22.2 *US projected demand for bio-based plastics in million kg*

Polymer type	2003	2007	2011	07–11 CAGR %
Polylactic acid (PLA)	20.41	68.04	136.08	8.62
Polyhydroxyalkonates (PHA,PHB)	0.00	<1	49.90	22.45
Starch-based	22.68	45.36	81.65	7.26
Bio-polyester	0.00	4.54	22.68	0.22
Cellulosic plastic	NA	NA	NA	NA
Total	43.09	117.93	290.30	0.11

Adapted from Table B of Sarnacke and Wildes (2008). Million lbs have been converted into million kgs.

Market Worldwide 2010/11-2015-2020-2025, 2012). European bio-plastics have revealed that the market for bio-based plastics in 2011 was around 1.2 million tonnes but the rapid scale of development in this area could see the market grow to nearly 6 million tonnes by 2016 (Aylott, 2012). Another study of the global biodegradable plastics market by Chemical Market Associates reported that the PLA and its blends will grow to 204 million kg in 2011. Growth in bio-plastics is fueled by several factors, some of which are listed below (Sarnacke and Wildes, 2008):

- Large retailers, such as Wal-Mart and Target, have introduced a "sustainability scorecard," which forces their suppliers to adopt measures that reduce the carbon footprint of products.
- In September 2010 consumer products giant P&G announced sweeping sustainability initiatives such as replacing 25% of all petroleum derived materials with sustainably sourced renewables and making all packaging from renewable or recyclable materials. More and more plastic consumers are expected to follow suit.
- Manufacturers are increasingly required to display "green labels" to quantify the carbon footprints of their products. Biodegradable plastics go a long way in improving such green scorecards.
- There is global concern over the rapid depletion of petroleum and natural gas resources, and efforts are on to look at more renewable feedstock.
- There is a growing public concern over environmental hazards caused by non-degradable plastic waste.
- Government support is increasing for use of bio-based products through regulations and subsidies.
- An increase in production capacity of biodegradable plastics in the worldwide can be expected to drive costs down. The price of crude oil is also expected to increase steadily.

Among numerous kinds of degradable polymers, PLA, sometimes called polylactide, an aliphatic polyester and biocompatible thermoplastic, is currently a promising and popular material with good development prospects. It has been considered as a "green" eco-friendly material. In 2010, total production capacity of PLA was about 180 000 t/a in the world, which is 14.6% higher than that in 2009. According to Thailand's National Innovation Agency (NIA) and the Germany-based Nova Institute GmbH, the worldwide polylactic acid (PLA) capacity will rise from 182,000 metric tons a year in 2011 to 721,000 metric tons in 2020. Asian capacity is expected to reach more than 350,000 metric tons (ChemistryViews, 2012). In 2011, the global consumption of PLA was about 120 000 tonnes, mainly in Western Europe and North America, and the consumption of PLA in Asia also keeps increasing. Currently, PLA consumption for packaging material is about 65% of total consumption and in the biomedical field is 26%. It is estimated that the compound annual growth rate of global lactic acid and PLA market will reach 18.7%, and the market value of PLA is expected to reach USD 3.831 billion in 2016 (CCM International, 2011).

Polylactic acid can be prepared by direct condensation of lactic acid and by the ring-opening polymerization of the cyclic lactide dimer. The direct condensation route is an equilibrium reaction and hence removal of water in the late stages of polymerization generally limits the ultimate molecular weight of polymer. Most work has focused on the ring-opening polymerization of lactide, although other approaches such as azeotropic distillation to remove water in the direct esterification process have also been practiced (Drumright *et al.*, 2000). Cargill Dow LLC has developed an environmentally benign, low-cost, continuous process for the production of lactic acid-based polymers. In this process both lactide and PLA are synthesized by melt polymerization rather than solution polymerization. They have, for the first time, shown a commercially viable biodegradable commodity polymer made from renewable resources. Nature Works LLC is the leader in lactic acid polymer technology and markets. Initially Dow-Cargill formed a joint venture for the development of downstream technologies for manufacturing LA, PLA and various consumer products from PLA by melt processing techniques. However Dow withdrew from the joint venture in 2005. Cargill went ahead alone to form NatureWorks LLC to take the PLA technology forward. The feedstock used by Cargill for making polymer grade lactic acid is corn. This company has done extensive work on the development of lactic acid-based products, which are of two types: the polydilactide-based resins (NatureWorks PLA®), used for plastics or packaging applications, and the Ingeo™ polydilactide-based fibers that are used in specialty textiles and fiber applications. Nature Works LLC has constructed a major lactic acid plant in Blair, NE, United States, with a capacity of 136 million kg/year for the production of lactic acid and PLA, and it began operating in late 2002. Later, Nature Works LLC wholly owned by Cargill Inc., and in 2007 the company started a joint venture between Cargill Inc., and Teijin Ltd (Nampoothiri *et al.*, 2010). NatureWorks is currently in the process of selecting the location of the 150 000 tonne PLA plant. Brazil is among the candidates while other potential locations are Thailand, Malaysia, and Singapore.

Purac is one of the major players in biodegradable materials; they have already developed and marketed various products like resorbable polymers (compatible biopolymer) and monomers worldwide under the PURASORB® brand name. PURASORB has been used commercially for several decades in applications as diverse as resorbable surgical sutures, orthopedic implant devices, cosmetic surgery products and drug delivery systems. Purac, has started a 100 000 t/a lactic acid plant in Thailand with an integrated process to make lactide. Purac has also entered into cooperation with Sulzer and Synbra to develop polymerization technology and associated products. Purac, a subsidiary of the food ingredients producer CSM, has started construction of its first full-scale lactide plant at Rayon, Thailand (Lindner, 2011). The First issue of *Biomaterials China News* by CCM International reveals that Zhejiang Hisun Biomaterials Co., Ltd. (Zhejiang Hisun) will expand its Polylactic acid (PLA) capacity to 50 000 t/a in 2013 from 5000 t/a in 2011. Mitsui Toatsu (Japan) developed a polymer from lactic acid under the brand name LACEA. Nets and non-woven products used in agriculture, fishing, and medical application developed by Lactron (Kanebo Goshen, Japan). Similarly, LACTY, PLA pellet for film and fiber extrusion was commercialized by Shimadzu, Japan. Galactic (Belgium) also commercialized multipurpose PLA pellets for the production of film, fibers, and so forth. Solanyl (Rodenburg Biopolymers, Holland) produce PLA from potatoes, and production capacity was about 40 000 t/a.

In India, Reliance Life Sciences has developed a complete process for the production of PLA from renewable feedstock by bacterial fermentation (Ramakrishna *et al.*, 2009). They were also developing co-polymers of lactic and glycolic acid of varying ratios and other biodegradable polymers including polyvinylcaprolactums, polyurethanes for biomedical applications and drug delivery systems based on nanotechnology using PLA/ PolyLactic-co-Glycolic Acid (PLGA) for generic drugs as well as therapeutic proteins. They have also started the preparation of fibers of PLA/PLGA/poly glycolic acid (PGA) by melt / wet spinning for sutures, scaffolds, and so forth.

Another SPC Biotech Private Limited (Hyderabad India) company has proposed a 5000 t/a pilot plant to make lactic acid and its polymer. The PLA so produced is proposed to be used as absorbable medical implants (Waltz, 2008). The technology for making lactic acid from corn and rice starch has been supplied to them by the Shanghai Institute of Industrial Microbiology, China. The license for making the implants comes from Vichy Biomaterials, France. Presently a small fraction of lactic acid is used for PLA production, and thus this biopolymer has high potential for further development. The PLA's higher cost (2–5 euros/kg) as compared to conventional plastic obtained from petroleum industry hydrocarbons (PET costs around 1 euro) hinders commercial large-scale utilization. Today, the manufacture of low-cost PLA from lactic acid is a major challenge to scientists and engineers. The cost of PLA depends upon the cost of lactic acid. Hence to compete with synthetic polymers it is necessary to have efficient and cost-effective technology for the manufacture of lactic acid.

22.3 Lactic acid recovery processes

Purification of dilute lactic acid obtained from bacterial fermentation is difficult due to its low vapor pressure, tendency to undergo self-esterification, and the presence of troublesome impurities. The downstream processes are very important in the production of lactic acid. The economics of lactic acid production process mainly depends upon the downstream process. It has been reported that 50% of the total cost of production is involved in the separation and purification processes. A number of techniques, like electrodialysis, solvent extraction, adsorption, reverse osmosis, and reactive distillation, have been studied and reported in the literature for the recovery of lactic acid. Each process is associated with some advantages and disadvantages. The details of each method are summarized as follows.

22.3.1 Electrodialysis

Electrodialysis employs ion-exchange membranes to remove ions from an aqueous solution under the driving force of electrical field. Instead of anion exchange membranes in desalting, special bipolar membranes are used to separate lactic acid from other impurities. Previous work in this area indicates the feasible production of lactic acid from lactate salts in two steps: (i) conventional electrodialysis (ED) for concentration and purification, and (ii) bipolar electrodialysis for conversion of lactate salts into lactic acid with recovery of alkali. However, electrodialysis requires pretreatment of the fermentation broth, which is performed by microfiltration. The purification of lactic acid from fermentation broth using electrodialysis has been extensively studied by various researchers (Hongo *et al.*, 1986; Herlban *et al.*, 1993; Lee *et al.*, 1998; Xuemei *et al.*, 1999; Madzingaidzo *et al.*, 2002; Habova *et al.*, 2004; Hirata *et al.*, 2005). Many studies have been carried out to improve the efficiency of this process, but commercialization of electrodialysis has not been reported (Joglekar *et al.*, 2006; Pal *et al.*, 2009). The major drawbacks associated with electrodialysis are as follows:

- The pretreatment of feed is required.
- Electrodialysis has the problem of corrosion and membrane fouling, which requires frequent cleaning of the dialyzer.
- Very expensive dialysis units, sometimes costing even more than the fermenter vessel, are required for a commercial-scale operation. Electrodialysis gives a higher extent of lactic acid separation but with increased power and energy consumption.
- The control of fermentation pH is very crucial for electrodialysis.
- Large amounts of byproduct salts from the ion-exchange regeneration are formed.

22.3.2 Adsorption

A large number of adsorbents have been used for the selective removal of lactic acid from aqueous solution. The critical assessment of various polymeric adsorbents (polymeric compound with Pyridine skeletal structure and cross linked structure, Amberlite IRA-400, Amberlite IRA-420, weak anion-exchange resin, D354 and Amberlite IR-120) were used for recovery of lactic acid has been studied by many researchers

(Osaka *et al*., 1982 ; Srivastava *et al*., 1992; Dai and King, 1996; Monteagudo *et al*., 1999; Raya-Tonetti *et al*., 1999; Sun *et al*., 1999; Sosa *et al*., 2001). It was found that IRA-400 was suitable for the recovery of lactic acid. Further investigation is needed to improve its efficiency. From these studies the following drawbacks were found to be associated with the adsorption technique:

- The adsorption of LA on solid sorbents under a certain pH range is feasible. The corrosion of material may prevent large-scale operation.
- The reuse of resin is difficult due to accumulations of inhibitory substances.
- Regeneration of an ion-exchange resin and adjustment of the feed pH requires large amount of chemicals and the need to dispose of large quantities of salts/effluents.

22.3.3 Reactive extraction

Lactic acid, being hydrophilic, is difficult to extract using common organic solvents; hence reactive extraction is employed. Amine extractant has been found to be prospective medium of separation of carboxylic acids from aqueous solution. Lactic acid can be readily extracted in to a number of organic solvents with high molecular mass aliphatic amines and phosphorous bounded oxygen donor solvents, exhibiting particularly good selectivity (Jarvinen *et al*., 2000; Wang *et al*., 1991; Yang *et al*., 1991; Kertes and King, 2004; Kyuchoukov *et al*., 2004). Wasewar *et al*., (2004)) have shown the comparison of extraction capacity and selectivity of different solvents used for recovery of LA. The main parameters for selection of a diluent-extractant system for extraction of lactic acid are distribution coefficient, toxicity, complexation constant, and feasibility for back extraction. It was found that Alamine 336 in proper diluents is the best single extractant in terms of distribution coefficient, toxicity and feasibility for back extraction; it however forms a third layer which is undesirable (Yang *et al*., 1991; Wasewar, 2005). The following drawbacks are associated with reactive extraction:

- The addition of extractants introduces an additional component and hence further purification issues. In many cases the system also requires diluents, which add to the separation and material costs.
- The toxicity of the extracting solvents may affect microbial strains.
- The process is strongly pH dependent. Most of the extractant works efficiently at low pH while most microbes give higher productivity at higher pH.
- The regeneration of lactic acid from the loaded organic phase by a suitable method (temperature swing and a swing of the diluent composition) is needed. This indicates that complete back extraction is required.

22.3.4 Reverse osmosis

The separation and purification of lactic acid from the fermentation broth using reverse osmosis and combination of nanofiltration and reverse osmosis has been reported by various researchers (Liew *et al*., 1995; Freedman and Shaban, 2007; Li *et al*., 2008). It has been found that the reverse osmosis could effectively concentrate lactic acid from the solution of 10 to 120 gm/l at 6.9 MPatrans membrane pressure at lower energy consumptions. From the literature it was found that the following drawbacks are associated with reverse osmosis:

- the reverse osmosis method does seem promising but very little research has been done on it and scale-up of this technique may be difficult;
- the cost of the membrane and fouling of the membrane are important issues;
- the overall energy cost is also an area of concern.

22.3.5 Reactive distillation

Purification of lactic acid by simple distillation is difficult due to its low volatility, highly solubility in water, and its polymerizing tendency at higher temperature. To overcome these limitations, the lactic acid is esterified by reacting it with alcohol, yielding corresponding lactate ester. The lactate ester is purified by distillation, and then hydrolyzed to obtain pure lactic acid. The esterification process is an equilibrium controlled reaction with low reaction rate and requires more time. Equilibrium reactions can be shifted to the right by the continuous removal of one of the products from the reaction mixture. In order to enhance the rate of reaction (esterification of lactic acid and hydrolysis of lactic esters), different types of polymeric cation exchange resins (DOWEX-50W and Amberlyst-15 and Amberlyst CSP2; Macroporous-D001, D002 and Gel-002) have been used by various researchers (Seo *et al.*, 1999; Kim *et al.*, 2000, 2002; Zhang *et al.*, 2004; Kumar *et al.*, 2006, 2007; Cho *et al.*, 2008).

From the above discussions it was found that each technique has some advantages and disadvantages, but reactive distillation was found to be quite promising at the industrial scale, with some cost incentive (Kumar *et al.*, 2006; Kumar and Mahajani, 2007). Hence in order to minimize the overall cost of downstream processing of lactic a promising novel autocatalytic, cost-effective, eco-friendly process concept for the recovery of high-quality lactic acid have been suggested. The detail of this concept is explained in the subsequent Section 22.4.

22.4 Separation performance and results of autocatalytic counter current reactive distillation of lactic acid with methanol and hydrolysis of methyl lactate into highly pure lactic acid using 3-CSTRs in series

The autocatalytic countercurrent reactive distillation of lactic acid (LA) with methanol (MeOH) using a packed column followed by hydrolysis of methyl lactate (MLA) with LA catalyst using three CSTRs in series has been studied for purification of lactic acid at pilot plant scale. The advantage of autocatalytic reaction is the absence of a separation step and the regeneration of the catalyst.

The autocatalytic continuous counter current esterification of crude lactic acid was performed in packed column (i.d. 50 mm dia. and length 5000 mm). The columns were packed with glass Raschig rings (i.d. 4/5 mm, 5 mm length) and have dry surface area of 525 m^2/m^3 and packed volume of 9.5 L. The esterification reaction was performed at 100 °C and the entire process of continuous esterification of crude lactic acid using methanol was controlled using supervisory control and data acquisition (SCADA) system. The block diagram for the purification of lactic acid is shown in Figure 22.1. The detailed explanation of complete process is reported elsewhere by Kamble *et al.* (2011). The esterification of lactic acid was performed under different experimental conditions such as with different residence times, or molar ratios (lactic acid to methanol). The operating parameters were finalized after conducting a number of pilot plant trials. The experimental data obtained during the pilot plant trials at steady state for five different runs are given in Table 22.3. Typically, each run takes about 0.5-2 hr to attain the steady state. The flow rate of lactic acid was varied from 1.6 to 1.9 kg/h, while methanol flow rate was varied from 1.4 to 2.4 kg/h. Table 22.3 also shows the effect of molar ratio on the esterification reaction. It was found that, with an increase in the molar ratio of LA: MeOH, the conversion of LA was found to increase. Table 22.4 shows the analysis of feed, distillate and bottom product (column overflow) during the continuous counter current esterification of crude lactic acid. It was observed that only 0.81 to 15% unreacted lactic acid was found in the overflow of column (product), while 66 to 85% of MLA was formed at different molar ratios of reactants. From the distillate analysis, it was found that about 1.3 to 6.88% MLA comes along with

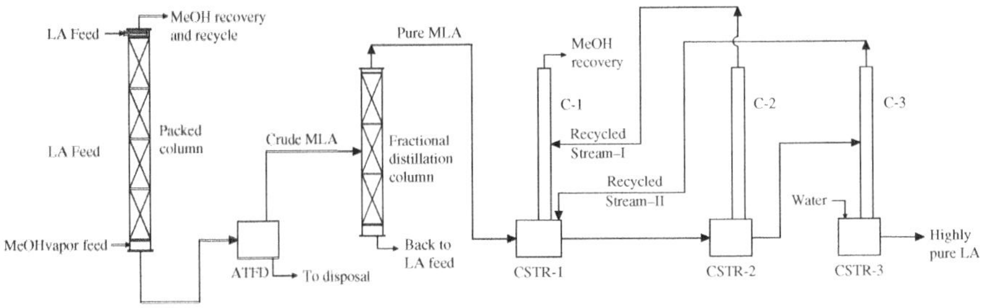

Figure 22.1 *Block diagram for the purification of lactic acid*

Table 22.3 *Continuous counter current autocatalytic esterification of crude concentrated lactic acid using packed column*

Run no.	Feed rate (kg/hr)		Output (kg/hr)		Rate of formation of MLA Kg/m^2-hr @ 373 K
	Crude LA	MeOH	Crude MLA	Distillate (MeOH and water)	
RN-01	1.850	2.282	2.056	2.019	0.261
RN-02	1.915	2.410	2.141	2.130	0.270
RN-03	1.890	1.890	1.960	1.805	0.257
RN-04	1.616	1.408	1.496	1.517	0.209
RN-05	1.744	1.472	1.645	1.558	0.229

Table 22.4 *The analysis of various streams generated during continuous counter current autocatalytic esterification of crude concentrated lactic acid*

Run no.		RN-01 (%)	RN-02 (%)	RN-03 (%)	RN-04 (%)	RN-05 (%)
Composition of streams		Analysis (wt%)				
Crude LA (feed)	LA	76.1	76.1	74.8	74.8	76.1
	DiLA	8.3	8.3	9.7	9.7	8.3
	Moisture	11.5	11.5	10.3	10.3	11.5
MeOH (Feed)	Moisture	0.52	0.52	0.52	0.56	0.56
	LA	3.1	0.81	6.5	11.51	14.56
Overflow of column C-1 (product)	MLA	81.2	85.8	83.5	68.13	66.4
	Moisture	1.5	1.1	2.28	3.79	3.3
	MeOH	9.8	9.58	10.2	10.01	9.18
Distillate	MeOH	72.5	73.1	71.3	79.9	85.07
	MLA	3.6	3.8	4.3	6.88	1.3
	Moisture	21.3	22.3	22.8	26.56	21.20

Source: Reprinted with permission from [39] © 2011, American Chemical Society.

methanol and water. This is due to the contribution of MLA vapor pressure at the boiling point of water. After establishing various process parameters such as flow rates of methanol, crude concentrated lactic acid, and feed temperatures, further emphasis was put on optimization of the process conditions so as to obtain the maximum conversion of crude lactic acid to methyl lactate with a good yield. The methyl lactate obtained at the bottom of packed column was further purified using agitated thin film dryer (ATFD) and fractional distillation column. Pure methyl lactate (99.93%) was obtained using fractional distillation. Around 30 stages are required for the separation of reaction mixture.

Generally, in a conventional co-current process, crude lactic acid, methanol and small amount of sulfuric acid are added in the reboiler of reactive distillation column. The temperature of the reboiler is kept around 398 to 423 K and the product (methyl lactate, water and excess methanol) is taken out of the reactive distillation column by maintaining the still at a higher temperature. This leads to the accumulation of the acidity in the reboiler and gives rise to the undesired by-products like hydroxyl methyl furfural, 2-pentene-1-ol, and pentyl lactate. Also in co-current reactive distillation, methyl lactate forms an azeotrope with water. The isolation of methyl lactate in pure anhydrous form from the mixture of methanol, water, and methyl lactate is difficult. If the methyl lactate is not purified further, then the lactic acid generated in hydrolysis cycle of methyl lactate becomes contaminated with other impurities like hydroxyl methyl furfural, 2-pentene-1-ol, pentyl lactate (Barve *et al*., 2010; Kamble *et al*., 2011).

In the present process a countercurrent mode of operation was employed for the esterification of crude lactic acid where crude methyl lactate is produced in anhydrous form, which is further separately purified using ATFD followed by a fractional distillation column. The major advantages of countercurrent reactive distillation are that the acidity buildup in the bottom of reactive distillation column is absent and the temperature of the reaction mixture across the reactive distillation column is maintained at a constant 373 K, which prevents the formation of side products. Hence, in the present process, a countercurrent mode of operation was employed for the esterification of crude lactic acid using the methanol and autocatalyst approach.

The continuous hydrolysis of MLA was performed at pilot scale using reactive distillation assembly consist of glass columns connected to a jacketed reboilers which work as CSTRs in series (capacity of each CSTR is 7 L), as shown in Figure 22.1. CSTR-1 and CSTR-2 were associated with fractional distillation column C-1 and C-2 respectively while CSTR-3 was having a stripping column C-3 attached to it. The hydrolysis of methyl lactate was performed at 373 K. Initially MLA, water, and LA in the ratios of 50%: 45%:5% and 50%, 35% and 15% were charged in CSTR-1 and CSTR-2 respectively, while CSTR-3 was charged with the mixture of LA and DM water in the ratio of 50%: 50%. All the CSTR-1, 2 and 3 were heated at 373 K. Initially column C-1 was put on total reflux. Once the steady state was reached then the desired flow rate of MLA and DM water was started. The distillate of column C-2 was recycled (recycle stream-I) to the middle of column C-1. The distillate obtained from column C-3 was recycled (recycle stream-II) to CSTR-1. The overflow of CSTR-2 was fed to the middle of stripper column C-3, where the contents of CSTR-2 were treated in countercurrent mode with the rising steam from reboiler of CSTR-3. The periodic samples of column C-1 distillate and CSTR-3 bottom were collected and analyzed for methanol, moisture, MLA, LA content, and so forth. The hydrolysis of methyl lactate was performed under different experimental conditions, such as with different residence time, and molar ratio (MLA to water). The operating parameters were obtained after conducting a number of pilot plant trials. The experimental data obtained during the pilot plant trials at steady state for four different runs are given in Table 22.5. Table 22.5 also shows the effect of the molar ratio on the hydrolysis of MLA and the rate of formation of lactic acid with respect to molar ratio was found to be nearly constant. The analytical details of different runs (RN No 1–4) are given in Table 22.6. It shows the concentration (wt%) of MLA in feed, overflow of CSTR-3 and distillate of column C-1. From the analysis of the overflow of CSTR-3, it was found that the complete conversion of MLA is possible and in most of the experiments the residual concentration of MLA was found to be practically zero. The purity of synthesized lactic acid was found

Table 22.5 *Hydrolysis MLA to highly pure lactic acid using three CSTRs in series*

Run no.	Feed rate (kg/hr)		Output rate (kg/hr)		Recycled stream (kg/hr)		Formation of lactic acid kg/L hr @ 373 K
	MLA	Water	LA	MeOH	Recycle stream R_1	Recycle stream R_2	
RN-001	3.12	2.67	4.81	0.97	1.10	3.85	0.128
RN-002	3.66	2.66	5.14	1.13	1.20	3.90	0.150
RN-003	3.74	3.11	5.67	1.15	1.25	3.95	0.153
RN-004	3.24	2.59	4.79	0.99	1.20	3.90	0.133

Table 22.6 *The analysis of various streams generated during hydrolysis of MLA using three CSTRs in series*

Run no. Compositions		RN-001 Analysis in wt%	RN-002	RN-003	RN −004
MLA	MLA	99.86	99.86	99.86	99.86
	Moisture	0.1	0.1	0.1	0.1
Overflow of CSTR-3	LA	56.10	61.50	57.3	58.5
	Moisture	43.60	37.92	41.9	41.2
	MLA	Nil	Nil	Nil	Nil
	MeOH	Nil	Nil	Nil	Nil
Distillate of C-1	MeOH	98.1	97.6	98.20	98.2
	Moisture	1.8	2.3	1.4	1.6

Source: Reprinted with permission from [39] © 2011, American Chemical Society.

to be 99.81% (wt on water free basis). The distillate of column C-1 mostly contains methanol and a small amount of water. This methanol is dehydrated using a fractional distillation column and recycled to esterification section. The advantage of this process is that the complete hydrolysis of methyl lactate is achieved without using catalyst and thus avoiding contamination of lactic acid. Optical activity of lactic acid directly affects the physiochemical properties of PLA such as mechanical strength, thermal plasticity, and ease of fabrication. Hence the optical activity of lactic acid prepared in the pilot plant and commercial lactic acid purchased from Malladi Specialties Ltd, Tamilnadu, India was measured and found to be 99.5% and 100% respectively. It indicates that during the purification of lactic acid via esterification of lactic acid and hydrolysis of MLA the optical activity remains practically the same (Kamble *et al*., 2011).

22.5 Economic importance and industrial challenges

Generally, all plastics are manufactured using crude oil as raw material, and about more than 270 million tonnes of oil and gas are consumed every year worldwide (Pimentel *et al*., 1999). There is ever increasing interest in the production and use of bio-renewable chemicals, because of rising global crude oil prices, and a growing desire to reduce dependence on petroleum. There has therefore been an emergent need for shifting the policies toward minimizing the resource consumption, and the application of alternative renewable resources for fuels and chemicals including synthetic plastics. Environmentally benign PLA synthesized from lactic acid is a prime candidate for replacing traditional petroleum-based polymers.

In 2008, NatureWorks use less than 0.05% of the annual global corn crop today for production of lactic acid; therefore there is little or no impact on food prices or supply. In 2007–8 world corn production was 790.91 MM MT and NatureWorks uses 0.997 kg of corn per kg of Ingeo manufactured. Any alternative sugar source such as sugar beet, sugar cane, wheat, and others, can be used as raw material. Recently, research on the production of lactic acid and PLA from other food grain such as from cellulosic raw materials, agricultural wastes, and non-food plants, has been going on in many industries, universities and institutions across the world. Despite these apparent advantages, very few industrial applications have been reported so far. Several drawbacks were shown to hinder implementation—mainly complexity of operation and higher cost as compared to synthetic plastic. There is therefore an emergent need to shift policies toward minimizing resource consumption, and toward the application of alternative renewable resources for fuels and chemicals, including synthetic plastics.

The main concern of PLA is the price of this polymer. On an industrial scale, the manufacturing cost of lactic acid monomer will be targeted to less than 0.8 US$/kg because the selling price of PLA should decrease roughly by half from its present price of 2.2 US$/kg in 2011. According to the cost analysis, the base manufacturing cost of lactic acid was estimated to be 0.55 US$/kg. In this case study the cost of purification of 1000 kg of lactic acid (>99%) was calculated in 2011 for both autocatalytic and ion-exchange resin processes based on operating cost. It was found that the cost of purification of lactic acid using autocatalysis is 1178 US $/ton, while for ion exchange resin process cost is about 1223 US$/ton. This clearly indicates that the final cost of purification of lactic acid using the current process is less (45 US$/ton) as compared to conversional heterogeneous catalyst.

22.6 Conclusions and future trends

It is vital to have a proficient and sustainable process for the separation of lactic acid from the fermentation broth. There are several issues that need to be addressed in order to produce lactic acid using a biochemical approach within the targeted cost, such as the development of high-performance lactic acid-producing microorganisms and the lowering of the costs of raw materials and fermentation processes. The biotechnological processes for the production of lactic acid from cheap raw materials should be improved further to make them competitive with the chemically derived approach. Research is being undertaken into the development of more cost-effective, environmentally clean processes for the synthesis of lactic acid, such as the direct esterification of calcium lactate using carbon dioxide and alcohol to give lactic ester and further hydrolysis to give highly pure lactic acid. The process route has the advantage that the synthesized byproduct (calcium carbonate) can be recycled into the fermenter to make the corresponding alkali metal lactate or can be used for various other applications (Barve *et al*., 2011)

Although a key driver in developing such products is the degradability of PLA, it is important to appreciate that PLA is fully biodegradable only when composted under controlled conditions. The primary mechanism of degradation of PLA occurs by a two-step process (i) hydrolysis, followed by (ii) bacterial attack. During the initial phases of degradation, the high molecular weight polyester chains hydrolyze to lower molecular weight oligomers. The rate of hydrolysis is accelerated by acids or bases and is dependent on moisture content, temperature etc. (Drumright *et al*., 2000; Nampoothiri *et al*., 2010). Polylactic acid cannot be reused easily or recycled along with other plastic waste. Applications will therefore have to be chosen carefully based on the value that PLA can bring in terms of its physical and degradability properties while keeping in mind its limitations in terms of recyclability.

Acknowledgements

Financial support from, NIMITLI, CSIR, Government of India for this work is gratefully acknowledged. The authors take the opportunity to thank to our research scholar Mr. Rishabh Chaudhary and Mr. Pushraj Chaudhary for their cooperation during the literature search and the analysis of data.

References

Aylott, M., Analysis: Bioplastics market set for fivefold growth by 2016, NNFCC, Press release, http://www.nnfcc.co.uk/news/bioplastics-market-set-for-fivefold-growth-by-2016 (2012).

Barve, P.P., Kamble, S.P., Joshi, J.B., Gupte, M.Y. and Kulkarni, B.D., Preparation of pure methyl esters from corresponding alkali metal salts of carboxylic acids using carbon dioxide and methanol, *Ind. Eng. Chem. Res.*, 51, 1498–1505 (2012).

Barve, P.P., Kulkarni, B.D., Nene, S.N., Shinde, R.W., Gupte, M.Y., Joshi, C.N., Thite, G.A., Chavan, V.B., Deshpande, T.R., Process for preparing L(+) lactic acid. Patent No US7820859 B2 (2010).

Bioplastics Market Worldwide 2010/11-2015-2020-2025; Applications, methods, competition, materials, technologies, development, recycling, renewable energy, production, consumption, Helmut Kaiser Consultancy, hkc22.com, ACON AG, 8004 Zurich Switzerland, http://www.hkc22.com/bioplastics.html (2012).

CCM International Ltd, Zhejiang Hisun to expand its PLA capacity in 2013, http://www.cnchemicals.com/PressRoom/PressRoomDetail_c_924.html (accessed September 30, 2012) (2011).

ChemistryViews, Future of Worldwide Polylactic Acid Capacity, Plastics News, http://www.chemistryviews.org/details/news/2782191/Future_of_Worldwide_Polylactic_Acid_Capacity.html (2012).

Cho, Y., Kim, B., Kim, D. and Han, M., Recovery of lactic acid by reactive dividing wall column, *Intern. Confer. Control, Automation Systems, in COEX*, Seoul, Korea, 14–17, 2596–2599 (2008).

Dai, Y. and King, C.J., Selectivity between lactic acid and glucose during recovery of lactic acid with basic extractants and polymeric sorbents, *Ind. Eng. Chem. Res.*, 35(4), 1215–1224 (1996).

Drumright, R.E., Gruber, P.R. and Henton, D.E., Polylactic acid technology, *Adv. Mater.*, 12(23), 1841–1846 (2000).

Farrington, D.W., Davies, J.L. and Blackburn, R.S., Poly(lactic acid) fibers, in biodegradable and sustainable fibers. In: Blackburn, R.S. (Ed.), Woodhead Publishing, Cambridge, England, pp. 191–220 (2005).

Freedman, A.M. and Shaban, H.I., Protein, lactose and lactic acid separation from cheese whey using reverse osmosis dynamically formed, *Intern. J. Food Sci. Tech.*, 6(3), 309–315 (2007).

Habova, V., Melzoch, K., Rychtera, M. and Sekavova, B., Electrodialysis as a useful technique for lactic acid separation from a model solution and a fermentation broth, *Desalination*, 163, 361–372 (2004).

Herlban, V., Skara, J., Sturdik, E. and Ilavsky, J., Isolation of free lactic acid using electrodialysis, *Biotechnol. Techniques*, 7(1), 63–68 (1993).

Hirata, M., Gao, M., Toorisaka, E., Takanashi, H. and Hano, T., Production of lactic acid by continuous electrodialysis fermentation with a glucose concentration controller, *Biochem. Eng. J.*, 25, 159–163 (2005).

Hongo, M., Nomura, Y. and Iwahara, M., Novel method of lactic acid production by electrodialysis fermentation, *Appl. Environ. Microbiol.*, 52(2), 314–319 (1986).

Jarvinen, M., Myllykoski, L., Keiski, R. and Sohlo, J., Separation of lactic acid from fermented broth by reactive extraction, *Bioseparation*, 9(3), 163–166 (2000).

Joglekar, H.G., Rahman, I., Babu, S., Kulkarni, B.D. and Joshi, A., Comparative assessment of downstream processing options for lactic acid, *Sep. Purif. Technol.*, 52, 1–17 (2006).

Kamble, S.P., Barve, P.P., Rahman, I., Joshi, J.B. and Kulkarni, B.D., Purification of lactic acid via esterification of lactic acid using packed column followed by hydrolysis of methyl lactate using 3-CSTRs in series: A continuous pilot plant study, *Ind. Eng. Chem. Res.*, 51, 1506–1514 (2012).

Kertes, A.S. and King, C.J., Extraction chemistry of fermentation product carboxylic acid, *Biotechnol Bioeng.*, 28(2), 269–282 (2004).

Kim, J.Y., Kim, Y.J., Hong, W.H. and Wonzy, G., Recovery process of lactic acid using two distillation columns, *Biotechnol. Bioprocess. Eng.*, 5, 196–201 (2000).

Kim, Y.J., Hong, H. and Wozny, G., Effect of recycle and feeding method on batch reactive recovery system of lactic acid, *Korean J. Chem. Eng.*, 19(5), 808–814 (2002).

Kumar, R., Mahajani, S.M., Esterification of lactic acid with n-butanol by reactive distillation, *Ind. Eng. Chem. Res.*, 46, 6873–6882 (2007).

Kumar, R., Nanavati, H., Noronha, S.B., Mahajani, S.M. A continuous process for the recovery of lactic acid by reactive distillation, *Chem. Technol. Biotechnol.*, 81,1767–1777 (2006).

Kyuchoukov, G., Marinova, M., Albet, J., Molinier, J. and Malmary, G., Separation of tartaric and lactic acids by means of solvent extraction, *Sep. Purif. Tech.*, 37, 199–207 (2004).

Lee, E.G., Moon, S., Chang, Y.K., Yoo, I. and Chang, H.N., Lactic acid recovery using two-stage electrodialysis and its modeling, *J. Membrane Sci.*, 145, 53–66 (1998).

Li, Y., Shahbazi, A., Williams, K. and Wan, C., Separate and concentrate lactic acid using combination of nanofiltration and reverse osmosis membranes, *Appl. Biochem. Biotechnol.*, 147, 1–9 (2008).

Liew, M.K.H., Tanaka, S. and Morita, M., Separation and purification of lactic acid: Fundamental studies on the reverse osmosis down-stream process, *Desalination*, 101(3), 269–277 (1995).

Lindner, E., CSM and Indorama Ventures in Alliance Discussion for PLA, Press Release http://www.purac.com/_sana_/handlers/getfile.ashx/2fbfe30a-9fec-4e2e-a7ee-c67d335937c8 /Press++release+–+CSM+and+Indorama+ Ventures +in+ Alliance+ Discussion +for+PLA.pdf (accessed September 30, 2012) (2011).

Lunt, J., Large-scale production, properties and commercial applications of polylactic acid polymers, *Polym. Degrad. Stabil.*, 59, 145–152 (1998).

Madzingaidzo, L., Danner, H. and Braun, H. Process development and optimization of lactic acid purification using electrodialysis, *J. Biotechnol.*, 96, 223–239 (2002).

Monteagudo, J.M. and Aldavero, M., Production of L-lactic acid by Lactobacillus delbrueckii in chemostat culture using an ion exchange resins system, *Chem. Technol. Biotechnol.*, 74, 627–634 (1999).

Nampoothiri, K.M., Nair, N.R. and John, R.P., An overview of the recent developments in polylactide (PLA) research, *Bioresource Technol.* 101, 8493–8501 (2010).

Osaka, K.N., Otsu, S.Y. and Sakai, T.Y., Process for recovering a carboxylic acid, USP 4323702 (1982).

Pal, P., Sikder, J., Roy, S. and Giorno, L., Process intensification in lactic acid production: A review of membrane based processes. *Chem. Eng. Process*, 48, 1549–1559 (2009).

Pimentel, D., Bailey, O., Kim, P., Mullaney, E., Calabrese, J., Walman, L., Nelson, F. and Yao, X., Will limits of the Earth's resources control human numbers? *Environ. Develop. Sustainability* 1, 19–39 (1999).

Ramakrishna, S.V., Rangaswamy, V., Jain, D., Jagdambalal, R.K., Patel, P.S., Kar, D., Ramachandran, S., Ganeshpure, P.A. and Satpathy, U.S. Process for the production of pylactic acid (PLA) from renewable feedstocks, Patent No. US750561 B2 (2009).

Raya-Tonetti, G., Cordoba, P., Bruno-Barcena, J., Siñeriz, F. and Perotti, N., Fluidized bed ion exchange for improving purification of lactic acid from fermentation, *Biotechnol. Techniques*, 13, 201–205 (1999).

Sarnacke, P. and Wildes, S., *Disposable Bioplastics, A Market Opportunity Study*, Omni Tech International, Ltd, Michigan, USA, 1–35 (2008).

Sauer, M., Porro, D., Mattanovich, D., Branduardi, P., 16 years research on lactic acid production, with yeast—ready for the market?, *Biotechnol. Genetic Eng. Reviews* 27, 229–256 (2010).

Seo, Y., Hong, W.H. and Hong, T.H., Effects of operation variables on the recovery of lactic acid in a batch distillation process with chemical reactions, *Korean J. Chem. Eng.*, 16(5), 556–561 (1999).

Sosa, A.V., Ochoa, J. and Perotti, N.I., Modeling of direct recovery of lactic acid from whole broths by ion exchange adsorption, *Bioseparation*, 9, 283–289 (2001).

Srivastava, A., Roychoudhury, P.K. and Roychoudhury, V., Extractive bioconversion of lactic acid, *Advances Biochem. Eng./Biotechnol.*, 53, 61–87 (1992).

Sun, Y., Li, Y., Bai, S. and Hu, Z.D., Modeling and simulation of an in situ product removal process for lactic acid production in an airlift bioreactor, *Ind. Eng. Chem. Res.* 38, 3290–3295 (1999).

Vaidya, A.N., Pandey, R.A., Mudliar, S., Suresh Kumar, M., Chakrabarti, T. and Devotta, S., Production and recovery of lactic acid for polylactide-an overview, *Critical Reviews Environ. Sci. Tech.*, 35, 429–467 (2005).

Vijayakumar, J., Aravindan, R. and Viruthagiri, T., Recent trends in the production, purification and application of lactic acid, *Chem. Biochem. Eng. Q.*22(2), 245–264 (2008).

Waltz, E., Do biomaterials really mean business? *Nature Biotech.*, 26(8), 851–853 (2008).

Wang, C.J., Bajpai, R.K. and Iannotti, E.L., Nondispersive extraction for recovering lactic acid, *Appl. Biochem. Biotechnol.*, 28-29(1), 589–603 (1991).

Wasewar, K.L., Separation of lactic acid: recent advances, *Chem. Biochem. Eng. Q.* 19(2), 159–172 (2005).

Wasewar, K.L., Yawalkar, A.A., Moulijn, J.A. and Pangarkar, V.G., Fermentation of glucose to lactic acid coupled with reactive extraction: A Review, *Ind. Eng. Chem. Res.*, 43, 5969–5982 (2004).

Williams, J. NNFCC Renewable Polymers Factsheet: Bioplastics, Press Release, http://www.nnfcc.co.uk/publications/nnfcc-renewable-polymers-factsheet-bioplastics (accessed September 30, 2012) (2010).

Xuemei, L., Jianping, L., Mo'e, L. and Peil, C., L-lactic acid production using immobilized Rhizopusoryzae in a three-phase fluidized-bed with simultaneous product separation by electrodialysis, *Bioprocess Eng.*, 20, 231–237 (1999).

Yang, S.T., White, S.A. and Hsu, S.T., Extraction of carboxylic acids with tertiary and quaternary amines: Effect of pH, *Ind. Eng. Chem. Res.*, 30(6), 1335–1342 (1991).

Zhang, Y., Ma, L. and Yang, J., Kinetics of esterification of lactic acid with ethanol catalyzed by cation-exchange resins, *React. Funct. Polym.*, 61, 101–114 (2004).

Zhao, W., Sun, X., Wang, Q., Ma, H., Teng, Y., Lactic acid recovery from fermentation broth of kitchen garbage by esterification and hydrolysis method, *Biomass Bioenergy* 33, 21–25 (2009).

Index

Separation and Purification Technologies in Biorefineries, First Edition.
Edited by Shri Ramaswamy, Hua-Jiang Huang, and Bandaru V. Ramarao.

Printed and bound by CPI Group (UK) Ltd, Croydon, CR0 4YY
18/09/2026
14964993-0002